CMOS ANALOG INTEGRATED CIRCUITS

HIGH-SPEED AND POWER-EFFICIENT DESIGN

CMOS ANALOG INTEGRATED CIRCUITS

HIGH-SPEED AND POWER-EFFICIENT DESIGN

Tertulien Ndjountche

CRC Press
Taylor & Francis Group
Boca Raton London New York

CRC Press is an imprint of the
Taylor & Francis Group, an **informa** business

CRC Press
Taylor & Francis Group
6000 Broken Sound Parkway NW, Suite 300
Boca Raton, FL 33487-2742

First issued in paperback 2017

ISBN 13: 978-1-138-07685-3 (pbk)
ISBN 13: 978-1-4398-5491-4 (hbk)

Contents

Preface

Hardware developments have been a major vehicle in popularizing the applications of signal processing theory in both science and engineering. The book describes the important trends of designing high-speed and power-efficient front-end analog circuits, which can be used alone or to interface modern digital signal processors and micro-controllers in various applications such as multimedia, communication, instrumentation, and control systems.

The book contains resources to allow the reader to design complementary metal oxide semiconductor (CMOS) analog integrated circuits with improved electrical performance. It offers a complete understanding of architectural- and transistor-level design issues of analog integrated circuits. It provides a comprehensive, self-contained, up-to-date, and in-depth treatment of design techniques, with an emphasis on practical aspects relevant to integrated circuit implementations.

Starting from an understanding of the basic physical behavior and modeling of MOS transistors, we review design techniques for more complex components such as amplifiers, comparators, and multipliers. The book details all aspects from specifications to the final chip related to the development and implementation process of filters, analog-to-digital converters (ADCs) and digital-to-analog converters (DACs), phase-locked loops (PLLs) and delay locked loops (DLLs). It provides the analysis of architectures and performance limitation issues affecting the circuit operation. The focus is on designing and verifying analog integrated circuits.

The book is intended to serve as a valuable guide and reference resource for analog circuit designers and graduate students in electrical engineering programs. It provides balanced coverage of both theoretical and practical issues in hierarchically organized format. With easy-to-follow mathematical derivations of all equations and formulas, the book also contains graphical plots, and a number of open-ended design problems to help determine the most suitable circuit architecture satisfying a given set of performance specifications. To appreciate the material in this book, it is expected that the reader has a rudimentary understanding of semiconductor physics, electronics, and signal processing.

Content overview

The book contains thirteen chapters and three appendices.

Chapter 1
Introduction
The use of CMOS technologies with a low device geometry and new architectures has accelerated the trend toward the system on a chip design, which merges analog, digital, and radio-frequency (RF) sections on a single integrated structure. While the manufacturing technology appears to be fundamentally limited by the material characteristics, the computer-aided design tools have to face the computational intractability of design optimizations. In this context, design techniques should be concerned with the automated conception, synthesis, and testing of microelectronic systems.

Chapter 2
MOS Transistors
A description of CMOS technology and transistor is provided. Different equivalent transistor models, including SPICE representations, are covered. The shrinking process of transistors results in an increase in the device physics complexity. Thus, advanced models, which can accurately describe the different electrical characteristics, are required to meet the circuit design specifications.

Chapter 3
Physical Design of MOS Integrated Circuits
Physical design and fabrication considerations for high-density integrated-circuits (ICs) in deep-submicrometer processes are reviewed. In addition to considering RLC models for the interconnect and package parasitic components, it appears also necessary to take into account the coupling through the common substrate during the IC design. Advances in packaging technology will be required to support high-performance ICs.

Chapter 4
Bias and Current Reference Circuits
Circuit structures for the design of current mirrors (current sources and sinks) and voltage references are reviewed. Generally, the effects of IC process and power supply variations on these basic blocks should be minimized to improve the overall performance of a device. Current mirrors are used in a variety of circuit building blocks to copy or scale a reference current, while voltage references are required to set an accurate and stable voltage for biasing circuits irrespective of fluctuations of the supply voltage and changes in operating temperature.

Chapter 5
CMOS Amplifiers

Topologies of amplifiers, which are suitable for the design of analog circuits, are described. The factors determining the nonideal behavior of an amplifier circuit are considered. To be tailored for a given application, an architecture has to meet the trade-off requirement among the different specifications, such as gain, bandwidth, phase margin, signal swing, noise, and slew rate. Design methods, which result in the optimization of specific performance characteristics, are summarized.

Chapter 6
Nonlinear Analog Components

Circuit architectures for comparators and multipliers are reviewed. Theoretical analysis is carried out for design and optimization purposes. The performances of comparators are essentially limited by the switching speed and mismatches of transistor characteristics, resulting in voltage offsets, while the main limitations affecting the operation of multipliers are nonlinear distortions. The design challenge is to meet the requirements of low-voltage and low-power circuits.

Chapter 7
Continuous-Time Circuits

Continuous-time circuits are required to interface digital signal processors to real-world signals. They are based on components such as transistors, resistors, capacitors, and inductors. The choice of an architecture and design technique depends on the performance parameters and application frequency range. The use of inductors, which can only be integrated with moderate efficiency (low quality factor, parasitic elements) in CMOS processes, is restricted to high-frequency building blocks with tuned characteristics. Using active components (transistor and operational amplifier) and capacitors, MOSFET-C and G_m-C structures have proven reliable for the design of integrated circuits in the video frequency (or MHz) range.

Chapter 8
Switched-Capacitor Circuits

Switched-capacitor circuits are used in the design of large-scale integrated systems. They are based on basic building blocks such as sample-and-hold, integrator, and gain stage, which can be optimized to meet the requirements of low power consumption and chip area. Design techniques, which result in the minimization of the circuit sensitivity to component imperfections, are described. Accurate switched-capacitor filters are obtained by performing the synthesis in the z-domain, and using stray insensitive circuits for the implementation of the resulting signal-flow graph.

Chapter 9
Data Converter Principles

The interface between real-world signals and digital-signal processors can be realized by data converters (analog-to-digital converters and digital-to-analog converters). An insight into the mathematical definitions of characteristics (quantization noise, component imperfections), which can affect the performance of data converters is provided. Depending on the sampling frequency, Nyquist and oversampling data converters can be distinguished. Generally, Nyquist converters based on a parallel operation exhibit a high speed. On the other hand, digital filtering is combined with oversampling, which relies on using a sampling rate which is several times higher than two times the signal bandwidth, to improve the converter resolution. For a given dynamic range, the reduced sensitivity of oversampling structures to component imperfections is the result of a trade-off between speed and accuracy.

Chapter 10
Nyquist Digital-to-Analog Converters

Digital-to-analog converters can be designed using various architectures, each with its distinctive advantages and limitations. A review of various Nyquist converter architectures illustrates the system level trade-offs and performance issues associated with the circuit design. For a given resolution, the difference between converter architectures can be an important factor for the selection of a specific application.

Chapter 11
Nyquist Analog-to-Digital Converters

A basic understanding of various ADC architectures is useful to meet the design challenges at the transistor level of high resolution converters. There are various ADC architectures, each with its peculiar advantages and limitations. The description of Nyquist converters are presented along with their performance modeling. Applications are key in selecting a given ADC architecture even though some overlap can exist between the characteristics of various architectures.

Chapter 12
Oversampling Data Converters

Oversampling data converters generally consist of a delta-sigma modulator and digital (decimation or interpolation) filter. By combining the noise shaping and oversampling, which is similar to the sampling of a signal at a rate higher than twice the maximum frequency in the input signal, their quantization noise is removed from the signal band and spread over a larger range of frequencies. Various modulator architectures will be reviewed, the effects of circuit nonidealities on the converter performance are analyzed, and the digital filters used to remove the out-of-band noise will be presented. An evaluation and a comparison of the different delta-sigma modulation-based approaches used to improve the linearity of Nyquist converters are also provided. Another application area for delta-sigma modulators is the test and instrumentation where a precise test signal is required.

Chapter 13
Circuits for Clock Signal Generation and Synchronization
Due to the increase of the IC clock frequency and data rate, circuits (phase-locked loop, delay-locked loop) for the clock signal generation and synchronization are generally included in electronic systems to avoid data read and transmission failures. They should be designed to operate with a low voltage, feature a low timing jitter, and be less sensitive to process and temperature variations. A tutorial survey of timing circuit and frequency synthesis architectures is presented.

Appendices
Three appendices cover the following topics:
Logic building blocks
Transistor sizing in building blocks
Signal flow graph

Feedback

The author welcomes feedback on any aspect of this book. He can be reached at the e-mail address: tndjountche@gmail.com.

Acknowledgments

I am grateful for the support of colleagues and students whose remarks helped refine the content of this book.

I would like to thank Prof. Dr.-Ing. h.c. R. Unbehauen (Erlangen-Nuremberg University, Germany). His continuing support, the discussions I had with him, and the comments he made have been very useful.

I express my sincere gratitude for all the support and spontaneous help I received from Dr. Fa-Long Luo (Element CXI, USA).

I wish to acknowledge the suggestions and comments provided by Prof. Avebe Zibi (UY-I, CM) and Prof. Emmanuel Tonye (ENSP, CM) during the early phase of this project.

While doing this work, I received much spontaneous help from some international experts: Prof. Ramesh Harjani (University of Minnesota, Minneapolis, Minnesota), Prof. Antonio Petraglia (Universidade Federal do Rio de Janeiro, Brazil), Dr. Schmid Hanspeter (Institute of Microelectronics, Windisch, Switzerland), Prof. Sanjit K. Mitra (University of California, Santa Babara, California), and Prof. August Kaelin (Siemens Schweiz AG, Zurich, Switzerland). I would like to express my thanks to all of them.

I am also indebted to the publisher, Nora Konopka, the project coordinator, Jessica Vakili, the project editor, Karen Simon, the editorial assistant, Brittany Gilbert, Stephany Wilken, Christian Munoz, Shashi Kumar, and all the staff at CRC Press for their valuable comments and reviews at various stages of the manuscript preparation, and their quality production of the book.

Finally, I would like to truly thank all members of my family and friends for the continual love and support they have given during the writing of this book.

List of Figures

l

List of Tables

1

Mixed-Signal Integrated Systems: Limitations and Challenges

CONTENTS

The objective of designing a complete system on a single chip has resulted in the complexity increase of application-specific integrated circuits (ASICs), application-specific standard parts (ASSPs), and very-large scale integrated circuits. The system on a chip (SoC), as shown in Figure 1.1, generally possesses complex signal paths through both analog devices and digital components (nonvolatile memory (NVM), random access memory (RAM), and digital signal processor (DSP)).

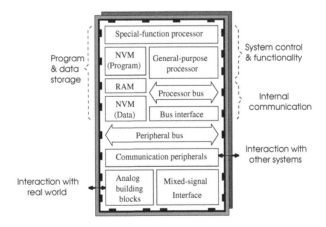

FIGURE 1.1
Example of an SoC floorplan.

Examples include multimedia devices, wireless transceivers, sensor and actuator controllers, instrumentation systems, and biomedical devices. The functions, which are realized in the analog domain, include

1

☐ Biasing

☐ Sensor and actuator signal conditioning

☐ Driver and buffer

☐ Signal down-conversion and up-conversion

☐ Mixed-signal DSP interfaces

☐ Clock signal generation and frequency synthesis

They are implemented using basic building blocks such as,

- Voltage and current references

- Low-noise and low-power amplifiers

- Variable-gain amplifier and automatic gain control circuit

- Filter

- Oscillator

- Mixer

- Sample-and-hold circuit

- Analog-to-digital converter

- Digital-to-analog converter

- Phase-locked loop (PLL) and delay-locked loop (DLL)

The SoC digital section essentially requires microprocessors, digital signal processors, memories, and control logics. The most important issues are then related to the integration of analog and digital sections. A fully monolithic chip appears to be limited, for instance, by the problematic isolation of analog sections with high-gain bandwidth from the noise generated by the substrate and digital circuits. Furthermore, the device-level simulation of mixed-signal integrated circuits in a realistic environment remains a challenge and testing chips with several complex functions is a difficult task.

1.1 Integrated circuit design flow

The specification partition into subsystems is illustrated in Figure 1.2. Tools such as SDL (Specification and Description Language), UML (Unified Modeling Language), and SystemC AMS are used to analyze the design at the higher

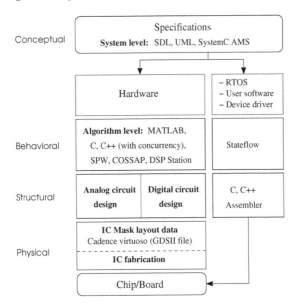

FIGURE 1.2
Specification partition into subsystems.

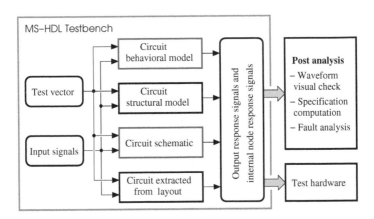

FIGURE 1.3
Circuit design verification.

level. At the system level, the design specifications are partitioned into hardware and software components. Note that SystemC AMS is particularly suited to provide functional modeling, architectural exploration, virtual prototyping, and integration validation for analog mixed-signal systems.

The development of signal processing algorithms can be performed using MATLAB®, SPW (Signal Processing Workbench), COSSAP (Communication System Simulation and Application Processor), DSP Station, C, and C++ (extended to handle concurrency). To help manage the complexity, the design implementation in hardware is supported by providing a link to hardware synthesis tools.

The functional description is realized by an analog circuit and a digital circuit, which can be designed and verified using various computer-aided design programs such Cadence, Synopsys, and Mentor Graphics tools. When the data processing in the digital domain may require processors or microcontrollers, the design of a real-time operating system (RTOS), application software and device driver is necessary. Besides allowing a modular and scalable programming approach, the desirable features of an RTOS include the ability to provide basic support for task scheduling, resource management, inter-task and input-output communication such that the processor functionality is available to application software in an optimized and predictable way. Stateflow is an interactive design and simulation tool that can be used to describe complex logic, such as an RTOS, in a form that can easily be coded using C/C++ language or assembler.

The functional description can also be refined to analog and digital models, which can be analyzed and verified using a simulator that can interpret mixed-signal hardware description languages (MS-HDLs). Verilog-AMS and VHDL-AMS are two examples in this category (VHDL stands for very high-speed integrated-circuit HDL). MS-HDLs are particularly well suited for the verification of very large and complex mixed-signal integrated circuit designs. An MS-HDL testbench, as shown in Figure 1.3, provides the stimulus required to drive various representations of a circuit while the response signals at nodes of interest are monitored. Specifications are checked by comparing the behavioral and structural models, while the implementation is verified by emphasizing the similarities between the circuit schematic and the circuit extracted from the layout.

The design flow of an integrated system is illustrated in Figure 1.4. The top-down synthesis process consists of the topology selection, specification translation or circuit sizing, and design verification (design rule check (DRC), electrical rule check (ERC), and layout versus schematic (LVS)). It is then followed by a bottom-up generation and verification of the circuit layout. The performance specifications are required at each step. Throughout the design flow, any change should be taken into account by propagating the associated constraints down the hierarchy, thus ensuring that the top-level block meets the target specifications.

Nowadays, the methodologies of top-down design and bottom-up verification are well-accepted standards in the digital domain. From bit true models of signal processing algorithms, C, Verilog, or VHDL code is generated or written for custom hardware or a DSP-based software solution. By defining a digital circuit at an architectural or behavioral level rather than at the gate level, hardware description languages, such as Verilog or VHDL, can help manage more large designs than tools based on schematic entry. An automated design flow is then adopted to convert the high-level description of the circuit into industry-standard output formats, such as GDSII, that can be integrated into chip layout tools.

In the analog domain, the current design approach — design, simulate, op-

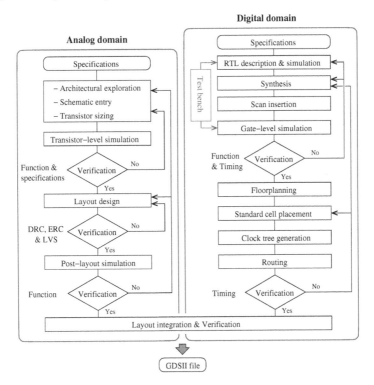

FIGURE 1.4
Design flow for an integrated system.

timize circuit specifications taking into account parasitic effects and process variations, repeat — can be very time consuming for large circuits and relies mostly on designer experience. This is due to the fact that second-order effects in analog circuits are difficult to model as the design evolves and automated tools are actually not available. Furthermore, the use of deep submicrometer CMOS processes contributes to making the verification of analog circuits substantially more difficult, as self-calibration or error cancelation schemes are often required to overcome the limitations of the components, thereby increasing the number of operating modes, behavioral complexity, and size of analog circuits.

The integration of circuit components takes place either during the physical design phase or after the fabrication. Physical design involves the floorplanning, timing optimization, placement and routing, or layout. The system timing and signal integrity verification is achieved by checking the electrical and design rules and comparing the layout and schematic. Detailed physical design information is required for accurate resistance, inductance, and capacitance parasitic extraction, and delay estimation. Note that the final system should include both hardware and software. The software platform binds the programmable cores and memories via the RTOS, the input-output interface via the device drivers, and the network connection via the communication subsystem.

1.2 Design technique issues

At the system level, an efficient solution is required for managing concurrency or assuring a real-time data flow. This is important for complex chips, which handle multiple tasks at the same time and in cases where the latency due to the interconnect delay dominates the signal bandwidth.

The design reuse results in a reduction of the cost and development time. A higher level of abstraction is necessary to create a library of subsystems that can be used for different designs.

The design of a complex chip should also include an adequate strategy for the verification of the functional blocks. The logic simulation is a suitable method for the functional verification. Here, the system is tested over a wide variety of operating conditions using simulated input patterns. However, the complexity of the chip can reduce the effectiveness of finding possible design errors. This is related to the higher number of likely events and the difficulty of determining whether the simulated behavior is correct. SoC testing suffers also from the lack of effective coverage metrics, that is, it is not always clear whether enough verification has been completed to confirm the reliability of a chip.

During the design, the circuit optimization is made with respect to timing, power, and area specifications for given values of interconnect load capacitances which can be different from the ones extracted from the final layout, specifically in deep submicrometer technologies. In addition, metal resistance effects are topology dependent and increase with the routing length, and the prediction of the delay propagation is not simple. That is, one-pass synthesis success becomes unlikely due to the requirement of physical design information. A possible solution can consist of using synthesis methods based on the delay equalization of all subsystems and the wire planning among blocks.

The speed-power performance of a design based on submicrometer integrated circuit (IC) process appears to be affected by the substrate and crosstalk-induced noises, signal delay, and parasitic inductance. The coupling effects can be controlled using low-swing differential pair structures, shield wires and repeater insertions, upper and lower bounding slew times, and increased spacing between wires. The increase in functionalities and operating frequencies results in more power dissipation. However, in addition to the supply voltage scaling, power consumption can be reduced by switching off unused subsystems via gated clocking modes.

It can be predicted that the use of IC process with low geometries will increase the impact of fabrication techniques on the design and verification.

The top-down design methodology provides a system-level model that can be used for the chip testing. But the mixed-signal nature of SoCs makes different test strategies suitable for each particular type of component, resulting in the requirement of a design for test and manufacturability across all abstraction levels.

1.3 Integrated system perspectives

The realization of integrated systems is influenced by several factors (IC process, circuit structure, package, software). Common goals such as performance optimization and development time reduction must be included in the suitable design framework.

Mixed-signal building blocks should be designed for reuse. Such a design is based on accurate high-level models, which can be used to evaluate the block suitability for a new design. The performance achievable in the hardware design reuse methodology seems to be limited in situations where the loading rules are highly complex or the circuit models exhibit interrelated features.

Due to the chip complexity, power and performance can be lost by using a single clocked-synchronous approach to manage the on-chip concurrency. An optimal implementation and verification of reliable communication among a collection of components may then be necessary. The programmable platform-based design emerges as a viable approach for the SoC implementation, and the optimal power/performance is dependent on the trade-off between hardware and software.

Methodologies for re-mapping and redesigning blocks based on physical information will be inessential only if the design approaches are either based on an improved nonideality (delay, noise, distortion) prediction or able to remove the requirement of predictability.

The performance of high-density ICs is mainly limited by noise and timing faults. For instance, the simultaneous switching of more devices increases the power supply noise. This can enlarge the timing delay by reducing the actual voltage that is applied to a device. The effect of capacitive coupling in submicrometer designs is also important and affects the signal integrity. Since the complexity of SoC makes a unified testing scheme difficult to implement, a self-test mechanism is required at the component and system levels. It can be developed based on new fault models with links to the layout and implemented as program executed by the processor core. This approach has the benefit of eliminating any additional built-in self-test (BIST) hardware, such as linear feedback shift registers. Another important issue in critical applications is related to self-repair techniques, which take advantage of the reconfigurability provided by adding coprocessors, appropriate instruction sets, and peripheral units to the embedded processor core.

To eliminate the need for external testers, the implementation of BIST for analog blocks should preferably exploit the capability of the digital processor for the signal generation and analysis and down-sampling techniques for the specific case of RF circuits.

1.4 Built-in self-test structures

With the increase in the density and complexity of mixed-signal integrated circuits (ICs), more complex measuring devices are required to meet ever more severe test specifications. The built-in self-test (BIST) appears as a suitable approach to resolve the problem related to the fact that mixed-signal circuits are verified by functionality, the number of which can be high in a single chip. Furthermore, a BIST section can facilitate the initialization and observation of the circuit nodes. BIST structures for digital circuits have reached a good level of maturity, and it can be expected that testing solutions for the analog section in mixed-signal systems will exploit the computation capability of logic gates and digital signal processors.

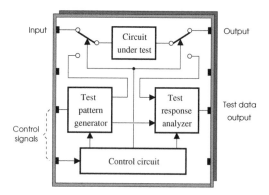

FIGURE 1.5
A chip including a built-in self-test structure.

The architecture of a chip including a BIST section is shown in Figure 1.5. It consists of a test pattern generator, a test response analyzer, and a control circuit, in addition to the circuit under test, which can be reconfigured by the control signals to support the test mode and the normal operation mode. Depending on the BIST flexibility, testing can be carried out while the circuit is in an idle state or during normal operation. A generator is used to provide the required test input signals and the analyzer features the capability of detecting various faults in the output response of the circuit under test.

1.5   Concluding remarks

The use of submicrometer process for the IC implementation extends the operating frequency range, but also results in the enhancement of nonideal effects related to the interconnect crosstalk and latency. The density of test data is growing as the complexity and the number of intellectual property cores required in a single chip is increasing. Thus, the viable development of SoCs should take into account aspects of the design, manufacturing and test at all abstraction levels.

1.6   To probe further

- Special issue on limits of semiconductor technology, *Proc. of the IEEE*, vol. 89, no. 3, March 2001.

- W. Müller, W. Rosenstiel, and J. Ruf, Eds., *SystemC Methodologies and Applications*, Dordrecht, The Netherlands: Kluwer Academic Publishers, 2003.

- D. Jansen et al, Eds., *The Electronic Design Automation Handbook*, Dordrecht, The Netherlands: Kluwer Academic Publishers, 2003.

- M. D. Birnbaum, *Electronic Design Automation*, Upper Saddle River, NJ: Prentice Hall, 2004.

- Special issue on system on chip: Design and integration, *Proc. of the IEEE*, vol. 94, no. 6, June 2006.

- Special issue on leading-edge computer aided design solutions for advanced digital and mixed-signal systems-on-chips, *Proc. of the IEEE*, vol. 59, no. 3, March 2007.

2

MOS Transistors

CONTENTS

In almost all modern electronic circuits, transistors are the key active element. By reducing the dimensions of MOS transistors and the wires connecting them in integrated circuits (ICs), it has been possible to increase the density and complexity of integrated systems. Figure 2.1 illustrates the reduction in feature size over time. It is expected that a chip designed in a 35-nm IC process will include more than 10^{11} transistors in a few years [1]. But up

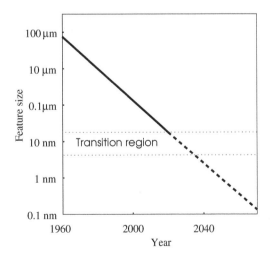

FIGURE 2.1
Transistor scaling advance: A plot of the feature size versus time.

to now, the scaling progress was essentially attributed to the improvements in manufacturing technology. However, as the physical limits are being met, some changes to the device structures and new materials will be necessary.

Next, we will describe the transistor structure and the different equivalent models that are generally used for simulations.

2.1   Transistor structure

Structures of MOS transistors are shown in Figure 2.2. The drain and source of the nMOS transistor (see Figure 2.2(a)) are realized by two heavily doped n-type semiconductor regions, which are implanted into a lightly doped p-type substrate or bulk. A thin layer of silicon dioxide (SiO_2) is thermally grown over the region between the source and drain, and is covered by a polycrystalline silicon (also shortly called, polysilicon or poly), which forms the gate of the transistor. The thickness t_{ox} of the oxide layer is on the order of a few angstroms. The useful charge transfer takes place in the induced channel of the transistor, which is the substrate region under the gate oxide. The length L and the width W of the gate are estimated along and perpendicularly to the drain-source path, respectively. The substrate connection is provided

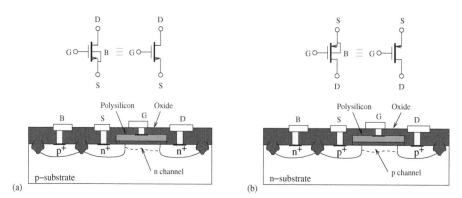

FIGURE 2.2
Model and structure of (a) nMOS transistor and (b) pMOS transistor.

by a doped p^+ regions. Generally, the substrate is connected to the most negative supply voltage of the circuit so that the source-drain junction diodes are reverse-biased.

In the case of the pMOS transistor (see Figure 2.2(b)), the drain and source are formed by p^+ diffusions in the n-type substrate. A doped n^+ region is required for the realization of the substrate connection. Otherwise, the cross-sections for the two transistor types are similar. Here, the substrate is to be connected to the most positive supply voltage.

The gate and substrate of an nMOS transistor can be assumed to form the plates of a capacitor using the silicon dioxide as the dielectric. As a positive voltage is applied to the gate, there is a movement of charges resulting in an augmentation of holes at the gate side and electrons at the substrate edge. Initially, the mobile holes are pushed away from the substrate surface, leaving behind a *depletion region* below the gate as shown in Figure 2.3, and the electron enhancement is first attributed to a *weak inversion*. By increasing the gate voltage, the concentration of electrons (minority carriers) can become larger than the one of holes (majority carriers) at the surface of the *p*-type substrate. This is known as the *strong inversion*, which occurs for voltages greater than two times the Fermi potential. When the gate voltage is negative, the concentration of holes will increase at the surface and an *accumulation region* is formed.

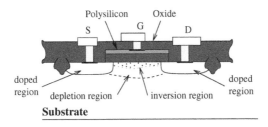

FIGURE 2.3
Localization of the inversion and depletion regions in a MOS transistor.

For a pMOS transistor, the current is carried by holes instead of electrons. The source should be the node biased at the most positive voltage. Similar results can then be obtained for the pMOS device connected to a gate voltage with the inverse polarity. Generally, the electrical characteristics of pMOS transistors can be obtained from the ones of nMOS transistors by reversing the sign of all currents and voltages.

2.1.1 I/V characteristics of MOS transistors

The static characteristics of a MOS device can be determined by solving numerically a set of differential equations governing the movement of electrical charges, given the relevant boundary conditions. The behavior of the internal electrostatic potential ϕ is given by the Poisson law,

$$\nabla^2 \phi = -\frac{q}{\epsilon}(N_d - N_a + p - n), \tag{2.1}$$

where q is the charge of the electron, ϵ is the dielectric constant of the semiconductor, and N_a and N_d are the concentrations of the acceptors (*n*-type dopant) and donors (*p*-type dopant), respectively. The electron and hole concentrations, n and p, can be derived respectively from the following conservation equations,

$$\frac{\partial n}{\partial t} = \frac{1}{q}\nabla \cdot \mathbf{J}_n - R_r + R_g \,, \tag{2.2}$$

$$\frac{\partial p}{\partial t} = -\frac{1}{q}\nabla \cdot \mathbf{J}_p - R_r + R_g \,, \tag{2.3}$$

where R_r and R_g denote, respectively, the recombination and generation rate of electrons and holes. The electron and hole current densities, $\mathbf{J}_n$ and $\mathbf{J}_p$, are, respectively, given by

$$\mathbf{J}_n = q\mu_n \left(-n\nabla\phi + \frac{kT}{q}\nabla n \right), \tag{2.4}$$

$$\mathbf{J}_p = q\mu_p \left(-p\nabla\phi - \frac{kT}{q}\nabla p \right), \tag{2.5}$$

where μ_n and μ_p are the electron and hole mobilities, respectively, k is Boltzmann's constant ($k = 1.38 \times 10^{-23}$ J/K), and T represents the absolute temperature (in K). It should be noted that simple and compact transistor models are generally used for the analysis and design at the circuit level.

2.1.2   Drain current in the strong inversion approximation

Based on Boltzmann's distribution [2, 3], the concentration of electrons and holes can be respectively computed as

$$n = n_{dp} \exp \left[\frac{q}{kT}(\phi - V - V_{SB}) \right], \tag{2.6}$$

$$p = p_{dp} \exp \left(-\frac{q\phi}{kT} \right), \tag{2.7}$$

where n_{dp} is the electron concentration within the diffusion region on the p side, p_{dp} is the corresponding concentration for holes, ϕ represents the potential of the field $\mathbf{E} = -\nabla\phi$, and V is the applied voltage. Given the relation, $p_{dp} \simeq N_a$, we can write $n_{dp} \simeq n_i^2/N_A$, where n_i is the intrinsic charge density.

With the assumption that the current flow is essentially one-dimensional from the source to drain and the mobility is constant throughout the channel, the current density becomes

$$J_n(x, y) = q\mu_n n(x) \frac{\partial V}{\partial y} \tag{2.8}$$

and using a spatial integration, the drain-source current for long channel transistors can be written as

$$I_{DS} \int_0^L dy = \int_0^W dz \int_0^{V_{DS}} dV \int_0^{x_W} q\mu_n n(x) dx \tag{2.9}$$

and

$$I_{DS} = \mu_n \left(\frac{W}{L}\right) \int_0^{V_{DS}} Q_n dV, \tag{2.10}$$

where

$$Q_n = q \int_0^{x_W} n(x)dx. \tag{2.11}$$

The mobile charge per unit area in the inversion region, Q_n, is related to the charge in the semiconductor, Q_s, and the charge in the depletion region, Q_d, that is,

$$Q_n = Q_s - Q_d. \tag{2.12}$$

To proceed further, we assume that

$$V_{GS} - V_{FB} = V_{GB} = V_{ox} + \phi_s, \tag{2.13}$$

where V_{FB} is the flat-band voltage, V_{ox} is the voltage drop across the oxide, and ϕ_s is the potential at the silicon-oxide interface referenced to the bulk. Furthermore,

$$\phi_s = \phi_s(0) + V(y) = -2\psi_F + V(y) \tag{2.14}$$

with ψ_F being the bulk Fermi potential, which is negative for a p-type substrate.

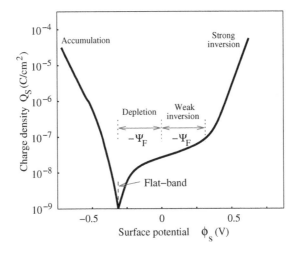

FIGURE 2.4
Plot of the charge density versus the surface potential.

Figure 2.4 shows the variation of the total charge per unit area, Q_s, as a function of the surface potential. The following relation can be written:

$$Q_s = -C_{ox}V_{ox} = -C_{ox}[V_{GB} + 2\psi_F - V(y)], \tag{2.15}$$

where $C_{ox} = \epsilon_{ox}/t_{ox}$ is the gate oxide capacitance per unit area, ϵ_{ox} is the oxide permittivity, and t_{ox} denotes the oxide thickness. As a small positive voltage is applied to the gate, holes are lessened from the vicinity of the oxide-silicon interface, and a space-charge region consisting of stationary acceptor ions is established. The depletion charge is then given by

$$Q_d = -qN_A W_d = -\sqrt{2q\epsilon N_A [V(y) - 2\psi_F]}, \tag{2.16}$$

where $W_d = \sqrt{2\epsilon[V(y) - 2\psi_F]/qN_A}$ is the width of the depletion region. In the small-signal analysis, the depletion capacitance can be obtained as $C_d = \partial Q_d/\partial V$. The expression of the drain current reads

$$I_{DS} = \mu_n C_{ox} \left(\frac{W}{L}\right)$$
$$\times \int_0^{V_{DS}} \left[V_{GB} + 2\psi_F - V(y) - \frac{1}{C_{ox}} 2q\epsilon N_A [V(y) - 2\psi_F]^{1/2} \right] dV. \tag{2.17}$$

Hence,

$$I_{DS} = \mu_n C_{ox} \left(\frac{W}{L}\right) \left[\left(V_{GS} - V_{FB} + 2\psi_F - \frac{V_{DS}}{2} \right) V_{DS} \right.$$
$$\left. - \frac{2}{3}\gamma \left((V_{DS} - 2\psi_F)^{3/2} - (-2\psi_F)^{3/2} \right) \right], \tag{2.18}$$

where

$$\gamma = \frac{\sqrt{2\epsilon q N_A}}{C_{ox}}. \tag{2.19}$$

It is generally assumed that $V_{DS} \ll -2\psi_F$ and the current I_{DS} in the triode region can be reduced to

$$I_{DS} = \mu_n C_{ox} \left(\frac{W}{L}\right)(V_{GS} - V_T - V_{DS}/2)V_{DS}, \tag{2.20}$$

where

$$V_T = V_{FB} - 2\psi_F + \gamma\sqrt{-2\psi_F}. \tag{2.21}$$

The current I_{DS} reaches its maximum at $V_{DS(sat)} = V_{GS} - V_T$, which can be obtained by solving the equation $\partial I_{DS}/\partial V_{DS} = 0$. Based on this result, the drain-source current in the saturation region is deduced from Equation (2.20) as follows:

$$I_{DS} = \frac{1}{2}\mu_n C_{ox} \left(\frac{W}{L}\right)(V_{GS} - V_T)^2. \tag{2.22}$$

Without the above simplifying assumption, the equation of the drain current can be derived as

$$I_{DS} = \mu_n C_{ox} \left(\frac{W}{L}\right) \left[\left(V_{GS} - V_T + \gamma\sqrt{-2\psi_F} - \frac{V_{DS}}{2} \right) V_{DS} \right.$$
$$\left. - \frac{2}{3}\gamma \left((V_{DS} - 2\psi_F)^{3/2} - (-2\psi_F)^{3/2} \right) \right]. \tag{2.23}$$

In the saturation region, the drain-source voltage can be obtained by solving $Q_n(L) = 0$ with $V(L) = V_{DS}$. Hence,

$$V_{DS(sat)} = V_{GS} - V_{FB} + 2\psi_F + \frac{\gamma^2}{2}\left[1 - \sqrt{1 + \frac{4(V_{GS} - V_{FB})}{\gamma^2}}\right]. \quad (2.24)$$

The triode and saturation regions are illustrated on the I/V characteristics shown in Figure 2.5. For a low drain-source voltage, V_{DS}, the charges in the inversion layer are uniformly induced along the channel, resulting in a current flowing from the source to the drain. The current, I_{DS}, grows proportionally to V_{DS} and the channel behaves as a voltage-controlled resistor in the *triode or linear region*. By increasing the drain voltage, there is a reduction of the charges at the drain boundary and the channel is *pinched off*. That is, $V_{DS} \geq V_{DS(sat)} = V_{GS} - V_T$, where V_{GS} denotes the gate-source voltage, the channel ceases to conduct electricity and its resistance becomes zero. The current is now due to the charge drift and I_{DS} remains practically constant. In this case, the transistor is considered to operate in the saturation region.

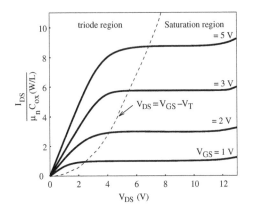

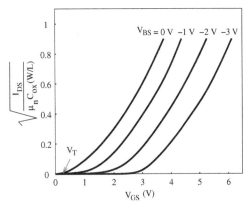

FIGURE 2.5
Plot of the drain current versus the drain-source voltage.

FIGURE 2.6
Plot of the drain current versus the gate-source voltage.

A plot of the square root of the normalized drain current versus the gate-source voltage is shown in Figure 2.6 for several values of the bulk-source voltage, V_{BS}. Remark that the threshold voltage changes with V_{BS}, provided that this latter is different from zero.

2.1.3 Drain current in the subthreshold region

For a gate voltage, V_G, greater than 0 and less than V_T, the drain-source current still exhibits a magnitude different from zero, which decreases exponentially. This corresponds to the subthreshold region. Here, the current I_{DS} is due to the diffusion [4] instead of the drift process, as is the case in the strong inversion region. Thus,

$$I_{DS} = -\mu_n qA \frac{kT}{q} \frac{\triangle n}{\triangle y} \tag{2.25}$$

$$= \mu_n qA \frac{kT}{q} \frac{n(0) - n(L)}{L}, \tag{2.26}$$

where $A = W \cdot \triangle d$ is the cross-section area of the current flow; $\triangle d$ is the channel depth, which is defined as the distance from the silicon-oxide interface at which the potential is lowered by kT/q, i.e., $\triangle d = kT/qE_s$; and E_s is the surface field given by

$$E_s = -\frac{Q_d}{\epsilon} = \sqrt{\frac{2qN_A\phi_s}{\epsilon}}. \tag{2.27}$$

The electron densities near the source, $n(0)$, and drain, $n(L)$, are, respectively, given by

$$n(0) = n_{dp} \exp\left[\frac{q}{kT}(\phi_s + 2\psi_F - V_{SB})\right], \tag{2.28}$$

$$n(L) = n_{dp} \exp\left[\frac{q}{kT}(\phi_s + 2\psi_F - V_{DB})\right], \tag{2.29}$$

where $n_{dp} \simeq N_a$, as the concentration of acceptors is the most significant. The drain-source current can be written as

$$I_{DS} = \mu_n q \left(\frac{W}{L}\right)\left(\frac{kT}{q}\right)^2$$
$$\times \sqrt{\frac{\epsilon N_a}{2q\phi_s}} \exp\left[\frac{q}{kT}(\phi_s + 2\psi_F - V_{SB})\right]\left[1 - \exp\left(-\frac{qV_{DS}}{kT}\right)\right]. \tag{2.30}$$

From Gauss' law of charge balance applied to the silicon-oxide interface [5], the gate-source voltage for a given surface potential, ϕ_s, is of the form

$$V_{GS} = V_{FB} + \phi_s - V_{SB} - \frac{Q_d + Q_i}{C_{ox}}, \tag{2.31}$$

where V_{FB} represents the flat-band voltage, V_{SB} is the source-substrate bias voltage, Q_d denotes the depletion charge, Q_i is the inversion charge at the silicon-oxide interface, and C_{ox} is the oxide capacitance per unit area. Due to the nonlinear dependence of the charge on ϕ_s, an expression of ϕ_s is easily obtained from the next first-order Taylor series of V_{GS} around $\phi_{so} + V_{SB}$ [6],

$$V_{GS} = V_{GS}^* + \eta(\phi_s - \phi_{so} - V_{SB}), \tag{2.32}$$

where

$$V_{GS}^* = V_{GS}|_{\phi_s = \phi_{so} + V_{SB}} \tag{2.33}$$

and

$$\eta = \left. \frac{dV_{GS}}{d\phi_s} \right|_{\phi_s = \phi_{so} + V_{SB}} . \tag{2.34}$$

Thus,

$$\phi_s = \frac{V_{GS} - V_{GS}^*}{\eta} + \phi_{so} + V_{SB} \tag{2.35}$$

and the subthreshold current becomes

$$\begin{aligned}
I_{DS} = & \mu_n \left(\frac{W}{L} \right) \left(\frac{kT}{q} \right)^2 C_d \\
& \times \exp \left[\frac{q}{kT} \left(\frac{V_{GS} - V_{GS}^*}{\eta} + \phi_{so} + 2\psi_F \right) \right] \left[1 - \exp \left(-\frac{qV_{DS}}{kT} \right) \right],
\end{aligned} \tag{2.36}$$

where C_d denotes the depletion capacitance given by

$$C_d = \sqrt{\frac{\epsilon q N_a}{2\phi_s}} . \tag{2.37}$$

In the weak inversion region, $-\psi_F + V_{SB} < \phi_s < -2\psi_F + V_{SB}$, and great accuracy can be ensured by choosing the reference point $\phi_{so} = -3\psi_F/2$. However, with $\phi_{so} = -2\psi_F$, the voltage V_{GS}^* is reduced to V_T. The current I_{DS} depends on the gate-source and drain-source voltages. But, its dependence on the voltage, V_{DS}, is considerably reduced as V_{DS} becomes greater than a few kT/q.

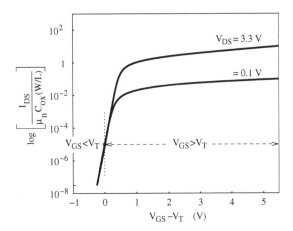

FIGURE 2.7
I/V characteristics of an nMOS transistor.

The I/V characteristics of an nMOS transistor is shown in Figure 2.7. For $V_{DS} > 0.1$ V, the current I_{DS} is almost independent of V_{DS}, and we have

$$I_{DS} \simeq \mu_n \left(\frac{W}{L} \right) \left(\frac{kT}{q} \right)^2 C_d \exp \left[\frac{q}{kT} \left(\frac{V_{GS} - V_{GS}^*}{\eta} + \phi_{so} + 2\psi_F \right) \right] . \tag{2.38}$$

It should be noted that nMOS and pMOS devices feature the characteristics that are the mirror of each other. In addition, pMOS transistors generally feature the lower holes mobility, $\mu_p \simeq \mu_n/4$. This can result in a lower current drive, transconductance and output resistance.

2.1.4 MOS transistor capacitances

A model for MOS transistor capacitances is required to accurately predict the *ac* behavior of circuits. The silicon oxide, which provides the isolation of the gate from the channel, can be considered the dielectric of a capacitor with the value C_{ox}. Due to lateral diffusion, the effective channel length, L_{eff}, is shorter than the drawn length, L, as shown in Figure 2.8, and overlap capacitors are formed between the gate and drain/source. The expression of the overlap capacitance can be given by

$$C_{gso} = C_{gdo} = C_{ox} W \triangle L, \qquad (2.39)$$

where $\triangle L = (L - L_{eff})/2$. Note that the accurate determination of the overlap capacitance per unit width, C_{ov}, can require more precise calculations. The capacitors related to the silicon oxide, $C_g = C_{ox} W L$, and the depletion region, C_d, exist between the gate and channel and between the channel and substrate. In addition, junction capacitors are present between the source/drain and substrate. They consist of two geometry-dependent components related respectively to the bottom-plate and side-wall of the junction.

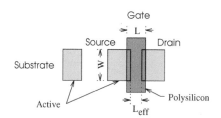

FIGURE 2.8
MOS transistor layout.

A transistor can then be represented as shown in Figure 2.9 with parasitic capacitors between every two of the four output nodes [7]. It is assumed that the capacitance between the source and drain is negligible.

Let Q_G, Q_B, Q_S, and Q_D be the gate, bulk, source, and drain charges, respectively. The charge conservation principle results in the following relation:

$$Q_G + Q_B + Q_S + Q_D = 0. \qquad (2.40)$$

The charges Q_G and Q_B can be computed from the Poisson equation. The sum of Q_D and Q_S is equal to the channel charge, and the drain and source charge partition changes uniformly from the Q_D/Q_S ratio of 50/50 in the triode

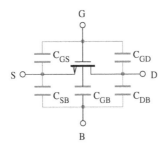

FIGURE 2.9
MOS transistor capacitance model.

region to 40/60 in the saturation region [8]. However, the charge partition will become closer to the ratio 0/100 when the transistor is switched at a speed greater than the channel charging time. Based on the charge model, the capacitances can be defined as

$$C_{GS} = \frac{\partial Q_G}{\partial V_S}, \qquad (2.41)$$

$$C_{GD} = \frac{\partial Q_G}{\partial V_D}, \qquad (2.42)$$

and

$$C_{GB} = \frac{\partial Q_G}{\partial V_B}. \qquad (2.43)$$

There are normally two capacitors between a couple of nodes. Here, the capacitors are nonlinear and the condition of reciprocity is not fulfilled. That is, $C_{SG} = \partial Q_S/\partial V_G$, $C_{DG} = \partial Q_D/\partial V_G$, and $C_{BG} = \partial Q_B/\partial V_G$. The three capacitances associated to the gate are represented in Figures 2.10 and 2.11.

The capacitors between the source/drain and the substrate, C_{SB} and C_{DB}, are caused by the charge in the depletion region of the *pn* regions. They can be considered passive, but dependent on the reverse voltage across the junction. Note that the gate-source, gate-drain, and parasitic source/drain-bulk capacitance per unit gate width of pMOS and nMOS transistors, which are on the order of 1 fF/μm, 0.5 fF/μm, and 1.5 fF/μm, respectively, remain almost unchanged across technology nodes [9].

2.1.5 Scaling effects on MOS transistors

Typical values of the threshold voltages and transconductance characteristics are provided in Table 2.1 for the 0.25 μm, 0.18 μm, and 0.13 μm CMOS process. Note that the transconductance is defined as half of the product of the charge carrier mobility and oxide capacitance.

While resulting in the improvement of the circuit performance (area, speed, power dissipation), the reduction of the transistor size is affected by limitations associated with the thickness and electrical characteristics of the gate

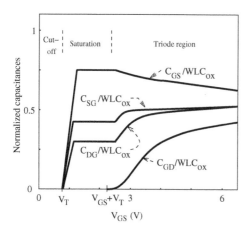

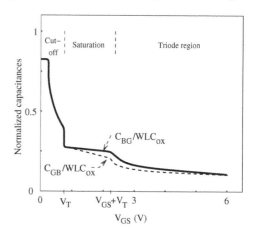

FIGURE 2.10

Plot of normalized capacitances (C_{DG}, C_{GD}, C_{GS}, and C_{SG}) versus the gate-source voltage. (Adapted from [7], ©1978 IEEE.)

FIGURE 2.11

Plot of normalized capacitances (C_{GB} and C_{BG}) versus the gate-source voltage. (Adapted from [7], ©1978 IEEE.)

TABLE 2.1

CMOS Process Characteristics

	0.25 µm $V_{DD} = 2.5$ V		0.18 µm $V_{DD} = 1.8$ V		0.13 µm $V_{DD} = 1.5$ V	
	nMOS	pMOS	nMOS	pMOS	nMOS	pMOS
V_T (V)	0.65	−0.51	0.49	−0.43	0.44	−0.42
K' (µA/V^2)	114.9	25.5	154.0	33.2	283.1	49.2

dielectric. With the constant-voltage scaling, the magnitude of the electric field in the channel can increase considerably, placing some limitations in the transistor miniaturization.

- Small-geometry effects

  When the effective channel length is on the order of the source/drain junction depletion width, the potential distribution in the channel becomes dependent on the lateral electric field component in addition to the normal one and features a two-dimensional representation. As a result, the threshold voltage is now dependent on the drain bias, channel length, and channel width. The threshold voltage is decreased as the channel length is reduced and also as the drain bias is raised. By reducing excessively the channel length, the depletion region of the drain junction can punch through the one of the source junction and the drain-source current ceases to be controlled by the gate-source voltage. The variation of the threshold voltage with the drain bias is caused by the drain-induced barrier lowering effect at the source junction.

- Hot carrier effects
  The lateral electric field increases quickly as the transistor is scaled down. During the displacement from the source to the drain, carriers can be accelerated by acquiring the energy related to the high electric field existing in the channel. They may collide with fixed atoms to generate additional electron/hole pairs. This process is repetitive and results in an abnormal increase in the drain current. The high-energy electrons may also be trapped at the silicon-oxide interface, and in turn cause the degradation of the device characteristics, such as the voltage threshold and transconductance. For the long-term operation, the transistor can exhibit reliability problems, due to the variation of the I/V characteristics.

- Gate-induced drain leakage current
  The gate-induced drain leakage current is observed in a transistor biased in the off-state. Due to the tunneling effect through the gate insulator, it is dependent on the drain-gate voltage and becomes important as the gate-dielectric thickness is decreased.

2.2 Transistor SPICE models

2.2.1 Electrical characteristics

The first transistor SPICE models are known as level 1, 2, and 3 [10]. They were developed based on the equations characterizing the physical behavior of MOS devices. (Note that SPICE stands for Simulation Program with Integrated Circuit Emphasis).

The accuracy of the level 1 model is adequate for transistors with channel length and width greater than 4 μm. The drain-source current, I_{DS}, in the cutoff, triode, and saturation regions is given by the next equation [11],

$$I_{DS} = \begin{cases} 0, & \text{for } V_{GS} \leq V_T \\ 2K\left((V_{GS} - V_T)V_{DS} - \dfrac{V_{DS}^2}{2}\right), & \text{for } V_{GS} > V_T \text{ and } V_{DS} < V_{GS} - V_T \\ K(V_{GS} - V_T)^2(1 + \lambda V_{DS}), & \text{for } V_{GS} > V_T \text{ and } V_{DS} \geq V_{GS} - V_T, \end{cases} \tag{2.44}$$

where

$$V_T = V_{T0} + \gamma(\sqrt{\psi_B + V_{SB}} - \sqrt{\psi_B}) \tag{2.45}$$

$$V_{T0} = V_{FB} + \psi_B + \gamma\sqrt{\psi_B} \tag{2.46}$$

and

$$\psi_B \simeq -2\psi_F \,. \tag{2.47}$$

The transconductance parameter K is defined as

$$\beta = 2K = \mu_n C_{ox} \frac{W}{L_{eff}} \,. \tag{2.48}$$

where L_{eff} is the effective channel length, and the saturation voltage is given by, $V_{DS(sat)} = V_{GS} - V_T$. The parameter γ represents the body-effect coefficient, ψ_B is the surface potential in the strong inversion for zero back-gate bias, and λ is the channel-length modulation coefficient. By introducing the parameter λ, the slight increase in the drain current in the saturation region is taken into account. However, the drain current in the triode region can be multiplied by the term $1 + \lambda V_{DS}$ to provide a continuous transition to the saturation current during the simulation. The drain-bulk and source-bulk currents, I_{DB} and I_{SB}, can be written as

$$I_{DB} = I_{D0} \left[\exp\left(\frac{qV_{DB}}{kT} \right) - 1 \right] \tag{2.49}$$

and

$$I_{SB} = I_{S0} \left[\exp\left(\frac{qV_{SB}}{kT} \right) - 1 \right], \tag{2.50}$$

where I_{D0} and I_{S0} are the saturation currents of the drain and source junctions, respectively.

A small-signal equivalent model of a MOS transistor [12] is shown in Figure 2.12. The gate-source and source-bulk (or simply bulk) transconductances,

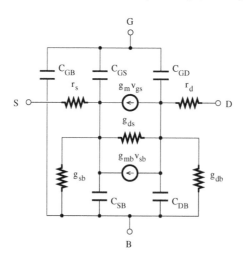

FIGURE 2.12
Small-signal equivalent model of a MOS transistor.

g_m and g_{mb}, and the drain-source conductance, g_{ds}, can be related to the I/V characteristic as follows:

$$g_m = \left.\frac{\partial I_{DS}}{\partial V_{GS}}\right|_{V_{DS}, V_{BS}} \tag{2.51}$$

$$g_{mb} = \left.\frac{\partial I_{DS}}{\partial V_{SB}}\right|_{V_{GS}, V_{DS}} \tag{2.52}$$

and

$$g_{ds} = \left.\frac{\partial I_{DS}}{\partial V_{DS}}\right|_{V_{GS}, V_{BS}}. \tag{2.53}$$

The five capacitances, C_{GD}, C_{GS}, C_{GB}, C_{DB}, and C_{SB}, are determined by the corresponding charges. The ohmic resistances, r_d and r_s, are not affected by the bias condition and are lower than the other transistor resistances. Due to the fact that the drain-bulk and source-bulk junctions are reverse-biased, the effect of the conductances $g_{db} = \partial I_{DB}/\partial V_{DB}$ and $g_{sb} = \partial I_{SB}/\partial V_{SB}$ on the device behavior is limited. It should be noted that additional substrate-coupling resistors and the gate resistor are to be included in the compact model of Figure 2.12 for an accurate description of the radio-frequency response.

Find the transconductances, g_m and g_{mb}, and the conductance, g_{ds}, for a transistor operating in the saturation region.

From the drain-source current in the saturation region

$$I_{DS} = K(V_{GS} - V_T)^2(1 + \lambda V_{DS}), \tag{2.54}$$

the threshold voltage

$$V_T = V_{T0} + \gamma(\sqrt{\psi_B + V_{SB}} - \sqrt{\psi_B}), \tag{2.55}$$

and $V_{BS} = -V_{SB}$, we can obtain the next relations

$$g_m = \left.\frac{\partial I_{DS}}{\partial V_{GS}}\right|_{V_{DS}, V_{BS}} = \frac{2I_D}{V_{GS} - V_T}, \tag{2.56}$$

$$g_{ds} = \left.\frac{\partial I_{DS}}{\partial V_{DS}}\right|_{V_{GS}, V_{BS}} = \lambda\frac{I_D}{1 + \lambda V_{GS}}, \tag{2.57}$$

and

$$g_{mb} = \left.\frac{\partial I_{DS}}{\partial V_{SB}}\right|_{V_{GS}, V_{DS}} = -g_m\frac{\partial V_T}{\partial V_{SB}} = -g_m\frac{\gamma}{\sqrt{2\psi_B - V_{BS}}}. \tag{2.58}$$

The level 1 model is adequate for initial design calculations and does not support the transistor operation in the subthreshold region.

The level 2 model includes the bulk-source voltage, V_{BS}, and the body-effect parameter, γ, in the I/V characteristics. By defining the ON voltage as, $V_{ON} = V_T + NkT/q$, the drain-source current in the subthreshold region can be written as

$$I_{DS} = I_{DS(ON)} \exp\left(q\frac{V_{GS} - V_{ON}}{NkT}\right), \tag{2.59}$$

where $N = 1 + qNFS/C_{ox} + C_d/C_{ox}$, NFS is an empirical parameter, and C_d is the depletion capacitance. During the normal operation of the transistor, the current I_{DS} takes the next form

$$I_{DS} = 2K_{eff}\left[(V_{GS} - V_{FB} - \psi_B - V_{DS}/2)V_{DS}\right.$$
$$\left. - \frac{2}{3}\gamma[(V_{DS} - V_{BS} + \psi_B)^{3/2} - (-V_{BS} + \psi_B)^{3/2}]\right], \tag{2.60}$$

where

$$K_{eff} = K\left(\frac{1}{1 - \lambda V_{DS}}\right)\left(\frac{E_{CRIT}\epsilon/C_{ox}}{V_{GS} - V_{ON}}\right)^{U_{EXP}}, \tag{2.61}$$

and

$$\gamma = \frac{\sqrt{2\epsilon q N_A}}{C_{ox}}. \tag{2.62}$$

Note that E_{CRIT} and U_{EXP} are empirical parameters, V_{ON} is the ON voltage, V_{FB} is the flat-band voltage, and N_A denotes the substrate doping concentration. The parameter $I_{DS(ON)}$ is the value of the above drain-source current at the subthreshold boundary, that is, for $V_{GS} = V_{ON}$. Here, the drain-source voltage in the saturation region becomes

$$V_{DS(sat)} = V_{GS} - V_{FB} - \psi_B + \frac{\gamma^2}{2}\left[1 - \sqrt{1 + 4\frac{(V_{GS} - V_{FB} - V_{BS})}{\gamma^2}}\right]. \tag{2.63}$$

In the level 2 representation, the slope of I_{DS} features a discontinuity between the subthreshold and strong inversion regions.

To improve the efficiency of the simulation results for transistors with channel length on the order of 1 μm, the level 3 model relies on empirical equations of the effective mobility and a correction factor, F_B, in order to take into account short-channel effects. The drain-source current is given by

$$I_{DS} = 2K\left[V_{GS} - V_T - \left(\frac{1 + F_B}{2}\right)V_{DS}\right]V_{DS}, \tag{2.64}$$

where

$$K = \mu_{eff}C_{ox}\frac{W_{eff}}{L_{eff}}. \tag{2.65}$$

W_{eff} and L_{eff} are the effective width and length, respectively. Let

$$V_C = \frac{V_{MAX} L_{eff}}{\mu_s}, \tag{2.66}$$

where μ_s is the surface mobility and V_{MAX} is used to steer the carrier velocity saturation. The expression of the effective mobility takes into account the degradation due to the lateral field and the carrier velocity saturation. Thus,

$$\mu_{eff} = \begin{cases} \dfrac{\mu_s}{1 + V_{DS}/V_C}, & \text{for} \quad V_{MAX} > 0 \\ \mu_s, & \text{otherwise} \end{cases} \tag{2.67}$$

and

$$\mu_s = \frac{U_0}{1 + \theta(V_{GS} - V_T)}, \tag{2.68}$$

where U_0 denotes the mobility constant value and θ is the mobility degradation coefficient.

By including the coefficients K_1, K_2, and η in the threshold voltage, which is now given by

$$V_T = V_{FB} + \psi_B + K_1\sqrt{\psi_B - V_{BS}} - K_2(\psi_B - V_{BS}) - \eta V_{DS} \tag{2.69}$$

the contributions due to the body, short-channel, narrow width, and drain-induced barrier lowering effects are considered.

The drain-source voltage in the saturation region, that is, where the displacement of carriers is governed by a constant mobility, is obtained as

$$V_{DS(sat)} = V_C + \frac{V_{GS} - V_T}{1 + F_B} - \sqrt{V_C^2 + \left(\frac{V_{GS} - V_T}{1 + F_B}\right)^2}. \tag{2.70}$$

The level 3 model exhibits some discontinuities (output conductance at $V_{DS} = V_{DS(sat)}$, transconductance at $V_{GS} = V_T$), which can result in the nonconvergence of simulations. The Berkeley short-channel IGFET models (BSIMs) were proposed to achieve an accurate description of transistors with submicrometer sizes [13] (IGFET stands for Insulated Gate Field Effect Transistor). They rely on empirical parameters obtained by data fitting to reduce the complexity of the equations of the device characteristics. In the BSIM4 transistor model, the continuity of the drain-source current and conductances is maintained using a single current equation for the different operating regions of the transistor.

2.2.2 Temperature effects

Due to the fact that a circuit can operate at a temperature different from the nominal one ($T_0 = 300\ K$) at which the model parameters are extracted, the temperature effects must be taken into account in the transistor representation. The mobility, μ, and the threshold voltage, V_T, are related to the absolute temperature, T, according to

$$\mu(T) = \mu(T_0) \left(\frac{T}{T_0}\right)^{3/2} \tag{2.71}$$

and

$$V_T(T) = V_T(T_0) + \text{VTC} \left(\frac{T}{T_0} - 1\right), \tag{2.72}$$

where VTC is a voltage temperature coefficient. The saturation current of the junction diodes (at the drain and source side) can be written as

$$J_s(T) = J_s(T_0) \exp\left\{\frac{1}{N}\left[\frac{q}{k}\left(\frac{E_g(T_0)}{T_0} - \frac{E_g(T)}{T}\right) + XT \cdot \ln\left(\frac{T}{T_0}\right)\right]\right\}, \tag{2.73}$$

where N and XT are two constant parameters, and the energy-band gap of the silicon, E_g, is given by

$$E_g(T) = 1.16 - \frac{7.02 \times 10^{-4} T^2}{T + 1108}. \tag{2.74}$$

The next equations can be used to express the temperature dependence of a resistor and capacitor:

$$R(T) = R(T_0) + \text{RTC}\left(\frac{T}{T_0} - 1\right) \tag{2.75}$$

and

$$C(T) = C(T_0) + \text{CTC}(T - T_0), \tag{2.76}$$

where RTC and CTC are the resistor and capacitor temperature coefficients, respectively.

2.2.3   Noise models

Different noise sources caused by fluctuations of the device characteristics affect the operation of MOS transistors. A simplified noise model of a transistor is depicted in Figure 2.13(a). It includes the flicker (or $1/f$ noise) and thermal channel noises, the spectrum density of which is represented in Figure 2.13(b). Note that $\overline{i_{1/f}^2} = g_m^2 \overline{v_{1/f}^2}$, where g_m is the transistor transconductance. The corner frequency, f_c, denotes the point at which both noise types intersect.

Specifically, the flicker noise, the spectral density of which is inversely proportional to the frequency, is given by an empirical equation of the form

$$\overline{i_d^2} = \frac{KF \cdot I_{DS}^{AF}}{C_{ox} L_{eff}^2} \frac{1}{f^{EF}}, \tag{2.77}$$

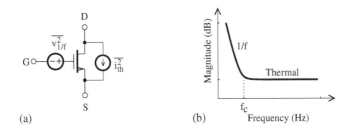

FIGURE 2.13
(a) Simplified noise model of a transistor; (b) plot of the noise spectrum density versus the frequency.

where KF is the flicker noise coefficient, I_{DS} is the drain-source current, AF is the flicker noise current exponent, C_{ox} is the oxide capacitance per unit area, L_{eff} is the effective length of the transistor, f is the operating frequency, and EF is the flicker noise frequency exponent. The other expressions are

$$\overline{i_d^2} = \frac{KF \cdot I_{DS}^{AF}}{C_{ox}W_{eff}L_{eff}}\frac{1}{f} \tag{2.78}$$

and

$$\overline{i_d^2} = \frac{KF \cdot g_m^2}{C_{ox}W_{eff}L_{eff}}\frac{1}{f^{EF}}, \tag{2.79}$$

where W_{eff} and g_m are the effective width and transconductance of the transistor, respectively. In the BSIM model used for submicrometer transistors, the flicker noise equation includes more parameters because it takes into account the fluctuations of the carrier number and surface mobility.

Let k represent the Boltzmann constant and T denote the absolute temperature. The drain channel thermal noise can be written as [14]

$$\frac{\overline{i_d^2}}{\triangle f} = 4kT\theta g_{do}, \tag{2.80}$$

where $\triangle f$ is the noise bandwidth, g_{do} is the channel conductance at $V_{DS} = 0$, i.e., $g_{do} = (\partial I_{DS}/\partial V_{DS})|_{V_{DS}=0}$, and θ is a bias-dependent noise coefficient, which, for long channel transistors, is unity at zero drain bias and about $2/3$ in the saturation region. However, the channel thermal noise can be accurately predicted for submicrometer transistors by the next model,

$$\frac{\overline{i_d^2}}{\triangle f} = \text{FP}\frac{4kT}{r_{ds} + \dfrac{L_{eff}^2}{\mu_{eff}|Q_{inv}|}}, \tag{2.81}$$

where r_{ds} is the drain-source resistance, μ_{eff} is the effective mobility, Q_{inv} is the inversion layer charge, and FP is a fitting parameter.

The induced gate current noise, $\overline{i_g^2}$, which is generally negligible, can become important for submicrometer devices operating at high frequencies or close to the transition frequency of the transistor. It is partially correlated with the drain current noise and is expressed as [15]

$$\frac{\overline{i_g^2}}{\triangle f} = 4kT\delta g_g(1 - |c|^2) + 4kT\delta g_g |c|^2 , \tag{2.82}$$

where c is a correlation coefficient given by

$$c = \frac{\overline{i_g i_d^*}}{\sqrt{\overline{i_g^2} \, \overline{i_d^2}}} \tag{2.83}$$

and δ is the gate noise coefficient. Because the conductance g_g increases with the square of the frequency, the power spectral density of the gate noise is different from that of a white noise.

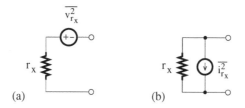

(a)   (b)

FIGURE 2.14
Representation of a resistor thermal noise: (a) voltage source, (b) current source.

Other noise sources include the thermal noise, which is related to various terminal resistances, and shot noise. For a resistor r_x, the thermal noise can be modeled either by a series voltage source with the power spectral density, $\overline{v_{r_x}^2}$, as shown in Figure 2.14(a), and given by

$$\frac{\overline{v_{r_x}^2}}{\triangle f} = 4kTr_x \tag{2.84}$$

or by a shunt current source with the power spectral density, $\overline{i_{r_x}^2}$, as illustrated in Figure 2.14(b), and of the form

$$\frac{\overline{i_{r_x}^2}}{\triangle f} = \frac{4kT}{r_x} , \tag{2.85}$$

where k is Boltzmann's constant, T is the absolute temperature, and $\triangle f$ denotes the measurement bandwidth in hertz (Hz).

Considering a resistor $r_x = 1$ kΩ, the power spectral densities of the thermal noise voltage and current at room temperature, (i.e., $T = 300$ K), are

$$\overline{v_{r_x}^2}/\triangle f \simeq 16 \times 10^{-18} \text{ V}^2/\text{Hz} \quad \text{and} \quad \overline{i_{r_x}^2}/\triangle f \simeq 16 \times 10^{-22} \text{ A}^2/\text{Hz}$$

or equivalently in terms of root-mean-square (rms) units,

$$\sqrt{\overline{v_{r_x}^2}/\triangle f} \simeq 4 \text{ nV}/\sqrt{Hz} \quad \text{and} \quad \sqrt{\overline{i_{r_x}^2}/\triangle f} \simeq 4 \text{ pA}/\sqrt{Hz},$$

respectively. Assuming a bandwidth of 1 kHz, we can obtain

$$\sqrt{\overline{v_{r_x}^2}} \simeq 4 \ \mu\text{V}_{rms} \quad \text{and} \quad \sqrt{\overline{i_{r_x}^2}} \simeq 4 \text{ nA}_{rms}.$$

Note that the contributions of independent noise sources should be added using mean squared quantities, instead of rms quantities.

The shot noise is related to the tunneling currents and can be written as

$$\frac{\overline{i_{sh}^2}}{\triangle f} = 2MqI, \tag{2.86}$$

where M is a multiplication factor, q is the electron charge, and I is the forward junction current. Note that the thermal noise is due to the random motion of charge carriers caused by an increase in the temperature, while the shot noise depends on the energy of carriers near a potential barrier or junction.

Using the transistor transconductance, g_m, the drain current noise, $\overline{i_d^2}$, can be related to the gate voltage noise, $\overline{v_{gs}^2}$, or the input referred noise. That is, $\overline{i_d^2} = g_m^2 \overline{v_{gs}^2}$. Note that the thermal noise generated by the transistor substrate is transmitted through the bulk transconductance, g_{mb}, as a noise current source at the drain. Its contribution to the input-referred voltage noise is $4kTr_b(g_{mb}/g_m)^2$, where r_b is the bulk resistance.

Typical values of the equivalent input voltage and current noises are generally expressed in pA/$\sqrt{\text{Hz}}$ and nV/$\sqrt{\text{Hz}}$ units, respectively.

2.3 Summary

Miniaturization of transistors has influenced almost all levels of the circuit design. While being subject to laws of physics, device technology development, and economic factors, the limits to this trend should also be application dependent due to power consumption issues. Accurate device models are essential for circuit design and analysis. They should exhibit a continuous and scalable electrical characteristics to meet the requirements of mixed-signal circuits based on submicrometer technologies. Recent modeling approaches rely on device physics and a suitable choice of empirical parameters to account for the different short-channel effects of transistors.

2.4 Circuit design assessment

1. **Capacitance due to the depletion charge**

   Consider a *pn* junction. The junction diffusion potential, ψ_0, is given by

   $$\psi_0 = \psi_n - \psi_p = \frac{kT}{q} \ln \frac{N_a N_d}{n_i^2}, \qquad (2.87)$$

   where $\psi_n = (kT/q) \ln(N_d/n_i)$ and $\psi_p = -(kT/q) \ln(N_a/n_i)$ are the potentials across the *n*-type and *p*-type regions, respectively; n_i is the intrinsic carrier concentration; and N_a and N_d are the acceptor and donor concentrations, respectively. The depletion width on the *n* and *p* regions, x_n and x_p, respectively, are related to the overall width of the depletion region, W_d, according to

   $$x_n = \frac{N_a}{N_a + N_d} W_d \qquad (2.88)$$

   and

   $$x_n = \frac{N_d}{N_a + N_d} W_d. \qquad (2.89)$$

   The Poisson equation can then be written as

   $$\frac{d^2 \phi(x)}{dx^2} = \begin{cases} \dfrac{qN_a}{\epsilon}, & \text{for} \quad -x_p \leq x < 0 \\[2mm] -\dfrac{qN_d}{\epsilon}, & \text{for} \quad 0 < x \leq x_n. \end{cases} \qquad (2.90)$$

   Show that the junction voltage is given by

   $$\phi(x) = \begin{cases} \dfrac{qN_a}{\epsilon} \left(\dfrac{x^2}{2} + x_p x + \dfrac{x_p^2}{2} \right), & \text{for} \quad -x_p \leq x < 0 \\[4mm] \dfrac{qN_d}{\epsilon} \left(-\dfrac{x^2}{2} + x_n x + \dfrac{x_n x_p}{2} \right), & \text{for} \quad 0 < x \leq x_n. \end{cases} \qquad (2.91)$$

   Based on the relation, $\psi_0 + V_r = \phi(x_n)$, determine x_n and x_p.

The depletion charge per unit area, Q_d, for a junction under reverse bias (i.e., the applied voltage V_r is negative) reads

$$Q_d = qN_d x_n = qN_a x_p = \sqrt{2q\epsilon \frac{N_a N_d}{N_a + N_d}(\psi_0 + V_r)}, \qquad (2.92)$$

where q is the electron charge and ϵ is the silicon dielectric.

Show that the junction capacitance can be obtained as

$$C_j = \frac{C_{j0}}{\sqrt{1 + \dfrac{V_r}{\psi_0}}}, \qquad (2.93)$$

where C_{j0} is the capacitance under the zero-bias condition to be determined.

Find the expression of C_{j0} in the case where $N_a \gg N_d$.

Use the following parameters, $C_{j0} = 0.47$ fF/μm^2, $\psi_0 = 0.65$ V, and $V_r = -3.3$ V to determine C_j.

2. **Analysis of common-source amplifier stages**
    Consider the inverting amplifier stages shown in Figure 2.15. Figure 2.16 shows the small-signal model of the transistor, where $g_m = \partial I_{DS}/\partial V_{GS}|_{V_{DS}}$ and $r_0 = \partial V_{DS}/\partial I_D$. Only the width, W, and length, L, of the transistors are assumed to be variable.

    Determine the small-signal voltage gain $A_v = v_0/v_i$.
    Compare the analytical analysis and SPICE simulations, and justify the dissimilarity between both results.

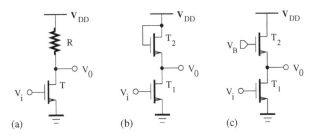

FIGURE 2.15
Circuit diagram of common-source amplifier stages.

3. **Analysis of source-degeneration amplifier stages**
    Repeat the previous exercise using the source-degeneration amplifier stages of Figure 2.17.

4. **Cascode amplifier**
    Using the transistor equivalent model of Figure 2.16, estimate the small-signal gain $G_m = v_0/i_i$ and the output resistance, r_{out}, of the cascode amplifier stage shown in Figure 2.18.

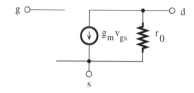

FIGURE 2.16
Small-signal model of the transistor.

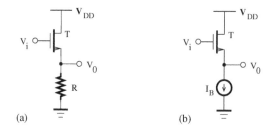

FIGURE 2.17
Circuit diagram of source-degeneration amplifier stages.

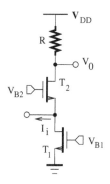

FIGURE 2.18
Circuit diagram of a cascode amplifier stage.

5. CMOS inverter

The circuit diagram of a CMOS inverter is depicted in Figure 2.19. When the input voltage is equal to V_{DD}, the transistor T_1 is on, while T_2 is off. On the other hand, the transistors T_1 and T_2 are, respectively off and on when the input voltage is equal to zero.

Determine the small-signal voltage gain $A_v = v_0/v_i$.
Let the charge stored on the capacitor C_L be

$$Q = C_L V_{DD} = I_0 T, \tag{2.94}$$

where I_0 is the output current and $T = 1/f$ is the signal period.

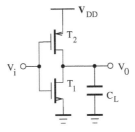

FIGURE 2.19
Circuit diagram of a CMOS inverter.

Show that

$$P = V_{DD}I_{DDQ} + C_L V_{DD}^2 f, \qquad (2.95)$$

where P is the power dissipated over a single signal cycle and I_{DDQ} denotes the quiescent leakage current, which flows through the transistor when it is off (V_{SB} or V_{DB} is different from zero).

6. **Transistor arrays**

Consider the transistor array shown in Figure 2.20(a). The transistors T_i ($i = 1, 2, 3$) are designed to have the same length but different widths. Verify that the structure of Figure 2.20(a) is equivalent to a single transistor (see Figure 2.20(b)) with the width equal to the sum of T_i widths and the same length as T_i.

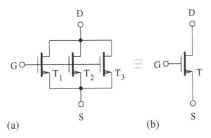

(a)   (b)

FIGURE 2.20
Circuit diagram of a transistor array.

For the circuit of Figure 2.21(a), the transistors T_i operate in the triode region and their drain-source voltage is fixed by the loop consisting of T_A, T_B, and the bias current I_B [16].

Find the current I_D and establish the equivalence between the structures of Figure 2.21.

Show that the transconductance of a transistor T_i is given by

$$g_{mi} = \frac{dI_{Di}}{dV_{GSi}} = K_i V_{DS}, \qquad (2.96)$$

where K_i is the transconductance parameter.

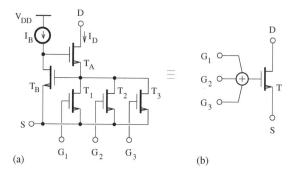

FIGURE 2.21

Circuit diagram of an array of weighted transistors.

7. **Simplified noise analysis of a single-stage amplifier**
    Verify that the output noise per unit bandwidth of the amplifier
    stage shown in Figure 2.15(a) is given by

    $$\overline{v_{n,o}^2} = (\overline{i_{th}^2} + \overline{i_{1/f}^2})R^2 + 4kTR, \qquad (2.97)$$

    where $\overline{i_{th}^2}$ and $\overline{i_{1/f}^2}$ denote the thermal and flicker noises of the tran-
    sistor, respectively. It should be assumed that the thermal noise of
    the resistor R is equivalent to a current source with the value $4kT/R$
    and $\overline{i_{1/f}^2} = g_m^2\overline{v_{1/f}^2}$, where g_m is the transistor transconductance.

Bibliography

[1] Semiconductor Industry Association (SIA), *International technology roadmap for semiconductor*, San Jose, CA: SIA, 1999.

[2] S. M. Sze, *Physics of Semiconductor Devices*, New York, NY: John Wiley & Sons, 1981.

[3] D. P. Foty, *MOSFET Modeling with SPICE: Principles and Practice*, Upper Saddle River, NJ: Prentice-Hall, 1996.

[4] R. J. Van Overstraeten, G. Declerck, and G. L. Broux, "Inadequacy of the classical theory of the MOS transistor operating in weak inversion," *IEEE Trans. on Electron Devices*, vol. 20, pp. 1150-1153, Dec. 1973.

[5] K. N. Ratnakumar and J. D. Meindl, "Short-channel MOST threshold voltage model," *IEEE J. of Solid-State Circuits*, vol. 17, pp. 937–948, Oct. 1982.

[6] T. Grotjohn and B. Hoefflinger, "A parametric short-channel MOS transistor model for subthreshold and strong inversion current," *IEEE Trans. on Electron Devices*, vol. 31, pp. 234–246, Feb. 1984.

[7] D. E. Ward and R. W. Dutton, "A charge-oriented model for MOS transistor capacitances," *IEEE J. of Solid-State Circuits*, vol. 17, pp. 703–708, Oct. 1978.

[8] B. J. Sheu, D. L. Scharfetter, C. Hu, and D. O. Pederson, "A compact IGFET charge model," *IEEE Trans. on Circuits and Syst.*, vol. 31, pp. 745–748, Aug. 1984.

[9] N.Weste and D. Harris, *CMOS VLSI Design*, Boston, MA: Addison-Wesley, 2005.

[10] T. Quarles, A. R. Newton, D. O. Pederson, and A. Sangiovanni-Vincentelli, *SPICE 3 version 3F5 user's manual*, University of California, Berkeley, CA, 1994.

[11] H. Shichman and D. A. Hodges, "Modeling and simulation of insulated field effect transistor switching circuits," *IEEE J. of Solid-State Circuits*, vol. 3, pp. 285–289, Sept. 1968.

37

[12] S. Liu and L. W. Nagel, "Small-signal MOSFET models for analog circuit design," *IEEE J. of Solid-State Circuits*, vol. 17, pp. 983–998, Oct. 1982.

[13] Y. Chen, M.-C. Jeng, Z. Liu, J. Huang, M. Chan, K. Chen, P. K. Ko, and C. Hu, "A physical and scalable *I-V* model in BSIM3v3 for analog/digital circuit simulation," *IEEE Trans. on Electron Dev.*, vol. 44, pp. 277–287, Feb. 1997.

[14] S. Tedja, J. Van der Spiegel, and H. H. Williams, "Analytical and experimental studies of thermal noise in MOSFET's," *IEEE Trans. on Electron Dev.*, vol. 41, pp. 2069–2075, Nov. 1994.

[15] A. van der Ziel, *Noise in Solid State Devices and Circuits*, New York, NY: John Wiley & Sons, 1986.

[16] Z. Czarnul, T. Iida, and K. Tsuji, "A low-voltage highly linear multiple weighted CMOS transconductor," *IEEE Trans. on Circuits and Systems*, vol. 42, pp. 362–364, May 1995.

3

Physical Design of MOS Integrated Circuits

CONTENTS

Modern lithography systems used in the integrated circuit (IC) fabrication employ an optical projection printing that operates almost at the diffraction limit. The mask of each IC layout layer is projected onto the wafer substrate, which has been coated with a photoresist material, also called resist. The structure of the resist is altered by the exposure to light so that, after the development, a silicon pattern can emerge. The next step can consist in implanting the dopant ions. Note that the small feature, which can be printed in this way, is about the wavelength of the light used.

Even with the great capabilities of lithography, it is impossible to produce completely identical devices over an entire wafer, or a much smaller die. As the size features are scaled down, the likelihood of process variations increases and the actual performance of a chip becomes more unpredictable. It is then necessary to take care of the process variation effects during the IC design phase, so that even worst-case fabrication conditions may provide a working circuit. The suitable design technique can consist of using worst-case (slow), nominal (typical), and best-case (fast) transistor models based on device measurements for SPICE simulations. However, to meet the demand for greater bandwidths and higher levels of integration, interconnects have to undergo a considerable reduction in size and currently represent an important limitation to the development of IC technology. To address this challenge, CAD and testing tools are required to link the interconnect characteristics to the circuit performance. Efficient interconnect models rely on simplifying assumptions, the validity of which is to be confirmed by on-chip measurements.

3.1 MOS Transistors

The structures of MOS transistors [1,2] are shown in Figure 3.1. The drain and source of the nMOS transistor (see Figure 3.1(a)) are realized by two heavily doped n-type semiconductor regions, which are implanted into a lightly doped p-type substrate or bulk. A thin layer of silicon dioxide (SiO_2) is thermally grown over the region between the source and drain, and is covered by the polycrystalline silicon (also shortly called, polysilicon or poly), which forms the gate of the transistor. The thickness t_{ox} of the oxide layer is on the order of a few angstroms. The length L and the width W of the gate are estimated along and perpendicularly to the drain-source path, respectively. Due to the lateral diffusion of the doped regions, the actual length is slightly smaller than L. The overlap region is symmetrical and is extended on the distance L_D on both sides. As a result, the effective length between the drain and source is given by, $L_{eff} = L_{drawn} - 2L_D$, where L_{drawn} is the dimension drawn in the layout. The width of the gate is also less than the drawn value due to a reduction in the active area by the field oxide growth. The substrate connection is

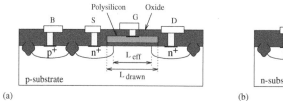 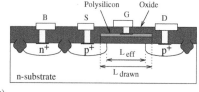

(a)   (b)

FIGURE 3.1
(a) nMOS transistor and (b) pMOS transistor structures.

provided by a doped p^+ region. Generally, the substrate is connected to the most negative supply voltage of the circuit so that the source-drain junction diodes are reverse-biased.

In the case of the pMOS transistor (see Figure 3.1(b)), the drain and source are formed by p^+ diffusions in the n-type substrate. A doped n^+ region is required for the realization of the substrate connection. Otherwise, the cross-sections for the two transistor types are similar. Here, the substrate is to be connected to the most positive supply voltage.

The layout of a transistor is shown in Figure 3.2. In addition to the active region representing the substrate, it is formed by the overlap of the active and polysilicon layers.

In practice, nMOS and pMOS transistors are fabricated on a wafer based on only one substrate, say of the p-type. In this case, the pMOS device is realized in an n-well as shown in Figure 3.3. Two transistor layouts are shown in Figures 3.4(a) and (b). The transistor can be fabricated in a well as is the

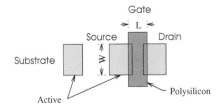

FIGURE 3.2
Transistor layout.

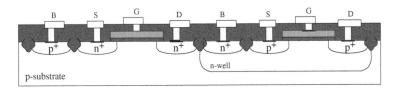

FIGURE 3.3
Transistor structure in a CMOS process with an *n*-well.

case in Figure 3.4(b). The active contact is a cut in the oxide, which allows the connection from the active and polysilicon regions to the first layer of metal.

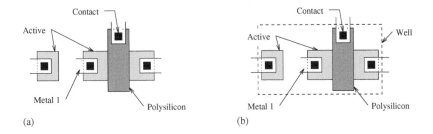

FIGURE 3.4
(a) Layout of a transistor realized on the substrate; (b) layout of a transistor fabricated in a well.

3.2 Passive components

The integrated-circuit (IC) performance is determined by the characteristics of basic components [1, 2] or elements, which can be fabricated in the MOS technology.

3.2.1 Capacitors

Let us consider a structure with a dominant parallel-plate capacitor having the length, L, width, W, and thickness, t_{ox}. The capacitance can be obtained as,

$$C = \frac{\epsilon_0 \epsilon_{ox} LW}{t_{ox}}, \tag{3.1}$$

where ϵ_0 is the vacuum dielectric constant and ϵ_{ox} represents the relative dielectric constant of the insulator (silicon dioxide). Accurate capacitances are the result of the reduction in the errors associated to the dielectric constants, also called oxide effects, and the variations due to geometrical parameters, known as edge effects.

A capacitor can be realized using the structures of Figure 3.5. The silicon dioxide dielectric layer can be deposited between a heavily doped crystalline silicon and polycrystalline silicon (see Figure 3.5(a)), two polycrystalline silicon layers (see Figure 3.5(b)), or polycrystalline silicon and metal layers (see Figure 3.5(c)). The capacitor value depends on the dielectric thickness, which

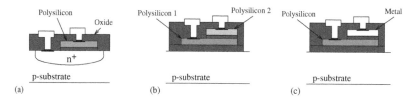

FIGURE 3.5
(a) Poly-diffusion capacitor; (b) poly-poly capacitor; (c) metal-poly capacitor.

can exhibit large variations in the case of the silicon dioxide. As a result, capacitance values can fluctuate by a few percent and the achievable device matching is limited. Specifically, the width of the depletion regions formed at the oxide contact-surface of the capacitor shown in Figure 3.5(a) depends on the applied voltage. As a result, the effective dielectric thickness is not constant and the capacitance is affected by the voltage variation and the bottom-plate parasitic capacitance. In the structures of Figure 3.5(b)-(c), the polysilicon region is isolated from the substrate by an oxide layer, which forms a parasitic capacitance to be included in the equivalent circuit model.

The layout of a capacitor based on two polysilicon layers is depicted in Figure 3.6. To minimize the contact resistance, contacts can be made everywhere it is possible on the polysilicon layer. It should be noted that a given capacitor consisting of two conducting planes separated by a thin insulator can occupy more area than the one implemented as a combination of several layers of conductors and insulators.

The transistor configuration of Figure 3.7(a) can also be used as a capacitor. The resulting capacitance, as shown in Figure 3.7(b) is a function of the voltage, $V = V_{GS}$, which should be chosen sufficiently greater than

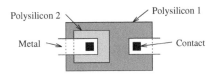

FIGURE 3.6
Layout of a poly-poly capacitor.

the threshold voltage, V_T, to take advantage of the linear capacitance-voltage characteristic of the transistor operating in the strong inversion region. The total capacitance, C, is given by the series connection of the gate-oxide capacitance, C_{ox}, and the depletion capacitance, C_d, between the channel induced under the oxide by increasing V_{GS}. That is,

$$C = \left(\frac{1}{C_{ox}} + \frac{1}{C_d} \right)^{-1}. \tag{3.2}$$

The capacitance C_d is a function of the gate-substrate voltage. Between the gate and drain/source, we simply have overlap capacitances. Note that a MOS capacitor is generally designed with a minimum length to reduce the influence of the channel resistance.

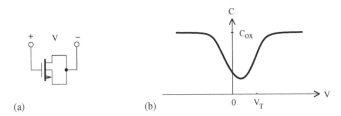

FIGURE 3.7
(a) A capacitor based on a MOS transistor and (b) its capacitance-voltage characteristic.

3.2.2 Resistors

The resistance of a uniformly doped structure of the length, L, width, W, and thickness, t, can be computed as

$$R = \frac{\rho}{t} \frac{L}{W} = R_\square \frac{L}{W}, \tag{3.3}$$

where ρ is the resistivity of the sample and $R_\square$ denotes the sheet resistance. The resistivity is determined by the type and concentration of impurity atoms. The temperature variation of the resistance can be modeled as

$$R = R(T_0[1 + (T - T_0)TC(R)], \tag{3.4}$$

where the first-order temperature coefficient is given by

$$TC(R) = \frac{1}{R}\frac{dR}{dT}. \tag{3.5}$$

The cross-section diagram of a polysilicon resistor is shown in Figure 3.8(a).

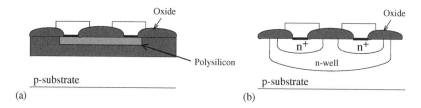

FIGURE 3.8

Structures of (a) polysilicon and (b) n-well resistors.

FIGURE 3.9

Layouts of (a) polysilicon and (b) n-well resistors.

The corresponding layouts are depicted in Figure 3.9. The resistor consists of the silicon dioxide thermally grown on a p-type substrate, undoped polysilicon settled by a low-pressure chemical-vapor deposition, and ohmic contact realized with aluminum electrodes. Experimental results show that the resistances vary depending on the polysilicon thickness. The resistor can feature a good linearity. However, for an accurate characterization of the high-frequency performance, the parallel-plate capacitance formed between the polysilicon and substrate must be taken into account.

The cross-section diagram of an n^+ diffusion resistor is shown in Figure 3.8(b). The diffusion layer is located within the n-well and isolated using the silicon dioxide. The sheet resistance of the device can be on the order of a kiloohm with a typical absolute value tolerance of a few percent.

3.2.3 Inductors

Although inductors are generally realized off-chip, there is an increasing interest in the integration all circuits elements in a single structure. To this end, MOS processes made up of lightly doped substrate and wells can be used for the inductor design. Due to the mutual magnetic coupling, the inductance value can be increased using a series connection of N wire segments in the

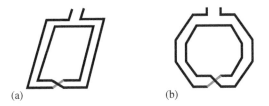

FIGURE 3.10
Layout of integrated inductors.

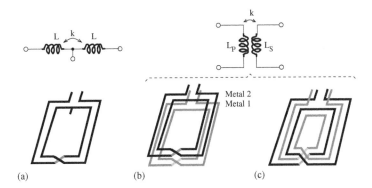

FIGURE 3.11
Layout of coupled inductors or transformers.

metal layer. This results in the inductance given by

$$L = \sum_{i=1}^{N} L_i + 2 \sum_{i=1}^{N} \sum_{i \neq j} M_{ij}, \tag{3.6}$$

where L_i is the inductance of the metal segment i, and M_{ij} is the mutual magnetic coupling between the segment i and j. The mutual inductance between two segments is determined by their length, separation distance, and intersection angle. The coupling between two perpendicular wires is negligible. However, the behavior of on-chip inductors is perturbed by dissipative mechanisms beyond the ones of a simple conductor loss. The planar spiral inductor has parasitic capacitors between the metal layer and substrate due to the insulating oxide. At high frequencies, a current can flow through these capacitors into the substrate, which can be modeled as RC networks. The achievable Q-factor is then about 5–20 for inductances in the range of a few nanohenries.

The inductors shown in Figure 3.10 consist of windings, which are fabricated with a metal layer patterned on the field oxide. The total self-inductance increases with the number of layers. Note that the difficulty in the realization of high-Q IC inductors is related to the creation of local insulating regions on the wafer for the device isolation.

A coupled inductor can be implemented using a center tapped inductor, as shown in Figure 3.11(a). The magnetic coupling coefficient is determined by the line width, spacing between conductors, and the substrate thickness.

A monolithic transformer can be designed using conductors overlaid as stacked metals or interwound in the same plane [3], as shown in Figures 3.11(b) and (c). The mutual inductance (and capacitance) is related to the common periphery between conductors. In practice, the achievable coupling coefficient can be as high as 0.9. Transformers using stacked conductors provide a slightly higher coupling coefficient, but their operating frequency can be limited by the lower self-resonance frequency associated with the large parallel-plate capacitance available between the windings.

3.3   Integrated-circuit (IC) interconnects

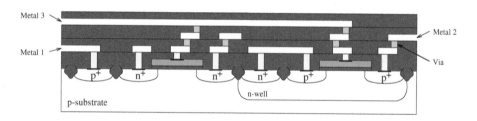

FIGURE 3.12
An IC cross-section with three metal layers.

The fabrication of an IC starts with the production of silicon wafers, which are then processed using various deposition and etching techniques, such as photolithography. The transistors are formed in wafer regions protected from oxidation and often referred to as moat regions. The area between adjacent transistors, which is generally covered with a thick-field oxide to prevent the formation of parasitic channel components, is called field region. Figure 3.12 shows the cross-section of an IC with three metal layers.

Unlike transistors, the interconnect performance is not improved through miniaturization. By scaling down the interconnect size, the crosstalk and latency effects are increased, while the inductance influence on the signal transients becomes dominant for large interconnect geometries. This is the case of power wires, which are connected to a large number of devices. Basically, a signal propagating on an interconnect is influenced by effects (delay, attenuation, reflection, and crosstalk) similar to the ones known in transmission lines. Depending on the IC process and signal operating frequency, the interconnects can be described by a lumped, or distributed (frequency independent/dependent parameters) model. All wires have a resistance, which is a function of the material resistivity. In the case of thin-film aluminum (Al) and copper (Cu), the resistivity is of 3.3 mΩ·cm and 2.2 mΩ·cm, respectively. Interconnections between metal layers, also called plugs or vias (see Figure 3.12), and made of tungsten (W) for aluminum wires, appear to be somewhat resistive (a few ohms). The current density in a metal wire is limited by the *electromigration* effect. At high current densities, the aluminum ions tend to migrate from one wire end, leaving voids, which can grow to a discontinuity after some time. At the other end, an accumulation of atoms in microscopic structures called hillocks can be observed. This failure is prevented by using via arrays for long wires and specifying design rules, which includes the minimum sizes of the wire to keep the current density less than 0.5-1 mA/μm. At the process level, alloying elements such as copper can be added to aluminum to avoid the ion displacement. Copper has also been introduced as a substitute to aluminum. In this case, the effect of electromigration is reduced because the IC process also permits vias made of copper.

The schematic of the equivalent model of a two-wire interconnect with the active line linked to a signal generator [4] is illustrated in Figure 3.13. To reduce the resistance, wires are generally designed to be taller than they

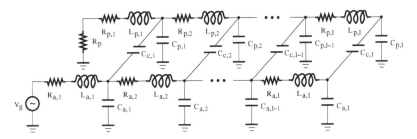

FIGURE 3.13
Equivalent circuit model of the interconnect.

are wide. Coupling noise can be important due to the high values of parasitic capacitance and inductance. The top and bottom capacitors are typically assumed to be grounded because they are associated with a set of orthogonally routed conductors that, averaged over the wire length, maintain a constant voltage. Due to the fact that a given wire is coupled to the neighbors by significant capacitances rather than ground, the signal integrity becomes dependent on the neighboring wire activity. At high frequencies, inductive effects become significant and the inductance can dominate the impedance of interconnect wires. Specifically, the coupling related to inductors is a nonlocal effect because the mutual inductance does not sink rapidly, as does the coupling capacitance. In fact, the inductance can be directly determined in cases where a time-varying current is supposed to flow in a known closed-loop and the flux of the resulting magnetic field is proportional to the loop area. But, the ambiguity in the determination of the current return path, which may include the ground, power lines, and neighboring wires, complicates the estimation of the interconnect inductance.

The computation of the interconnect resistance, inductance, and capacitance involves the solution of quasi-static electromagnetic systems. Due to the high amount of parasitic data to be extracted, CAD tools are required for high-density designs in a deep-submicrometer technology. The matching of impedances, insertion of repeaters (or inverters with large transistors), and the use of three-dimensional structures consisting of multiple levels of devices and layers can offer prospects for relief from the parasitic effects of interconnects.

3.4 Physical design considerations

Generally, the physical design involves steps of floor planning, timing optimization, placement, routing, and layout generation. The design process is increasingly interlaced with the verification, which consists of electrical rule checking (ERC), design rule checking (DRC), layout-versus-schematic (LVS) comparison, as well as resistance, inductance and capacitance parasitic extraction, and interconnect and signal integrity characterization.

Lay out the inverter shown in Figure 3.14(a) using the design

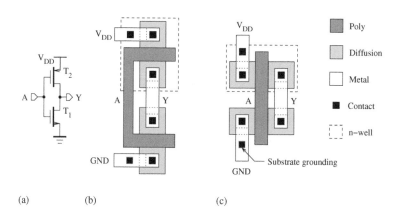

(a)   (b)   (c)

FIGURE 3.14
CMOS inverter: (a) Circuit diagram, (b) and (c) two possible layouts.

rules of a given CMOS process with n-well.

Two possible layouts can be derived, as illustrated in Figures 3.14(b) and (c). A metal layer is required between the n and p diffusion areas, as they should not be directly in contact. Note that contact openings can only be realized between metal and any other layer. To avoid the forward-biasing of pn junctions, the substrate and *n*-well are tied to the ground and the voltage V_{DD}, respectively.

ERC involves checking a circuit design for proper geometry and connectivity specifications. Examples of electrical rule violations include floating nodes, open circuits, improper supply voltage/ground and well/substrate connections, and limited power-carrying capacity on wires. The main objective of DRC is to provide a high overall die yield and reliability for a given design. Design rules specify certain geometric and connectivity restrictions (active-to-active spacing, metal-to-metal spacing, well-to-well spacing, minimum metal width, minimum channel length of transistors, ...) to ensure sufficient margins to account for common variations, and are specific to an integrated-circuit fabrication process. They can be based on the λ rule or micron rule. In the first case, all geometric specifications in a design are specified as integer multiples of the scalable parameter, λ, which is chosen such that the minimum gate length of a transistor is equal to 2λ, and micrometer units (or absolute measurements) are used in the second case. With the λ rule, the migration from one fabrication process to another is simplified. On the other hand, micron rules have the advantage of minimizing the resulting layout size. Computer-aided tools for DRC and ERC enable the verification of a circuit design prior to the fabrication. The ERC performs the syntax analysis on the circuit network, while the DRC verifies that the circuit layout does not have any specific errors.

Physical design results in a layout or geometric patterns, which are used for the IC manufacturing. Each fabrication step is based on a different layer mask extracted from the layout.

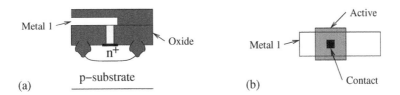

FIGURE 3.15
(a) Cross-section of an n^+ region and (b) the corresponding layout.

The fabrication of basic MOS structures requires several masking sequences, such as *n*-well, active, polysilicon, select, contact, metal, via, and overglass. The geometric size of features on every layer is defined by relevant design rules. Let us consider the n^+-doped region shown in Figure 3.15. It is realized by implanting ions into the substrate through the area described by the active mask. The minimum width of the active area is specified by the design rule. The active contact is shaped in the oxide to allow a connection between the first layer of metal and active region. Here, the minimum spacing between the active and contact area, vertical and horizontal sizes of the contact have to be defined. The metal, which is deposited after the oxide layer, is used as interconnect. It is subject to rules such as the metal-to-active contact minimum spacing and minimum width of the metal line.

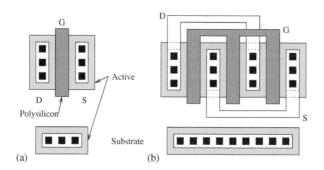

FIGURE 3.16
(a) Layout of a single transistor; (b) layout of a large transistor with multiple (three) fingers.

Different layout techniques are often adopted to minimize the parasitic effect and improve the device matching. As shown in the transistor layout of Figure 3.16(a), the overall contact resistance can be reduced by using as many contacts as the design rules allow. With N contacts, the resulting resistance is $1/N$ times the value of a single contact, as it is the case in a parallel configuration of resistors.

The design hierarchy permits the use of library cells to construct more complex circuits. The layout of a large transistor, which consists of a parallel connection of three minimum-size structures with the same length, is shown in Figure 3.16(b). The final transistor features the length of a single device and the sum of the individual widths. Note that by using a transistor with parallel fingers, the gate resistance can be reduced while the capacitance related to the source-to-drain areas increased. For this reason, excessively long geometries should preferably be avoided in the design of wide transistors.

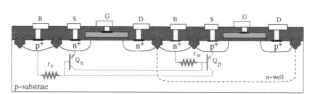

FIGURE 3.17
Parasitic resistors and bipolar transistors in a CMOS circuit.

In CMOS circuits in which a thyristor consisting of parasitic npn and pnp bipolar transistors, Q_n and Q_p, (see Figure 3.17) are formed, a *latch-up* effect can occur [5]. The parasitic circuit includes the substrate and well resistors,

r_s and r_w, and a positive feedback is formed around Q_n and Q_p. Due to transient noises, one of the two transistors can become forward biased and feeds the base of the other transistor. As a result, a current flows between the supply voltage lines and the circuit is unable to deliver a response to an input signal. Provided the feedback gain is greater than or equal to unity, this current will increase until the circuit burns out.

The latch-up effect can be mitigated by minimizing the current gains of the parasitic bipolar transistors, and the substrate and well resistances. This can be achieved by placing guard rings or n^+ and p^+ regions around MOS transistors. Suitable layout design rules, and appropriate selection of doping concentrations and profiles, can contribute to the reduction of latch-up susceptibility.

However, it should be noted that a substrate with a low resistivity can exhibit various parasitic paths between on-chip devices, thereby coupling the substrate noise to the signal of interest. The problem of *substrate coupling* is remarkable in mixed-signal circuits, where the analog components can be perturbed by the switching noise generated during the transitions of the clock signal used to control the switched devices. It is solved at the circuit level using a differential configuration, which is less sensitive to the common-mode noise. Another solution can consist of isolating sensitive circuit sections from the substrate noise.

Device matching can be enhanced by adopting layout techniques, which can reduce the statistical variations of the IC process. To this end, symmetry must be applied to the device layout as well as its nearby environment. In a set of transistors, for instance, a better matching is achieved by adopting the same orientation to place the transistors in the layout. This is due to the fact that the transconductances of MOS transistors depend on carrier mobilities, which are known to be sensitive to orientation-dependent stress.

Matched pairs of devices can be laid out with interdigitated or common-centroid structure [6]. Two layout examples of two capacitors X and Y are shown in Figure 3.18. In Figure 3.18(a), the construction is realized simply by cross-coupling the connection between the different cells. A common-centroid layout (see Figure 3.18(b)) features more symmetry and can be obtained by arranging elements as $XYYXXYYX$, or $XYXYYXYX$. It can also be obtained by using a two-dimensional configuration of the following form:

$$XYYX$$
$$YXXY$$
$$XYYX$$

As a result, the linear gradient in the oxide thickness for instance is equally distributed to the capacitor pairs and its effect is attenuated. Note that patterns of the form

$$XYXY$$

and

$$XYYXXY$$

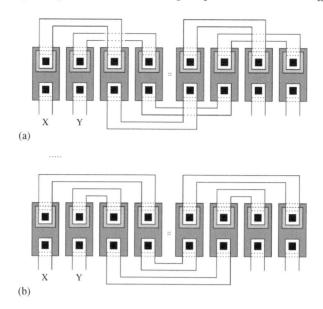

FIGURE 3.18
Poly-poly capacitor: (a) Interdigitated and (b) common-centroid layouts.

are often avoided because they exhibit a symmetry for each set of elements rather than for the whole pattern or are not uniformly distributed. The layout of Figure 3.19 can be used for the implementation of two capacitors C_1 and C_2. The unit cell located in the layout center is required for C_1, while C_2 is formed using eight unit capacitors. Dummy capacitors without an electrical role are added to similarly provide the same adjacent environment to both devices.

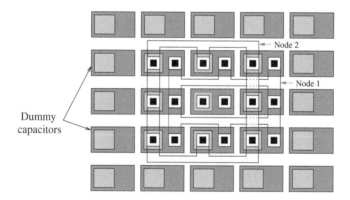

FIGURE 3.19
Layouts with dummy capacitors.

3.5 IC packaging

The IC package should provide a protection from external environments and facilitate the mounting of components in a test fixture without the risk of damage. In most demanding applications, multi-chip modules seem to be one of the few moderate-to-high lead count structures that can provide adequate electrical and thermal performance. The trend is toward thinner, smaller, and lighter packages with a high pin count. Advances in electronic packaging are based on improvements in the IC fabrication facilities, as well as in the deposited and laminated multi-chip processes. Generally, the interconnect non-ideal components together with the resistor, inductor, and capacitor of the package, pads, and bond wires form various parasitic RLC circuits that can add substantial noises to the signal of interest. The quality factor determines the amplitude and sharpness of the resonance. It can be tuned by steering the value of on-chip and off-chip passive devices. An accurate circuit model for packages is then indispensable for high-performance designs.

3.6 Summary

The design of high-density IC with deep-submicrometer process faces increasingly difficult challenges. Due to the scaling, transistors have become smaller and faster, while the chip power dissipation has increased. The size reduction of interconnects and packaging has led to an increase of parasitic effects. It is then necessary to adopt innovative approaches for the modeling, design, simulation, fabrication, and testing. Design technology, which includes software and hardware tools, and methodology, is thus the mean in approaching and realizing the limits imposed at the different levels of the IC development. Its quality determines the design time, performance, cost, and reliability of the final chip.

3.7 Circuit design assessment

1. **High-dynamic range MOS capacitor**
   The main disadvantage of capacitors implemented using a MOS transistor is the nonlinear capacitance-voltage characteristic due to different charge distributions in the accumulation, depletion, and inversion regions. Two gate-coupled transistors, T_1-T_2, can be config-

ured as shown in Figure3.20 to compensate the voltage dependence and provide a linear capacitor over a wide dynamic range [7,8].

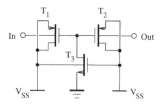

FIGURE 3.20
Circuit diagram of a high-dynamic range MOS capacitor.

Analyze the capacitance-voltage characteristic of the circuit using SPICE simulations.

2. **Inductor analysis**
Consider the equivalent circuit model of a MOS spiral inductor shown in Figure 3.21 [9]. Estimate the Q-factor defined as

$$Q = 2\pi \frac{E_m - E_e}{\triangle E},\tag{3.7}$$

where E_m and E_e are the peak magnetic and electric energies, respectively, and $\triangle E$ denotes the energy loss in one oscillation cycle. Verify that the maximum value of Q, $Q_{max} = \omega L / R_p$, is obtained at low frequencies.

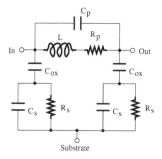

FIGURE 3.21
Equivalent circuit model of a MOS inductor.

3. **NAND gate and current mirror layouts**
• The circuit diagram and layout of a NAND gate are shown in Figures 3.22(a) and (b), respectively.

Verify the NAND gate layout and estimate the aspect ratios of transistors.

• Consider the circuit diagram and layout of a current mirror depicted in Figures 3.23(a) and (b), respectively.

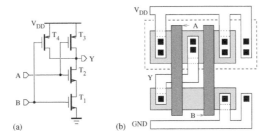

FIGURE 3.22
NAND gate: (a) Circuit diagram, (b) layout.

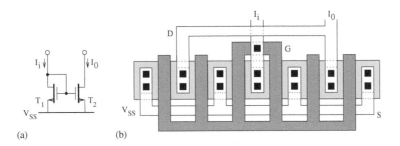

FIGURE 3.23
Current mirror: (a) Circuit diagram, (b) layout.

Determine the aspect ratios of the transistors from the layout.

Lay out the current mirror assuming that each transistor is now realized with a single gate structure.

Determine the errors introduced in the current ratio in each of the layouts.

4. **Differential amplifier layout**
To obtain a rectangular layout of a circuit, a transistor with the aspect ratio W/L can be realized as a single or a series stack structure.

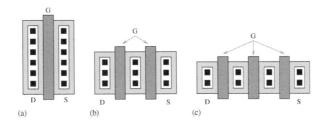

FIGURE 3.24
Layouts of a transistor and the equivalent series stack transistors.

Establish the equivalence between the characteristics (e.g., drain-

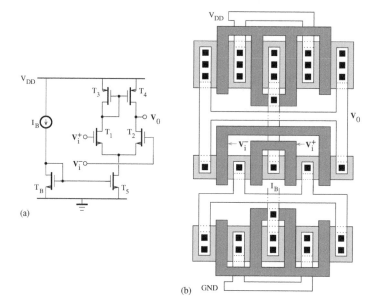

FIGURE 3.25
Differential amplifier layout.

substrate and source-substrate capacitances) of the transistor described by the layouts depicted in Figure 3.24, where the transistor width are (a) W, (b) $W/2$, and (c) $W/3$.

The differential amplifier of Figure 3.25(a) can be laid out as shown in Figure 3.25(b). The aspect ratios of transistors are as follows:

$$T_1\text{-}T_2 : W/L = 3 \quad T_3\text{-}T_4 : W/L = 9 \quad T_5\text{-}T_6 : W/L = 6$$

Assuming that the differential amplifier should be realized using a CMOS process with an n-well, complete the layout by placing all pMOS transistors in the same well to be tied to the voltage V_{DD}, and by connecting the substrate to the ground.

Use a layout tool to perform the design rule check.

Bibliography

[1] D. J. Allstot and W. C. Black, Jr., "Technological design considerations for monolithic MOS switched-capacitor filtering systems," *Proc. of the IEEE*, vol. 71, pp. 967–986, Aug. 1983.

[2] B. Razavi, *Design of Analog CMOS Integrated Circuits*, New York, NY: McGraw-Hill, 2001.

[3] J. R. Long, "Monolithic transformers for silicon RF IC design," *IEEE J. of Solid-State Circuits*, vol. 35, pp. 1368–1382, Sept. 2000.

[4] Y. Ismail, E. Friedman, and J. Neves, "Equivalent Elmore delay for RLC trees," *IEEE Trans. on Computer-Aided Design*, vol. 19, pp. 83–97, Jan. 2000.

[5] T. Cabara, "Reduced ground bounce and improved latch-up suppression through substrate conduction," *IEEE J. of Solid-State Circuits*, vol. 23, pp. 1224–1232, Oct. 1988.

[6] R. Jacob Baker, Harry W. Li, David E. Boyce, *CMOS Circuit Design, Layout, and Simulation*, Piscataway, NJ: IEEE Press, 1998.

[7] P. Larsson, "A 2-1600-MHz CMOS clock recovery PLL with low-V_{dd} capability," *IEEE J. of Solid-State Circuits*, vol. 34, pp. 1951–1960, Dec. 1999.

[8] T. Tille, J. Sauerbrey, and D. Schmitt-Landsiedel, "A 1.8-V MOSFET-only $\Sigma\Delta$ modulator using substrate biased depletion-mode MOS capacitors in series compensation," *IEEE J. of Solid-State Circuits*, vol. 36, pp. 1041–1047, Jul. 2001.

[9] T. H. Lee and S. S. Wong, "CMOS RF integrated circuits at 5 GHz and beyond," *Proc. of the IEEE*, vol. 88, pp. 1560–1571, Oct. 2000.

4

Bias and Current Reference Circuits

CONTENTS

As the power supply is scaled down, the operation of analog circuits becomes difficult because the threshold voltage of MOS transistors and dc operating points are not scaling proportionally. The bias and current reference circuits should be designed to feature a good stability over the IC process, and supply voltage and temperature variations. A review of different techniques used to achieve this goal is then necessary.

Current mirrors should desirably feature a high output resistance, which is required for an accurate replication of currents, and high output swing, which is essential for a low-voltage operation. There are many current mirror architectures, each with its advantages and inconveniences.

A voltage reference is commonly used in IC design for providing a constant voltage in spite of the IC process, supply voltage, and temperature variations. Various voltage reference structures are available with varying degrees of initial accuracy and drift over the temperature range of operation. Some voltage reference circuits provide an output voltage that is determined by a resistor or transistor. However, the resulting reference voltage is limited by IC process deviations. Band-gap voltage reference circuits generate an output voltage that is less sensitive to variations of the operating temperature. But often, external voltage references, which exhibit a much higher accuracy and lower drift than on-chip voltage references, are required in high-precision applications.

4.1 Current mirrors

Current mirrors find applications in analog ICs as biasing elements, which can set transistor bias levels such that the circuit characteristics are less affected by power supply and temperature variations. They are also used in the amplifier design as load devices, whose high impedance is exploited to increase the resulting gain. The operation of current mirrors relies on the principle that a reference current in the input stage is either sinked or sourced in such a way as to be reproduced with a given scaling factor in the output stage. Generally, the generated output current or bias current is used to drive a given load.

4.1.1 Simple current mirror

The circuit diagram of a simple current mirror is shown in Figure 4.1(a). Because the transistor T_1 is diode-connected, it operates in the saturation region, and T_2 is also assumed to operate in the saturation region.

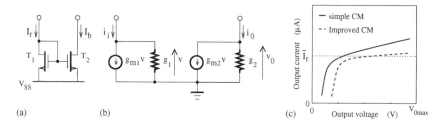

(a) (b) (c)

FIGURE 4.1
(a) Simple current mirror (CM); (b) small-signal equivalent model; (c) output characteristics of a simple CM and an improved CM.

The minimum voltage at the output node of the current mirror is

$$V_{bmin} = V_{DS(sat)} = V_{GS} - V_T,\qquad(4.1)$$

where $V_{DS(sat)}$ is the drain-source saturation voltage. The drain currents I_{D1} and I_{D2} are given by

$$I_{D_1} = K_1' \left(\frac{W_1}{L_1}\right)(V_{GS_1} - V_T)^2(1 + \lambda V_{DS_1})\qquad(4.2)$$

$$I_{D_2} = K_2' \left(\frac{W_2}{L_2}\right)(V_{GS_2} - V_T)^2(1 + \lambda V_{DS_2}),\qquad(4.3)$$

where $K_1' = K_2' = \mu C_{ox}/2$ is the transconductance parameter, μ denotes the electron mobility, C_{ox} is the gate oxide capacitance per unit area, and λ is the channel-length modulation coefficient. With $V_{GS_1} = V_{GS_2} = V_{GS}$, the current ratio can be obtained as

$$\frac{I_{D_2}}{I_{D_1}} = \frac{W_2/L_2}{W_1/L_1} \cdot \frac{1 + \lambda V_{DS_2}}{1 + \lambda V_{DS_1}}. \tag{4.4}$$

Neglecting the effect due to the channel-length modulation and assuming that $I_{D_1} = I_r$ and $I_{D_2} = I_b$, we can write

$$\frac{I_{D_2}}{I_{D_1}} = \frac{I_b}{I_r} \simeq \frac{W_2/L_2}{W_1/L_1}. \tag{4.5}$$

With reference to the small-signal equivalent model of Figure 4.1(b), where $i_i = 0$ and the transistor body transconductances are assumed to be negligible, it can be deduced that $v = 0$ and the output resistance is obtained as

$$r_0 = \frac{v_0}{i_0} = \frac{1}{g_2} = r_{DS_2} = \frac{1 + \lambda V_{DS_2}}{\lambda I_{D_2}}. \tag{4.6}$$

For typical values of the transistor parameters, the resistance r_0, which is about a few hundreds kiloohms, can appear to be too small in high-accuracy applications. Figure 4.1(c) shows the output characteristics of current mirrors. The effect of the higher output resistance exhibited by an improved current mirror can be clearly observed in these plots.

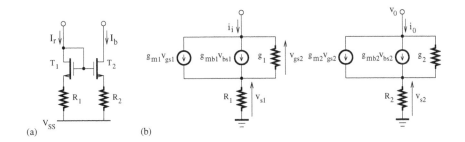

(a)   (b)

FIGURE 4.2
(a) Current mirror with source degeneration resistors and (b) its small-signal equivalent model.

A simple approach to increase the current mirror resistance can consist of using source degeneration resistors, as illustrated in Figure 4.2. Both transistors operate in the saturation region and it can be shown that

$$R_2 I_b + V_{GS_2} = R_1 I_r + V_{GS_1}, \tag{4.7}$$

where $V_{GS_1} = V_{DS_1(sat)} + V_{T_1}$ and $V_{GS_2} = V_{DS_2(sat)} + V_{T_2}$. Because the transistor threshold voltages are matched and $V_{DS_1(sat)} = \sqrt{I_b/[K_2'(W_2/L_2)]}$, we have

$$R_2 I_b + \sqrt{\frac{I_b}{K_2'(W_2/L_2)}} - (R_1 I_r + V_{DS_1(sat)}) = 0. \tag{4.8}$$

The square of the only positive root of this quadratic equation is given by

$$I_b = \frac{1}{4R_2^2} \left(-\sqrt{\frac{1}{K_2'(W_2/L_2)}} + \sqrt{\frac{1}{K_2'(W_2/L_2)} + 4R_2 \left(R_1 I_r + V_{DS_1(sat)} \right)} \right)^2,$$

(4.9)

where $V_{DS_2(sat)} = \sqrt{I_r/[K_1'(W_1/L_1)]}$. In the specific case of a Widlar MOS current mirror, $R_1 = 0$ and the source of T_1 is directly connected either to a supply voltage terminal or ground.

With reference to the small-signal equivalent model of Figure 4.2(b), we have

$$v_0 = [i_0 - (g_{m2}v_{gs2} + g_{mb2}v_{bs2})]/g_2 + v_{s2},$$

(4.10)

where $v_{s2} = R_2 i_0$. For the determination of the output resistance, the voltage at the input node is set to zero, that is, $v_{d1} = v_{g1} = v_{g2} = 0$. As a result, $v_{gs2} = -v_{s2}$. With $v_{bs2} \simeq -v_{s2}$, the output resistance can be computed as

$$r_0 = \frac{v_0}{i_0} = \frac{1}{g_2} + R_2 \left(1 + \frac{g_{m2} + g_{mb2}}{g_2} \right).$$

(4.11)

Hence, the use of source degeneration resistors leads to an increase in the current mirror output resistance. By reducing the sensitivity of the output current to variations of the voltage at the current mirror output, a high output resistance helps provide an accurate replication of currents.

4.1.2 Cascode current mirror

In order to obtain a high resistance, current mirrors can be designed using cascode transistor structures. However, an adequate transistor biasing is required to achieve a high output swing, which is desired for the low-voltage operation.

The cascode current mirror shown in Figure 4.3(a) has the advantage of reducing the effect of the channel-length modulation and increasing the output resistance. It is assumed that the effects due to the bulk of the transistors T_1 and T_2 are negligible, and all transistors are matched. To ensure the operation of all the transistors in the saturation region, the drain-source voltages of the diode-connected transistors, T_1 and T_3, should be at least $V_{DS_1} = V_{DS_3} = V_{DS(sat)} + V_T$, yielding the minimum voltages of the value

$$V_{rmin} = V_{DS_1} + V_{DS_3} = V_{G_2} = 2(V_{DS(sat)} + V_T)$$

(4.12)

and

$$V_{bmin} = V_{G_2} - V_T = 2(V_{DS(sat)} + V_T) - V_T = 2V_{DS(sat)} + V_T,$$

(4.13)

respectively, at the input and output nodes. Note that V_{bmin} represents the

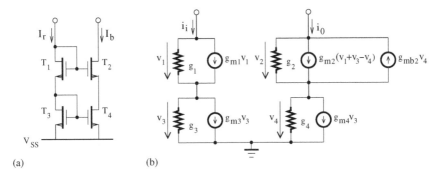

FIGURE 4.3
(a) Cascode current mirror and (b) its small-signal equivalent model.

sum of two overdrive voltages and one threshold voltage. The drain-source voltages of the transistors T_4 and T_2 are then given by

$$V_{DS_4} = V_{G_2} - V_{GS_2}$$
$$= 2(V_{DS(sat)} + V_T) - (V_{DS(sat)} + V_T) = V_{DS(sat)} + V_T \tag{4.14}$$

and

$$V_{DS_2} = V_{bmin} - V_{DS_4} = V_{DS(sat)}. \tag{4.15}$$

In general, the relation $V_{GS_1} + V_{DS_3} = V_{GS_2} + V_{DS_4}$ holds for the circuit of Figure 4.3(a). Thus, $V_{GS_1} = V_{GS_2}$ and $V_{DS_3} = V_{DS_4}$ provided $(W_2/L_2)(W_3/L_3) = (W_1/L_1)(W_4/L_4)$. In this case, the current I_b has a value close to the one of I_r.

The output resistance of the cascode current mirror can be derived using the small-signal equivalent circuit shown in Figure 4.3(b). Then, we can write

$$v_2 = [i_0 + g_{mb2}v_4 - g_{m2}(v_1 + v_3 - v_4)]/g_2 \tag{4.16}$$
$$v_4 = (i_0 - g_{m4}v_3)/g_4. \tag{4.17}$$

Assuming that $i_r = 0$, the values of the voltage v_1 and v_3 are reduced to zero and the output resistance is given by

$$r_0 = \frac{v_0}{i_0} = \frac{1}{g_2} + \frac{1}{g_4} + \frac{g_{m2} + g_{mb2}}{g_2 g_4}, \tag{4.18}$$

where $v_0 = v_2 + v_4$. Note that $v_{sb2} \neq 0$ in contrast to the source-substrate voltage of the other transistors. The body effect due to the g_{mb2} contribution tends to increase the output resistance.

In the cascode current mirror shown in Figure 4.4(a), the output current will follow I_r if V_B is chosen such that $V_{G2} = V_{D2}$. Applying Kirchhoff's current and voltage laws to the small-signal equivalent model of Figure 4.4(b),

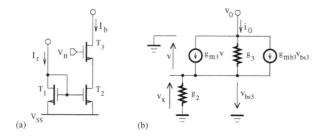

FIGURE 4.4

(a) Cascode current mirror with a single input transistor and (b) its small-signal equivalent model.

we obtain

$$i_0 = g_{m3}v + g_{mb3}v_{bs3} + g_3(v_0 - v_x) \qquad (4.19)$$

$$v_x = i_0/g_2 \qquad (4.20)$$

$$v_{bs3} = v = -v_x, \qquad (4.21)$$

where g_{mb3} denotes the transconductance due to the body of the transistor T_3. The output resistance r_0 is then given by

$$r_0 = \frac{v_0}{i_0} = \frac{1}{g_2} + \frac{1}{g_3} + \frac{g_{m3} + g_{mb3}}{g_2 g_3}. \qquad (4.22)$$

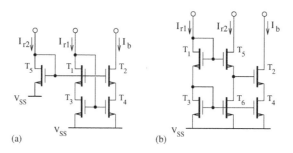

FIGURE 4.5

(a) High-swing cascode current mirror; (b) high-swing cascode current mirror using a source follower output stage.

A high-swing cascode current mirror is illustrated in Figure 4.5(a), where $I_{r1} = I_{r2} = I_r$. All transistors are assumed to operate in the saturation region where the $I - V$ characteristic is reduced to a square law for $V_{DS} \geq V_{DS(sat)} = V_{GS} - V_T$. The currents flowing through transistors T_5, T_4, and T_2 can be respectively expressed as

$$I_r = K_5'\left(\frac{W_5}{L_5}\right)(V_{GS_5} - V_{T_5})^2 \quad \text{or} \quad V_{GS_5} = \sqrt{\frac{I_r}{K_5'(W_5/L_5)}} + V_{T_5}, \quad (4.23)$$

$$I_r = K_4' \left(\frac{W_4}{L_4}\right) (V_{GS_4} - V_{T_4})^2 \quad \text{or} \quad V_{GS_4} = \sqrt{\frac{I_r}{K_4'(W_4/L_4)}} + V_{T_2}, \quad (4.24)$$

and

$$I_r = K_2' \left(\frac{W_2}{L_2}\right) (V_{GS_2} - V_{T_2})^2 \quad \text{or} \quad V_{GS_2} = \sqrt{\frac{I_r}{K_2'(W_2/L_2)}} + V_{T_2}. \quad (4.25)$$

With the assumption that the size of transistors T_1 and T_3 is identical to the one of T_2 and T_4, the current I_b should be a replica of I_r. The drain-source voltage of T_4 is given by

$$V_{DS_4} = V_{GS_5} - V_{GS_2} \qquad (4.26)$$
$$= V_{DS_4(sat)} = V_{GS_4} - V_{T_4}. \qquad (4.27)$$

Upon substitution of V_{GS_5}, V_{GS_2}, and V_{GS_4} into Equations (4.26) and (4.27), we obtain

$$\left(\sqrt{\frac{I_r}{K_5'(W_5/L_5)}} + V_{T_5}\right) - \left(\sqrt{\frac{I_r}{K_2'(W_2/L_2)}} + V_{T_2}\right) = \sqrt{\frac{I_r}{K_4'(W_4/L_4)}} \quad (4.28)$$

or, equivalently,

$$\sqrt{\frac{1}{K_4'(W_4/L_4)}} + \sqrt{\frac{1}{K_2'(W_2/L_2)}} - \sqrt{\frac{1}{K_5'(W_5/L_5)}} = \frac{V_{T_5} - V_{T_2}}{\sqrt{I_r}}. \quad (4.29)$$

Assuming that transistors are designed with identical parameters K_i' and V_{T_i} ($i = 1, 2, 3, 4, 5$), the requirement, $V_{DS_4} = V_{DS_4(sat)}$, is met provided

$$\sqrt{\frac{1}{(W_5/L_5)}} = \sqrt{\frac{1}{(W_4/L_4)}} + \sqrt{\frac{1}{(W_2/L_2)}}. \quad (4.30)$$

A proper operation of the current mirror then relies on the insensitivity of the achievable transistor matching to the effects of process variations. Furthermore, the use of long channel transistors whose substrates are connected to the corresponding sources may be necessary to reduce current variations related to the effects of the channel length modulation.

In the particular cases where the transistors $T_1 - T_4$ are designed with the same W/L ratio while the ratio of T_5 is $W/(4L)$, the minimum voltages needed at the input and output nodes can be expressed as

$$V_{r1min} = V_{GS_3} = V_{GS_4} = V_{GS_2} = \sqrt{\frac{I_r}{K'(W/L)}} + V_T \qquad (4.31)$$
$$= V_{DS(sat)} + V_T$$

and

$$V_{r2min} = V_{GS_5} = \sqrt{\frac{I_r}{K'(W/4L)}} + V_T = 2V_{DS(sat)} + V_T, \qquad (4.32)$$

where

$$V_{bmin} = V_{DS_4} + V_{DS_2} = (V_{GS_4} - V_T) + (V_{GS_2} - V_T) = 2V_{DS(sat)} \qquad (4.33)$$

and $K' = \mu C_{ox}/2$. Here, the minimum output voltage or compliance voltage is reduced to the sum of two overdrive voltages.

Consider the high-swing cascode current mirror using a source follower output stage, as shown in Figure 4.5(b). It is assumed that $I_{r1} = I_{r2} = I_b$ and all transistors operate in the saturation region. The use of Kirchhoff's voltage law can yield

$$V_{DS_4} = V_{GS_3} + V_{GS_1} - V_{GS_5} - V_{GS_2} \qquad (4.34)$$

and

$$V_{GS_3} = V_{GS_6} = V_{GS_4} . \qquad (4.35)$$

Because $V_{DS_4} = V_{DS_4(sat)} = V_{GS_4} - V_{T_4}$ and $V_{GS_4} = V_{GS_3}$, Equation (4.34) becomes

$$-V_{T_4} = V_{GS_1} - V_{GS_5} - V_{GS_2} . \qquad (4.36)$$

Based on the simplified $I - V$ characteristic for MOS transistors operating in the saturation region, we obtain

$$I_r = K_5' \left(\frac{W_5}{L_5}\right) (V_{GS_5} - V_{T_5})^2 \quad \text{or} \quad V_{GS_5} = \sqrt{\frac{I_r}{K_5'(W_5/L_5)}} + V_{T_5}, \qquad (4.37)$$

$$I_r = K_2' \left(\frac{W_2}{L_2}\right) (V_{GS_2} - V_{T_2})^2 \quad \text{or} \quad V_{GS_2} = \sqrt{\frac{I_r}{K_2'(W_2/L_2)}} + V_{T_2}, \qquad (4.38)$$

and

$$I_r = K_1' \left(\frac{W_1}{L_1}\right) (V_{GS_1} - V_{T_1})^2 \quad \text{or} \quad V_{GS_1} = \sqrt{\frac{I_r}{K_1'(W_1/L_1)}} + V_{T_1}. \qquad (4.39)$$

Upon substitution of V_{GS_5}, V_{GS_2}, and V_{GS_1} into Equation (4.36), we arrive at

$$\sqrt{\frac{1}{K_4'(W_4/L_4)}} - \sqrt{\frac{1}{K_2'(W_2/L_2)}} - \sqrt{\frac{1}{K_5'(W_5/L_5)}} = \frac{V_{T_2} + V_{T_5} - (V_{T_1} + V_{T_4})}{\sqrt{I_r}}.$$
$$(4.40)$$

In the case where the parameters K_i' and V_{T_i} ($i = 1, 2, 3, 4, 5, 6$) are identical for all transistors, it can be deduced that

$$\sqrt{\frac{1}{K_4'(W_4/L_4)}} = \sqrt{\frac{1}{K_2'(W_2/L_2)}} + \sqrt{\frac{1}{K_5'(W_5/L_5)}}. \qquad (4.41)$$

The choice of the same W/L ratio for transistors $T_2 - T_6$ results in a ratio of $W/(4L)$ for T_1. Hence, the gate-source voltages for transistors T_5, T_3, T_2, and T_1 are given by

$$V_{GS_3} = V_{GS_5} = V_{GS_2} = \sqrt{\frac{I_r}{K'(W/L)}} + V_T = V_{DS(sat)} + V_T \qquad (4.42)$$

$$V_{GS_1} = \sqrt{\frac{I_r}{K'(W/(4L))}} + V_T = 2V_{DS(sat)} + V_T , \qquad (4.43)$$

where $K' = \mu C_{ox}/2$. Because the transistors T_6 and T_2 are biased at the boundary of the saturation region, we have

$$V_{DS_6} = V_{DS_2} = V_{DS(sat)} . \qquad (4.44)$$

The minimum voltages required at the input and output nodes can then be obtained as

$$V_{r1min} = V_{GS_1} + V_{GS_3} = 3V_{DS(sat)} + 2V_T , \qquad (4.45)$$

$$V_{r2min} = V_{DS_5} + V_{DS_6} = (V_{r1min} - V_{GS_5}) + V_{DS_6} = 3V_{DS(sat)} + V_T , \qquad (4.46)$$

and

$$V_{bmin} = V_{DS_4} + V_{DS_2} = (V_{DS_6} - V_{GS_2}) + V_{DS_2} = 2V_{DS(sat)} . \qquad (4.47)$$

Although the aforementioned cascode current mirror requires a minimum output voltage of only two overdrive voltages, its accuracy can be limited by the difference in the body effects of transistors T_2 and T_5, and the discrepancy between the drain-source voltages of transistors T_4 and T_6.

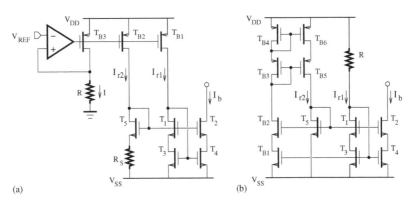

(a) (b)

FIGURE 4.6
(a) High-swing cascode current mirror with voltage-controlled reference currents; (b) high-swing cascode current mirror with a reference current duplicated by a complementary circuit.

In general, variations in the IC process and supply voltage can lead to changes in the saturation voltage margin so that the output resistance and swing of the current mirror are reduced. The use of an adequate bias circuit may then be required to maintain the cascode connected transistors in the saturation region regardless of changes in the current and voltage levels due to nonideal effects.

The circuit diagram of a high-swing cascode current mirror with a bias circuit driven by a reference voltage is shown in Figure 4.6(a). An accurate control of the bias voltage applied between the drain and source of the transistor T_3 is achieved by inserting the resistor R_S in series with the source of T_5 [1]. With the assumption that the operational amplifier is ideal, the reference current I is of the form $I = V_{REF}/R$. For transistors T_{B3}, T_{B2}, and T_{B1} designed with identical threshold voltage and process transconductance parameters, and operating with the same biasing condition, the drain current is proportional to the ratio of the width to the length, and we have

$$\frac{I}{W_{B3}/L_{B3}} = \frac{I_{r2}}{W_{B2}/L_{B2}} = \frac{I_{r1}}{W_{B1}/L_{B1}}. \tag{4.48}$$

Because $V_{GS_5} = V_{GS_1}$, the drain-source voltage of T_3 is given by

$$V_{DS_3} = V_{S_5} = R_S I_{r2} = \frac{(W_{B2}/L_{B2})}{(W_{B3}/L_{B3})} \frac{R_S}{R} V_{REF}. \tag{4.49}$$

In order to reduce the dependence of the current mirror performance to variations in the IC process, supply voltage, and temperature, the voltage V_{REF} should be generated by a bandgap reference source and the resistor ratio should be accurately matched.

Another cascode bias circuit is illustrated in Figure 4.6(b) [2]. The transistors T_{B1} and T_{B2} of the bias circuit dynamically detect the actual gate voltages of the current mirror transistors $T_1 - T_4$ and this information is used to appropriately set the value of the current I_{r2}. As a result, the current mirror output exhibits a low swing regardless of variations in the IC process and supply voltage, and the operating range or headroom available to the output load is maximized.

Note that degeneration resistors may be included between the transistor sources and V_{SS} to help maintain a constant current flowing through transistors over a wider range of the supply voltages.

The regulated cascode structure [8], which is based on a transistor arrangement featuring a very high output resistance, can be used to improve the performance of current mirrors, as shown in Figure 4.7. In the circuit of Figure 4.7(a), the input and output currents are denoted by I_{r1} and I_b, respectively, while I_{r2} is used to bias the transistors T_1 and T_2. With reference to the practical implementation shown in Figure 4.7(b), the input and biasing currents delivered, respectively, by transistors T_{B1} and T_{B2} should nominally be identical to achieve $V_{GS_1} = V_{GS_4}$ for any level of the input current, which is mirrored through T_3 [3]. If T_1 and T_3 have the same size, $V_{GS_1} = V_{DS_3} = V_{GS_3}$

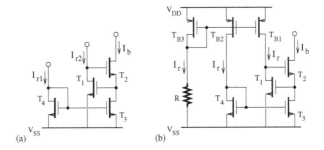

FIGURE 4.7
(a) Active-feedback cascode current mirror; (b) active-feedback cascode current mirror with a duplicated reference current.

and T_3 will be saturated because $V_{DS_3} \geq V_{GS_3} - V_T$. The minimum output voltage is then primarily determined by the drain-source saturation voltage required to keep T_2 in the saturation region. However, the accuracy of the equality between the input and output currents can be limited by the achievable matching between the gate-source voltages of T_1 and T_4.

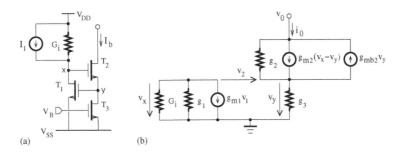

FIGURE 4.8
(a) Equivalent circuit of the active-feedback cascode current mirror and (b) its small-signal equivalent model.

Consider the equivalent circuit of the active-feedback cascode current mirror shown in Figure 4.8(a) [8], where V_B is a bias voltage, and G_i is the conductance of the current source I_1. From its small-signal equivalent model depicted in Figure 4.8(a), the output current can be computed as

$$i_0 = g_{m2}(v_x - v_y) - g_{mb2}v_y + g_2(v_0 - v_y), \tag{4.50}$$

Because

$$v_y = v_1 = i_0/g_3 \tag{4.51}$$

and

$$v_x - v_y = -\frac{g_{m1}v_1}{g_1 + G_i} - v_1 = -\left(1 + \frac{g_{m1}}{g_1 + G_i}\right)\frac{i_0}{g_3}, \tag{4.52}$$

we obtain

$$i_0 = -\left[\frac{g_{m2}}{g_3}\left(1 + \frac{g_{m1}}{g_1 + G_i}\right) + \frac{g_{mb2}}{g_3} + \frac{g_2}{g_3}\right]i_0 + g_2 v_0. \tag{4.53}$$

Hence, the output resistance is given by

$$r_0 = \frac{v_0}{i_0} = \frac{1}{g_2}\left[1 + \frac{g_{m2}}{g_3}\left(1 + \frac{g_{m1}}{g_1 + G_i}\right) + \frac{g_{mb2}}{g_3} + \frac{g_2}{g_3}\right]. \tag{4.54}$$

Assuming that the drain-source voltage of the transistor T_3 is kept constant by the structure consisting of $T_1 - T_2$ and the current source I_1 to reduce the effect of the body transconductance, g_{mb2}, and that the ratio g_2/g_3 is relatively negligible, we can write

$$r_0 = \frac{v_0}{i_0} \simeq \frac{g_{m1}g_{m2}}{g_2 g_3 (g_1 + G_i)}. \tag{4.55}$$

With identical transconductances and all conductances being equal to the inverse of the drain-source resistance, the output resistance of the active-feedback cascode current mirror is approximately $g_m^2 r_{ds}^3/2$. On the other hand, the transistors should operate in the saturation region to ensure proper operation of the current mirror, leading to the minimum output voltage of

$$V_{bmin} = V_{GS_1} + V_{DS_2} = 2V_{DS(sat)} + V_T. \tag{4.56}$$

It was assumed that the transistors feature the same drain-source saturation voltage and threshold voltage.

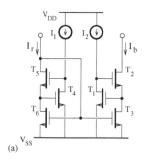 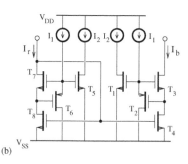

FIGURE 4.9
(a) Regulated cascode current mirror; (b) regulated cascode current mirror with the optimum output swing.

The regulated cascode current mirror of Figure 4.9(a) consists of the same transistor structure in the input and output paths and the biasing current sources I_1 and I_2. Its overall symmetric structure results in the reduction of the systematic matching error in the current ratio at *dc*.

For low-voltage applications, the minimum output voltage of the regulated cascode current mirror can be reduced using source followers operating as level shifters to lower the transistor bias voltages [4], as shown in Figure 4.9(b). It is assumed that the pMOS and nMOS transistors are designed with the process transconductance and threshold voltage of the form $K'_n = K'_p = K'$ and $V_{T_n} = -V_{T_p} = V_T$, and that the value of the bias current I_2 is four times that of I_1. Under these conditions, we can have

$$V_{G_3} = V_{GS_1} = \sqrt{\frac{I_2}{K'(W/L)}} + V_T = 2V_{DS(sat)} + V_T, \qquad (4.57)$$

where $I_2 = 4I_1$ and $V_{DS(sat)} = \sqrt{I_1/[K'(W/L)]}$. The drain-source voltage of the transistor T_4 is computed as

$$V_{DS_4} = V_{GS_1} + V_{GS_2} = V_{DS(sat)}, \qquad (4.58)$$

where $V_{GS_2} = -V_{DS(sat)} - V_T$, and the gate-source voltage of the transistor T_3 is given by

$$V_{GS_3} = V_{GS_1} - V_{DS_4} = V_{DS(sat)} + V_T. \qquad (4.59)$$

The minimum output voltage is reduced to

$$V_{bmin} = V_{DS_3} + V_{DS_4} = 2V_{DS(sat)}, \qquad (4.60)$$

where $V_{DS_3} = V_{GS_3} - V_T = V_{DS(sat)}$. The current mirror then features a high output impedance over a wide output swing, but this performance can be affected by mismatches between the parameters of the nMOS and pMOS transistors.

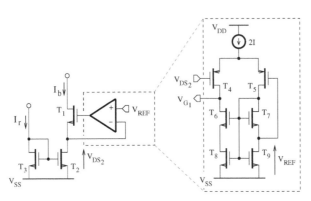

FIGURE 4.10
Active cascode current mirror with a wide output swing.

Another architecture for a cascode current mirror with a high output resistance and a wide output swing is shown in Figure 4.10 [5]. In the ideal case, the amplifier should have a high *dc* gain and a zero offset voltage. This makes the voltages at both input terminals of the amplifier to be equal, that is, $V_{DS_2} = V_{REF}$.

The structure of the amplifier is chosen such that the transistor is biased to operate in the saturation region with $V_{DS_1} = V_{DS_1(sat)}$ and the reference voltage, V_{REF}, is determined by the drain-source voltage of T_9. Hence, the minimum output voltage is given by

$$V_{bmin} = V_{DS_1(sat)} + V_{REF} . \tag{4.61}$$

The amplifier is designed such that the level of V_{REF} is appropriate for the biasing of the transistor T_2 in the saturation region, while maintaining the minimum output voltage as low as possible. With reference to the amplifier of Figure 4.10, we have

$$V_{GS_9} = V_{DS_7} + V_{DS_9} , \tag{4.62}$$

where $V_{REF} = V_{DS_9}$. The transistors $T_6 - T_9$ all have equal threshold voltages. Because the transistor T_7 operates in the saturation region, its I/V characteristic leads to

$$V_{DS_7} = V_{GS_7} = \sqrt{\frac{I}{K_7'(W_7/L_7)}} + V_T . \tag{4.63}$$

In the case of the transistor T_9, which operates in the triode region, we can write

$$V_{GS_9} = \frac{1}{2V_{DS_9}} \left(\frac{I}{K_9'(W_9/L_9)} + V_{DS_9}^2 \right) + V_T . \tag{4.64}$$

Substituting Equations (4.63) and (4.64) into Equation (4.62) gives

$$V_{DS_9}^2 + 2V_{DS_7}V_{DS_9} - \frac{I}{K_9'(W_9/L_9)} = 0. \tag{4.65}$$

The only positive solution of this quadratic equation is of the form

$$V_{DS_9} = \sqrt{\frac{I}{K_7'(W_7/L_7)}} \left(\sqrt{1 + \frac{K_7'(W_7/L_7)}{K_9'(W_9/L_9)}} - 1 \right) . \tag{4.66}$$

The transistor parameters can then be sized such that the reference voltage is on the order of the drain-source saturation voltage of T_2.

The small-signal equivalent model depicted in Figure 4.11 can be considered for the determination of the output resistance. With the assumption that the input voltage is equal to zero, $v_{gs3} = v_{gs2} = 0$, we have

$$v_0 = [i_0 - (g_{m1}v_{gs1} + g_{mb1}v_{bs1})]/g_1 + i_0/g_2 , \tag{4.67}$$

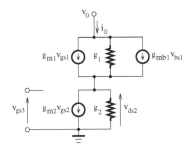

FIGURE 4.11
Equivalent model of the active cascode current mirror.

where $v_{bs1} = -v_{ds2}$. The effect of the body transconductance of T_2 is considered negligible because $v_{bs2} = 0$. Noting that the gate voltage of the transistor T_1 is $v_{g1} = -Av_{ds2}$, we can write

$$v_{gs1} = v_{g1} - v_{s1} = -(A+1)v_{ds2}, \qquad (4.68)$$

where $v_{ds2} = i_0/g_2$ and A denotes the amplifier gain. The output resistance can then be derived as

$$r_0 = \frac{v_0}{i_0} = \frac{1}{g_1} + \frac{1}{g_2} + \frac{1}{g_1 g_2}[(A+1)g_{m1} + g_{mb1}]. \qquad (4.69)$$

In addition to a low output compliance, the active cascode current mirror can also feature a high output resistance.

The aforementioned cascode current mirrors do provide a high output resistance, but at the cost of a large power consumption due to the additional circuit branches or components required for adequate biasing. Self-biased cascode current mirrors, which require only one reference current source, may be used to reduce power consumption.

With reference to Figure 4.12(a), a self-biased high-swing cascode current mirror with the bias voltage fixed by a resistor [6] is shown in schematic form. Assuming that all transistors are matched and operate in the saturation region, we have $V_{GS_3} = V_{GS_4}$ and it follows that $I_b = I_r$. The transistors T_3 and T_1, which are in series, are traversed by the same current. Their gate-source voltages are then equal, that is,

$$V_{GS_3} = V_{GS_1} = V_T + V_{DS(sat)}. \qquad (4.70)$$

The current I_r flowing through the resistor R establishes a bias voltage at the gate of T_2, which is by an amount RI_r greater than the one at the gate of T_4. With an appropriate selection of R, $V_{DS(sat)} = RI_r$ and we obtain

$$V_{G_1} = V_{G_2} = V_T + 2V_{DS(sat)}. \qquad (4.71)$$

Hence, a minimum output voltage of $V_{bmin} = 2V_{DS(sat)}$ is required to maintain a high resistance at the output of the self-biased cascode current mirror.

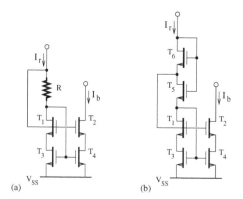

FIGURE 4.12
(a) Self-biased high-swing cascode current mirror with the bias voltage fixed by a resistor; (b) self-biased high-swing cascode current mirror with the bias voltage fixed by an active load.

An alternative self-biased cascode current mirror is depicted in Figure 4.12(b) [7], where the difference between the gate voltages of transistors T_4 and T_2 is equal to the drain-source voltage of the transistor T_5. It is assumed that all transistors are designed with the same process transconductance parameter, K', and threshold voltage, V_T. With $V_{DS6} \geq V_{DS6(sat)} = V_{GS6} - V_T$, the transistor T_6 operates in the saturation region and we can write

$$I_r = K'(W_6/L_6)(V_{GS6} - V_T)^2, \qquad (4.72)$$

where $K' = \mu C_{ox}/2$. The characteristic of T_5, which is assumed to operate in the triode region, is given by

$$I_r = K'(W_5/L_5)[2(V_{GS5} - V_T)V_{DS5} - V_{DS5}^2] \qquad (4.73)$$

when $V_{DS5} \leq V_{DS5(sat)} = V_{GS5} - V_T$. For the optimum biasing condition,

$$V_{DS5} = V_{DS(sat)} \qquad (4.74)$$

and

$$V_{GS6} = V_{DS(sat)} + V_T. \qquad (4.75)$$

Applying Kirchhoff's voltage law around the loop including transistors T_6 and T_5, we obtain

$$V_{GS5} = V_{GS6} + V_{DS5} = 2V_{DS(sat)} + V_T. \qquad (4.76)$$

Because the same current flows through transistors T_6 and T_5, it can be deduced that

$$K'(W_6/L_6)V_{DS(sat)}^2 = K'(W_5/L_5)[2(2V_{DS(sat)})V_{DS(sat)} - V_{DS(sat)}^2]. \qquad (4.77)$$

Hence,

$$W_5/L_5 = (1/3)(W_6/L_6). \qquad (4.78)$$

All the transistors except T_6 operate in the saturation region and can be designed with the same width-to-length ratio as the transistor T_5. The currents I_r and I_b are matched provided the gate-source voltages of transistors T_3 and T_4 are identical, that is,

$$V_{GS_3} = V_{GS_4} = V_{DS(sat)} + V_T. \tag{4.79}$$

The voltage at the gates of transistors T_1 and T_2 is

$$V_{G_1} = V_{G_2} = 2V_{DS(sat)} + V_T, \tag{4.80}$$

and the minimum output voltage for which a high resistance is ensured is $2V_{DS(sat)}$.

4.1.3 Low-voltage active current mirror

Cascode-based current mirrors generally feature a high output resistance with a minimum output voltage of about $2V_{DS(sat)}$. This, however, may still too large for low-voltage applications. In the current mirror of Figure 4.13(a), a sensing amplifier is used to force the similar transistors T_1 and T_2 to have the same drain-source voltage, and hence the same drain current. The minimum output voltage is reduced to $V_{DS(sat)}$ when the transistor operates in the saturation region.

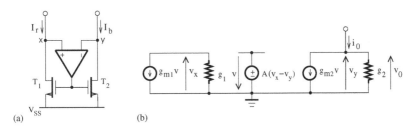

FIGURE 4.13
(a) Current mirror using a sensing amplifier and (b) its small-signal equivalent circuit.

Consider the small-signal equivalent circuit shown in Figure 4.13(b). Applying the current and voltage laws, we have

$$i_0 = g_{m2}v + g_2 v_y, \tag{4.81}$$

$$v_x = -g_{m1}g_1 v, \tag{4.82}$$

$$v = A(v_x - v_y). \tag{4.83}$$

The output resistance, r_0, of the current mirror can then be computed as

$$r_0 = \frac{v_0}{i_0} = \frac{1 + Ag_{m1}/g_1}{g_2(1 + Ag_{m1}/g_1 - Ag_{m2}/g_2)}, \tag{4.84}$$

where $v_0 = v_y$ and A denotes the amplifier gain. It should be noted that a more accurate analysis taking into account the parasitic impedances may be necessary in order to predict the stability and high-frequency behavior of the circuit.

4.2 Current and voltage references

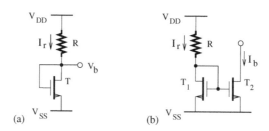

FIGURE 4.14
Simple (a) voltage and (b) current references.

A simple voltage reference is shown in Figure 4.14(a). It operates with a current level set by the resistor R. The transistor T, whose gate and drain are linked, operates in the saturation region provided that $V_{GS} \geq V_T$. That is, we have

$$I_r = K(V_b - V_{SS} - V_T)^2 \tag{4.85}$$

$$= \frac{V_{DD} - V_b}{R}, \tag{4.86}$$

where $K = (1/2)\mu C_{ox}(W/L)$ is the transconductance parameter. The equation for the output voltage can be written as

$$KRV_b^2 + [1 - 2KR(V_{SS} + V_T)]V_b + KR(V_{SS} + V_T)^2 - V_{DD} = 0. \tag{4.87}$$

Because $V_{GS} = V_b - V_{SS} \geq V_T$, the only valid root of this quadratic equation is given by

$$V_b = V_{SS} + V_T - \frac{1}{2KR} + \sqrt{\frac{1}{KR}\left[V_{DD} - (V_{SS} + V_T) + \frac{1}{4KR}\right]}. \tag{4.88}$$

The bias voltage, V_b, is a function of the resistance; the threshold voltage, V_T, of the transistor; and the supply voltages. Thus, the accuracy of the current reference will be affected by the supply-voltage variations and the changes in V_T due to the temperature and IC process fluctuations.

A current reference based on a simple current is shown in Figure 4.14(b). The reference current I_r is defined by the resistor R. Let

$$I_{D1} = I_r = \frac{V_{DD} - V_{GS} - V_{SS}}{R} ; \qquad (4.89)$$

the current I_{D2} reads

$$I_{D2} = I_b = \frac{(W_2/L_2)}{(W_1/L_1)} \frac{V_{DD} - V_{GS} - V_{SS}}{R} . \qquad (4.90)$$

Thus, the bias current I_b is affected by the variations in the supply voltages.

The sensitivity of I_b to the supply voltage, $V_{sup} = V_{DD} - V_{SS}$, can be defined by

$$S_{V_{sup}}^{I_b} = \lim_{\triangle V_{sup} \to 0} \frac{\triangle I_b/I_b}{\triangle V_{sup}/V_{sup}} = \frac{V_{sup}}{I_b} \frac{\partial I_b}{\partial V_{sup}} . \qquad (4.91)$$

Assuming that V_{GS} is constant, we can write

$$S_{V_{sup}}^{I_b} = \frac{1}{1 - \dfrac{V_{GS}}{V_{sup}}} . \qquad (4.92)$$

Given the percentage variation in V_{sup}, the change in the bias current can be computed as

$$\frac{\triangle I_b}{I_b} = S_{V_{sup}}^{I_b} \frac{\triangle V_{sup}}{V_{sup}} . \qquad (4.93)$$

Generally, the reference circuit must be designed to feature a sensitivity to the supply-voltage change less than 1%.

In the case of the temperature dependence of I_b, the fractional temperature coefficient, which is given by

$$TC(I_b) = \frac{1}{I_b} \frac{\partial I_b}{\partial T} , \qquad (4.94)$$

appears to be useful. Using the expression of the bias current, we obtain

$$TC(I_b) = -\frac{1}{I_b} \left[\frac{(W_2/L_2)}{(W_1/L_1)} \frac{1}{R} \frac{\partial V_{GS}}{\partial T} + \frac{I_b}{R} \frac{\partial R}{\partial T} \right] . \qquad (4.95)$$

The voltage V_{GS} is related to the parameters, such as the transistor transconductance parameter and threshold voltage, and the resistor R, whose values depend on the temperature. Note that the fractional temperature coefficient is a function of the temperature, that is, it is specified only at a given temperature. Ideally, a reference circuit can be temperature-independent due to the cancelation of the temperature coefficients of the individual components.

4.2.1 Supply-voltage independent current reference

A supply-voltage independent current reference is depicted in Figure 4.15(a) [9]. It is based on two current mirrors interconnected into a closed loop. A resistor R is linked to the source of the transistor T_1. The requirement, $I_1 = I_1 = I_r$, is met because the p-channel transistors are designed to have the same size. The next equations can be written as

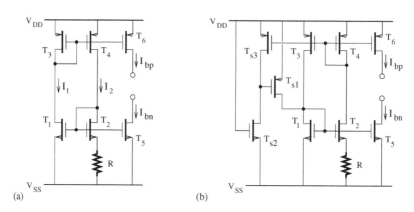

FIGURE 4.15
(a) Supply-voltage independent current reference; (b) current reference including a start-up circuit.

$$V_{GS1} = V_{GS2} + RI_r \, , \tag{4.96}$$

$$V_{GS1} = \sqrt{\frac{2I_r}{\mu_n C_{ox}(W_1/L_1)}} + V_{T1} \, , \tag{4.97}$$

$$V_{GS2} = \sqrt{\frac{2I_r}{\mu_n C_{ox}(W_2/L_2)}} + V_{T2} \, . \tag{4.98}$$

With the assumption that $V_{T1} = V_{T2}$ and $(W_1/L_1) = \kappa(W_2/L_2)$, the current I_r is given by

$$I_r = \frac{2}{\mu_n C_{ox}(W_1/L_1)} \left[\frac{1}{R}\left(1 - \frac{1}{\sqrt{\kappa}}\right) \right]^2 . \tag{4.99}$$

The versions of the reference current, I_{bn} and I_{bp}, which are respectively suitable for the biasing of n- and p-channel transistors, are generated by T_5 and T_6.

The start-up circuit $T_{s1} - T_{s3}$ of the reference circuit of Figure 4.15(b) injects a current into one branch during the initial power-on to prevent the occurrence of the zero current state. It is designed so that T_{s1} will turn off (i.e., $V_{SG} < V_T$) during normal operation of the current reference.

4.2.2 Bandgap references

The bandgap reference is designed to provide a voltage, that is independent of supply voltage and temperature variations. The temperature insensitivity is practically realized by adding a voltage with a positive temperature coefficient to a voltage with an equal but negative temperature coefficient. The first type of voltage can be obtained by amplifying the difference between the base-emitter voltages of two forward-biased bipolar transistors operating at different current densities, while the second one is proportional to the base-emitter voltage of a forward-biased bipolar transistor. Numerous variations can be introduced in the bandgap reference circuitry to improve some performance characteristics, such as the temperature coefficient, initial accuracy, noise, and stability.

The collector current of a bipolar transistor operating in the forward-active region, that is, $V_{BC} \leq 0$, may be expressed as

$$I_c = I_s(e^{V_{BE}/\eta V_T} - 1) \simeq I_s e^{V_{BE}/\eta V_T}, \qquad (4.100)$$

where I_s is the saturation current, $V_T = kT/q$ is the thermal voltage, $k = 1.38 \cdot 10^{-23}$ J/K is Boltzmann's constant, $q = 1.6 \cdot 10^{-19}$ C is the electronic charge, and η is an empirical scaling constant, which depends on the geometry, material, and doping levels. The current I_s can be defined by

$$I_s = CT^{1/2}e^{-E_g/kT}, \qquad (4.101)$$

where $E_g \simeq 1.12$ eV is the silicon bandgap energy and C is a proportionality constant. With

$$V_{BE} = \eta V_T \ln\left(\frac{I_c}{I_s}\right) \qquad (4.102)$$

we can compute

$$\frac{\partial V_{BE}}{\partial T} = \eta\left[\frac{\partial V_T}{\partial T}\ln\left(\frac{I_c}{I_s}\right) - \frac{\partial I_s}{\partial T}\frac{V_T}{I_s}\right]. \qquad (4.103)$$

Therefore,

$$\frac{\partial V_{BE}}{\partial T} = \eta\frac{V_{BE} - V_T/2 - E_g/q}{T}. \qquad (4.104)$$

Let $\eta = 1$, $T = 300$K or $27°$C (room temperature), and $V_{BE} \simeq 0.75$ V, $\partial V_{BE}/\partial T$ is on the order of -2.2 mV/$°$C, while V_T, which is proportional to absolute temperature (PTAT), exhibits a temperature coefficient of $+0.086$ mV/$°$C. The base-emitter voltage of a bipolar transistor possesses a negative temperature coefficient and has a complementary to absolute temperature (CTAT) dependency. In general, the bandgap voltage, V_{REF}, is generated as a linear combination of V_T and V_{BE}, that is,

$$V_{REF} = V_{BE} + \alpha V_T, \qquad (4.105)$$

where the constant parameter α is set by the circuit component such that V_{REF} is almost independent of the temperature variations.

In the bandgap reference of Figure 4.16(a), the voltage drop across R_3 is equal to the difference, $\triangle V_{BE}$, between the V_{BE}'s of the bipolar transistors. The reference voltage is then a combination of the V_{BE} of one transistor and a version of $\triangle V_{BE}$ scaled by a factor determined by the resistances, which may be chosen so that the opposite temperature coefficients of V_{BE} and $\triangle V_{BE}$ counterbalance. We have

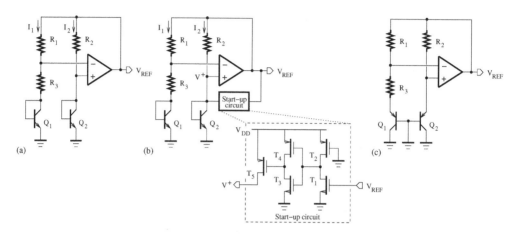

FIGURE 4.16
(a) Bandgap reference using *npn* transistors; (b) bandgap reference with a start-up circuit; (c) modified bandgap reference with grounded collector *pnp* transistors suitable for *n*-well fabrication process.

$$V_{REF} = V_{BE1} + (R_1 + R_3)I_1 . \tag{4.106}$$

For a high-gain amplifier, the voltage drop between the input nodes is reduced to zero, and the next relationships can be written:

$$I_1 = \frac{V_{BE2} - V_{BE1}}{R_3} , \tag{4.107}$$

$$R_1 I_1 = R_2 I_2 . \tag{4.108}$$

The reference voltage V_{REF} is then given by

$$V_{REF} = V_{BE1} + \left(1 + \frac{R_1}{R_3}\right) V_T \ln\left(\frac{R_1}{R_2}\frac{I_{S1}}{I_{S2}}\right) , \tag{4.109}$$

where I_{S1} and I_{S2} are the saturation currents of transistors Q_1 and Q_2. The temperature coefficient $\partial V_T / \partial T$ is about $+0.087$ mV/K. For $(1 + R_1/R_3) \ln(R_1 I_{S1}/R_2 I_{S2}) \simeq 17.2$, the reference voltage will feature a zero temperature coefficient. Furthermore, by using the relationship $\partial V_{REF}/\partial T = 0$, the dependence of V_{REF} to the bandgap voltage, E_g, can be established.

In order to provide an output voltage that is independent of the variations in the supply voltage, the bandgap reference circuit is designed to be self-biased. As a consequence, the operating point obtained when all currents in the circuit are equal to zero can also be stable. A start-up circuit is required to force the bandgap reference circuit to operate with nonzero bias currents. Figure 4.16(b) shows a bandgap reference with a start-up circuit [10] connected between the output and the noninverting input of the amplifier. In normal operation, the output level of the bandgap reference circuit is greater than the threshold voltage of the transistor T_1, assumed to be the start-up voltage threshold, and the start-up circuit is disabled so as not to interfere with the normal operation of the band-gap reference circuit. On the contrary, if the bandgap reference circuit is in the zero-current state, the output of the inverter including transistors T_3 and T_4 will generate a logic high voltage and an initial current will be injected into the bandgap reference circuit by the start-up circuit. Note that the p-channel transistor T_2 is always activated because its gate is connected to ground.

In practice, the accurate prediction of the temperature coefficient should rely on simulations due to the nonideal effects such as

- The temperature dependence of the collector current

- The amplifier offset voltage and output impedance

Furthermore, note that the temperature compensation achieved in a bandgap reference is limited to first order.

Due to the fact that the transistor collectors are not connected to the most negative supply voltage, the circuit of Figure 4.16(a) can be incompatible with implementations using the CMOS process where the p-type substrate has to serve as the collector. This problem is solved in the bandgap reference shown in Figure 4.16(c).

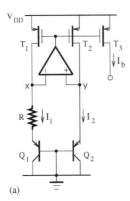
(a)

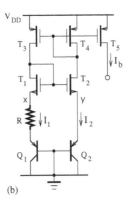
(b)

FIGURE 4.17

Bandgap reference generating a bias current: Implementations based on (a) amplifier and (b) transistor loads.

Bandgap reference implementations using amplifier and transistor loads are shown in Figures 4.17(a) and (b), respectively. These structures generate a bias current, I_b, which is proportional-to-absolute temperature. With the assumption that all MOS transistors are identical, I_b is similar to the currents I_1 and I_2, which are given by $I_1 = I_2 = (V_T/R)\ln(I_{S1}/I_{S2})$, provided $V_x = V_y$. A supply voltage dependence can be observed in the circuit of Figure 4.17(b) due to the channel-length modulation of the MOS transistors. A solution to this problem can consist of using cascode structures.

4.2.2.1 Low-voltage bandgap voltage reference

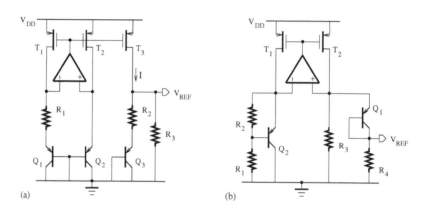

(a)   (b)

FIGURE 4.18
(a) Current mode low-supply voltage bandgap reference; (b) low-supply voltage bandgap reference based on the reverse bandgap voltage principle.

Generally, the normal operation of conventional bandgap references requires the use of a power supply voltage higher than the resulting reference voltage. For instance, a 2.5-V supply voltage is needed for the generation of an output voltage of about 1.25 V. To satisfy the requirements for low-voltage applications, the current-mode operation principle, which relies on the combination of current sources with positive and negative temperature coefficients to create a temperature-independent current, can be exploited, as illustrated in Figure 4.18(a) [11]. The resulting current is transferred by a current mirror to a network operating as a current divider, and the output voltage level can be scaled to a given value of the form

$$V_{REF} = \frac{R_3}{R_2 + R_3}(V_{EB3} + R_2 I).\tag{4.110}$$

Hence,

$$V_{REF} = \frac{R_3}{R_2 + R_3}\left[V_{EB3} + \frac{R_2}{R_1}V_T \ln\left(\frac{I_{S1}}{I_{S2}}\right)\right].\tag{4.111}$$

It should be noted that a low-voltage amplifier structure may be required to reduce the influence of the input common-mode range on the minimum supply voltage of the bandgap reference. Furthermore, the accuracy of the reference voltage is limited by the nonideal characteristics of the current mirror, resistor mismatches, and noises.

To reduce the nonideal effect introduced by the active current mirror, the reversed bandgap voltage principle can be exploited for the voltage reference design, as illustrated in Figure 4.18(b) [14]. Assuming that the transistors T_1 and T_2 are identical and $R_3 = R_1 + R_2$, the collector currents of transistors Q_1 and Q_2 should be equal, provided the base current of Q_2 is negligible. According to the voltage divider principle, the voltage applied to the inverting node of the amplifier is the sum of V_{EB_2} and $V_{EB_2}(R_1/R_2)$. Because the voltages at the input nodes of an ideal amplifier are equal, we have

$$V_{EB_2}\left(1 + \frac{R_1}{R_2}\right) = V_{REF} + V_{EB_1}. \qquad (4.112)$$

Hence,

$$V_{REF} = \frac{R_1}{R_2} V_{EB_2} + \triangle V_{EB}, \qquad (4.113)$$

where $\triangle V_{EB} = V_{EB_2} - V_{EB_1}$. For bipolar transistors, the collector currents are of the form

$$I_{c_1} = I_{s_1} e^{V_{EB_1}/V_T}, \qquad (4.114)$$

$$I_{c_2} = I_{s_2} e^{V_{EB_2}/V_T}. \qquad (4.115)$$

Here $I_{c_1} = I_{c_2}$, and we can obtain

$$\triangle V_{EB} = V_{EB_2} - V_{EB_1} = V_T \ln\left(I_{s_1}/I_{s_2}\right). \qquad (4.116)$$

The reference voltage can then be expressed as

$$V_{REF} = \frac{R_1}{R_2} V_{EB_2} + V_T \ln\left(\frac{I_{s_1}}{I_{s_2}}\right). \qquad (4.117)$$

With the voltage reference architecture based on the reversed bandgap voltage, the supply voltage can be on the order of 1 V while the value of V_{REF} is set as low as 200 mV.

4.2.2.2 Curvature-compensated bandgap voltage reference

In practice, the base-emitter voltage of a bipolar transistor exhibits a non-linear temperature relationship. However, the temperature compensation of the reference voltage in classical structures of the bandgap voltage reference is realized only in a vicinity of the reference temperature value and is limited to the first order. Due to this nonlinearity, the reference voltage is generated with a temperature curvature error, which is proportional to $T \ln(T)$, where T

is the absolute temperature. Uncompensated bandgap references have a temperature coefficient[1] greater than 20 ppm/°C (parts-per-million per degree Celsius) over a typical industrial temperature range from $-40°C$ to $85°C$.

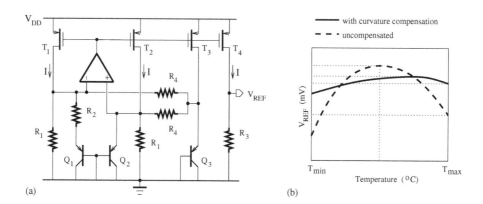

(a)   (b)

FIGURE 4.19
(a) Low-supply voltage bandgap reference with the temperature curvature compensation; (b) variations of V_{REF} as a function of the temperature.

The bandgap voltage reference structure shown in Figure 4.19(a) is designed to exhibit an output voltage that is almost independent of the temperature. This is achieved by compensating for the curvature introduced by the nonlinear temperature dependence of bipolar transistor characteristics on the output voltage.

The base-emitter voltage can be expressed as [15]

$$V_{BE} = V_{G0} - [V_{G0} - V_{BE}(T_R)]\frac{T}{T_R} - (\eta - x)V_T \ln\left(\frac{T}{T_R}\right), \qquad (4.118)$$

where V_{G0} is the energy-bandgap voltage at zero degree Kelvin, T is the absolute temperature in degrees Kelvin, T_R is the reference temperature, $V_{BE}(T_R)$ is the base-emitter voltage at the temperature T_R, n denotes a temperature constant depending on the IC process, η represents a process-dependent constant, x is the exponential order of the temperature dependence of the collector current (i.e., $I_C \propto T^x$), and V_T is the thermal voltage. Typical values for these parameters are $V_{G0} = 1.17$ V, $\eta = 3.5$, and at $T_R = 300K$, $V_{BE}(T_R) = 0.65$

---

[1] The temperature coefficient TC can be defined using the difference in the maximum and minimum values of the reference voltage over the entire temperature range, i.e.,

$$TC \text{ (ppm/°C)} = \frac{10^6}{T_{max} - T_{min}}\left(\frac{V_{REF,max} - V_{REF,min}}{V_{REF,25°C}}\right)$$

where $V_{REF,25°C}$ is the value of the reference voltage at the room temperature, and T_{max} and T_{min} are respectively the maximum and minimum temperatures.

V. A high-order compensation of the bandgap voltage reference is required to reduce the effect of the nonlinearity related to the logarithmic temperature ratio. Its principle can be based on a proper combination of the base-emitter voltage across a junction with a temperature-independent current ($x = 0$) and the one across a junction with a PTAT current ($x = 1$). The transistor Q_2 is biased by a PTAT current so that

$$V_{EB_2} = V_{G0} - [V_{G0} - V_{EB_2}(T_R)]\frac{T}{T_R} - (\eta - 1)V_T \ln\left(\frac{T}{T_R}\right), \qquad (4.119)$$

while Q_3 is biased by a temperature-independent current delivered by a p-channel MOS transistor, and

$$V_{EB_3} = V_{G0} - [V_{G0} - V_{EB_3}(T_R)]\frac{T}{T_R} - \eta V_T \ln\left(\frac{T}{T_R}\right). \qquad (4.120)$$

Because the voltage across the resistor R_4 is equal to the difference of the base-emitter voltages of Q_2 and Q_3, the curvature correction current is given by

$$I' = \frac{V_{EB_2} - V_{EB_3}}{R_4} = \frac{V_T}{R_4}\ln\left(\frac{T}{T_R}\right). \qquad (4.121)$$

The reference voltage can then be computed as

$$V_{REF} = R_3 I, \qquad (4.122)$$

where

$$
\begin{aligned}
I &= \frac{V_{EB_2} - V_{EB_1}}{R_2} + \frac{V_{EB_2}}{R_1} + I', \\
&= \frac{V_T}{R_2}\ln\left(\frac{I_{s_1}}{I_{s_2}}\right) + \frac{V_{EB_2}}{R_1} + \frac{V_T}{R_4}\ln\left(\frac{T}{T_R}\right). \qquad (4.123)
\end{aligned}
$$

Hence,

$$V_{REF} = \frac{R_3}{R_1}\left[\frac{R_1}{R_2}V_T \ln\left(\frac{I_{s_1}}{I_{s_2}}\right) + V_{EB_2} + \frac{R_1}{R_4}V_T \ln\left(\frac{T}{T_R}\right)\right]. \qquad (4.124)$$

The curvature compensation will be realized if the following requirement [16] is met:

$$R_4 = R_1/(\eta - 1) \qquad (4.125)$$

and the value of the ratio R_1/R_2 should be determined to minimize the drift of V_{REF} due to the linear variation of the temperature. Ideally, the resulting voltage reference is reduced to

$$V_{REF} = R_3 V_{G0}/R_1. \qquad (4.126)$$

However, the output voltage reference, as illustrated in Figure 4.19(b), may not remain constant with respect to the temperature due to component mismatches and the ignored temperature coefficients of resistors.

4.2.3 Floating-gate voltage reference

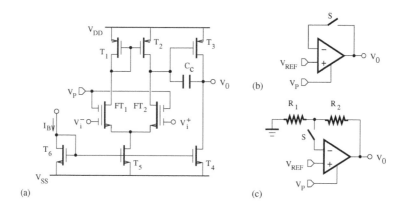

FIGURE 4.20
Floating-gate voltage reference.

A typical bandgap voltage reference, which consists of bipolar transistors combined with either the resistive feedback network of an amplifier or MOS transistor load, can exhibit a number of shortcomings. This is the case, for instance, for high circuit complexity, large silicon area, and high power consumption. Furthermore, the reference voltage precision depends on bipolar transistor matching and may be affected by the offset voltage introduced by the amplification circuit required to appropriately scale the base-emitter voltage difference. For applications requiring a reference voltage that is less than 1.25 V, the reference voltage must be scaled to values other than the one set by the silicon bandgap by using additional circuit components, which not only increase the circuit size and power consumption but also can constitute a source of distortions affecting the output accuracy.

One method to overcome the problems related to the implementation of bandgap voltage references using CMOS processes with parasitic bipolar devices and integrated resistors is to design the voltage reference using floating-gate transistors. Generally employed for its capability to store an electrical charge for extended periods of time even without a connection to a power supply, a floating-gate transistor can be fabricated by inserting a layer of oxide to electrically isolate the gate of a conventional MOS transistor and depositing extra secondary gates or input nodes. The control signal applied to each input is capacitively coupled to the floating gate, and then appears to be attenuated by a factor C_i/C_T, where C_i is the coupling capacitance for the *i-th* input and C_T is the total load capacitance seen from the gate. Programming involves adding charges to the floating gate to raise the threshold voltage, while erasing is achieved by removing charge from the floating gate to lower the threshold voltage. This is generally realized under high applied voltages (> 10 V) that can create electric fields with sufficient strength to allow electrons to gain kinetic energy to overcome the potential barrier, or to tunnel through the potential barrier. Essentially, the performance characteristics of physical mechanisms that may be used to vary the amount of charge on the floating gate, such as hot electron injection and Fowler-Nordheim tunneling, are affected by the programming voltage and time.

The design of precise and stable voltage references in a standard CMOS process without using parasitic bipolar devices and integrated resistors makes use of floating-gate transistors, whose charge storage capability can be exploited for the post-fabrication programmability. An amplifier based on two floating-gate transistors in the differential configuration, as shown in Figure 4.20(a), can constitute the main building block required for the voltage reference design [17]. It is needed to provide a reproduction of the programmed voltage level at a low impedance node. Figure 4.20(b) illustrates the operation as a unity-gain voltage follower, where $V_0 = V_{REF}$. In the noninverting amplifier configuration in Figure 4.20(c), the output voltage is of the form $V_0 = (1 + R_2/R_1)V_{REF}$. Here, the nature of the feedback forces V_0 to be dependent on V_{REF}, which is a function of the difference in the amount of charge stored on the floating gates of the transistor differential pairs. The switch S is in open state during the reference voltage programming, which may require many iterations to obtain the target, depending on the adopted technique.

4.3 Summary

Basic building blocks necessary for the design of active components such as amplifiers, comparators, and multipliers, were presented. They include current mirrors, voltage, and current references. The accuracy of a current mirror is improved by increasing the output resistance. This objective can be met by using cascode or feedback structures. On the other hand, the techniques that can be used to reduce the dependence of the supply voltage and temperature on the output signal provided by the reference circuit were reviewed.

4.4 Circuit design assessment

1. **Improved cascode current mirror**
   Analyze each of the improved cascode curent mirrors of Figure 4.21 with its respective transistor sizes given in Table 4.1 and show that the minimum output voltage is about $2V_{DS(sat)}$.

2. **Low-voltage current mirror**
   Determine the small-signal transfer function, $a_i = i_0/i_i$, of the current mirror shown in Figure 4.22.

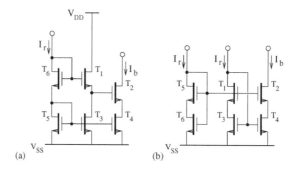

FIGURE 4.21

Circuit diagram of improved cascode current mirrors.

TABLE 4.1

Transistor Sizes

	$T_1 - T_5$	T_6
Figure 4.21(a):	(W/L)	$(1/4)(W/L)$
Figure 4.21(b):	(W/L)	$(1/3)(W/L)$

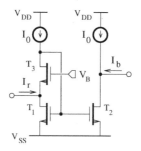

FIGURE 4.22

Circuit diagram of a low-voltage current mirror.

3. **Self-biasing current reference**

   The self-biasing current reference of Figure 4.23 is designed with $(W_2/L_2) = \kappa(W_1/L_1)$ and $(W_4/L_4) = (W_3/L_3)$. Here, the body effect is eliminated because the transistors, T_1 and T_2, have the same source potential.

   Show that

   $$I = \frac{1}{KR^2}\left(1 - \frac{1}{\sqrt{\kappa}}\right)^2, \qquad (4.127)$$

   where K is the transistor transconductance parameter.

   Propose a start-up circuit for the current reference.

4. **Analysis of an active cascode current mirror**

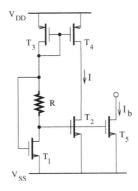

FIGURE 4.23

Circuit diagram of a self-biasing current reference.

Consider the current mirror shown in Figure 4.24 [18]. The difference between the drain-source voltages of transistors T_2 and T_3 is detected by the amplifier and used to control the gate voltage of T_1. Assuming that all transistors feature the same drain-source satura-

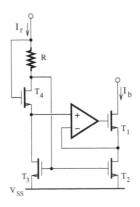

FIGURE 4.24

Active cascode current mirror.

tion voltage, $V_{DS(sat)}$, determine the resistance R as a function of I_r.

Estimate the minimum output voltage of the current mirror.

Derive an expression for the output resistance r_0.

5. **Voltage-controlled current source**

   For the voltage-controlled current source/sink of Figure 4.25, determine the current I as a function of the resistor R.

   Show that $I_r = I$, provided the transistors T_{B1} and T_{B2} are matched.

   Consider the circuit structures shown in Figure 4.26 [12]. The cur-

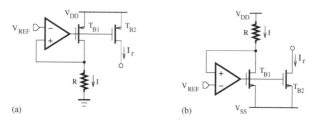

FIGURE 4.25
Voltage-controlled current (a) source and (b) sink.

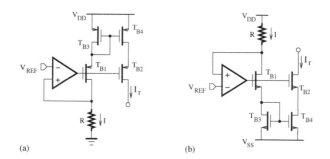

FIGURE 4.26
Improved voltage-controlled current (a) source and (b) sink.

rent mirror transistors T_{B3} and T_{B4} are used to improve the accuracy of the relation between the current I and the output current, I_r.

Assuming that the width-to-length ratio of T_{B1} and T_{B3} is equal to W_1/L_1, and that for T_{B2} and T_{B4} is equal to W_2/L_2, determine the current I_r as a function of I.

With $V_{DS_{B1}} + V_{DS_{B3}}$ being either equal to $V_{DD} - V_{REF}$ or $V_{REF} - V_{SS}$, find the minimum value of V_{REF} to maintain the transistors in the saturation region.

6. **Bandgap reference for low supply-voltage applications**
   The implementation of a low-voltage bandgap reference [13] is depicted in Figure 4.27(a). To reduce the minimum supply voltage to about 1 V, MOS transistors with a low threshold voltage are required in the amplifier implementation. The transistors are assumed to be identical and $R_1 = R_3$.

   Taking into account the amplifier offset voltage, V_{off}, show that

   $$V_{REF} = \frac{R_4}{R_1} \left[V_{EB2} + \frac{R_1}{R_2} V_T \ln\left(\frac{I_{S1}}{I_{S2}}\right) - \left(1 + \frac{R_1}{R_2}\right) V_{off} \right].$$
   (4.128)

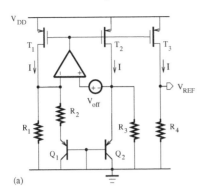
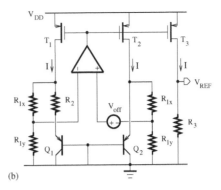

(a) (b)

FIGURE 4.27
Low supply-voltage bandgap references.

Estimate the contribution of the offset dispersion, $\triangle V_{off}$, on the variation of V_{REF}.

The requirement of using a CMOS fabrication process with low threshold voltages can be avoided in the case of the low-voltage bandgap reference shown in Figure 4.27(b) [19]. To maintain the same voltage at the drain of transistors T_1 and T_2, it is necessary to have $R_{1x} = R_{1y}$ and $R_1 = R_{1x} + R_{1y}$.

Verify that

$$V_{REF} = \frac{R_3}{R_1} \left[V_{EB2} + \frac{R_1}{R_2} \left(V_T \ln \left(\frac{I_{S1}}{I_{S2}} \right) + 2V_{off} \right) \right]. \qquad (4.129)$$

Compare these circuits for use as bandgap voltage references with a low supply-voltage.

Bibliography

[1] C. Abel, "Bias circuit for high-swing cascode current mirrors," U.S. Patent 7,208,998, filed April 12, 2005; issued April 24, 2007.

[2] J. B. Hughes and I. C. MacBeth, "Circuit arrangement for processing sampled analogue electrical signals," U.S. Patent 4,897,596, filed December 16, 1988; issued January 30, 1990.

[3] P. C. Kwong, "Cascode current mirror with amplifier," U.S. Patent 6,124,705, filed August 20, 1999; issued September 26, 2000.

[4] S. K. Hoon and J. Chen, "Regulated cascode current source with wide output swing," U.S. Patent 6,903,539, filed November 19, 2003; issued June 7, 2005.

[5] B. R. Gregoire, Jr., "Low voltage enhanced output impedance current mirror," U.S. Patent 6,707,286, filed February 24, 2003; issued March 16, 2004.

[6] T. L. Brooks and M. A. Rybicki, "Self-biased cascode current mirror having high voltage swing and low power consumption," U.S. Patent 5,359,296, filed September 10, 1993; issued October 25, 1994.

[7] N. S. Sooch, "MOS Cascode current mirror," U.S. Patent 4,550,284, filed May 16, 1984; issued October 29, 1985.

[8] E. Säckinger and W. Guggenbühl, "A high-swing, high-impedance MOS cascode circuit," *IEEE J. of Solid-State Circuits*, vol. 25, pp. 289–298, Feb. 1990.

[9] E. Vittoz and J. Fellrath, "CMOS analog circuits based on weak inversion operation," *IEEE J. of Solid-State Circuits*, vol. 12, pp. 224–231, June 1977.

[10] D. Y. Yu, "Startup circuit for band-gap reference circuit," U.S. Patent 5,867,013, filed November 20, 1997; issued February 2, 1999.

[11] H. Neuteboom, B. M. J. Kup, and M. Janssens, "A DSP-based hearing instrument IC," *IEEE J. of Solid-State Circuits*, vol. 32, pp. 1790–1806, Nov. 1997.

[12] C.-J. Yen and C.-H. Cheng, "Voltage-controlled current source," U.S. Patent 7,417,415, filed June 10, 2005; issued August 26, 2008.

[13] H. Banba, H. Shiga, A. Umezawa, T. Miyaba, T. Tanzawa, S. Atsumi, and K. Sakui, "A CMOS bandgap reference circuit with sub-1-V operation," *IEEE J. of Solid-State Circuits*, vol. 34, pp. 670–674, May. 1999.

[14] V. V. Ivanov and K. E. Sanborn, "Precision reversed bandgap voltage reference circuits and method," U.S. Patent 7,411,443, filed June 22, 2006; issued August 12, 2008.

[15] Y. P. Tsividis, "Accurate analysis of temperature effects in I_C-V_{BE} characteristics with application to bandgap reference sources," *IEEE J. Solid-State Circuits*, vol. 15, pp. 1076–1084, Dec. 1980.

[16] M. Gunawan, G. C. M. Meijer, J. Fonderie, and J. H. Huijsing, "A curvature-corrected low-voltage bandgap reference," *IEEE J. Solid-State Circuits*, vol. 28, pp. 667–670, June 1993.

[17] K. C. Adkins , "Structure and method for programmable and non-volatile analog signal storage for a precision voltage reference," U.S. Patent 6,791,879, filed September 23, 2002; issued September 14, 2004.

[18] T. Itakura and Z. Czarnul, "Current mirror circuit," U.S. Patent 5,986,507, filed September 12, 1996; issued November 16, 1999.

[19] K. N. Leung and P. K. T. Mok, "A sub-1-V 15-ppm/°C CMOS bandgap voltage reference without requiring low threshold Voltage Device," *IEEE J. of Solid-State Circuits*, vol. 37, pp. 526–530, April 2002.

5

CMOS Amplifiers

CONTENTS

The performance of mixed-signal circuits are generally determined by the characteristics of active components. Analog cells used inside the chip have to drive well-defined loads that are often purely capacitive, while the output buffer may demonstrate the capability of driving variable resistive and capacitive loads with low distortion. For this reason, the choice of the architecture and the design approach are mainly determined by the function to be realized. Furthermore, the realization of an amplifier, which features both high dc gain and bandwidth, can lead to conflicting demands. The high gain requirement is met by multistage designs with long-channel transistors biased at a low current level, while the high bandwidth specification is achieved in single-stage structures using short-channel transistors biased by a high-level current source.

Many architectures are available for the design of amplifiers [1]. However, these integrated circuits (ICs) would only benefit marginally from the decrease in transistor feature sizes and power supply voltages due to noise and offset requirements. Efficient solutions should be able to associate the low-voltage operation with high power efficiency and result in simple structures, that is, with a low die area.

Various performance specifications can be taken into consideration for the choice of an amplifier structure. They include dc characteristics such as input common-mode range, output swing, and offset voltage. In addition, there are specifications that can be described in the frequency domain (gain, bandwidth, phase margin) and in the time domain (slew rate, settling time).

5.1   Differential amplifier

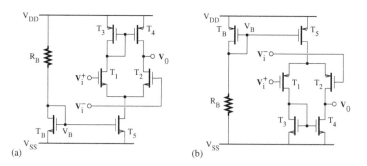

FIGURE 5.1
Circuit diagram of the single-stage differential amplifier based on (a) n-channel and (b) p-channel transistor pair.

The circuit diagram of a basic differential amplifier is shown in Figure 5.1. Note that, to achieve an accurate biasing of the amplifier in the presence of process, voltage, and temperature variations, the resistor R_B is generally implemented off-chip. The input differential transistor pair is biased by a constant current source and loaded by a current mirror. In the n-well process, a better threshold voltage matching can be achieved with the differential stage using n-channel transistors (see Figure 5.1(a)). On the other hand, the body effect can be removed by implementing the amplifier input stage with p-channel transistors as shown in Figure 5.1(b). Furthermore, the output noise is minimized because a p-channel transistor is less affected by $1/f$ noise than an n-channel transistor. In general, a large transconductance is obtained by designing the input transistors with a high W/L ratio. The parameter L is chosen to keep the channel length modulation effects low and W is sized for a given gate-source voltage.

5.1.1 Dynamic range

In general, the minimum supply voltage is determined based on the input common-mode range and output swing specifications. The input common-mode range is the range of the dc input voltage for which the specified gain of the differential stage is maintained unchanged. The output swing is the range over which the output voltage can be driven without being distorted. These specifications are estimated by assuming that the input voltage applied to both inputs is reduced to its common-mode dc component.

A differential stage is designed to amplify the voltage difference between both inputs and reject the common-mode input voltage. Let us consider the differential stage shown in Figure 5.1(a). In the worst case, the common-mode input voltage can be given by

$$V_{ICM} = V_{DD} - V_{SG_3} - V_{DS_1} + V_{GS_1} \qquad (5.1)$$

or

$$V_{ICM} = V_{SS} + V_{DS_5} + V_{GS_1} . \qquad (5.2)$$

Because $V_{DS_1} = V_{DS_1(sat)} = V_{GS_1} - V_{T_n}$, $V_{SG_3} = V_{SD_3} = V_{SD_3(sat)} - V_{T_p}$, and $V_{DS_5} = V_{DS_5(sat)} = V_{GS_5} - V_{T_n}$, the maximum and minimum values of the common-mode input voltage can be found as

$$V_{ICM} = V_{DD} - V_{SD_3(sat)} + V_{T_p} + V_{T_n} = V_{ICM,max} \qquad (5.3)$$

and

$$V_{ICM} = V_{SS} + V_{DS_5(sat)} + V_{DS_1(sat)} + V_{T_n} = V_{ICM,min} . \qquad (5.4)$$

Thus,

$$V_{SS} + V_{DS_5(sat)} + V_{DS_1(sat)} + V_{T_n} \leq V_{ICM} \leq V_{DD} - V_{SD_3(sat)} + V_{T_p} + V_{T_n} . \qquad (5.5)$$

The limits of the output voltage swing are set by the requirement of maintaining the transistors T_2 and T_4 in the saturation region. For the transistor T_2, this results in the condition

$$V_{DS_2} \geq V_{DS_2(sat)} = V_{GS_2} - V_{T_n} . \tag{5.6}$$

Because $V_{DS_2} = V_0 - V_{S_2}$ and $V_{GS_2} = V_{ICM} - V_{S_2}$, we can obtain

$$V_0 \geq V_{ICM} - V_{T_n} . \tag{5.7}$$

In the case of the transistor T_4, we have

$$V_{SD_4} \geq V_{SD_4(sat)} = V_{SG_4} + V_{T_p} , \tag{5.8}$$

where $V_{SD_4} = V_{DD} - V_0$. It can be shown that

$$V_0 \leq V_{DD} - V_{SG_4} - V_{T_p} = V_{DD} - V_{SD_4(sat)} . \tag{5.9}$$

Hence,

$$V_{ICM} - V_{T_n} \leq V_0 \leq V_{DD} - V_{SD_4(sat)} . \tag{5.10}$$

An analogous analysis can be performed for the differential stage with p-channel input transistors of Figure 5.1(b). The input common-mode voltage can be expressed as

$$V_{ICM} = -V_{SG_1} + V_{SD_1} + V_{GS_3} + V_{SS} \tag{5.11}$$

and

$$V_{ICM} = V_{DD} - V_{SD_5} - V_{SG_1} , \tag{5.12}$$

where $V_{GS_3} = V_{DS_3}$. Assuming that $V_{GS_3} = V_{DS_3(sat)} + V_{T_n}$, the input common-mode range is then given by

$$V_{T_p} + V_{T_n} + V_{DS_3(sat)} + V_{SS} \leq V_{ICM} \leq V_{DD} - V_{SD_5(sat)} - V_{SD_1(sat)} + V_{T_p} . \tag{5.13}$$

The transistors T_4 and T_2 will remain in the saturation region if the output swing is

$$V_{SS} + V_{DS_4(sat)} \leq V_0 \leq V_{ICM} + V_{T_p} . \tag{5.14}$$

It was assumed that $V_{T_n} > 0$, $V_{T_p} < 0$, $V_{DD} > 0$, and $V_{SS} \leq 0$. With the difference between the threshold voltages of n-channel and p-channel transistors not exceeding the drain-source saturation voltage, whose typical value is on the order of 0.3 V, it can be noted that the n-channel input stage exhibits a wide positive input common-mode swing, while the p-channel structure has a wide negative input common-mode swing. This suggests the use of a parallel combination of n-channel and p-channel differential pairs to achieve an almost rail-rail input range. On the other hand, the basic differential stage can only provide a limited output swing.

5.1.2 Source-coupled differential transistor pair

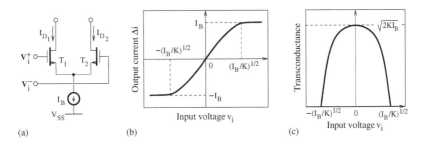

(a) (b) (c)

FIGURE 5.2
Source-coupled differential transistor pair: (a) Circuit diagram, (b) I/V transfer characteristic, (c) transconductance plot.

Let the n-channel transistor be described by the next equation,

$$I_D = \begin{cases} K(V_{GS} - V_{T_n})^2 & \text{if} \quad V_{GS} \geq V_{T_n}, \quad \text{saturation region} \\ 0 & \text{if} \quad V_{GS} < V_{T_n}, \quad \text{cutoff region,} \end{cases} \tag{5.15}$$

where $K = \mu_n(C_{ox}/2)(W/L)$ is the transconductance parameter; μ_n is the effective surface carrier mobility; C_{ox} is the gate-oxide capacitance per unit area; W and L are the channel width and length, respectively; and V_{T_n} denotes the threshold voltage. The bias current, I_B, of the source-coupled transistor structure of Figure 5.2(a) can be expressed as

$$I_{D_1} + I_{D_2} = I_B. \tag{5.16}$$

Applying Kirchhoff's voltage law to the loop involving the noninverting and inverting input nodes, the differential signal, V_i, is derived as

$$V_i = V_i^+ - V_i^- = V_{GS_1} - V_{GS_2} \tag{5.17}$$

Because the transistors T_1 and T_2 are identical and operate in the saturation region, we have

$$V_{GS_1} = \sqrt{\frac{I_{D_1}}{K}} + V_{T_n} \tag{5.18}$$

and

$$V_{GS_2} = \sqrt{\frac{I_{D_2}}{K}} + V_{T_n}. \tag{5.19}$$

Hence, Equation (5.17) becomes

$$\sqrt{I_{D_1}} - \sqrt{I_{D_2}} = \sqrt{K}V_i. \tag{5.20}$$

Solving the system of Equations (5.16) and (5.20) gives

$$I_{D_1} = \frac{I_B}{2} \pm \sqrt{2KI_B} \frac{V_i}{2} \sqrt{1 - \frac{V_i^2}{2I_B/K}}. \tag{5.21}$$

Because the drain current should be greater than $I_B/2$ when the input voltage increases, the only acceptable solution is given by

$$I_{D_1} = \frac{I_B}{2} + \sqrt{2KI_B} \frac{V_i}{2} \sqrt{1 - \frac{V_i^2}{2I_B/K}}. \tag{5.22}$$

Using Equation (5.16), the drain current of T_2 can then be computed as

$$I_{D_2} = \frac{I_B}{2} - \sqrt{2KI_B} \frac{V_i}{2} \sqrt{1 - \frac{V_i^2}{2I_B/K}}. \tag{5.23}$$

For input voltages within the linear range, the current flowing through one transistor will increase while the current in the other transistor will decrease. The maximum value of the input voltage is reached when T_1 operates in the saturation region and T_2 turns off, resulting in $I_{D_1} = I_B$ and $I_{D_2} = 0$, while the minimum value of the input voltage is attained when T_1 turns off and T_2 operates in the saturation region, yielding $I_{D_1} = 0$ and $I_{D_2} = I_B$. Using these values together with Equation (5.20), the input linear range can be expressed as

$$|V_i| \leq \sqrt{\frac{I_B}{K}}. \tag{5.24}$$

The differential output current, Δi, is derived as follows:

$$\Delta i = I_{D_1} - I_{D_2} = \begin{cases} \sqrt{2KI_B}\, V_i \sqrt{1 - \dfrac{V_i^2}{2I_B/K}}, & \text{if } |V_i| \leq \sqrt{\dfrac{I_B}{K}} \\ I_B \,\text{sign}(V_i), & \text{if } |V_i| > \sqrt{\dfrac{I_B}{K}}. \end{cases} \tag{5.25}$$

This last expression is valid provided the transistors do not turn off. The linear region can be made large by increasing the bias current or transistor lengths. The transconductance of the differential transistor pair represents the slope of the I/V characteristic depicted in Figure 5.2(b). Taking the derivative of Δi with respect to V_i gives

$$g_m = \frac{d\Delta i}{dV_i} = \begin{cases} \sqrt{2KI_B} \left(\sqrt{1 - \dfrac{V_i^2}{2I_B/K}} - \dfrac{\dfrac{V_i^2}{2I_B/K}}{\sqrt{1 - \dfrac{V_i^2}{2I_B/K}}} \right), & \text{if } |V_i| \leq \sqrt{\dfrac{I_B}{K}} \\ \\ 0, & \text{if } |V_i| > \sqrt{\dfrac{I_B}{K}}. \end{cases} \tag{5.26}$$

Figure 5.2(c) shows the plot of the transconductance versus the input voltage. At $V_i = 0$, the transconductance is reduced to $\sqrt{2KI_B}$. The square root term is associated with nonlinear distortions that can affect the transconductance for high values of the input voltages.

5.1.3 Current mirror

The circuit diagram of a current mirror using p-channel transistors is depicted in Figure 5.3(a). The currents flowing through the transistors T_1 and T_2 are given by

$$I_i = K_1(V_{SG_1} + V_{T_p})^2 \tag{5.27}$$

and

$$I_0 = K_2(V_{SG_2} + V_{T_p})^2, \tag{5.28}$$

respectively, where $K_1 = \mu_p(C_{ox}/2)(W_1/L_1)$ and $K_2 = \mu_p(C_{ox}/2)(W_2/L_2)$. Assuming that $V_{SG_1} = V_{SG_2}$, the current ratio can be expressed as

$$\frac{I_0}{I_i} = \frac{W_2/L_2}{W_1/L_1}. \tag{5.29}$$

The output current is represented in Figure 5.3(b) in the case where the input

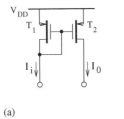
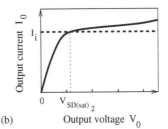

(a)   (b)

FIGURE 5.3
(a) Circuit diagram and (b) output characteristic of a simple current mirror.

current is constant. The output signal swing is limited by the voltage required to maintain the transistor in the saturation region, that is,

$$V_{SD(sat)} = V_{SG} + V_{T_p}. \tag{5.30}$$

Ideally, the input and output current should have the same value provided that the transistors are matched. However, the actual output resistance of the current mirror is limited instead of being infinite, and the input and output currents are equal only when $V_{SG_1} = V_0 = V_{SD_2}$. As a result, the achievable linearity range of a simple current mirror is reduced.

5.1.4 Slew-rate limitation

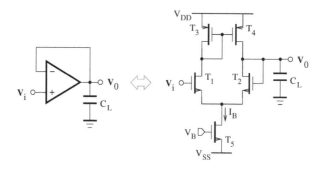

FIGURE 5.4
Differential stage in unity-gain configuration with a capacitive load.

The output slew rate of a differential stage is defined as the maximum rate of change of the output voltage. Because the current required to charge the output capacitor is given by

$$i = C_L \frac{dV_0}{dt}, \tag{5.31}$$

the slew rate of the differential stage shown in Figure 5.4 can be expressed as

$$SR = \max\left(\left|\frac{dV_0}{dt}\right|\right) = \frac{I_B}{C_L}. \tag{5.32}$$

The slew rate is generally expressed in units of V/µs. For example, the slew rate of a differential stage with $I_B = 20$ µA and $C_L = 2$ pF is 10 V/µs. In the case of a differential stage configured, as shown in Figure 5.4, to operate as a unity-gain amplifier with a sinusoidal input signal of the form

$$V_i = V_{max} \sin(2\pi f t), \tag{5.33}$$

the derivative of the output voltage is

$$\frac{dV_0}{dt} = 2\pi f V_{max} \cos(2\pi f t), \tag{5.34}$$

where V_{max} is the peak voltage and f denotes the input frequency. To avoid distortions due to slewing, the slew rate must satisfy the following constraint:

$$SR \geq 2\pi f_{max} V_{max}, \tag{5.35}$$

where f_{max} is the maximum input frequency.

5.1.5 Small-signal characteristics

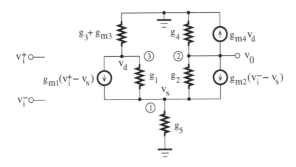

FIGURE 5.5
Small-signal equivalent model of the differential stage.

The small-signal equivalent model of the differential stage of Figure 5.1(a) is depicted in Figure 5.5. The body effects of transistors are assumed to be negligible. The transistor T_5, which generates the bias current, is represented by its output resistance. The diode-connected transistor, T_3, is modeled by a resistor with the conductance $g_3 + g_{m3}$. Applying Kirchhoff's current law to the nodes 1, 2, and 3, we obtain

$$g_{m1}(v_i^+ - v_s) + g_1(v_d - v_s) + g_{m2}(v_i^- - v_s) + g_2(v_0 - v_s) - g_5 v_s = 0, \quad (5.36)$$
$$g_{m2}(v_i^- - v_s) + g_2(v_0 - v_s) + g_{m4}v_d + g_4 v_0 = 0, \quad (5.37)$$

and

$$(g_3 + g_{m3})v_d + g_{m1}(v_i^+ - v_s) + g_1(v_d - v_s) = 0, \quad (5.38)$$

respectively. Under the assumption that the transistors, T_1 and T_2, and T_3 and T_4, are matched, that is, $g_{m1} = g_{m2}$, $g_1 = g_2$, $g_{m3} = g_{m4}$, and $g_3 = g_4$, replacing Equation (5.37) by the difference of Equations (5.38) and (5.37), and Equation (5.38) by the sum of Equations (5.38) and (5.37) gives

$$2g_{m1}v_{ic} + g_1 v_0 + g_1 v_d - [g_5 + 2(g_1 + g_{m1})]v_s = 0, \quad (5.39)$$
$$g_{m1}v_{id} - (g_1 + g_3)v_0 + (g_1 + g_3)v_d = 0, \quad (5.40)$$

and

$$2g_{m1}v_{ic} + (g_1 + g_3)v_0 + (g_1 + g_3 + 2g_{m3})v_d - 2(g_1 + g_{m1})v_s = 0, \quad (5.41)$$

where

$$v_{id} = v_i^+ - v_i^- \quad (5.42)$$

and

$$v_{ic} = \frac{v_i^+ + v_i^-}{2}. \quad (5.43)$$

Solving this last system of three equations gives the small-signal output voltage of the form

$$v_0 = A_d v_{id} + A_c v_{ic}, \tag{5.44}$$

where

$$A_d = \frac{g_{m1}\{g_1 g_5 + (g_3 + 2g_{m3})[g_5 + 2(g_1 + g_{m1})]\}}{2(g_1 + g_3)\{g_1 g_5 + (g_3 + g_{m3})[g_5 + 2(g_1 + g_{m1})]\}} \tag{5.45}$$

and

$$A_c = \frac{-g_5 g_{m1}}{g_1 g_5 + (g_3 + g_{m3})[g_5 + 2(g_1 + g_{m1})]}. \tag{5.46}$$

The common-mode rejection ratio (CMRR) is defined by

$$CMRR = \left| \frac{A_d}{A_c} \right|, \tag{5.47}$$

where A_d and A_c are the differential and common-mode gains, respectively.

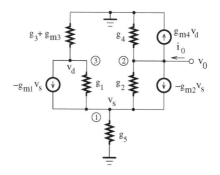

FIGURE 5.6
Small-signal equivalent model of the differential stage for the output resistance determination.

To determine the output resistance, the input nodes of the differential stage are short-circuited to the ground and a test generator is connected to the output node. The corresponding small-signal equivalent model is illustrated in Figure 5.6. Applying Kirchhoff's current law to the nodes 1, 2, and 3, we get

$$-g_{m1}v_s + g_1(v_d - v_s) - g_{m2}v_s + g_2(v_0 - v_s) - g_5 v_s = 0 \tag{5.48}$$

$$-g_{m2}v_s + g_2(v_0 - v_s) + g_{m4}v_d + g_4 v_0 = i_0 \tag{5.49}$$

and

$$(g_3 + g_{m3})v_d - g_{m1}v_s + g_1(v_d - v_s) = 0, \tag{5.50}$$

where $g_{m1} = g_{m2}$, $g_{m3} = g_{m4}$, $g_1 = g_2$, and $g_3 = g_4$. These equations can be solved for the output resistance given by

$$r_0 = \frac{v_0}{i_0} = \frac{g_1 g_5 + g_1(g_1 + g_{m1}) + (g_3 + g_{m3})[g_5 + 2(g_1 + g_{m1})]}{(g_1 + g_3)\{g_1 g_5 + (g_3 + g_{m3})[g_5 + 2(g_1 + g_{m1})]\}}. \tag{5.51}$$

In practice, g_{m1}, $g_{m3} \gg g_1$, g_3, g_5, and it can then be shown that

$$A_d \simeq \frac{g_{m1}}{g_1 + g_3}, \tag{5.52}$$

$$A_c \simeq \frac{-g_5}{2g_{m3}}, \tag{5.53}$$

$$r_0 \simeq \frac{1}{g_1 + g_3}, \tag{5.54}$$

and

$$CMRR \simeq 2\frac{g_{m1}g_{m3}}{g_5(g_1 + g_3)}, \tag{5.55}$$

where $g_1 = g_{ds_1}$, $g_3 = g_{ds_3}$, and $g_5 = g_{ds_5}$. The transconductance and conductance are proportional to the W/L ratio of the transistor, and a large differential gain is realized for a given bias current provided the aspect ratio of differential transistors is much greater than the one of load transistors. To obtain a high CMRR, the value of g_5 must be small, as is the case when a bias current source with a high output resistance is used.

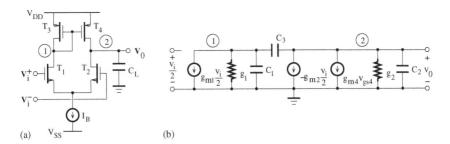

FIGURE 5.7
(a) Single-ended transconductance amplifier; (b) small-signal equivalent model of the transconductance amplifier including parasitic capacitances.

At high frequencies, the effect of parasitic capacitances can no longer be neglected. The amplifier gain and impedances then become dependent on the signal frequency. The circuit diagram of a single-ended transconductance amplifier and its small-signal equivalent model are respectively shown in Figures 5.7(a) and (b). Using Kirchhoff's current law, the equations for the node 1 and node 2 can be written as

$$g_{m1}V_i/2 + (g_1 + SC_1)V_1 - SC_3(V_0 - V_1) = 0 \tag{5.56}$$

and

$$-g_{m2}V_i/2 + SC_3(V_0 - V_1) + g_{m4}V_{gs4} + (g_2 + SC_2)V_0 = 0, \tag{5.57}$$

respectively. The system of Equations (5.56) and (5.57) can be solved for the transfer function given by

$$A(s) = \frac{V_0(s)}{V_i(s)} = \frac{g_{m1}(g_1 + g_{m4} + sC_1)/2}{(g_1 + sC_1)(g_2 + sC_2) + sC_3[g_1 + g_2 + g_{m4} + s(C_1 + C_2)]},$$
(5.58)

where $C_1 = C_{gd1} + C_{db1} + C_{db3} + C_{gs3} + C_{gs4}$, $C_2 = C_{gd2} + C_{db2} + C_{db4} + C_L$, $C_3 = C_{gd4}$, $g_1 = g_{ds1} + g_{ds3} + g_{m3}$, and $g_2 = g_{ds2} + g_{ds4}$. Due to the relative low value of C_{gd4}, the capacitance C_3 can be considered negligible. Furthermore, because $g_{ds1} + g_{ds3} + g_{m3} \simeq g_{m3}$, and the transistors T_3 and T_4 are matched ($g_{m3} = g_{m4}$), we obtain

$$A(s) = \frac{V_0(s)}{V_i(s)} \simeq A_0 \frac{1 - s/\omega_{z_1}}{(1 - s/\omega_{p_1})(1 - s/\omega_{p_2})},$$
(5.59)

where $A_0 = g_{m1}/(g_{ds2} + g_{ds4})$, $\omega_{z_1} = -2g_{m3}/C_1$, $\omega_{p_1} = -(g_{ds2} + g_{ds4})/C_2$ and $\omega_{p_2} = -g_{m3}/C_1$. The main contribution to the value of C_1 can be attributed to the sum of equal capacitances C_{gs3} and C_{gs4}. The second pole is then located at approximately half the transition frequency of the transistor T_3, while the zero occurs almost at the transition frequency, which is generally in the range of few hundred megahertz. Therefore, the zero and second pole are rejected at very high frequencies, and the amplifier frequency response can be described by a transfer function with a single dominant pole.

(a)   (b)

FIGURE 5.8
(a) Symbol of a single-ended transconductance amplifier; (b) small-signal equivalent model of a single-ended transconductance amplifier.

When the differential stage is used with a capacitive load, it operates as an operational transconductance amplifier (OTA), which is a differential voltage-controlled current source. Figure 5.8(a) shows the symbol of a single-ended OTA. A single-stage transconductance amplifier essentially features a dominant-pole behavior and can be modeled as a first-order system. Its small-signal equivalent model is depicted in Figure 5.8(b), where the current delivered to the overall output load, or the parallel combination of the output resistor and capacitor, is directly related to the input differential voltage. The proportionality factor, g_m, between the output current and input differential voltage is known as the transconductance. In the frequency domain, the voltage-gain transfer function can be written as

$$A(s) = \frac{V_0(s)}{V_i(s)} = g_{m1}Z_0,$$
(5.60)

where $Z_0 = r_0 \parallel 1/(sC_0)$, $s = j\omega$ and $\omega = 2\pi f$. It can then be found that

$$A(s) = \frac{V_0(s)}{V_i(s)} = \frac{A_0}{1 + s/\omega_c}, \qquad (5.61)$$

where the dc gain is of the form, $A_0 = g_{m1}r_0$, and $\omega_c = 1/r_0C_0$ is the 3-dB cutoff frequency. The *gain-bandwidth product* is defined as

$$\text{GBW} = A_0\omega_c = g_{m1}/C_0. \qquad (5.62)$$

In practice, ω_c is generally very small and $\omega \gg \omega_c$. Hence,

$$A(s) = \frac{V_0(s)}{V_i(s)} \simeq \frac{\omega_t}{s}, \qquad (5.63)$$

where $\omega_t = \text{GBW}$. Because $|A(j\omega_t)| = 1$, the parameter ω_t is called the *unity-gain frequency* or *transition frequency*. The basic differential stage can exhibit a gain-bandwidth product in the megahertz range, but its gain is generally less than 40 dB.

Hand analysis of the differential stage in the frequency domain using a high-frequency equivalent model of transistors is quite tedious when no simplifying assumption is made. Thus, an insight into the frequency response is usually gained using computer-aided design programs, such as SPICE and Spectre.

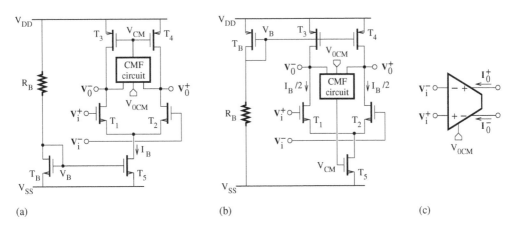

(a) (b) (c)

FIGURE 5.9
Circuit diagram of fully differential amplifiers with (a) output and (b) input common-mode control; (c) symbol of a fully differential amplifier.

The output swing and linearity can be further improved using fully differential amplifier structures, as shown in Figure 5.8(a) and (b), which feature differential input and output voltages. The transistors T_1 and T_2, and T_3 and T_4, are assumed to be matched. The common-mode gain of the differential input stage is approximately $A_c \simeq -g_5/(2g_l)$, where g_l is the conductance of the load transistor T_3 or T_4. In general, the conductance, g_5, of the transistor T_5, or say the output conductance of the bias current source, is small and A_c has a very low value. Therefore, only limited control can be exerted by the common-mode input voltage on the common-mode output voltage through the local feedback for common-mode signals. A common-mode feedback (CMF) circuit is then required to regulate the output common-mode voltage to the desired value, V_{OCM}. Furthermore, the CMF circuit can suppress the common-mode variations at the output, thereby preventing a change in the operation region for the load transistors due to IC process variations and loading conditions. Ideally, the differential output voltage is not affected by common-mode signals and the small-signal differential gain is approximately given by

$$A_d = \frac{v_0}{v_i} \simeq g_{m1}r_0 , \qquad (5.64)$$

where $r_0 \simeq 1/(g_1 + g_3)$, $v_i = v_i^+ - v_i^-$, and $v_0 = v_0^+ - v_0^-$. On the other hand, the voltage at each output node can swing symmetrically around the output common-mode level set by the CMF circuit.

The symbol of a fully differential amplifier is shown in Figure 5.8(c). For simplification, the V_{OCM} node is generally omitted. A fully differential amplifier has the advantage over its single-ended counterpart of rejecting the common-mode noise. It should also be noted that even-order nonlinearities are canceled at the outputs of a differential circuit that is balanced or whose both sides are electrically similar and symmetrical with respect to the ground.

5.1.6   Offset voltage

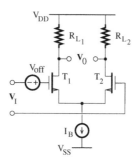

FIGURE 5.10
Circuit diagram of a differential transistor pair with the input offset voltage.

Amplifier stages may exhibit an input offset voltage, V_{off}, in the range of µV to mV due to mismatches of the transistor characteristics. Figure 5.10 shows a differential stage, where the offset voltage is modeled by a voltage source, V_{off}, in series with one of the input nodes. The differential input voltage can be expressed as

$$V_{ICM} = V_{GS_1} - V_{GS_2} . \qquad (5.65)$$

By exploiting the square-law I/V characteristic of MOS transistors in the saturation region, it can be shown that

$$V_{GS_1} = V_{T_1} + \sqrt{\frac{I_{D_1}}{K_1}} \tag{5.66}$$

and

$$V_{GS_2} = V_{T_2} + \sqrt{\frac{I_{D_2}}{K_2}}. \tag{5.67}$$

To proceed further, the drain currents of transistors T_1 and T_2 can be expressed as

$$\triangle I_D = I_{D_1} - I_{D_2} \tag{5.68}$$

and

$$I_B = I_{D_1} + I_{D_2}, \tag{5.69}$$

where I_B is the bias current. Solving this last system of equations gives

$$I_{D_1} = (I_B + \triangle I_D)/2 \tag{5.70}$$

and

$$I_{D_2} = (I_B - \triangle I_D)/2. \tag{5.71}$$

The output shift from zero experienced by the differential stage when the inputs are short-circuited to the ground can be described by the input offset voltage, which is the value of the input differential voltage, V_{ICM}, needed to reset the differential output voltage to zero. That is,

$$V_{off} = V_{T_1} - V_{T_2} + \sqrt{\frac{I_B}{2K_1}\left(1 + \frac{\triangle I_D}{I_B}\right)} - \sqrt{\frac{I_B}{2K_2}\left(1 - \frac{\triangle I_D}{I_B}\right)} \tag{5.72}$$

and

$$V_0 = R_{L_1}I_{D_1} - R_{L_2}I_{D_2} = 0. \tag{5.73}$$

Considering the output load resistances of the form $R_{L_1} = (R_L + \triangle R_L)/2$ and $R_{L_2} = (R_L - \triangle R_L)/2$, it can be shown that

$$\frac{\triangle I_D}{I_B} = -\frac{\triangle R_L}{R_L}. \tag{5.74}$$

The input offset voltage of the differential stage is then given by

$$V_{off} = V_{T_1} - V_{T_2} + \sqrt{\frac{I_B}{2K_1}\left(1 - \frac{\triangle R_L}{R_L}\right)} - \sqrt{\frac{I_B}{2K_2}\left(1 + \frac{\triangle R_L}{R_L}\right)}. \tag{5.75}$$

In practice, the mismatch errors are small so that the contribution of the square root terms is almost negligible, and V_{off} is primarily due to differences in the threshold voltage, V_T, caused by variations in the width, length, thickness, and doping levels of the transistor channels.

5.1.7 Noise in a differential transistor pair

A differential stage including the input-referred noise, which is modeled by a voltage source and a current source, is depicted in Figure 5.11. In this way, the effect of the source finite input impedance on the output noise is accurately described. The output noise is due to only the current source provided the input is open, while the voltage source will be the only contribution to the output noise if the input is set to zero. To proceed further, the different noise sources in the circuit are assumed to be uncorrelated.

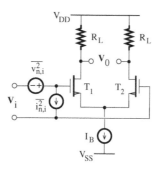

FIGURE 5.11

Circuit diagram of a differential MOS transistor stage with input-referred noises.

By shorting the input nodes together, the total output noise voltage can be written as

$$\overline{v_{n,0}^2} = (\overline{i_{th1}^2} + \overline{i_{th2}^2})R_L^2 + 2(4kTR_L), \tag{5.76}$$

where $\overline{i_{th1}^2}$ and $\overline{i_{th2}^2}$ are the mean-square values of the thermal noise currents of T_1 and T_2, respectively, k is the Boltzmann constant, and T is the absolute temperature. Let g_m be the transistor transconductance. The mean-square value of the input-referred noise voltage is then given by

$$\overline{v_{n,i}^2} = \frac{\overline{v_{n,0}^2}}{g_m^2 R_L^2}, \tag{5.77}$$

where $g_m R_L$ denotes the gain of the differential transistor pair. The $1/f$ noise is generally represented by a voltage source, $\overline{v_{1/f}^2}$, in series with the transistor gate. In this case, we have

$$\overline{v_{n,i}^2} = \frac{\overline{v_{n,0}^2}}{g_m^2 R_L^2} + 2\overline{v_{1/f}^2}. \tag{5.78}$$

Here, the common unit for the input-referred noise voltage is usually nV/$\sqrt{\text{Hz}}$.

The input-referred noise current of the differential MOS transistor stage can be determined by opening the input nodes and expressing the ratio of the spectral density of the total noise voltage at the output node and the square of the transimpedance gain in terms of the circuit noise contributions. Typically, it is primarily dependent on the shot noise of the input bias current and can be assumed to be negligible due to its value, which is generally in the fA/√Hz range.

In practice, the level of the input-referred noise signal in the passband can be related to the smallest level of the signal that the amplifier circuit can adequately process.

5.1.8 Operational amplifier

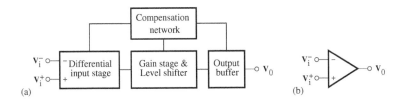

FIGURE 5.12
(a) Block diagram and (b) symbol of a single-ended operational amplifier.

Operational transconductance amplifiers, which are equivalent to a voltage-controlled current source, are generally based on single-stage structures, while additional gain stages and an output buffer are required for the design of operational amplifiers operating as a voltage-controlled voltage source. A high-performance operational amplifier should provide a high input impedance, a high open-loop gain, a large CMRR, a low dc offset voltage, a low noise, and a low output impedance. Figure 5.12 shows the block diagram and symbol of a single-ended operational amplifier. The output of the differential stage is supplied to additional gain stages in order to meet the high gain requirement. A level shifter is needed whenever the dc voltage difference introduced between the input and output voltages of a stage should be canceled. An output buffer with unity gain will be necessary if the amplifier is supposed to drive a resistive load. The use of a compensation network is necessary to avoid the conditions leading to instability when the amplifier operates with a feedback.

A differential-to-single-ended converter can be implemented, as depicted in Figure 5.13. Using the superposition theorem and voltage division principle, it can be shown that

$$V^+ = \frac{R_4}{R_3 + R_4}V_i^+ \tag{5.79}$$

and

$$V^- = \frac{R_2}{R_1 + R_2}V_i^- + \frac{R_1}{R_1 + R_2}V_0, \tag{5.80}$$

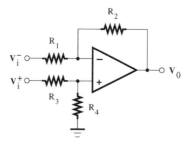

FIGURE 5.13
Circuit diagram of a differential-to-single-ended converter.

where $V_0 = A(V^+ - V^-)$. Solving the system of Equations (5.79) and (5.80) gives

$$V_0 = \frac{\dfrac{R_1 + R_2}{R_1} \dfrac{R_4}{R_3 + R_4} V_i^+ - \dfrac{R_2}{R_1} V_i^-}{1 + \dfrac{1}{A} \dfrac{R_1 + R_2}{R_1}}. \tag{5.81}$$

Assuming that $R_2/R_1 = R_4/R_3 = k$, we obtain

$$V_0 = \frac{k(V_i^+ - V_i^-)}{1 + \dfrac{1 + k}{A}}. \tag{5.82}$$

In practice, the amplifier gain, A, is very high and does not affect the output voltage. However, the achievable common-mode rejection ratio may be limited by the resistor mismatches.

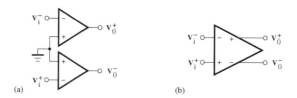

FIGURE 5.14
Symbols of (a) pseudo-differential and (b) fully differential operational amplifiers.

For low-voltage applications, the signal swing may be increased using a pseudo-differential structure based on two single-ended amplifiers with their noninverting inputs connected to the ground, as illustrated in Figure 5.14(a). However, due to the lack of a common-mode feedback path, the amplifier linearity may be affected by any imbalance between the signal paths. Another solution is to design the operational amplifier with a differential configuration. Figure 5.14(b) shows the symbol of a fully differential operational amplifier.

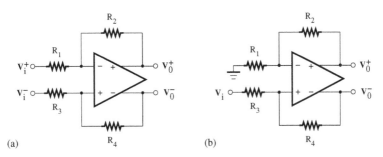

FIGURE 5.15
(a) Differential amplifier with negative feedback; (b) single-ended-to-differential converter.

More insight into the operation of a fully differential amplifier is gained by analyzing the structure shown in Figure 5.15(a). Here, the amplifier output voltage can be expressed as

$$V_0^+ - V_0^- = A(V^+ - V^-), \tag{5.83}$$

where A is the amplifier gain, V^+ and V^- are the voltage at the noninverting and inverting input nodes, respectively. Using the principles of superposition and voltage division, we find

$$V^- = \frac{R_2}{R_1 + R_2}V_i^+ + \frac{R_1}{R_1 + R_2}V_0^+ \tag{5.84}$$

and

$$V^+ = \frac{R_4}{R_3 + R_4}V_i^- + \frac{R_3}{R_3 + R_4}V_0^-. \tag{5.85}$$

Let $\alpha = R_3/(R_3 + R_4)$ and $\beta = R_1/(R_1 + R_2)$. Combining Equations (5.84) and (5.85) with Equation (5.83), we obtain

$$(1 + \beta A)V_0^+ - (1 + \alpha A)V_0^- = A[(1 - \alpha)V_i^- - (1 - \beta)V_i^+]. \tag{5.86}$$

The voltage at each output node is independently defined by setting the output common-mode voltage to a given level, V_{0CM}, that is,

$$\frac{V_0^+ + V_0^-}{2} = V_{0CM} \tag{5.87}$$

Solving Equations (5.86) and (5.87) gives

$$V_0^- = \frac{-(1 - \alpha)V_i^- + (1 - \beta)V_i^+ + 2(\beta + 1/A)V_{0CM}}{(\alpha + \beta)\{1 + 2/[(\alpha + \beta)A]\}} \tag{5.88}$$

and

$$V_0^+ = \frac{(1-\alpha)V_i^- - (1-\beta)V_i^+ + 2(\alpha + 1/A)V_{0CM}}{(\alpha + \beta)\{1 + 2/[(\alpha + \beta)A]\}}. \tag{5.89}$$

The differential output voltage is then given by

$$V_0 = \frac{2[(1-\alpha)V_i^- - (1-\beta)V_i^+ + (\alpha - \beta)V_{0CM}]}{(\alpha + \beta)\{1 + 2/[(\alpha + \beta)A]\}}, \tag{5.90}$$

where $V_0 = V_0^+ - V_0^-$. The differential amplifier is generally designed with identical resistors R_1 and R_3, and R_2 and R_4. Hence, $\alpha = \beta$ and the differential output voltage is reduced to

$$V_0 = \frac{-(1-\alpha)V_i}{\alpha[1 + 1/(\alpha A)]}, \tag{5.91}$$

where $V_i = V_i^+ - V_i^-$. In the ideal case, the gain of the amplifier is very high so that $1 + 1/(\alpha A) \simeq 1$. The differential input voltage is then amplified by the factor $(1 - \alpha)/\alpha = R_2/R_1$.

Singled-ended signals can be converted to differential signals using the circuit shown in Figure 5.15(b). It can be shown that

$$V_0 = \frac{2[(1-\alpha)V_i + (\alpha - \beta)V_{0CM}]}{(\alpha + \beta)\{1 + 2/[(\alpha + \beta)A]\}}, \tag{5.92}$$

where $V_i^- = V_i$, $V_i^+ = 0$, and $V_0^+ - V_0^- = V_0$. To prevent the output common-mode voltage, V_{0CM}, from affecting the output differential voltage, the resistors R_1 and R_3, and R_2 and R_4, should be matched so that $\alpha = \beta$.

5.2 Linearization techniques for transconductors

In order to increase the input dynamic range of the basic differential stage shown in Figure 5.16(a), various linearization techniques can be used. Consider a differential stage consisting of two source-connected transistors operating in the saturation region, where the lowest internal channel resistance can be achieved. Assuming that the transistors are matched, the drain currents are given by

$$I_{D_1} = K(V_{GS_1} - V_T)^2 \tag{5.93}$$

and

$$I_{D_2} = K(V_{GS_2} - V_T)^2. \tag{5.94}$$

The output current of the differential stage can be expressed as

$$\Delta i = I_{D_1} - I_{D_2} = K(V_{GS_1} + V_{GS_2} - 2V_T)(V_{GS_1} - V_{GS_2}), \tag{5.95}$$

where the differential input voltage is equal to the difference of the gate-source voltages, while the linearity of the transconductance is achieved by maintaining the term $V_{GS_1} + V_{GS_2} - 2V_T$ or the sum of the gate-source voltages constant. Approaches used to linearize the transconductance can then be implemented using a source degeneration resistor or floating dc voltages. In the case of a transconductance realized by transistors operating in the triode region, where the drain current is of the form

$$I_D = K[2(V_{GS} - V_T)V_{DS} - V_{DS}^2], \qquad (5.96)$$

the drain-source voltage should be maintained constant to improve the linear range of the differential stage.

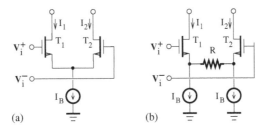

(a)   (b)

FIGURE 5.16
(a) Basic differential stage; (b) differential stage with resistor source degeneration.

A differential stage with resistor source degeneration is depicted in Figure 5.16(b). The currents flowing through the transistors T_1 and T_2 can be written as

$$I_{D_1} = I_B + \triangle i = K(V_{GS_1} - V_T)^2 \qquad (5.97)$$

and

$$I_{D_2} = I_B - \triangle i = K(V_{GS_2} - V_T)^2. \qquad (5.98)$$

Applying Kirchhoff's voltage law to the input loop involving T_1, R, and T_2 gives

$$V_i^+ - V_{GS_1} - R\triangle i + V_{GS_2} - V_i^- = 0, \qquad (5.99)$$

where

$$V_{GS_1} = V_T + \sqrt{\frac{I_B + \triangle i}{K}} \qquad (5.100)$$

and

$$V_{GS_2} = V_T + \sqrt{\frac{I_B - \triangle i}{K}}. \qquad (5.101)$$

Hence,

$$V_i - R\triangle i = \sqrt{\frac{I_B}{K}} \left[\left(1 + \frac{\triangle i}{I_B}\right)^{1/2} - \left(1 - \frac{\triangle i}{I_B}\right)^{1/2} \right], \qquad (5.102)$$

where $V_i = V_i^+ - V_i^-$. Assuming that $\Delta i/I_B \ll 1$, we get $(1 + \Delta i/I_B)^{1/2} \simeq 1 + \Delta i/2I_B$ and $(1 - \Delta i/I_B)^{1/2} \simeq 1 - \Delta i/2I_B$. Thus,

$$\Delta i \simeq \frac{g_m}{1 + g_m R} V_i ,\tag{5.103}$$

where $g_m = \sqrt{KI_B}$.

The linearity can be further improved using amplifiers to sense the input voltages, as shown in Figure 5.17(a). Ideally, the difference between the voltages at the noninverting and inverting nodes of the amplifier is zero, and the input voltages are directly applied across the resistor R due to the feedback path. Hence,

$$\Delta i = V_i/R.\tag{5.104}$$

Note that it is generally required to optimize the bandwidth of these extra amplifiers so as not to reduce the speed of the overall circuit. Furthermore, the accuracy of the transconductance, which is inversely proportional to the resistance R, may be affected by process variations.

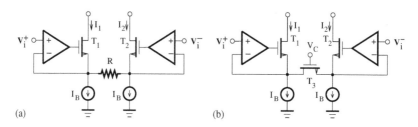

(a) (b)

FIGURE 5.17
(a) Improved differential stage with a resistor source degeneration; (b) differential stage with a single transistor source degeneration.

In the case of the differential stage shown in Figure 5.17(b), a transistor operating in the triode region is used to realize the source degeneration resistor [2], whose value can be controlled by means of the drain voltage. The transistors T_1 and T_2 are assumed to operate in the saturation region. They play the role of source followers driving T_3, which is biased in the triode region. Let V_{ICM} and V_C be the dc component of the input voltages and the gate voltage of T_3, respectively. Assuming that $V_i^+ = V_{ICM} + V_i/2$ and $V_i^- = V_{ICM} - V_i/2$, the current flowing through T_3 can be expressed as

$$\Delta i = K[2(V_{GS_3} - V_T)V_{DS_3} - V_{DS_3}^2],\tag{5.105}$$

where

$$V_{GS_3} = V_C - V_{ICM} - \frac{V_i}{2},\tag{5.106}$$

$$V_{DS_3} = V_i ,\tag{5.107}$$

and $V_i = V_i^+ - V_i^-$ is the differential input voltage. Hence,

$$\triangle i = 2K[(V_C - V_{ICM} - V_T)V_i - V_i^2]. \qquad (5.108)$$

For large values of the differential input voltage, the transconductance linearity can be degraded due to the significant contribution associated with the V_i^2 term.

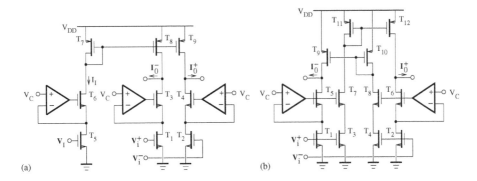

FIGURE 5.18
Differential stages with an active cascode configuration: (a) Basic structure and (b) implementation with improved linearity.

A linear transconductance characteristic can also be provided by transistors operating in the triode region with constant drain-source voltages. Figure 5.18(a) shows the basic structure of a differential stage with an active cascode configuration [3]. The amplifier feedback helps maintain constant the drain-source voltage of transistors T_1 and T_2, thereby guaranteeing the accuracy of the resulting transconductance. Assuming that $V_i^+ = V_{ICM} + V_i/2$ and $V_i^- = V_{ICM} - V_i/2$, the drain currents of the identical transistors T_1, T_2, and T_5 in the triode region can be expressed as

$$I_{D_1} = K[2(V_{GS_1} - V_T)V_{DS_1} - V_{DS_1}^2], \qquad (5.109)$$

$$I_{D_2} = K[2(V_{GS_2} - V_T)V_{DS_2} - V_{DS_2}^2], \qquad (5.110)$$

and

$$I_I = K[2(V_{GS_5} - V_T)V_{DS_5} - V_{DS_5}^2], \qquad (5.111)$$

where $V_{GS_1} = V_{ICM} + V_i/2$, $V_{GS_2} = V_{ICM} - V_i/2$, $V_{GS_5} = V_{ICM}$, $V_{DS_1} = V_{DS_2} = V_{DS_5} = V_C$, and $K = (1/2)\mu C_{ox}(W/L)$. The output currents are respectively given by

$$I_0^- = I_I - I_{D_1} = -KV_C V_i \qquad (5.112)$$

and

$$I_0^+ = I_I - I_{D_2} = KV_C V_i. \qquad (5.113)$$

The resulting transconductance is of the form

$$g_m = \triangle i/V_i = 2KV_C, \qquad (5.114)$$

where $\Delta i = I_0^+ - I_0^-$, and it can be tuned by varying the level of the control voltage V_C. For operation in the triode region, it is required that $V_{DS} \leq V_{GS} - V_T$, or in the worst case,

$$V_C \leq V_{ICM} - V_T - V_i/2. \tag{5.115}$$

In practice, the transconductance linearity can be degraded due to variations in the mobility, μ, which exhibits a nonlinear dependence with respect to the gate-source voltage, that is, $\mu = f(V_{GS})$, where f is a nonlinear function. To overcome this drawback, the input transistor pairs can be duplicated as shown in Figure 5.18(b) [4]. Assuming that the transistors $T_1 - T_4$ are matched and operate in the triode region, we have

$$I_{D_1} = I_{D_3} = K[2(V_{ICM} + V_i/2 - V_T)V_C - V_C^2] \tag{5.116}$$

and

$$I_{D_2} = I_{D_4} = K[2(V_{ICM} - V_i/2 - V_T)V_C - V_C^2]. \tag{5.117}$$

The current mirrors T_9-T_{10} and $T_{11}-T_{12}$ are required to respectively combine the currents I_{D_4} and I_{D_1}, and I_{D_3} and I_{D_2}, resulting in

$$I_0^- = I_{D_4} - I_{D_1} = -KV_CV_i \tag{5.118}$$

and

$$I_0^+ = I_{D_3} - I_{D_2} = KV_CV_i. \tag{5.119}$$

Here, the solution adopted for the common-mode rejection at the output nodes relies on using an extra feed-forward transconductor, which can only provide a dc signal equal in magnitude to the common-mode component to be subtracted. Note that the effect of IC process variations affecting the drain currents of the input transistors can also be canceled in the same way.

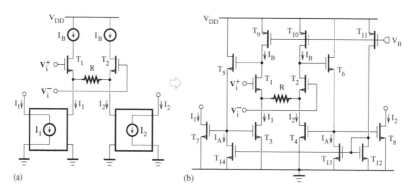

FIGURE 5.19

Differential stage with a resistor source degeneration: (a) Operation principle, (b) implementation.

An alternative structure for the realization of a differential stage with resistor source degeneration is derived by maintaining the same voltage between the gate and source of the input transistors, irrespective of the input voltage magnitude [5]. The principle and implementation of this approach are illustrated in Figures 5.19(a) and (b). The transistors operate in the saturation region. The biasing of each of the differential input transistors, T_1 and T_2, with the identical current source of the value, I_B, and the feedback path realized by T_5 and T_6, force the gate-source voltages of transistors T_1 and T_2 to be constant and equal, that is,

$$V_{GS_1} = V_{GS_2}. \tag{5.120}$$

The transistors $T_{11} - T_{14}$ are required to set the bias current, I_A, for the transistors T_5 and T_6. The input voltage is level-shifted and applied across the resistor R, and the current flowing through the resistor R is then given by

$$\triangle i = V_i/R, \tag{5.121}$$

where $V_i = V_i^+ - V_i^-$. The drain currents of the transistors T_3 and T_4 are respectively of the form

$$I_{D_3} = I_1 = I_B - \triangle i \tag{5.122}$$

and

$$I_{D_4} = I_2 = I_B + \triangle i. \tag{5.123}$$

By connecting the gates of transistors T_7 and T_8 to the gates of transistors T_3 and T_4, respectively, we get $I_1 = I_{D_3}$ and $I_2 = I_{D_4}$.

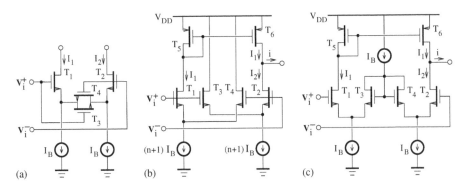

FIGURE 5.20
Differential stage with active source degeneration: (a) Transistor in the triode region, (b) cross-coupled configuration, (c) transistor in the saturation region.

A differential stage with active source degeneration is shown in Figure 5.20(a) [32]. Let V_{ICM} and V_C be the dc component of the input voltages and the gate voltage of T_3, respectively; $V_i^+ = V_{ICM} + V_i/2$; and $V_i^- = V_{ICM} - V_i/2$. The transistors T_3 and T_4, which are identical and biased in the triode region, are equivalent to a parallel configuration of resistors, whose conductances are of the form

$$g_{ds_3} = I_{D_3}/V_{DS_3} = K[2(V_{GS_3} - V_T) - V_{DS_3}] \tag{5.124}$$

and

$$g_{ds_4} = I_{D_4}/V_{DS_4} = K[2(V_{GS_4} - V_T) - V_{DS_4}], \qquad (5.125)$$

where

$$V_{GS_3} = V_i + V_{GS_2}, \qquad (5.126)$$
$$V_{GS_4} = -V_i + V_{GS_1}, \qquad (5.127)$$
$$V_{DS_3} = V_i - V_{GS_1} + V_{GS_2} = -V_{DS_4}, \qquad (5.128)$$

and $V_i = V_i^+ - V_i^-$ is the differential input voltage. The transconductance of the differential stage can then be written as

$$g_m = \frac{g_{m_1}}{1 + g_{m_1}/(g_{ds_3} + g_{ds_3})}, \qquad (5.129)$$

where

$$g_{m_1} = \sqrt{K_1 I_B} = K_1 \sqrt{\frac{I_B}{K_1}} \qquad (5.130)$$

and

$$g_{ds_3} + g_{ds_4} = 2K_3(V_{GS_1} + V_{GS_2} - 2V_T). \qquad (5.131)$$

Assuming that the transistors T_1 and T_2 are matched and operate in the saturation region, and that $\triangle i \ll I_B$, we have

$$V_{GS_1} = V_T + \sqrt{\frac{I_B + \triangle i}{K_1}} \simeq V_T + \sqrt{\frac{I_B}{K_1}}\left(1 + \frac{\triangle i}{2I_B}\right), \qquad (5.132)$$

$$V_{GS_2} = V_T + \sqrt{\frac{I_B - \triangle i}{K_1}} \simeq V_T + \sqrt{\frac{I_B}{K_1}}\left(1 - \frac{\triangle i}{2I_B}\right), \qquad (5.133)$$

and Equation (5.131) becomes

$$g_{ds_3} + g_{ds_4} \simeq 4K_3\sqrt{\frac{I_B}{K_1}}. \qquad (5.134)$$

Hence,

$$g_m \simeq \frac{g_{m_1}}{1 + \dfrac{K_1}{4K_3}} = \frac{g_{m_1}}{1 + \dfrac{(W_1/L_1)}{4(W_3/L_3)}}, \qquad (5.135)$$

where W_1/L_1 and W_3/L_3 denote the gate width-to-length ratios of the transistors T_1 and T_3, respectively.

The input dynamic range of a differential stage can also be improved using two floating dc voltage sources to keep constant the sum of gate-source voltages of input transistors. The differential stage implementation shown in Figure 5.20(b) is based on cross-coupled transistors [7]. Using Kirchhoff's voltage law, the input voltage can be related to the gate-source voltages of transistors $T_1 - T_4$ as

$$V_i = V_{GS_1} - V_{GS_4} = V_{GS_3} - V_{GS_2}, \quad (5.136)$$

where

$$V_{GS_j} = V_T + \sqrt{\frac{I_{D_j}}{K_j}} \quad (5.137)$$

and $V_i = V_i^+ - V_i^-$ is the differential input voltage. The transistor pairs T_1 and T_4, and T_3 and T_2, behave as differential stages. Assuming that $K_1 = K_2 = K$ and $K_1 = K_2 = K/n$, and taking into account the fact that

$$I_{D_1} = I_1 = I_B + i_1, \quad I_{D_2} = I_2 = I_B - i_2, \quad I_{D_3} = nI_B + i_2, \quad I_{D_4} = nI_B - i_1, \quad (5.138)$$

Equations (5.136) and (5.137) can be solved for i_1 and i_2. That is,

$$i_1 = \gamma K V_i^2 + \frac{\alpha}{2}\sqrt{K I_B} V_i \sqrt{1 - \eta K V_i^2 / I_B} \quad (5.139)$$

and

$$i_2 = -\gamma K V_i^2 + \frac{\alpha}{2}\sqrt{K I_B} V_i \sqrt{1 - \eta K V_i^2 / I_B}. \quad (5.140)$$

The output current i is then given by

$$i = i_1 + i_2 = \alpha\sqrt{K I_B} V_i \sqrt{1 - \eta K V_i^2 / I_B}, \quad (5.141)$$

where $\alpha = 4n/(n+1)$, $\eta = n/(n+1)^2$, and $\gamma = n(n-1)/(n+1)^2$. The input range is characterized by $|V_i| \leq \sqrt{I_B/(\eta K)}$. Provided the transistors are accurately matched, the linearity is improved as the value of η is reduced by increasing n, or correspondingly, the bias current.

In the case of the differential stage depicted in Figure 5.20(c), the transconductance linearization is achieved using two differential transistor pairs, T_1 and T_3, and T_2 and T_4, connected in series [8]. The voltages V_i^+ and V_i^- are applied to the gates of transistors T_1 and T_2, respectively. Because each transistor pair is driven by the bias current I_B and the diode-connected transistors T_3 and T_4 are wired to the current source I_B, we can write

$$I_B = I_{D_1} + I_{D_3} \quad (5.142)$$
$$= I_{D_2} + I_{D_4} \quad (5.143)$$
$$= I_{D_3} + I_{D_4}. \quad (5.144)$$

Thus,

$$I_{D_1} = I_{D_4} \quad \text{and} \quad I_{D_2} = I_{D_3}. \quad (5.145)$$

Assuming that the transistors $T_1 - T_4$ are identical and operate in the saturation region, the voltage applied to each transistor pair is $V_i/2$, where $V_i = V_i^+ - V_i^-$. On the basis of Kirchhoff's voltage law, it can be easily shown that

$$V_i/2 = V_{GS_1} - V_{GS_3} \tag{5.146}$$
$$= V_{GS_4} - V_{GS_2}. \tag{5.147}$$

The drain currents of transistors T_1 and T_2 can then be obtained as

$$I_{D_1} = \frac{1}{2}\left[I_B + K\frac{V_i}{2}\sqrt{\frac{2I_B}{K} - \frac{V_i^2}{4}}\right] \tag{5.148}$$

and

$$I_{D_2} = \frac{1}{2}\left[I_B - K\frac{V_i}{2}\sqrt{\frac{2I_B}{K} - \frac{V_i^2}{4}}\right], \tag{5.149}$$

where $|V_i| \le 2\sqrt{2I_B/K}$. Note that $I_1 = I_{D_1}$ and $I_2 = I_{D_2}$. Therefore, the output current i is given by

$$i = I_{D_1} - I_{D_2} = K\frac{V_i}{2}\sqrt{\frac{2I_B}{K} - \frac{V_i^2}{4}}. \tag{5.150}$$

In comparison with the case of the basic differential stage, the input dynamic range is extended by a factor of 2 while the transconductance is decreased by a factor of 2.

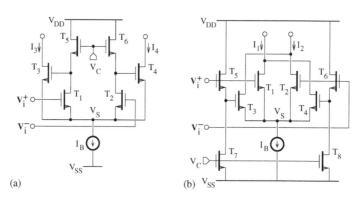

(a)   (b)

FIGURE 5.21
Circuit diagram of differential stages using voltage-controlled floating dc sources and featuring an output current (a) dependent on and (b) independent of the threshold voltage.

Differential stages can also be designed using voltage-controlled floating dc sources to maintain constant the sum of gate-source voltages of input transistors, as illustrated in Figure 5.21. This approach is known as the bias offset technique. The transistors are assumed to be identical and operate in the saturation region. In the case of the circuit depicted in Figure 5.21(a) [9], the drain currents of transistors T_1 and T_2 can be written as

$$I_{D_1} = K(V_{ICM} + V_i/2 - V_S - V_T)^2 \qquad (5.151)$$

and

$$I_{D_2} = K(V_{ICM} - V_i/2 - V_S - V_T)^2, \qquad (5.152)$$

where $V_i^+ = V_{ICM} + V_i/2$ and $V_i^- = V_{ICM} - V_i/2$. Because $I_{D_1} = I_{D_5}$ and $I_{D_2} = I_{D_6}$, we have $V_{GS_1} = V_{GS_5}$ and $V_{GS_2} = V_{GS_6}$. The gate voltages of transistors T_3 and T_4 are then of the form

$$V_{G_3} = V_C - V_{GS_5} = V_C - V_{GS_1} = V_C - V_{ICM} - V_i/2 \qquad (5.153)$$

and

$$V_{G_4} = V_C - V_{GS_6} = V_C - V_{GS_2} = V_C - V_{ICM} + V_i/2. \qquad (5.154)$$

Hence,

$$I_{D_3} = K(V_C - V_{ICM} - V_i/2 - V_S - V_T)^2 \qquad (5.155)$$

and

$$I_{D_4} = K(V_C - V_{ICM} + V_i/2 - V_S - V_T)^2. \qquad (5.156)$$

The difference in the output currents is given by

$$\triangle i = I_3 - I_4 = I_{D_3} - I_{D_4} = 2K(V_{ICM} - V_C - V_T)V_i. \qquad (5.157)$$

To remove the effect of the threshold voltage on the output voltage, the differential stage of Figure 5.21(b) [10] can be used. The transistors are identical and biased in the saturation region. Assuming that $V_i^+ = V_{ICM} + V_i/2$ and $V_i^- = V_{ICM} - V_i/2$, the drain currents of transistors T_1 and T_2 can be expressed as

$$I_{D_1} = K(V_{ICM} + V_i/2 - V_S - V_T)^2 \qquad (5.158)$$

and

$$I_{D_2} = K(V_{ICM} - V_i/2 - V_S - V_T)^2. \qquad (5.159)$$

With the same current flowing through T_5 and T_7, and T_6 and T_8, it can be shown that $V_{GS_5} = V_{GS_7}$ and $V_{GS_6} = V_{GS_8}$. The gate voltages of transistors T_3 and T_4 are then given by

$$V_{G_3} = V_i^+ - V_{GS_5} = V_{ICM} + V_i/2 - V_{GS_7} = V_{ICM} + V_i/2 - V_B \quad (5.160)$$

and

$$V_{G_4} = V_i^- - V_{GS_6} = V_{ICM} - V_i/2 - V_{GS_8} = V_{ICM} - V_i/2 - V_B, \quad (5.161)$$

where $V_B = V_C - V_{SS}$. Hence,

$$I_{D_3} = K(V_{ICM} + V_i/2 - V_B - V_S - V_T)^2 \quad (5.162)$$

and

$$I_{D_4} = K(V_{ICM} - V_i/2 - V_B - V_S - V_T)^2. \quad (5.163)$$

The difference in the output currents is of the form

$$\triangle i = I_1 - I_2 = (I_{D_1} + I_{D_4}) - (I_{D_2} + I_{D_3}) = 2KV_BV_i. \quad (5.164)$$

Applying Kirchhoff's current law at the common source node of transistors $T_1 - T_4$, we have

$$(I_{D_1} + I_{D_4}) + (I_{D_2} + I_{D_3}) = I_1 + I_2 = I_B. \quad (5.165)$$

Solving Equations (5.164) and (5.165) gives

$$I_1 = I_{D_1} + I_{D_4} = I_B/2 + KV_BV_i \quad (5.166)$$

and

$$I_2 = I_{D_2} + I_{D_3} = I_B/2 - KV_BV_i. \quad (5.167)$$

The range of input voltages over which the transistors still operate in the saturation region can be determined by the next worst-case requirement,

$$V_{GS_4} = V_{ICM} - V_i/2 - V_B - V_S \geq V_T. \quad (5.168)$$

Substituting Equations (5.158) and (5.163) into Equation (5.166), we obtain

$$(V_{ICM} - V_S - V_T)^2 - V_B(V_{ICM} - V_S - V_T) + (V_i^2 + 2V_B^2 - I_B/K)/4 = 0. \quad (5.169)$$

This quadratic equation can be solved for $V_{ICM} - V_S - V_T$. That is,

$$V_{ICM} - V_S - V_T = \frac{V_B}{2} \pm \frac{1}{2}\sqrt{\frac{I_B}{K} - V_B^2 - V_i^2}, \quad (5.170)$$

where $I_B/K \geq V_B^2 + V_i^2$. Because the condition set by Equation (5.168) can only be met by the solution with the sign + between both terms, combining Equations (5.170) and (5.168) gives

$$V_i^2 + V_B V_i + V_B^2 - \frac{I_B}{2K} \leq 0, \qquad (5.171)$$

or equivalently,

$$|V_i| \leq -\frac{V_B}{2} + \sqrt{\frac{I_B}{2K} - \frac{3V_B^2}{4}}, \quad |V_B| \leq \sqrt{\frac{2I_B}{3K}}. \qquad (5.172)$$

Note that the other possible solution of Equation (5.171) provides a negative bound for the magnitude of the differential input voltage and is not suitable.

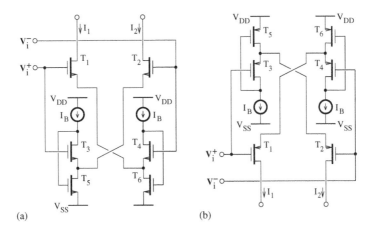

FIGURE 5.22
Circuit diagram of differential stages using floating voltage follower sources: (a) n-channel and (b) p-channel input transistors.

An alternative approach that can be adopted to increase the linearity range consists of using the differential stages shown in Figure 5.22 [11, 12]. Here, two floating dc voltage sources consisting of two transistors biased by a constant current are used to maintain constant the sum of the gate-source voltages associated with the differential transistor pair. One terminal of the input voltage is connected to the gate of the transistors T_1 and T_3, while the other is connected to the gate of transistors T_2 and T_4. The dc voltages available at the source-drain junctions of transistors T_3 and T_5, and T_4 and T_6, are applied to the sources of the input transistors T_2 and T_1, respectively.

Let us consider the differential stage with n-channel input transistors. Assuming that the transistors have the same threshold voltage and operate in the saturation region, we can write

$$I_1 = K_1(V_{GS_1} - V_T)^2 \qquad (5.173)$$

$$I_2 = K_2(V_{GS_2} - V_T)^2 \qquad (5.174)$$

and

$$I_B = K_3(V_{GS_3} - V_T)^2 \qquad (5.175)$$
$$= K_4(V_{GS_4} - V_T)^2. \qquad (5.176)$$

Because $V_{G_1} = V_{G_3} = V_i^+$ and $V_{G_2} = V_{G_4} = V_i^-$, it can be shown that

$$V_{GS_1} = V_{G_1} - V_{S_1} = V_i^+ - V_{S_4} = V_i + \sqrt{\frac{I_B}{K_4}} + V_T \qquad (5.177)$$

and

$$V_{GS_2} = V_{G_2} - V_{S_2} = V_i^- - V_{S_3} = -V_i + \sqrt{\frac{I_B}{K_3}} + V_T, \qquad (5.178)$$

where $V_i = V_i^+ - V_i^-$. Hence,

$$I_1 = K_1 \left(\sqrt{\frac{I_B}{K_4}} + V_i \right)^2 \qquad (5.179)$$

and

$$I_2 = K_2 \left(\sqrt{\frac{I_B}{K_3}} - V_i \right)^2. \qquad (5.180)$$

With $K_1 = K_2$ and $K_3 = K_4$, the difference in the output currents can be computed as

$$i = I_1 - I_2 = 4K_1 \sqrt{\frac{I_B}{K_3}} V_i. \qquad (5.181)$$

To maintain the input transistors in the saturation region, the currents I_1 and I_2 should not be equal to zero. That is,

$$-\sqrt{\frac{I_B}{K_3}} < V_i < \sqrt{\frac{I_B}{K_3}}. \qquad (5.182)$$

This differential input stage structure is capable of a class AB operation and is suitable for low-voltage applications.

Another approach for the transconductor linearization consists of using two current-controlled floating dc voltage sources to keep constant the sum of gate-source voltages of input transistors, as illustrated in Figures 5.23(a) and (b) for n-channel and p-channel input transistors, respectively [13]. All transistors operate in the saturation region and it is assumed that the transistors with the same channel type are matched. Let us consider the structure of Figure 5.23(a). Applying Kirchhoff's voltage law to both input loops gives

$$V_i = V_{GS_1} + V_{GS_5} - (V_{GS_7} + V_{GS_4}) \qquad (5.183)$$

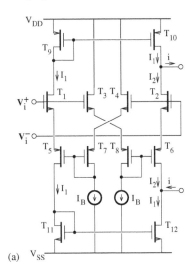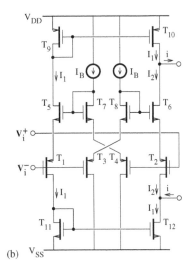

FIGURE 5.23
Circuit diagram of differential stages using current-controlled floating voltage sources: (a) *n*-channel and (b) *p*-channel input transistors.

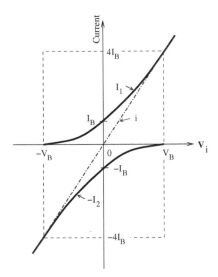

FIGURE 5.24
I/V transfer characteristic of a differential stage using current-controlled floating voltage sources.

and

$$-V_i = V_{GS_2} + V_{GS_6} - (V_{GS_8} + V_{GS_3}), \tag{5.184}$$

where

$$V_{GS_1} = \sqrt{\frac{I_1}{K_n}} + V_{T_n}, \tag{5.185}$$

$$V_{GS_2} = \sqrt{\frac{I_2}{K_n}} + V_{T_n} \,, \tag{5.186}$$

$$V_{GS_3} = V_{GS_4} = \sqrt{\frac{I_B}{K_n}} + V_{T_n} \,, \tag{5.187}$$

$$V_{GS_5} = \sqrt{\frac{I_1}{K_p}} + V_{T_p} \,, \tag{5.188}$$

$$V_{GS_6} = \sqrt{\frac{I_2}{K_p}} + V_{T_p} \,, \tag{5.189}$$

and

$$V_{GS_7} = V_{GS_8} = \sqrt{\frac{I_B}{K_p}} + V_{T_p} \,. \tag{5.190}$$

To proceed further, it can be shown that

$$V_i = \sqrt{\frac{I_1}{K_{eq}}} - V_B \tag{5.191}$$

and

$$-V_i = \sqrt{\frac{I_2}{K_{eq}}} - V_B \,, \tag{5.192}$$

where

$$V_B = \sqrt{\frac{I_B}{K_{eq}}} \tag{5.193}$$

and

$$K_{eq} = \frac{K_n K_p}{(\sqrt{K_n} + \sqrt{K_p})^2} \,. \tag{5.194}$$

The currents I_1 and I_2 are, respectively, given by

$$I_1 = K_{eq}(V_B + V_i)^2 \tag{5.195}$$

and

$$I_2 = K_{eq}(V_B - V_i)^2 \,. \tag{5.196}$$

The output current is then obtained as

$$i = I_1 - I_2 = 4K_{eq}V_B V_i = 4\sqrt{K_{eq}I_B}\, V_i \,. \tag{5.197}$$

As illustrated in Figure 5.24, the differential stage remains in the linear range provided the currents I_1 and I_2 are different from zero. From Equations (5.195) and (5.196), it can be deduced that

$$-\sqrt{I_B/K_{eq}} < V_i < \sqrt{I_B/K_{eq}} \tag{5.198}$$

or equivalently,

$$-4I_B < i < 4I_B \,. \tag{5.199}$$

This differential stage design has the advantages of featuring a source node and a sink node for the output current.

5.3 Single-stage amplifier

The circuit diagram of a single-stage amplifier [14] is shown in Figure 5.25. The differential transistor pair, $T_1 - T_2$, converts the input voltage into currents, which are directed to the output stage by stacked current mirrors. The tail current of the input stage, the transconductance of which is assumed to be g_m, is set by $T_{15} - T_{16}$. Using Kirchhoff's voltage law, we can obtain

$$V_i^+ = V_{DD} - V_{SG_5} - V_{SG_3} - V_{DS_1} + V_{GS_1} \tag{5.200}$$
$$= V_{GS_1} + V_{DS_{15}} + V_{SS} \,. \tag{5.201}$$

Assuming that the dc voltage applied to both amplifier inputs is V_{ICM}, it can be shown that

$$\begin{aligned} V_{DS_1(sat)} + & V_{DS_{15}(sat)} + V_{T_n} + V_{SS} \\ & \leq V_{ICM} \leq V_{DD} - V_{SD_5(sat)} - V_{SD_3(sat)} + 2V_{T_p} + V_{T_n} \,, \end{aligned} \tag{5.202}$$

where V_{ICM} denotes the input common-mode voltage. To maintain the transistors T_8, T_{10}, T_{12}, and T_{14} in the saturation region, the signal swing at the output node should be bounded as follows:

$$V_{DS_{12}(sat)} + V_{DS_{14}(sat)} + V_{T_n} + V_{SS} \leq V_0 \leq V_{DD} - V_{SD_{10}(sat)} - V_{SD_8(sat)} + V_{T_p} \tag{5.203}$$

Note that the available output swing can be increased using high-swing cascode current mirrors, which are able to cancel the voltage threshold term.

Let a_i be the amplification factor provided by the load devices, T_3-T_6 and T_7-T_{10}, and r_0 be the output resistance. The dc voltage gain of the amplifier can be written as

$$A_0 = g_m a_i r_0 \,, \tag{5.204}$$

where g_m is the transconductance of each of the input transistor T_1 or T_2. Assuming that $W_3/L_3 = W_5/L_5$, $W_4/L_4 = W_6/L_6$, $W_7/L_7 = W_9/L_9$, and $W_8/L_8 = W_{10}/L_{10}$, it can be shown that $a_i = (W_6/L_6)/(W_5/L_5) = (W_{10}/L_{10})/(W_9/L_9)$. The amplifier is compensated by output capacitive loads. It can be modeled by a single-stage, small-signal equivalent circuit, provided that C_L is sufficiently large so that the pole associated to the output

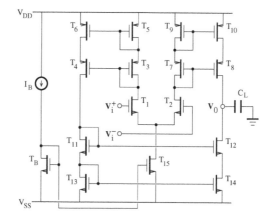

FIGURE 5.25
Circuit diagram of a single-stage amplifier.

node is dominant. The gain-bandwidth product, GBW, and slew rate, SR, are then given by

$$GBW = \frac{g_m a_i}{C_0} \simeq \frac{g_m a_i}{C_L} \tag{5.205}$$

and

$$SR = \frac{a_i I_B}{C_0} \simeq \frac{a_i I_B}{C_L}, \tag{5.206}$$

where I_B is the bias current applied to the input differential stage, C_0 is the total output capacitance, and C_L denotes the external load capacitance connected to the output node.

5.4 Folded-cascode amplifier

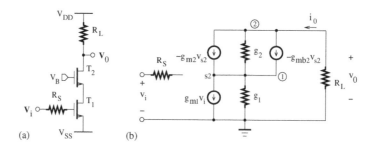

FIGURE 5.26
(a) Circuit diagram and (b) small-signal equivalent model of a cascode amplifier.

An amplifier with a cascode configuration has the advantage of exhibiting a higher bandwidth due its improved isolation between the input and output nodes. Figure 5.26(a) shows the circuit diagram of a cascode amplifier, whose input stage is a common-source transistor driving a common-gate transistor loaded by a resistor. By replacing each transistor with its equivalent model, the small-signal circuit shown in Figure 5.26(b) can be derived. It is assumed that $v_{gs2} = -v_{s2}$ and $v_{bs2} = -v_{s2}$. Using Kirchhoff's current law for the nodes 1 and 2, we may write

$$g_{m_1}v_i + g_1 v_{s2} + (g_{m_2} + g_{mb_2})v_{s2} + g_2(v_{s2} - v_0) = 0, \tag{5.207}$$

$$-(g_{m_2} + g_{mb_2})v_{s2} - g_2(v_{s2} - v_0) + v_0/R_L = 0. \tag{5.208}$$

Solving the system of Equations (5.207) and (5.208), we get

$$A_v = \frac{v_0}{v_i} = \frac{-g_{m_1}(g_{m_2} + g_{mb_2} + g_2)}{(g_1 + g_2)/R_L + g_2(g_1 + g_{m_2} + g_{mb_2})}. \tag{5.209}$$

For the determination of the output resistance, the input voltage is set to zero and a voltage source is applied at the amplifier output node. The output node equation is

$$i_0 = -(g_{m_2} + g_{mb_2})v_{s2} + g_2(v_0 - v_{s2}), \tag{5.210}$$

where

$$v_{s2} = i_0/g_1 . \tag{5.211}$$

It can then be shown that

$$r_0 = \frac{v_0}{i_0} = 1/g_1 + 1/g_2 + (g_{m_2} + g_{mb_2})/g_1 g_2 , \tag{5.212}$$

where $g_1 = g_{ds_1}$ and $g_2 = g_{ds_2}$. The output resistance is derived as

$$R_0 = r_0 \parallel R_L . \tag{5.213}$$

When R_L is implemented by a passive resistor, it is much lower than r_0, so that $R_0 \simeq R_L$ and the voltage gain is reduced to $A_v \simeq -g_{m_1} R_L$. On the other hand, an active load is used for the implementation of R_L in the folded-cascode amplifier in order to meet the high gain specification.

The folded-cascode amplifier [15, 16] schematic with a single-ended output is shown in Figure 5.27. The voltage-to-current converter based on transistors $T_1 - T_2$ is connected to an output stage with the folded-cascode configuration consisting of $T_5 - T_{12}$. Using Kirchhoff's voltage law, it can be found that

$$V_i^- = V_{DD} - V_{DS_6} - V_{DS_1} + V_{GS_1} \tag{5.214}$$

$$= V_{GS_1} + V_{DS_3} + V_{DS_4} + V_{SS} \tag{5.215}$$

and

$$V_0^- = V_{DD} - V_{SD_6} - V_{SD_8} \tag{5.216}$$

$$= V_{DS_{10}} + V_{DS_{12}} + V_{SS} . \tag{5.217}$$

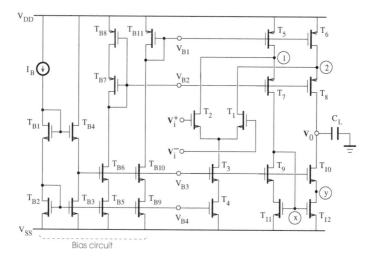

FIGURE 5.27
Circuit diagram of a folded-cascode amplifier.

To ensure normal operation of the amplifier, the transistors should be biased slightly above the saturation region. In the worst case, the input common-mode range is given by

$$V_{DS_1(sat)} + V_{T_n} + V_{DS_3(sat)} + V_{DS_4(sat)} + V_{SS} \leq V_{ICM} \leq V_{DD} - V_{SD_6(sat)} + V_{T_n}, \quad (5.218)$$

where V_{ICM} is the dc voltage that can be applied to both inputs. To keep the transistors in the saturation region, the output voltage swing should be bounded as follows:

$$V_{DS_{12}(sat)} + V_{DS_{10}(sat)} + V_{SS} \leq V_0 \leq V_{DD} - V_{SD_6(sat)} - V_{SD_8(sat)}. \quad (5.219)$$

The biasing circuit [6] consisting of transistors $T_{B1} - T_{B11}$ should be designed to set the quiescent points of the amplifier transistors, such that they can operate in the saturation region. Note that the stability of transistor quiescent points may be affected by variations of the current I_B and fluctuations of the IC process.

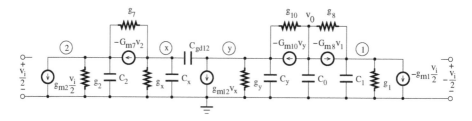

FIGURE 5.28
Equivalent circuit model of a folded-cascode amplifier.

The small-signal equivalent model of the folded-cascode amplifier is depicted in Figure 5.28. Applying Kirchhoff's current law, the equations for the node 1, node 2, node x, node y, and output node can be written as

$$-g_{m_1}V_i/2 + (G_{m_8} + g_1 + sC_1)V_1 + g_8(V_1 - V_0) = 0, \tag{5.220}$$

$$g_{m_2}V_i/2 + (g_2 + sC_2 + G_{m_7})V_2 + g_7(V_2 - V_x) = 0, \tag{5.221}$$

$$-G_{m_7}V_2 - g_7(V_2 - V_x) + (g_x + sC_x)V_x = 0, \tag{5.222}$$

$$g_{m_{12}}V_x + (G_{m10} + g_y + sC_y)V_y + g_{10}(V_y - V_0) = 0, \tag{5.223}$$

and

$$-G_{m10}V_y - g_{10}(V_y - V_0) - G_{m_8}V_1 - g_8(V_1 - V_0) + sC_0V_0 = 0, \tag{5.224}$$

respectively, where $G_{m_k} = g_{m_k} + g_{mb_k}$, $(k = 7, 8, 10)$, $g_1 = g_{ds_1} + g_{ds_6}$, $g_2 = g_{ds_2} + g_{ds_5}$, $g_x = g_{m11}$, and $g_y = g_{ds_{12}}$. Note that g_{ml} and g_{ds_l} denote the transconductance and the drain-source conductance of the transistor T_l (l is an integer), respectively. For the determination of node capacitances, it can be assumed that the scaling factor of the capacitor $C_{gd_{12}}$ provided by the Miller effect is approximately equal to unity due to the low amplification gain available at the source of the transistor T_{10}. Hence,

$$C_1 = C_{gd_1} + C_{db_1} + C_{gd_6} + C_{db_6} + C_{gs_8} + C_{sb_8}, \tag{5.225}$$
$$C_2 = C_{gd_2} + C_{db_2} + C_{gd_5} + C_{db_5} + C_{gs_7} + C_{sb_7}, \tag{5.226}$$
$$\begin{aligned} C_x = C_{gd_7} + C_{db_7} + C_{gd_9} + C_{db_9} \\ + C_{gs_{11}} + C_{gd_{11}} + C_{gb_{11}} + C_{gs_{12}} + C_{gd_{12}} + C_{gb_{12}}, \end{aligned} \tag{5.227}$$
$$C_y = C_{gs_{10}} + C_{sb_{10}} + C_{gd_{12}} + C_{db_{12}}, \tag{5.228}$$

and

$$C_0 = C_L + C_{gd_8} + C_{db_8} + C_{gd_{10}} + C_{db_{10}}. \tag{5.229}$$

Assuming that the transconductances are much greater than the conductances, the system of Equations (5.221–5.224) can be solved for a transfer function of the form

$$A(s) = \frac{V_0(s)}{V_i(s)} = A_0 \frac{(1 - s/\omega_{z_1})(1 - s/\omega_{z_2})}{(1 - s/\omega_{p_1})(1 - s/\omega_{p_2})(1 - s/\omega_{p_3})(1 - s/\omega_{p_4})}. \tag{5.230}$$

For practical component values, the first pole is dominant, that is, $|\omega_{p_1}| \ll |\omega_{p_2}|, |\omega_{p_3}|, |\omega_{p_4}|$. The amplifier transfer function can then be approximated as

$$A(s) \simeq A_0 \frac{(1 - s/\omega_{z_1})(1 - s/\omega_{z_2})}{D(s)}, \tag{5.231}$$

where

$$D(s) = 1 - \frac{s}{\omega_{p1}} + \frac{s^2}{\omega_{p1}} \left(\frac{1}{\omega_{p2}} + \frac{1}{\omega_{p3}} + \frac{1}{\omega_{p4}} \right)$$
$$- \frac{s^3}{\omega_{p1}} \left(\frac{1}{\omega_{p2}\omega_{p3}} + \frac{1}{\omega_{p2}\omega_{p4}} + \frac{1}{\omega_{p3}\omega_{p4}} \right) + \frac{s^4}{\omega_{p1}\omega_{p2}\omega_{p3}\omega_{p4}}. \tag{5.232}$$

Assuming that the transistor T_1 and T_2 are matched, the small-signal dc gain of the amplifier is written as

$$A_0 \simeq \frac{g_{m1}}{\dfrac{(g_{ds1} + g_{ds6})g_{ds8}}{G_{m8}} + \dfrac{g_{ds10}g_{ds12}}{G_{m10}}}. \tag{5.233}$$

The poles of the transfer function are given by

$$\omega_{p1} = -\left(\frac{(g_{ds1} + g_{ds6})g_{ds8}}{G_{m8}} + \frac{g_{ds10}g_{ds12}}{G_{m10}} \right) \frac{1}{C_0} = \frac{g_{m1}}{A_0 C_0}, \tag{5.234}$$

$$\omega_{p2} = -\frac{g_{m11}}{C_x}, \tag{5.235}$$

$$\omega_{p3} = -\frac{G_{m10}}{C_y}, \tag{5.236}$$

and

$$\omega_{p4} = -\frac{G_{m8}}{C_1}, \tag{5.237}$$

while for the zeros, we have

$$\omega_{z1}, \omega_{z2} = -\frac{\omega_{p2} + \omega_{p3}}{2} \pm \sqrt{\frac{\omega_{p2}^2 + \omega_{p3}^2}{4} - \frac{3\omega_{p2}\omega_{p3}}{2}}. \tag{5.238}$$

Note that the dc gain is of the form, $A_0 \simeq g_{m1}R_0$, where R_0 is the output resistance. The frequency response shows that the folded-cascode amplifier has two left-half plane zeros, a dominant pole associated with the output node, and nondominant poles introduced by the current mirror, n-channel and p-channel cascode transistors.

The zeros, the second and third poles, which are closely located, form two doublets. Generally, pole-zero doublets have less influence on the frequency response, but can degrade the settling response. They should be located at frequencies greater than the unity-gain frequency of the amplifier in order to achieve the optimum settling time.

That is, the frequency behavior of the folded-cascode amplifier can be described by a two-pole system with the next transfer function

$$A(s) = \frac{V_0(s)}{V_i(s)} \simeq A_0 \frac{1}{(1 - s/\omega_{p_1})(1 - s/\omega_{p_4})}. \tag{5.239}$$

The phase margin can be obtained as

$$\phi_M = 180^o - \angle A[j(GBW)] \tag{5.240}$$

$$= 180^o - \arctan(GBW/\omega_{p_1}) - \arctan(GBW/\omega_{p_4}), \tag{5.241}$$

where GBW denotes the gain-bandwidth product or unity-gain frequency. Because $GBW = A_0\omega_{p_1}$ and the dc gain, A_0, is very high, the first arctan term tends to 90°. The expression of the phase margin then becomes

$$\phi_M = 90^o - \arctan(GBW/\omega_{p_4}). \tag{5.242}$$

In practice, the nondominant pole ω_{p_4} is located at a high frequency and the frequency response of the folded-cascode amplifier is primarily determined by the single dominant pole[1]. The frequency compensation, which should affect only the output node pole, is implemented by the load capacitor C_L.

5.5 Fully differential amplifier architectures

Differential amplifier architectures offer many design advantages (e.g., improved dynamic range, availability of inverting and noninverting functions on the same structure).

5.5.1 Fully differential folded-cascode amplifier

Depending on the trade-off to be achieved between the gain and speed specifications in a given application, fully differential folded-cascode amplifiers can be designed using either a basic or gain-enhanced structure.

5.5.1.1 Basic structure

The single-stage amplifier with cascode structure [15, 17], as shown in Figure 5.29, can provide an acceptable gain without degrading the high-frequency performance in most applications. It consists of a differential input gain stage $(T_1 - T_4)$ followed by a cascode loading structure $(T_5 - T_{12})$. The common-mode feedback (CMF) circuit is used in order to constrain the common-mode

---

[1]Assuming a dominant pole model for the amplifier, the gain-bandwidth product is approximately equal to the unity-gain or transition frequency.

(CM) output voltage to a desired dc operating point so that the output voltage swing can be maximized. Assuming that the transistors are biased to exhibit identical saturation voltages, the differential output swing is about $2V_{sup} - 4V_{DS(sat)} - 4V_{SD(sat)}$, where $V_{sup} = V_{DD} - V_{SS}$ represents the total supply voltage.

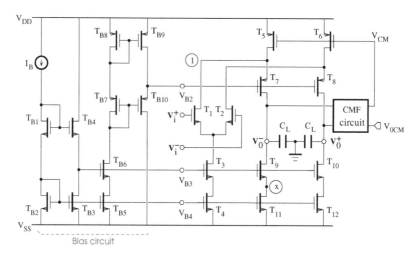

FIGURE 5.29
Fully differential folded-cascode amplifier.

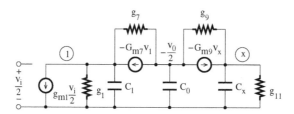

FIGURE 5.30
Equivalent circuit model of the differential amplifier half circuit.

The small-signal equivalent circuit of the differential amplifier half circuit, as shown in Figure 5.30, is used for the frequency domain analysis. Because both circuit sections around the axis of symmetry are matched, the axis of symmetry can be considered the ac ground. For the node 1, node x, and output node, the equations obtained using Kirchhoff's current law can be written as

$$g_{m1}V_i/2 + (g_1 + sC_1 + G_{m7})V_1 + g_7(V_1 + V_0/2) = 0, \quad (5.243)$$
$$G_{m9}V_x + g_9(V_x + V_0/2) + (g_{11} + sC_x)V_x = 0, \quad (5.244)$$

and

$$G_{m7}V_1 + g_7(V_1 + V_0/2) + G_{m9}V_x + g_9(V_x + V_0/2) + sC_0V_0/2 = 0, \quad (5.245)$$

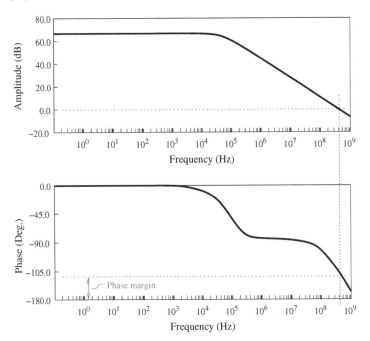

FIGURE 5.31
(a) Magnitude and (b) phase response (output v_0^-) of a folded-cascode amplifier.

respectively, where $G_{mk} = g_{mk} + g_{mbk}$, $(k = 7, 9)$, $g_1 = g_{ds_1} + g_{ds_5}$, $g_7 = g_{ds_7}$, $g_9 = g_{ds_9}$, and $g_{11} = g_{ds_{11}}$. Solving the system of Equations (5.243)–(5.245), the transfer function of the amplifier can be derived in the form

$$A(s) = \frac{V_0(s)}{V_i(s)} = A_0 \frac{1 - s/\omega_{z_1}}{(1 - s/\omega_{p_1})(1 - s/\omega_{p_2})(1 - s/\omega_{p_3})}, \qquad (5.246)$$

where ω_{z_1} represents the frequency of the zero z_1, and ω_{p_1}, ω_{p_2}, and ω_{p_3} are the frequencies of the poles p_1, p_2, and p_3, respectively. Generally, the amplifier is designed to feature the behavior of a dominant pole system, whose poles are widely spaced, that is, $|\omega_{p_1}| \ll |\omega_{p_2}|, |\omega_{p_3}|$ and $|\omega_{p_2}| \ll |\omega_{p_3}|$. Hence,

$$A(s) = \frac{V_0(s)}{V_i(s)} \simeq A_0 \frac{1 - s/\omega_{z_1}}{1 - s/\omega_{p_1} + s^2/\omega_{p_1}\omega_{p_2} - s^3/\omega_{p_1}\omega_{p_2}\omega_{p_3}}. \qquad (5.247)$$

Assuming that the transconductances are much higher than the conductances, we arrive at

$$A_0 \simeq \frac{g_{m1}}{\dfrac{(g_{ds_1} + g_{ds_5})g_{ds_7}}{G_{m7}} + \dfrac{g_{ds_9}g_{ds_{11}}}{G_{m9}}}. \qquad (5.248)$$

As the zero and third pole, which are located at about $-G_{m9}/C_x$, cancel each other out, the amplifier transfer function is reduced to the one of a second-order system, whose poles are given by

$$\omega_{p_1} = -g_{m1}/A_0 C_0 \tag{5.249}$$

and

$$\omega_{p_2} = -G_{m7}/C_1 , \tag{5.250}$$

where $C_0 = C_L + C_{gd_7} + C_{db_7} + C_{gd_9} + C_{db_9} + C'_i$, $C_1 = C_{gd_1} + C_{db_1} + C_{gd_5} + C_{db_5} + C_{gs_7} + C_{sb_7}$, and C'_i is the input capacitance of the CMF circuit. The compensation of the folded-cascode amplifier is achieved by the load capacitor C_L (see Figure 5.29), which primarily determines the frequency location of the dominant pole. In this case, the gain-bandwidth product, ω_{GBW}, can be expressed as

$$\omega_{GBW} \simeq \frac{g_{m1}}{C_L} . \tag{5.251}$$

The slew rate is given by

$$SR \simeq \frac{I_B}{C_L} , \tag{5.252}$$

where I_B is the bias current of the differential input stage. But, due to the amplifier stability condition, which imposes that all nondominant poles occur at frequencies past the unity-gain bandwidth frequency, the amplifier speed is limited by the position of the first nondominant pole. This pole comes from the G_{m7}/C_1 time constant of the cascode transistor (here T_7) and is specified by a pole frequency approximately at the transition frequency of this transistor. Note that the capacitor C_1 represents the total capacitive load at the source of the related transistor (or node 1). Of practical importance is the parameter K, expressed as

$$K = \tan(\phi_M) \simeq \frac{\omega_{p_2}}{\omega_{GBW}} , \tag{5.253}$$

where ω_{p_2} characterizes the first nondominant pole of the amplifier. The phase margin ϕ_M is commonly used for the definition of the stability. It should be noted that the minimum settling time at 0.1% is obtained for a value of ϕ_M around 76°.

Frequency responses of the fully differential amplifier are shown in Figures 5.31(a) and (b), where a phase difference of 180° exists between the two output voltages v_0^+ and v_0^-. For these plots, $V_{DD} = 3.3$ V, $V_{SS} = 0$, $C_L = 0.75$ pF, and a common-mode dc voltage of 1.5 V was added to the input voltages.

5.5.1.2 Gain-enhanced structure

Typically, amplifiers should exhibit a high open-loop gain and a high bandwidth to minimize errors in the output voltage. The high gain requirement can be met by cascading gain stages, the number of which is limited by the need for frequency compensation to enable stable feedback. The gain enhancement technique can also be implemented by inserting each cascode transistor in an amplifier feedback path to increase the overall output resistance.

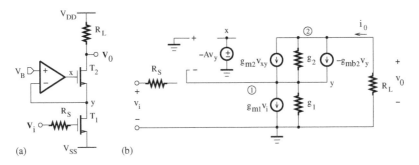

FIGURE 5.32
(a) Circuit diagram and (b) small-signal equivalent model of a gain-enhanced cascode amplifier.

The circuit diagram of a gain-enhanced cascode amplifier is shown in Figure 5.32(a). The feedback amplifier has a voltage gain, A, and V_B is a constant voltage. A small-signal equivalent model for this amplifier is illustrated in Figure 5.32(b), where $v_{gs2} = v_{xy}$ and $v_{bs2} = -v_{ds1} = -v_y$. Applying Kirchhoff's current law at the nodes 1 and 2 gives

$$g_{m_1} v_i + g_1 v_y - g_{m_2} v_{xy} + g_{mb_2} v_y + g_2(v_y - v_0) = 0, \tag{5.254}$$

$$g_{m_2} v_{xy} - g_{mb_2} v_y - g_2(v_y - v_0) + v_0/R_L = 0, \tag{5.255}$$

where

$$v_{xy} = v_x - v_y = -(A + 1)v_y. \tag{5.256}$$

Combining Equations (5.254) and (5.255), the voltage gain is obtained as

$$A_v = \frac{v_0}{v_i} = \frac{-g_{m_1}[(1 + A)g_{m_2} + g_{mb_2} + g_2]}{[(1 + A)g_{m_2} + g_{mb_2} + g_1 + g_2]/R_L + g_1 g_2}. \tag{5.257}$$

To find the output resistance, a test generator is connected to the amplifier output and the input node is short-circuited to ground. For the transistor T_1, $g_{m_1} v_i = 0$. The output node current equation can be written as

$$i_0 = g_{m_2} v_{xy} - g_{mb_2} v_y + g_2(v_0 - v_y), \tag{5.258}$$

where

$$v_y = i_0/g_1. \tag{5.259}$$

By solving Equations (5.258) and (5.259), we get

$$r_0 = \frac{v_0}{i_0} = 1/g_1 + 1/g_2 + [(1 + A)g_{m_2} + g_{mb_2}]/g_1 g_2. \tag{5.260}$$

Therefore, the overall output resistance is given by

$$R_0 = r_0 \parallel R_L. \tag{5.261}$$

Provided the resistance R_L is sufficiently high, the output resistance is enhanced by a factor on the order of the gain of the feedback amplifier or auxiliary amplifier. This principle can be exploited to meet the high gain and fast settling requirements in the design of amplifiers.

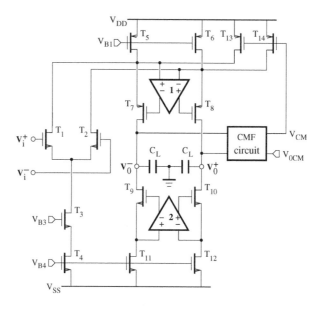

FIGURE 5.33

Fully differential folded-cascode amplifier with gain enhancement.

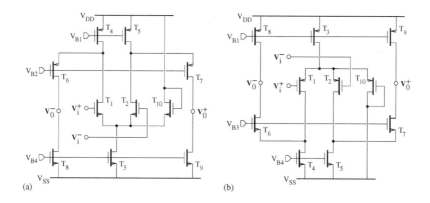

FIGURE 5.34

(a) Differential gain enhancement amplifier 1 and (b) its complementary version for the gain enhancement amplifier 2.

A single-stage amplifier based on an active cascode output stage is shown in Figure 5.33. The feedback created around the output transistors helps increase the output impedance of the main amplifier. As a result, its gain is enhanced without altering the bandwidth due to the scaling effect provided by the auxiliary amplifier gain [18–20]. The resulting dc gain, A_0, of the active cascode amplifier is given approximately by

$$A_0 \simeq A_0'(1 + A_0''), \qquad (5.262)$$

where A_0' denotes the dc gain of the main amplifier and A_0'' is the dc gain of the auxiliary amplifier. The auxiliary amplifiers, as shown in Figure 5.34, are fully differential and use a single transistor CMF circuit. In this way, only two auxiliary amplifiers are required, resulting in a reduction in area and power consumption in comparison with an architecture based on four single-ended auxiliary amplifiers. To provide a high open-loop gain, the output stage of the auxiliary amplifier must be implemented using transistors with minimum channel length.

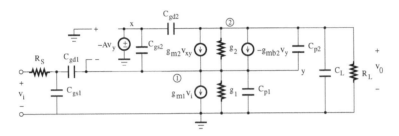

FIGURE 5.35

Small-signal equivalent model of the gain-enhanced cascode amplifier with parasitic and output capacitors.

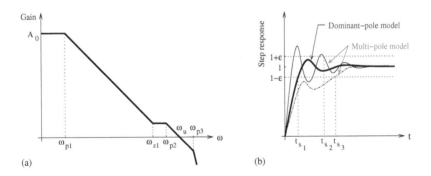

FIGURE 5.36

(a) Bode plot of the amplifier gain magnitude in open-loop configuration; (b) amplifier step response.

In order to analyze the high-frequency response of the gain-enhanced amplifier, the effects of parasitic and output capacitors are taken into account in the amplifier of Figure 5.32(a) to derive the small-signal equivalent model depicted in Figure 5.35, where C_L is the output load capacitor, C_{p_1} and C_{p2} represent all parasitic capacitors seen at the output nodes of transistors T_1 and T_2, respectively. An intuitive analysis can show that the behavior of the

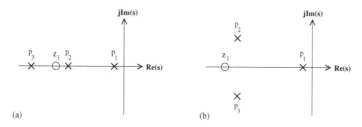

FIGURE 5.37
(a) Pole-zero locations of a nonoptimized gain-enhanced amplifier; (b) pole-zero locations of an optimized gain-enhanced amplifier.

gain-enhanced amplifier is determined by the first three poles and the first zero [20–22]. Figure 5.36(a) shows the Bode plot of the gain-enhanced amplifier gain. The first pole, which is located at the lowest frequency and is well separated from the second and third poles, is dominant. Its frequency is on the order of $g_1 g_2 / A_0'' g_{m1} g_{m2} C_L$ and is related to the gain-bandwidth product. When the closely spaced pole and zero (or doublet) characterized by ω_{p2} and ω_{z1} are located near the unity-gain frequency of the auxiliary amplifier, the settling time is increased due to slow components associated with a multi-pole system [23]. Because this is achieved even though the 60° phase margin criterion for the amplifier stability is met, a closed-loop configuration should be considered for the requirement specification to ensure a fast settling response similar to the one of an amplifier with a single dominant pole. In Figure 5.36(b), the settling responses to a voltage step input are shown as a function of time. Note that each small-signal settling time, t_{s_j} ($j = 1, 2, 3$), is defined as the minimum time required for the amplifier output voltage to settle to within an error tolerance, ϵ, of its final steady-state value.

For high-speed applications, the amplifier settling time should be optimized. It can be related to the characteristics of the closed-loop frequency response. For instance, an increase of the 3-dB bandwidth leads to a reduction in the settling time, while any oscillation or ringing in the closed-loop response may increase the settling time. By including the gain-enhanced amplifier in a closed loop with a constant feedback factor β, the closed-loop gain is given by

$$A_{CL}(s) = \frac{V_0(s)}{V_i(s)} = \frac{A(s)}{1 + \beta A(s)}, \tag{5.263}$$

where $A(s)$ is the open-loop gain of the amplifier. With the assumption that the unity-gain frequency, ω_u'', of the auxiliary amplifier is greater than the first pole frequency, ω_{p_1}', of the main amplifier and is lower than the unity-gain frequency, ω_u', of the main amplifier, that is,

$$\omega_{p_1}' < \omega_u'' < \omega_u', \tag{5.264}$$

the first pole of the gain-boosted amplifier is moved at a lower frequency than the remaining poles and zero and can then be considered dominant. Hence, the transfer function of the gain-boosted amplifier is approximately given by

$$A(s) \simeq \frac{A_0}{1 + s/\omega_{p_1}}. \tag{5.265}$$

This implies that the dominant-pole frequency response is obtained by designing the auxiliary amplifier to be slower than the main amplifier. Note that the gain-boosted amplifier and the main amplifier exhibit the same unity-gain frequency. Combining Equations (5.265) and (5.263) gives

$$A_{CL}(s) \simeq \frac{A_0/(1 + \beta A_0)}{1 + s/\omega_{p_1}(1 + \beta A_0)}. \tag{5.266}$$

In the feedback configuration, the first pole is moved at the frequency $\omega_{p_1}(1 + \beta A_0)$, which can be approximated by $\beta A_0 \omega_{p_1}$, or equivalently, $\beta \omega_u$, where ω_u is the unity-gain frequency.

To reduce the effect of slow-settling components on the transient response, the pole-zero doublet should be moved to higher frequency. This can be achieved by keeping the unity-gain frequency, ω_u'', of the auxiliary amplifier greater than $\beta \omega_u$. On the other hand, the stability requirement is met provided ω_u'' remains lower than the second-pole frequency, ω_{p2}', of the main amplifier. Therefore,

$$\beta \omega_u < \omega_u'' < \omega_{p2}'. \tag{5.267}$$

This is satisfied in practical implementations, where the load or compensation capacitor of the auxiliary amplifier is generally chosen to be much smaller than the one of the main amplifier. By increasing ω_u'', while satisfying the condition defined by Equation (5.267), the pole-zero doublet is pushed to higher frequencies until it merges with the third pole to generate a complex-conjugate pole pair and a real zero. Because the real part of this complex-conjugate pole pair determines the decrease rate in the output response, its optimum value can be found by further increasing ω_u'' to obtain a fast settling response. As a result, the phase margin of the auxiliary amplifier can be greater than $60°$. Figure 5.37 shows the pole-zero locations of the nonoptimized and optimized gain-enhanced amplifiers.

5.5.2 Telescopic amplifier

The circuit diagram of a telescopic amplifier is depicted in Figure 5.38. A common-mode feedback stage sets the bias voltage V_{CM}. The different bias currents are derived from a master current source using mirrors. An internal bias circuit is adopted for transistors $T_3 - T_4$. The proper common-mode rejection is obtained by maintaining at less a voltage of $V_{DS(sat)}$ on $T_1 - T_2$, $T_7 - T_8$, and T_9. From the amplifier circuit, we can obtain the following

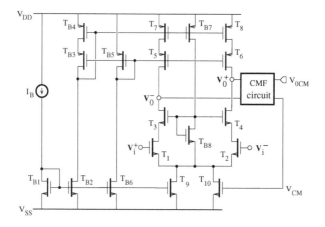

FIGURE 5.38
Circuit diagram of a telescopic amplifier.

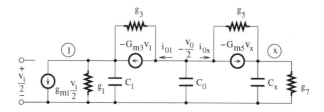

FIGURE 5.39
Equivalent circuit model of the telescopic amplifier.

relations:

$$V_i^+ = V_{DD} - V_{SD_7} - V_{SD_5} - V_{DS_3} - V_{DS_1} + V_{GS_1} \qquad (5.268)$$
$$= V_{GS_1} + V_{DS_9} + V_{SS}. \qquad (5.269)$$

Due to the amplifier symmetry, the transistors T_1 and T_2, T_3 and T_4, T_5 and T_6, and T_7 and T_8 are matched. Assuming that both input nodes are connected to the same dc voltage V_{ICM}, it can be shown that

$$V_{DS_1(sat)} + V_{DS_9(sat)} + V_{T_n} + V_{SS}$$
$$\leq V_{ICM} \leq V_{DD} - V_{SD_7(sat)} - V_{SD_5(sat)} - V_{DS_3(sat)} + V_{T_n}. \qquad (5.270)$$

On the other hand, the expression of V_0^- can be written as

$$V_0^- = V_{DD} - V_{SD_7} - V_{SD_5} \qquad (5.271)$$
$$= V_{DS_3} + V_{DS_1} + V_{DS_9} + V_{SS}. \qquad (5.272)$$

To maintain the transistors in the saturation region, the voltage swing at the inverting output node should be of the form

$$V_{DS_3(sat)} + V_{DS_1(sat)} + V_{DS_9(sat)} + V_{SS} \leq V_0^- \leq V_{DD} - V_{SD_7(sat)} - V_{SD_5(sat)}. \qquad (5.273)$$

Similar equations can be derived for the voltage swing, V_0^+, at the noninverting output node. To proceed further, we exploit the fact that the voltage V_0^- can also be expressed as

$$V_0^- = V_{DS_3} + V_{DS_1} - V_{GS_1} + V_i^+ . \tag{5.274}$$

For a given input common-mode voltage, V_{ICM}, the transistors T_1 and T_3 should be biased slightly above the saturation region to optimize the output swing. That is,

$$V_0^- \le V_{ICM} + V_{DS_3(sat)} - V_{T_n} . \tag{5.275}$$

Therefore, we can conclude that the maximum output voltage is dependent on the input common-mode voltage, whose variations may cause a premature clipping of the output voltage.

The telescopic amplifier, whose number of current paths between the supply voltages is two in comparison to four in the folded-cascode architecture, should consume the smaller static power. It can provide the superior speed but features the smaller differential output swing of about $2[V_{sup} - (3V_{DS(sat)} + 2V_{SD(sat)})]$, where $V_{sup} = V_{DD} - V_{SS}$ and the transistors of the same type are assumed to operate with identical saturation voltages.

The equivalent circuit model of the telescopic amplifier is depicted in Figure 5.39. Based on a dominant-pole model, the frequency response of the amplifier is primarily determined by the dc gain, the first and the second pole frequencies. Because the parasitic and output capacitors act like open circuits at dc, the dc gain is given by

$$A_0 = v_0/v_i \simeq -g_{m1}(r_{01} \parallel r_{0x}), \tag{5.276}$$

where

$$r_{01} = -\frac{v_0}{2i_{01}}\bigg|_{v_i=0} = \frac{1}{g_1} + \frac{1}{g_3}\left(1 + \frac{G_{m3}}{g_1}\right) \tag{5.277}$$

and

$$r_{0x} = -\frac{v_0}{2i_{0x}} = \frac{1}{g_7} + \frac{1}{g_5}\left(1 + \frac{G_{m5}}{g_7}\right). \tag{5.278}$$

The first pole of the amplifier transfer function can be obtained as

$$\omega_{p_1} = -g_{m1}/A_0 C_0 , \tag{5.279}$$

where $C_0 = C_L + C_{gd_3} + C_{db_3} + C_{gd_5} + C_{db_5} + C_i'$. Here, C_0 is the total load capacitance at the output node, C_L denotes the external load capacitance connected to the output node, and C_i' represents the input capacitance of the CMF circuit. The second pole of the amplifier is due to the parasitic capacitance at the source of the n-channel cascode transistor, T_3 or T_4, and is of the form

$$\omega_{p_2} \simeq -G_{m3}/C_1 , \tag{5.280}$$

where $C_1 = C_{gd_1} + C_{db_1} + C_{gs_3} + C_{sb_3}$ and $G_{m3} = g_{m3} + g_{mb3}$. The slew rate is given by

$$SR \simeq \frac{I_{D_9}}{C_L}, \tag{5.281}$$

where $I_{D_9} = I_B$, and I_B is the bias current.

Note that the gain-boosting technique can also be used with the transistors $T_3 - T_4$ and $T_5 - T_6$, to further increase the achievable amplification gain.

5.5.3 Common-mode feedback circuits

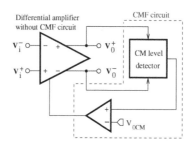

FIGURE 5.40
Block diagram of a differential amplifier.

In fully differential amplifiers, a common-mode circuit is used to control the common-mode voltage level and cancel the undesirable common-mode components of signals. Figure 5.40 shows the block diagram of a differential amplifier. The amplification is achieved by a fully differential stage. The common-mode circuit includes a common-mode level detector and a sense amplifier. Due to the negative feedback loop including the common-mode circuit, the output common-mode voltage is forced to be equal to V_{0CM} in the steady state. In the case where $V_{0CM} = (V_0^+ + V_0^-)/2$, the amplifier should exhibit a large and symmetric output dynamic range. To meet the stability requirement, all poles introduced in the common-mode loop should be well above the ones of the differential loop.

5.5.3.1 Continuous-time common-mode feedback circuit

A continuous-time common-mode feedback circuit can be realized as shown in Figure 5.41(a) [24]. The CM detection is performed by an RC network with two equal-valued resistors and capacitors. The role of the capacitors is to stabilize the common-mode feedback loop. Using the principle of superposition and voltage division, we obtain

$$V_{0CM}' = \frac{V_0^+ + V_0^-}{2}. \tag{5.282}$$

The voltages V_{0CM}' and V_{0CM} are applied at the inputs of a sense amplifier and the CM error detection and amplification are accomplished by the differential

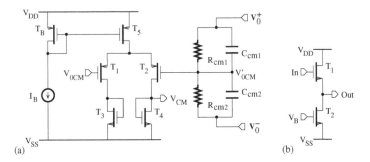

FIGURE 5.41

(a) Circuit diagram of a CMF using an RC network; (b) circuit diagram of a source follower.

pairs consisting of $T_1 - T_2$. The CM level can be held at a reference potential by increasing or decreasing V_{CM}, which is assumed to be connected to a high resistance node. The implementation of the CMF loop then allows the voltage V_{CM} to be adjusted until $V_{0CM} = V'_{0CM}$. The gain of the CMF circuit should not be excessively high to eliminate any undesirable waveform oscillation at the amplifier output.

Because the resistors used in the CMF circuit can severely degrade the resulting dc gain of the differential amplifier, a source follower, as shown in Figure 5.41(b), is often inserted between each of the V_0^+ and V_0^- terminals and the corresponding node of the RC network. But, source followers can limit the signal swing and increase the noise level. It should also be noted that the use of resistors with high value is limited by the required large silicon area.

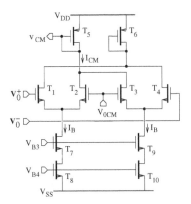

FIGURE 5.42

Circuit diagram of a CMF based on two differential transistor pairs.

Another structure of the common-mode feedback circuit is shown in Figure 5.42 [15,25]. The input transistors $(T_1 - T_4)$ are assumed to operate in the saturation region. Ideally, they have to sense and amplify only the common-mode signal. Through the use of a feedback between the CMF circuit and the amplifier, the CM voltage and V_{0CM} are made equal. This is done by applying the CMF circuit output voltage, V_{CM}, to a suitable internal node of the amplifier. Let the current I_{CM} be expressed as

$$I_{CM} = I_{D_2} + I_{D_3}. \tag{5.283}$$

Because each of the transistor pairs T_1 and T_2, and T_3 and T_4, operate as a differential stage biased by the tail current I_B, it can be found that the drain currents of T_2 and T_3 are respectively given by

$$I_{D_2} = \frac{I_B}{2} - \sqrt{2KI_B}\frac{V_0^+ - V_{0CM}}{2}\sqrt{1 - \frac{(V_0^+ - V_{0CM})^2}{2I_B/K}} \tag{5.284}$$

and

$$I_{D_3} = \frac{I_B}{2} - \sqrt{2KI_B}\frac{V_0^- - V_{0CM}}{2}\sqrt{1 - \frac{(V_0^- - V_{0CM})^2}{2I_B/K}}. \tag{5.285}$$

Assuming that

$$V_0^+ = V'_{0CM} + V_0/2, \tag{5.286}$$
$$V_0^- = V'_{0CM} - V_0/2, \tag{5.287}$$

and

$$|V'_{0CM} - V_{0CM}| \ll |V_0|, \tag{5.288}$$

we can obtain

$$I_{CM} \simeq I_B - \sqrt{2KI_B}\frac{V'_{0CM} - V_{0CM}}{2}\sqrt{1 - \frac{(V_0/2)^2}{2I_B/K}}. \tag{5.289}$$

For the normal operation, it is required to have $V'_{0CM} = V_{0CM}$. As a result, the current I_{CM} becomes equal to I_B, which is identical to the constant tail current of the input differential stage of the amplifier requiring the CMF circuit.

The aforementioned CMF circuit has the advantage of exhibiting a high input impedance, which can be useful in the design of operational transconductance amplifiers with a reduced sensitivity to the loading effect.

For differential amplifiers based on transistors operating in triode region, the CMF circuit can be designed as shown in Figure 5.43(a) [26,27]. Here, the transistors $T_1 - T_4$ operate in the triode region, while the remaining transistors are biased in the saturation region. The control current I_{CM} is generated by comparing the currents I'_{0CM} and I_{0CM}, which are related to the output CM voltage, V'_{0CM}, and the reference CM voltage, V_{0CM}, respectively. We can write

$$I_{CM} = I'_{0CM} - I_{0CM}, \tag{5.290}$$

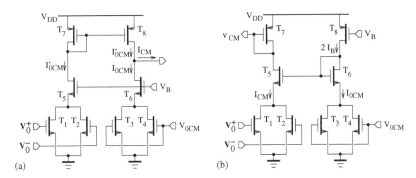

FIGURE 5.43
Circuit diagram of a CMF using two differential transistor pairs operating in the triode region: (a) Voltage output, (b) current output.

where

$$I'_{0CM} = I_{D_1} + I_{D_2} \tag{5.291}$$

and

$$I_{0CM} = I_{D_3} + I_{D_4}. \tag{5.292}$$

Assuming that the transistors $T_1 - T_4$ are matched and the same voltage is maintained at the source of T_5 and T_6, we have

$$I_{D_1} = K[2(V'_{0CM} + V_0/2 - V_T)V_D - V_D^2], \tag{5.293}$$

$$I_{D_2} = K[2(V'_{0CM} - V_0/2 - V_T)V_D - V_D^2], \tag{5.294}$$

and

$$I_{D_3} = I_{D_4} = K[2(V_{0CM} - V_T)V_D - V_D^2], \tag{5.295}$$

where $V_D = V_{DS_1} = V_{DS_2} = V_{DS_3} = V_{DS_4}$. Hence,

$$I_{CM} = 4K(V'_{0CM} - V_{0CM}). \tag{5.296}$$

The variations of I_{CM} are exploited to adjust the CM level. When $I'_{0CM} = I_{0CM}$, the voltage V_{0CM} is reduced to

$$V_{0CM} = \frac{V_0^+ + V_0^-}{2}. \tag{5.297}$$

In the case where the CM control signal should be applied to a high resistance node, a modified version of the CMF circuit, as shown in Figure 5.43(b) [28], may be suitable. The transistors $T_1 - T_6$ are assumed to be identical. Let the drain current of transistors T_1 and T_2, which operate in the triode region, be given by

$$I_{D_1} = K[2(V'_{0CM} + V_0/2 - V_T)V_D - V_D^2] \tag{5.298}$$

and

$$I_{D_2} = K[2(V'_{0CM} - V_0/2 - V_T)V_D - V_D^2], \tag{5.299}$$

where $V_D = V_{DS_1} = V_{DS_2}$. The current I_{CM} is of the form

$$I_{CM} = I_{D_1} + I_{D_2} = 2K[2(V'_{0CM} - V_T)V_D - V_D^2]. \tag{5.300}$$

On the other hand, the voltage equation of the loop including T_1, T_5, T_6, and T_2 can be written as

$$V_{DS_1} + V_{GS_5} = V_{DS_3} + V_{GS_6}. \tag{5.301}$$

Because the currents flowing through T_5 and T_6 are almost equal, it can be assumed that $V_{GS_5} \simeq V_{GS_6}$ and $V_{DS_1} \simeq V_{DS_3}$. The transistors T_3 and T_4 operate in the triode region, and their drain currents are given by

$$I_{D_3} = I_{D_4} = K[2(V_{0CM} - V_T)V_D - V_D^2] = I_B, \tag{5.302}$$

where $V_D = V_{DS_3} = V_{DS_4}$. Taking into account the fact that the triode region is characterized by $0 < V_{DS} < V_{GS} - V_T$, or equivalently $0 < V_D < V_{0CM} - V_T$, the second-order equation (5.302) can be solved for V_D. Therefore,

$$V_D = V_{0CM} - V_T - \sqrt{(V_{0CM} - V_T)^2 - I_B/K}, \tag{5.303}$$

where $(V_{0CM} - V_T)^2 - I_B/K \geq 0$. Substituting Equation (5.303) into (5.300) gives

$$I_{CM} = 2K\left[2(V'_{0CM} - V_{0CM})\left(V_{0CM} - V_T - \sqrt{(V_{0CM} - V_T)^2 - I_B/K}\right) + I_B/K\right]. \tag{5.304}$$

When $V'_{0CM} = V_{0CM}$, the current I_{CM} is reduced to $2I_B$. The variations in I_{CM} take place around $2I_B$, depending on the difference $V'_{0CM} - V_{0CM}$.

The drawbacks of this approach are essentially due to the fact that the input transistors are required to operate in the triode region, instead of the saturation region. As a consequence, the amplifier output swing and the small-signal gain of the CMF circuit can be reduced.

5.5.3.2 Switched-capacitor common-mode feedback circuit

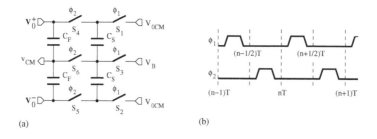

(a)   (b)

FIGURE 5.44
Circuit diagram of a switched-capacitor CMF circuit.

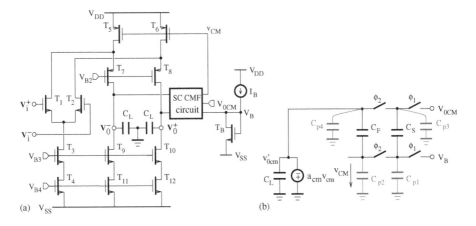

FIGURE 5.45
(a) Folded-cascode fully differential amplifier; (b) equivalent circuit model of
the SC CMF loop ($v'_{0cm} = (v_0^+ + v_0^-)/2$ and $v_{CM} = V_{CM} + v_{cm}$).

Switched-capacitor (SC) common-mode feedback (CMF) circuits have several advantages over continuous-time counterparts, including improved output
range and linearity, low power dissipation and silicon area. The circuit diagram
and clock waveforms of an SC CMF circuit are shown in Figure 5.44 [29].

Capacitors C_F establish an ac feedback path between the amplifier output and CM nodes. To set the dc level of the CM voltage, the capacitors C_S
are charged to the difference between the bias voltage, V_B, and the desired
CM voltage, V_{0CM}, during the first phase, ϕ_1, and then connected across the
capacitors C_F during the clock phase ϕ_2. Because both terminals of the capacitors C_S are periodically switched, a nonoverlapping two-phase clock signal is
required to control the switches so as to minimize errors due to charge injection and clock feedthrough. The detection of the actual output CM level is
performed using the ac voltage divider composed of capacitors C_F to determine the average of both amplifier outputs, which is periodically compared to
the desired CM output voltage available across the switched capacitors C_S.
The CM control voltage supplied to the amplifier is then adjusted based on
the difference between the actual and desired CM output voltages. When the
steady state is reached, there is no longer a charge transfer between the capacitors C_F and C_S. Ideally, the difference between the amplifier output CM
voltage and the CM control voltage should then be equal to the difference
between the desired CM voltage and the bias voltage.

With reference to the fully differential amplifier depicted in Figure 5.45(a),
the equivalent model of the common-mode feedback loop [30] can be derived as shown in Figure 5.45(b). During the clock phase, ϕ_2, or say, for
$(n - 1/2)T < t \le nT$, where T is the clock signal period, the equation for the

charge conservation at the v_{CM} node is of the form

$$
\begin{aligned}
C_S(V_B - V_{0CM}) + C_{p1}V_B & \\
+ C_F[v_{CM}(n - 1/2) - v'_{0cm}(n - 1/2)] + C_{p2}v_{CM}(n - 1/2) & \\
= (C_{p1} + C_{p2})v_{CM}(n) + (C_S + C_F)[v_{CM}(n) - v'_{0cm}(n)],
\end{aligned}
\tag{5.305}
$$

where $v'_{0cm} = (v_0^+ + v_0^-)/2$, and C_{p1} and C_{p2} are parasitic capacitances. When the clock phase ϕ_1 is high, that is, for $(n-1)T \le t < (n-1/2)T$, we have

$$
v_{CM}(n - 1/2) = v_{CM}(n - 1)
\tag{5.306}
$$

and

$$
v'_{0cm}(n - 1/2) = v'_{0cm}(n - 1).
\tag{5.307}
$$

Combining Equations (5.307), (5.306), and (5.305) gives

$$
\begin{aligned}
C_S(V_B - V_{0CM}) + C_{p1}V_B & \\
+ C_F[v_{CM}(n - 1) - v'_{0cm}(n - 1)] + C_{p2}v_{CM}(n - 1) & \\
= (C_{p1} + C_{p2})v_{CM}(n) + (C_S + C_F)[v_{CM}(n) - v'_{0cm}(n)].
\end{aligned}
\tag{5.308}
$$

Because $v_{CM} = V_{CM} + v_{cm}$, it can be deduced that

$$
v'_{0cm}(n) = -a_{cm}v_{cm}(n) = a_{cm}(V_{CM} - v_{CM}),
\tag{5.309}
$$

where V_{CM} and v_{cm} denote the dc and ac components of the CM control voltage, respectively. The system of Equations (5.308) and (5.309) can be solved for v_{CM} and v'_{0cm}. Hence,

$$
v_{CM}(n) = r v_{CM}(n - 1) + p
\tag{5.310}
$$

and

$$
v'_{0cm}(n) = r v'_{0cm}(n - 1) + q,
\tag{5.311}
$$

where

$$
r = \frac{(1 + a_{cm})C_F + C_{p2}}{(1 + a_{cm})(C_S + C_F) + C_{p1} + C_{p2}},
\tag{5.312}
$$

$$
p = \frac{C_S(a_{cm}V_{CM} - V_{0CM}) + (C_S + C_{p1})V_B}{(1 + a_{cm})(C_S + C_F) + C_{p1} + C_{p2}},
\tag{5.313}
$$

and

$$
q = a_{cm}[(1 - r)V_{CM} - p].
\tag{5.314}
$$

Using the iterative method, the solutions of the first-order linear autonomous difference equation, (5.310), can be computed as,

$$
v_{CM}(n) = p \sum_{k=0}^{n-1} r^k + r^n v_{CM}(0)
\tag{5.315}
$$

$$
= p\left(\frac{1 - r^n}{1 - r}\right) + r^n v_{CM}(0),
\tag{5.316}
$$

where $r \neq 1$. Similarly, for Equation (5.311), it can be shown that

$$v'_{0CM}(n) = q\left(\frac{1 - r^n}{1 - r}\right) + r^n v'_{0CM}(0). \qquad (5.317)$$

The sequences $v_{CM}(n)$ and $v'_{0CM}(n)$ can then be expressed as

$$v_{CM}(n) = \overline{v}_{CM} + r^n[v_{CM}(0) - \overline{v}_{CM}] \qquad (5.318)$$

and

$$v'_{0CM}(n) = \overline{v'}_{0CM} + r^n[v'_{0CM}(0) - \overline{v'}_{0CM}], \qquad (5.319)$$

where $v_{CM}(0)$ and $v'_{0CM}(0)$ represent initial values, and the steady-state values are given by

$$\overline{v}_{CM} = \lim_{n \to \infty} v_{CM}(n) = \frac{p}{1 - r} = \frac{V_{CM} + \left[\left(1 + \dfrac{C_{p1}}{C_S}\right)V_B - V_{0CM}\right]\dfrac{1}{a_{cm}}}{1 + \left(1 + \dfrac{C_{p1}}{C_S}\right)\dfrac{1}{a_{cm}}}$$

$$(5.320)$$

and

$$\overline{v'}_{0CM} = \lim_{n \to \infty} v'_{0CM}(n) = \frac{q}{1 - r} = \frac{V_{0CM} + \left(1 + \dfrac{C_{p1}}{C_S}\right)(V_{CM} - V_B)}{1 + \left(1 + \dfrac{C_{p1}}{C_S}\right)\dfrac{1}{a_{cm}}}. \qquad (5.321)$$

Because $r < 1$, it appears that $v_{CM}(n)$ and $v'_{0CM}(n)$ will converge to steady state. To proceed further, we can write

$$\overline{v'}_{0CM} - \overline{v}_{CM} = \frac{(V_{0CM} - V_B)\left(1 + \dfrac{1}{a_{cm}}\right) + \left(V_{CM} - V_B\right)\dfrac{C_{p1}}{C_S} - V_B \dfrac{C_{p1}}{a_{cm}C_S}}{1 + \left(1 + \dfrac{C_{p1}}{C_S}\right)\dfrac{1}{a_{cm}}}.$$

$$(5.322)$$

Assuming that the CM gain a_{cm} is very high, we arrive at

$$\overline{v'}_{0CM} - \overline{v}_{CM} \simeq V_{0CM} - V_B + (V_{CM} - V_B)C_{p1}/C_S. \qquad (5.323)$$

For accurate control of the output CM, the values of C_S and V_B should be chosen to make the effect of the last term negligible.

Ideally, the SC CMF circuit should exhibit a high gain and bandwidth. For a fast and accurate response, the value of C_S is chosen to be greater than the one of C_F, which can be determined by making the bandwidth of the CM loop at least equal to the one of the differential loop. Note that the amplifier load due to the aforementioned SC CMF circuit is different from one clock phase to another, and may affect the resulting settling time and bandwidth.

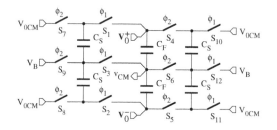

FIGURE 5.46

Circuit diagram of a switched-capacitor CMF circuit with symmetric loading.

For applications where the output voltage is valid during the whole clock period, the SC CMF circuit structure shown in Figure 5.46 [31] may be suitable. Additional capacitors and switches are used to improve the circuit symmetry. The capacitors are switched such that the effect of the SC CMF circuit on the amplifier loading remains the same for both clock phases.

5.5.4 Pseudo fully differential amplifier

In the cases where the variations of the common-mode voltage at the amplifier output remain small, fully differential amplifiers can also be implemented without an extra CMF circuit, which has the drawback of increasing noise and limiting the output swing and speed.

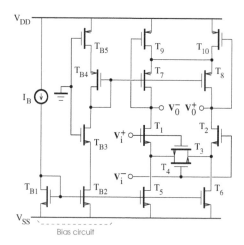

FIGURE 5.47

Circuit diagram of a pseudo fully differential transconductance amplifier using a voltage-controlled output load.

In the single-stage pseudo fully differential amplifier depicted in Figure 5.47 [32, 33], the solution adopted for the stabilization of the common-mode output voltage, $V_{0CM} = (V_0^+ + V_0^-)/2$, defined by the biasing circuit is based on a voltage-controlled output load. The input signal is applied to the

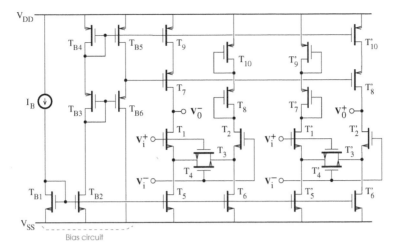

FIGURE 5.48
Circuit diagram of a pseudo fully differential transconductance amplifier with duplicated differential stages.

gates of transistors $T_1 - T_4$. The bias currents are provided by transistors T_5 and T_6, and the output loads consist of $T_7 - T_{10}$. The linearity of the amplifier transconductance, g_m, is improved by the source degeneration provided by $T_3 - T_4$, which should operate in the triode region. The resulting transfer characteristic is similar to the one of a source-coupled transistor pair, but with a different biasing condition.

An identical dc current flows through T_7 and T_8, which are biased by the same gate voltage chosen such that T_9 and T_{10} operate in the triode region with source-drain voltages set to a stable value. Any variation in the common-mode voltage first affects the resistance of T_9 and T_{10} and then the source-gate voltages of T_7 and T_8. Because the dc currents flowing through the transistors remain constant, this induces a modification of the source-drain voltages of T_7 and T_8, thereby forcing the compensation of the common-mode voltage variations.

The common-mode output voltage can also be defined by duplicating the differential stage [34], as shown in Figure 5.48. In this approach, each differential stage operates as a single-ended circuit and the dc components of the output voltage can be set to a desired level, making the use of a CMF circuit unnecessary. The currents available at the noninverting and inverting output nodes can be expressed as $I_0^+ \simeq I_B/2 + \triangle i$ and $I_0^- \simeq I_B/2 - \triangle i$, respectively. By taking the difference of output currents, the common-mode component is canceled. Hence, $I_0^+ - I_0^- \simeq 2\triangle i$, where $\triangle i$ represents the output contribution associated to the differential mode. But here, the cancelation of the common-mode component can be limited by the achievable transistor matching. It should also be noted that the variations of the common-mode signals must remain low enough to prevent the transistors from moving outside their operating region.

In general, transconductance amplifiers are designed to have a sufficient g_m tuning range for the correction of IC process variations. The aforementioned transconductance amplifiers are suitable for high-frequency applications due to their single-stage structure. However, they are limited by the achievable gain-bandwidth product.

5.6 Multi-stage amplifier structures

In the multistage amplifier design approach, the requirement of a high gain is met by combining a differential stage and extra output stages. Various architectures are available for the implementation of output stages, which should be designed to provide an adequate level of the signal power to the amplifier load. The output stage should exhibit a lower output resistance to drive the load as if it were an ideal voltage source.

Output stages are amplification circuits, which can be described in terms of the conduction angle, or the portion of the input signal cycle for which there is an output signal. The output stage can be of the class A, class B, or class AB type, depending on the transistor's conduction angle which typically varies between $0°$ and $360°$.

For class A amplifiers, the conduction angle is $360°$. Because the transistors are biased to never reach the cutoff region during the normal operation, a linear amplification of the input signal is performed. However, class A amplifiers typically have a low power efficiency. This is due to the fact that an important part of the supply power is required to bias the transistors during the whole cycle of the input signal such that the ratio of the output signal power to the total input power is relatively small.

Class B amplifiers are generally based on push-pull configuration, consisting of complementary transistors, each of which conducts on alternating half-cycles of the input signal. Each transistor has a conduction angle of about $180°$. It is then biased to operate in the linear range during approximately one-half of the input signal cycle and is turned off for the other half cycle. Due to the fact that a gate-source voltage greater than the threshold voltage is required to initiate the transistor conduction, there is a transition region where both transistors of the push-pull amplifier are turned off, thereby producing crossover distortions in the output waveform. Note that in a push-pull configuration, the p-channel transistor sources (or pushes) current to the output load while the n-channel transistor sinks (or pulls) current from the output load.

Class AB amplifiers have the same configuration as class B gain stages, but both transistors are biased slightly above the cutoff region when the value of the input signal is around zero. The conduction angle of each transistor is slightly greater than 180° to reduce the crossover distortions.

5.6.1 Output stage

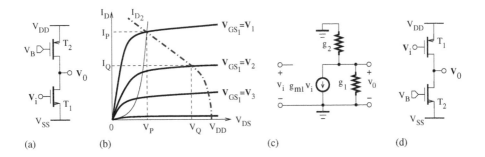

(a) (b) (c) (d)

FIGURE 5.49
(a) Circuit diagram of a common-source gain stage with n-channel input transistor; (b) transistor I/V characteristics; (c) small-signal equivalent model; (d) circuit diagram of a common-source gain stage with p-channel input transistor.

A common-source amplifier with an n channel input transistor is shown in Figure 5.49(a). Assuming that the transistors are biased in the saturation region, we have

$$V_{DS_1} \geq V_{DS_1(sat)} = V_{GS_1} - V_{T_n} \qquad (5.324)$$

and

$$V_{SD_2} \geq V_{SD_2(sat)} = V_{SG_2} + V_{T_p}, \qquad (5.325)$$

where $V_{DS_1} = V_0 - V_{SS}$, $V_{GS_1} = V_i - V_{SS}$, $V_{SD_2} = V_{DD} - V_0$, and $V_{SG_2} = V_{DD} - V_B$. The output voltage should then remain in the range

$$V_{SS} + V_{DS_1(sat)} \leq V_0 \leq V_{DD} - V_{SD_2(sat)}. \qquad (5.326)$$

For the normal operation of the output stage, it is required that

$$V_i - V_{T_n} \leq V_0 \leq V_B - V_{T_p}. \qquad (5.327)$$

The I/V characteristics of the input and load transistors are depicted in Figure 5.49(b). The input voltage, which is directly related to the gate-source voltage, V_{GS_1}, determines the curves of the characteristics to be considered in the case of T_1 for the determination of the operating point. Because the voltage V_B is constant, only the curve of the I/V characteristic associated with $V_{SG_2} = V_{DD} - V_B$ is retained for the load transistor T_2. To achieve a class A amplification, the operating point should be set half way between the points (I_P, V_P) and (I_Q, V_Q). The small-signal equivalent model of the amplifier is shown in Figure 5.49(c). Applying Kirchhoff's current law gives

$$g_{m_1} v_i + g_1 v_0 + g_2 v_0 = 0. \qquad (5.328)$$

The voltage gain can then be computed as

$$A = \frac{v_0}{v_i} = -g_{m_1} r_0, \quad (5.329)$$

where

$$r_0 = 1/(g_1 + g_2). \quad (5.330)$$

Here, $g_1 = g_{ds_1}$, $g_2 = g_{ds_2}$, and r_0 represents the output resistance.

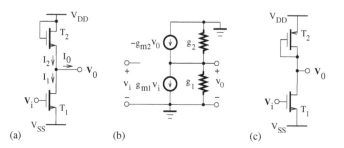

(a)   (b)   (c)

FIGURE 5.50
(a) Common-source gain stage with n-channel input transistor and diode-connected n-channel transistor load; (b) small-signal equivalent model; (c) common-source gain stage with n-channel input transistor and diode-connected p-channel transistor load.

A diode-connected transistor can also be used as a load for the common-source gain stage, as shown in Figure 5.50(a) for the case of an n-channel transistor. It is assumed that both transistors operate in the saturation region. For the transistor T_1, we can obtain

$$V_{DS_1} \geq V_{DS_1(sat)} = V_{GS_1} - V_{T_n}, \quad (5.331)$$

where $V_{DS_1} = V_0 - V_{SS}$ and $V_{GS_1} = V_i - V_{SS}$. Therefore, the output voltage is such that $V_0 \geq V_i - V_{T_n}$. The transistor T_2 is saturated because its gate and drain are connected together. The currents flowing through the transistors T_1 and T_2 are respectively given by

$$I_1 = K_1(V_{GS_1} - V_{T_n})^2 \quad (5.332)$$

and

$$I_2 = K_2(V_{GS_2} - V_{T_n})^2, \quad (5.333)$$

where $V_{GS_2} = V_{DD} - V_0$. Applying Kirchhoff's current law at the output node gives

$$I_2 = I_1 + I_0. \quad (5.334)$$

Substituting Equations (5.332) and (5.333) into Equation (5.334), we obtain

$$V_0 = V_{DD} - V_{T_n} - \sqrt{[K_1(V_i - V_{SS} - V_{T_n})^2 + I_0]/K_2}. \quad (5.335)$$

When the amplifier stage is connected to a high resistance load, the output current can be considered negligible. Hence, $I_0 \simeq 0$ and

$$V_0 = V_{DD} - V_{T_n} - \sqrt{K_1/K_2}(V_i - V_{SS} - V_{T_n}) \qquad (5.336)$$

The dc input voltage is then amplified by a factor equal to the square root of transconductance parameters.

The small-signal equivalent model is depicted in Figure 5.50(b). Using Kirchhoff's current law, we can obtain

$$g_{m_1} v_i + g_{m_2} v_0 + g_1 v_0 + g_2 v_0 = 0. \qquad (5.337)$$

Hence,

$$A = \frac{v_0}{v_i} = \frac{-g_{m_1}}{g_{m_2}} \left(\frac{1}{1 + \dfrac{g_1 + g_2}{g_{m_2}}} \right). \qquad (5.338)$$

The gain is lower than the one of the aforementioned amplifier. Assuming that $(g_1 + g_2)/g_{m_2} \ll 1$, and taking into account the fact that T_1 and T_2 are biased by the same current I_D, $g_{m_1} = 2\sqrt{K_1 I_D}$ and $g_{m_2} = 2\sqrt{K_2 I_D}$, the voltage gain is reduced to

$$A = \frac{v_0}{v_i} \simeq \frac{-g_{m_1}}{g_{m_2}} = -\sqrt{\frac{W_1/L_1}{W_2/L_2}}. \qquad (5.339)$$

Due to the gain dependence on the width-to-length ratios of transistors, the requirement of a high gain is limited by the available amplifier area. Figure 5.50(c) shows the circuit diagram of a common-source gain stage with a p-channel transistor load.

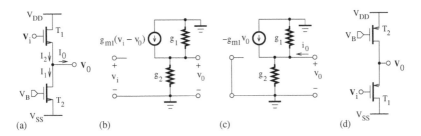

(a)   (b)   (c)   (d)

FIGURE 5.51
(a) nMOS source follower; (b) small-signal equivalent model; (c) small-signal equivalent model for the determination of the output resistance; (d) pMOS source follower.

A source follower, as shown in Figure 5.51(a) can be used as a buffer amplifier or dc level shifter. Both the input and load transistors are biased in the saturation region. Hence,

$$V_{DS_1} \geq V_{DS_1(sat)} = V_{GS_1} - V_{T_n} \qquad (5.340)$$

and

$$V_{DS_2} \geq V_{DS_2(sat)} = V_{GS_2} - V_{T_n} , \qquad (5.341)$$

where $V_{GS_1} = V_i - V_0$ and $V_{GS_2} = V_B - V_{SS}$. Because $V_{DS_2} = V_0 - V_{SS}$, the input range can be obtained as

$$V_{SS} + V_{DS_2(sat)} + V_{DS_1(sat)} + V_{T_n} \leq V_i \leq V_{DD} + V_{T_n} . \qquad (5.342)$$

The normal operation also requires that $V_0 \geq V_B - V_{T_n}$. The currents that flow through the transistors T_1 and T_2 can respectively be written as

$$I_1 = K_1 (V_{GS_1} - V_{T_n})^2 \qquad (5.343)$$

and

$$I_2 = K_2 (V_{GS_2} - V_{T_n})^2 . \qquad (5.344)$$

From Kirchhoff's current law at the output node, we obtain

$$I_2 = I_1 + I_0 . \qquad (5.345)$$

To find the output voltage, Equation (5.345) can be rewritten using Equations (5.343) and (5.344) as

$$V_0 = V_i - V_{T_n} - \sqrt{[K_2 (V_B - V_{SS} - V_{T_n})^2 + I_0]/K_1} . \qquad (5.346)$$

The output voltage is simply equal to the input voltage minus a term determined by the amplifier and transistor characteristics. The source follower then provides a dc voltage gain of unity.

The small-signal equivalent model is depicted in Figure 5.51(b). Applying Kirchhoff's current law at the output node, we find

$$g_{m_1}(v_i - v_0) - g_1 v_0 - g_2 v_0 = 0. \qquad (5.347)$$

The voltage gain is then obtained as

$$A = \frac{v_0}{v_i} = \frac{g_{m_1}/(g_1 + g_2)}{1 + g_{m_1}/(g_1 + g_2)} . \qquad (5.348)$$

For typical values of the transistor characteristics, $g_{m_1} \gg (g_1 + g_2)$ and $A \simeq 1$. To derive the output resistance, the input node is connected to the ground and the amplifier is driven by a voltage source applied to the output node. With reference to the small-signal equivalent model of Figure 5.51(c), it can be shown that

$$i_0 = g_{m_1} v_0 + g_1 v_0 + g_2 v_0 . \qquad (5.349)$$

Hence,

$$r_0 = \frac{v_0}{i_0} = \frac{1}{g_{m_1} + g_1 + g_2} \simeq \frac{1}{g_{m_1}} . \qquad (5.350)$$

Because the small-signal output resistance, r_0, is generally less than 1 kΩ, it is

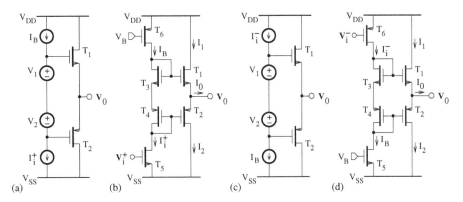

FIGURE 5.52

(a) Principle and (b) circuit diagram of the first version of the complementary source follower; (c) principle and (d) circuit diagram of the second version of the complementary source follower.

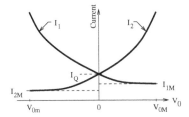

FIGURE 5.53

Plot of the currents I_1 and I_2.

considered small. Figure 5.51(d) shows the circuit diagram of a pMOS source follower.

The aforementioned gain stages are suitable for class A operation, but output stages can also be designed using class AB circuit configuration. A class AB output stage should preferably provide a high maximum output current, while requiring only a low quiescent current. The principle and circuit diagram of complementary source followers are illustrated in Figures 5.52(a) and (b) in the case where the input voltage is applied to an n-channel transistor, and in Figures 5.52(c) and (d) for designs requiring a p-channel input transistor. Two floating dc voltages are used to set the gate-source voltages of transistors T_1 and T_2, thereby defining the quiescent currents flowing through T_1 and T_2 so that crossover distortions are eliminated [35]. Applying Kirchhoff's voltage law to the translinear loop including $T_1 - T_4$, we obtain

$$V_{GS_3} + V_{SG_4} = V_{GS_1} + V_{SG_2}, \qquad (5.351)$$

where

$$V_{GS_1} = V_{T_n} + \sqrt{\frac{I_1}{K_{n_1}}}, \qquad (5.352)$$

$$V_{SG_2} = -V_{T_p} + \sqrt{\frac{I_2}{K_{p2}}}, \tag{5.353}$$

$$V_{GS_3} = V_{T_n} + \sqrt{\frac{I_B}{K_{n3}}}, \tag{5.354}$$

and

$$V_{SG_4} = -V_{T_p} + \sqrt{\frac{I_B}{K_{p4}}}. \tag{5.355}$$

Hence,

$$\sqrt{\frac{I_1}{K_{n1}}} + \sqrt{\frac{I_2}{K_{p2}}} = \left(\sqrt{\frac{1}{K_{n3}}} + \sqrt{\frac{1}{K_{p4}}} \right) \sqrt{I_B}. \tag{5.356}$$

Note that the current I_B is set by the bias voltage V_B. The output current can be written as

$$I_0 = I_1 - I_2. \tag{5.357}$$

Considering that only dc voltages are applied to the output stage, so that $V_0 = 0$ and $I_0 = 0$, a current with the same value now flows through the transistors $T_1 - T_2$ and $T_3 - T_4$. That is, $I_1 = I_2 = I_Q$, where the quiescent current, I_Q, is given by

$$I_Q = I_B \frac{\left(\sqrt{\frac{1}{K_{n3}}} + \sqrt{\frac{1}{K_{p4}}} \right)^2}{\left(\sqrt{\frac{1}{K_{n1}}} + \sqrt{\frac{1}{K_{p2}}} \right)^2}. \tag{5.358}$$

A class AB operation is achieved because both transistors T_1 and T_2 are biased to conduct when the output voltage is reduced to zero. Using the square-law characteristic of MOS transistors, the currents I_1 and I_2 can be expressed as

$$I_1 = K_{n1}(V_{GS_1} - V_{T_n})^2 \tag{5.359}$$

and

$$I_2 = K_{p2}(V_{SG_2} + V_{T_p})^2, \tag{5.360}$$

where

$$V_{GS_1} = V_{G_1} - V_0 = V_{DD} - V_{SD_6} - V_0 \tag{5.361}$$

and

$$V_{SG_2} = V_0 - V_{G_2} = V_0 - V_{DS_5} - V_{SS}. \tag{5.362}$$

The swing of the output voltage, V_0, is limited due to the requirement of maintaining the transistors in the saturation region. Considering the path from the output node to the positive supply voltage, we can write

$$V_0 = V_{DD} - V_{SD_6} - V_{GS_1}. \tag{5.363}$$

The maximum value of the output voltage is then given by

$$V_{0M} = V_{DD} - V_{SD_6(sat)} - V_{DS_1(sat)} - V_{T_n} . \tag{5.364}$$

In the case of the path from the output node to the negative supply voltage, it can be shown that

$$V_0 = V_{SS} + V_{DS_5} + V_{SG_2} . \tag{5.365}$$

Therefore, the maximum value of the output voltage is of the form

$$V_{0m} = V_{SS} + V_{DS_5(sat)} + V_{SD_2(sat)} + V_{T_p} . \tag{5.366}$$

The graphical representations of currents I_1 and I_2 are shown in Figure 5.53, where the minimum values of I_1 and I_2 are respectively related to V_{0M} and V_{0m}.

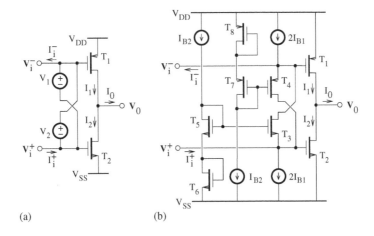

(a) (b)

FIGURE 5.54
(a) Principle and (b) circuit diagram of a complementary common source gain stage.

An improved class AB operation can be achieved by an output stage based on the principle illustrated in Figure 5.54(a), where an independent loop is used to set each floating voltage required for the biasing of the output transistor. In the class AB output stage implementation shown in Figure 5.54(b) [36, 37], the common-source connected output transistors, T_1 and T_2, are driven by in-phase input signals. The quiescent current flowing through the output transistors is determined by two independent translinear loops. Let the transistors of the same channel type be designed with identical threshold voltages and mobility parameters. Using Kirchhoff's voltage law for the loop including T_1, T_4, T_7, and T_8, we obtain

$$V_{SG_1} + V_{SG_4} = V_{SG_7} + V_{SG_8} . \tag{5.367}$$

Based on the square law characteristic of transistors, the current I_1 can be written as

$$I_1 = K_{p_1} \left(\sqrt{\frac{I_{B_2}}{K_{p_7}}} + \sqrt{\frac{I_{B_2}}{K_{p_8}}} - \sqrt{\frac{I_{B_1}}{K_{p_4}}} \right)^2 . \tag{5.368}$$

Considering the loop formed by T_2, T_3, T_5, and T_6, the voltage equation is of the form

$$V_{GS_2} + V_{GS_3} = V_{GS_5} + V_{GS_6} . \tag{5.369}$$

It can then be shown that

$$I_2 = K_{n_2} \left(\sqrt{\frac{I_{B_2}}{K_{n_5}}} + \sqrt{\frac{I_{B_2}}{K_{n_6}}} - \sqrt{\frac{I_{B_1}}{K_{n_3}}} \right)^2 . \tag{5.370}$$

To proceed further, we assume that

$$\frac{W_2/L_2}{W_1/L_1} = \frac{W_6/L_6}{W_8/L_8} = \frac{W_5/L_5}{W_7/L_7} = \frac{W_3/L_3}{W_4/L_4} = \frac{\mu_p}{\mu_n} \tag{5.371}$$

and

$$I_{B_1} = I_{B_2} . \tag{5.372}$$

When the output current I_0 is set to zero, each of the currents I_1 and I_2 is reduced to the quiescent current, I_Q, flowing through the transistors T_1 and T_2. Hence,

$$I_Q = \frac{W_1/L_1}{W_8/L_8} I_{B_2} = \frac{W_2/L_2}{W_6/L_6} I_{B_2} . \tag{5.373}$$

The maximum value of the current I_1 is obtained when T_4 is forced to operate in the cutoff region, that is, $V_{SG_4} \leq -V_{T_p}$, and the overall current $2I_{B_1}$ flows through T_3. As a result,

$$I_1 = I_{max} = K_{p_1} \left(\sqrt{\frac{I_{B_2}}{K_{p_7}}} + \sqrt{\frac{I_{B_2}}{K_{p_8}}} \right)^2 = 4\frac{W_1/L_1}{W_8/L_8} I_{B_2} = 4I_Q . \tag{5.374}$$

The current I_2 is reduced to the minimum value given by

$$I_2 = I_{min} = K_{n_2} \left(\sqrt{\frac{I_{B_2}}{K_{n_5}}} + \sqrt{\frac{I_{B_2}}{K_{n_6}}} - \sqrt{\frac{2I_{B_1}}{K_{n_3}}} \right)^2$$

$$= (2 - \sqrt{2})^2 \frac{W_2/L_2}{W_6/L_6} I_{B_2} = (2 - \sqrt{2})^2 I_Q . \tag{5.375}$$

Similarly, when the current I_1 becomes equal to its minimum value, due to the fact that T_3 is forced to operate in the cutoff region, that is, $V_{GS_3} \leq V_{T_n}$, the overall current $2I_{B_1}$ flows through T_4. The current I_2 is then set to its

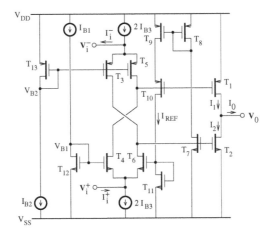

FIGURE 5.55
Circuit diagram of a low-voltage class AB output stage.

maximum value. Thus, this output stage configuration has the advantage of maintaining a minimum current in the inactive output transistor.

With reference to Figure 5.55 [38], a class AB output stage that can still operate with a low supply voltage is shown. In general, the minimum supply voltage is limited by the output stage, which uses transistors operating with sufficiently high gate-source voltages in order to drive high output currents. To reduce the minimum value of the supply voltage to the sum of one gate-source voltage and two saturation voltages, the gate voltages of the output transistors T_1 and T_2 are set by the folded mesh loop consisting of $T_3 - T_6$, which, together with the minimum current selector realized by $T_7 - T_{10}$, also regulates the minimum current flowing through the output transistors.

The transistors T_{12}, T_4, T_6, and T_{11} form a translinear loop, which defines the current I_{REF}. Applying Kirchhoff's voltage law around this loop gives

$$V_{GS_{12}} + V_{GS_4} = V_{GS_6} + V_{GS_{11}}, \tag{5.376}$$

where

$$V_{GS_{12}} = V_{T_n} + \sqrt{I_{B1}/K_{n_{12}}}, \tag{5.377}$$

$$V_{GS_4} = V_{T_n} + \sqrt{I_{B3}/K_{n_4}}, \tag{5.378}$$

$$V_{GS_6} = V_{T_n} + \sqrt{I_{B3}/K_{n_6}}, \tag{5.379}$$

and

$$V_{GS_{11}} = V_{T_n} + \sqrt{I_{REF}/K_{n_{11}}}. \tag{5.380}$$

Hence,

$$I_{REF} = K_{n_{11}} \left(\sqrt{\frac{I_{B_1}}{K_{n_{12}}}} + \sqrt{\frac{I_{B_3}}{K_{n_4}}} - \sqrt{\frac{I_{B_3}}{K_{n_6}}} \right)^2. \tag{5.381}$$

When $K_{n_4} = K_{n_6}$, the expression of the current I_{REF} is reduced to

$$I_{REF} = \frac{K_{n_{11}}}{K_{n_{12}}} I_{B_1} \, . \tag{5.382}$$

Let

$$I_1 = K_{p_1}(V_{SG_1} - V_{T_p})^2 \tag{5.383}$$

and

$$I_{REF} = K_{p_{10}}(V_{SG_{10}} - V_{T_p})^2 \tag{5.384}$$

$$= K_{p_9}[2(V_{SG_9} - V_{T_p})V_{SD_9} - V_{SD_9}^2], \tag{5.385}$$

where $0 < V_{SD_9} < V_{SG_9} - V_{T_p}$. For the loop including transistors T_1, T_{10}, and T_9, Kirchhoff's voltage law equation can be written as

$$V_{SG_1} = V_{SG_{10}} + V_{SD_9} \, , \tag{5.386}$$

where

$$V_{SG_1} = V_{T_p} + \sqrt{I_1/K_{p_1}} \, , \tag{5.387}$$

$$V_{SD_9} = V_{SG_9} - V_{T_p} - \sqrt{(V_{SG_9} - V_{T_p})^2 - I_{REF}/K_{p_9}} \, , \tag{5.388}$$

and

$$V_{SG_{10}} = V_{T_p} + \sqrt{I_{REF}/K_{p_{10}}} \, . \tag{5.389}$$

Because $V_{GS_2} = V_{GS_7}$, we have $I_{D_7} = I_2 = I_{D_8}$. Using the fact that

$$V_{SG_9} = V_{SG_8} = V_{T_p} + \sqrt{I_2/K_{p_8}} \, , \tag{5.390}$$

we obtain

$$\sqrt{I_1/K_{p_1}} = \sqrt{I_{REF}/K_{p_{10}}} + \sqrt{I_2/K_{p_8}} - \sqrt{I_2/K_{p_8} - I_{REF}/K_{p_{10}}} \, , \tag{5.391}$$

where $I_2 \geq (K_{p_8}/K_{p_{10}})I_{REF}$. Assuming that $I_1 = I_2 = I_Q$ and $K_{p_1} = K_{p_8}$, it can be deduced from Equation (5.391) that

$$I_Q = 2\frac{K_{p_1}}{K_{p_{10}}} I_{REF} \, , \tag{5.392}$$

where I_Q is the quiescent current flowing through the output transistors.

During normal operation, the transistor T_9 operates in the linear region, where its drain current is a function of both the source-gate voltage set by the transistor T_8 and the source-drain voltage adjusted via the transistor T_{10}. The source-drain voltage of the transistor T_9, or the source voltage of the transistor T_{10}, can then be maintained sufficiently low such that the variations in the current I_1 can be tracked by the transistor T_{10}. The transistor T_7, which operates with the same gate-source voltage as the transistor T_2, is used to detect the current I_2. The minimum selector circuit $T_7 - T_{10}$ then evaluates the magnitudes of the currents I_1 and I_2 to help set a minimum current flowing through each of the output transistors as a function of the current I_{REF}.

However, as the drain current of the transistor T_1 increases such that its source-gate voltage becomes sufficiently high to provide enough headroom for the operation of T_9 in the saturation region, the transistors $T_8 - T_{10}$ realize a cascoded current mirror. When the current I_2 reaches its minimum value, $I_Q/2$, the maximum value of the current I_1 derived from Equation (5.391) is $2I_Q$. With $V_{SD_9} = V_{SG_9} - V_{T_p}$ and $V_{SG_9} = V_{SG_{10}}$, Equation (5.386) is reduced to $V_{SG_1} = 2V_{SG_9}$ and the drain current of T_9 is equal to $I_Q/2$. Because $V_{SG_8} = V_{SG_9}$, the bias current of the transistor T_7 is also set to $I_Q/2$. On the other hand, an increase in the current I_2 produces an augmentation of the current flowing through T_7 and T_8, and a decrease in the current I_1 leading to a reduction in the source-gate voltage of the transistor T_1. The source-gate voltage of the transistor T_1 can then be reduced until the source-drain voltage of the transistor T_9 becomes negligible. Hence, $V_{SG_1} \simeq V_{SG_{10}}$ and the current I_1 takes the minimum value $I_Q/2$, while the current I_2 is maximum.

It should be noted that the stability can be affected by poles associated with the folded mesh loop, the current mirror $T_8 - T_9$ and cascode transistor T_{10}. The frequency stabilization is achieved in practical implementations by using a pole-splitting compensation network. Furthermore, the control of the quiescent current can be limited by mismatches of transistors $T_8 - T_{10}$.

5.6.2 Two-stage amplifier

The two-stage amplifiers of Figures 5.56 and 5.57 consist of a differential input stage, an inverting output stage, and a biasing circuit. The single-ended signal is provided by the current mirror used as load of the transistor differential pair and the amplifier output swing is $V_{sup} - 2V_{DS(sat)}$, where V_{sup} denotes the supply voltage. The Miller frequency compensation with a pole-zero cancelation is adopted to overcome the gain-bandwidth trade-off. A one-pole frequency response of the amplifier is obtained by including a zero in the left-half plane of the s-domain to cancel the first nondominant pole. The compensation section [35, 40, 43] can be implemented using a resistor in series with a capacitor, as shown in Figure 5.56, or a series connection of a MOS transistor operating in the triode region and a capacitor, as illustrated in Figure 5.57 [35].

An equivalent circuit of the RC-compensated two-stage amplifier is shown in Figure 5.58. Assuming that the parasitic coupling capacitances C_{p1} and C_{p2} can be neglected, the next nodal equations can be written

$$g_{m_1} V_i(s) + V_1(s)(g_1 + sC_1) - (V_0(s) - V_1(s))/(R_c + 1/sC_c) = 0 \quad (5.393)$$

and

$$g_{m_2} V_1(s) + V_0(s)(g_2 + sC_2) + (V_0(s) - V_1(s))/(R_c + 1/sC_c) = 0 \quad (5.394)$$

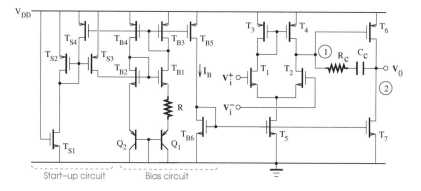

FIGURE 5.56
Circuit diagram of a two-stage amplifier with RC compensation.

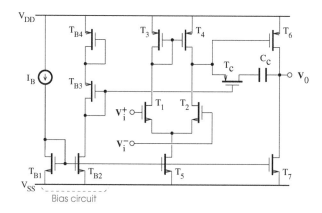

FIGURE 5.57
Circuit diagram of a two-stage amplifier with the compensation resistor implemented by a transistor.

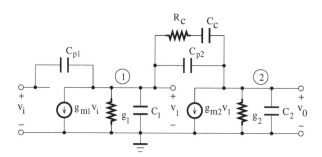

FIGURE 5.58
Small-signal equivalent of the RC-compensated two-stage amplifier.

for the node 1 and node 2, respectively, where g_{m_k}, g_k, and C_k $(k = 1, 2)$ denote the transconductance, output conductance, and output capacitor of the k stage, respectively, and R_c and C_c are the compensation resistor and capacitor, respectively. Solving the above system of equations gives

$$A(s) = \frac{V_0(s)}{V_i(s)} = A_0 \frac{1 + (R_c - 1/g_{m_2})\, sC_c}{1 + cs + bs^2 + as^3}, \tag{5.395}$$

where

$$A_0 = \frac{g_{m_1} g_{m_2}}{g_1 g_2}, \tag{5.396}$$

$$a = \frac{C_1 C_2 R_c C_c}{g_1 g_2}, \tag{5.397}$$

$$b = \frac{C_1 C_c + C_1 C_2 + C_2 C_c}{g_1 g_2} + \left(\frac{C_1}{g_1} + \frac{C_2}{g_2}\right) R_c C_c, \tag{5.398}$$

$$c = \frac{C_1 + C_c}{g_1} + \frac{C_2 + C_c}{g_2} + \frac{g_{m_2} C_c}{g_1 g_2} + R_c C_c. \tag{5.399}$$

The capacitances C_1 and C_2 can be related to the transistor capacitances by equations of the form $C_1 = C_{gd_4} + C_{db_4} + C_{gd_2} + C_{db_2} + C_{gs_2}$ and $C_2 = C_{db_6} + C_{db_7} + C_{gd_7} + C_L$, where C_L represents the load capacitance at the amplifier output and the contribution due to C_{gd_6} is assumed to be negligible because $C_{gd_6} \ll C_c$. The third-order transfer function, $A(s)$, can be put into the form

$$A(s) = \frac{V_0(s)}{V_i(s)} = A_0 \frac{1 - s/\omega_{z_1}}{(1 - s/\omega_{p_1})(1 - s/\omega_{p_2})(1 - s/\omega_{p_3})}, \tag{5.400}$$

where ω_{z_1} is the frequency of the zero z_1, and ω_{p_1}, ω_{p_2}, and ω_{p_3} are the frequencies of the poles p_1, p_2, and p_3, respectively. In practice, $A_0 \gg 1$, $g_1 R_c \gg 1$, $g_2 R_c \gg 1$, $C_1/C_c \gg 1$, and $C_1/C_2 \gg 1$, and it is assumed that the poles are widely spaced. Thus, we can write

$$A(s) = \frac{V_0(s)}{V_i(s)} \simeq A_0 \frac{1 - s/\omega_{z_1}}{1 - s/\omega_{p_1} + s^2/\omega_{p_1}\omega_{p_2} - s^3/\omega_{p_1}\omega_{p_2}\omega_{p_3}}, \tag{5.401}$$

where

$$\omega_{z_1} = \frac{1}{C_c(1/g_{m_2} - R_c)}, \tag{5.402}$$

$$\omega_{p_1} = -\frac{g_1 g_2}{g_{m_2} C_c} = -\frac{g_{m_1}}{A_0 C_c}, \tag{5.403}$$

$$\omega_{p_2} = -\frac{g_{m_2} C_c}{C_1 C_2 + C_c(C_1 + C_2)} \simeq -\frac{g_{m_2}}{C_2}, \tag{5.404}$$

$$\omega_{p_3} = -\frac{1}{R_c C_1}. \tag{5.405}$$

Note that p_1 represents the dominant pole. The compensation leads to a splitting of the initially close poles as depicted in Figure 5.59.

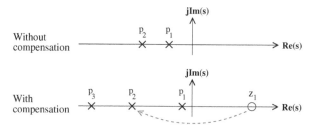

FIGURE 5.59
Pole-zero representations of a two-stage amplifier without and with frequency compensation.

By setting, $R_c = 1/g_{m_2}$, the zero is rejected at infinity and the stability of the amplifier is guaranteed provided that p_2 and p_3 are located far beyond the unity-gain frequency, that is, $|\omega_{p_3}| > |\omega_{p_2}| > A_0|\omega_{p_1}|$. In this case, $C_c > g_{m_1}C_2/g_{m_2}$ and a large value of C_c or g_{m_2} is required to split the poles. As a result, the die area or power consumption must be increased.

It is also possible to cancel the pole p_2 using the zero z_1, which is moved from the right-half plane to the left-half plane (see Figure 5.59). Hence,

$$\omega_{p_2} = \omega_{z_1} \tag{5.406}$$

and the value of the compensation resistor is derived as

$$R_c = \frac{C_2 + C_c}{g_{m_2}C_c}. \tag{5.407}$$

The compensation of the amplifier, whose frequency response is now determined by the two remaining poles, requires that $|\omega_{p_3}| > A_0|\omega_{p_1}|$. By using a small compensation capacitor to realize the pole-zero cancelation, the resulting amplifier can exhibit a wide bandwidth and a high slew rate.

In the case of the amplifier shown in Figure 5.56, the bias circuit is designed to be less sensitive to temperature and process variations. Initially, the start-up circuit [39] triggers the flow of a current used to activate the bandgap bias circuit. Due to the conduction of the transistor T_{S1}, a start-up current is mirrored into the loop $T_{B1} - T_{B4}$ of the bias circuit, via $T_{S2} - T_{S3}$. When the stationary state of the bias current is reached, the transistor T_{S4} starts conducting. The voltage level at the gates of transistors $T_{S2} - T_{S3}$ is approximately equal to the positive supply voltage, rendering T_{S2} and T_{S3} nonconducting. The generation of the start-up current is then interrupted.

The biasing circuit of the amplifier can be designed to ensure that the pole and zero track over process, voltage, and temperature variations. With reference to Figure 5.56, we can write

$$R_B I_B = V_{GS3} - V_{GS4}, \tag{5.408}$$

and the bias current, I_B, is given by

$$I_B = \frac{1}{K_p R_B^2} \left(\sqrt{\frac{1}{W_3/L_3}} - \sqrt{\frac{1}{W_4/L_4}} \right)^2, \qquad (5.409)$$

where $K_p = \mu_p C_{ox}/2$. If the transistors match accurately, the transconductance will be independent of K_p and determined by R_B.

The transistor T_c used in the circuit of Figure 5.57 operates in the triode region. Its drain-source resistance is given by

$$R_{DS} = \left. \frac{\partial I_D}{\partial V_{SD}} \right|_{V_{SG}} = [2K_p(V_{SG} - V_T - V_{SD})]^{-1} \qquad (5.410)$$

and is continuously adjusted to allow an adequate settling response.

The RC-based compensation network also introduces some limitations on the performance of the aforementioned two-stage amplifiers. The amplifier is stable in a feedback configuration only when the value of the load capacitor does not exceed the one of the compensation capacitor. Due to the extra path from the power supply voltage through the compensation network to the amplifier output, the positive and negative power supply rejections are degraded for frequencies greater than the pole frequency associated with the compensation capacitor in the case of n-channel and p-channel differential input transistor pairs, respectively.

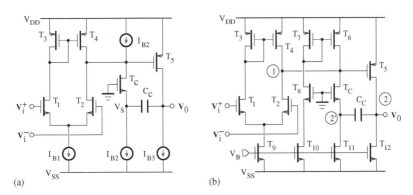

FIGURE 5.60
(a) Principle and (b) circuit diagram of a two-stage amplifier with a current buffer.

In the amplifier structures illustrated in Figure 5.60, the feed-forward path is removed by inserting a current buffer between the input and output stages [40–42]. The small-signal equivalent model is depicted in Figure 5.61, where C_{p2} denotes the gate-source capacitor of T_c. To simplify the circuit analysis, the feedback current source can be replaced by its equivalent Y network consisting of the current source i_f connected between the node 1 and

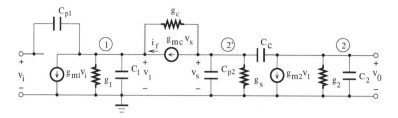

FIGURE 5.61
Small-signal equivalent model of the two-stage amplifier with a current buffer.

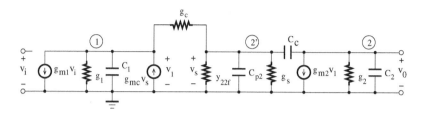

FIGURE 5.62
Shunt-shunt version of the amplifier small-signal equivalent model.

the ground, and a resistor with the admittance y_{22f} and connected between the node 2' and the ground. Because $i_f = g_{m_c} v_s$, we can obtain

$$y_{22f} = \left. \frac{i_f}{v_s} \right|_{v_i=0} = g_{m_c} . \qquad (5.411)$$

Assuming that the parasitic capacitor C_{p_1} is negligible, a shunt-shunt version of the small-signal equivalent model can be derived as shown in Figure 5.62. The use of Kirchhoff's current law at the circuit nodes 1, 2', and 2 gives

$$g_{m_1} V_i(s) + g_1 V_1(s) + sC_1 V_1(s) + g_c(V_1(s) - V_s(s)) = g_{m_c} V_s(s), \qquad (5.412)$$

$$g_c(V_1(s) - V_s(s)) = g_{m_c} V_s(s) + sC_{p2} V_s(s) + g_s V_s(s) + sC_c(V_s(s) - V_0(s)), \qquad (5.413)$$

and

$$sC_c(V_s(s) - V_0(s)) = g_{m_2} V_1(s) + g_2 V_0(s) + sC_2 V_0(s), \qquad (5.414)$$

respectively, where

$$C_1 = C_{db_2} + C_{gd_2} + C_{db_4} + C_{gd_4} \\ + C_{gs_5} + C_{db_6} + C_{gd_6} + C_{db_c} + C_{gd_c} , \qquad (5.415)$$

$$C_{p2} = C_{gs_c} + C_{sb_c} + C_{db_{11}} + C_{gd_{11}} , \qquad (5.416)$$

$$C_2 = C_{db_5} + C_{gd_5} + C_{db_{12}} + C_{gd_{12}} + C_L , \qquad (5.417)$$

$$g_1 = g_{ds_2} + g_{ds_4} , \qquad (5.418)$$

$$g_2 = g_{ds_5} + g_{ds_{12}} , \qquad (5.419)$$

and g_s can be reduced to the output conductance of the current source I_{B_2}. By solving the system of Equations (5.412), (5.413), and (5.414), the transfer function can obtained as

$$A(s) = \frac{V_0(s)}{V_i(s)} = A_0 \frac{1 + \dfrac{g_c + g_s}{g_{m_c}} + \left(\dfrac{C_c + C_{p2}}{g_{m_c}} - \dfrac{C_c g_c}{g_2 g_{m_c}} \right) s}{1 + \gamma s + \beta s^2 + \alpha s^3}, \tag{5.420}$$

where

$$A_0 = \frac{g_{m_1} g_{m_2}}{g_1 g_2}, \tag{5.421}$$

$$\alpha = C_1 \frac{C_2 C_c + C_2 C_{p2} + C_c C_{p2}}{g_{m_c} g_1 g_2}, \tag{5.422}$$

$$\begin{aligned} \beta = {} & C_1 \frac{C_2 + C_c}{g_1 g_2} \left(1 + \frac{g_c + g_s}{g_{m_c}} \right) \\ & + C_1 \frac{C_c + C_{p2}}{g_{m_c} g_1} + \frac{C_2 C_c + C_2 C_{p2} + C_c C_{p2}}{g_{m_c} g_2} \left(1 + \frac{g_c}{g_1} \right), \end{aligned} \tag{5.423}$$

$$\begin{aligned} \gamma = {} & \frac{g_{m_2} C_c}{g_1 g_2} \left(1 + \frac{g_c}{g_{m_c}} \right) + \frac{C_2 + C_c}{g_2} \left(1 + \frac{g_c}{g_{m_c}} + \frac{g_s}{g_{m_c}} + \frac{g_c g_s}{g_{m_c} g_1} \right) \\ & + \frac{C_1}{g_1} \left(1 + \frac{g_c + g_s}{g_{m_c}} \right) + \frac{C_c + C_{p2}}{g_{m_c}} \left(1 + \frac{g_c}{g_1} \right). \end{aligned} \tag{5.424}$$

In general, it can be shown that

$$A(s) = \frac{V_0(s)}{V_i(s)} = A_0 \frac{1 - s/\omega_{z_1}}{(1 - s/\omega_{p_1})(1 - s/\omega_{p_2})(1 - s/\omega_{p_3})}. \tag{5.425}$$

where the locations of the poles p_1, p_2, and p_3 can be determined using a computer program for symbolic analysis. The compensation capacitor can then be chosen to achieve a given settling time or phase margin specification. Assuming a dominant pole model, or say $|\omega_{p_1}| \ll |\omega_{p_2}|, |\omega_{p_3}|$, and $|\omega_{p_2}| \ll |\omega_{p_3}|$, the transfer function can be approximated as

$$A(s) = \frac{V_0(s)}{V_i(s)} \simeq A_0 \frac{1 - s/\omega_{z_1}}{1 - s/\omega_{p_1} + s^2/\omega_{p_1}\omega_{p_2} - s^3/\omega_{p_1}\omega_{p_2}\omega_{p_3}}. \tag{5.426}$$

Assuming that $g_{m_2}, g_{m_c} \gg g_1, g_2, g_c, g_s$; $C_c, C_2 \gg C_1, C_{p2}$; $g_{m_c}/g_c \gg g_s/g_1$, and $g_{m_2}/g_1 \gg C_2/C_c$, we obtain

$$\omega_{z_1} = -\frac{g_{m_c}}{C_c + C_{p2}}, \tag{5.427}$$

$$\omega_{p_1} = -\frac{g_1 g_2}{C_c g_{m_2}}, \tag{5.428}$$

$$\omega_{p_2} = -\frac{g_{m_2} C_c}{C_1(C_2 + C_c)}, \tag{5.429}$$

$$\omega_{p_3} = -\frac{g_{m_c}(C_2 + C_c)}{C_2 C_c + C_2 C_{p2} + C_c C_{p2}} = -\frac{g_{m_c}}{\dfrac{C_2 C_c}{C_2 + C_c} + C_{p2}}. \tag{5.430}$$

A cancelation between p_3 and z_1 is achieved only if $C_c/C_2 \ll 1$. It is incomplete in practice due to component mismatches, and p_3 and z_1 then form a pole-zero doublet. If now $|\omega_{p_2}| < |\omega_{z_1}| < |\omega_{p_3}|$, for instance, the stability requirement of the amplifier in feedback loop will be met.

The magnitude of the nondominant pole p_2 in this case is greater than the one of the conventional Miller compensated amplifier, which is $g_{m_2} C_c/[C_1 C_2 + C_c(C_1 + C_2)]$. To achieve the same frequency response, the use of a current buffer then offers the advantage of reducing the required component values. Furthermore, the value of g_{m_c} can be determined by appropriately biasing and sizing the transistor T_c to set p_3 and z_1 well beyond p_2. However, the mismatch of the biasing currents of T_c may increase the amplifier offset voltage.

Note that if the current buffer is modeled as an ideal current source with a zero input resistance [40], or equivalently, g_{mc} is considered to be infinite, the right-half plane zero is canceled and the transfer function only exhibits two poles.

5.6.3 Optimization of a two-pole amplifier for fast settling response

Amplifiers used in high-speed applications should preferably be designed to achieve the minimum settling time [45, 46]. In general, the settling time, t_s, is defined as the time required for the amplifier output voltage to settle to within an error tolerance, ϵ, around the final steady-state value.

Consider a two-stage amplifier characterized by an open-loop transfer function of the form

$$A(s) = \frac{A_0}{(1 - s/\omega_{p_1})(1 - s/\omega_{p_2})}, \tag{5.431}$$

where A_0 is the dc gain, and ω_{p_1} and ω_{p_2} are the frequency locations of the first and second poles (p_1 and p_2), respectively. In a unity-gain configuration,

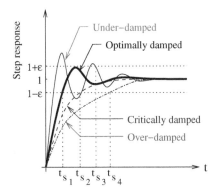

FIGURE 5.63
Step responses of a two-stage amplifier.

the closed-loop transfer function is

$$A_{CL}(s) = \frac{A(s)}{1 + A(s)} = \frac{A_0'}{(s/\omega_0)^2 + 2k(s/\omega_0) + 1}, \qquad (5.432)$$

where

$$A_0' = \frac{A_0}{1 + A_0}, \qquad (5.433)$$

$$\omega_0 = [\omega_{p_1}\omega_{p_2}(1 + A_0)]^{1/2}, \qquad (5.434)$$

and

$$k = -\frac{\omega_{p_1} + \omega_{p_2}}{2\omega_0}. \qquad (5.435)$$

With $\beta = \omega_{p_2}/\omega_{p_1}$ being the pole separation factor, the damping factor k can be rewritten as

$$k = \frac{1 + \beta}{2[\beta(1 + A_0)]^{1/2}}. \qquad (5.436)$$

For a unit step input, $u(t)$, defined as

$$u(t) = \begin{cases} 1, & \text{for} \quad t \geq 0 \\ 0, & \text{for} \quad t < 0, \end{cases} \qquad (5.437)$$

the closed-loop response in the time domain is given by

$$v_0(t) = \mathcal{L}^{-1}[H(s)], \qquad (5.438)$$

where $H(s) = A_{CL}(s)/s$ and $\mathcal{L}^{-1}$ denotes the inverse Laplace transform. Depending on the value of k, there are three possible responses.

For $k > 1$, the closed-loop response is said to be over-damped. It shows no oscillation and is slow to settle. The output voltage is given by

$$v_0(t) = \left\{ 1 - \frac{1}{2(k^2 - 1)^{1/2}} \left[\frac{1}{k_1} \exp(-k_1\omega_0 t) - \frac{1}{k_2} \exp(-k_2\omega_0 t) \right] \right\} u(t), \quad (5.439)$$

where $k_1 = k - (k^2 - 1)^{1/2}$ and $k_2 = k + (k^2 - 1)^{1/2}$.

For $k = 1$, the closed-loop response is critically damped. It quickly converges to the steady-state value, and without oscillating. We then have

$$v_0(t) = [1 - (1 + \omega_0 t) \exp(-\omega_0 t)] u(t). \quad (5.440)$$

For $0 < k < 1$, the closed-loop response is under-damped and exhibits some oscillations before approaching the steady-state value. The output voltage can be written as

$$v_0(t) = \left\{ 1 - \left(\frac{k \sin[(1 - k^2)^{1/2}\omega_0 t]}{(1 - k^2)^{1/2}} + \cos[(1 - k^2)^{1/2}\omega_0 t] \right) \exp(-k\omega_0 t) \right\} u(t). \quad (5.441)$$

The over-damped, critically damped, and under-damped step responses are depicted in Figure 5.63. Also shown in Figure 5.63 is the optimally damped response. It is apparent that the amplifier exhibits a minimum settling time when the first peak of the step response just touches the upper settling error bound.

Let us consider the case of an under-damped response with the peak of the first overshoot occurring at the instant t_p. By setting the derivative of $v_0(t)$ equal to zero, we have

$$t = t_p = \frac{\pi}{\omega_0(1 - k^2)^{1/2}}. \quad (5.442)$$

The level of the first peak of the step response is of the form, $v_0(t_p) = 1 + \epsilon$, where

$$\epsilon = \exp[-k\pi/(1 - k^2)^{1/2}]. \quad (5.443)$$

With $\beta \gg 1$, we can combine the expression derived from Equation (5.436), that is, $k \simeq (1/2)\sqrt{\beta/(1 + A_0)}$, and Equation (5.443) to get

$$\beta = \beta_{mst} \simeq \frac{4(1 + A_0)}{1 + (\pi/\ln \epsilon)^2}, \quad (5.444)$$

where β_{mst} is the pole separation factor associated to the minimum settling time for a given error tolerance, ϵ. However, it is common to characterize an amplifier using the phase margin, ϕ_M, instead of the pole separation factor. When the amplifier poles are sufficiently separated, we have

$$\phi_M = 180^\circ - \angle A(j\omega_u)$$
$$= 180^\circ - \arctan(\omega_u/|\omega_{p_1}|) - \arctan(\omega_u/|\omega_{p_2}|), \quad (5.445)$$

where ω_u is the unity-gain frequency. Assuming that $\omega_u \simeq A_0|\omega_{p_1}|$, the term $\arctan(\omega_u/|\omega_{p_1}|)$ is on the order of $90°$ as the dc gain A_0 is very high. Because $\omega_{p_2} = \beta\omega_{p_1}$, the substitution of Equation (5.444) into (5.445) then gives [46]

$$\phi_M = \phi_{M,mst} \simeq 90° - \arctan\left[\frac{1 + (\pi/\ln\epsilon)^2}{4}\right]. \qquad (5.446)$$

For very small values of ϵ, $(\pi/\ln\epsilon)^2$ becomes negligible and $\phi_{M,mst}$ is about $76°$, which is in the same order as the phase margin of a critically damped system. Here, the minimum settling time is then achieved when the amplifier response is critically damped.

In the case of a two-stage amplifier with the pole-splitting frequency-compensation network, the transfer function is given by

$$A(s) = A_0 \frac{1 - s/\omega_{z_1}}{(1 - s/\omega_{p_1})(1 - s/\omega_{p_2})(1 - s/\omega_{p_3})}. \qquad (5.447)$$

To obtain a two-pole frequency response and minimize the settling time, we need to have

$$\omega_{p_2} = \omega_{z_1} \qquad (5.448)$$

and

$$\omega_{p_3} = \beta_{mst}\omega_{p_1}. \qquad (5.449)$$

The values of the components used in the compensation network should then be chosen such that the above requirements are met.

5.6.4 Three-stage amplifier

The circuit diagram of a three-stage amplifier with the nested Miller frequency compensation [47,48] is shown in Figure 5.64. The first stage is implemented by the differential input transistors, $T_1 - T_2$, biased by T_{15}, and loaded by $T_3 - T_8$, while the second and third ones consist of $T_9 - T_{12}$ and T_{13}, respectively. The compensation capacitors C_{c1} and C_{c2} are used to stabilize the amplifier.

The equivalent model of the three-stage amplifier with the nested Miller frequency compensation is shown in Figure 5.65. The first and third gain stages are of the inverting type. The noninverting second amplification section is required to ensure the negative feedback around the nested compensation paths. Applying Kirchhoff's current law at nodes 1, 2, and 3, we obtain

$$g_{m_1}V_i(s) + V_1(s)(g_1 + sC_1) - (V_0(s) - V_1(s))sC_{c_1} = 0, \qquad (5.450)$$
$$-g_{m_2}V_1(s) + V_2(s)(g_2 + sC_2) - (V_0(s) - V_2(s))sC_{c_2} = 0, \qquad (5.451)$$

and

$$g_{m_3}V_2(s) + V_0(s)(g_3 + sC_3) + (V_0(s) - V_1(s))sC_{c_1} + (V_0(s) - V_2(s))sC_{c_2} = 0, \qquad (5.452)$$

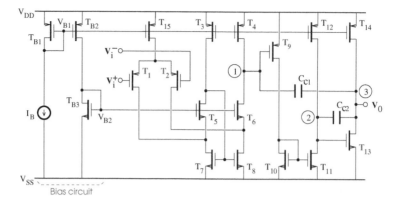

FIGURE 5.64

Circuit diagram of a three-stage amplifier with a nested Miller compensation.

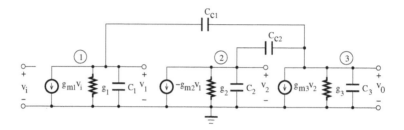

FIGURE 5.65

Small-signal equivalent circuit of the three-stage amplifier with a nested Miller compensation.

where g_{m_k}, g_k, and C_k ($k = 1, 2, 3$) denote the transconductance, output conductance, and output capacitor of the k stage, respectively. The node capacitances are given by

$$C_1 = C_{gd_4} + C_{gb_4} + C_{gd_6} + C_{gb_6} + C_{gs_9}, \tag{5.453}$$

$$C_2 = C_{gd_{11}} + C_{gb_{11}} + C_{gd_{12}} + C_{gb_{12}}, \tag{5.454}$$

and

$$C_3 = C_{gb_{13}} + C_{gd_{14}} + C_{gb_{14}} + C_L, \tag{5.455}$$

where C_L is the external load capacitance that can be connected to the output node. Assuming that $g_{m_k} \gg g_k$ ($k = 1, 2, 3$) and $C_{c1}, C_{c2}, C_3 \gg C_1, C_2$, the amplifier small-signal voltage gain can be computed as [49]

$$A(s) = \frac{V_0(s)}{V_i(s)} \simeq A_0 \frac{1 + ds + cs^2}{(1 + s/\omega_{p_1})(1 + bs + as^2)}, \tag{5.456}$$

where

$$A_0 = \frac{g_{m_1} g_{m_2} g_{m_3}}{g_1 g_2 g_3}, \tag{5.457}$$

$$\omega_{p_1} = \frac{g_1 g_2 g_3}{g_{m_2} g_{m_3} C_{c1}}, \tag{5.458}$$

$$a = \frac{C_{c2} C_3}{g_{m_2} g_{m_3}}, \tag{5.459}$$

$$b = \frac{C_{c2}(g_{m_3} - g_{m_2})}{g_{m_2} g_{m_3}}, \tag{5.460}$$

$$c = -\frac{C_{c1} C_{c2}}{g_{m_2} g_{m_3}}, \tag{5.461}$$

and

$$d = -\frac{C_{c2}}{g_{m_3}}. \tag{5.462}$$

The zero frequencies are usually much greater than the unity-gain frequency of the amplifier. With the assumptions that $g_{m_3} \gg g_{m_1}, g_{m_2}$, and $b \simeq C_{c2}/g_{m_2}$, the voltage transfer function can be further simplified to

$$A(s) = \frac{V_0(s)}{V_i(s)} \simeq A_0 \frac{1}{(1 + s/\omega_{p_1})(1 + bs + as^2)}. \tag{5.463}$$

The dominant pole is related to the output node (node 3). The nested capacitor C_{c1} splits the poles at nodes 2 and 3, while C_{c1} splits the ones at nodes 1 and 3 (see Figure 5.66). With the poles being sufficiently separated, we can obtain

$$A(s) = \frac{V_0(s)}{V_i(s)} \simeq \frac{1}{(s/A_0 \omega_{p_1})(1 + bs + as^2)}. \tag{5.464}$$

The stabilization will be achieved if the amplifier exhibits a third-order Butterworth frequency response in the unity-gain feedback configuration [50]. That is,

$$A_{CL}(s) = \frac{A(s)}{1 + A(s)} = \frac{1}{1 + 2\dfrac{s}{\omega_c} + 2\dfrac{s^2}{\omega_c^2} + \dfrac{s^3}{\omega_c^3}}, \tag{5.465}$$

where ω_c is the cutoff frequency. Consequently, the gain $A(s)$ should be of the form

$$A(s) = \frac{1}{2\dfrac{s}{\omega_c}\left(1 + \dfrac{s}{\omega_c} + \dfrac{1}{2}\dfrac{s^2}{\omega_c^2}\right)}. \tag{5.466}$$

Comparing Equations (5.464) and (5.466) gives

$$\frac{2}{\omega_c} = \frac{1}{A_0 \omega_{p_1}} \simeq \frac{C_{c1}}{g_{m_1}}, \tag{5.467}$$

$$\frac{1}{\omega_c} = b \simeq \frac{C_{c2}}{g_{m_2}}, \tag{5.468}$$

$$\frac{1}{2\omega_c^2} = a = \frac{C_{c2} C_3}{g_{m_2} g_{m_3}}. \tag{5.469}$$

This leads to the values of the compensation capacitors given by

$$C_{c1} = 4 \left(\frac{g_{m_1}}{g_{m_3}} \right) C_3, \tag{5.470}$$

$$C_{c2} = 2 \left(\frac{g_{m_2}}{g_{m_3}} \right) C_3. \tag{5.471}$$

Because the Butterworth response is said to be under-damped, it can also yield the minimum settling time. It is characterized by a unity-gain phase margin given by

$$\phi_M = 180^o - \angle A[j(A_0 \omega_{p_1})] \tag{5.472}$$

$$= 180^o - \angle(2s/\omega_c) - \arctan\left(\frac{A_0 \omega_{p_1}/\omega_c}{1 - (A_0 \omega_{p_1}/\sqrt{2}\omega_c)^2} \right) \tag{5.473}$$

$$= 180^o - 90^o - \arctan(4/7) \simeq 60^o. \tag{5.474}$$

With $A_0 \omega_{p_1}$ and ω_c representing the amplifier unity-gain frequency and the cutoff frequency of the Butterworth response, respectively, it was assumed that $A_0 \omega_{p_1}/\omega_c = 1/2$.

Note that the locations of the nondominant poles depend on C_3 and the amplifier stability can be affected by the output load capacitor. The gain-bandwidth product is given by

$$GBW \simeq A_0 \omega_{p_1} = \frac{g_{m_1}}{C_{c1}} = \frac{1}{4} \frac{g_{m_3}}{C_3}. \tag{5.475}$$

The resulting GBW is four times smaller than the one of an uncompensated structure. The slew rate (SR) of the amplifier can be obtained as

$$SR = \min \left\{ \frac{I_{B1}}{C_{c1}}, \frac{I_{B2}}{C_{c2}}, \frac{I_{B3}}{C_3} \right\}, \tag{5.476}$$

where I_{B1}, I_{B2}, and I_{B3} are the bias current of the first, second, and third stage, respectively. The amplifier should exhibit a poor SR for large compensation capacitors.

Generally, an N-stage amplifier needs $N - 1$ compensation capacitors

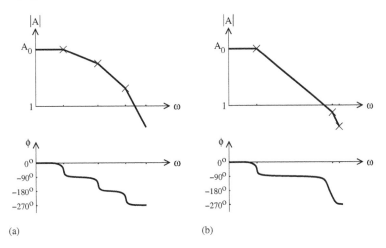

(a)   (b)

FIGURE 5.66
Magnitude and phase responses of the three-stage amplifier (a) before and (b) after the nested Miller compensation.

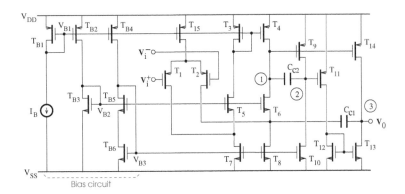

FIGURE 5.67
Circuit diagram of a three-stage RAFFC amplifier.

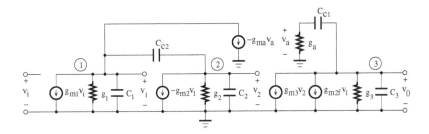

FIGURE 5.68
Small-signal equivalent circuit of the three-stage RAFFC amplifier.

driven by the output stage. In this case, the requirement of a large bandwidth can be fulfilled only at the price of a high-power consumption.

Due to the fact that both compensation capacitors affect the amplifier loading, the bandwidth achievable with the nested Miller compensation is limited. An improvement can be obtained using amplifiers based on the reversed nested Miller compensation, especially when driving large capacitive loads.

The circuit diagram of a three-stage reversed active feedback frequency compensation (RAFFC) amplifier [51] is depicted in Figure 5.67. It consists of an input stage with the folded-cascode structure, a common-source inverting stage, and a noninverting output stage. The feed-forward transconductance is realized by linking the gate of the load transistor T_{14} to the first stage output, while the feedback transconductance is obtained by connecting C_{c1} to the source of T_6 instead of the first stage output. With reference to the small-signal equivalent circuit shown in Figure 5.68, the equations for nodes 1, 2, and 3 can be written as

$$g_{m_1} V_i(s) + V_1(s)(g_1 + sC_1) - g_{ma} V_a(s) - (V_2(s) - V_1(s))sC_{c_2} = 0, \quad (5.477)$$
$$-g_{m_2} V_1(s) + V_2(s)(g_2 + sC_2) + (V_2(s) - V_1(s))sC_{c_2} = 0, \quad (5.478)$$

and

$$g_{m_3} V_2(s) + g_{m2f} V_1(s) + V_0(s)(g_3 + sC_3) + (V_0(s) - V_a(s))sC_{c_1} = 0, \quad (5.479)$$

respectively, where g_{m2f} is the feed-forward transconductance, and g_{ma} is the feedback transconductance. The node capacitances can be expressed as

$$C_1 = C_{gd_4} + C_{gb_4} + C_{gd_6} + C_{gb_6} + C_{gs_9}, \quad (5.480)$$
$$C_2 = C_{gd_9} + C_{gb_9} + C_{gd_{10}} + C_{gb_{10}} + C_{gs_{11}}, \quad (5.481)$$

and

$$C_3 = C_{gd_{13}} + C_{gb_{13}} + C_{gd_{14}} + C_{gb_{14}} + C_L, \quad (5.482)$$

where C_L is the external output load capacitance. Using the principle of voltage division, it can be shown that

$$V_a(s) = \frac{sC_{c_1} V_0(s)}{g_a + sC_{c_1}}. \quad (5.483)$$

Assuming that $g_{m_k} \gg g_k$ $(k = 1, 2, 3)$ and $C_{c1}, C_{c2}, C_3 \gg C_1, C_2$, the amplifier small-signal voltage gain can be computed as

$$A(s) = \frac{V_0(s)}{V_i(s)} \simeq A_0 \frac{1 + ds + cs^2}{(1 + s/\omega_{p_1})(1 + bs + as^2)}, \quad (5.484)$$

where

$$A_0 = \frac{g_{m_1} g_{m_2} g_{m_3}}{g_1 g_2 g_3}, \tag{5.485}$$

$$\omega_{p_1} = \frac{g_1 g_2 g_3}{g_{m_2} g_{m_3} C_{c1}}, \tag{5.486}$$

$$a = \frac{C_{c2} C_3}{g_{ma} g_{m_3}}, \tag{5.487}$$

$$b = \frac{(g_{m_2} + g_{m2f} - g_{m_3}) C_{c1} C_{c2} + g_{m_2} C_{c2} C_3}{g_{m_2} g_{m_3} C_{c1}}, \tag{5.488}$$

$$c = \frac{(g_{m2f} - g_{m_3}) C_{c1} C_{c2}}{g_{m_2} g_{m_3} g_a}, \tag{5.489}$$

and

$$d = \frac{C_{c1}}{g_a} + \frac{(g_{m2f} - g_{m_3}) C_{c2}}{g_{m_2} g_{m_3}}. \tag{5.490}$$

To proceed further, it is assumed that $g_{m2f} = g_{m_3}$. As a result, the coefficient c of the transfer function is reduced to zero. The amplifier transfer function now exhibits one left-hand plane zero and three poles. By choosing g_a appropriately, the zero is at a much higher frequency than the dominant pole, and its effect can be neglected. The stability of the amplifier can then be ensured by considering that the denominator of the transfer function in the unity-gain closed-loop configuration is a third-order Butterworth polynomial with a cutoff frequency of ω_c. That is,

$$\frac{2}{\omega_c} = \frac{1}{A_0 \omega_{p_1}} \simeq \frac{C_{c1}}{g_{m_1}}, \tag{5.491}$$

$$\frac{1}{\omega_c} = b \simeq \frac{(C_{c1} + C_3) C_{c2}}{g_{m_3} C_{c1}}, \tag{5.492}$$

and

$$\frac{1}{2\omega_c^2} = a = \frac{C_{c2} C_3}{g_{ma} g_{m_3}}. \tag{5.493}$$

By solving the system of Equations (5.491), (5.492), and (5.493), the values of the compensation capacitors can be obtained as

$$C_{c_1} = \frac{1}{2} \left(\frac{4 g_{m_1}}{g_{ma}} - 1 \right) C_3 \tag{5.494}$$

and

$$C_{c_2} = \frac{g_{ma} g_{m_3}}{8} \left[\frac{1}{g_{m_1}} \left(\frac{4 g_{m_1}}{g_{ma}} - 1 \right) \right]^2 C_3, \tag{5.495}$$

where $g_{m_1} > g_{ma}/4$. The gain-bandwidth product of the amplifier is of the form

$$GBW \simeq \frac{g_{m_1}}{C_{c_1}} = N \frac{1}{4} \frac{g_{m_3}}{C_3}, \tag{5.496}$$

where

$$N = \frac{2}{g_{m_3} \left(\dfrac{1}{g_{ma}} - \dfrac{1}{4} \dfrac{1}{g_{m_1}} \right)}. \tag{5.497}$$

Because $N > 1$, the GBW of the RAFFC amplifier is much greater than the one of the NMC amplifier. The phase margin of the RAFFC amplifier can be computed as

$$\phi_M = 180^o - \angle A[j(A_0 \omega_{p_1})] \tag{5.498}$$

$$\simeq 60^o + \arctan[A_0 \omega_{p_1}/(g_a/C_{c1})] \simeq 60^o + \arctan(g_{m1}/g_a) > 74^o. \tag{5.499}$$

Due to the contribution of the left-hand plane zero, the phase margin is greater than the value of 60° obtained in the case of the NMC amplifier [51,52]. The slew rate depends upon the value of compensation capacitors and the bias currents.

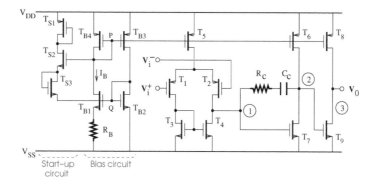

FIGURE 5.69
Circuit diagram of a three-stage amplifier with a single RC compensation.

The performance improvement provided by three-stage amplifiers using nested compensation is obtained at the expense of power dissipation. To overcome this limitation, the amplifier architecture shown in Figure 5.69 can be adopted [43]. The start-up circuit [44] consisting of transistors $T_{S1} - T_{S3}$ is used to prevent the bias circuit, $T_{B1} - T_{B4}$, from entering the zero-current state. When the circuit is powered up, the voltage V_P at the gates of transistors T_{B3} and T_{B4} begins to increase, while the voltage V_Q at the gates of transistors T_{B1} and T_{B2} may remain at zero. Because the difference voltage $V_P - V_Q$ is greater than the sum of two threshold voltages, $2V_{T_n}$, both transistors T_{S2} and T_{S3} are conducting and the start-up current can be injected into

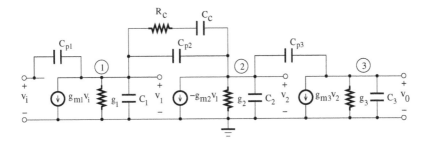

FIGURE 5.70

Small-signal equivalent circuit of the three-stage amplifier with a single RC compensation.

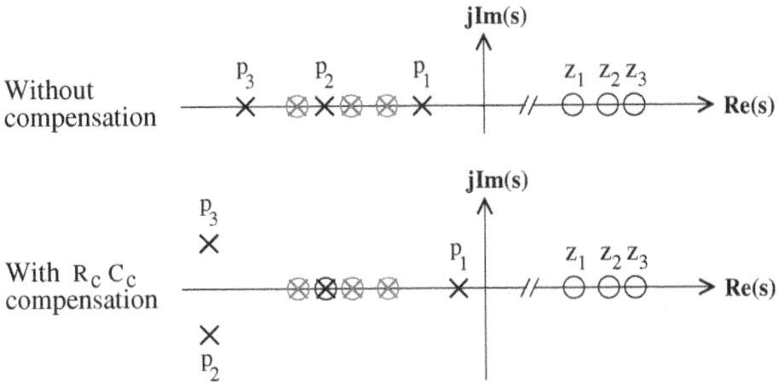

FIGURE 5.71

Pole-zero representations of the uncompensated and compensated three-stage amplifier in the s-domain.

the bias circuit. When the normal operation state is subsequently reached, the voltage $V_P - V_Q$ becomes less than $2V_{T_n}$, such that at least the transistor T_{S3} is biased in the cutoff region. The start-up circuit is then deactivated. Considering the bias circuit, it can be found that $V_{GS_{B2}} = V_{GS_{B1}} + R_B I_B$, and the current used to bias each of the amplifier stages is of the form

$$I_B = \frac{2}{\mu_n C_{ox} R_B^2} \left(\sqrt{\frac{1}{W_2/L_2}} - \sqrt{\frac{1}{W_1/L_1}} \right)^2 . \tag{5.500}$$

The small-signal equivalent circuit of the three-stage amplifier with a single RC compensation is depicted in Figure 5.70, where C_{p1}, C_{p2}, and C_{p3} denote the parasitic coupling capacitors. Without taking into account the $R_c C_c$ compensation network, the amplifier transfer function can be computed as

$$\frac{V_0(s)}{V_i(s)} = \frac{V_0(s)}{V_2(s)} \times \frac{V_2(s)}{V_1(s)} \times \frac{V_1(s)}{V_i(s)} = -\frac{(g_{m_1} - sC_1)(g_{m_2} - sC_2)(g_{m_3} - sC_3)}{D(s)}, \tag{5.501}$$

where

$$D(s) = \{[g_1 + s(C_1 + C_{p1})][g_2 + s(C_2 + C_{p2})] + sC_{p2}(g_{m_2} + g_2 + sC_2)\}$$
$$\times [g_3 + s(C_3 + C_{p3})] + [g_1 + s(C_1 + C_{p2} + C_{p1})](g_{m_3} + g_3 + sC_3)sC_{p3}.$$
$$(5.502)$$

Usually $C_c \gg C_{gd_7}$ and assuming that $R_cC_c \ll 1$, the node capacitances can be computed as

$$C_1 = C_{gd_2} + C_{gb_2} + C_{gd_4} + C_{gb_4} + C_{gs_7}, \qquad (5.503)$$
$$C_2 = C_{gd_6} + C_{gb_6} + C_{gb_7} + C_{gs_9}, \qquad (5.504)$$

and

$$C_3 = C_{gd_8} + C_{gb_8} + C_{gd_9} + C_{gb_9} + C_L, \qquad (5.505)$$

where C_L is the external output load capacitance. It can be observed that the pole introduced by the third stage is canceled by the zero of the second stage, while the pole associated with the second stage is eliminated by the zero of the first stage. Because the transfer function of the uncompensated amplifier exhibits three poles and three zeros located in the left- and right-half planes, respectively, a compensation network consisting of a resistor R_c in series with a capacitor C_c is required to improve the amplifier stability.

Due to the capacitor C_c, the poles and zeros are shifted to lower or higher frequencies. The resistor R_c should move the zero at the lower frequency back to almost its initial location, create a new zero to cancel the first nondominant pole and a pole that can merge with the second nondominant pole to form a complex-conjugate pole pair. By appropriately choosing the values of R_c and C_c, the nondominant poles appear at frequencies beyond the unity-gain frequency and do not affect the amplifier behavior. Therefore, the unity-gain frequency is reduced to the product of the dominant pole frequency and the dc gain. Figure 5.71 shows the pole-zero representations of the uncompensated and compensated three-stage amplifier in the s-domain.

Note that the bias circuit is designed such that the required pole-zero cancelation is accurate over process, voltage and temperature variations. With the pole frequency determined by the g_m/C ratio and the zero frequency related to $1/R_cC_c$, the transistor transconductance g_m is made proportional to $1/R_B$ so that R_B can track R_c.

5.7 Rail-to-rail amplifiers

Generally, rail-to-rail amplifiers are useful in low-voltage applications, where it is necessary to efficiently use the limited span offered by the power supply. While conventional amplifiers are capable of linear operation only for signals with a small excursion around the common-mode levels, rail-to-rail amplifiers are designed to allow signals to swing within millivolts of either power supply rail. The input and output voltage ranges are dependent on the amplifier topology, and the rail-to-rail operation can be achieved for either the input or the output, or both input and output.

5.7.1 Amplifier with a class AB input stage

The circuit diagram of the amplifier is depicted in Figure 5.72. It uses an input stage consisting of source cross-coupled transistors [53] as shown in Figure 5.73.

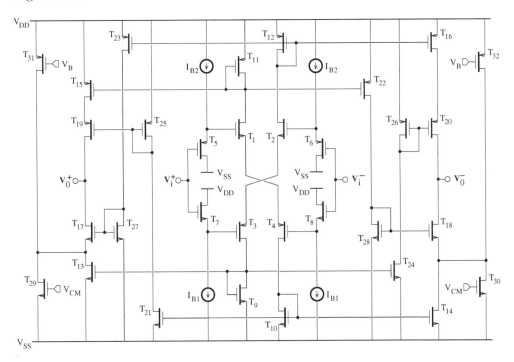

FIGURE 5.72
Cross-coupled transistor-based differential amplifier.

Using the square-law model for MOS devices in the saturation region, the drain currents are given by

$$i_D = \begin{cases} K_n(v_{GS} - V_{Tn})^2, & n\text{-channel transistor} \\ K_p(v_{GS} - V_{Tp})^2, & p\text{-channel transistor.} \end{cases} \tag{5.506}$$

Applying Kirchhoff's voltage law on the left and right sides of the transconductor, we have

$$v_i = v_i^+ - v_i^- = v_{GS7} + v_{GS3} - (v_{GS2} + v_{GS6}), \tag{5.507}$$

$$-v_i = v_i^- - v_i^+ = v_{GS8} + v_{GS4} - (v_{GS1} + v_{GS5}). \tag{5.508}$$

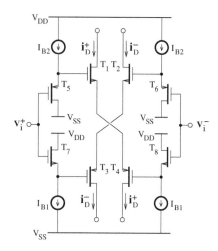

FIGURE 5.73

Source cross-coupled transistor pair.

With $i_{D5} = i_{D7} = i_{D6} = i_{D8} = I_{SS} = I_{DD} = I_B$, the drain currents are written as

$$i_D^+ = K_{eq} \left(v_i + \sqrt{\frac{I_B}{K_{eq}}} \right)^2 , \tag{5.509}$$

$$i_D^- = K_{eq} \left(v_i - \sqrt{\frac{I_B}{K_{eq}}} \right)^2 , \tag{5.510}$$

where

$$i_D^+ = i_{D1} = i_{D4} , \tag{5.511}$$

$$i_D^- = i_{D2} = i_{D3} , \tag{5.512}$$

and

$$K_{eq} = \frac{K_n K_p}{\left(\sqrt{K_n} + \sqrt{K_p} \right)^2} . \tag{5.513}$$

The differential output current is given by

$$\Delta i = i_D^+ - i_D^- = 4\sqrt{K_{eq} I_B}\, v_i . \tag{5.514}$$

The transfer characteristic of the differential stage is plotted in Figure 5.74. The corresponding transconductance is $g_m = 4\sqrt{K_{eq} I_B}$. The transistors remain in the saturation region provided that

$$|v_i| \leq \sqrt{\frac{I_B}{K_{eq}}} . \tag{5.515}$$

The class AB operation is achieved because the transconductor can generate an output current which is larger than the dc quiescent current flowing in the circuit.

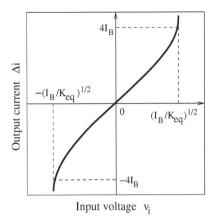

FIGURE 5.74
Transfer characteristic of the source cross-coupled transistor pair.

Due to the biasing condition realized by source followers $T_5 - T_8$, a current can flow through the input stage even for zero differential input. Given an increase in the voltage on the positive input and a corresponding decrease on the negative input, the drain currents of T_1 and T_4, and the ones of T_2 and T_3 increase and decrease from their initial values, as a result of a rise and reduction of their gate-source voltages, respectively. That is, the current in one side of the differential stage increases monotonically with the applied voltage and is limited by the power supply level, while the one in the other side decreases until a transistor turns off. The input currents are then directed to the output branches by current mirrors. The cascode transistors, $T_{17} - T_{20}$, are used to increase the amplifier gain. The conflicting requirements of high output current during the slewing period and large output swing during the settling are met by dynamically biasing the gates of cascode transistors so that the common-source transistors $T_{13} - T_{16}$ remain in the saturation region.

The common-mode feedback is realized by controlling the bias current of T_{29} and T_{30}. Transistors T_{31} and T_{32} are connected to the bias voltage, V_B, in order to deliver constant currents.

5.7.2 Two-stage amplifier with class AB output stage

The circuit diagram of an amplifier based on the four-transistor class AB output stage is shown in Figure 5.75. The signal provided by the differential input stage is applied to the output stage consisting of common drain complementary transistors, whose quiescent point is set by a translinear loop. As a result, the conduction of the output transistors is maintained even in the absence of an incoming ac signal. Furthermore, the output stage has the capability to source or sink the output current. However, the available output dynamic range may be affected by fluctuations in the transistor threshold voltages. The amplifier stability is maintained by inserting a compensation network

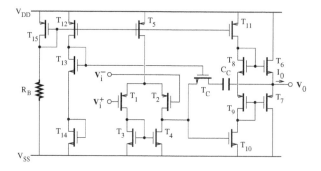

FIGURE 5.75
Circuit diagram of a two-stage amplifier with a class AB output stage.

consisting of T_C and C_C between the input and output of the second stage. The transistor T_C is biased such that a proper compensation resistance can be maintained over IC process, temperature, and supply voltage variations.

5.7.3 Amplifier with rail-to-rail input and output stages

In low-voltage designs, a rail-to-rail input stage is required to improve the signal dynamic range for a given supply voltage [54, 55]. It is generally implemented, as shown in Figure 5.76, by a parallel connection of n-channel and p-channel transistor pairs. Due to the different dc behavior of each pair, the resulting transconductance shown in Figure 5.77 varies with the common-mode input voltage and is not constant.

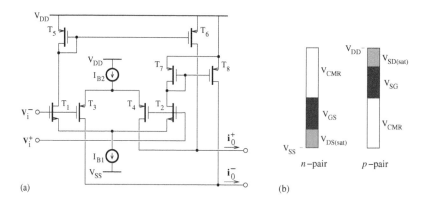

FIGURE 5.76
(a) Circuit diagram of a rail-to-rail input stage; (b) voltage swings of n- and p-channel transistor pairs.

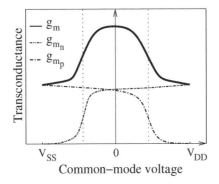

FIGURE 5.77

Transconductance variations versus the common-mode voltage.

The output current of the differential stage can be written as

$$i^+ = \frac{I_B}{2} + g_m \frac{v_i}{2}, \qquad (5.516)$$

$$i^- = \frac{I_B}{2} - g_m \frac{v_i}{2}. \qquad (5.517)$$

The transconductance is either

$$g_m = \frac{I_B}{\alpha U_T}, \qquad \text{for} \quad v_i < 2U_T \qquad (5.518)$$

in the weak inversion region, where α is the week inversion slope and U_T is the thermal potential, or

$$g_m = \sqrt{2KI_B}, \qquad \text{for} \quad |v_i| < (2I_B/K)^{1/2} \qquad (5.519)$$

in the saturation region. The tail current I_B assumes the value of I_{B1} for the n-channel transistor pair while it is I_{B2} for the stage using p-channel transistors. The parameter K is given by

$$K = \begin{cases} K_n = \frac{1}{2}\mu_n C_{ox} \frac{W}{L}, & n\text{-channel transistor} \\ K_p = \frac{1}{2}\mu_p C_{ox} \frac{W}{L}, & p\text{-channel transistor.} \end{cases} \qquad (5.520)$$

When the common-voltage, v_{CM}, is at mid-rail, both n- and p-channel transistor stages operate normally, as a result, the total transconductance g_{mT}, which is about two times greater than that obtained in the cases where v_{CM} is close to either of the supply voltages, V_{SS} or V_{DD}, and only one pair type remains operational. This transconductance variation can be efficiently reduced by using differential pairs based on improved biasing circuits.

A constant transconductance can be obtained by inserting a constant voltage, V_Z, between the tails of the complementary input pairs, as shown in Figure 5.78. It is the result of stabilizing the gate-source voltages of the transistors.

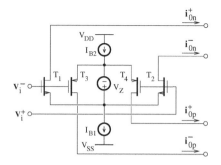

FIGURE 5.78

Circuit diagram of a rail-to-rail input stage with a constant voltage regulation.

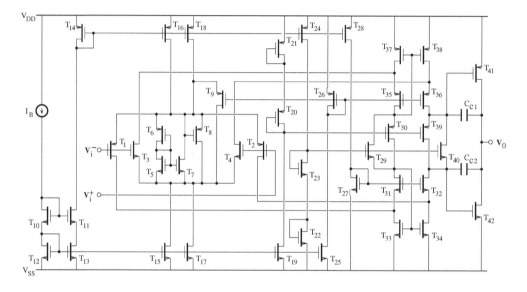

FIGURE 5.79

Circuit diagram of a rail-to-rail amplifier with the constant voltage regulation of the input stage.

The circuit diagram of the overall rail-to-rail amplifier is depicted in Figure 5.79. Transistors $T_5 - T_8$ and $T_{17} - T_{18}$ implement the voltage source V_Z. The parameters W/L of $T_5 - T_8$ can be sized to match the ones of the corresponding input transistors and $T_{15} - T_{16}$ should drive a current eight times greater than the one of $T_{17} - T_{18}$. The gate voltage of T_9 is used to limit the drain voltage of T_{18}, reducing in this way any additional variation of the current delivered by T_{15}.

The class AB output stage has the advantage of operating with a high speed and low supply voltages. The quiescent current flowing through the complementary output transistors, T_{41} and T_{42}, is minimized by the biasing structures established by $T_{19} - T_{21}$, T_{39} and $T_{22} - T_{24}$, T_{40}, respectively. For a negative output slew, the gate voltage of T_{42} increases. Because T_{40} is biased at a fixed voltage, it will turn off and the whole bias current now flows through T_{39}. As a result, the gate-source voltages of T_{39} and T_{41} decrease. A similar operation will be observed when the bias of T_{41} is reduced due to a positive change at the output. The floating current sources $T_{29} - T_{30}$ have the same structure as the ones used for the control of the output transistors, whose quiescent current is made less sensitive to supply voltage variations by using two current mirrors biased independently. The input signals of the current mirrors $T_{31} - T_{34}$ and $T_{35} - T_{38}$ are obtained from the p- and n-channel transistor pairs, $T_1 - T_2$ and $T_3 - T_4$, respectively. Assuming identical saturation and threshold voltages, the minimum supply voltage is about $3V_{DS(sat)} + 2V_T$.

The amplifier phase margin is determined by the two compensation capacitors, C_{c1} and C_{c2}, which split apart the amplifier poles to provide a first-order gain characteristic.

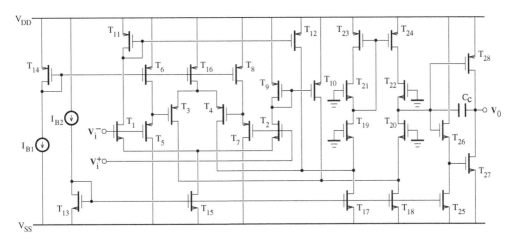

FIGURE 5.80
Circuit diagram of a rail-to-rail amplifier with dc level shifters.

The amplifier architecture of Figure 5.80 is based on another simple technique, which can be adopted to achieve a constant transconductance. The dc level shifters, realized by the source followers $T_5 - T_6$ and $T_7 - T_8$, are required to overlap the transition region of the tail currents for the n- and p-channel transistor pairs $T_1 - T_2$ and $T_3 - T_4$. The input stage is loaded by the folded cascode structure, $T_{17} - T_{24}$, followed by a class AB output stage compensated by C_c. Note that the deviation of the transconductance due to layout mismatches and IC process variations can be reduced by tuning the bias currents I_{B1} and I_{B2}.

5.8 Amplifier characterization

The ideal model of an amplifier features an infinite gain and bandwidth. Furthermore, the input and output impedances are assumed to be infinite or zero. However, the performance characteristics of amplifiers are generally limited by various nonideal effects (finite gain and bandwidth, common-mode range, power supply rejection, input and output impedance, slew rate, offset) in practice.

5.8.1 Finite gain and bandwidth

The frequency response of an amplifier is shown in Figure 5.81. The first pole of the transfer function, which is characterized by ω_1, is generally made dominant using a suitable compensation technique to have a first-order frequency response. In this case, the 3-dB frequency of the amplifier is reduced to ω_1. The parasitic effects can then be modeled by a pole at ω_2. The parameter A_0 is the dc gain and ω_t represents the unity-gain frequency.

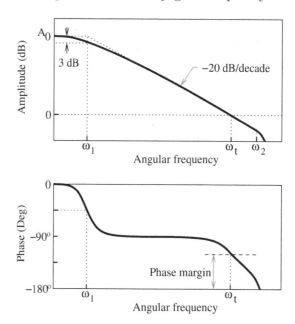

FIGURE 5.81
Frequency response of an amplifier.

The output voltage of an amplifier can be written as

$$v_0 = A_{dm}v_d + A_{cm}v_c \,, \tag{5.521}$$

where

$$v_d = v^+ - v^- \tag{5.522}$$

and

$$v_c = \frac{v^+ - v^-}{2} \qquad (5.523)$$

are the differential and common-mode voltages, respectively. The common-mode rejection ratio (CMRR) is defined by

$$\text{CMRR} = \frac{A_{dm}}{A_{cm}}, \qquad (5.524)$$

where A_{dm} and A_{cm} denote the small-signal differential mode and common mode gains, respectively. Ideally, the CMRR should be zero. It is often expressed in dB.

5.8.2 Phase margin

The phase margin, ϕ_M, is the amount by which the phase of the amplifier transfer function exceeds $-180°$ at the unity-gain frequency. It can be measured in degrees and indicates the relative stability of an amplifier in a closed-loop configuration. Typically, the minimum value of the phase margin is on the order of $45°$.

5.8.3 Input and output impedances

Ideally, the input and output impedances of an amplifier can be either infinite or zero. However, they are determined in a practical amplifier by the resistors and capacitors associated with the input stage and output buffer, and can be finite and frequency dependent.

5.8.4 Power supply rejection

The power supply rejection ratio (PSSR) is used to characterize the amplifier sensitivity to variations in the supply voltage. It can be expressed as

$$\text{PSSR} = \frac{V_0/V_i}{V_0/V_{sup}}, \qquad (5.525)$$

where V_{sup} can denote the positive or negative supply voltage. An ideal amplifier would feature an infinite PSSR. Due to the use of a compensation structure, the PSSR will be degraded as the number of amplification stages is increased.

5.8.5 Slew rate

The slew rate represents the maximum rate of change of the amplifier output in response to a step input signal. It is given by

$$SR = \max\left(\left|\frac{dv_0(t)}{dt}\right|\right) = \frac{I_B}{C_L}, \qquad (5.526)$$

where v_0 denotes the amplifier output voltage, I_B is the bias current of the input differential transistor pair, and C_L is the output load capacitor. The SR worst-case value is obtained when the amplifier is in the noninverting unity gain configuration.

The slew rate can be improved by using an amplifier biased dynamically, as shown in Figure 5.82 [57]. This structure is based on the current subtractor depicted in Figure 5.83. By applying a signal at the amplifier input, the currents i_1 and i_2 become different. If i_2 is greater than i_1, a current $i = \alpha(i_2 - i_1)$, where $\alpha < 1$, will be generated and the overall bias current of the input stage will evolve into $I_B + i$. Note that in the cases where i_2 is less or equal to i_1, no current is mirrored in $T_3 - T_4$.

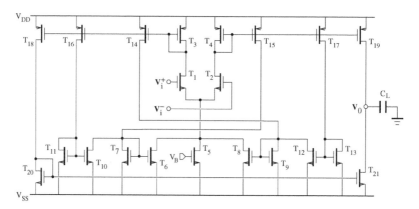

FIGURE 5.82
Single-stage amplifier with dynamic biasing.

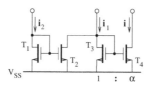

FIGURE 5.83
Current subtractor.

5.8.6   Low-frequency noise and dc offset voltage

The noise voltage spectrum of an amplifier is shown in Figure 5.84(a). It is dominated at low frequencies by the $1/f$ noise, which varies almost inversely with the frequency. The magnitude of the $1/f$ noise is dependent on the size of the input transistors and the IC process used. For high frequencies, the main contribution is due to the thermal noise, the spectrum of which is assumed to be flat. The frequency at which the thermal and $1/f$ noise components are equal corresponds to the knee frequency, f_k.

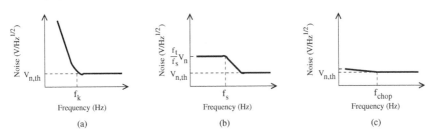

FIGURE 5.84

(a) CMOS amplifier, (b) auto-zeroed amplifier, and (c) chopper amplifier noise voltage spectra.

The straightforward approach to reduce the $1/f$ noise is to design the amplifier input differential pair using pMOS transistors, whose noise contribution is generally lower than the one of nMOS transistors. In a multi-stage amplifier, the gain of the input stage should be made as high as possible.

Let us consider the differential pair (see Figure 5.85) consisting

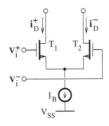

FIGURE 5.85

Differential transistor pair.

of transistors $T_1 - T_2$, which operate in the saturation region, but under different biasing conditions. Neglecting the variations of the effective surface carrier mobility, μ_n, and the gate oxide capacitance per unit area, C_{ox}, the offset voltage is obtained as

$$V_{off} = V_{GS_1} - V_{GS_2}, \qquad (5.527)$$

where

$$V_{GS_1} = V_{T_1} + \sqrt{\frac{2I_D^+}{\mu_n C_{ox}\left(\dfrac{W_1}{L_1}\right)}} \tag{5.528}$$

$$V_{GS_2} = V_{T_2} + \sqrt{\frac{2I_D^-}{\mu_n C_{ox}\left(\dfrac{W_2}{L_2}\right)}}. \tag{5.529}$$

The main contribution to V_{off} is due to the mismatches between the threshold, length and width of the transistors.

Auto-zeroing and chopper techniques can be used to reduce the effect of the $1/f$ noise and dc offset in CMOS amplifiers [58,60]. They operate with two clock phases. While in the auto-zero method, the low-frequency noise is first estimated and then subtracted from the corrupted signal in the next phase, it is modulated to higher frequencies in the chopper approach. Figures 5.84(b) and (c) show the resulting amplifier noise voltage spectra provided by both techniques. The noise is reduced for the auto-zeroing sampling frequency, f_s, or chopping frequency, f_{chop}, greater than f_k. If the noise is stationary, it will be completely eliminated by the auto-zeroing; otherwise, the residual noise at low frequencies will be above the thermal noise floor. This is due to the noise components, which have a magnitude proportional to f_t/f_s, where f_t is the amplifier transition frequency, and are aliased by the sampling process into the signal baseband. In contrast to an amplifier using the auto-zero calibration, the one based on the chopper technique features a residual noise approximately equal to the thermal noise. Typically, the residual input noise level can be scaled from the millivolt range to the microvolt range, as a result of the use of either the auto-zeroing or chopper technique.

5.8.6.1 Auto-zero compensation scheme

The objective of the auto-zero scheme as shown in Figure 5.86 is to compensate the amplifier dc offset voltage induced by transistor mismatches.

The zeroing amplifier is implemented using a single-ended version of the main amplifier structure. The amplifier differential input stages ($T_1 - T_4$) have been modified to include auxiliary inputs for dc offset voltage correction. Its realization, as depicted in Figure 5.87, includes an additional transistor pair connected in parallel with the main differential input stage. If the potential differences V_i and V_i' exist between the primary and auxiliary inputs, respectively, the amplifier output voltage can be written as

$$V_0 = A_0(V_i + V_i'/\alpha + V_{off}), \tag{5.530}$$

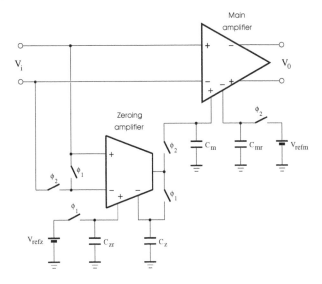

FIGURE 5.86
Circuit diagram of an auto-zero amplifier.

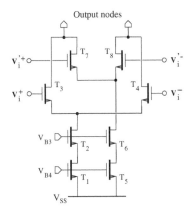

FIGURE 5.87
Differential input stage for the auto-zeroed amplifier.

where V_{off} is the amplifier offset voltage and the ratio of the open-loop dc gains between the primary and auxiliary inputs is defined as $\alpha = A_0/A'_0$. Basically, the role of the zeroing amplifier is to sense the dc offset voltage and to generate a correction voltage that is stored periodically on the capacitors and is used to drive the auxiliary input nodes of the amplifiers [61].

During the clock phase ϕ_1, the inputs of the zeroing amplifier are shorted and a feedback path is created between its output and inverting auxiliary input. The differential voltage applied to the auxiliary inputs of the zeroing amplifier can be expressed as

$$V_{C_{zr}} - V_{C_z} = V_{refz} - V_{0z} . \qquad (5.531)$$

According to Equation (5.530), we can obtain

$$V_{0z} = A_{0z}\left[\frac{V_{refz} - V_{0z})}{\alpha_z} + V_{offz}\right] \tag{5.532}$$

or equivalently,

$$V_{0z} = \frac{1}{1 + A_{0z}/\alpha_z}(A_{0z}V_{refz}/\alpha_z + A_{0z}V_{offz}), \tag{5.533}$$

where V_{offz} is the zeroing amplifier dc offset voltage, A_{0z} and $A'_{0z} = A_{0z}/\alpha_z$ are the open-loop dc gains of the zeroing amplifier with respect to the primary and auxiliary inputs, respectively. Combining Equations (5.531) and (5.533) gives

$$V_{C_{zr}} - V_{C_z} = \frac{1}{1 + A_{0z}/\alpha_z}(V_{refz} - A_{0z}V_{offz}) \tag{5.534}$$

In the clock phase ϕ_2, the main inputs of the amplifiers are connected together and the zeroing amplifier output is switched to the noninverting auxiliary input of the main amplifier. Due to the error voltage, $\triangle V_{C_{zr}} - \triangle V_{C_z}$, introduced by the opening of switches, the differential voltage maintained between the capacitors C_{zr} and C_z is

$$V_{C_{zr}} - V_{C_z} = \frac{1}{1 + A_{0z}/\alpha_z}(V_{refz} - A_{0z}V_{offz}) + (\triangle V_{C_{zr}} - \triangle V_{C_z}). \tag{5.535}$$

The output voltage of the main amplifier is given by

$$V_0 = A_{0m}(V^+ - V^- + V_{offm}) + A'_{0m}(V_{C_m} - V_{C_{mr}}), \tag{5.536}$$

where V_{offm} is the main amplifier offset voltage, A_{0m} and $A'_{0m} = A_{0m}/\alpha_m$ are the open-loop dc gains of the main amplifier with respect to the primary and auxiliary inputs, respectively. The differential voltage driving the auxiliary inputs of the main amplifier can be written as

$$V_{C_m} - V_{C_{mr}} = V_{0z} - V_{refm} \tag{5.537}$$

$$= A_{0z}(V^+ - V^- + V_{offz}) + A'_{0z}(V_{C_{zr}} - V_{C_z}) - V_{refm}. \tag{5.538}$$

Note that the value of the above differential voltage stored on the capacitors during the next phase is affected by the switch error voltages, $\triangle V_{C_{mr}} - \triangle V_{C_m}$, according to the next equation:

$$V_{C_m} - V_{C_{mr}} = A_{0z}(V^+ - V^- + V_{offz}) \\ + A'_{0z}(V_{C_{zr}} - V_{C_z}) - V_{refm} + (\triangle V_{C_m} - \triangle V_{C_{mr}}). \tag{5.539}$$

A first-order compensation of the switch error voltages is achieved by the differential structure and the substitution of Equation (5.535) into (5.539) can be reduced to

$$V_{C_m} - V_{C_{mr}} \simeq A_{0z}(V^+ - V^-) + \frac{A_{0z}V_{offz} + A'_{0z}V_{refz} - (1 + A'_{0z})V_{refm}}{1 + A'_{0z}}.$$
(5.540)

Combining Equations (5.536) and (5.540), the output voltage, V_0, can then be expressed as

$$V_0 = (A_{0m} + A'_{0m}A_{0z})(V^+ - V^-) + A_{0m}V_{offm}$$
$$+ A'_{0m}\frac{A_{0z}V_{offz} + A'_{0z}V_{refz} - (1 + A'_{0z})V_{refm}}{1 + A'_{0z}}.$$
(5.541)

By choosing the reference voltages such that $A'_{0z}V_{refz} = (1 + A'_{0z})V_{refm}$, and assuming that $A'_{0m}A_{0z} \gg A_{0m}$ and $A'_{0z} \gg 1$, Equation (5.541) can be put into the form

$$V_0 \simeq A'_{0m}A_{0z}(V^+ - V^- + V_{off,res}),$$
(5.542)

where the residual offset voltage, $V_{off,res}$, of the compensated amplifier is given by

$$V_{off,res} \simeq \frac{1}{A'_{0z}}\left[V_{offz} + \left(\frac{\alpha_m}{\alpha_z}\right)V_{offm}\right].$$
(5.543)

For a very low value of $V_{off,res}$, the gain A'_{0z} should be made very high. The effect of the offset voltage is similar to the one of the low-frequency or $1/f$ noise, which is also expected to be reduced in auto-zeroed amplifier. Generally, the residual offset can be estimated to be in the range of 100 µV.

It is necessary that the zeroing amplifier has the structure of a transconductor. Then, it can implement together with the load capacitor C_m a lowpass filter, whose cutoff frequency can be very low for large values of C_m. The advantage of this structure is to reduce the effect of the parasitic signal caused by the sampling process.

5.8.6.2 Chopper technique

The principle of the chopper technique [56] is illustrated in Figure 5.88, where ϕ_1 and ϕ_2 are two square signals with nonoverlapping phases. Each array of cross-coupled switches, which is inserted at the input and output of the first amplifier stage, is then considered to be steered by a chopping square wave taking $+1$ or -1 values. Let us assume that the signal has a spectrum limited to half of the chopping frequency so that no aliasing occurs. After the first multiplication by V_{chop}, the input signal, V_i, is modulated and translated to odd harmonic frequencies. It is then amplified and translated back to the baseband, while the low-frequency noise voltage (offset voltage, $1/f$ noise), V_n, which is modulated only by the second multiplication stage, appears at odd harmonic frequencies of V_{chop} (see Figure 5.88(c)). The high-frequency

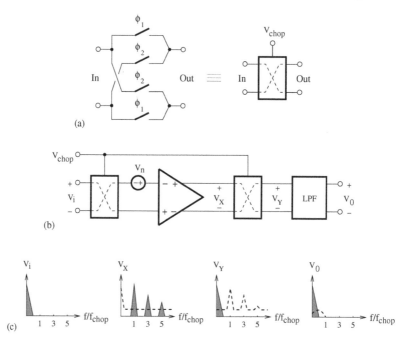

FIGURE 5.88
Principle of the chopper technique: (a) circuit symbol of the chopper section, (b) chopper amplifier, (c) signal spectra.

components are eliminated by the lowpass filter (LPF) included in the output stage of the chopper amplifier.

Let A be the amplifier gain. Taking into account the chopper effect, the amplifier output voltage can be written as

$$V_0(t) = [V_i(t) \cdot m(t) + V_n]A \cdot m(t), \tag{5.544}$$

where $m(t)$ represents the chopping signal alternating between 1 and -1 with a frequency f_{chop}, and V_n is the equivalent input noise of the amplifier. The signal multiplication by $m(t)$ induces a polarity change at the input nodes and a polarity restoration the output nodes. That is,

$$V_0(t) = A \cdot V_i(t) + A \cdot V_n \cdot m(t) \tag{5.545}$$

The chopping signal $m(t)$ can be represented as Fourier series given by

$$m(t) = \frac{4}{\pi} \sum_{k=1}^{\infty} \frac{1}{2k+1} \sin(2(2k+1)\pi f_{chop}t). \tag{5.546}$$

Using Fourier transform, the spectrum of the signal $m(t)$ can be obtained as

$$M(f) = \frac{2}{\pi} \sum_{k=1}^{\infty} \frac{1}{j(2k+1)}[\delta(f - (2k+1)f_{chop}) - \delta(f + (2k+1)f_{chop})]. \tag{5.547}$$

It can then be shown that the output power spectral density due to noise source is of the form

$$S_{0,v_n}(f) = S_{v_n}(f) * |M(f)|^2 , \tag{5.548}$$

where $*$ denotes the convolution operation, S_{v_n} is the power spectral density of the equivalent input noise, and

$$|M(f)|^2 = \left(\frac{2}{\pi}\right)^2 \sum_{k=1}^{\infty} \frac{1}{(2k+1)^2}[\delta(f + (2k+1)f_{chop}) - \delta(f - (2k+1)f_{chop})]. \tag{5.549}$$

The low-frequency noise is then transposed to odd harmonic frequencies of the chopping signal $m(t)$, where it can be attenuated by a lowpass filter. Due to the $1/(2k+1)^2$ scaling factor, the contribution to the baseband of noise replicas is greatly attenuated.

Simulation results show that the output power spectral density increases with the ratio of the amplifier cutoff frequency to the chopper frequency, but always remains smaller than the power spectral density of the white noise inherent to the original amplifier. Due to the fact that each signal component is not sampled and held, but periodically inverted, the chopper technique has the advantage of not aliasing the broadband noise, in contrast to the auto-zero approach [59]. In practice, the frequency of the modulating signal should be greater than the sum of the signal bandwidth and the $1/f$ corner frequency to minimize the noise contribution in the baseband and relax the filtering specifications. As a result, the achievable frequency range may appear to be limited by the required high-frequency chopping signals.

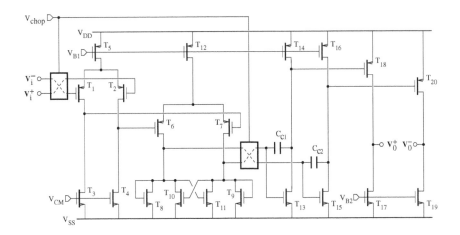

FIGURE 5.89
Circuit diagram of a low-noise amplifier using the chopper technique.

The circuit diagram of an amplifier using the chopper technique is shown in Fig 5.89. The first differential stage consisting of $T_1 - T_5$ should exhibit a low gain and noise. Transistors $T_6 - T_{12}$ form the second stage, which features a low unity-gain frequency and a high gain. These features can be useful for the reduction of the effect of the modulated offset and the offset due to the output stage comprising a gain inverting section and a source follower. The amplifier is compensated by the Miller capacitors, C_{c1} and C_{c2}. The voltage V_{CM} is generated by a CMF circuit and the cross-coupled connection of transistors $T_8 - T_{11}$ is used to define the CM level for the second differential stage. In this case, the residual offset voltage is about 10 µV [60]. It is generally due to spikes generated by mismatches between the charge injections of switches.

5.9 Summary

The available amplifier circuits can be divided into two main groups: single- and multi-stage architectures. Generally, the existing trade-off between speed and gain makes it difficult for CMOS amplifiers to exhibit a high bandwidth and dc gain simultaneously. Then, the need of a high gain leads to multi-stage designs with long-channel transistors biased at low current levels, whereas the requirement of a high unity-gain frequency is fulfilled by a single-stage topology with short-channel transistors biased at high current levels.

The choice of the amplifier architecture plays an important role in the design of low-voltage circuits. In addition to the amplifier characteristics such as the gain, bandwidth, and slew rate, the output swing also becomes critical. Generally, the power dissipation increases as the supply voltage is reduced. The two-stage amplifier appears to be suitable for low-voltage operation due to its larger output swing, while the telescopic structure achieves superior speed and power consumption.

5.10 Circuit design assessment

1. **Amplifier design challenges**

   Show that the transfer function of the RC circuit of Fig 5.90(a) can be written as

   $$H(j\omega) = \frac{V_0(j\omega)}{V_i(j\omega)} = \frac{1}{1 + j\omega RC}. \tag{5.550}$$

   Let k and T be the Boltzmann's constant and absolute temperature, respectively, and $\overline{v_{nf}^2}$ denote the output noise of the RC circuit over the frequency range from dc to infinity. The thermal noise can be modeled by adding a current source with the root mean squared value of $\sqrt{\overline{i_R^2}} = \sqrt{(4kT)/R}$ in parallel with the resistor R. Use the formula

$$\sqrt{\overline{v_{nf}^2}} = \left(\int_0^{+\infty} \overline{v_n^2} df \right)^{1/2}, \qquad (5.551)$$

where $\overline{v_n^2} = R \overline{i_R^2} |H(j\omega)|^2$, to verify that

$$\sqrt{\overline{v_{nf}^2}} = \sqrt{\frac{kT}{C}}. \qquad (5.552)$$

Note that

$$\int \frac{\mathrm{d}x}{x^2 + a^2} = \frac{1}{a} \arctan\left(\frac{x}{a}\right) + c, \qquad (5.553)$$

where a and c are two real constants.

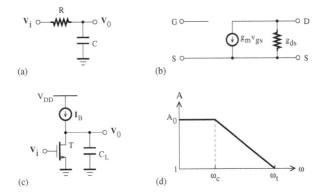

(a) (b) (c) (d)

FIGURE 5.90
(a) First-order RC circuit; (b) MOS transistor model; (c) basic single-stage amplifier; (d) amplifier frequency response.

The transistor in the saturation region can be described by the square law given by

$$I_D = K(v_{GS} - V_T)^2 (1 + \lambda v_{DS}), \qquad (5.554)$$

where $K = \mu(C_{ox}/2)(W/L)$ and λ is the channel-length modulation parameter. The transconductance is defined by

$$g_m = \left. \frac{\partial I_D}{\partial v_{GS}} \right|_{v_{DS}} \simeq 2K(v_{GS} - V_T) = 2\sqrt{KI_D}, \qquad (5.555)$$

while the conductance can be written as

$$g_{ds} = \left. \frac{\partial I_D}{\partial v_{DS}} \right|_{v_{GS}} = K\lambda(v_{GS} - V_T)^2 = I_D\lambda. \qquad (5.556)$$

Show that the gain of the amplifier of Figure 5.90(c) can be written as

$$A(j\omega) = \frac{v_o(j\omega)}{v_i(j\omega)} = A_0 \frac{1}{1 + j\dfrac{\omega}{\omega_c}}, \tag{5.557}$$

where $A_0 = g_m/g_{ds}$ and $\omega_c = g_{ds}/C_L$ denote the dc gain and bandwidth, respectively.

Add the asymptotic bode representation of a gain A', which corresponds to $A'_0 > A_0$, to the graph of Figure 5.90(d). Can we meet simultaneously the high dc gain and bandwidth requirements?

Explain why the power consumption increases by α^2, as a result of the scaling by a factor α of the amplifier signal swing and thermal noise such that the dynamic range and bandwidth are maintained constant.

Hint: The dynamic range (DR), which is limited by the thermal noise, is given by

$$DR = \frac{S_S}{S_{kT/C}} = \frac{V_{max}^2}{kT/C}, \tag{5.558}$$

where V_{max} is the maximum signal swing. By reducing the supply voltage by α, the DR and bandwidth become

$$DR' = \frac{V_{max}^2/\alpha^2}{kT/\alpha^2 C} \tag{5.559}$$

and

$$\omega_c' = \frac{\alpha^2 g_m}{\alpha^2 C_L}, \tag{5.560}$$

respectively. The transconductance is given by

$$g_m = \sqrt{2\mu C_{ox}\frac{W}{L}I_D} \tag{5.561}$$

while C_{ox} is converted to αC_{ox} and I_D to $\alpha^3 I_D$. As a result, the power consumption contribution, $P = I_D V_{max}$, is magnified by α^2.

2. **Comparison of folded-cascode and two-stage amplifiers**
   Consider the folded-cascode and two-stage amplifiers shown in Figures 5.91(a) and (b), respectively. With the assumption that symmetrical transistors are matched, and all transistors operate in the saturation region and exhibit the same drain-source saturation voltage, compare both amplifier structures and verify the results (slew rate, unity gain frequency, lowest nondominant pole, output swing, input-referred thermal noise, minimum supply voltage) summarized

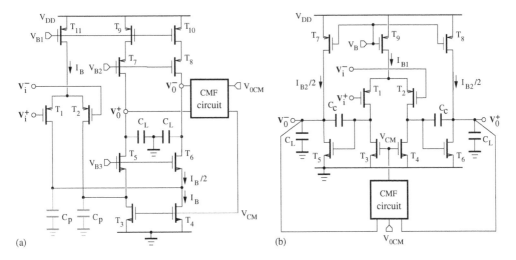

FIGURE 5.91

Circuit diagrams of (a) folded-cascode and (b) two-stage amplifiers.

TABLE 5.1

Performance Characteristics of Folded-Cascode and Two-Stage Amplifiers

	Folded-Cascode Amplifier	Two-Stage Amplifier
Slew rate	$\dfrac{I_B}{C_L}$	$\min\left(\dfrac{I_{B1}}{C_C}, \dfrac{I_{B2}}{C_C + C_L}\right)$
Unity-gain frequency	g_{m1}/C_L	g_{m1}/C_C
Nondominant pole	g_{m5}/C_p	g_{m5}/C_L
Output swing	$2V_{DD} - 8|V_{DS(sat)}|$	$2V_{DD} - 4|V_{DS(sat)}|$
Thermal noise	$\dfrac{8\gamma kT}{g_{m1}}\left(1 + \dfrac{g_{m1}}{g_{m3}} + \dfrac{g_{m1}}{g_{m5}}\right)$	$\dfrac{8\gamma kT}{g_{m1}}\left(1 + \dfrac{g_{m1}}{g_{m3}}\right)$
Minimum supply	$|V_T| + 2|V_{DS(sat)}|$	$|V_T| + 2|V_{DS(sat)}|$

in Table 5.1, where C_p represents a parasitic capacitance, and γ is the transistor noise coefficient.

<u>Hint:</u> Let $\overline{v_{n,i}^2}$ be the input-referred noise power spectral density of the amplifier, $\overline{i_{n,0}^2}$ be the output noise power spectral density of the amplifier, and $\overline{v_{nk}^2}$ be the input-referred noise power spectral density of the transistor T_k. It is convenient to assume that

$$\overline{v_{n,i}^2} = \overline{i_{n,0}^2}/g_{m1}^2 \tag{5.562}$$

and to model $\overline{v_{nk}^2}$ as the sum the thermal noise and the $1/f$ noise, that is,

$$\overline{v_{nk}^2} = \left(\frac{2}{3}\right)\frac{4kT}{g_{m,k}} + \frac{KF}{C_{ox}W_kL_kf}. \tag{5.563}$$

− Folded-cascode amplifier

Assuming that the noise due to the transistor T_{11} is canceled by the circuit symmetry and input transistor matching, and the noise contribution of cascode transistors is negligible because they are source degenerated and exhibit a small effective transconductance, verify that the equivalent input-referred noise power spectral density, $\overline{v_{n,i}^2}$, in V^2/Hz is of the form

$$\overline{v_{n,i}^2} = 2\left[\overline{v_{n1}^2} + \left(\frac{g_{m3}}{g_{m1}}\right)^2 \overline{v_{n3}^2} + \left(\frac{g_{m9}}{g_{m1}}\right)^2 \overline{v_{n9}^2}\right]. \tag{5.564}$$

− Two-stage amplifier

With the assumption that the input-referred noise is mainly determined by the first stage, verify that

$$\overline{v_{n,i}^2} = 2\left[\overline{v_{n1}^2} + \left(\frac{g_{m3}}{g_{m1}}\right)^2 \overline{v_{n3}^2}\right]. \tag{5.565}$$

3. **Design of a low-voltage amplifier**

For the amplifier shown in Figure 5.92, verify the following statements:

• The input common-mode is limited by
$$V_{SS} \quad \text{and} \quad V_{DD} - |V_T| - 2V_{DS(sat)}.$$
• The minimum supply voltage can be expressed as
$$\max\{V_{DS3(sat)} + V_{DS5(sat)} + V_{T5}, V_{DS3(sat)} + V_{DS5(sat)} + V_{DS7(sat)}\}.$$

Size of the amplifier components using a submicrometer IC process to meet the following specifications: 50 dB dc gain and 4 MHz unity-gain bandwidth.

Estimate the slew rate for a load capacitance of 10 pF and the power dissipation of the amplifier.

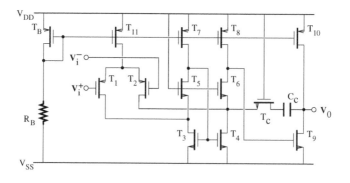

FIGURE 5.92
Circuit diagram of a low-voltage amplifier.

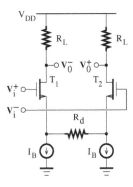

FIGURE 5.93
Differential amplifier with a source degeneration resistance.

4. **Amplifier stage with a source degeneration resistance**
   Use the small signal equivalent model of the transistors to determine the voltage gain of the amplifier structure shown in Figure 5.93.

5. **Amplifier stage based on a series of differential pairs**
   The linearity of a differential pair can be improved by using the amplifier stage shown in Figure 5.94 [8]. All transistors operate in the saturation region. Find the relationship between the output current $\triangle i_0 = i_0^+ - i_0^-$ and the input voltage $V_i = V_i^+ - V_i^-$.

   For a given CMOS technology, verify your calculations using SPICE simulations.

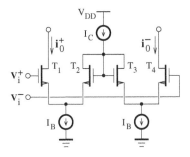

FIGURE 5.94
Differential amplifier based on a series of differential pairs.

6. **Fully differential folded-cascode amplifier with a common-mode feedback (CMF) circuit**
   In a 0.13 μm CMOS technology, design the differential folded-cascode amplifier shown in Figure 5.95 to meet the specifications given in Table 5.2.

   Use SPICE simulations to adequately adjust the transistor aspect ratios.

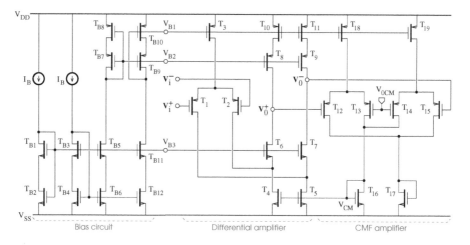

FIGURE 5.95

Differential folded-cascode amplifier with a common-mode feedback (CMF) circuit.

TABLE 5.2

Amplifier Specifications

Supply voltage	± 1.25 V
Capacitive load	2 pF
Slew rate	100 V/μs
DC voltage gain	60 dB
Gain-bandwidth product	100 MHz
Phase margin	> 60°
Power consumption	< 15 mW)

Provide a table including the simulated dc results.

Plot the transient step response and frequency response of the resulting amplifier in the unity feedback configuration.

Determine the settling time achieved by the resulting amplifier.

7. **Fully differential amplifier without a common-mode feedback circuit**

   Due to the loading of the common-mode feedback circuit on the signal path, a fully differential amplifier can exhibit a limited speed and require more power. The amplifier structure of Figure 5.96 [62] can serve for the implementation of differential switched-capacitor (SC) circuits without a common-mode feedback stage. The SC gain stage shown in Figure 5.97, where V_{icm} and V_{0cm} are the input and output common-mode voltages, respectively, uses the clock phase 2 to define a common-mode voltage reference for the amplifier nodes.

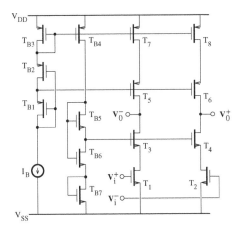

FIGURE 5.96
Circuit diagram of a pseudo fully differential amplifier.

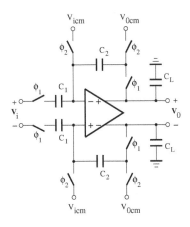

FIGURE 5.97
Circuit diagram of an SC gain stage.

Use simulation results to analyze the effect of the common-mode signal at the amplifier input on the circuit behavior.

8. **Stability of a differential gain stage**

   Consider the equivalent circuit of a differential gain stage shown in Figure 5.98, where $C_1 = 2$ pF, $C_2 = 1$ pF, and $C_L = 1$ pF.

   Use the small-signal equivalent model of a fully differential amplifier depicted in Figure 5.99, where Z_i and Z_{cm} denote the impedance of a parallel combination of a resistor and a capacitor, to determine the transfer function V_0/V_i, where $V_0 = V_0^+ - V_0^-$ and $V_i = V_i^+ - V_i^-$.

   Starting with the following initial component values, $R_i = R_0 = R_0' = R_{cm} = 1$ MΩ, $C_i = C_0 = C_0' = C_{cm} = C_p = 1$ pF, $g_m = g_m' = 1$ mS, analyze the stability of the gain stage using SPICE simulations.

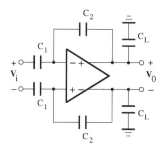

FIGURE 5.98
Equivalent circuit of a differential gain stage.

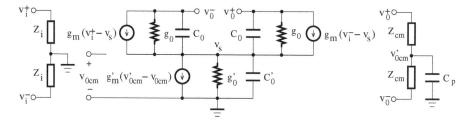

FIGURE 5.99
Small-signal equivalent model of a fully differential amplifier.

9. **Low-voltage amplifier with a class AB output**
   Consider the low-voltage class AB amplifier shown in Figure 5.100.
   The transistor T_{11} operates in the triode region, while the remaining
   transistors are biased in the saturation region.

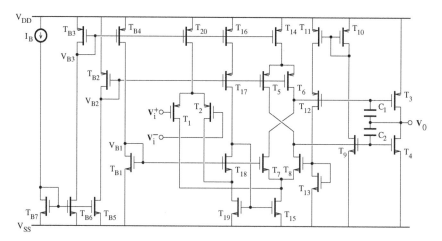

FIGURE 5.100
Circuit diagram of a low-voltage class AB amplifier.

Assuming that $V_{DD} = -V_{SS} = 1$ V and $I_B = 200$ µA, estimate
the values of the width and length of each transistor to achieve a

voltage gain of 10^4, a gain-bandwidth product of 3.5 MHz, and a phase margin of $70°$ in a given CMOS process. The capacitive and resistive loads of the amplifier are on the order of 2 pF and 10 kΩ, respectively.

10. **Two-stage amplifier with a current-buffer compensation**
To improve the capacitive load range of a two-stage amplifier, the compensation can be implemented using a current buffer as shown in Figure 5.101 [40]. A virtual ground is realized for the compensation capacitor as a result of the biasing of the transistor T_5 at a fixed dc potential by one of the two current sources, I_c.

Use the small-signal model of Figure 5.102 with the assumption that the controlled current connected to the first stage is given by $I_c(s) = sC_cV_0(s)$ and show that the resulting transfer function of the compensated amplifier reads

$$A(s) = \frac{V_0(s)}{V_i(s)} = -A_0\frac{1}{1 + bs + as^2}, \tag{5.566}$$

where

$$A_0 = \frac{g_{m_1}g_{m_2}}{g_1g_2}, \tag{5.567}$$

$$a = \frac{C_1(C_2 + C_c)}{g_1g_2}, \tag{5.568}$$

$$b = \frac{C_1}{g_1} + \frac{C_2 + C_c}{g_2} + \frac{g_{m_2}C_c}{g_1g_2}. \tag{5.569}$$

Verify that

$$A(s) = \frac{V_0(s)}{V_i(s)} = -A_0\frac{1}{(1 + s/p_1)(1 + s/p_2)}, \tag{5.570}$$

where the poles p_1 and p_2 are computed as

$$\frac{1}{p_1}, \frac{1}{p_1} = \frac{1}{2}(-b \pm \sqrt{\Delta}) \tag{5.571}$$

and $\Delta = b^2 - 4a$ should be positive.

Let p_2 be much greater than p_1. The gain-bandwidth product, GBW, and phase margin, ϕ_M, are defined by

$$GBW \simeq g_{m_1}/C_c \tag{5.572}$$

and

$$\tan \phi_M = p_2/GBW, \tag{5.573}$$

respectively. Relate C_c to ϕ_M and analyze the amplifier stability.

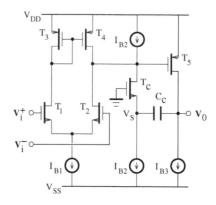

FIGURE 5.101
Circuit diagram of a two-stage amplifier with a current buffer compensation.

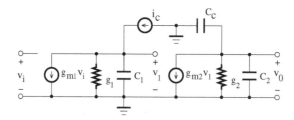

FIGURE 5.102
Small-signal equivalent of the two-stage amplifier.

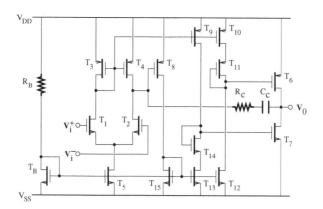

FIGURE 5.103
Circuit diagram of a two-stage amplifier with a class AB output stage.

11. **Two-stage amplifier with a class AB output**

Consider the circuit diagram of a two-stage amplifier with a class AB output shown in Figure 5.103 [63]. To minimize cross-over distortion, the output transistors should be kept at the boundary of the conduction region in the absence of an incoming ac signal. This can be achieved through unbalanced current mirroring, thereby resulting in transistors sizes of the form

$W_1/L_1 = W_2/L_2$	$W_3/L_3 = W_4/L_4 = W_8/L_8$
$W_7/L_7 = \alpha(W_6/L_6)$	$W_{10}/L_{10} = \beta(W_9/L_9)$
$W_{12}/L_{12} = W_{13}/L_{13} = W_{15}/L_{15}$	$W_{11}/L_{11} = \gamma(W_{14}/L_{14})$

where α, β, and γ are constant numbers. A frequency compensation network consisting of R_c and C_c is used to ensure the closed-loop stability when the amplifier operates with a load.

Based on the equations

$$V_{SG_6} = V_{SG_{11}} + V_{SD_{10}} \tag{5.574}$$

and

$$V_{GS_7} = V_{GS_{14}} + V_{DS_{13}}, \tag{5.575}$$

determine the quiescent current, I_Q, flowing through the output transistors.

Assuming that $V_{DD} = -V_{SS} = 2.5$ V and $R_B = 100$ kΩ, estimate the values of the width and length of each transistor to achieve a voltage gain of 10^3 and a gain-bandwidth product of 2 MHz in a given CMOS process.

12. Three-stage RNMC amplifier with a feed-forward transconductance

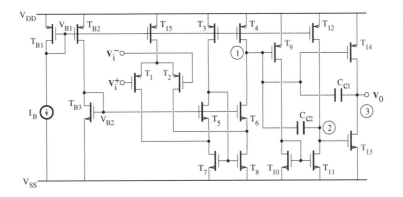

FIGURE 5.104
Circuit diagram of a three-stage RNMC amplifier with a feed-forward transconductance.

Consider the circuit diagram of a three-stage reversed nested Miller compensation (RNMC) amplifier with a feed-forward transconductance shown in Figure 5.104. With reference to the small-signal

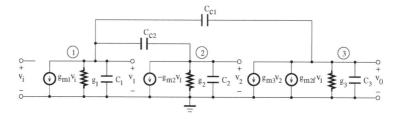

FIGURE 5.105
Small-signal equivalent circuit of the three-stage RNMC amplifier with a feed-forward transconductance.

equivalent circuit shown in Figure 5.105, verify that the equations for the node 1, node 2, and node 3, can be written as

$$g_{m_1} V_i + V_1(g_1 + sC_1) - (V_0 - V_1)sC_{c_1} - (V_2 - V_1)sC_{c_2} = 0, \tag{5.576}$$

$$-g_{m_2} V_1 + V_2(g_2 + sC_2) + (V_2 - V_1)sC_{c_2} = 0, \tag{5.577}$$

and

$$g_{m_3} V_2 + g_{m2f} V_1 + V_0(g_3 + sC_3) + (V_0 - V_1)sC_{c_1} = 0, \tag{5.578}$$

respectively. Assuming that $g_{m_k} \gg g_k$ ($k = 1, 2, 3$) and $C_{c1}, C_{c2}, C_3 \gg C_1, C_2$, show that the amplifier small-signal voltage gain can be put into the form

$$A(s) = \frac{V_0(s)}{V_i(s)} \simeq A_0 \frac{1 + ds + cs^2}{(1 + s/\omega_{p_1})(1 + bs + as^2)}, \tag{5.579}$$

where

$$A_0 = \frac{g_{m_1} g_{m_2} g_{m_3}}{g_1 g_2 g_3}, \tag{5.580}$$

$$\omega_{p_1} = \frac{g_1 g_2 g_3}{g_{m_2} g_{m_3} C_{c1}}, \tag{5.581}$$

$$a = \frac{C_{c2} C_3}{g_{m_2} g_{m_3}}, \tag{5.582}$$

$$b = \frac{(g_{m_3} + g_{m2f} - g_{m_2})C_{c1}C_{c2} - g_{m_2}C_{c2}C_3}{g_{m_2} g_{m_3} C_{c1}}, \tag{5.583}$$

$$c = -\frac{C_{c1} C_{c2}}{g_{m_2} g_{m_3}}, \tag{5.584}$$

and

$$d = \frac{(g_{m3} + g_{m2f})C_{c2}}{g_{m2}g_{m3}}. \tag{5.585}$$

Determine the gain-bandwidth product and slew rate of the amplifier.

13. **Three-stage amplifier with a feed-forward compensation**
For the three-stage amplifier shown in Figure 5.106, verify that the small-signal equivalent model can be derived as depicted in Figure 5.107.

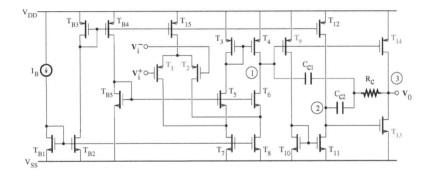

FIGURE 5.106
Circuit diagram of a three-stage amplifier.

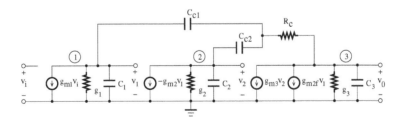

FIGURE 5.107
Small-signal equivalent model of the three-stage amplifier.

Show that

$$A(s) = \frac{V_0(s)}{V_i(s)} = A_0 \frac{1 + ds + cs^2}{(1 + s/\omega_{p_1})(1 + bs + as^2)}, \tag{5.586}$$

where

$$A_0 = \frac{g_{m_1} g_{m_2} g_{m_3}}{g_1 g_2 g_3}, \tag{5.587}$$

$$\omega_{p_1} = \frac{g_1 g_2 g_3}{g_{m_2} g_{m_3} C_{c1}}, \tag{5.588}$$

$$a = \frac{C_3 C_{c2}}{g_{m_2} g_{m_3}}, \tag{5.589}$$

$$b = \frac{C_{c2}(g_{m_3} + g_{m2f} - g_{m_2})}{g_{m_2} g_{m_3}}, \tag{5.590}$$

$$c = \frac{C_{c1} C_{c2}[(g_{m2f} + g_{m_3})R_c - 1]}{g_{m_2} g_{m_3}}, \tag{5.591}$$

and

$$d = (C_{c1} + C_{c2})R_c + \frac{C_{c2}(g_{m2f} - g_{m_2})}{g_{m_2} g_{m_3}}. \tag{5.592}$$

With the assumption that $R_c = 1/(g_{m2f} + g_{m_3})$, compute the gain-bandwidth product and slew rate of the amplifier.

Bibliography

[1] P. R. Gray and R. G. Meyer, "MOS operational amplifier design — a tutorial overview," *IEEE J. of Solid-State Circuits*, vol. 22, pp. 969–982, Dec. 1987.

[2] Y. Tsividis, Z. Czarnul, and S. C. Fang, "MOS transconductors and integrator with high linearity," *Electronics Letters*, vol 22. pp. 245–246, 1986.

[3] J. L. Pennock, P. Frith, and R. G. Barker, "CMOS triode transconductor continuous time filters," *Proc. of the IEEE CICC*, pp. 378–381, 1986.

[4] A. A. Fayed and M. Ismail, "A low-voltage, highly linear voltage-controlled transconductor," *IEEE Trans. on Circuits and Systems–II*, vol. 52, pp. 831–835, Dec. 2005.

[5] I. Mehr and D. R. Welland, "A CMOS continuous-time $G_m - C$ filter for PRML read channel applications at 150 Mb/s and beyond," *IEEE J. of Solid-State Circuits*, vol. 32, pp. 499–513, April 1997.

[6] J. N. Babanezhad, "A rail-to-rail CMOS op amp," *IEEE J. of Solid-State Circuits*, vol. 23, pp. 1414–1417, Dec. 1988.

[7] A. Nedungadi and T. R. Viswanathan, "Design of linear CMOS transconductance elements," *IEEE Trans. on Circuits and Systems*, vol. 31, pp. 891–894, Oct. 1984.

[8] R. R. Torrance, T. R. Viswanathan, and J. V. Hanson, "CMOS voltage to current transducers," *IEEE Trans. on Circuits and Systems*, vol. 32, pp. 1097–1104, Nov. 1985.

[9] K. Bult and H. Wallinga, "A CMOS four-quadrant analog multiplier," *IEEE J. of Solid-State Circuits*, vol. 21, pp. 430–435, June 1986.

[10] Z. Wang and W. Guggenbühl, "A voltage-controllable linear MOS transconductor using bias offset technique," *IEEE J. of Solid-State Circuits*, vol. 25, pp. 315–317, Feb. 1990.

[11] V. Peluso, P. Vancorenland, A. M. Marques, M. S. J. Steyaert, and W. Sansen, "A 900 mV low-power $\Delta - \Sigma$ A/D converter with 77-dB dynamic range," *IEEE J. of Solid-State Circuits*, vol. 33, pp. 1887–1897, Dec. 1998.

[12] A. J. López-Martín, S. Baswa, J. Ramirez-Angulo, and R. G. Carvajal, "Low-voltage super class AB CMOS OTA cells with very high slew rate and power efficiency," *IEEE J. of Solid-State Circuits*, vol. 40, pp. 1068–1077, May 2005.

[13] E. Seevinck and R. F. Wassenaar, "A versatile CMOS linear transconductor/square-law function circuit," *IEEE J. of Solid-State Circuits*, vol. 22, pp. 366–377, June 1987.

[14] M. Milkovic, "Current gain high-frequency CMOS operational amplifiers," *IEEE J. of Solid-State Circuits*, vol. 20, pp. 845–851, Aug. 1985.

[15] C.-C. Shih, and P. R. Gray, "Reference refreshing cyclic analog-to-digital and digital-to-analog converters," *IEEE J. of Solid-State Circuits*, vol. 21, pp. 544–554, Aug. 1986.

[16] J. N. Babanezhad and R. Gregorian, "A programmable gain/loss circuit," *IEEE J. of Solid-State Circuits*, vol. 22, pp. 1082–1090, Dec. 1987.

[17] S. M. Mallya and J. H. Nevin, "Design procedures for a fully differential folded-cascode CMOS operational amplifier," *IEEE J. of Solid-State Circuits*, vol. 24, pp. 1737–1740, Dec. 1989.

[18] K. Bult and G. J. G. M. Geelen, "A fast-settling CMOS op-amp for SC circuits with 90dB DC gain," *IEEE J. of Solid-State Circuits*, vol. 25, pp. 1379–1384, Dec. 1990.

[19] J. Lloyd and H.-S. Lee, "A CMOS op-amp with fully differential gain-enhancement," *IEEE Trans. on Circuits and Systems*, vol. 41, pp. 241–243, March 1994.

[20] D. Flandre, A. Viviani, J.-P. Eggermont, B. Gentine, and P. G. A. Jespers, "Improved synthesis of gain-boosted regulated-cascode CMOS stages using symbolic analysis and gm/ID methodology," *IEEE J. of Solid-State Circuits*, vol. 32, pp. 1006–1012, July 1997.

[21] M. Das, "Improved design criteria of gain-boosted CMOS OTA with high-speed optimizations," *IEEE Trans. on Circuits and Systems–II*, vol. 49, pp. 204–207, March 2002.

[22] M. M. Ahmadi, "A new modeling and optimization of gain-boosted cascode amplifier for high-speed and low-voltage applications," *IEEE Trans. on Circuits and Systems–II*, vol. 53, pp. 169–173, March 2006.

[23] B. Y. Kamath, R. G. Meyer, and P. R. Gray, "Relationship between frequency response and settling time of operational amplifiers," *IEEE J. Solid-State Circuits*, vol. SC-9, pp. 347–352, Dec. 1974.

[24] M. Banu, J. M. Khoury, and Y. Tsividis, "Fully differential operational amplifiers with accurate output balancing," *IEEE J. Solid-State Circuits*, vol. 23, pp. 1410–1414, Dec. 1988.

[25] R. A. Whatley, "Fully differential operational amplifier with D.C. common-mode feedback," U.S. Patent 4,573,020, filed Dec. 18, 1984; issued Feb. 25, 1986.

[26] T. C. Choi, R. T. Kaneshiro, R. W. Brodersen, P. R. Gray, W. B. Jett, and M. Wilcox, "High-frequency CMOS switched-capacitor filters for communications application," *IEEE J. Solid-State Circuits*, vol. SC-18, pp. 652–664, Dec. 1983.

[27] Z. Czamul, S. Takagi, and N. Fuji, "Common-mode feedback circuit with differential-difference amplifier," *IEEE Trans. on Circuits and Systems–I*, vol. 41, pp. 243–246, March 1994.

[28] X. Zhang, and E. I. El-Masry, "A novel CMOS OTA based on body-driven MOSFETs and its applications in OTA-C filter," *IEEE Trans. on Circuits and Systems–I*, vol. 54, pp. 1204–1212, June 2007.

[29] D. Senderowicz, S. F. Dreyer, J. H. Huggins, C. F. Rahim and C. A. Laber, "A family of differential NMOS analog circuits for a PCM codec filter chip," *IEEE J. of Solid-State Circuits*, vol. 17, pp. 1014–1023, Dec. 1982.

[30] O. Choksi and L. R. Carley, "Analysis of switched-capacitor common-mode feedback circuit," *IEEE Trans. on Circuits and Systems–II*, vol. 50, pp. 906–917, Dec. 2003.

[31] D. A. Garrity and P. L. Rakers, "Common-Mode Output Sensing Circuit," U.S. Patent 5,894,284, filed Dec. 2, 1996; issued April 13, 1999.

[32] F. Krummenacher and N. Joehl, "A 4-MHz CMOS continuous-time filter with on-chip automatic tuning," *IEEE J. of Solid-State Circuits*, vol. 23, pp. 750–758, June 1988.

[33] H. Khorramabadi and P. R. Gray, "High-frequency CMOS continuous-time filters," *IEEE J. of Solid-State Circuits*, vol. SC-19, pp. 939–948, Dec. 1984.

[34] P. D. Walker and M. M. Green, "An approach to fully differential circuit design without common-mode feedback," *IEEE Trans. on Circuits and Systems–II*, vol. 43, pp. 752–762, Nov. 1996.

[35] W. C. Black, Jr., D. J. Allstot, and R. A. Reed, "A high performance low power filter," *IEEE J. of Solid-State Circuits*, vol. SC-15, pp. 929–938, Dec. 1980.

[36] D. M. Monticelli, "A quad CMOS single-supply opamp with rail-to-rail output swing," *IEEE J of Solid-State Circuits*, vol. SC-21, pp. 1026–1034, Dec. 1986.

[37] W.-C. S. Wu, W. J. Helms, J. A. Kuhn, and B. E. Byrkett, "Digital-compatible high-performance operational amplifier with rail-to-rail input and output ranges," *IEEE J. of Solid-State Circuits*, vol. SC-29, pp. 63–66, Jan. 1994.

[38] K. J. de Langen, and J. H. Huijsing, "Compact low-voltage power-efficient operational amplifier cells for VLSI," *IEEE J. of Solid-State Circuits*, vol. 33, pp. 1482–1496, Oct. 1998.

[39] P. Migliavacca, "Start-up aid circuit for a plurality of current sources," U.S. Patent 6,002,242, filed Aug. 4, 1998; issued Dec. 14, 1999.

[40] B. K. Ahuja, "An improved frequency compensation technique for CMOS operational amplifiers," *IEEE J. of Solid-State Circuits*, vol. 18, pp. 629–633, Dec. 1983.

[41] G. Palmisano and G. Palumbo, "A compensation strategy for two-stage CMOS opamps based on current buffer," *IEEE Trans. on Circuits and Systems–I*, vol. 44, pp. 257–262, March 1997.

[42] P. J. Hurst, S. H. Lewis, J. P. Keane, F. Aram, and K. C. Dyer, "Miller compensation using current buffers in fully differential CMOS two-stage operational amplifiers," *IEEE Trans. on Circuits and Systems–I*, vol. 51, pp. 275–285, Feb. 2004.

[43] H-T. Ng, R. M. Ziazadeh, and D. J. Allstot, "A multistage amplifier with embedded frequency compensation," *IEEE J. of Solid-State Circuits*, vol. 34, pp. 339–347, March 1999.

[44] Y.-F. Chou, "Low-power start-up circuit for a reference voltage generator," U.S. Patent 6,201,435, filed Aug. 26, 1999; issued March 13, 2001.

[45] C. T. Chuang, "Analysis of the settling behavior of an operational amplifier," *IEEE J. Solid-State Circuits*, vol. SC-17, pp. 74–80. Feb. 1982.

[46] H. C. Yang and D. J. Allstot, "Considerations for fast settling operational amplifiers," *IEEE Trans. on Circuits and Systems*, vol. 37, pp. 326–334, March 1990.

[47] S. Pernici, G. Nicollini, and R. Castello, "A CMOS low-distortion fully differential power amplifier with double nested Miller compensation," *IEEE J. of Solid-State Circuits*, vol. 28, pp. 758–763, July 1993.

[48] R. G. H. Eschauzier, R. Hogervorst, and J. H. Huijsing, "A programmable 1.5 V CMOS class-AB operational amplifier with hybrid nested Miller compensation for 120 dB gain and 6 MHz UGF," *IEEE J. of Solid-State Circuits*, vol. 29, pp. 1497–1504, Dec. 1994.

[49] K. N. Leung and P. K. T. Mok, "Analysis of multistage amplifier-frequency compensation," *IEEE Trans. on Circuits Systems–I*, vol. 48, pp. 1041–1056, Sept. 2001.

[50] R. G. H. Eschauzier, L. P. T. Kerklaan, and J. H. Huijsing, "A 100-MHz 100-dB operational amplifier with multipath nested Miller compensation structure," *IEEE J. Solid-State Circuits*, vol. 27, pp. 1709-1717, Dec. 1992.

[51] A. D. Grasso, G. Palumbo, and S. Pennisi, "Advances in reversed nested miller compensation," *IEEE Trans. on Circuits Systems–I*, vol. 54, pp. 1459-1470, July 2007.

[52] H. Lee, and P. K. T. Mok, "Active-feedback frequency-compensation technique for low-power multistage amplifiers," *IEEE J. of Solid-State Circuits*, vol. 38, pp. 511–520, March 2003.

[53] S. H. Lewis and P. R. Gray, "A pipelined 5-Msamples/s 9-bit analog-to-digital converter," *IEEE J. of Solid-State Circuits*, vol. 22, pp. 954–961, Dec. 1987.

[54] R. Hogervorst, J. P. Tero, and J. H. Huijsing, "Compact CMOS constant-g_m rail-to-rail input stage with g_m-control by an electronic Zener diode," *IEEE J. of Solid-State Circuits*, vol. 31, pp. 1035–1040, July 1996.

[55] M. Wang, T. L. Mayhugh, Jr., S. H. K. Embabi and E. Sànchez-Sinencio, "Constant-g_m rail-to-rail CMOS op-amp input stage with overlapped transition regions," *IEEE J. of Solid-State Circuits*, vol. 34, pp. 148–156, Feb. 1999.

[56] K-C. Hsieh, P. R. Gray, D. Senderowicz, and D. G. Messerschmitt, "A low-noise chopper-stabilized differential switched-capacitor filtering technique," *IEEE J. of Solid-State Circuits*, vol. 16, pp. 708–715, Dec. 1981.

[57] M. G. Degrauwe, J. Rijmenants, E. A. Vittoz, and H. J. De Man, "Adaptive biasing CMOS amplifiers," *IEEE J. of Solid-State Circuits*, vol. 17, pp. 522–528, June 1982.

[58] M. C. W. Coln, "Chopper stabilization of MOS operational amplifiers using feed-forward techniques," *IEEE J. of Solid-State Circuits*, vol. 16, pp. 745–748, Dec. 1981.

[59] C. Enz and G. C. Temes, "Circuit techniques for reducing the effects of op-amp imperfections: Autozeroing, correlated double sampling, and chopper stabilization," *Proceedings of the IEEE*, vol. 84, pp. 1584–1614, Nov. 1996.

[60] C. C. Enz, E. A. Vittoz, and F. Krummenacher, "A CMOS chopper amplifier," *IEEE J. of Solid-State Circuits*, vol. 22, pp. 335–342, June 1987.

[61] I. G. Finvers, J. W. Haslett, and F. N. Trofimenkoff, "A high temperature precision amplifier," *IEEE J. of Solid-State Circuits*, vol. 30, pp. 120–128, Feb. 1995.

[62] G. Nicollini, F. Moretti, and M. Conti, "High-frequency fully differential filter using operational amplifiers without common-mode feedback," *IEEE J. of Solid-State Circuits*, vol. 24, pp. 803–813, June 1989.

[63] D. K. Su, "Class AB CMOS output amplifier," U.S. Patent 5,039,953, filed May 18, 1990; issued Aug. 13, 1991.

6

Nonlinear Analog Components

CONTENTS

Nonlinear analog components in MOS technology essentially include comparators and multipliers. They can find applications in communication and instrumentation systems.

A comparator provides an output signal to indicate whether the input signal is higher or lower than a reference voltage, or equivalently, determines the sign of a voltage difference. Its output then assumes either the high or low level, depending on the relative magnitude of the input signals. In most applications, comparators are required to exhibit a high speed, a high gain, a high common-mode rejection, and a low offset voltage.

Multipliers or mixers are used in a variety of electronic systems to provide the product of two signals. The basic idea of the multiplier implementation relies on applying the input signals to a nonlinear device and canceling the undesired output components. High linearity, or say, low intermodulation distortion, and low noise can be considered important performance characteristics in a multiplier, because they can affect the resulting dynamic range.

The next sections deal with the analysis and design of nonlinear active components (comparators and multipliers). Generally, the objective is to achieve a sufficient dynamic range and bandwidth using submicrometer IC process with a low supply voltage.

6.1 Comparators

Comparators are used in applications such as data conversion and interfacing, where a decision regarding the relative value of the input signals is required. They can be realized using high-gain differential amplifiers, charge balancing techniques, and open-loop structures with a positive feedback [1, 2].

In general, the speed and accuracy of comparators are enhanced by reducing the probability of metastability. A metastable state occurs, for instance, when the input voltage difference is too small to be resolved unambiguously. Although the metastability cannot be completely avoided, its rate of occurrence is reduced by maximizing the available settling time, as by employing a cascade of output latches.

6.1.1 Amplifier-based comparator

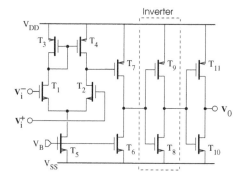

FIGURE 6.1
Circuit diagram of an amplifier-based comparator.

Amplifiers may appear suitable for comparing two signals. Figure 6.1 shows a comparator based on an uncompensated two-stage transconductance amplifier driving two inverters in series. The input and reference voltages are applied to either the positive node or negative node of the first stage of the comparator, which consists of a differential transistor pair loaded by a current mirror. The output of the second stage, which is a source follower, is used to drive the inverters. The compensation network of the amplifier is removed to somewhat increase the operation speed while the use of the inverter stage helps drive output capacitive loads with optimum speed.

In the noninverting configuration, the comparator output voltage can

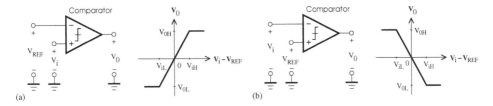

FIGURE 6.2
Practical transfer characteristic: (a) noninverting and (b) inverting comparators.

be written as,

$$V_0 = \begin{cases} V_{OH}, & \text{if } V_i - V_{REF} > V_{iH} \\ A_v(V_i - V_{REF}), & \text{if } V_{iL} \leq V_i - V_{REF} \leq V_{iH} \\ V_{OL}, & \text{if } V_i - V_{REF} < V_{iL} \end{cases} \tag{6.1}$$

while, in the inverting configuration, it is given by

$$V_0 = \begin{cases} V_{OL}, & \text{if } V_i - V_{REF} > V_{iH} \\ -A_v(V_i - V_{REF}), & \text{if } V_{iL} \leq V_i - V_{REF} \leq V_{iH} \\ V_{OH}, & \text{if } V_i - V_{REF} < V_{iL}, \end{cases} \tag{6.2}$$

where A_v is the amplification gain; V_{iL} and V_{iH} denote the low and high level of the input thresholds, respectively; and V_{OL} and V_{OH} represent the lower and higher limits of the output thresholds, respectively.

The smallest voltage difference, which can be resolved by the comparator, is dependent on the gain, A_v, and

$$A_v = \frac{V_{OH} - V_{OL}}{\triangle V}, \tag{6.3}$$

where $\triangle V$ is the resolution. In the ideal case, A_v is assumed to be infinite and $V_{iL} = V_{iH}$.

Due to transistor mismatches, the comparator exhibits an input-referred offset voltage that is equal to the value of the output signal when $V_i - V_{REF}$ is set to zero. Offsets, that are generally in the range of ± 10 mV, can be reduced to a few microvolts using trimming or auto-zero techniques.

In high-speed applications, the comparator performance is also dependent on the delay between the instant where the signals are applied at the inputs and the one where the output voltage reaches the steady state. In contrast to amplifiers, the speed requirement is achieved in multistage comparators by not using the frequency compensation.

Ideally, the transition between logic levels at the output node will occur when both input signals have the same magnitude, as shown in Figure 6.3(a).

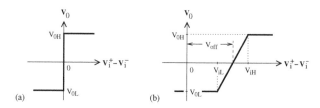

(a) (b)

FIGURE 6.3

(a) Ideal transfer characteristic of the comparator; (b) comparator transfer characteristic showing the effect of the finite gain and offset voltage.

This is not the case in practice due the offset voltage caused by mismatches between nMOS and pMOS transistor characteristics. The comparator transfer characteristic showing the effect of the finite gain, A_v, and offset voltage, V_{off}, is illustrated in Figure 6.3(b). Furthermore, the transition between the output logic levels will occur with a time delay, which increases for large voltage swings on the inputs.

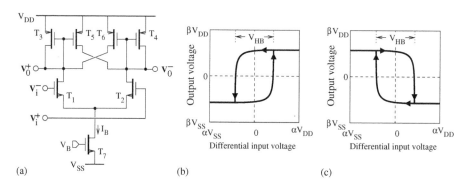

(a) (b) (c)

FIGURE 6.4

(a) Comparator circuit with hysteresis; comparator characteristics in the (b) noninverting and (c) inverting configurations.

The operation with a high gain can contribute to increasing the comparator sensitivity to the effect of the noise, which can corrupt the input signal. In the presence of noise or in the case where the comparator speed is much higher than the variation rate of the input signal around the reference signal level, the switching of the comparator output can become erratic. This problem is overcome by using a comparator with hysteresis. A common technique to generate the hysteresis is based on the use of a positive feedback, which forces the threshold levels for the output switching to depend not only on the difference voltage at the input nodes, but also on the previous states of the input signals. The comparator will then exhibit a characteristic with hysteresis if the switching thresholds for low-to-high and high-to-low transitions of the input voltages are different from each other.

A hysteresis circuit is depicted in Figure 6.4(a) [4,5]. It consists of a differential transistor pair, $T_1 - T_2$, loaded by current mirrors, $T_3 - T_6$, with cross-coupled outputs. The current flowing through T_3 and T_4 are of the form $i_1 = i_3 + i_6$ and $i_2 = i_4 + i_5$. The extra current contributions of T_5 and T_6, which form the positive feedback, help increase not only the transconductance of the differential stage, but also the output slew rate.

Based on the small-signal assumption, the current flowing through each of the transistors T_1 and T_2 forming the differential input stage of the comparator shown in Figure 6.4(a) can be computed as

$$i_{d_1} \simeq \frac{I_B}{2} + g_{m_1} \frac{V_i}{2} \tag{6.4}$$

and

$$i_{d_2} \simeq \frac{I_B}{2} - g_{m_2} \frac{V_i}{2}, \tag{6.5}$$

respectively, where $V_i = V_i^+ - V_i^-$ and I_B is the bias current. For the transistors T_5 and T_6, we have

$$i_{d_5} \simeq g_{m_5} \frac{V_0}{2} \tag{6.6}$$

and

$$i_{d_6} \simeq -g_{m_6} \frac{V_0}{2}, \tag{6.7}$$

respectively, where $V_0 = V_0^+ - V_0^-$. By replacing the diode connected transistor T_3 by its equivalent resistance, $(1/g_{m3}) \parallel (1/g_3)$, the output voltages can be expressed as

$$V_0^+ = [(1/g_{m_3}) \parallel (1/g_3) \parallel (1/g_1)]i_{d_3} \tag{6.8}$$
$$V_0^- = [(1/g_{m_4}) \parallel (1/g_4) \parallel (1/g_2)]i_{d_4}, \tag{6.9}$$

where $i_{d_3} = i_{d_1} - i_{d_6}$ and $i_{d_4} = i_{d_2} - i_{d_5}$. In practice, $g_{m_1} = g_{m_2}$, $g_{m_3} = g_{m_4}$, $g_{m_5} = g_{m_6}$, $g_1 = g_3$, and $g_2 = g_4$, while $1/g_1$ and $1/g_3$ are assumed to be negligible in comparison with $1/g_{m_3}$. The differential output voltage is given by

$$V_0 = \frac{1}{g_{m_3}}[(i_{d_1} - i_{d_2}) + (i_{d_5} - i_{d_6})], \tag{6.10}$$

where $i_{d_1} - i_{d_2} = g_{m_1} V_i$ and $i_{d_5} - i_{d_6} = g_{m_5} V_0$. The differential gain can then be written as

$$A_d = \frac{V_0}{V_i} = \frac{1}{1 - \eta} \frac{g_{m_1}}{g_{m_3}}, \tag{6.11}$$

where $\eta = g_{m_5}/g_{m_3}$. Note that the drain current for a transistor operating in the saturation region is of the form

$$i_{d_j} = \frac{1}{2} K' \left(\frac{W_j}{L_j} \right) (V_{gs_j} - V_T)^2, \tag{6.12}$$

where $K' = \mu_n C_{ox}$, and the transconductance around the bias point can be derived as

$$g_{m_j} = \frac{\partial i_{d_j}}{\partial V_{gs_j}} = \sqrt{\frac{2K'(W_j/L_j)}{i_{d_j}}}, \tag{6.13}$$

where W_j and L_j denote the width and length of the transistor, respectively.

The operation mode of the comparator is determined by the value of the positive feedback factor, η. If $\eta < 1$, the comparator will operate as an improved gain stage [3]. If $\eta = 1$, the comparator will exhibit the behavior of a positive feedback latch. If $\eta > 1$, the operation of the comparator will be similar to the one of a Schmitt trigger circuit with the hysteresis width related to the value of the positive feedback factor.

In the case where $\eta > 1$, the comparator characteristics in the noninverting and inverting configurations are respectively shown in Figures 6.4(b) and (c), where $\alpha < 1$ and $\beta < 1$. To derive the positive trigger point, it is assumed that $V_i^+ = 0$. By connecting the inverting input node to a negative voltage, the transistor T_1 is turned on while T_2 is turned off. Because the current flowing through T_1 and T_3 is almost equal to the bias current, V_0^+ is set to the high logic level and V_0^- takes the low logic level. Hence, the current flowing through the transistors T_4, T_5, and T_6 is nearly zero.

By increasing the voltage at the inverting input node, T_2 can start to conduct and the load of the differential stage is reduced to the current mirror formed by T_3 and T_5. This electrical conduction is developed gradually until the drain currents of T_2 and T_5 are equal. It can be found that

$$i_{d_5} = \frac{W_5/L_5}{W_3/L_3} i_{d_3}, \tag{6.14}$$

$$i_{d_2} = i_{d_5}, \tag{6.15}$$

where $i_{d_3} = i_{d_1}$. Because the input differential pair consisting of transistors T_1 and T_2 is biased by the constant current I_B, we have

$$i_{d_1} + i_{d_2} = I_B . \tag{6.16}$$

The drain current of T_1 and T_2 can then be given by

$$i_{d_1} = \frac{I_B}{1 + (W_5/L_5)/(W_3/L_3)} \tag{6.17}$$

and

$$i_{d_2} = \frac{(W_5/L_5)/(W_3/L_3)I_B}{1 + (W_5/L_5)/(W_3/L_3)} . \tag{6.18}$$

The positive trigger level is defined as

$$V_{trig+} = v_{gs_2} - v_{gs_1} , \tag{6.19}$$

where

$$v_{gs_1} = \sqrt{\frac{i_{d_1}}{2K'(W_1/L_1)}} + V_T \tag{6.20}$$

and

$$v_{gs_2} = \sqrt{\frac{i_{d_2}}{2K'(W_2/L_2)}} + V_T . \tag{6.21}$$

Assuming that the transistors T_1 and T_2 are matched, we obtain

$$V_{trig+} = \sqrt{\frac{I_B}{2K'(W_1/L_1)}} \frac{\sqrt{(W_5/L_5)/(W_3/L_3)} - 1}{\sqrt{1 + (W_5/L_5)/(W_3/L_3)}} . \tag{6.22}$$

As soon as the input voltage becomes greater than V_{trig+}, the switching of comparator outputs will occur. As a result, V_0^+ changes to the low logic level and V_0^- now assumes the high logic level. With the transistor T_1 being off and T_2 on, the bias current primarily flows through T_2 and T_4 and there is no current flowing through transistors T_3, T_5, and T_6.

For the derivation of the negative trigger point, the voltage at the inverting input node is decreased to initiate the conduction of T_1. This is associated with a reduction in the current through T_2 and continues until the drain currents of T_1 and T_6 become equal. This is achieved at the trigger point. In a similar manner as previously, we find that

$$i_{d_6} = \frac{W_6/L_6}{W_4/L_4} i_{d_4} , \tag{6.23}$$

$$i_{d_1} = i_{d_6} , \tag{6.24}$$

$$i_{d_1} + i_{d_2} = I_B , \tag{6.25}$$

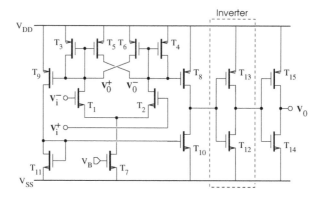

FIGURE 6.5
Circuit diagram of a comparator with hysteresis.

where $i_{d_4} = i_{d_2}$. This set of equations can then be solved for

$$i_{d_2} = \frac{I_B}{1 + (W_6/L_6)/(W_4/L_4)} \tag{6.26}$$

and

$$i_{d_1} = \frac{(W_6/L_6)/(W_4/L_4)I_B}{1 + (W_6/L_6)/(W_4/L_4)}. \tag{6.27}$$

The negative trigger level is obtained as

$$V_{trig-} = v_{gs_2} - v_{gs_1}, \tag{6.28}$$

where

$$v_{gs_1} = \sqrt{\frac{i_{d_1}}{2K'(W_1/L_1)}} + V_T \tag{6.29}$$

and

$$v_{gs_2} = \sqrt{\frac{i_{d_2}}{2K'(W_2/L_2)}} + V_T. \tag{6.30}$$

With $W_1/L_1 = W_2/L_2$, the expression for V_{trig-} becomes

$$V_{trig-} = \sqrt{\frac{I_B}{2K'(W_1/L_1)}} \frac{1 - \sqrt{(W_6/L_6)/(W_4/L_4)}}{\sqrt{1 + (W_6/L_6)/(W_4/L_4)}}. \tag{6.31}$$

Due to the circuit symmetry, transistors T_3, T_4, T_5, and T_6 are matched. The difference between the positive and negative trigger points is the hysteresis band, which is given by

$$V_{HB} = V_{trig^+} - V_{trig^-} = 2\sqrt{\frac{I_B}{2K'(W_1/L_1)} \frac{\sqrt{(W_5/L_5)/(W_3/L_3)} - 1}{\sqrt{1 + (W_5/L_5)/(W_3/L_3)}}}. \quad (6.32)$$

The complete circuit diagram of a comparator based on the hysteresis circuit is depicted in Figure 6.5 [4,5]. It also includes an output stage composed of $T_8 - T_{11}$ and two buffer inverters.

Unfortunately, the hysteresis introduced in the comparator characteristic has the inconvenience of limiting the speed of operation and the minimum level of the input voltage difference that can be accurately resolved. Due to the limitations of amplifier-based comparator structures, it may be better in practice to use comparators, which are designed to drive logic circuits, and operate in open-loop and with a high speed even when overdriven.

6.1.2 Comparator using charge balancing techniques

In switched capacitor circuits, the comparator can consist of an input stage based on charge balancing techniques [6] followed by a latch. Figure 6.6 shows the comparator input stages in the case of the single-ended and fully differential structures.

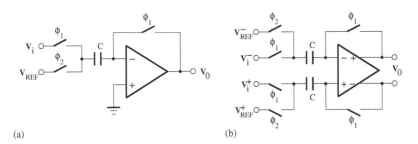

(a)   (b)

FIGURE 6.6
Circuit diagram of the comparator input stage: (a) single-ended and (b) differential configurations.

Let us consider the single-ended circuit of Figure 6.6(a), where V_i and V_{REF} are the input and reference voltages, respectively. The amplifier is in a unity-gain feedback loop during the clock phase ϕ_1, that is, $(n-1)T < t \leq (n-1/2)T$, while it operates in an open loop during the clock phase ϕ_2, that is, $(n-1/2)T < t \leq nT$. With $V_0 = A_0(V^+ - V^-)$, $V^+ = V_{off}$ and $V^- = V_0$, the voltage at the inverting node of the amplifier can be written as

$$V^-((n-1/2)T) = \frac{A_0 V_{off}}{1 + A_0} \quad (6.33)$$

in the clock phase 1, and

$$V^-(nT) = \frac{C}{C+C_p}[V_{REF}(nT) - V_i((n-1/2)T)] + \frac{q_{inj}}{C+C_p} + \frac{A_0 V_{off}}{1+A_0} \quad (6.34)$$

in the clock phase ϕ_2, where q_{inj} represents the error due to the charge injection, C_p is the total parasitic capacitance at the inverting node of the amplifier, V_{off} is the offset voltage, and A_0 is the amplifier dc gain. The output signal in the clock phase ϕ_2 is given by

$$V_0(nT) = -V_{sup} \cdot \text{sign}\{V^-(nT) - V^-((n-1/2)T)\}, \quad (6.35)$$

where V_{sup} denotes the supply voltage of the amplifier. The resolution and settling speed of the comparator based on charge balancing techniques is limited by the bottom-plate parasitic of the capacitor and charge injections of switches.

6.1.3 Latched comparators

A latched comparator is configured with clock signals to sample the input signals and to generate the corresponding output signal. Its amplification gain and propagation delay time are improved by using a positive feedback loop for the signal regeneration into the full-scale logic level. The main advantage of a latched comparator is its high sensitivity to a small input difference.

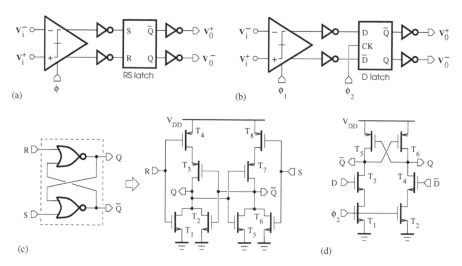

FIGURE 6.7
Circuit diagram of a latched comparator based on (a) an RS latch and (b) a D latch; circuit diagrams of (c) an RS latch and (d) a D latch.

The circuit diagrams of a latched comparator based on an RS latch and a D latch are depicted in Figure 6.7(a) and (b), respectively. These structures consist of a regenerative comparator and a latch driven and loaded by two inverters, which provide the buffer function. The latch is used to store the result of the comparison. Figs. 6.7(c) and (d) show the circuit diagrams of a RS latch and D latch, respectively.

The overall power consumption of a comparator can be divided into static, dynamic, and short-circuit components. The static power is caused by the leakage current and dc current needed for the comparator operation. The dynamic power is associated with the charging and discharging of various switched capacitors. The short-circuit power is caused by the dc current flow from power rails as a result of the time delay to switch from one state to the other. Due to the relative negligible value of the short-circuit power, circuit-level techniques are essentially exploited to reduce the static and dynamic power contributions. The regenerative comparator can then be implemented using either the dynamic or static logic circuits. The difference between both structures is basically related to the fact that the output state of a static logic circuit can change at any time in accordance with the input signal.

In a static comparator, there is always a low-impedance path between the output and both the supply voltage and ground, yielding an uninterrupted power consumption during the whole time that the circuit is powered up. Because the regeneration speed is typically proportional to the magnitude of this current, the power consumption can be considerable at high speeds.

A dynamic comparator uses a quiescent current to bias the output stage only during the time required for the generation of the output signals. As a result, it will quite likely dissipate less power than a static comparator. However, the speed of a dynamic comparator may be limited by the time delay required to charge capacitors before each comparison. Furthermore, the kickback charge injection, which is caused by large voltage transitions associated with the regeneration mechanism, can be transmitted back to the input stage through capacitive coupling, thereby affecting the comparator accuracy.

6.1.3.1 Static comparator

The circuit diagram of a static comparator, which includes a preamplifier stage and a regenerative stage based on cross-coupled transistors, is depicted in Figure 6.8(a). The operation of the comparator is controlled by two nonoverlapping clock signals. During the clock phase ϕ_1, a bias current is supplied to the differential transistor pair of the preamplifier while the regenerative stage is disabled. Because the differential output voltage is an amplified version of the input voltage difference, the comparator is in the *tracking* mode. During the clock phase, ϕ_2, the preamplifier is disabled and the regenerative stage, which is now connected to a bias current, is enabled. The amplification of the tracked voltage difference to valid logic levels can then be achieved without being further affected by the actual signal levels at the input nodes. On the falling edge of the clock signal or at the end of the *latching* mode, the

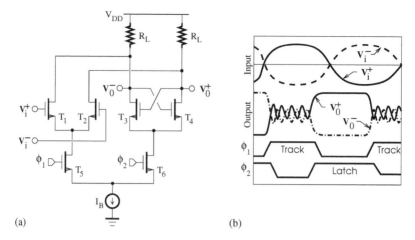

(a)   (b)

FIGURE 6.8

(a) Circuit diagram of a latched comparator; (b) input, output and clock waveforms.

FIGURE 6.9

Small-signal equivalent model of the latched comparator.

comparator outputs are reset, and the comparison process can start again on the next clock rising edge. Figure 6.8(b) shows the input, output, and clock waveforms explaining the comparator operation.

For small-signal analysis, a latched comparator is often represented as two identical back-to-back inverting gain stages driving equal impedance loads [12, 13]. Figure 6.9 shows the small-signal equivalent model of the aforementioned latched comparator.

During the tracking mode, g_{m_k} is equal to g_{m_i}, which is the transconductance of the input differential transistor pair, R_{0k} and C_{0k} respectively denote the total resistance, R_{0i}, and capacitance, C_{0i}, at the output nodes. The output voltages, V_0^+ and V_0^-, satisfy the following differential equations

$$g_{m_i} V_i^- + \frac{V_0^+}{R_{0i}} + C_{0i} \frac{dV_0^+}{dt} = 0, \tag{6.36}$$

$$g_{m_i} V_i^+ + \frac{V_0^-}{R_{0i}} + C_{0i} \frac{dV_0^-}{dt} = 0. \tag{6.37}$$

Subtracting Equation (6.37) from (6.36), we obtain

$$\frac{dV_0}{dt} + \frac{V_0}{R_{0i} C_{0i}} = \frac{g_{m_i} V_i}{C_{0i}}, \tag{6.38}$$

where $V_i = V_i^+ - V_i^-$ and $V_0 = V_0^+ - V_0^-$. Assuming that the initial condition is determined by the comparator state at the end of the previous latching mode, the differential output voltage can be computed as

$$V_0 = \overline{V}_{0r}\, e^{-t/\tau_i} + g_{m_i} R_{0i} V_i (1 - e^{-t/\tau_i}), \tag{6.39}$$

where $\overline{V}_{0r}$ is the steady-state output voltage in the latching mode, and $\tau_i = R_{0i} C_{0i}$ is the preamplification time constant.

During the latching mode, the preamplifier is disabled and the operation of the comparator is determined by the regenerative stage. Hence, g_{m_k} is identical to the transconductance, g_{m_r}, of the cross-coupled transistors, and R_{0k} and C_{0k} respectively represent the total resistance, R_{0r}, and capacitance, C_{0r}, at the output nodes. The dynamic behavior of the output voltages, V_0^+ and V_0^-, can be modeled by the next differential equations,

$$g_{m_r} V_0^- + \frac{V_0^+}{R_{0r}} + C_{0r} \frac{dV_0^+}{dt} = 0, \tag{6.40}$$

$$g_{m_r} V_0^+ + \frac{V_0^-}{R_{0r}} + C_{0r} \frac{dV_0^-}{dt} = 0. \tag{6.41}$$

The difference between Equations (6.40) and (6.41) can be put into the form

$$\frac{dV_0}{dt} - \frac{(g_{m_r} - 1/R_{0r})V_0}{C_{0r}} = 0, \tag{6.42}$$

where $V_0 = V_0^+ - V_0^-$. Solving for the differential output voltage gives

$$V_0 = \overline{V}_{0i}\, e^{t/\tau_r}, \tag{6.43}$$

where the regenerative time constant can be written as

$$\tau_r = \frac{C_{0r}}{g_{m_r} - 1/R_{0r}} \tag{6.44}$$

and $\overline{V}_{0i}$ is the steady-state output voltage in the tracking mode. When g_{m_r} is greater than $1/R_{0r}$, the time constant is positive, and the output voltage can increase exponentially until it reaches the valid logic level.

For high-frequency applications, the comparator should be designed to exhibit a short signal propagation delay. Hence, the operation speed of the comparator can be increased by minimizing the preamplification and regeneration time constants.

• The circuit diagram of the comparator with coupling capacitors is shown in Figure 6.10 [14]. During the resetting phase, that is, when the clock phase 2 is high, the difference of the input signal is amplified by $T_1 - T_2$ and applied to the parasitic capacitors connected at the drains of these transistors, and the comparator output nodes are shorted by the switch T_6. When the clock

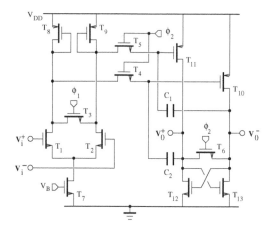

FIGURE 6.10
Circuit diagram of a static comparator with coupling capacitors.

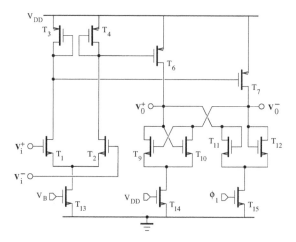

FIGURE 6.11
Circuit diagram of a static comparator with diode-connected transistor-based output initialization.

phase 1 is high, the preamplifier is disabled and the regeneration is achieved in the output stage based on the signal acquired during the previous phase. By establishing a cross-coupling between $T_{10}-T_{11}$ and $T_{12}-T_{13}$, the positive feedback realized by the capacitors C_1 and C_2 increases the regeneration speed. However, because the kickback noise can also be transmitted through this connection, switches T_3-T_4 are required to uncouple the input and output stages.

• The circuit diagram of a static comparator with diode connected transistors is depicted in Figure 6.11 [15]. This structure is based on a preamplifier and latch output stage. The clock phase 1 is high during the resetting period and the transistors $T_{11}-T_{12}$, which are now activated, affect the amplification gain. The latch output resets at a speed, which can be optimized by sizing the biasing transistors. When the clock signal goes low, transistors $T_{11} - T_{12}$ are disconnected and the comparator operation is determined by the positive feedback due to the cross-coupled transistors $T_9 - T_{10}$. The delay required by the output to reach the low and high level is shortened because the connection between T_9 and T_{10} is maintained during both phases. However, this also results in more power dissipation. The kickback noise is reduced by the isolation of the input stage provided by the current mirrors.

6.1.3.2 Dynamic comparator

In general, the operation of a dynamic comparator can be divided into successive phases determined by clock signals. After each comparison of voltage levels periodically supplied to the regenerative stage by the preamplifier connected to the input signals, the comparator outputs should be reset.

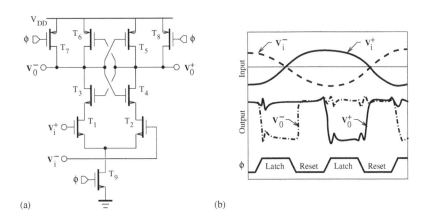

(a)　　　　　　　　　　　　　　　　　　　(b)

FIGURE 6.12
(a) Circuit diagram of a dynamic comparator based on the current-controlled sense-amplifier; (b) input, output, and clock waveforms.

• The circuit diagram of a dynamic comparator based on the current-controlled sense-amplifier is shown in Figure 6.12(a) [11]. When the input differential transistor pair is disabled, the comparator outputs are pulled up to the positive supply voltage, V_{DD}, by switch transistors, $T_7 - T_8$. During the clock phase, ϕ, the sense amplifier is enabled, and the current generated by the input differential transistor pair, $T_1 - T_2$, is used to drive the serially connected regenerative stage, $T_3 - T_6$. Due to the positive feedback provided by cross-coupled transistor pairs, a small input difference can be translated to full logic output voltages. If V_i^+ is greater than V_i^-, V_0^+ will remain at V_{DD} and V_0^- will be set to the ground. Conversely, if V_i^+ is smaller than V_i^-, V_0^+ will be set to the ground and V_0^- will remain at V_{DD}. Because the dc current flow is restricted to the relatively short period that is allocated to the transistor switching, the static power consumption of the comparator is

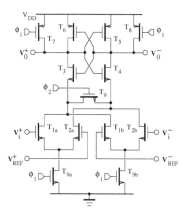

FIGURE 6.13

Circuit diagram of a fully differential dynamic comparator based on the current-controlled sense-amplifier.

almost negligible. Figure 6.12(b) shows the input, output, and clock waveforms during normal operation.

A fully differential dynamic comparator based on the current-controlled sense-amplifier is depicted in Figure 6.13 [7, 8]. The input stage consists of two differential pairs connected respectively to the input and reference voltages with the same polarity. The regenerative output stage is driven by the sum of currents caused by the voltages at the input nodes. The transistor switch, T_9, is used to reset the input nodes of the regenerative stage, so that each comparison is made independent of the previous one. It is controlled by the clock signal ϕ_2, which can be of the form $\phi_2 = \overline{\phi}_1$ and should be designed with minimum sizes to keep the discharging time of parasitic capacitors as low as possible.

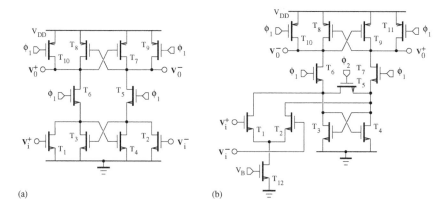

FIGURE 6.14

Circuit diagram of a dynamic comparator with two cross-coupled transistor pairs and input transistor pair (a) without and (b) with a bias current.

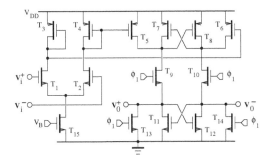

FIGURE 6.15
Circuit diagram of a dynamic comparator including a preamplifier.

• The dynamic comparator with two cross-coupled transistor pairs, as shown in Figure 6.14(a) [9, 10], can detect the difference between the input signals within a few millivolts and regeneratively switch to the appropriate output levels. The outputs rise simultaneously to the same logic state related to the supply-voltage level when the clock signal goes low. The cross-coupled transistors $T_7 - T_8$ are used to speed the pull-up of the outputs on the rising edge of the clock signal. When one of the output nodes drops less than $V_{DD} - |V_T|$, the transistor whose gate is connected to this node turns on, linking the other node to V_{DD}. The offset voltage can be minimized by designing the cross-coupled transistors with non-minimum gate lengths. The power dissipation is reduced because the static current flows only during the transition phase of the outputs. This is due to the fact that the nMOS and pMOS transistors are turned off at the end of the pull-up and pull-down operations, respectively.

In the dynamic comparator implementation of Figure 6.14(b), the differential transistor pair is biased by a current to ensure a high transconductance and, consequently, a high gain required to detect a small voltage difference between the inputs. When the transistor switch T_5 is closed by the high logic level of the clock signal ϕ_2, the charges previously stored on parasitic capacitors at the input of the regenerative stage are equalized and made independent of the input voltages because the low logic level of the clock signal ϕ_1 further turns off the pass transistors, T_6 and T_7, and turns on the pre-charge transistor switches, T_{10} and T_{11}.

A modified version of the above comparator is depicted in Figure 6.15. It includes an input preamplifier, which can isolate the input signal from the latch kickback noise. The output current provided by the differential pair $T_1 - T_2$ is mirrored from T_3 and T_4 to T_5 and T_6, respectively, and fed to the output latch.

6.2   Multipliers

Analog multipliers generally produce an output signal that is proportional in magnitude to the product of the two input signals. They can be designed to perform either a four-quadrant multiplication, when both input signals can be bipolar (i.e., there is no restriction on the sign of the input signals), or a two-quadrant multiplication in the case where one of the input signals is unipolar and the other can be bipolar.

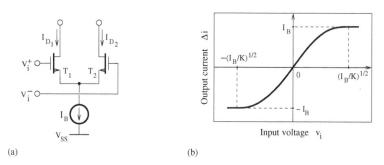

(a)   (b)

FIGURE 6.16
(a) Differential source-coupled transistor pair; (b) transfer characteristic.

Various techniques can be used for the implementation of multipliers. Let the drain current, i_D, of an nMOS transistor be expressed as,

$$I_D = \begin{cases} 0, & \text{if } V_{GS} \leq V_T \\ K[2(V_{GS} - V_T)V_{DS} - V_{DS}^2], & \text{if } V_{GS} > V_T;\ 0 < V_{DS} \leq V_{DS(sat)} \\ K(V_{GS} - V_T)^2, & \text{if } V_{GS} > V_T;\ V_{DS} \geq V_{DS(sat)} \end{cases}$$

$$(6.45)$$

in the cutoff, triode, and saturation regions, respectively, where the drain-source saturation voltage is of the form, $V_{DS(sat)} = V_{GS} - V_T$, V_{GS} is the gate-source voltage, V_T is the threshold voltage, $K = \mu_n(C_{ox}/2)(W/L)$ is related to the transconductance parameter, μ_n is the effective surface carrier mobility, C_{ox} is the gate oxide capacitance per unit area, and W and L are the channel width and length, respectively. Figure 6.16(a) shows a differential source-coupled transistor pair, which can be used in the multiplier design. Assuming that the transistors are matched and operate in the saturation region, we have

$$V_{GS_1} = V_T + \sqrt{\frac{I_{D_1}}{K}}$$

$$(6.46)$$

and

$$V_{GS_2} = V_T + \sqrt{\frac{I_{D_2}}{K}}.$$

$$(6.47)$$

Kirchhoff's voltage law equation for the input loop can be written as

$$V_i^+ - V_{GS_1} + V_{GS_2} - V_i^- = 0. \tag{6.48}$$

Substituting Equations (6.46) and (6.47) into Equation (6.48), we can obtain

$$\sqrt{I_{D_1}} - \sqrt{I_{D_2}} = \sqrt{K}V_i, \tag{6.49}$$

where $V_i = V_i^+ - V_i^-$. Using Kirchhoff's current law at the source node, we find

$$I_{D_1} + I_{D_2} = I_B, \tag{6.50}$$

where I_B is the bias current. The system involving Equations (6.49) and (6.50) can be solved for I_{D_1}. That is,

$$I_{D_1} = \frac{I_B}{2} \pm K\frac{v_i}{2}\sqrt{\frac{2I_B}{K} - V_i^2}, \tag{6.51}$$

where $|V_i| \leq \sqrt{I_B/K}$. For positive values of V_i, the current I_{D_1} should be greater than $I_B/2$. Hence,

$$I_{D_1} = \frac{I_B}{2} + K\frac{V_i}{2}\sqrt{\frac{2I_B}{K} - V_i^2}. \tag{6.52}$$

The substitution of Equation (6.52) into (6.50) yields

$$I_{D_2} = \frac{I_B}{2} - K\frac{V_i}{2}\sqrt{\frac{2I_B}{K} - V_i^2}. \tag{6.53}$$

The difference between I_{D_1} and I_{D_2} is then given by

$$\triangle i = I_{D_1} - I_{D_2} = KV_i\sqrt{\frac{2I_B}{K} - V_i^2}. \tag{6.54}$$

A plot of the transfer characteristic is shown in Figure 6.16(b). As illustrated by the characteristic curvature, the linear range is limited because it can only be extended by increasing the magnitude of the bias current, or equivalently, the power consumption.

In general, the multiplication of two voltages can be achieved using the transistor characteristic in the saturation and triode regions. Various circuit techniques have been proposed to implement multipliers in CMOS technology, which is known to feature a high integration density. For instance, they can exploit the variation in the transconductance of differential transistor stages or be based on the subtraction of the sum-squared and difference-squared of two input signals. The most critical design specification is the linear dynamic range, which is limited in some multiplier structures even for high supply voltages.

6.2.1 Multiplier cores

6.2.1.1 Multiplier core based on externally controlled transconductances

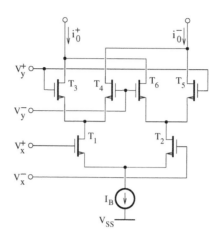

FIGURE 6.17
Multiplier core based on cross-coupled transconductance stages.

Various analog multiplier architectures based on externally controlled transconductances are available. In general, they consist of differential voltage-to-current converters and provide an output proportional to the product of two input signals.

- The multiplier core shown in Figure 6.17 [16,17] is based on cross-coupled transconductance stages. Assuming that all transistors operate in the saturation region, the difference between the output currents is given by

$$\triangle i_0 = i_0^+ - i_0^- \tag{6.55}$$
$$= (I_{D_3} + I_{D_6}) - (I_{D_4} + I_{D_5})$$
$$= (I_{D_3} - I_{D_4}) - (I_{D_5} - I_{D_6}), \tag{6.56}$$

where

$$I_{D_3} - I_{D_4} = KV_y \sqrt{\frac{2I_{D_1}}{K} - V_y^2} \tag{6.57}$$

and

$$I_{D_5} - I_{D_6} = KV_y \sqrt{\frac{2I_{D_2}}{K} - V_y^2} \tag{6.58}$$

To proceed further, it can be shown that

$$I_{D_1} = \frac{I_{SS}}{2} + K \frac{V_x}{2} \sqrt{\frac{2I_B}{K} - V_x^2} = \frac{K}{4} \left(\sqrt{\frac{2I_B}{K} - V_x^2} + V_x \right)^2 \qquad (6.59)$$

and

$$I_{D_2} = \frac{I_B}{2} - K \frac{V_x}{2} \sqrt{\frac{2I_B}{K} - V_x^2} = \frac{K}{4} \left(\sqrt{\frac{2I_B}{K} - V_x^2} - V_x \right)^2. \qquad (6.60)$$

Substituting Equations (6.60), (6.59), (6.58), and (6.57) into Equation (6.56), we obtain

$$\triangle i_0 = KV_y \left[\sqrt{\frac{1}{2} \left(\sqrt{\frac{2I_B}{K} - V_x^2} + V_x \right)^2 - V_y^2} \right.$$

$$\left. - \sqrt{\frac{1}{2} \left(\sqrt{\frac{2I_B}{K} - V_x^2} - V_x \right)^2 - V_y^2} \right]. \qquad (6.61)$$

Note that $\triangle i_0$ has a nonlinear relationship with V_x and V_y. Nevertheless, when V_x and V_y are small, and

$$|V_y| \leq \sqrt{\frac{I_B}{K} - \frac{1}{2}V_x^2} - \frac{1}{\sqrt{2}}V_x, \qquad (6.62)$$

it can be shown that

$$\triangle i_0 \simeq KV_y \left[\sqrt{\frac{1}{2} \left(\sqrt{\frac{2I_B}{K} - V_x^2} + V_x \right)^2} - \sqrt{\frac{1}{2} \left(\sqrt{\frac{2I_B}{K} - V_x^2} - V_x \right)^2} \right] \qquad (6.63)$$

and finally,

$$\triangle i_0 \simeq \sqrt{2} K V_x V_y. \qquad (6.64)$$

Because the multiplier core consists of differential transistor pairs, it operates in the linear region only if $|V_x| \leq (I_B/K)^{1/2}$ and $|V_y| \ll (2\min(I_{D_1}, I_{D_2})/K)^{1/2}$, where I_{D_1} and I_{D_2} are the drain currents of the transistors T_1 and T_2, respectively.

The circuit diagram of a high-linearity multiplier core is depicted in Figure 6.18 [18]. It is composed of nMOS transistors, T_1, T_2, T_3, T_4, T_5, and T_6, and pMOS transistors, T_7 and T_8, which can be partitioned into two sections. A constant current sink, I_{B1}, is connected to the sources of T_1, T_2, and T_5, and the drain of T_7, while the drain of T_5, which is connected to the gate of

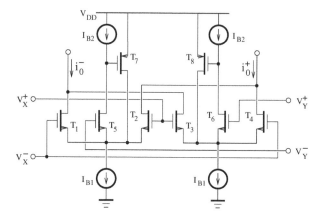

FIGURE 6.18

High-linearity multiplier core.

T_7, is driven by a constant current source, I_{B2}. Similarly, a constant current sink, I_{B1}, is connected to the sources of T_3, T_4, and T_6 and the drain of T_8, while the drain of T_6, which is connected to the gate of T_8, is driven by a constant current source, I_{B2}.

All transistors operate in the saturation region. Let V_I be the dc component associated with the input voltages. Assuming that $V_x^+ = V_I + V_x/2$, $V_x^- = V_I - V_x/2$, $V_y^+ = V_I + V_y/2$, and $V_y^- = V_I - V_y/2$, we can obtain

$$V_{GS_5} = V_{G_5} - V_{S_5} = V_I - V_y/2 - V_{S_5} \tag{6.65}$$
$$V_{GS_6} = V_{G_6} - V_{S_6} = V_I + V_y/2 - V_{S_6} \tag{6.66}$$

and

$$I_{D_5} = K(V_{GS_5} - V_T)^2 = K(V_I - V_y/2 - V_{S_5} - V_T)^2 = I_{B2} \tag{6.67}$$
$$I_{D_6} = K(V_{GS_6} - V_T)^2 = K(V_I + V_y/2 - V_{S_6} - V_T)^2 = I_{B2}. \tag{6.68}$$

Hence,

$$V_{S_5} = V_I - \frac{V_y}{2} - V_T - \sqrt{\frac{I_{B2}}{K}}, \tag{6.69}$$

$$V_{S_6} = V_I + \frac{V_y}{2} - V_T - \sqrt{\frac{I_{B2}}{K}}. \tag{6.70}$$

The gate-source voltages and the drain currents of transistors $T_1 - T_4$ can also be expressed as

$$V_{GS_1} = V_{G_1} - V_{S_1} = V_I - V_x/2 - V_{S_5} \tag{6.71}$$
$$V_{GS_2} = V_{G_2} - V_{S_2} = V_I + V_x/2 - V_{S_5} \tag{6.72}$$
$$V_{GS_3} = V_{G_3} - V_{S_3} = V_I + V_x/2 - V_{S_6} \tag{6.73}$$
$$V_{GS_4} = V_{G_4} - V_{S_4} = V_I - V_x/2 - V_{S_6} \tag{6.74}$$

and

$$I_{D_1} = K(V_{GS_1} - V_T)^2 = K(V_I - V_{\mathrm{x}}/2 - V_{S_5} - V_T)^2 \tag{6.75}$$

$$I_{D_2} = K(V_{GS_2} - V_T)^2 = K(V_I + V_{\mathrm{x}}/2 - V_{S_5} - V_T)^2 \tag{6.76}$$

$$I_{D_3} = K(V_{GS_3} - V_T)^2 = K(V_I + V_{\mathrm{x}}/2 - V_{S_6} - V_T)^2 \tag{6.77}$$

$$I_{D_4} = K(V_{GS_4} - V_T)^2 = K(V_I - V_{\mathrm{x}}/2 - V_{S_6} - V_T)^2 . \tag{6.78}$$

Note that $I_{B1} = I_{D_1} + I_{D_2} + I_{D_5} + I_{D_7}$ and $I_{B1} = I_{D_3} + I_{D_4} + I_{D_6} + I_{D_8}$. Combining Equations (6.69), (6.70), and (6.75) through (6.78) and simplifying, we find

$$I_{D_1} = K \left(-\frac{V_{\mathrm{x}}}{2} + \frac{V_{\mathrm{y}}}{2} + \sqrt{\frac{I_{B2}}{K}} \right)^2 \tag{6.79}$$

$$I_{D_2} = K \left(\frac{V_{\mathrm{x}}}{2} + \frac{V_{\mathrm{y}}}{2} + \sqrt{\frac{I_{B2}}{K}} \right)^2 \tag{6.80}$$

$$I_{D_3} = K \left(\frac{V_{\mathrm{x}}}{2} - \frac{V_{\mathrm{y}}}{2} + \sqrt{\frac{I_{B2}}{K}} \right)^2 \tag{6.81}$$

$$I_{D_4} = K \left(-\frac{V_{\mathrm{x}}}{2} - \frac{V_{\mathrm{y}}}{2} + \sqrt{\frac{I_{B2}}{K}} \right)^2 . \tag{6.82}$$

The difference of the output currents is given by

$$\triangle i_0 = i_0^+ - i_0^- \tag{6.83}$$
$$= (I_{D_2} + I_{D_4}) - (I_{D_1} + I_{D_3})$$
$$= -(I_{D_1} - I_{D_2}) - (I_{D_3} - I_{D_4}) \tag{6.84}$$
$$= 2KV_{\mathrm{x}}V_{\mathrm{y}} , \tag{6.85}$$

where

$$I_{D_1} - I_{D_2} = -2KV_{\mathrm{x}} \left(\frac{V_{\mathrm{y}}}{2} + \sqrt{\frac{I_{B2}}{K}} \right) , \tag{6.86}$$

$$I_{D_3} - I_{D_4} = 2KV_{\mathrm{x}} \left(-\frac{V_{\mathrm{y}}}{2} + \sqrt{\frac{I_{B2}}{K}} \right) . \tag{6.87}$$

The output dc characteristic, $\triangle i_0$, which is proportional to the product of the differential input voltages, is linear over a wide input range.

- The circuit diagram of a multiplier core based on transconductance stages

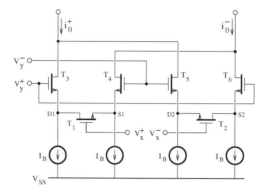

FIGURE 6.19

Multiplier core based on transconductance stages with source degeneration.

with source degeneration is depicted in Figure 6.19 [19]. The currents I_1 and I_2 flowing respectively through the transistors T_1 and T_2, which are assumed to be identical and operate in the triode region, are given by

$$I_1 = K[2(V_x^+ - V_{S1} - V_T)V_y - V_y^2] \tag{6.88}$$

$$I_2 = K[2(V_x^- - V_{S2} - V_T)V_y - V_y^2], \tag{6.89}$$

where $V_{DS_1} = V_{DS_2} = V_y$. The transistors $T_3 - T_6$ have the same geometry and operate in the saturation region. The next relation can be written

$$V_{GS_4} = \sqrt{\frac{I_B}{K}} + V_T = V_y^- - V_{S1} \tag{6.90}$$

$$V_{GS_6} = \sqrt{\frac{I_B}{K}} + V_T = V_y^+ - V_{S2}, \tag{6.91}$$

where I_B is the bias current of the transistor. The currents i_0^+ and i_0^- can be obtained as

$$i_0^+ = 2I_B - (I_1 - I_2) = 2I_B + 2KV_xV_y \tag{6.92}$$

$$i_0^- = 2I_B + (I_1 - I_2) = 2I_B - 2KV_xV_y \tag{6.93}$$

and the resulting differential current is

$$\triangle i_0 = 4KV_xV_y. \tag{6.94}$$

The dynamic rage is limited by the fact that T_1 and T_2 should be in the triode region, that is, $V_{DS} \leq V_{DS(sat)} = V_{GS} - V_T$.

- The circuit diagram of a multiplier core based on transconductance stages with active cascode structure is depicted in Figure 6.20. Transistors $T_1 - T_4$ operate in the triode region and the currents i_0^+ and i_0^- can be written as

$$i_0^+ = K[2(V_x^+ - V_T)V_{D1} - V_{D1}^2] + K[2(V_x^- - V_T)V_{D3} - V_{D3}^2] \tag{6.95}$$

$$i_0^- = K[2(V_x^+ - V_T)V_{D4} - V_{D4}^2] + K[2(V_x^- - V_T)V_{D2} - V_{D2}^2]. \tag{6.96}$$

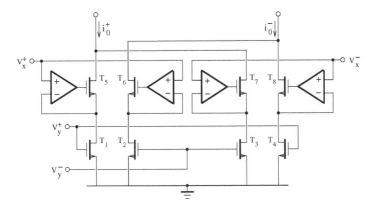

FIGURE 6.20
Multiplier core based on transconductance stages with active cascode structure.

Using the gate voltages given by

$$V_{Gi} = A(V_x^+ - V_{D(i-4)}), \quad \text{if} \quad i = 5, 6 \tag{6.97}$$

$$V_{Gj} = A(V_x^- - V_{D(j-4)}), \quad \text{if} \quad j = 7, 8, \tag{6.98}$$

where A represents the gain of the amplifier, and the fact that $T_5 - T_8$ operate in the saturation region, we obtain

$$V_{Dk} = \frac{A}{1+A}V_x^+ - \frac{1}{1+A}\left(\sqrt{\frac{I_{Dk}}{K}} + V_T\right) \simeq V_x^+, \quad \text{if} \quad k = 1, 2 \tag{6.99}$$

$$V_{Dl} = \frac{A}{1+A}V_x^- - \frac{1}{1+A}\left(\sqrt{\frac{I_{Dl}}{K}} + V_T\right) \simeq V_x^-, \quad \text{if} \quad l = 3, 4. \tag{6.100}$$

The difference of the output currents can be expressed as

$$\triangle i_0 = i_0^+ - i_0^- = 2KV_xV_y. \tag{6.101}$$

It should be noted that the effective transconductance of $T_5 - T_8$ is increased due to the presence of amplifiers. As a result, the linearity of the multiplier is improved.

6.2.1.2 Multiplier core based on the quarter-square technique

Multipliers based on stacking transconductance stages have poor linearity when they operate with low supply voltages. A way to circumvent this problem is to use the quarter-square technique.

• Using two squarer stages whose output nodes are cross-coupled, a multiplier based on the quarter-square technique can be implemented as

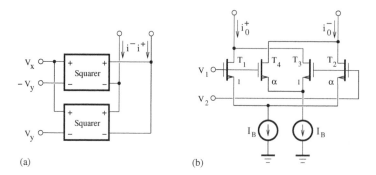

FIGURE 6.21
(a) Block diagram of a multiplier based on the quarter-square technique; (b) circuit diagram of a squarer.

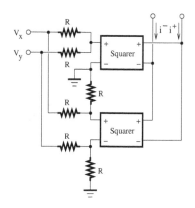

FIGURE 6.22
Quarter-square multiplier core with an input-resistive network.

shown in Figure 6.21(a). The multiplication of the input voltages, V_x and V_y, is achieved by taking the difference between the output currents due to $(V_x + V_y)^2$ and $(V_x - V_y)^2$. The circuit diagram of a squarer is depicted in Figure 6.21(b) [24,25]. It consists of two unbalanced source-coupled transistor pairs with the W/L ratio of α. The drain currents of transistors $T_1 - T_4$ are respectively given by

$$I_{D_1} = K'\frac{W}{L}(V_{GS_1} - V_T)^2 \tag{6.102}$$

$$I_{D_2} = K'\alpha\frac{W}{L}(V_{GS_2} - V_T)^2 \tag{6.103}$$

$$I_{D_3} = K'\frac{W}{L}(V_{GS_3} - V_T)^2 \tag{6.104}$$

$$I_{D_4} = K'\alpha\frac{W}{L}(V_{GS_4} - V_T)^2, \tag{6.105}$$

where $K' = \mu_n C_{ox}/2$. Using Kirchhoff's voltage law, we can obtain

$$V_{GS_1} - V_{GS_2} = V_{GS_4} - V_{GS_3} = V_1 - V_2. \tag{6.106}$$

Combining Equations (6.102) through (6.105) and (6.106), it can be shown that

$$\sqrt{I_{D_1}} - \sqrt{I_{D_2}/\alpha} = \sqrt{K}(V_1 - V_2) \tag{6.107}$$

and

$$\sqrt{I_{D_3}} - \sqrt{I_{D_4}/\alpha} = -\sqrt{K}(V_1 - V_2). \tag{6.108}$$

By applying Kirchhoff's current law, we find

$$I_{D_1} + I_{D_2} = I_{D_3} + I_{D_4} = I_B. \tag{6.109}$$

Solving the system of Equations (6.107), (6.108), and (6.109) gives

$$I_{D_1}, I_{D_3} = \frac{1}{\left(1+\dfrac{1}{\alpha}\right)^2}\left[\left(1+\frac{1}{\alpha}\right)\frac{I_B}{\alpha} + \left(1-\frac{1}{\alpha}\right)K(V_1-V_2)^2\right.$$
$$\left. \pm 2K(V_1-V_2)\sqrt{\left(1+\frac{1}{\alpha}\right)\frac{I_B}{K\alpha} - \frac{(V_1-V_2)^2}{\alpha}}\right], \tag{6.110}$$

where the upper plus sign is for I_{D_1}, while the lower minus sign is for I_{D_3}, and

$$I_{D_2}, I_{D_4} = \frac{1}{\left(1+\dfrac{1}{\alpha}\right)^2}\left[\left(1+\frac{1}{\alpha}\right)I_B - \left(1-\frac{1}{\alpha}\right)K(V_1-V_2)^2\right.$$
$$\left. \mp 2K(V_1-V_2)\sqrt{\left(1+\frac{1}{\alpha}\right)\frac{I_B}{K\alpha} - \frac{(V_1-V_2)^2}{\alpha}}\right], \tag{6.111}$$

where the upper minus sign is for I_{D_2} while the lower plus sign is for I_{D_4}. The difference in the output currents can then be expressed as

$$\triangle I_0 = I_0^+ - I_0^- \tag{6.112}$$
$$= (I_{D_1} + I_{D_3}) - (I_{D_2} + I_{D_4}) \tag{6.113}$$
$$= (I_{D_1} - I_{D_2}) + (I_{D_3} - I_{D_4}) \tag{6.114}$$
$$= \frac{-2\left(1+\dfrac{1}{\alpha}\right)\left[\left(1-\dfrac{1}{\alpha}\right)I_B - 2K(V_1-V_2)^2\right]}{\left(1+\dfrac{1}{\alpha}\right)^2}. \tag{6.115}$$

For the normal operation of the squarer circuit, the input voltage, $V_1 - V_2$, should be in the range $|V_1 - V_2| \leq \sqrt{I_B/(K\alpha)}$.

The aforementioned quarter square can also be implemented as shown in Figure 6.22. Here, the inversion of one of the inputs is not required. The input resistive network operates as a voltage divider with a factor of 2 to produce the voltages $(V_x + V_y)/2$ and $(V_x - V_y)/2$, which are applied at the input nodes of the multiplier core.

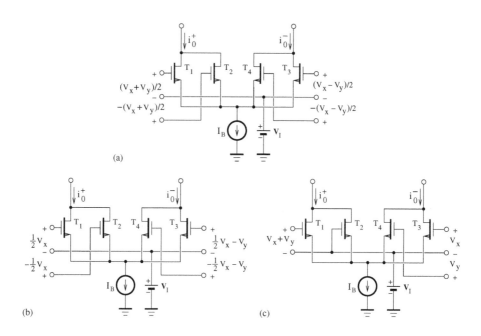

FIGURE 6.23

Multiplier core based on the quarter-square technique.

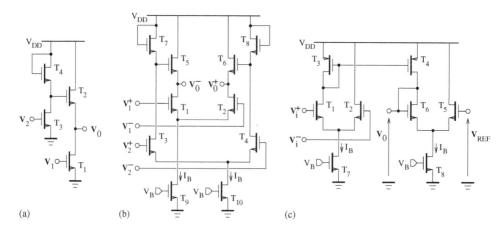

FIGURE 6.24

Summer circuit diagrams.

• Other structures of the multiplier core based on the quarter-square technique are depicted in Figure 6.23(a) [27], (b) [28,29], and (c) [30]. They consist of four source-coupled transistors biased by a constant current source, but differ in the required combination of the input voltages. The voltage V_I represents the *dc* component of the input voltages.

Let us consider the structure of Figure 6.23(a). Assuming the current-voltage characteristic of MOS transistors operating in the saturation region, the drain currents are given by

$$I_{D_1} = K \left(V_I + \frac{V_x + V_y}{2} - V_S - V_T \right)^2 , \tag{6.116}$$

$$I_{D_2} = K \left(V_I - \frac{V_x + V_y}{2} - V_S - V_T \right)^2 , \tag{6.117}$$

$$I_{D_3} = K \left(V_I + \frac{V_x - V_y}{2} - V_S - V_T \right)^2 , \tag{6.118}$$

and

$$I_{D_4} = K \left(V_I - \frac{V_x - V_y}{2} - V_S - V_T \right)^2 . \tag{6.119}$$

The difference in the output currents can be written as

$$\triangle i_0 = i_0^+ - i_0^- = (I_{D_1} + I_{D_2}) - (I_{D_3} + I_{D_4}) = 2KV_xV_y . \tag{6.120}$$

Applying Kirchhoff's current law at the source node, we have

$$I_B = (I_{D_1} + I_{D_2}) + (I_{D_3} + I_{D_4}) = i_0^+ + i_0^- . \tag{6.121}$$

Combining Equations (6.120) and (6.121), it can be shown that

$$i_0^+ = I_{D_1} + I_{D_2} = \frac{I_B}{2} + KV_xV_y \tag{6.122}$$

and

$$i_0^- = I_{D_3} + I_{D_4} = \frac{I_B}{2} - KV_xV_y . \tag{6.123}$$

The range of input voltages over which the transistors still operate in the saturation region can be determined by the next worst-case requirement,

$$V_{GS_2} = V_I - \frac{V_x + V_y}{2} - V_S \geq V_T . \tag{6.124}$$

Substituting Equations (6.116) and (6.117) into Equation (6.122), we obtain

$$V_I - V_S - V_T = \sqrt{\frac{I_B}{4K} - \frac{V_x^2 + V_y^2}{4}} . \tag{6.125}$$

Combining Equations (6.125) and (6.124) gives

$$V_x^2 + V_y V_x + V_y^2 - \frac{I_B}{2K} \leq 0 \qquad (6.126)$$

or equivalently,

$$|V_x| \leq -\frac{V_y}{2} + \sqrt{\frac{I_B}{2K} - \frac{3V_y^2}{4}}, \quad |V_y| \leq \sqrt{\frac{2I_B}{3K}}. \qquad (6.127)$$

A similar analysis of the multiplier cores shown in Figures 6.23(b) and (c) also results in the output current characteristic and dynamic range given by Equations (6.120) and (6.126), respectively.

Depending on the implementation of the input stage, the resulting multiplier can have the advantage of exhibiting the same transfer characteristic with respect to any of the differential input voltages.

A single-ended summer structure is depicted in Figure 6.24(a) [20]. It is based on two nMOS inverting stages. Using Kirchhoff's voltage law for the loop including the transistors T_2 and T_2, the output voltage can be expressed as

$$V_0 = V_{DD} - V_{GS_2} - V_{GS_4}. \qquad (6.128)$$

The transistors are assumed to have the same threshold voltage, V_T, and to operate in the saturation region. Because $I_{D_1} = I_{D_2}$ and $I_{D_3} = I_{D_4}$, we obtain

$$V_{GS_2} = \sqrt{\frac{(W_1/L_1)}{(W_2/L_2)}}(V_{GS_1} - V_T) + V_T \qquad (6.129)$$

and

$$V_{GS_4} = \sqrt{\frac{(W_3/L_3)}{(W_4/L_4)}}(V_{GS_3} - V_T) + V_T, \qquad (6.130)$$

where $V_{GS_1} = V_1$ and $V_{GS_3} = V_2$. Assuming that $(W_1/L_1) = (W_3/L_3)$ and $(W_2/L_2) = (W_4/L_4)$ and combining Equations (6.128), (6.129), and (6.130), we find

$$V_0 = V_0' - \sqrt{\frac{(W_1/L_1)}{(W_2/L_2)}}(V_1 + V_2), \qquad (6.131)$$

where

$$V_0' = V_{DD} + 2V_T \left(\sqrt{\frac{(W_1/L_1)}{(W_2/L_2)}} - 1 \right). \qquad (6.132)$$

The dc component, V_0', of the output voltage can be cancelled by adopting the differential summer structure of Figure 6.24(b). Performing a similar analysis as previously, it can be shown that

$$V_0^+ = V_0' - \sqrt{\frac{(W_1/L_1)}{(W_2/L_2)}}(V_1^- + V_2^-) \qquad (6.133)$$

and

$$V_0^- = V_0' - \sqrt{\frac{(W_1/L_1)}{(W_2/L_2)}}(V_1^+ + V_2^+). \qquad (6.134)$$

The differential output voltage is then given by

$$V_0 = V_0^+ - V_0^- = \sqrt{\frac{(W_1/L_1)}{(W_2/L_2)}}(V_1 + V_2), \qquad (6.135)$$

where $V_1 = V_1^+ - V_1^-$ and $V_2 = V_2^+ - V^-$.

An alternative summer circuit is shown in Figure 6.24(c) [21]. It is composed of two differential stages, which are coupled by a current mirror in such a way that $I_{D_1} = I_{D_6}$. Hence, the gate-source voltages of transistors T_1 and T_6 have the same value. Because both differential stages are assumed to be biased by a constant current, I_B, we can write

$$I_{D_1} + I_{D_2} = I_B = I_{D_5} + I_{D_6}. \qquad (6.136)$$

The drain currents of transistors T_2 and T_5 should also be identical, thereby forcing the gate-source voltages of transistors T_2 and T_5 to be matched. It can then be shown that

$$V_0 - V_{REF} = V_{GS_6} - V_{GS_5} = V_{GS_1} - V_{GS_2} = V_i, \qquad (6.137)$$

where $V_i = V_i^+ - V_i^-$.

6.2.1.3 Design issues

In practice, the accuracy of the output voltage can be affected by the input offset voltage due to mismatches between transistor characteristics, and can become degraded at high frequencies. Without resorting to special matching methods, it may be difficult to preserve the advantage of the quarter-square technique, such as the symmetry of the output characteristic.

For low-voltage applications, the aforementioned multiplier core can be designed without the bias current I_B. In this case, the transistor sources are directly connected to either the ground or a supply voltage terminal. A disadvantage of the resulting circuit is that the dc components of the input voltages are simultaneously the bias voltages. As a result, it may be impossible to independently control the common-mode levels at the inputs.

6.2.2 Design examples

The circuit diagram of a multiplier with single-ended output is shown in Figure 6.25. The transistors operate in the saturation region and the output

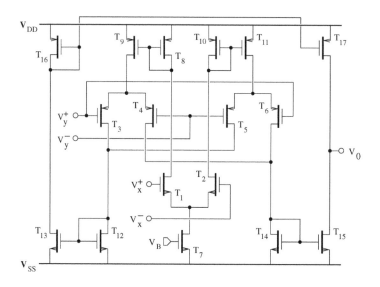

FIGURE 6.25
Four-quadrant multiplier circuit with a single-ended output.

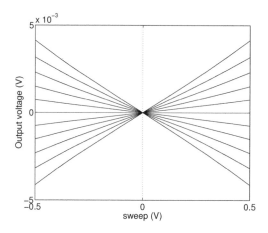

FIGURE 6.26
Plot of dc transfer characteristics of the four-quadrant multiplier.

current is given by

$$i_0 = \sqrt{2mK_nK_p}v_xv_y , \qquad (6.138)$$

where m is the scaling ratio of the current mirrors $T_8 - T_9$ and $T_{10} - T_{11}$, and K_n and K_p are the transconductance of the n-channel and p-channel transistors, respectively.

The structure of a multiplier with differential outputs is depicted in Figure 6.27. It is based on the multiplier core of Figure 6.17. The common-mode

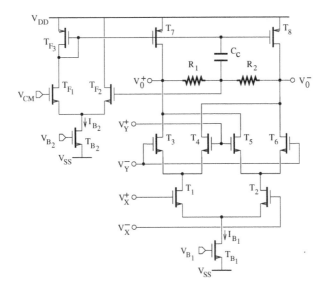

FIGURE 6.27

Four-quadrant multiplier circuit with differential outputs.

feedback is implemented by the transistors $T_{F1} - T_{F3}$ and resistors R_1 and R_2. The reference level is defined by the voltage V_{CM} and the frequency compensation is provided by C_c.

The dc characteristics of the multiplier are computed based on 0.5-µm CMOS transistors parameters, by performing SPICE [26] simulations and are shown in Figure 6.26.

Another multiplier implementation is shown in Figure 6.28 [27]. It is based on the quarter-square technique. A current I_B generated by a transistor current source is used to bias each set of transistors, $T_1 - T_4$ and $T_5 - T_8$, whose sources are connected together. Assuming that transistors $T_1 - T_{12}$ are identical and operate in the saturation region, we can write

$$V_{G_1} = V_{DD} - V_{DS_9} = V_{DD} - (V_{GS_9} - V_T), \qquad (6.139)$$

$$V_{G_2} = V_{DD} - V_{DS_{10}} = V_{DD} - (V_{GS_{10}} - V_T), \qquad (6.140)$$

$$V_{G_3} = V_{DD} - V_{DS_{11}} = V_{DD} - (V_{GS_{11}} - V_T), \qquad (6.141)$$

and

$$V_{G_4} = V_{DD} - V_{DS_{12}} = V_{DD} - (V_{GS_{12}} - V_T), \qquad (6.142)$$

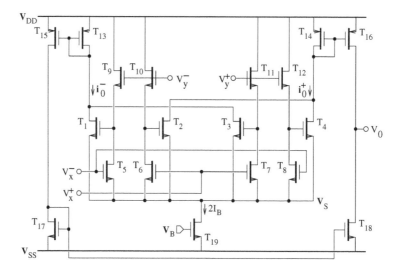

FIGURE 6.28
Four-quadrant multiplier based on the quarter-square technique.

where

$$V_{GS_9} = V_y^- - (V_{DS_5} + V_S) = V_y^- - (V_{GS_5} - V_T) - V_S, \tag{6.143}$$

$$V_{GS_{10}} = V_y^- - (V_{DS_6} + V_S) = V_y^- - (V_{GS_6} - V_T) - V_S, \tag{6.144}$$

$$V_{GS_{11}} = V_y^+ - (V_{DS_7} + V_S) = V_y^+ - (V_{GS_7} - V_T) - V_S, \tag{6.145}$$

and

$$V_{GS_{12}} = V_y^+ - (V_{DS_8} + V_S) = V_y^+ - (V_{GS_8} - V_T) - V_S, \tag{6.146}$$

where V_S is the common source voltage. Also, it can be shown that

$$V_{GS_5} = V_x^- - V_S, \tag{6.147}$$

$$V_{GS_6} = V_x^+ - V_S, \tag{6.148}$$

$$V_{GS_7} = V_x^+ - V_S, \tag{6.149}$$

and

$$V_{GS_8} = V_x^- - V_S. \tag{6.150}$$

Combining Equations (6.147) through (6.150), (6.143) through (6.146), and (6.139) through (6.142) gives

$$V_{G_1} = V_{DD} + (V_x^- - V_y^-), \tag{6.151}$$

$$V_{G_2} = V_{DD} + (V_x^+ - V_y^-), \tag{6.152}$$

$$V_{G_3} = V_{DD} + (V_x^+ - V_y^+), \tag{6.153}$$

and

$$V_{G_4} = V_{DD} + (V_\mathrm{x}^- - V_\mathrm{y}^+). \tag{6.154}$$

Let $V_\mathrm{x}^+ = V_I + v_\mathrm{x}/2$, $V_\mathrm{x}^- = V_I - v_\mathrm{x}/2$, $V_\mathrm{y}^+ = V_I + v_\mathrm{y}/2$, and $V_\mathrm{y}^- = V_I - v_\mathrm{y}/2$, where V_I is the common dc component of the input voltage. The drain currents of transistors $T_1 - T_4$ are respectively given by

$$I_{D_1} = K[V_{DD} - (v_\mathrm{x} - v_\mathrm{y})/2 - V_S - V_T]^2, \tag{6.155}$$

$$I_{D_2} = K[V_{DD} + (v_\mathrm{x} + v_\mathrm{y})/2 - V_S - V_T]^2, \tag{6.156}$$

$$I_{D_3} = K[V_{DD} + (v_\mathrm{x} - v_\mathrm{y})/2 - V_S - V_T]^2, \tag{6.157}$$

and

$$I_{D_4} = K[V_{DD} - (v_\mathrm{x} + v_\mathrm{y})/2 - V_S - V_T]^2. \tag{6.158}$$

The difference of output currents can be expressed as

$$\triangle i_0 = i_0^+ - i_0^- \tag{6.159}$$

$$= (I_{D_2} + I_{D_4}) - (I_{D_1} + I_{D_3}) \tag{6.160}$$

$$= 2K v_\mathrm{x} v_\mathrm{y}. \tag{6.161}$$

Transistors $T_{13} - T_{18}$ are required for the generation of a single-ended version of the output. The linear operation of the multiplier is achieved only for the range of input voltages over which the transistors remain in the saturation region.

6.3 Summary

Several circuit structures are available for the implementation of nonlinear functions, such as the comparison and multiplication. They can be used in a given application depending on the design requirement to be met, that is, low supply voltage, low power consumption, high bandwidth, and low sensitivity to component nonidealities. In most cases, a trade-off exists between the performance characteristics.

6.4 Circuit design assessment

1. **Analysis of a comparator circuit**
   The comparator block diagram, as shown in Figure 6.29(a), consists

of an input amplifier followed by the dynamic and storage latches. Let g_{mn} and g_{mp} be the transconductance of the n-channel and p-channel transistors, respectively. Verify that the dynamic latch of Figure6.29(b) can be described by equations of the form

$$C_{pi}\frac{dV_0(t)}{dt} = g_m V_0(t) \tag{6.162}$$

$$V_0(t) = (V_i \pm V_{off}) \exp\left[\frac{g_m}{C_{pi}}(t - T_S)\right], \tag{6.163}$$

where $g_m = g_{mn} + g_{mp}$, g_m is the equivalent transconductance of a single inverter, C_{pi} is the parasitic capacitance at the input node, T_S is the sampling period, and V_{off} denotes the offset voltage. Estimate the response time of the latch.
Determine the equivalent bit error rate of the comparator.

Hint: Assuming that LSB is the voltage level corresponding to the least-significant bit of a converter, the bit error rate (BER) can be defined as

$$\text{BER} = \frac{\text{SLMR}}{\text{Input amplifier gain} \times (1 \text{ LSB}/2)}, \tag{6.164}$$

SLMR being the input referred metastable region of the storage latch given by

$$\text{SLMR} = \triangle V_0 \exp\left(-\frac{T_S/2 - t_d}{\tau_{dlatch}}\right), \tag{6.165}$$

where $\triangle V_0$ is the initial output metastable region, τ_{dlatch} is the time constant of the dynamic latch, and t_d is the delay between the output signals of the dynamic and storage latches.

2. **Comparator design**
   Complete the design of the comparator shown in Figure 6.30.
   Use SPICE simulations to obtain the time response of the comparator. The signals are connected to the input nodes during the clock phase $\overline{\phi}_1$.

3. **Squaring circuit**
   Consider the circuit shown in Figure 6.31(a) [22], where the transistor T_1 operates in the triode region, while T_2 is biased in the saturation region.
   Assuming that $V_{GS1} = V + V_{GS2}$, show that

$$I - 2K_1V\sqrt{\frac{I}{K_2}} - K_1V^2 = 0 \tag{6.166}$$

   and

$$I = \left(1 + 2\frac{K_1}{K_2} \pm 2\sqrt{\frac{K_1}{K_2}}\sqrt{1 + \frac{K_1}{K_2}}\right)K_1V^2, \tag{6.167}$$

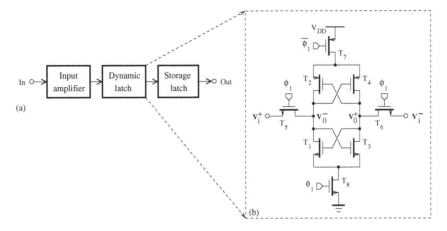

FIGURE 6.29

(a) Comparator block diagram; (b) circuit diagram of a dynamic latch.

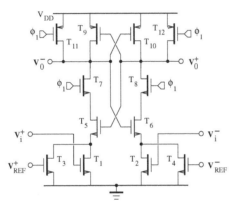

FIGURE 6.30

Circuit diagram of a fully differential comparator.

where $K_1 = \mu_n C_{ox} W_1/(2L_1)$ and $K_2 = \mu_n C_{ox} W_2/(2L_2)$.

The circuit of Figure 6.31(b) can provide the square of a positive input signal. Let $0 < V \le V_m$ and show that

$$V_m = \frac{\sqrt{(1/K_1) + (1/K_2)} - \sqrt{(1/K_2)}}{\sqrt{(1/K_1) + (1/K_2)} + \sqrt{(1/K_3)}}(V_{DD} - V_{Tn} - |V_{Tp}|), \quad (6.168)$$

where $K_3 = \mu_p C_{ox} W_3/(2L_3)$, V_{Tn}, and V_{Tp} are the threshold voltages of the n-channel and p-channel transistors, respectively.

The operating range can be defined by noting that the transistor T_3 will be shut down if

$$V_{GS1} = V_{SG3} = V_{DD}, \quad (6.169)$$

where

$$V_{GS1} = V_{Tn} + \sqrt{(I/K_1) + (I/K_2)} \qquad (6.170)$$

and

$$V_{SG3} = |V_{Tp}| + \sqrt{\frac{I}{K_3}} \,. \qquad (6.171)$$

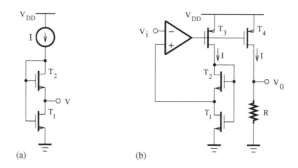

(a)   (b)

FIGURE 6.31
(a) Two transistor circuit; (b) squaring circuit.

4. **Multiplier design**

In the circuit of Figure 6.32 [27], the transistors operate in the triode region. Let I_{D_i} $(i = 1, 2, 3, 4)$ be the current flowing through the

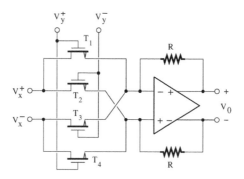

FIGURE 6.32
Multiplier circuit based on four transistors.

transistor T_i. Write the current $\triangle i = (I_{D_1} + I_{D_3}) - (I_{D_2} + I_{D_4})$ in the form

$$\triangle i = 2K[(V_x^+ - V_x^-)(V_y^+ - V_y^-)$$
$$- (V_P - V_N)(V_y^+ + V_y^-) + 2V_T(V_P - V_N) + V_P^2 - V_N^2], \qquad (6.172)$$

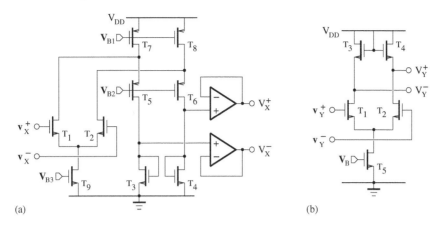

(a) (b)

FIGURE 6.33
(a) Level shifter with a buffer for the input nodes X; (b) level shifter for the input nodes Y.

where V^+ and V^- are the voltages at the noninverting and inverting nodes of the amplifier, respectively.

What condition must be met to realize a multiplier?

Show that

$$V_0 = \frac{R\Delta i}{1 + 1/A},\qquad(6.173)$$

where A is the amplifier dc gain.

The common-mode characteristic of the structure shown in Figure 6.32 can be improved by connecting two level shifters (see Figure 6.33) at its inputs.

Explain why buffers are required to drive the X inputs.

Verify that the gain of level shifters will be unity if the design requirement, $V_{GS_1} - V_{GS_2} = \pm(V_{GS_4} - V_{GS_3})$, is met.

5. **Multiplier circuit**

   The circuit shown in Figure 6.34 [30] is the generalized structure for the implementation of multiplier based on the quarter-square technique. The bias current is represented by I_B and V_I is the dc component of the input voltage. Transistors $T_1 - T_4$ operate in the saturation region and are designed with the same transconductance parameter, $K = \mu_n(C_{ox}/2)(W/L)$, and threshold voltage, V_T.

   Show that $\Delta i = i_0^+ - i_0^- = 2KV_\mathrm{x}V_\mathrm{y}$, for any numbers p, q, and r.

6. **Multiplier based on two squaring circuits**

   The squaring circuit of Figure 6.35 is based on unbalanced differential transistor pairs. The transistors operate in the saturation region and are sized so that $(W_2/L_2) = (W_4/L_4) = \alpha(W_3/L_3) = \alpha(W_3/L_3)$.

   Estimate the differential output current $\Delta i_0 = i_0^+ - i_0^-$ as a function

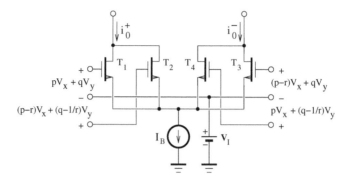

FIGURE 6.34
Generalized multiplier core based on the quarter-square technique.

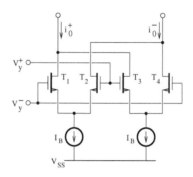

FIGURE 6.35
Squaring circuit based on four transistors.

of the input voltage $V_i = V_i^+ - V_i^-$.

Based on the relation $(V_x + V_y)^2 - (V_x - V_y)^2 = 4V_x V_y$, design a multiplier circuit.

<u>Hint:</u> The output current, $\triangle i$, of a single unbalanced differential pair connected to the input voltage, V_i, is given by

$$\triangle i = \begin{cases} i - \dfrac{1-\alpha}{1+\alpha} I_B, & \text{for} \quad -\sqrt{\dfrac{I_B}{K}} < V_i < \sqrt{\dfrac{I_B}{\alpha K}} \\ I_B \operatorname{sign}(V_i), & \text{for} \quad V_i \leq -\sqrt{\dfrac{I_B}{K}}, \quad \sqrt{\dfrac{I_B}{\alpha K}} \leq V_i, \end{cases} \tag{6.174}$$

where

$$i = \frac{2\alpha(1-\alpha)KV_i^2 + 4\alpha\sqrt{KI_B}V_i\sqrt{1+\alpha-\alpha KV_i^2/I_B}}{(1+\alpha)^2} \tag{6.175}$$

and K is the transconductance parameter of the transistors.

Bibliography

[1] R. Gregorian, *Introduction to CMOS op-amps and comparators*, New York, NY: John Wiley & Sons, 1999.

[2] P. E. Allen and D. R. Holberg, *CMOS analog circuit design*, 2nd ed., New York, NY: Oxford University Press, 2002.

[3] R. Wang ang R. Harjani, "Partial positive feedback for gain enhancement of low-power CMOS OTAs," *Analog Integrated Circuits and Signal Processing*, vol. 8, no 1, pp. 21–35, July 1995.

[4] K. B. Ohri and M. J. Callahan, Jr., "Integrated PCM codec," *IEEE J. of Solid-State Circuits*, vol. 14, pp. 38–46, Feb. 1979.

[5] D. J. Allstot, "A precision variable-supply CMOS comparator," *IEEE J. of Solid-State Circuits*, vol. 17, pp. 1080–1087, Dec. 1982.

[6] B. J. Hosticka, W. Brockherde, U. Kleine, and R. Schweer, "Design of nonlinear analog switched-capacitor circuits using building blocks," *IEEE Trans. on Circuits and Systems*, vol. 31, pp. 354–368, April 1984.

[7] M. Matsui, H. Hara, Y. Uetani, L.-S. Kim, T. Nagamatsu, Y. Watanabe, A. Chiba, K. Matsuda, and T. Sakurai, "A 200 MHz 13 mm^2 2-D DCT macrocell using sense-amplifying pipeline flip-flop scheme," *IEEE J. of Solid-State Circuits*, vol. 29, pp. 1482–1490, Dec. 1994.

[8] B. Nikolić, V. G. Oklobdžija, V. Stojanović, W. Jia, J. K.-S. Chiu, and M. M.-T. Leung, "Improved sense-amplifier-based flip-flop: Design and measurements," *IEEE J. of Solid-State Circuits*, vol. 35, pp. 876–884, June 2000.

[9] A. Yukawa, "A CMOS 8-bit high-speed A/D converter IC," *IEEE J. of Solid-State Circuits*, vol. 22, pp. 775–779, June 1985.

[10] B. P. Brandt, D. E. Wingard, and B. A. Wooley, "Second-order sigma-delta modulation for digital-audio signal acquisition," *IEEE J. of Solid-State Circuits*, vol. 26, pp. 618–627, April 1991.

[11] T. Kobayashi, K. Nogami, T. Shirotori, and Y. Fujimoto, "A current-controlled latch sense amplifier and a static power-saving input buffer for low-power architecture," *IEEE J. of Solid-State Circuits*, vol. 28, pp. 523–527, April 1993.

[12] G. M. Yin, F. Opt Eynde, and W. Sansen, "A high-speed CMOS comparator with 8-b resolution," *IEEE J. of Solid-State Circuits*, vol. 27, pp. 208–211, Feb. 1992.

[13] S. Park and M. P. Flynn, "A regenerative comparator structure with integrated inductors," *IEEE Trans. on Circuits and Systems*, vol. 53, pp. 1704-1711, Aug. 2006.

[14] J. Robert, G. C. Temes, V. Valencic, R. Dessoulavy, and P. Deval, "A 16-bit low-voltage CMOS A/D converter," *IEEE J. of Solid-State Circuits*, vol. 22, pp. 157–163, April 1987.

[15] B.-S. Song, S.-H. Lee, and M. F. Tompsett, "A 10-b 15-MHz CMOS recycling two-step A/D converter," *IEEE J. of Solid-State Circuits*, vol. 25, pp. 1328–1338, Dec. 1990.

[16] B. Gilbert, "A precise four-quadrant multiplier with subnanosecond response," *IEEE J. of Solid-State Circuits*, vol. 3, pp. 365–373, Dec. 1968.

[17] J. N. Babanezhad and G. C. Temes, "A 20-V four-quadrant CMOS analog multiplier," *IEEE J. of Solid-State Circuits*, vol. 20, pp. 1158–1168, Dec. 1985.

[18] Y. H. Kim and S. B. Park, "Four-quadrant CMOS analogue multiplier," *Electronics Letters*, vol. 28, no. 7, pp. 649–650, March 1992.

[19] C. W. Kim and S. B. Park, "New four-quadrant CMOS analog multiplier," *Electronics Letters*, vol. 23, no. 24, pp. 1268–1270, Nov. 1987.

[20] J. S. Peña-Finol and J. A. Connelly, "A MOS four-quadrant analog multiplier using the quarter-square technique," *IEEE J. of Solid-State Circuits*, vol. 22, pp. 1064–1073, Dec. 1987.

[21] R. R. Torrance, T. R. Viswanathan, and J. V. Hanson, "CMOS voltage to current transducers," *IEEE Trans. Circuits and Systems*, vol. CAS-32, pp. 1097–1104, Nov. 1985.

[22] I. M. Filanovsky and H. P. Baltes, "Simple CMOS square-rooting and squaring circuits," *IEEE Trans. on Circuits and Systems*, vol. 39, pp. 312–315, April 1992.

[23] B.-S. Song, "CMOS RF circuits for data communications applications," *IEEE J. of Solid-State Circuits*, vol. 21, pp. 310–317, April 1986.

[24] A. Nedungadi and T. R. Viswanathan, "Design of linear CMOS transconductance elements," *IEEE Trans. Circuits and Systems*, vol. CAS-31, pp. 891–894, Sept. 1984.

[25] K. Kimura, "Some circuit design techniques for low-voltage analog functional elements using squaring circuits," *IEEE Trans. on Circuits and Systems*, vol. 43, pp. 559–576, July 1996.

[26] *HSPICE Users' Guide*, Campbell, CA: Meta-Software, Inc., Feb. 1996.

[27] K. Bult and H. Wallinga, "A CMOS four-quadrant analog multiplier," *IEEE J. of Solid-State Circuits*, vol. 21, pp. 430–435, June 1986.

[28] K. Bult, "Analog CMOS square-law circuits," Ph.D. dissertation, University of Twente, The Netherlands, 1988.

[29] Z. Wang, "A CMOS four-quadrant analog multiplier with single-ended voltage output and improved temperature performance," *IEEE J. of Solid-State Circuits*, vol. 26, pp. 1293–1301, Sept. 1991.

[30] K. Kimura, "An MOS four-quadrant analog multiplier based on the multitail technique using a quadritail cell as a multiplier core," *IEEE Trans. on Circuits and Systems*, vol. 42, pp. 448–454, Aug. 1995.

7

Continuous-Time Circuits

CONTENTS

Real-world data are generally converted by sensors into analog electrical signals corrupted by noise. In this case, the extraction of useful information, which can be displayed by actuators, requires the use of complex algorithms, whose efficient implementation exploits the programmability and flexibility of digital circuits. The suitable level and representation of sensor output information are provided by signal conditioning and interface structures. In communication systems, the tasks of signal processing generally encountered in the front-end and back-end sections of a digital signal processor, namely, amplification of the desired channel to the full-scale of the data converters, automatic gain control, and filtering to remove the interference of the adjacent channels are preferably implemented using continuous-time (CT) circuits, which offer the advantages of high-speed operation and low-power consumption. Other applications include anti-aliasing and smoothing filters, channel equalization in magnetic disk drives and high-speed data links.

Due to factors such as the fabrication tolerance, temperature variation and aging, the values of components can drift to about 10% to 50% from their nominal specifications. The automatic tuning scheme included in CT filters provides a means to solve this problem. When the requirement of a high dynamic range results in a large power consumption and the noise level can be reduced only at the price of a large chip, it is common to adopt a solution based on multiple integrated circuits (ICs) and discrete components instead of a monolithic IC. With the down-scaling of MOS transistors into the sub-micrometer regime, analog and digital circuits may share the same die on a single chip in mixed-signal design. The result is the onset of several problems related to the signal integrity, substrate noise, crosstalk, interconnect parasitic impedances, and electromagnetic interference. By increasing the dynamic range, differential architectures can reduce the sensitivity to some of these effects and supply voltage variations.

In addition to a survey of various CT building blocks (integrator, summer, gain stage), conventional design techniques of CT circuits will be addressed.

7.1   Wireless communication system

A transceiver consists of a receiver and transmitter. It includes the antenna, and radio frequency (RF), intermediate frequency (IF), and baseband and bit-stream processing functions. This partition is justified by the large change in bandwidth due to the decimation or interpolation within each section and the flexibility and cost of the hardware implementation. The RF section, which is generally implemented with gallium arsenide (GaAs), silicon germanium (SiGe), indium phosphide (InP), and bi-complementary metal-oxide semiconductor (BiCMOS), performs the frequency conversion.

In the receiver, the RF signal delivered by a low-noise amplifier (LNA) is down-converted into an IF waveform, which is then sent to the demodulator. It should be noted that the incoming signal at the antenna can be lower than a half microvolt root-mean square (μV_{rms}) and is modulated in the gigahertz (GHz) frequency range.

The transmitter follows the reverse path. The IF of the modulator output signal is up-converted into RF and a power amplifier (PA) is required to adjust the level of the signal to be emitted by the antenna.

Basically, a communication system can be reduced to a channel decoder, source decoder, source encoder, and channel encoder.

The trend in software defined radios [1] is to move the analog-to-digital converter (ADC) and digital-to-analog converter (DAC) closer to the antenna so that more functions can be realized digitally. Such an approach is necessary to meet the requirements (high degree of adaptation, easy reconfiguration) of a multi-standard communication environment based on multi-band antenna and RF conversion. It relies on the use of high-speed and high-accuracy data converters. Figure 7.1 shows the distribution of converter resolutions versus the signal bandwidth for some common applications.

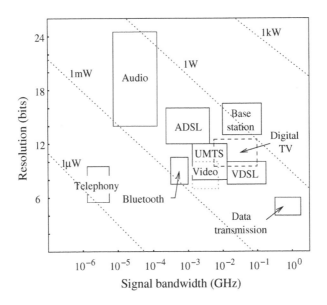

FIGURE 7.1
Specifications of CMOS data converters.

The high-speed attribute of submicrometer CMOS process can offer the opportunity for a single-chip implementation of transceivers [2,3], which actually require costly RF discrete components. To achieve the complete integration of such a system, suitable architectures and circuit techniques are required to overcome the limitations inherent in low-Q components available in CMOS IC technology. Specifically, RF devices involve complex noise, intermodulation, and electromagnetic interference effects.

7.1.1 Receiver and transmitter architectures

RF transceivers find applications in bidirectional communication systems. They include a receiver section, which down-converts the detected RF signal to a baseband frequency (or a low frequency around 0 Hz), and a transmitter section, which up-converts the baseband signal to a radio (or very high) frequency. Transceivers should be designed to achieve a low power consumption and to feature a high IC density in order to be suitable for portable devices. Other criteria such as the number of off-chip components and cost may play an important role in the transceiver architecture choice.

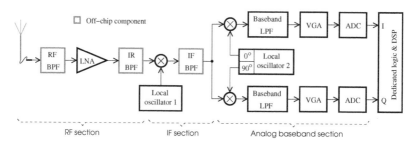

FIGURE 7.2

Block diagram of a superheterodyne receiver.

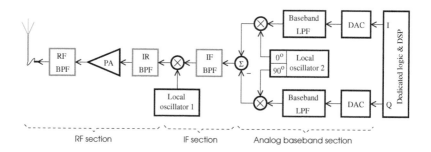

FIGURE 7.3

Block diagram of a superheterodyne transmitter.

The superheterodyne architecture [4,5] is commonly used due to its capacity to detect a low-level RF signal in the presence of strong interferers. Note that the term "hetero" means "different," while "dyne" stands for "power." Hence, "heterodyne" is related to the fact that the conversion to an intermediate frequency is achieved by mixing a signal frequency with a locally generated frequency; "super" comes from the fact that these frequencies may be above the audible spectrum.

The block diagram of a superheterodyne receiver is shown in Figure 7.2. The RF bandpass filter immediately after the antenna is used to select the desired RF band. The low-noise amplifier (LNA) is designed to boost the signal level while introducing as little of its own noise as possible. The resulting signal is passed through an image-reject bandpass filter, which also attenuates the LNA noise contribution present in the image band. The down-conversion to an intermediate frequency, which is equal to the difference between the frequency of the incoming RF signal and the frequency of the local oscillator, is then achieved by the first mixer and IF bandpass filter. The small transition band of each of the IF bandpass filtersr needed for channel selection results in the requirement of a high Q-factor usually satisfied by using an off-chip surface acoustic wave or ceramic circuits. The choice of the IF is determined by the trade-off to be achieved between image rejection and channel selection, or say, between the sensitivity and selectivity of the receiver. The down-conversion from the IF to the baseband frequency is realized by quadrature mixers followed by lowpass filters. The frequency of the local oscillator that feeds the second mixer section should be equal to the IF. The last section of the receiver front end includes variable gain amplifiers (VGAs) and ADCs. It is placed just before the digital signal processor (DSP), which performs the digital demodulation and decoding.

At the system level, the superheterodyne transmitter is configured to realize the reverse operation of the corresponding receiver. The block diagram of a superheterodyne transmitter is depicted in Figure 7.3. A quadrature mixer followed by an IF filter is used to translate the baseband signal, which is provided by the DAC, to an intermediate frequency. The IF-to-RF conversion, which then follows, is achieved with the help of another mixer and filter stage. Finally, the power amplifier (PA) increases the signal level and the RF filter transmits only the signal components satisfying the spectral mask requirements.

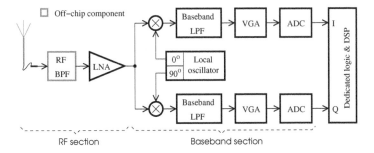

FIGURE 7.4
Block diagram of a homodyne receiver.

Homodyne architectures [6] for receivers and transmitters, as illustrated in Figures 7.4 and 7.5, respectively, have the advantage of eliminating many off-chip components in the signal paths. They are sometimes referred to as direct-conversion or zero-IF receivers and transmitters. Here, the conversion

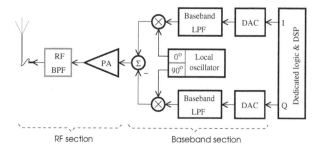

FIGURE 7.5
Block diagram of a homodyne transmitter.

from the radio frequency to the baseband frequency, and vice versa, is achieved using only one mixer stage. The word "homo" implies "same;" this is equivalent to having an identical frequency for the signal of interest and local oscillator signal. Even though the IF frequency is zero, the mirrored image of the desired signal will be superimposed on the down-converted signal, which is generated by the mixer. The image problem is solved by performing the frequency conversion in quadrature. The amplitude and phase of the desired signal can readily be determined from the in-phase and quadrature components. In homodyne architectures, the channel select filtering at baseband is simply performed by lowpass filters, suitable to monolithic IC integration. Hence, the requirements associated with different communication standards can be met using programmable filters and high-dynamic range data converters. Overall, this approach is excellent at saving cost, die area, and power consumption. However, the effects of dc offset, device flicker noise, even-order harmonics, and local-oscillator leakage can critically limit the accuracy of the signal detection.

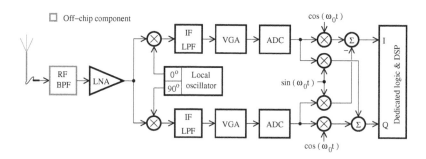

FIGURE 7.6
Block diagram of a low-IF receiver.

Low-IF architectures exploit the advantages of both heterodyne and homodyne structures. They include quadrature frequency conversions that can be both done in the analog domain or be split into analog and digital stages.

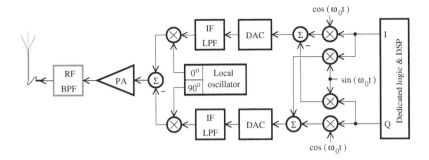

FIGURE 7.7
Block diagram of a low-IF transmitter.

However, the realization of data conversions at the IF instead of the baseband may result in a power consumption increase and could make the overall performance sensitive to clock jitter, distortion, and noise errors. Typically, the IF should be a half to a few channel spacings from dc.

In the low-IF receiver of Figure 7.6, the RF signal is first translated to a low IF, before being down-converted to baseband frequency in the digital domain. Conversely, the low-IF transmitter, as depicted in Figure 7.7, uses both digital and analog steps to perform the up-conversion from the baseband frequency to RF. The image rejection is achieved by summing the output signals provided by a pair of quadrature mixers such that image-band signals ideally cancel out while the desired signals add together coherently. The use of harmonic rejection mixers excludes the need for discrete IF filters, thus making low-IF architectures well suited for single-chip integration than superheterodyne structures. In contrast to direct-conversion architectures, low-IF structures use local oscillators operating at frequencies that are lower than that of the incoming RF signal, to reduce the LO re-transmission, thereby attenuating the dc offset level. It should be noted that the static errors associated with the baseband section can generally be canceled by suitable calibration techniques.

7.1.2 Frequency translation and quadrature multiplexing

Frequency translations are generally used in transmitters to reduce the cost associated with the processing of radio-frequency signals and to meet the transmission specifications. Basically, the frequency translation along with sideband modulation and demodulation relies on the multiplication of two signals. Given a carrier $\cos(\omega_c t)$, with the angular frequency ω_c, the modulated signal can be derived from the waveform $m(t)$ containing the message to be transmitted, as

$$x(t) = m(t)\cos(\omega_c t). \tag{7.1}$$

The Fourier transform of $x(t)$ is

$$X(\omega) = \frac{1}{2}[M(\omega - \omega_c) + M(\omega + \omega_c)], \tag{7.2}$$

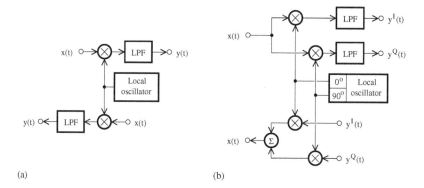

FIGURE 7.8
Functional block diagram of (a) basic and (b) quadrature mixers.

where $M(\omega)$ is the spectrum of the message signal $m(t)$. Assuming that $\omega = 0$, the spectrum of $m(t)$ is shifted to the left and right of the origin by ω_c. This process, known as double-sideband modulation (see Figure 7.8), exhibits a transmission bandwidth that is two times the bandwidth ω_B of $m(t)$. To recover the message signal by the demodulation, the spectra located at $\pm\omega_c$ should not overlap as it is the case for $\omega_c \geq \omega_B$. The multiplication of $x(t)$ by the carrier provides a signal of the form

$$y(t) = x(t)\cos(\omega_c t)$$
$$= \frac{1}{2}m(t)[1 + \cos(2\omega_c t)]. \tag{7.3}$$

The spectrum of $y(t)$ is given by

$$Y(\omega) = \frac{1}{2}M(\omega) + \frac{1}{4}[M(\omega - 2\omega_c) + M(\omega + 2\omega_c)]. \tag{7.4}$$

A version of the desired signal, that is, $m(t)/2$, is obtained by eliminating the spectral components with the frequency $2\omega_c$ using a lowpass filter.

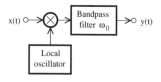

FIGURE 7.9
Block diagram of a frequency translation mixer.

The quadrature multiplexing scheme (see Figure 7.8(b)) makes use of the orthogonality of sine and cosine waves to transmit and receive two different signals simultaneously on the same carrier frequency. The transmission bandwidth remains $2\omega_B$, and the modulated and demodulated signals can be

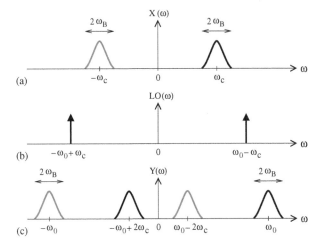

(a)

(b)

(c)

FIGURE 7.10
Plot of spectra of (a) the message, (b) LO signal, and (c) frequency-translated signals for the down-conversion.

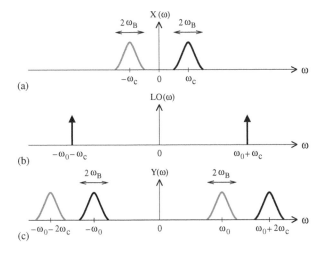

(a)

(b)

(c)

FIGURE 7.11
Plot of spectra of (a) the message, (b) LO signal, and (c) frequency-translated signals for the up-conversion.

written as

$$x(t) = Q(t)\sin(\omega_c t) + I(t)\cos(\omega_c t), \tag{7.5}$$

and

$$y^I(t) = x(t)\cos(\omega_c t)$$
$$= \frac{1}{2}I(t) + \frac{1}{2}[I(t)\cos(2\omega_c t) + Q(t)\sin(2\omega_c t)] \tag{7.6}$$

$$y^Q(t) = x(t) \sin(\omega_c t)$$

$$= \frac{1}{2}Q(t) + \frac{1}{2}[I(t) \sin(2\omega_c t) - Q(t) \cos(2\omega_c t)], \qquad (7.7)$$

respectively. The terms at $2\omega_c$ are suppressed by a lowpass filter, yielding $y^I(t) = I(t)/2$ and $y^Q(t) = Q(t)/2$ for the in-phase and quadrature channels, respectively. In general, the local carriers at the demodulator and modulator should be synchronized in frequency and phase for efficient demodulation.

The frequency translation, which is the process of shifting a signal from one frequency to another, can be realized using mixer. It is then also referred to as frequency mixing. Figure 7.9 shows the block diagram of a frequency translation mixer. Let

$$x(t) = m(t) \cos(\omega_c t). \qquad (7.8)$$

First, the multiplication of the signal $x(t)$ by a locally generated sine wave, $2 \cos(\omega_{LO} t)$, is realized. Thus,

$$y(t) = 2x(t) \cos(\omega_{LO} t)$$

$$= m(t) \cos[(\omega_c - \omega_{LO})t] + m(t) \cos[(\omega_c + \omega_{LO})t]. \qquad (7.9)$$

With the assumption that $\omega_{LO} = \omega_c \pm \omega_0$, we can obtain

$$y(t) = m(t) \cos(\omega_0 t) + m(t) \cos[(2\omega_c \pm \omega_0)t]. \qquad (7.10)$$

A bandpass filter, whose center frequency is ω_0 and bandwidth is equal to or greater than $2\omega_B$, is used to detect the signal component at the angular frequency ω_0. The carrier frequency is then translated from ω_c to ω_0. The plus and minus signs in the ω_{LO} expression correspond to an up-conversion and a down-conversion, respectively. The spectra of signals are shown in Figures 7.10 and 7.11 for a message signal featuring a bandwidth of $2\omega_B$.

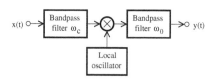

FIGURE 7.12
Block diagram of a frequency translation mixer with the bandpass pre-filter.

By multiplying the input signal and a reference oscillator signal, spectral images can be produced at either the sum or difference frequencies. In particular, interference signals or harmonic components at $\omega_c \pm 2\omega_0$ are also translated to the frequency ω_0, due to the fact that

$$2 \cos[(\omega_c \pm 2\omega_0)t] \cos[(\omega_c \pm \omega_0)t] = \cos(\omega_0 t) + \cos[(2\omega_c \pm 3\omega_0)t] \qquad (7.11)$$

The problem of spectral images can be addressed by placing a high-quality pre-filter at the mixer input, as shown in Figure 7.12. In order to pass only the signal of interest, this filter should have a bandwidth equal to or greater than the bandwidth of the message signal and a center frequency ω_c. However, for wideband applications, the pre-filter may become complex due to the required tunable range.

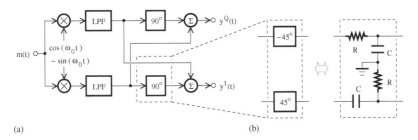

(a)   (b)

FIGURE 7.13
(a) Block diagram of Hartley image-reject down-converter; (b) phase shifter implementation.

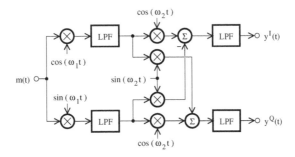

FIGURE 7.14
Block diagram of Weaver image-reject down-converter.

Other mechanisms of dealing with the image rejection problem involve the use of special image-rejection architectures, such as Hartley architecture [7] shown in Figure 7.13, and Weaver architecture [8] depicted in Figure 7.14. The exploitation of quadrature conversions in the separation of the signal of interest from image components makes these architectures more suitable for IC implementations than conventional structures requiring a high-quality filter with sharp cutoff characteristics. In Hartley and Weaver architectures, an opposite phase difference is introduced between the image band signals, which are then canceled out by a summing operation, while the desired signals, whose phase difference remains equal to zero, adds up coherently. The main difference between both architectures is the implementation of the 90° phase shift required on the signal path. Hartley architectures typically use a passive phase shifter, while a second mixing stage is used in the Weaver architecture.

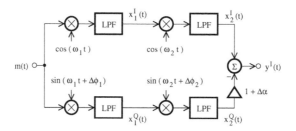

FIGURE 7.15
Equivalent model of a Weaver converter with mismatches.

The image rejection ratio (IRR) of the Weaver converter can be determined in accordance with the equivalent model shown in Figure 7.15. The message signal is assumed to be of the form

$$m(t) = \cos \omega t. \tag{7.12}$$

After each mixer stage, the signal component at the frequency $\omega + \omega_1$ is eliminated by the lowpass filter. The in-phase and quadrature signals, $x_1^I(t)$ and $x_1^Q(t)$, can be respectively expressed as

$$x_1^I(t) = \frac{1}{2}\cos(\omega - \omega_1)t \tag{7.13}$$

and

$$x_1^Q(t) = -\frac{1}{2}\sin[(\omega - \omega_1)t - \triangle\phi_1], \tag{7.14}$$

where $\triangle\phi_1$ denotes the phase error of the first local oscillator. The converter output signal is given by

$$y^I(t) = x_2^I(t) - (1 + \triangle\alpha)x_2^Q(t) \tag{7.15}$$

where

$$x_2^I(t) = \frac{1}{4}\cos(\omega - \omega_1 - \omega_2)t \tag{7.16}$$

and

$$x_2^Q(t) = -\frac{1}{4}\{\cos[(\omega - \omega_1 - \omega_2)t]\cos(\triangle\phi_1 - \triangle\phi_2) \\ - \sin[(\omega - \omega_1 - \omega_2)t]\sin(\triangle\phi_1 - \triangle\phi_2)\}, \tag{7.17}$$

where $\triangle\phi_2$ represents the phase error of the second local oscillator and $\triangle\alpha$ is the gain mismatch. Assuming that $\omega = \omega_{RF}$ for the message of interest, and $\omega = \omega_{IM}$, where $\omega_{IM} = 2\omega_1 - \omega_{RF}$, for the image signal, the IRR (in dB) can be computed as

$$\text{IRR} = 10\log_{10}\left[\frac{|y^I(t)|^2_{\omega=\omega_{RF}}}{|y^I(t)|^2_{\omega=\omega_{IM}}}\right] \tag{7.18}$$

$$= 10\log_{10}\left[\frac{1+(1+\triangle\alpha)^2+2(1+\triangle\alpha)\cos(\triangle\phi_1+\triangle\phi_2)}{1+(1+\triangle\alpha)^2-2(1+\triangle\alpha)\cos(\triangle\phi_1-\triangle\phi_2)}\right]. \tag{7.19}$$

In this case, the image rejection performance can then be improved by making both phase errors equal. Note that an expression identical to Equation (7.19) can also be derived for the IRR of the Hartley converter.

Various nonideal effects are present in a practical transceiver. In the following, we review the most important of them.

- The dc offset is mainly due to transistor mismatches, self-mixing of a strong interference signal coupled to LO nodes by parasitic capacitors, and the in-band signal associated with the leakage of the LO signal into the antenna. It can be mitigated by ac coupling or self-calibration.

- The mismatch between the signal path and the phase error of the LO signals results in undesired spectral components in the signal spectrum. Differential circuit structures and adaptive algorithms can be required to improve the image rejection capability of the transceiver.

7.1.3 Architecture of a harmonic-rejection transceiver

Basic frequency translation techniques (see Figure 7.8) can be combined in different ways to optimize the transceiver performance. Figure 7.16 shows the block diagram of a harmonic-rejection transceiver [9–11]. The transmitter and receiver can operate simultaneously using separate channels, as in a duplex system. Their isolation from the undesirable noise, which can arise at the antenna, is achieved by RF bandpass filters.

A low-noise amplifier (LNA) adjusts the level of the incoming RF signal. It is followed by a quadrature down-conversion stage, which includes two mixers and the lowpass filters necessary to deliver a wideband IF signal and remove all up-converted frequency components. The conversion from IF to baseband is achieved by an image-reject structure. After the signal is processed by the mixers, the useful frequency components are added, while the undesired interference is canceled by subtraction. The first local oscillator (LO) can operate at a fixed frequency, while a tunable frequency characteristic is required for the second LO so that all channel information can be accurately transferred

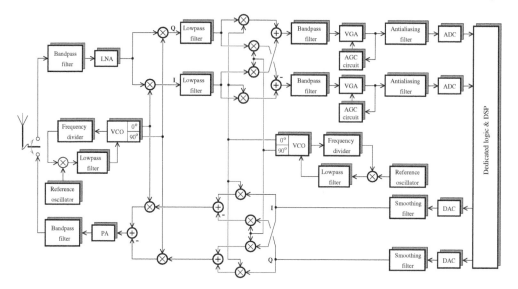

FIGURE 7.16
Block diagram of a harmonic-rejection transceiver.

to the baseband. The use of digitally programmable filters for channel selection provides the flexibility, which is essential in multi-standard receivers. The suitable signal level is fixed by the following variable gain amplifier (VGA), which is steered by an automatic gain control (AGC) circuit. The resulting I and Q signals are then digitized by the ADC and demodulated in the digital domain using dedicated logic circuits and digital signal processor (DSP).

In the transmit path, the baseband signal provided by the DAC is up-converted in two steps. The signal is first translated from the baseband to IF, and the channel tuning is performed using four mixers and a quadrature LO based on a voltage-controlled oscillator (VCO) structure. The next stage achieved the IF-to-RF conversion using an LO operating at a fixed frequency. The specifications of RF filter, which follows the power amplifier (PA), are relaxed due to the image rejection feature of the frequency translation.

In general, any mismatches in the I and Q signal paths and fluctuations of the LO signals limit the image reject capability and therefore the sensitivity of the transceiver. However, a total interference rejection of at least 60 dB can still be achieved provided the IF is chosen high enough and well-suited RF filters are employed.

7.1.4 Amplifiers

Amplifiers are used to boost the amplitude level of a signal. A broad range of circuit architectures is available to meet various amplification requirements. Nevertheless, there are common performance characteristics for almost all amplifiers, such as the linearity, bandwidth, power efficiency, noise figure, and impedance matching. The appropriate amplifier structure for a given application is determined by the requirement of optimizing a specific characteristic and performance trade-offs (e.g., power amplifier, low-noise amplifier).

7.1.4.1 Power amplifier

• Principle and architectures

Amplifiers [12,13] are required in the transmitter to scale the signal amplitude or power to the desired level. For wireless systems, the power is on the order of tens to hundreds of milliwatts (mW) and can be expressed in dBm. That is,

$$P(\text{in dBm}) = 10 \log_{10} \left(\frac{P(\text{in W})}{10^{-3}} \right). \tag{7.20}$$

The efficiency, η, which characterizes the useful power consumption during the amplifier operation, is defined as

$$\eta = \frac{\text{Power delivered to load}}{\text{Power drawn from supply}}. \tag{7.21}$$

The power-added efficiency (PAE) is defined as the difference between the output power, $P_{RF,out}$, and the input power, $P_{RF,in}$, divided by the supply power, P_{DC}, that is,

$$\text{PAE} = \frac{P_{RF,out} - P_{RF,in}}{P_{DC}}. \tag{7.22}$$

An amplifier is then suitable for a given application due to its high PAE.

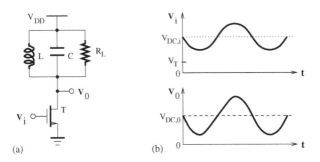

FIGURE 7.17
(a) Circuit diagram of a class A amplifier; (b) input and output voltages.

The operating point of the transistor determines whether an amplifier belongs to class A, B, AB, or C. Given a sinusoidal input voltage, the transistor conduction in a class A amplifier (see Figure 7.17(a)) should be guaranteed for the overall signal period. The value of the input bias, $V_{DC,i}$, as shown in Figure 7.17(b), is chosen so that the maximum swing of the input signal is kept above the threshold voltage necessary to maintain the transistor on. The average value of the output voltage, $V_{DC,0}$, is generally near V_{DD} and the output peak-to-peak swing is on the order of V_{DD}. Because the maximum power delivered to the load is expressed as $V_{DD}I_D/2$, and the power drawn from the supply voltage is of the form, $V_{DD}I_D$, where I_D is the transistor drain current, the maximum efficiency is 50%. This value is reduced in practical implementations due to the nonzero drain-source saturation voltage and the extra power dissipated by parasitic resistors. By consuming dc power regardless of the input signal level, class A amplifiers may appear to be inefficient for some applications.

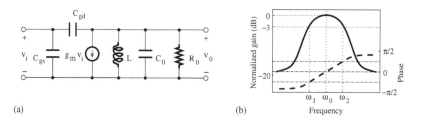

(a)   (b)

FIGURE 7.18
(a) Small-signal equivalent model of the class A amplifier; (b) frequency response characteristics.

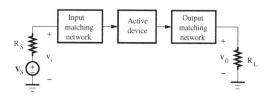

FIGURE 7.19
Generic block diagram of a power amplifier.

A class A tuned amplifier, as shown in Figure 7.17(a), is often used in applications where a narrow band of frequencies around a center frequency is of interest. The small-signal equivalent model is depicted in Figure 7.18(a). It is assumed that R_0 and C_0 denote the total resistance and capacitance at the output node, respectively. Applying Kirchhoff's current law at the output node gives

$$(V_0 - V_i)sC_{gd} + g_m V_i + V_0(G_0 + sC_0 + 1/sL) = 0, \qquad (7.23)$$

where $G_0 = 1/R_0 = G_L + g_{ds}$, $C_0 = C_L + C_{db}$, and G_L and C_L are the load conductance and capacitance, respectively. This last equation can be solved for the voltage gain, which is expressed as

$$A(s) = \frac{V_0(s)}{V_i(s)} = \frac{sC_{gd} - g_m}{G_0 + s(C_0 + C_{gd}) + 1/sL}. \qquad (7.24)$$

Assuming that $g_m \gg \omega C_{gd}$ and $C_0 \gg C_{gd}$, it can be shown that

$$A(s) = \frac{V_0(s)}{V_i(s)} \simeq -g_m Z_0(s), \qquad (7.25)$$

where the output impedance, Z_0, is given by

$$Z_0(s) = \frac{1}{G_0 + sC_0 + 1/sL} = \frac{1}{G_0 + C_0(s + \omega_0^2/s)}. \qquad (7.26)$$

At the center (or resonant) frequency, $\omega_0 = 1/\sqrt{LC_0}$, the gain is reduced to

$$A_0 = -g_m R_0. \qquad (7.27)$$

The frequencies at which $|A(\omega)| = |A_0|/\sqrt{2}$, or the half-power frequencies, can be found as

$$\omega_1 = -\frac{1}{2R_0 C_0} + \sqrt{\omega_0^2 + \frac{1}{4R_0^2 C_0^2}} \qquad (7.28)$$

and

$$\omega_2 = \frac{1}{2R_0 C_0} + \sqrt{\omega_0^2 + \frac{1}{4R_0^2 C_0^2}}. \qquad (7.29)$$

Note that $\omega_0 = \sqrt{\omega_1 \omega_2}$. The -3 dB bandwidth of the amplifier is given by

$$BW = \omega_2 - \omega_1 = \frac{1}{R_0 C_0}. \qquad (7.30)$$

The quality factor of the amplifier is defined as

$$Q = \frac{\omega_0}{BW} = \omega_0 R_0 C_0 = \frac{R_0}{L\omega_0}. \qquad (7.31)$$

The frequency response characteristics are illustrated in Figure 7.18(b). It can be observed that the phase is zero for $\omega = \omega_0$, and equal to $-\pi/4$ and $\pi/4$ for $\omega = \omega_1$ and $\omega = \omega_2$, respectively.

Considering a situation in which a class A amplifier with 1.5-V supply voltage should transmit a power of 500 mW, the output resistance will be given by

$$R_0 = \frac{V_{DD}^2}{2P_{max}} = \frac{1.5^2}{2 \times 0.5} = 2.25 \, \Omega \qquad (7.32)$$

Because the antenna resistance is generally chosen to be 50 Ω, an impedance matching network should be inserted between the amplifier and antenna to transform the load resistance to the required value. Hence, a high output power can be delivered without having to increase the value of the supply voltage. Figure 7.19 shows the generic block diagram of a power amplifier. A matching network is ideally lossless, and it is composed of inductors and capacitors.

For the untuned and tuned amplifiers shown in Figures 7.20(a) and (b), respectively, the transistor can be biased for operation in class A, B, or C mode. In a class B amplifier, the transistor is biased such that it operates in the linear region for a half signal cycle and is in the cutoff region during the next half cycle. The output signal of a class AB amplifier is available for a duration between the half and full period. The class C operation is achieved by biasing the transistor to have a conduction for less than a half signal period. Due to the distortion of the output waveform, this mode of operation is mainly used in tuned amplifiers (see Figure 7.20), where a pulsed current can be filtered to extract the fundamental frequency component. In both cases, the amplification is sustained by the current drawn from the supply voltage using a radio-frequency choke (RFC), which is an inductor designed to block high-frequency (or RF) signals while passing signal components at the low frequency (or dc). Generally, the output power can be increased by reducing the conduction time of the transistor for a given input power level. However, the device matching constraint required to get a linear reproduction of the input waveform can become difficult to meet.

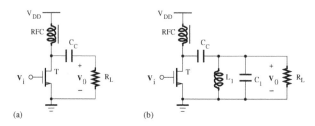

FIGURE 7.20
Circuit diagrams of (a) untuned and (b) tuned amplifiers.

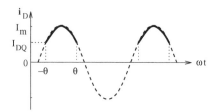

FIGURE 7.21
Representation of the drain current waveform.

Assuming that the conduction angle of the transistor, as shown in Figure 7.21, is such that $-\theta < \omega t < \theta$, the *dc* component of the drain current can be obtained as

$$I_{DC} = \frac{1}{2\pi} \int_{-\theta}^{\theta} i_D d\omega t, \tag{7.33}$$

where

$$i_D = I_m \cos \omega t - I_{CQ} = I_m \cos \omega t - I_m \cos \theta. \tag{7.34}$$

Because the cosine function is even, it can be shown that

$$
\begin{aligned}
I_{DC} &= \frac{I_m}{2\pi} \int_{-\theta}^{\theta} (\cos \omega t - \cos \theta) d\omega t \\
&= \frac{I_m}{\pi} \int_{0}^{\theta} (\cos \omega t - \cos \theta) d\omega t \\
&= \frac{I_m}{\pi} (\sin \theta - \theta \cos \theta).
\end{aligned}
\tag{7.35}
$$

The dc power consumption can be written as

$$
P_{DC} = V_{DD} I_{DC} = \frac{V_{DD} I_m}{\pi} (\sin \theta - \theta \cos \theta).
\tag{7.36}
$$

The fundamental component of the drain current is given by

$$
\begin{aligned}
I_{D0} &= \frac{1}{2\pi} \int_{-\theta}^{\theta} i_D \cos \omega t \, d\omega t \\
&= \frac{I_m}{\pi} \int_{0}^{\theta} (\cos \omega t - \cos \theta) \cos \omega t \, d\omega t = \frac{I_m}{2\pi} (2\theta - \sin 2\theta).
\end{aligned}
\tag{7.37}
$$

The maximum power delivered to the output load can then be expressed as

$$
P_{0,max} = \frac{1}{2} V_{DD} I_{D0} = \frac{V_{DD} I_m}{4\pi} (2\theta - \sin 2\theta).
\tag{7.38}
$$

Thus, the maximum efficiency of the amplifier is of the form

$$
\eta_{max} = \frac{P_{0,max}}{P_{DC}} = \frac{2\theta - \sin 2\theta}{4(\sin \theta - \theta \cos \theta)}.
\tag{7.39}
$$

Equation (7.39) can be used for amplifiers of the class A, B, AB, or C. Note that $\theta = \pi$ for a class A amplifier, $\theta = \pi/2$ for a class B amplifier, $\pi/2 < \theta < \pi$ for a class AB amplifier, and $\theta < \pi/2$ for a class C amplifier. The efficiency increases as the conduction angle decreases.

In the case where the input signal is constant-envelope modulated such that the amplitude remains fixed and the frequency or phase is time varying (e.g., pulse-width modulated signals, sigma-delta modulated signals, etc.), a better approach to achieve an optimal drain voltage waveform is to rely on a transistor operating as a switch. The input signal then synchronizes the switching times of the transistor so that the frequency and phase information is transferred to the output. Switching-mode power amplifiers of class D, E, and F can feature a higher efficiency than structures based on transistors operating in the linear region. They can exhibit an efficiency almost equal to 100% provided the switching devices are ideal.

The circuit diagram of an ideal class D amplifier is shown in Figure 7.22(a).

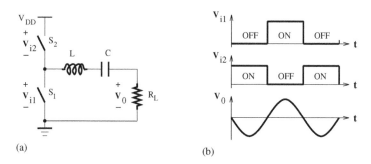

(a)

(b)

FIGURE 7.22
(a) Circuit diagram of an ideal class D amplifier; (b) voltage waveforms.

The voltages V_{i1} and V_{i2} (see Figure 7.22(b)) are assumed to be square wave-forms. The transistors used for the switch implementation operate either in the cutoff region or in the saturation region. They conduct on alternate half periods of the input signal. A series LC circuit tuned at a desired frequency is necessary to recover the output signal, which will be a sinusoid. Due to the nonzero on-resistance and the parasitic capacitances of the transistor, the maximum attainable efficiency can be less than 100%. A power loss occurs at the switching transitions due to the discharge of the capacitors connected at the transistor nodes. In practice, this power dissipation is mitigated by intro-ducing a dead time of a few nanoseconds between the turn-on and turn-off of the transistors. The gate voltages are now two sinusoids, which are out of phase by 180°.

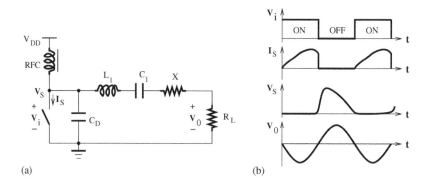

(a)

(b)

FIGURE 7.23
(a) Circuit diagram of an ideal class E amplifier; (b) voltage and current waveforms.

The efficiency can be improved by reducing the power dissipation in the transistor, as is the case in a class E amplifier. The circuit diagram of an ideal class E amplifier is shown in Figure 7.23(a). The switch, whose implementation is based on transistors, is controlled by the input signal and operates with a duty cycle of 50%. The $L_1 C_1$ circuit is supposed to resonate at the frequency of the first harmonic of the input signal. The suitable phase shift between the voltage across the switch and the output voltage is introduced by the reactive element, X. Ideally, the switch power dissipation is zero, because the voltage and current waveforms of the switch, as shown in Figure 7.23(b), do not overlap. In order to avoid the power dissipation due to the discharging of the capacitor C_D when it is connected to the ground, the circuit must be designed such that the voltage across the switch returns to zero with zero slope (that is, $V_s = 0$ and $dV_S/dt = 0$) right before the switch is turned on. All the power from the dc supply voltage can then be transmitted to the output load.

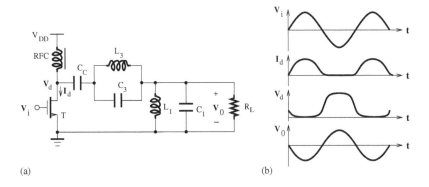

(a) (b)

FIGURE 7.24
(a) Circuit diagram of a class F1 amplifier; (b) voltage and current waveforms.

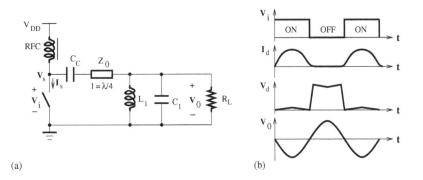

(a) (b)

FIGURE 7.25
(a) Circuit diagram of a class F2 amplifier; (b) voltage and current waveforms.

A class F PA enhances the efficiency by using harmonic resonators in the output network to shape the output waveforms in such a way to minimize the device power dissipation. There are two types of class F amplifiers.

With reference to the class F1 amplifier shown in Figure 7.24, the drain current and voltage of the transistor are processed by two LC circuits, which are connected in series and are tuned to the first and third harmonics of the output signal, respectively. The transistor should be biased in the same way as in a class B PA. However, the drain voltage swing of the class F1 PA is compressed with respect to the fundamental component because the ac component of the drain voltage is equal to the difference between the fundamental and third components. As a result, the class F1 PA exhibits a better power efficiency than a PA of the class B type due to the fact that the compressed drain voltage waveform has less overlap with the drain current and thus less power is dissipated in the device. By increasing the number of harmonics used in the PA, the efficiency can be further improved.

In the case of the class F2 PA depicted in Figure 7.25, the input device is assumed to operate as a switch rather than a transconductor. Here, a quarter wavelength ($\lambda/4$) transmission line acting as an impedance transformer is used instead of a third-order harmonic resonator. At the carrier frequency or the resonance frequency of the LC network, the transmission line loaded by the impedance Z_L exhibits an input impedance of the form, Z_0^2/Z_L, where Z_0 is the characteristic impedance, because its length is equal to $\lambda/4$. The ac contributions superposed to the dc and fundamental components of the current I_s and voltage V_s are associated with even and odd harmonics, respectively, as the resistance at the switch output node is finite for even harmonics and infinite for odd harmonics. Ideally, the current I_s and voltage V_s should be insensitive to the input signal amplitude. This requires that the input signal must be large enough to turn on the transistor. By sizing the switch such that its on-resistance is much smaller than the input impedance of the matching network, the waveforms of the current I_s and voltage V_s are essentially determined by the matching network termination, which can then be designed to minimize the overlapping time between the current I_s and voltage V_s for improved power efficiency.

• Design examples

The circuit diagram of a class AB PA is depicted in Figure 7.26. The PA isolation from the mixer is ensured by the cascode transistors ($T_3 - T_4$) included in the pre-amplification stage, whose outputs are ac coupled to a class AB differential pair. The inductively degenerated output stage, shunt inductors, and coupling capacitors contribute to the fulfillment of the 50-Ω impedance matching requirement. A balun using a pair of two short coupled microstrip lines and tuning capacitors may be required for the conversion of the differential signal into the single-ended one, which can be processed by the following filter and antenna.

The circuit diagram of a class E PA is shown in Figure 7.27. It is based on the differential configuration, which has the advantage of increasing the signal swing and reducing the substrate coupling effects. The overall amplifier consists of a driver and class E amplification stages with cross-coupled switching transistors. The current flowing through an inductive load is also used to control the switching of the other circuit half. Ideally, the class E amplifier should operate with sharp input pulses having a 50% duty cycle, and the peak voltage on the switch can be higher than three times V_{DD}, putting a limit on

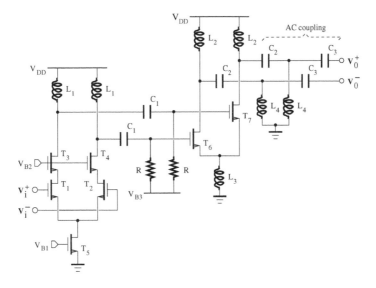

FIGURE 7.26
Circuit diagram of a class AB power amplifier.

the maximum possible supply voltage. To overcome this limitation, the input capacitors of the class E amplifier are tuned out by the dc-feed inductors. The driver is designed to deliver sinusoid switching signals with a peak-to-peak voltage on the order of two times V_{DD}. The 50-Ω output impedance is provided by an output matching network, which involves off-chip capacitors and inductors realized with the bond wires. Note that the self-oscillation of the amplifier in the absence of the input signal can be prevented by switching the common source to the supply voltage. An integrated PA can deliver 1 W with up to 48% PAE.

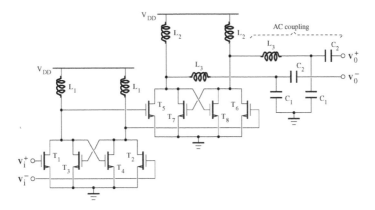

FIGURE 7.27
Circuit diagram of a class E power amplifier.

For the design of efficient integrated PAs, the inductor and its associated parasitic components are dependent upon each other and should be included in an iterative optimization process. A CAD program is then necessary to reduce the effect of loss in the matching network.

7.1.4.2 Low-noise amplifier

• Fundamentals

In the receiver, the low-noise amplifier (LNA) should increase by 10 to 100 times the level of the signal coming from the antenna. It is generally designed to feature a low additive noise and to maintain linearity under large-signal conditions. An estimation of the amplifier contribution to the overall noise is provided by the noise figure, which is given by

$$\text{NF} = 10 \log_{10}(\text{F}). \tag{7.40}$$

where F is the noise factor defined as

$$F = \frac{\text{Overall output noise}}{\text{Output noise due to the source resistor}}. \tag{7.41}$$

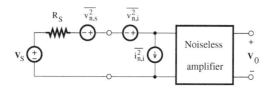

FIGURE 7.28
Amplifier equivalent model including noise sources.

The amplification stage, as shown in Figure 7.28, is assumed to be driven by a voltage source with the resistance R_s, whose noise at the output nodes of a noiseless amplifier model should be proportional to the gain factor. The source resistor gives rise to a thermal noise per unit bandwidth with the mean-square voltage of $\overline{v_{n,s}^2} = 4kTR_s$, where k is the Boltzmann constant and T is the absolute temperature. Let A_v be the amplifier gain. The noise factor, F, can be computed as

$$F = \frac{\overline{v_{n,0}^2}}{A_v \overline{v_{n,s}^2}}, \tag{7.42}$$

where $\overline{v_{n,0}^2}$ is the mean-square output noise voltage estimated with $v_s = 0$. To proceed further,

$$F = 1 + \frac{\overline{v_{n,i}^2}}{4kTR_s} + \frac{\overline{i_{n,i}^2}R_s}{4kT}, \tag{7.43}$$

where $\overline{v_{n,i}^2}$ and $\overline{i_{n,i}^2}$ are the input-referred voltage and current noises per unit bandwidth, respectively.

The amplifier will exhibit a nonlinear transfer function when it processes

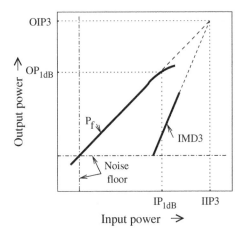

FIGURE 7.29
Plot of P_f and $IMD3$ versus the input power.

input voltages with a large magnitude. That is, two close frequencies, f_1 and $f_2 = f_1 + \triangle f$, contained in a signal can interact and produce third-order intermodulation distortion (IMD3) at $2f_1 - f_2$ and $2f_2 - f_1$. These interference products cannot be easily eliminated by a filter. The output power, P_f, of the signal component at the fundamental frequency and IMD3 versus the input power are represented in Figure 7.29. The IMD3 characteristic increases with a higher slope than P_f. The input-referred intercept point (IIP3) and output-referred intercept point (OIP3) correspond to the location, where the extrapolated IMD3 and the tangent to the P_f curve meet each other. These figures of merit provide a relative comparison of P_f and IMD3 up to the start of the power compression. The output compression point (OP_{1dB}) is defined as the output power level, which is 1 dB lower than it should be by applying the equivalent input level to an amplifier assumed to be linear. The input compression point (IP_{1dB}) corresponds to the input level that produces an output power 1 dB lower than it should be in an amplifier operating linearly. The spurious free dynamic range (SFDR) specifies the range over which the input level must be increased over the minimum detectable signal (MDS) level so that the third-order intermodulation products and the MDS take the same magnitude. The blocking dynamic range (BDR) lies between the MDS and the power compression point. It is determined by the noise and large-signal limitations.

In the design of LNAs, the cascode structure is generally adopted because it exhibits better isolation, improved bandwidth, and higher gain even at millimeter-wave frequencies when compared to architectures based on a single transistor. LNA circuits that are based on the inductively degenerated common source topology [19–21] have demonstrated a relatively low noise figure, and adequate gain, power consumption, and impedance matching for narrowband and wideband applications. The circuit diagram of an LNA based on

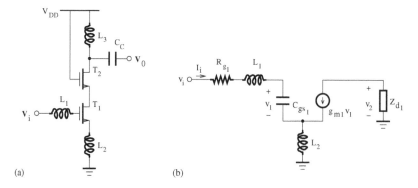

(a)   (b)

FIGURE 7.30
(a) Circuit diagram of a cascode LNA with source degeneration inductor; (b) small-signal equivalent circuit for the calculation of the input impedance.

such a topology is shown in Figure 7.30(a). The coupling capacitor C_C should be sized such that only the ac signal is transferred to the output load while the dc voltage component is blocked. Because the Miller effect of the gate-drain capacitance of T_1 is reduced by the cascode transistor T_2, the small-signal equivalent circuit for the calculation of the input impedance can be obtained as shown in Figure 7.30(b). Using Kirchhoff's voltage law, the input voltage can be derived as

$$V_i = \left(R_{g_1} + sL_1 + \frac{1}{sC_{gs_1}} \right) I_i + sL_2(I_i + +g_{m1}V_1), \qquad (7.44)$$

where $V_1 = I_i/sC_{gs_1}$ and R_{g_1} is the equivalent series resistance available at the gate of T_1. The input impedance is then given by

$$Z_i = \frac{V_i}{I_i} \simeq R_{g_1} + g_{m1}\frac{L_2}{C_{gs_1}} + s(L_1 + L_2) + \frac{1}{sC_{gs_1}}. \qquad (7.45)$$

At the series resonance of the input circuit, $\omega^2 = \omega_0^2 = 1/(L_1 + L_2)C_{gs_1}$, the impedance becomes purely real and equal to $R_{g_1} + g_{m1}L_2/C_{gs_1}$. By carefully dimensioning the inductors L_1 and L_2, and the transistor T_1, the real term of the input impedance can be made equal to 50 Ω.

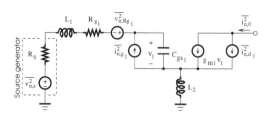

FIGURE 7.31
Small-signal equivalent circuit for the noise analysis.

The small-signal equivalent model of the source degenerated LNA shown in Figure 7.31 is considered for the noise analysis. The noise factor, F, is defined as the ratio of the total output noise power to the output noise power due to the signal source. In the case where the noise contribution due to the load of transistor T_1 can be ignored, we have

$$F = \frac{\overline{i_{n,0s}^2} + \overline{i_{n,0Rg_1}^2} + \overline{i_{n,0g_1}^2} + \overline{i_{n,0d_1}^2} + \overline{i_{n,0T_2}^2}}{\overline{i_{n,0s}^2}},$$

(7.46)

where $\overline{i_{n,0s}^2}$ is the output noise power contribution of the signal source, $\overline{i_{n,0Rg_1}^2}$ denotes the output noise power due to the overall resistance at the gate of T_1, and $\overline{i_{n,0g_1}^2} + \overline{i_{n,0d_1}^2}$ is the output noise power contribution due to the gate and drain noise currents of T_1. To proceed further,

$$F = 1 + F_1,$$

(7.47)

where

$$F_1 = \frac{\overline{i_{n,0Rg_1}^2} + \overline{i_{n,0g_1}^2} + \overline{i_{n,0d_1}^2}}{\overline{i_{n,0s}^2}}.$$

(7.48)

Let G, F, and E be the transfer functions relating the noise sources, $\overline{v_{n,Rg_1}^2}$, $\overline{i_{n,g_1}^2}$, and $\overline{i_{n,d_1}^2}$, respectively, to the output noise current. It can be shown that

$$F_1 = \frac{1}{\overline{v_{n,s}^2}|G(\omega)|^2}\left[\overline{v_{n,Rg_1}^2}|G(j\omega)|^2 + \overline{i_{n,g_1}^2}|F(j\omega)|^2 + \overline{i_{n,d_1}^2}|E(j\omega)|^2 \right.$$

$$\left. + \overline{i_{n,g_1}i_{n,d_1}^*}F(j\omega)E^*(j\omega) + \overline{i_{n,g_1}^*i_{n,d_1}}F^*(j\omega)E(j\omega)\right].$$

(7.49)

For the determination of G, F, and E, it is assumed that the output noise contribution is solely caused by the noise source of interest. Applying the principle of voltage division to determine the ratio of V_1 to either V_s or V_{Rg_1}, we get

$$\frac{V_1}{V_{Rg_1}} = \frac{V_1}{V_s} = \frac{1/sC_{gs_1}}{R_s + R_{g_1} + g_{m1}\dfrac{L_2}{C_{gs_1}} + s(L_1 + L_2) + \dfrac{1}{sC_{gs_1}}}.$$

(7.50)

The output noise current is given by

$$I_{n,0} = g_{m_1}V_1.$$

(7.51)

Assuming that the input impedance matching condition is fulfilled, Equations (7.50) and (7.51) can be solved for

$$G(j\omega_0) = \frac{I_{n,0}(j\omega_0)}{V_s(j\omega_0)} = \frac{I_{n,0}(j\omega_0)}{V_{Rg_1}(j\omega_0)} = \frac{g_{m1}}{\omega_0 C_{gs_1}(R_s + R_{g_1} + g_{m1}L_2/C_{gs_1})}.$$

(7.52)

Similarly, for the noise source i_{n,d_1}, it can be shown that

$$(R_s + R_{g_1} + sL_1 + 1/sC_{gs_1})sC_{gs_1}V_1 = -sL_2(sC_{gs_1}V_1 + g_{m1}V_1 + I_{n,d_1}) \quad (7.53)$$

and

$$I_{n,0} = g_{m_1}V_1 + I_{n,d_1}. \quad (7.54)$$

Combining Equations (7.53) and (7.54) gives

$$E(j\omega_0) = \frac{I_{n,0}(j\omega_0)}{I_{n,d_1}(j\omega_0)} = \frac{R_s + R_{g_1}}{R_s + R_{g_1} + g_{m1}L_2/C_{gs_1}}. \quad (7.55)$$

The equations considered in the case of the noise source i_{n,g_1} are

$$(R_s + R_{g_1} + sL_1)(I_{n,g_1} - sC_{gs_1}V_1) = V_1 + sL_2(sC_{gs_1}V_1 + g_{m1}V_1 - I_{n,g_1}) \quad (7.56)$$

and

$$I_{n,0} = g_{m_1}V_1. \quad (7.57)$$

Using Equations (7.56) and (7.57), the next transfer function can be derived:

$$F(j\omega_0) = \frac{I_{n,0}(j\omega_0)}{I_{n,g_1}(j\omega_0)} = \frac{g_{m1}[R_s + R_{g_1} + j\omega_0(L_1 + L_2)]}{\omega_0 C_{gs_1}(R_s + R_{g_1} + g_{m1}L_2/C_{gs_1})}. \quad (7.58)$$

The input source has a power spectral density

$$\overline{v_{n,s}^2} = 4kTR_s\triangle f, \quad (7.59)$$

and the power spectral density of the noise due to the parasitic resistance at the gate of the transistor T_1 is of the form

$$\overline{v_{n,Rg_1}^2} = 4kTR_{g_1}\triangle f, \quad (7.60)$$

where $\triangle f$ is the noise bandwidth (in Hz), k is Boltzmann's constant, and T denotes the absolute temperature. For MOS transistors, the power spectral density of the gate-induced noise can be written as

$$\overline{i_{n,g_1}^2} = 4kT\delta g_g\triangle f, \quad (7.61)$$

and the power spectral density of the channel noise is

$$\overline{i_{n,d_1}^2} = 4kT(\gamma/\alpha)g_{m1}\triangle f = 4kT\gamma g_{d10}\triangle f, \quad (7.62)$$

where $g_g = \omega_0^2 C_{gs_1}^2/5g_{d10}$, g_{d10} is the conductance of the transistor T_1 when the drain-source voltage is equal to zero, and δ and γ are the coefficients of channel and gate induced noises, respectively. Furthermore,

$$\overline{i_{n,g_1}i_{n,d_1}^*} = c\sqrt{\overline{i_{n,g_1}^2}}\sqrt{\overline{i_{n,d_1}^2}}. \quad (7.63)$$

The parameter F_1 can then be expressed as

$$F_1 = \frac{R_{g_1}}{R_s} + \frac{\gamma}{\alpha}\frac{\chi}{Q_L}\frac{\omega_0}{\omega_T}, \tag{7.64}$$

where

$$\chi = 1 - 2|c|\sqrt{\frac{\delta\alpha^2}{5\gamma}} + \frac{\delta\alpha^2}{5\gamma}(1 + Q_L^2), \tag{7.65}$$

$$Q_L = \frac{\omega_0(L_1 + L_2)}{R_s + R_{g_1}} \simeq \frac{\omega_0(L_1 + L_2)}{R_s} \simeq \frac{1}{\omega_0 C_{gs_1} R_s}, \tag{7.66}$$

and

$$\omega_T = \frac{g_{m1}}{C_{gs_1}}. \tag{7.67}$$

For long channel transistors, $\gamma = 2/3$, $\delta = 4/3$, $\alpha = 1$, and $c = 0.395j$. It can be observed that the gate noise contribution is reduced by decreasing Q_L, while the channel noise contribution is attenuated by increasing Q_L. Therefore, there is an optimum value of Q_L that minimizes the noise figure. That is, $\partial F/\partial Q_L|_{Q_L=Q_{L,opt}} = 0$ at fixed ω_T, and

$$Q_{L,opt} = \sqrt{1 + 2|c|\sqrt{\frac{5\gamma}{\delta\alpha^2}} + \frac{5\gamma}{\delta\alpha^2}}. \tag{7.68}$$

A typical value of $Q_{L,opt}$ is about 1.5 to 3. Generally, a low-Q input-matching network is preferred to make the design less sensitive to variations in the inductance and parasitic capacitances. Furthermore, the noise figure can also be minimized with respect to the width of T_1, which is related to g_{m1} and g_{d10}.

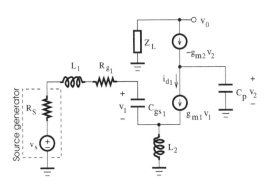

FIGURE 7.32
Small-signal equivalent circuit of the LNA.

With reference to the small-signal equivalent model of the LNA depicted in Figure 7.32, it can be shown that

$$V_2 = sC_p(-g_{m2}V_2 - I_{d_1}) \tag{7.69}$$

and

$$g_{m2}V_2 = V_0/Z_L,\qquad(7.70)$$

where $C_p \simeq C_{gs_2} + C_{sb_2} + C_{db_1}$. The output voltage is then given by

$$V_0 = \frac{-sC_p}{1 + sg_{m2}C_p}g_{m2}Z_L I_{d_1},\qquad(7.71)$$

where $I_{d_1} = G(j\omega_0)V_s$. In practice, the parasitic capacitances connected to the common lower-drain-to-upper-source node can reduce the gain of the input transistor T_1 in such a way that the noise contribution of T_2 becomes relatively significant. They may be lowered by utilizing the dual-gate layout technique [22, 23] based on a transistor with two gates sharing one channel disposed between a source and a drain. Specifically, when T_1 and T_2 are of an identical size, the cascode structure is realized using the primary gate as the input node and biasing the second gate appropriately. Because the gate nearest the drain is ac grounded, it acts as a shield between the input gate and the output drain, thereby minimizing the parasitic coupling capacitance.

Note that a more accurate expression of the noise figure can be derived by taking into account the parasitic capacitances and the noise contribution of T_2. In general, an LNA should be designed with a 50-Ω input impedance, minimum noise contribution, and maximum gain and IIP3. The inductances used in the LNA implementation are generally lower than a few hundred picohenries, making the amplification characteristics sensible to package parasitic components as the frequency is increased.

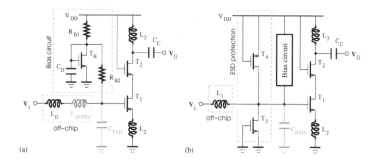

FIGURE 7.33
(a) Implementation of a cascode LNA with source degeneration inductor; (b) LNA implementation having ESD protection.

To provide low-cost devices, RF circuits are realized using packages that are generally designed for low-frequency analog and digital ICs. At high frequencies, the parasitic components of these packages may be the cause of signal attenuation, poor isolation of signal and ground. The effect of the pad capacitance (a few hundred femtofarads) and the lead and bonding wire inductance (a few nanohenries) should then be taken into account in the matching requirement for the input impedance. Figure 7.33(a) shows an implementation of a cascode LNA with source degeneration inductor. The bias circuit for the transistor T_1 includes the resistors, R_{B_1} and R_{B_2}, the decoupling capacitor C_D, and the transistor T_B. The sum of the off-chip inductance, L_G, and the bond wire inductance, L_{BOND}, is equal to the inductance L_1. To overcome the lack of precise on-chip passive elements, the noise figure of the LNA is improved by using a high Q off-chip inductor in the input matching network.

Transistors become more sensitive to electrostatic discharge (ESD) as the oxide thickness is lowered as the result of the CMOS process scaling toward short channel lengths. Figure 7.33(b) shows an LNA implementation having ESD protection. A low resistance path is associated to the input pin such that the ESD current mainly flows through the protection device (e.g., a power clamp). The transistors T_3 and T_4 should have thicker gate oxide than the transistors of the LNA core to exhibit higher threshold and breakdown voltages [14]. However, an increase in the LNA noise level can be caused by the parasitic components introduced by the ESD protection transistors. It is then important to keep the size of these parasitic components very small.

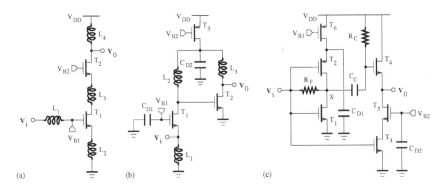

FIGURE 7.34
Circuit diagrams of low-noise amplifiers with (a) middle inductor and inductively degenerated, (b) common gate, and (c) resistive feedback input stages.

The impact of the parasitic capacitances connected to the common lower-drain-to-upper-source node on the cascode LNA performance at high frequencies can also be attenuated using an extra series inductor [15]. With reference to Figure 7.34(a), the inductor L_3 is inserted to create the appropriate series resonance for the compensation of the parasitic capacitance pole. In contrast to the approach where an inductor should be connected in parallel with the parasitic capacitance, this compensation method has the advantage of preventing the use of a large and accurate bypass capacitor. However, the inductor L_3 may provoke a reduction of the real part of the input impedance [16]. Assuming that $g_{m1} = g_{m2} = g_m$, $L_2 = L_3 = L$, $C_{gs_1} \gg C_{gd_1}$, and $g_m^2 L/C_{gs_1}$ is approximately equal to unity, the real part of the input impedance remains positive provided $L < 4C_{gs_1}/g_m^2$.

A common-source LNA exhibits relatively higher gain and a lower noise figure than the common-gate counterpart, which is limited by the absence of any degree of freedom in the choice of the input transistor transconductance that is essentially defined by the impedance-matching condition. However, the common-gate architecture has the highest potential to achieve a wide-band input matching, especially at frequencies greater than 5 GHz. The circuit diagram of an LNA with a common gate input stage [3, 17] is shown in Figure 7.34(b). Because the resistance looking into the source terminal of the input transistor is on the order of $1/g_{m_1}$, the input matching condition, which is determined at the resonance of L_1 with $C_{gs_1} + C_{sb_1}$ and the pad capacitance, is reduced to $R_s \simeq 1/g_{m_1}$. The first stage of the LNA is inductively loaded by L_2. Transistor T_2, along with the inductor L_3, constitutes the second amplification stage, which also improves the output drive capability. The first stage is biased such that ac coupling can be avoided between the two stages of the LNA, thereby improving the amplifier response at higher frequencies. Note that C_{D1} and C_{D2} are decoupling capacitors.

To reduce the chip area, a wideband LNA can replace several LC-tuned LNAs typically used in multi-band and multi-mode narrowband receivers. This design solution can be adopted for analog cable (50 MHz to 850 MHz), satellite system (950 MHz to 2150 MHz), terrestrial digital video broadcasting (450 MHz to 850 MHz), and any application requiring reconfigurability for agile service switching. By using a feedforward noise-canceling technique, which can attenuate the noise and distortion contributions of the input (or matching) transistors, the noise and impedance-matching requirements can be met simultaneously. This allows the circuit components to be sized such that the resulting wideband impedance-matching LNA can exhibit a sufficiently large gain, adequate linearity, and a noise figure well below 3 dB. Figure 7.34(c) shows the circuit diagram of an LNA with a resistive feedback input stage [18]. With the assumption that $g_{m1} + g_{m2} \gg 1/R_{L_1}$ and $1 \gg R_F/R_{L_1}$, the shunt feedback around the CMOS inverter makes the input impedance equal to about $1/(g_{m1} + g_{m2})$. Here, R_{L_1} is the load resistance seen at the inverter output node. By adopting a current source to bias the inverter and connecting the source of T_2 to the ground via the decoupling capacitor C_{D_1}, the effects of supply voltage variations on the gain and input impedance is attenuated. To improve the overall noise figure, a highpass filter, $C_C - R_C$, is inserted between the inverter output and the gate of the transistor T_4. In the output stage, the transistor T_4 operates as a source follower while T_3, which is configured as a common-source stage, provides the gain. The transistor T_5, whose gate is ac coupled by the decoupling capacitor C_{D_2} to the ground, attenuates the Miller effect due to the gate-drain capacitance of T_3 and thus improves input-output isolation. Hence, two feed-forward paths leading to the output node are implemented such that the signal contributions are combined in phase while the noise contribution from each input matching transistor equally counterbalances another. The common-source output stage including the transistors $T_3 - T_4$ provides a gain of about g_{m3}/g_{m4}. Considering the LNA with a source voltage v_s and a source resistance R_S, the superposition principle is used to show that

$$v_0 = v_x - (g_{m3}/g_{m4})v_i \,, \tag{7.72}$$

where $v_i = v_s - R_S i_d$ and $v_x = v_s - (R_S + R_F)i_d$. Hence,

$$v_0 = \left(1 - \frac{g_{m3}}{g_{m4}}\right)v_s + \left[R_S\frac{g_{m3}}{g_{m4}} - (R_S + R_F)\right]i_d \,. \tag{7.73}$$

Ideally, the fluctuations of i_d due to noise is canceled provided $g_{m3}/g_{m4} = 1 + R_F/R_S$. Note that the aforementioned noise canceling approach can also be applied to other LNA architectures.

• Differential LNA architectures

The circuit diagram of an LNA with inductively degenerated input stage [19] is shown in Figure 7.35(a). It exhibits a pseudo-differential configuration. The low noise requirement is met using large input transistors with bias currents on the order of a few milliamperes. The LNA input impedance depends on L_2 and the gate-source capacitance of the transistor, and its value at the resonance is to be matched to 50 Ω. The cascode transistors T_3 and T_4 reduce the Miller effect of the gate-drain capacitors, and improve the amplifier reverse isolation. The output load inductors should resonate with the capacitors connected at

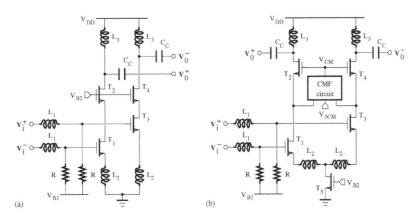

FIGURE 7.35
(a) Circuit diagram of pseudo-differential LNA with inductively degenerated input stage; (b) circuit diagram of a fully differential LNA with inductively degenerated input stage.

the transistor drains. Furthermore, the network consisting of L_3 and C_C is sized to ensure an optimal power transfer to the following stages.

For a better rejection of on-chip interferences and a lower sensitivity to the substrate and supply voltage noises, a differential LNA architecture can be selected. Figure 7.35(b) shows the circuit diagram of a fully differential LNA with inductively degenerated input stage [24]. In order to stabilize the dc bias voltage at the output nodes, the actual common-mode voltage is detected and compared to a given reference voltage by the common-mode feedback circuit. In comparison with a single-ended structure, a differential LNA offers a twofold increase in dynamic range. However, its power consumption is somewhat increased.

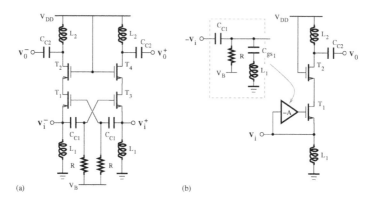

(a)   (b)

FIGURE 7.36
(a) Circuit diagram of pseudo-differential, low-noise amplifier with capacitor cross-coupled input stage; (b) single-ended equivalent model.

A common-source LNA is known to feature better noise performance than a common-gate LNA at operating frequencies well below the transistor transition frequency, but at the cost of a higher power consumption and use of off-chip matching components. The noise figure of the common-gate LNA can be improved without affecting the power consumption by using a pseudo-differential cross-coupled LNA, as shown in Figure 7.36(a). The achieved improvement is due to the transistor transconductance boosting, which is realized by inserting an inverting amplification stage between the source and gate nodes of the input transistors [25], as illustrated in Figure 7.36(b) using a single-ended equivalent model. The effective transconductance looking into the source node is now $(1 + A)g_m$, where A is the gain of the amplification stage inserted between the source and gate and g_m denotes the transconductance of each of the matched transistors T_1 and T_2. Hence, the input matching condition is of the form $R_s \simeq 1/g_m(1 + A)$. The contribution of the drain-induced noise to the amplifier noise factor is then reduced by the factor $1 + A$ when compared to the case of a conventional common-gate LNA.

7.1.5 Mixer

Mixers are important components for transceivers. They are necessary for the RF-to-IF down-conversion and IF-to-RF up-conversion of the signals. Figure 7.37(a) shows the circuit diagram of a single-balanced mixer. Inductive loads are used to minimize the noise level. Transistors T_1 and T_2 should be

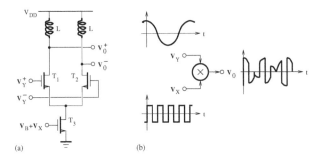

FIGURE 7.37
(a) Circuit diagram of a single-balanced mixer; (b) operation principle of a switching mixer.

sized based on the trade-off to be achieved between the switching time and the conversion gain, which can be degraded by parasitic capacitors. The third intercept point is determined by the overdrive voltage of the transistor T_1.

In communication applications, the single-balanced mixer [26] operates as a switching device to perform the signal multiplication, as shown in Figure 7.37(b). Let V_X be the input RF signal and V_Y be a square wave with amplitude of ± 1 and 50% duty cycle. The RF signal is a sinusoid of the form

$$V_X = A\cos(\omega_{RF}t), \tag{7.74}$$

where A is the signal amplitude. The operating point is set by a bias voltage superposed to V_X. The Fourier series representation of the square wave generated by the local oscillator (LO) is given by

$$V_Y = \sum_{n=1}^{+\infty} B_n \cos(n\omega_{LO}t), \tag{7.75}$$

where $\omega_{LO} = 2\pi/T$ denotes the LO fundamental frequency, T is the waveform period, and

$$B_n = \frac{4}{T}\int_0^{T/2} \cos(n\omega_{LO}t)\mathrm{d}t = \frac{\sin(n\pi/2)}{n\pi/4} \quad \text{for} \quad n \neq 0. \tag{7.76}$$

The output signal, which is obtained by periodically switching the drain current of T_3 to either the noninverting or inverting output node, can be expressed as

$$V_0 = KV_XV_Y = \frac{KA}{2}\sum_{n=1}^{+\infty} B_n[\cos(n\omega_{LO} + \omega_{RF})t + \cos(n\omega_{LO} - \omega_{RF})t], \tag{7.77}$$

where K is the mixer gain. Due to the spectral contents of the square waveform, the spectrum of the mixer output exhibits components around the

LO fundamental frequency and its odd harmonics. Ideally, these components should represent only the sum and difference frequencies of the two input signals.

To determine the mixer gain, it is assumed that the transistors T_1 and T_2 operate as ideal switches. The transistor T_3, whose transconductance is equal to g_m, is biased such that the input voltage is converted into a current, $g_m V_X$, which, together with the bias current, is transferred to either the noninverting node for one half of the switching signal period or the inverting node for the other half. Hence, the mixer gain is of the form $K = g_m R_L$, where R_L is the equivalent output load resistance. Because the important contribution to the mixer output is associated with the fundamental Fourier coefficient, B_1, of the switching signal, the conversion gain can be reduced to

$$K_{CG} = \frac{g_m R_L}{2} B_1 = \frac{2 g_m R_L}{\pi}. \tag{7.78}$$

However, the nonlinearity of the transconductor and mismatches of transistor switches have the effect of producing intermodulation products and dc offset at the mixer output. Double-balanced mixers feature the advantage of suppressing output spurs caused by the local oscillator as well as some high-order products.

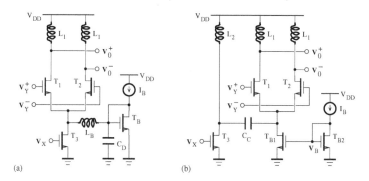

(a)   (b)

FIGURE 7.38
Circuit diagram of a single-balanced mixer (a) using auxiliary bias current path and (b) with capacitive coupling.

For the conventional single-balanced mixer, the frequency response is limited by the pole due to the parasitic capacitance at the drain of the transistor T_3, and the switching speed remains low because all the bias current has to flow through either T_1 or T_2, and the noise contribution of the switching transistors may be folded into the frequency domain of the desired signal. Furthermore, to achieve an acceptable level of linearity, a drain-source voltage well above the saturation voltage is necessary for T_3. The choice of operating T_1 and T_2 in the saturation region requires a significant voltage headroom, thereby limiting the dynamic range available for the load and hence, the conversion gain [26].

The performance characteristics can be improved by adopting the mixer structure shown in Figure 7.38(a). This is achieved with the help of the inductor L_B which can resonate with the parasitic capacitance at the drain of the transistor T_3 and can drive almost half of the bias current. Although the aforementioned technique can help to improve the switching speed, the accuracy of the bias current can be degraded due to the impedance variations of the network consisting of C_D, L_B, and T_3.

Another design approach consists of using the mixer topology depicted in Figure 7.38(b). The coupling capacitor, C_C, provides an isolation between the bias current of the switching transistors, T_1 and T_2, and that of the input transistor, T_3. As result, an increase of the mixer gain and a reduction of the noise factor can be observed.

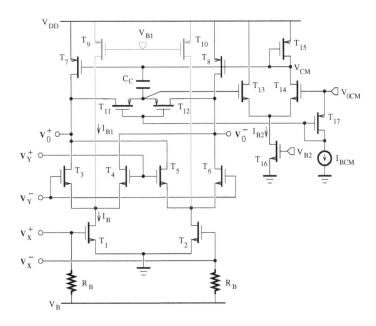

FIGURE 7.39

Circuit diagram of a double-balanced mixer with the common-mode feedback circuit. (Adapted from [10], ©1997 IEEE.)

Double-balanced mixers are widely used because they can reject the noise from the LO circuit and can minimize the LO signal feed-through, which can affect the performance of single-balanced structures. The circuit diagram [27] of a double balanced mixer is shown in Figure 7.39. Transistors $T_1 - T_2$ realize a transconductor, while $T_3 - T_6$ operate as a switch controlled by the signal provided by the LO. The actual output common-mode (CM) voltage is detected using two matched transistors, T_{11} and T_{12}, which are biased by the diode-connected transistor T_{17} to operate in the triode region, and compared to the desired common-mode reference by the amplifier $T_{13} - T_{16}$. The common-mode voltage is then set by varying the bias voltage of transistors

$T_7 - T_8$ in order to reduce the comparison difference. The compensation capacitor C_C is used to provide an appropriate bandwidth to the CM feedback loop. The contribution of the switching transistors to the mixer noise is dominant during the time period in which both transistors in the source-coupled pair are conducting. It can be reduced using an LO waveform with sharpen transitions and minimizing the parasitic capacitors of transistors. Transistors driven by a sine wave, whose amplitude is raised to sharpen its transitions, can be forced to operate in the triode region. As a result, an increase in the mixer distortion can be observed due to the nonlinear output resistance of the transistors. The reduction in the overdrive of the LO signals necessary for the switching of $T_3 - T_6$ and the increase in the conversion gain are simultaneously achieved using the current sources I_{B_1}. This latter is set to about $3I/4$ to preserve the linearity of the mixer, which can be deteriorated for a value of I_{B_1} approaching the one of the bias current I_B of T_1 or T_2.

Mixers with a grounded-source differential transistor pair feature a better linearity than the ones based on a differential stage biased by a constant tail current. Furthermore, the current consumption, which is on the order of a few milliamperes, should be reduced as the transistor is scaled down.

7.1.6   Voltage-controlled oscillator

• Noise

The signal generated by an ideal oscillator can be written as

$$v_0(t) = A\cos(\omega_0 t + \phi), \tag{7.79}$$

where A denotes the amplitude, ω_0 is the angular oscillation frequency, and ϕ represents the phase. Due to the different noise sources, the amplitude and phase can become a time-varying function. As a result, the spectrum of a practical oscillator exhibits sidebands close to ω_0. The noise caused by the amplitude and phase fluctuations can be characterized by

$$\mathcal{L}\{\triangle\omega\} = 10\log_{10}\left(\frac{P_{sideband}(\omega_0 + \triangle\omega)}{P_{carrier}}\right), \tag{7.80}$$

where $P_{sideband}(\omega_0 + \triangle\omega)$ is the power per unit bandwidth of a single sideband at a frequency offset of $\triangle\omega$ from the carrier, and $P_{carrier}$ is the carrier power. The spectral density $\mathcal{L}$ is expressed in units of decibels below the carrier per hertz (dBc/Hz). Note that the amplitude noise can be minimized by an appropriate limiter and the overall oscillator noise is then dominated by the phase noise. An oscillator based on a lossless LC network should have no phase noise and exhibit a noise factor equal to one.

• Differential LC oscillator

Although various oscillator configurations can be used in wireless communication systems, a differential architecture is more suitable for integrated-circuit implementations, especially when the effects of power supply noise and substrate noise coupling are to be minimized. A voltage-controlled oscillator (VCO) consisting of a cross-coupled differential pair of transistors loaded by inductors is shown in Figure 7.40(a). To sustain the oscillation, the resistive losses in the passive elements are compensated by the negative resistance provided by the cross-coupled transistors. The tuning of the oscillation frequency is achieved by varying the control voltage, which determines the capacitances, C, realized by MOS transistors in the inversion or accumulation mode. In addition to the reduction in parasitic resistances in the inductors and MOS variable capacitors, an acceptable level of phase noise can be achieved by sizing the transistors appropriately. Note that the minimum supply voltage is $V_{DS(sat),T_B} + V_{GS,T_1}$. A small value of $V_{GS} - V_T$ for T_1 and T_2 can then provide a large transconductance and a small power consumption. However, in this case, the transistor sizes can become very large, resulting in large parasitic capacitances.

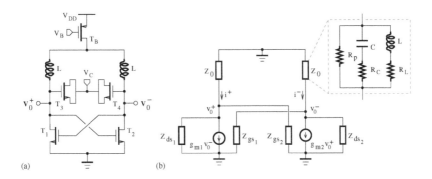

FIGURE 7.40
(a) Circuit diagram and (b) small-signal equivalent model of a voltage-controlled oscillator.

With reference to the small-signal equivalent model of the VCO depicted in Figure 7.40(b), the nodal equations at both outputs can be written as

$$i^+ = \frac{v_0^+}{Z_{ds1}} + \frac{v_0^+}{Z_{gs2}} + g_{m1}v_0^- \qquad (7.81)$$

and

$$i^- = \frac{v_0^-}{Z_{ds2}} + \frac{v_0^-}{Z_{gs1}} + g_{m2}v_0^+ . \qquad (7.82)$$

In practice, the transistors T_1 and T_2 are matched, that is, $g_{m1} = g_{m2} = g_m$, $Z_{ds2} = Z_{ds1} = Z_{ds}$, and $Z_{gs2} = Z_{gs1} = Z_{gs}$. Assuming that $v_0^+ = v_0/2$ and

$v_0^- = -v_0/2$, the impedance provided by the cross-coupled section is given by

$$Z_i = \frac{v_0}{i^+} = -\frac{v_0}{i^-} = \frac{-2}{g_m - \dfrac{Z_{ds} + Z_{gs}}{Z_{ds}Z_{gs}}}. \tag{7.83}$$

Because the impedance Z_{ds} is very high, it can be shown that $(Z_{ds} + Z_{gs})/Z_{ds}Z_{gs} \simeq 1/Z_{gs}$. For $Z_{gs} = 1/(sC_{gs})$, we obtain

$$Z_i = \frac{v_0}{i^+} = -\frac{v_0}{i^-} \simeq \frac{-2}{g_m(1 - s/\omega_t)}, \tag{7.84}$$

where $\omega_t = g_m/C_{gs}$. When $\omega \ll \omega_t$, $Z_i = R_i = -2/g_m$ and the cross-coupled section realizes a negative resistance that can sustain oscillations by compensating loss in the LC tank. The use of Kirchhoff's current law leads to

$$(Z_i + Z_0)i^+ = 0 \tag{7.85}$$

and

$$(Z_i + Z_0)i^- = 0. \tag{7.86}$$

For the VCO to oscillate, the currents i^+ and i^- must be nonzero. Hence,

$$Z_i + Z_0 = 0. \tag{7.87}$$

To proceed further, it can be assumed that the impedance Z_0 is realized by the parallel connection of the parasitic resistance, R_p, in parallel with the LC-tank, the capacitor C with its parasitic resistance R_C, and the inductor L with its parasitic resistance R_L. The resonant frequency at which oscillations will occur is derived from $\text{Im}(Z_0) = -\text{Im}(Z_i)$ to be

$$\omega_0 = \frac{1}{\sqrt{LC}}\sqrt{\frac{L/C - R_L^2}{L/C - R_C^2}}. \tag{7.88}$$

At the oscillation frequency, the magnitude of the output voltage is either equal to the product of the tail current and the LC tank equivalent resistance (current-limited operation regime) or the minimum value between the supply voltage and the voltage at which there is a change in the operating regions of transistors (voltage-limited operation regime). On the other hand, by setting $\text{Re}(Z_i) = -\text{Re}(Z_0)$, the value of transconductance, g_m, for each of the transistors T_1 and T_2 is obtained as

$$g_m = \frac{2}{R_p} + \frac{2R_C}{R_C^2 + (1/\omega_0 C)^2} + \frac{2R_L}{R_L^2 + \omega_0^2 L^2} \tag{7.89}$$

$$= \frac{2}{R_p} + \frac{2}{R_C(1 + Q_C^2)} + \frac{2}{R_L(1 + Q_L^2)}, \tag{7.90}$$

where $Q_C = 1/(\omega_0 C R_C)$ and $Q_L = \omega_0 L/R_L$ denote the quality factor of the capacitor and inductance, respectively. To guarantee the start-up of the oscillator, the transconductance g_m should be sufficiently high to overcome all resistive losses in the oscillator circuit. Ideally, $L/C \gg R_C, R_L$, and the oscillation frequency is reduced to $\omega_0 = 1/\sqrt{LC}$.

Due to the signal phase fluctuations caused by the noise contribution of oscillator components, the oscillation criterion may not hold perpetually. The phase noise then appears to be a performance characteristic of the oscillator. It provides a measure of the stability of the output frequency over a given duration. In the case where R_C and R_L are negligible and the overall resistance, R, at the output node is compensated by the negative resistance of the cross-coupled transistors, the current noise flows through a lossless LC network with the impedance given by

$$Z(\omega_0 + \triangle\omega) = \left[j(\omega_0 + \triangle\omega)C + \frac{1}{j(\omega_0 + \triangle\omega)L} \right]^{-1} \quad (7.91)$$

$$= \frac{j(\omega_0 + \triangle\omega)L}{1 - (\omega_0^2 + 2\omega_0\triangle\omega + \triangle\omega^2)LC} \quad (7.92)$$

$$= \frac{j\omega_0(1 + \triangle\omega/\omega_0)L}{2(\triangle\omega/\omega_0) + (\triangle\omega/\omega_0)^2}, \quad (7.93)$$

where $\omega_0^2 = 1/LC$. Because $\omega_0 \gg \triangle\omega$, it can be shown that

$$Z(\omega_0 + \triangle\omega) \simeq -j\frac{\omega_0 L}{2\triangle\omega/\omega_0} = -j\frac{R}{2Q\triangle\omega/\omega_0}, \quad (7.94)$$

where $Q = R/(\omega_0 L)$. In the absence of amplitude saturation, the power contribution due to the phase noise can be written as

$$P_{sideband} = \frac{1}{2}\frac{\overline{v_n^2}}{B \cdot R} = \frac{1}{2}\frac{\overline{i_n^2} \cdot |Z|^2}{B \cdot R} = 2kT\left(\frac{\omega_0}{2Q\triangle\omega}\right)^2 \quad (7.95)$$

where $\overline{i_n^2}$ and $\overline{v_n^2}$ are the mean-square noise current and voltage, respectively; B denotes the bandwidth; ω_0 is the oscillation frequency; Q is the effective quality factor of the LC tank; k is Boltzmann's constant; T is the temperature in degrees Kelvin; and $\triangle\omega$ represents the offset from the carrier. The use of the $1/2$ factor is justified by the equipartition theorem of thermodynamics, which predicts an even distribution of the noise power among all of the quadratic degrees of freedom (here, the amplitude and phase) in thermal equilibrium.

A first-order approximation of the single-sided noise spectral density, or phase noise, for the oscillator is of the form

$$\mathcal{L}\{\triangle\omega\} = 10\log_{10}\left[\frac{2kT}{P_{carrier}}\left(\frac{\omega_0}{2Q\triangle\omega}\right)^2 \right], \quad (7.96)$$

where $P_{carrier} = \overline{v_0^2}/R$. The phase noise is determined by considering only the resistor thermal noise and can be reduced by increasing the quality factor, Q, of the LC tank section. While a high Q is desirable, it should also be noted that Q is a function of L, and thereby the available circuit area. The values of the phase noise computed from Equation (7.96) exhibit a different behavior and are smaller than the ones measured between the output nodes of a practical circuit due to the effect of other noise sources.

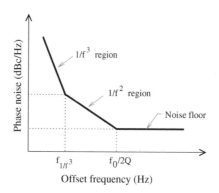

FIGURE 7.41
Plot of the phase noise according to the empirical model.

By considering the VCO as a linear, time-invariant system, a semi-empirical model [28, 29] can be adopted for the derivation of the phase noise as follows

$$\mathcal{L}\{\Delta\omega\} = 10\log_{10}\left\{\frac{2FkT}{P_{carrier}}\left[1+\left(\frac{\omega_0}{2Q\Delta\omega}\right)^2\right]\left(1+\frac{\omega_{1/f^3}}{\Delta\omega}\right)\right\}, \quad (7.97)$$

where F is an empirical parameter, which is also known as the device excess noise number, ω_{1/f^3} is the frequency of the corner between the $1/f^2$ and $1/f^3$ regions. Figure 7.41 shows the phase noise versus the offset frequency. It appears that the phase noise spectrum exhibits a $1/f^3$ region in addition to the $1/f^2$ region and a noise floor at large frequency offsets. The use of Equation (7.97) for the computation of the phase noise is limited by the fact that fitting parameters $(F, \omega_{1/f^3})$, which should be determined from measurements, are required.

In practice, the circuit parameters required for the phase noise computation change as the bias point of the transistors fluctuates during each oscillation cycle. For each noise source, the VCO can then be considered a linear, time variant system.

Let $h_\phi(t, \tau)$ be the impulse response at time τ for the excess phase ϕ, and $\Gamma(\omega_0\tau)$ be the 2π-periodic impulse sensitivity function (ISF). The excess phase can be obtained as the superposition of impulse responses for all τ [30]. That is,

$$\phi(t) = \int_{-\infty}^{\infty} h_\phi(t,\tau)d\tau = \frac{1}{q_{max}} \int_{-\infty}^{t} \Gamma(\omega_0\tau)i(\tau)d\tau, \tag{7.98}$$

where

$$h_\phi(t,\tau) = \frac{\Gamma(\omega_0\tau)}{q_{max}} u(t-\tau). \tag{7.99}$$

Here, $u(t)$ is the unit step function, and q_{max} is the maximum charge swing across the capacitor on the output node. The ISF is normalized to q_{max} in order to make $h_\phi(t,\tau)$ independent of the amplitude. For each noise source, the impulse response $h_\phi(t,\tau)$ can be determined by SPICE simulations based on the extracted VCO model. The ISF can be expressed as a Fourier series of the form

$$\Gamma(\omega_0\tau) = \frac{c_0}{2} + \sum_{n=1}^{\infty} c_n \cos(n\omega_0\tau + \theta_n). \tag{7.100}$$

Substituting Equation (7.100) into (7.98) gives

$$\phi(t) = \frac{1}{q_{max}} \left[\frac{c_0}{2} \int_{-\infty}^{t} i(\tau)\,d\tau + \sum_{n=1}^{\infty} c_n \int_{-\infty}^{t} i(\tau)\cos(n\omega_0\tau)d\tau \right]. \tag{7.101}$$

We observe that the excess phase $\phi(t)$ can be computed for an arbitrary input current injected into any circuit node using the corresponding Fourier coefficients of the ISF. Let the current to be injected in a given node be a sinusoidal function, for instance. Thus,

$$i(t) = I_M \cos[(m\omega_0 + \triangle\omega)t], \tag{7.102}$$

where I_M represents the maximum amplitude. Combining Equations (7.101) and (7.102), we obtain

$$\phi(t) = \frac{c_0 I_M}{2q_{max}} \int_{-\infty}^{t} \cos[(m\omega_0 + \triangle\omega)t]d\tau$$
$$+ \frac{I_M}{q_{max}} \sum_{n=1}^{\infty} c_n \int_{-\infty}^{t} \cos[(m\omega_0 + \triangle\omega)\tau]\cos(n\omega_0\tau)\,d\tau. \tag{7.103}$$

For causal signals, recall that

$$\int_{-\infty}^{t} \cos[(m\omega_0 + \triangle\omega)\tau]\cos(n\omega_0\tau)\,d\tau$$
$$= \frac{1}{2} \left[\frac{\sin(m\omega_0 + n\omega_0 + \triangle\omega)t}{m\omega_0 + n\omega_0 + \triangle\omega} + \frac{\sin(m\omega_0 - n\omega_0 + \triangle\omega)t}{m\omega_0 - n\omega_0 + \triangle\omega} \right] \tag{7.104}$$

is nonnegligible only when $n = m$. Because $\triangle\omega$ is close to any integer multiple of the oscillation frequency, and $\sin(x)/x$ decays to zero as x tends to infinity, has a maximum value of unity at $x = 0$, and is zero at $x = k\pi$ ($k = \pm 1, \pm 2, \ldots$), the excess phase can be approximated as

$$\phi(t) \simeq \frac{c_m I_M}{2q_{max}\Delta\omega}\sin(\Delta\omega\, t). \tag{7.105}$$

Combining Equation (7.105) with the oscillator output signal considered to be of the form

$$v_0(t) = A\cos[\omega_0 t + \phi(t)], \tag{7.106}$$

we have

$$v_0(t) = A\cos\left[\omega_0 t + \frac{c_m I_M}{2q_{max}\Delta\omega}\sin(\Delta\omega\, t)\right]. \tag{7.107}$$

If $c_m I_M/2q_{max}\Delta\omega \ll 1$, it can be found that

$$v_0(t) \simeq A\left[\cos(\omega_0 t) - \frac{c_m I_M}{2q_{max}\Delta\omega}\sin(\omega_0 t)\sin(\Delta\omega\, t)\right], \tag{7.108}$$

or equivalently,

$$v_0(t) \simeq A\cos(\omega_0 t) + \frac{c_m I_M A}{4q_{max}\Delta\omega}[\cos(\omega_0 + \Delta\omega)t - \cos(\omega_0 - \Delta\omega)t]. \tag{7.109}$$

Hence, the injection of a current at $n\omega_0 + \Delta\omega$ into an oscillator node produces a pair of equal sidebands at $\omega_0 \pm \Delta\omega$ with a sideband power relative to the carrier given by

$$\frac{P_{sideband}(\Delta\omega)}{P_{carrier}} = \left(\frac{c_m I_M}{4q_{max}\Delta\omega}\right)^2, \tag{7.110}$$

where $I_M^2/2 = \overline{i_n^2}/\Delta f$. To proceed further, Parseval's theorem can be used to show that

$$\sum_{m=0}^{\infty} c_m^2 = \frac{1}{\pi}\int_0^{\pi}|\Gamma(x)|dx = 2\Gamma_{rms}^2, \tag{7.111}$$

and the phase noise expression for the $1/f^2$ region can then be written as

$$\mathcal{L}(\Delta\omega) = 10\log\left[\frac{(\overline{i_n^2}/\Delta f)\sum_{m=0}^{\infty}c_m^2}{8q_{max}^2\Delta\omega^2}\right] = 10\log\left(\frac{\Gamma_{rms}^2}{q_{max}^2}\frac{\overline{i_n^2}/\Delta f}{4\Delta\omega^2}\right), \tag{7.112}$$

where $\overline{i_n^2}/\Delta f$ is the power spectral density of the current noise source, and Γ_{rms} is the root-mean-square value of the ISF. In the simple case where a noise-free sinusoid waveform is used for the ISF determination, we have $\Gamma_{rms}^2 = 1/2$, and the noise contribution due to the parasitic resistance R_p alone can be derived from Equation (7.112), assuming that $\overline{i_n^2}/\Delta f = 4kT/R_p$ and $q_{max} = CV_{max}$, where V_{max} is the maximum voltage swing across the

LC tank. The time-variant noise analysis approach is limited by the fact that the determination of the ISF can require complex simulations. In practice, simulation programs such as SpectreRF can be used to compute the phase noise directly.

For a 5-GHz oscillator, the computed phase noise, $\mathcal{L}\{\triangle f\}$, at $\triangle f = 100$ kHz is on the order of 103 dBc/Hz. The performance of various oscillator architectures can be compared using a figure of merit (FOM) defined as

$$FOM = \mathcal{L}(\triangle f) + 10 \log_{10}\left[\left(\frac{\triangle f}{f_0}\right)^2 P\right],\tag{7.113}$$

where $\mathcal{L}(\triangle f)$ is the single sideband phase noise at offset frequency $\triangle f$ from the oscillation frequency, f_0; and P represents the total power consumption of the oscillator in milliwatts (mW). In CMOS designs whose power consumption is to be minimized, the oscillator performance is improved as the absolute value of FOM is increased.

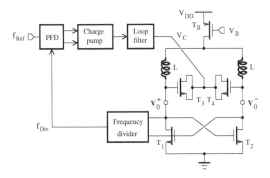

FIGURE 7.42
Circuit diagram of a voltage-controlled oscillator with the frequency control loop.

Due to IC process and temperature variations, the accuracy requirement of the oscillation frequency may no longer be satisfied. In practice, the oscillation frequency is tuned by making the capacitor C variable and controllable through a voltage. The circuit diagram of a VCO with the frequency control loop [31] is shown in Figure 7.42. The VCO output is connected to a frequency divider, whose output signal is compared to a reference signal, using a phase and frequency detector (PFD) followed by a charge pump circuit. If the frequency of the reference signal is higher than that of the feedback signal, the control voltage of the VCO available at the loop filter output will be increased; otherwise, it will be reduced.

Furthermore, because the oscillation amplitude changes with the biasing condition of transistors and the additive amplitude noise may be detected by varactors, resulting in the reduction of the tuning range and sensitivity, practical implementations must often include a loop for the automatic control of

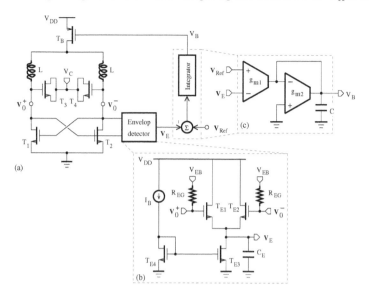

FIGURE 7.43

Circuit diagram of a voltage-controlled oscillator with the amplitude control loop.

the output level [32]. With reference to Figure 7.43(a), the oscillation amplitude is sensed by the peak detector circuit (rectifier and filter), whose output drives the error amplifier stage. The resulting signal is then processed by a lowpass filter (here, an integrator), before being used to control the magnitude of the VCO bias current. Note that, in [33], by combining discrete and continuous tuning, it was also possible to lower the phase noise sensitivity while maintaining a wide tuning range. In this case, the capacitor is realized using varactors, whose capacitance depends continuously on the control voltage, and a binary-weighted switched-capacitor array with a given number of control bits. First, the oscillator center frequency is digitally set to one of the possible discrete frequencies, and varactors are then interpolated continuously around this frequency.

Consider now the envelope detector circuit depicted in Figure 7.43(b). The input signal is sensed by a pair of transistors, T_{E1} and T_{E2}, which is biased by the current I_B flowing through the current mirror composed of T_{E3} and T_{E4}. If the input peak voltage is greater than $V_E + V_T$, where V_T is the transistor threshold voltage, the transistors T_{E1} and T_{E2} will operate in the strong inversion region, thereby supplying a charge current to the capacitor C_E so that the output voltage, V_E, is increased. In contrast, if the input peak voltage becomes less than $V_E + V_T$, transistors T_{E1} and T_{E2} will be biased in the subthreshold region, and the discharge of C_E induced by leakage currents yields a reduction in the output signal level. The time required to track the input signal amplitude depends on the value of the capacitor C_E and the bias current source I_B. In the subthreshold region, the $I-V$ characteristic of the n-channel transistor with the source connected to the substrate is approximated by

$$I_d = I_t \frac{W}{L} \exp\left(\frac{qV_{gs}}{nkT}\right), \tag{7.114}$$

where I_t is a technology-dependent positive constant; n is the subthreshold swing parameter, typically ranging from 1 to 2; and kT/q represents the thermal voltage. Assuming that the input signal whose amplitude is to be detected is sinusoidal, the drain current for each of the transistors T_{E1} and T_{E2} satisfies the next relation [34]

$$I_d = I_t \frac{W}{L} \exp\left(\frac{qV_{GS}}{nkT}\right) \exp\left(\frac{q}{nkT} v_i \cos \omega t\right), \tag{7.115}$$

where V_{GS} is the gate-source dc voltage. It is then possible to expand the drain current as follows,

$$I_d = I_t \frac{W}{L} \exp\left(\frac{qV_{GS}}{nkT}\right)\left[I_0\left(\frac{qv_i}{nkT}\right) + 2\sum_{k=1}^{\infty} I_j\left(\frac{qv_i}{nkT}\right)\cos(j\omega t)\right], \tag{7.116}$$

where I_j is the j-th order modified Bessel functions of the first kind. The average of the drain current is of the form

$$\overline{I_d} = I_t \frac{W}{L} \exp\left(\frac{qV_{GS}}{nkT}\right) I_0\left(\frac{qv_i}{nkT}\right), \tag{7.117}$$

where

$$I_0\left(\frac{qv_i}{nkT}\right) \simeq \frac{\exp\left(\dfrac{qv_i}{nkT}\right)}{\sqrt{2\pi \dfrac{qv_i}{nkT}}}. \tag{7.118}$$

This last approximation results in less than 1% error for $qv_i/nkT > 15$. Because $V_{GS} = V_G - V_S = V_G - V_0$ and $\overline{I_d} = I_B/2$, the output voltage can be written as

$$V_0 = v_i + V_G - \frac{nkT}{q}\log\left[\frac{I_B}{2I_t(W/L)}\sqrt{2\pi \frac{qv_i}{nkT}}\right]. \tag{7.119}$$

Note that the logarithmic term is generally assumed negligible, especially for large input signals.

One approach to realize the integrator section relies on the use of transconductors and capacitor. The circuit diagram of the integrator is illustrated in Figure 7.43(c). The output node equation can be written as

$$g_{m1}(V_E - V_{Ref}) + g_{m2}V_B + sCV_B = 0 \tag{7.120}$$

or equivalently,

$$V_B = -\frac{g_{m1}(V_E - V_{Ref})}{sC + g_{m2}}.$$ (7.121)

Equation (7.121) corresponds to that of a lossy integrator with the dc gain, g_{m1}/g_{m2}, and the 3-dB cutoff frequency, g_{m2}/C. However, the resulting frequency response may be limited by the parasitic capacitors and resistances associated with real transconductors.

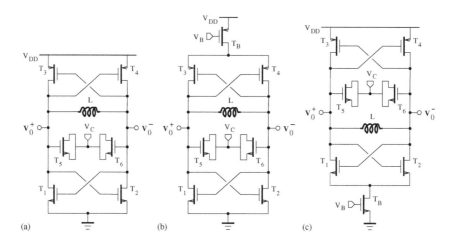

FIGURE 7.44
Circuit diagrams of complementary VCOs (a) without current source, (b) with pMOS current source, and (c) with nMOS current source.

The VCO performance is primarily determined by the phase noise, tuning range, and power dissipation. In addition to the aforementioned LC VCO, various other architectures can be used for the oscillator design. Figure 7.44 shows the circuit diagram of complementary VCOs using both nMOS and pMOS cross-coupled transistors to compensate for the losses in the LC tank. A reduction in the phase noise is achieved for these last structures because the symmetry of the oscillation waveform generally facilitates the cancelation of the phase noise dc component. The tail current source, which can significantly increase the phase noise, was omitted in the VCO structure of Figure 7.44(a) to maximize the signal swing. However, by using nMOS or pMOS current sources, as shown in Figures 7.44(b) and (c), either the ground or the power supply is isolated from the output nodes so that the contribution of the substrate or supply voltage to the phase noise is further reduced.

• Generation of I and Q signals

One simple approach for generating quadrature signals is to use a set of RC and CR networks. Figure 7.45(a) shows the circuit diagram of an RC-CR phase shift network. Using the voltage division principle gives the next transfer

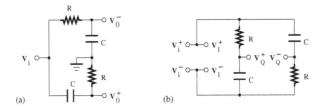

FIGURE 7.45
(a) Circuit diagram of an RC-CR phase shift network; (b) circuit diagram of an allpass-based phase shifter.

function

$$T^-(s) = \frac{V_0^-(s)}{V_i(s)} = \frac{1/sC}{R + 1/sC}. \tag{7.122}$$

The magnitude of the transfer function is of the form

$$|T^-(\omega)| = \frac{1}{\sqrt{1 + (\omega RC)^2}}, \tag{7.123}$$

and the phase can be expressed as

$$\angle T^-(\omega) = \arctan(-\omega RC) = -\arctan\left(\frac{1}{\omega RC}\right). \tag{7.124}$$

Similarly, it can be shown that

$$T^+(s) = \frac{V_0^+(s)}{V_i(s)} = \frac{R}{R + 1/sC}. \tag{7.125}$$

The magnitude of the transfer function is obtained as

$$|T^+(\omega)| = \frac{\omega RC}{\sqrt{1 + (\omega RC)^2}}, \tag{7.126}$$

and the phase is given by

$$\angle T^+(\omega) = \arctan\left(\frac{1}{\omega RC}\right). \tag{7.127}$$

At the 3-dB frequency, $\omega_c = 1/RC$, the magnitudes, $|T^-|$ and $|T^+|$, are attenuated by $1/\sqrt{2}$, while the phases, $\angle T^-$ and $\angle T^+$, are equal to $-45°$ and $45°$, respectively. Hence, a $90°$ phase shift is realized between the outputs V_0^+ and V_0^-.

The magnitude of signals can be maintained constant using the allpass-based phase shifter depicted in Figure 7.45(b). The I signal is derived directly from the input signal, while the Q signal is obtained at the output of the allpass filter. Using the superposition principle, we can obtain

$$V_Q^+ = \frac{V_i^+ + sRCV_i^-}{1 + sRC} \tag{7.128}$$

and

$$V_Q^- = \frac{sRCV_i^+ + V_i^-}{1 + sRC}. \tag{7.129}$$

Hence,

$$T(s) = \frac{V_Q^+ - V_Q^-}{V_i^+ - V_i^-} = \frac{1 - sRC}{1 + sRC}. \tag{7.130}$$

The transfer function, $T(s)$, is of the allpass type. Its magnitude is unity and its phase is of the form

$$\angle T(\omega) = \arctan(-\omega RC) - \arctan(\omega RC) = -2\arctan(\omega RC). \tag{7.131}$$

At the 3-dB cutoff frequency, $\omega_c = 1/RC$, the phase difference between the I and Q signals is 90°.

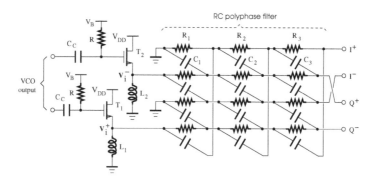

FIGURE 7.46
Circuit diagram of a VCO output buffer followed by a polyphase RC filter-based phase shifter for the generation of I and Q signals.

The allpass-based phase shifter is designed to feature a gain of unity at any frequency, while the RC-CR phase shift network should ideally provide a constant 90° phase shift at all frequencies. However, the performance of the aforementioned phase shifters is limited by component mismatches due to process variations. To overcome this problem, the RC-CR network is preferably configured in the form of a polyphase filter. Fully differential phase shifters are adopted to further attenuate the effect of component mismatches on the achievable accuracy. Figure 7.46 shows the circuit diagram of an amplifier buffer and a phase shifter based on a polyphase RC filter for the I and Q signal generation [10]. The VCO outputs are connected to source follower-based buffers using large dc-blocking capacitors, C_C. The inductive loads of the buffer are sized to resonate with the equivalent input capacitors of the polyphase filter at the frequency of interest, providing high ac impedances. In this way, transistors with small aspect ratios can be used to provide output currents, as high as required, while still loading the VCO core appropriately. The buffer output signals are applied to the polyphase RC filter, which is less sensitive to the absolute variations of the R and C values due to the adopted multi-stage filter structure.

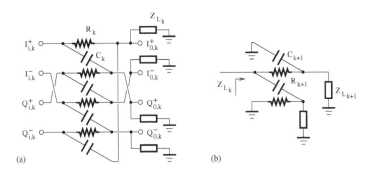

FIGURE 7.47
(a) Circuit diagram of the k-th stage of a phase shifter based on a polyphase RC filter; (b) circuit diagram of the Z_{L_k} loading network.

For the analysis [35, 36] of the phase shifter based on a polyphase RC filter, the k-th stage illustrated in Figure 7.47(a) can be considered. The load impedance, Z_{L_k}, of each stage except the last one, can be derived from the network depicted in Figure 7.47(b) Let $\mathbf{V}_{i,k} = [V_{I_{i,k}^+} \ V_{Q_{i,k}^+} \ V_{I_{i,k}^-} \ V_{Q_{i,k}^-}]^T$ and $\mathbf{V}_{0,k} = [V_{I_{0,k}^+} \ V_{Q_{0,k}^+} \ V_{I_{0,k}^-} \ V_{Q_{0,k}^-}]^T$. Using the division and superposition principles with the assumption that the input nodes, which are not driven by a signal source, are virtual grounds, we arrive at

$$\mathbf{V}_{0,k} = \mathbf{T}_k \mathbf{V}_{i,k}, \qquad (7.132)$$

where

$$\mathbf{T}_k = \begin{bmatrix} Z_k & 0 & 0 & Z'_k \\ Z'_k & Z_k & 0 & 0 \\ 0 & Z'_k & Z_k & 0 \\ 0 & 0 & Z'_k & Z_k \end{bmatrix}, \qquad (7.133)$$

$$Z_k = \frac{(1/sC_k) \parallel Z_{L_k}}{R_k + (1/sC_k) \parallel Z_{L_k}}, \qquad (7.134)$$

$$Z'_k = \frac{R_k \parallel Z_{L_k}}{R_k \parallel Z_{L_k} + (1/sC_k)}, \qquad (7.135)$$

and

$$Z_{L_k} = [(1/sC_{k+1}) + R_{k+1} \parallel Z_{L_{k+1}}] \parallel [R_{k+1} + (1/sC_{k+1}) \parallel Z_{L_{k+1}}] \quad (7.136)$$

$$= \frac{R_{k+1} + Z_{L_{k+1}} + C_{k+1}R_{k+1}Z_{L_{k+1}}s}{1 + C_{k+1}(R_{k+1} + 2Z_{L_{k+1}})s}. \quad (7.137)$$

The differential in-phase and quadrature voltages can then be obtained by putting Equation (7.132) into the form

$$\begin{bmatrix} V_{I_{0,k}} \\ V_{Q_{0,k}} \end{bmatrix} = \frac{Z_{L_k}}{R_k + Z_{L_k} + sR_kZ_{L_k}C_k} \begin{bmatrix} 1 & -sR_kC_k \\ sR_kC_k & 1 \end{bmatrix} \begin{bmatrix} V_{I_{i,k}} \\ V_{Q_{i,k}} \end{bmatrix}, \quad (7.138)$$

where $V_{Q_{0,k}} = V_{Q_{0,k}^+} - V_{Q_{0,k}^-}$, $V_{I_{0,k}} = V_{I_{0,k}^+} - V_{I_{0,k}^-}$, $V_{Q_{i,k}} = V_{Q_{i,k}^+} - V_{Q_{i,k}^-}$, and $V_{I_{i,k}} = V_{I_{i,k}^+} - V_{I_{i,k}^-}$. Applying Equation (7.138) successively to the first, second, and third stage of the RC polyphase filter shown in Figure 7.46, the ratio of the quadrature to the in-phase outputs is derived as

$$\frac{V_{Q_{0,3}}}{V_{I_{0,3}}} = \frac{(R_1C_1 + R_2C_2 + R_3C_3)s - R_1R_2R_3C_1C_2C_3s^3}{1 - (R_1R_2C_1C_2 + R_1R_3C_1C_3 + R_2R_3C_2C_3)s^2}. \quad (7.139)$$

The phase difference between the outputs is 90° regardless of the frequency, because $V_{Q_{0,3}}/V_{I_{0,3}}$ is purely imaginary. However, the magnitudes of the outputs are equal to unity at the resonant frequencies of each stage, $\omega_{c_k} = 1/R_kC_k$ ($k = 1, 2, 3$). A minimum image-reject ratio can be achieved provided the resonant frequency of the second stage is almost equal to the geometric mean of the other two resonant frequencies, that is, $\omega_{c_2} \simeq \sqrt{\omega_{c_1}\omega_{c_3}}$. For any input signal with a frequency within the band delimited by lowest and highest resonant frequencies, the quadrature outputs exhibit almost identical gains and phases, making the multi-stage RC polyphase filter suitable for wideband applications. The error resulting from all of the stages prior to the last stage is averaged out by the subsequent stages. Each additional stage then improves the matching accuracy of the signals to be generated. On the other hand, a single-stage *RC-CR* phase shifter, whose ratio of the quadrature to the in-phase outputs is reduced to sRC and the amplitudes of the quadrature outputs are equal only at the frequency $\omega_c = 1/RC$, is generally adopted for narrowband applications.

- Quadrature voltage-controlled oscillator

The use of a VCO driving an RC-CR polyphase filter to generate quadrature signals can be limited by mismatches of passive elements and the high power consumption of the required output buffers. An alternative approach, which consists of coupling two identical oscillators in such a way that their outputs are forced to oscillate either 90° or −90° out of phase, can be adopted to achieve some performance improvements. This can be implemented as shown in Figure 7.48. Each of both VCOs includes an additional transistor pair, which

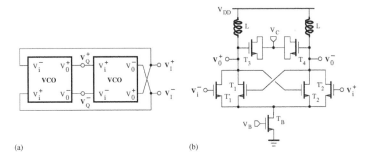

FIGURE 7.48
(a) Quadrature coupled LC VCO; (b) circuit diagram of the LC VCO with coupling transistors.

forms the coupling network [37]. In the steady state, the coupled oscillators are synchronized to the same frequency, which is proportional to the resonance frequency of the unloaded tank network, while their output phases are in quadrature. Quadrature coupled oscillators can exhibit a higher phase noise than the corresponding stand-alone VCO. In practice, coupling transistors can then be designed to be a few times smaller than other transistors to limit their noise contribution, while still providing the required coupling factor. Furthermore, a 90° phase shifter can be inserted in each coupling path [38–40] to reduce the effects of mismatches on the oscillator output phases.

7.1.7 Automatic gain control

In the presence of signals with variable amplitude, a variable gain amplifier (VGA) and automatic gain control (AGC) is required to maintain constant the dynamic range. The AGC loop shown in Figure 7.49 includes a VGA, a peak (or envelope) detector, and an integrator (or lowpass filter). The gain of the VGA is controlled by the integrator output signal based on the difference between the actual output amplitude and the reference voltage V_{REF}. The gain control feedback then forces the estimated peak amplitude to track the dc reference voltage V_{REF}.

A variable-gain amplifier [41] and its common-mode feedback amplifier can be implemented as shown in Figures 7.50(a) and (b), respectively. The amplification stage consists of a source-coupled input transistor pair and diode-connected transistors biased by the currents $I_B(1 + x)$ and $I_B(1 - x)$, respectively. It is assumed that the transistors $T_1 - T_4$ are matched. The drain currents of transistors T_1 and T_2 can be written as

$$I_{D_1} = K(V_{GS_1} - V_T)^2 \tag{7.140}$$

and

$$I_{D_2} = K(V_{GS_2} - V_T)^2 . \tag{7.141}$$

Applying Kirchhoff's current and voltage laws at the source node of transistors

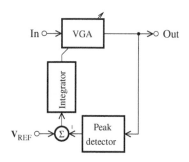

FIGURE 7.49
Block diagram of a variable-gain amplifier with automatic gain control.

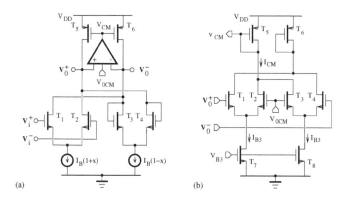

FIGURE 7.50
(a) Circuit diagram of a variable-gain fully differential amplifier; (b) circuit diagram of the common-mode feedback amplifier.

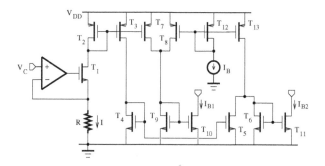

FIGURE 7.51
Circuit diagram of the control and bias generator.

T_1 and T_2, we have

$$I_{D_1} + I_{D_2} = I_B(1 + x) \tag{7.142}$$

and

$$V_{GS_1} - V_{GS_2} = V_i. \tag{7.143}$$

Solving Equations (7.140) through (7.143), we arrive at

$$I_{D_1} = \frac{1}{2}\left[I_B(1+x) + KV_i\sqrt{\frac{2I_B(1+x)}{K} - V_i^2}\right] \tag{7.144}$$

$$I_{D_2} = \frac{1}{2}\left[I_B(1+x) - KV_i\sqrt{\frac{2I_B(1+x)}{K} - V_i^2}\right]. \tag{7.145}$$

Similarly, for transistors T_3 and T_4, the drain currents are given by

$$I_{D_3} = K(V_{GS_3} - V_T)^2 \tag{7.146}$$

and

$$I_{D_4} = K(V_{GS_4} - V_T)^2. \tag{7.147}$$

To proceed further, it can be shown that

$$I_{D_3} + I_{D_4} = I_B(1-x) \tag{7.148}$$

and

$$V_{GS_3} - V_{GS_4} = V_0. \tag{7.149}$$

Combining Equations (7.140) through (7.143) gives

$$I_{D_3} = \frac{1}{2}\left[I_B(1-x) + KV_0\sqrt{\frac{2I_B(1-x)}{K} - V_0^2}\right] \tag{7.150}$$

$$I_{D_4} = \frac{1}{2}\left[I_B(1-x) - KV_0\sqrt{\frac{2I_B(1-x)}{K} - V_0^2}\right]. \tag{7.151}$$

Because $I_{D_1} + I_{D_3} = I_B$ and $I_{D_2} + I_{D_4} = I_B$, it can be found that

$$V_i\sqrt{\frac{2I_B(1+x)}{K} - V_i^2} = -V_0\sqrt{\frac{2I_B(1-x)}{K} - V_0^2}. \tag{7.152}$$

Assuming that $V_i \ll \sqrt{2I_B(1+x)/K}$ and $V_0 \ll \sqrt{2I_B(1-x)/K}$, the amplifier gain is derived as

$$A = \frac{V_0}{V_i} \simeq \frac{\sqrt{\dfrac{2I_B(1+x)}{K}}}{\sqrt{\dfrac{2I_B(1-x)}{K}}} = \sqrt{\frac{1+x}{1-x}}. \tag{7.153}$$

The circuit diagram of the control and bias generator, which can provide the currents, $I_B(1+x)$ and $I_B(1-x)$, is shown in Figure 7.51. With the assumption that the amplifier is ideal, that is, $V_C = V^+ = V^-$ and $I = V_C/R$, and that each of the current mirrors (T_2, T_3, T_7), (T_4, T_5), (T_6, T_{11}), (T_9, T_{10}), and (T_8, T_{12}, T_{13}) is realized using transistors of identical size, we can obtain

$$I_{B_1} = I_B + V_C/R = I_B(1 + x) \tag{7.154}$$

and

$$I_{B_2} = I_B - V_C/R = I_B(1 - x), \tag{7.155}$$

where $x = V_C/RI_B$. Because the approximation exploited here to obtain a linear variation of the amplifier gain in dB is of the form

$$e^x \simeq \sqrt{\frac{1 + x}{1 - x}} \tag{7.156}$$

the tuning range is limited to about 15 dB with a linearity error of less than 0.5 dB. A multi-stage amplifier configuration may then be needed in applications requiring a wide tuning range.

7.2 Continuous-time filters

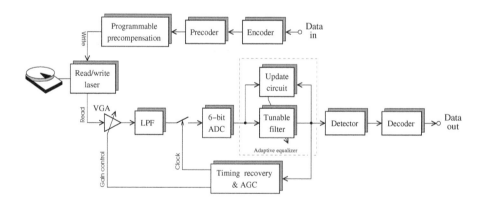

FIGURE 7.52
Block diagram of a read/write channel integrated circuit.

Continuous-time (CT) circuits find applications in the direct processing of analog signals and interfacing of digital signal processors. The block diagram of a read/write channel integrated circuit is shown in Figure 7.52. It is a mixed-signal system, which includes a variable gain amplifier (VGA), a lowpass filter (LPF), an analog-to-digital converter (ADC), and an automatic gain control (AGC) in addition to various digital building blocks. Due to the hysteresis effects in the magnetic disk, the number of signal levels injected through the channels can be limited to two (e.g., ± 1).

Error-correcting encodings are used to correct burst errors (or error patterns) associated with media defects or generated in the write or read process. During the write process, magnetic fields are converted into binary waveforms using modulation or encoding methods, known to reduce the timing uncertainty that can affect the decoding of the stored data. Due to noise errors still remaining in the stream of digital samples after the read sensing, amplification, and conversion and equalization operations, there is a need for a detector and decoder to efficiently recover the data bits.

The increase in the storage capacity is primarily related to the technological improvements in the design of magnetic media and heads. However, as the storage density increases, the performance and speed of the different signal processing blocks (encoder, modulator, equalizer, and decoder) play an important role in the reduction of inter-symbol interference and noise emerging in magnetic disk systems.

Various methods can be used for the synthesis of CT filters. In the cascade design, the filter transfer function is realized as a connection of first- and second-order sections. The sensitivity in the passband can be reduced using LC ladder-based design approaches. An active filter is derived either by simulating the equations of a passive LC prototype, or by replacing the inductors of a passive LC prototype with impedance converters.

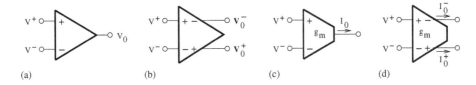

(a) (b) (c) (d)

FIGURE 7.53
(a) Single-ended and (b) fully differential voltage amplifier structures; (c) single-ended and (d) fully differential transconductance amplifier structures.

In general, integrated filters are implemented using resistors, capacitors, and amplifiers. The filter design is commonly based on two types of amplifiers. The first ones (see Figures 7.53(a) and (b)) operate as voltage-controlled voltage source and are described by

$$V_0 = A(V^+ - V^-) \qquad (7.157)$$

and

$$V_0^+ - V_0^- = A(V^+ - V^-), \qquad (7.158)$$

where A is the amplifier gain, while the second structures (see Figures 7.53(c) and (d)), known as operational transconductance amplifiers, are equivalent to a voltage-controlled current source of the form,

$$I_0 = g_m(V^+ - V^-) \qquad (7.159)$$

and

$$I_0^+ - I_0^- = g_m(V^+ - V^-), \tag{7.160}$$

respectively, where g_m is the amplifier transconductance. For differential structures, it was assumed that $V_0^+ = A(V^+ - V^-)/2$, $V_0^- = -A(V^+ - V^-)/2$, $I_0^+ = g_m(V^+ - V^-)/2$, and $I_0^- = -g_m(V^+ - V^-)/2$. Note that V^+ and V^- denote the voltages at the noninverting and inverting input nodes of the amplifier, respectively.

The integrator and gyrator are the fundamental building blocks used for a transfer function description and their different implementations will be reviewed in the next section.

7.2.1 RC circuits

RC circuits consist of operational amplifiers (OA), resistors, and capacitors. Applying Kirchhoff's current law at the inverting node of the single-ended

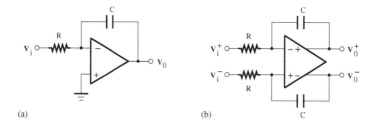

(a) (b)

FIGURE 7.54
(a) Single-ended and (b) fully differential RC integrator structures.

integrator shown in Figure 7.54(a) yields

$$\frac{V_i(s) - V^-(s)}{R} = sC[V^-(s) - V_0(s)], \tag{7.161}$$

where $V_0(s) = -A(s)V^-(s)$. The transfer function is then derived as

$$\frac{V_0(s)}{V_i(s)} = -\frac{1}{sRC} \frac{1}{1 + \frac{1}{A(s)}\left(1 + \frac{1}{sRC}\right)}. \tag{7.162}$$

Ideally, $|A(j\omega)| \gg 1$ and we arrive at $V_0(s)/V_i(s) \simeq -1/(sRC)$. Hence, the gain is inversely proportional to the frequency, while the phase remains equal to $90°$. Equation (7.162) can also be applied to the fully differential structure depicted in Figure 7.54(b), provided $V_0 = (V_0^+ - V_0^-)/2$ and $V_i = (V_i^+ - V_i^-)/2$. The noninverting version of the differential integrator is obtained by permutating the polarities of the input signal.

Generally, it is assumed that the integrator operates with ac signals. But,

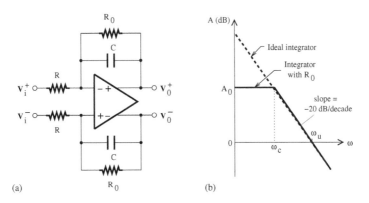

FIGURE 7.55
(a) Integrator with dc feedback; (b) plot of the magnitude of the integrator transfer functions.

in the case where the input also includes dc components, the feedback capacitors, which act like an open circuit at dc, should be periodically discharged to prevent saturation of the amplifier outputs. Alternatively, resistors with large values can be added in parallel with the feedback capacitor, as shown in Figure 7.55(a), to provide the necessary dc feedback. Figure 7.55(b) shows the plot of the magnitude of the integrator transfer functions. The dc gain of the integrator is now reduced to $A_0 = 20 \log_{10}(R_0/R)$, where $R_0 > R$. Assuming that the amplifier is ideal, the cutoff and transition frequencies are of the form $\omega_c = 1/R_0 C$ and $\omega_u = 1/RC$, respectively. At high frequencies, the integrator gain rolls off at a rate of -20 dB/decade.

The straightforward integration of these structures, however, suffers from the imprecise RC product realizable in MOS technology. Since both the resistors and capacitors are designed with random processing variations on the order of 10 to 20%, the overall accuracy of the RC time constant can be as high as 20%. This imprecision can also be observed in the frequency response of the integrator and is unacceptable in most applications.

7.2.2 MOSFET-C circuits

In this approach, the metal oxide semiconductor field-effect transistors (MOSFET) operating in triode region are used as variable resistors that are automatically adjusted to provide accurate RC products by an on-chip control circuit. According to the n-channel transistor shown in Figure 7.56, if V_{GS}, V_{DS}, V_T, and V_B are the gate-source, drain-source, threshold, and substrate voltages, respectively, the drain current, I, in the triode region will be given by

$$I = K[2(V_{GS} - V_T)V_{DS} - V_{DS}^2], \tag{7.163}$$

where $K = \mu_n C_{ox}(W/2L)$, $V_{GS} > V_T$, $V_{DS} < V_{DS(sat)} = V_{GS} - V_T$, and $V_B \leq V_S$. Assuming that $V_{GS} = V_C$, the current, I, can be written as [42]

$$I = \frac{V_{DS}}{R} + f(V_{DS}), \qquad (7.164)$$

where $f(V_{DS}) = -KV_{DS}^2$ and the tunable resistor, R, is of the form

$$R = \frac{1}{2K(V_C - V_T)} = \frac{1}{\mu_n C_{ox}(W/L)(V_C - V_T)}, \qquad (7.165)$$

where W and L are the channel width and length, respectively; C_{ox} is the gate capacitance per unit area; and μ_n is the effective mobility of the transistor. The MOSFET behaves as a voltage-controlled resistor only for small signals,

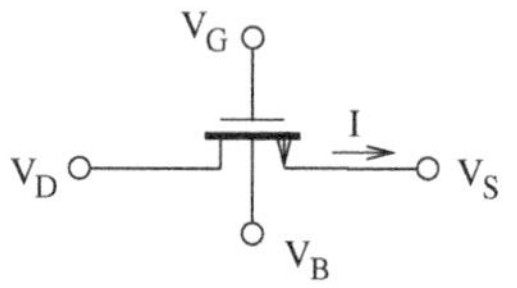

FIGURE 7.56
Symbol of an *n*-channel transistor.

and the higher-order terms introduced by the nonlinear function $f(V_{DS})$ must be considered in the context of large signals.

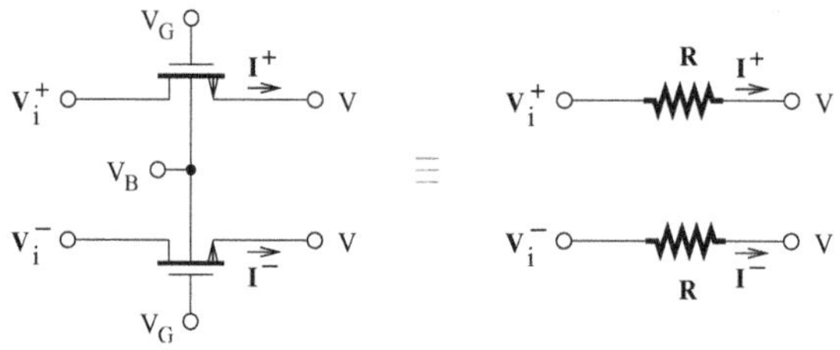

FIGURE 7.57
Resistor based on a balanced transistor configuration.

By driving two transistor devices in balanced form by $V_i^+ = V_i/2$ and $V_i^- = -V_i/2$ at the input terminals and the same voltage V at the other terminals [43] (see Figure 7.57), the effects of transistor nonlinearities are considerably reduced especially when the voltage V is almost equal to zero, and the difference of the output currents is reduced to

$$I^+ - I^- = \frac{V_1^+ - V_1^-}{R}, \qquad (7.166)$$

where $R = 1/[2K(V_C - V_T)]$. The magnitude of the control voltage, V_C, can be changed to adjust the resistor value. Thus, the differential configuration has the advantage of canceling even nonlinearities in terms of the input signal.

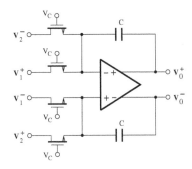

FIGURE 7.58
Differential MOSFET-C integrator.

Resistors based on a balanced transistor configuration find applications in the design of gain stages and integrators. The analysis of the differential integrator shown in Figure 7.58 results in the following expressions in the time and frequency domain:

$$v_0(t) = -\frac{1}{RC} \int_{-\infty}^{t} [v_1(\tau) - v_2(\tau)] \, d\tau \Rightarrow V_0(s) = -\frac{1}{sRC} [V_1(s) - V_2(s)],$$
(7.167)

where $V_1 = V_1^+ - V_1^-$, $V_2 = V_2^+ - V_2^-$, and $V_0 = V_0^+ - V_0^-$. The fact that the allowable signal swing is limited by the triode region of transistors can be a drawback for the low-voltage operation.

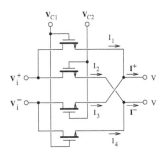

FIGURE 7.59
Four-transistor implementation of the resistor.

Typically, the remaining nonlinearities arising from device mismatches can be estimated to be about 0.1% for 1 V peak-to-peak signals. If a high linearity is needed, the MOSFET pair will be replaced by the four MOSFET cross-coupled structure of Figure 7.59 [44–46]. In this case, the difference of the output currents can be written as

$$I^+ - I^- = (I_1 + I_3) - (I_2 + I_4)$$
(7.168)

$$= (I_1 - I_2) + (I_3 - I_4)$$

$$= 2K(V_{C1} - V_{C2})(V_i^+ - V_i^-).$$
(7.169)

The resulting resistance is then given by

$$R = \frac{V_i^+ - V_i^-}{I^+ - I^-} = \frac{1}{2K(V_{C1} - V_{C2})}. \tag{7.170}$$

All transistors remain in the triode region provided $V_{DS} < V_{DS(sat)} = V_{GS} - V_T$, or equivalently, $V_i^+, V_i^- < \min[V_{C1} - V_T, V_{C2} - V_T]$. Ideally, the nonlinear contributions to the output current are canceled, and the equivalent resistor is controlled by the differential voltage, $V_{C1} - V_{C2}$, such that its value remains independent of the threshold voltage, V_T.

7.2.3 g_m-C circuits

The active component of a g_m-C structure is a transconductor. This latter is a voltage-to-current converter, which gives out a current proportional to a frequency-dependent transconductance, $g_m(s)$. Generally, its value can be regarded as constant, that is, $g_m(s) = g_m$, as long as the transconductor bandwidth is assumed to be sufficiently large. For the single-ended integrator shown in Figure 7.60(a), a transconductor and a capacitor are used to provide the amplification and the integration, respectively [47, 48].

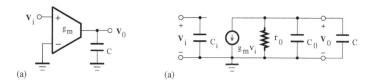

(a)   (a)

FIGURE 7.60
(a) Single-ended g_m-C integrator; (b) small-signal equivalent model.

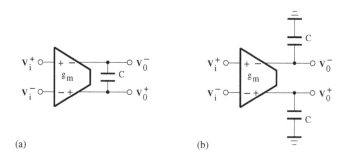

(a)   (b)

FIGURE 7.61
Fully differential g_m-C integrators.

  With reference to the small-signal equivalent model of Figure 7.60(b), $V_0(s) = Z I(s)$, where $I(s) = g_m V_i(s)$ and $Z = r_0 \parallel C_0 \parallel C$. The transfer function of the integrator can then be written as

$$\frac{V_0(s)}{V_i(s)} = \frac{\omega_u}{s\left(1 + \dfrac{C_0}{C}\right) + \dfrac{1}{r_0 C}} . \tag{7.171}$$

Ideally, the output capacitance, C_0, should be small, and the output resistance, r_0, should be large, that is, $C_0/C \ll 1$ and $r_0 C \gg 1$. Therefore, we obtain

$$\frac{V_0(s)}{V_i(s)} = \frac{\omega_u}{s} , \tag{7.172}$$

where $\omega_u = g_m/C$ is the unity-gain frequency. Although the transconductance amplifier can be designed to exhibit a sufficiently high output resistance, r_0, the output parasitic capacitance, C_0, can generally not be assumed to be negligible. Theoretically, parasitic capacitances should be absorbed into a total integrating capacitance value, but their effect becomes dominant in circuits for high-frequency applications. In this case, the integrator unity-gain frequency is found to be $\tilde{\omega}_u = \omega_u/(1 + C_0/C)$. Because the value of C_0 is not accurately known, the value of $\tilde{\omega}_u$ can be scaled only by adjusting the parameters g_m and C. However, this solution implies an augmentation or a reduction of device dimensions, and thus an increase in C_0 or in mismatches and distortions.

The total integrating capacitance of the fully differential integrator shown in Figure 7.61(a) is two times smaller than the one needed in the integrator configuration of Figure 7.61(b), which, however, has the advantage of exhibiting a low sensitivity to parasitic capacitances in the case where C_0 becomes large relative to C.

Practically, due to parasitic capacitances, g_m-C circuits are only suitable for low or medium linearity applications and their high-frequency performance tends to be limited.

7.2.4 g_m-C operational amplifier (OA) circuits

In g_m-C OA circuits, the problem of parasitic capacitances at the transconductor output is minimized by using an OA. These capacitors are connected between the OA inputs (virtual ground) and the ground and thus do not carry any charge. In addition, the OA plays the role of a buffer at the outputs of the transconductor and the structure of this latter can be very simple. But the use of two active components can still mean significant power consumption and large chip area.

Let C_p be the total equivalent capacitance connected at each of transconductance amplifier outputs. Figure 7.62 shows the circuit diagram of g_m-C OA inverting and noninverting integrators. Applying Kirchhoff's current law to the integrator of Figure 7.62(a) [31] gives

$$I^-(s) = sC_p V^+(s) + sC(V^+(s) - V_0^-(s)) \tag{7.173}$$

and

$$I^+(s) = sC_pV^-(s) + sC(V^-(s) - V_0^+(s)), \qquad (7.174)$$

where $I^+(s) - I^-(s) = g_m[V_i^+(s) - V_i^-(s)]$ and $V_0^+(s) - V_0^-(s) = A(s)[V^+(s) - V^-(s)]$. The transfer function can then be computed as

$$\frac{V_0(s)}{V_i(s)} = -\frac{\omega_u}{s} \frac{1}{1 + (1 + C_p/C)/A(s)}, \qquad (7.175)$$

where $\omega_u = g_m/C$, $V_0(s) = [V_0^+(s) - V_0^-(s)]/2$, and $V_i(s) = [V_i^+(s) - V_i^-(s)]/2$.

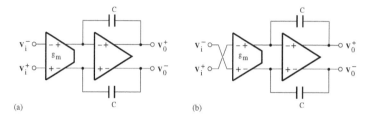

(a) (b)

FIGURE 7.62
(a) g_m-C OA inverting integrator; (b) g_m-C OA noninverting integrator.

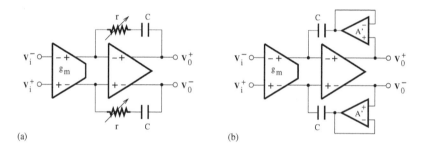

(a) (b)

FIGURE 7.63
(a) g_m-C OA integrator with passive compensation; (b) g_m-C OA integrator with active compensation.

When the OA gain is modeled as $A(s) = \omega_t/s$, it can be shown that a pole at $s = -\omega_t/(1 + C_P/C)$, where ω_t is the OA unity-gain frequency, has been introduced in the integrator transfer function. The price to be paid in order to reduce the error caused by this parasitic pole is an increase in the integrator's dc gain. However, for high-frequency applications, a special compensation technique may be needed. This can be achieved by inserting tunable resistors in series with the capacitors, as shown in Figure 7.63(a). The resistor r can be realized, for instance, by a MOS transistor operating in the triode region. The application of Kirchhoff's current law to the input nodes of the OA yields

$$I^-(s) = sC_pV^+(s) + \frac{V^+(s) - V_0^-(s)}{r + \dfrac{1}{sC}} \qquad (7.176)$$

and

$$I^+(s) = sC_pV^-(s) + \frac{V^-(s) - V_0^+(s)}{r + \dfrac{1}{sC}}. \tag{7.177}$$

The voltage transfer function can readily be written as

$$\frac{V_0(s)}{V_i(s)} = -\frac{\omega_u}{s} \frac{1 + srC}{1 - \dfrac{\omega^2 rC_p}{\omega_t} + \dfrac{s}{\omega_t}\left(1 + \dfrac{C_p}{C}\right)} \simeq -\frac{\omega_u}{s} \frac{1 + srC}{1 + \dfrac{s}{\omega_t}\left(1 + \dfrac{C_p}{C}\right)}. \tag{7.178}$$

The pole-zero cancelation can be achieved provided

$$r = \frac{1}{\omega_t C}\left(1 + \frac{C_p}{C}\right). \tag{7.179}$$

Taking into consideration the value of r, the assumption of neglecting the term $\omega^2 rC_p/\omega_t$ may be justified by the fact that, in practice, $C_p/C \ll 1$ and $\omega/\omega_t \ll 1$. The passive compensation appears to be limited by matching errors and process variations affecting the value of the resistor r.

An improved tracking performance can be achieved by adopting the integrator structure depicted in Figure 7.63(b). In this approach, the frequency responses of two unity-gain buffers used to isolate the integrating capacitors from the OA outputs are exploited to minimize the integrator losses [49]. The equations for the OA input nodes are given by

$$I^-(s) = sC_pV^+(s) + sC[V^+(s) - V_0'^-(s)] \tag{7.180}$$

and

$$I^+(s) = sC_pV^-(s) + sC[V^-(s) - V_0'^+(s)], \tag{7.181}$$

where $V_0'^-(s) = V_0^-(s)/[1 + 1/A'(s)]$ and $V_0'^+(s) = V_0^+(s)/[1 + 1/A'(s)]$. It can then be shown that

$$\frac{V_0(s)}{V_i(s)} = -\frac{\omega_u}{s} \frac{1}{\dfrac{1}{1 + 1/A'(s)} + \dfrac{s}{\omega_t}\left(1 + \dfrac{C_p}{C}\right)}. \tag{7.182}$$

Because $1/|A'(\omega)| \ll 1$, we have

$$\frac{1}{1 + 1/A'(\omega)} \simeq 1 - \frac{1}{A'(\omega)} + \frac{1}{[A'(\omega)]^2} - \cdots \tag{7.183}$$

This suggests that the integrator quality factor, which is computed as the ratio of the imaginary part to the real part of the transfer function denominator, can be maximized by making the gain, A', of the buffer amplifier identical to A. That is,

$$\frac{V_0(j\omega)}{V_i(j\omega)} \simeq -\frac{\omega_u}{j\omega\left(1 - \dfrac{\omega^2}{\omega_t^2}\right)} \frac{1}{1 + j\dfrac{1}{Q_I}}, \tag{7.184}$$

where

$$Q_I = \frac{C}{C_p} \frac{\omega_t}{\omega} \left(1 - \frac{\omega^2}{\omega_t^2}\right) = \frac{C}{C_p} |A(\omega)| \left(1 - \frac{1}{|A(\omega)|^2}\right). \tag{7.185}$$

Here, the matching can be efficiently realized since the involved components are both of the same type, that is, active, and the integrator quality factor, Q_I, may be made high. Note that, Q_I is infinite for an integrator with no losses.

Structures based on fully differential amplifiers are easily derived from single-ended circuits consisting of amplifiers with one input node connected to the ground. When this requirement is not met (see Figure 7.64(a)), the differential input and output are realized by coupling two single-ended circuits. With $V_0(s) = I(s)/sC$ and $I(s) = g_m[V_i(s) - V_0(s)]$, the circuit of Figure 7.64(a) exhibits the next transfer function,

$$H(s) = \frac{V_0(s)}{V_i(s)} = \frac{g_m}{sC + g_m}, \tag{7.186}$$

and its differential version (see Figure 7.64(b)) should have a common-mode rejection ratio (CMRR) given by

$$\text{CMRR} = \frac{1}{2} \frac{H_1(s) + H_2(s)}{H_1(s) - H_2(s)}, \tag{7.187}$$

where $H_j(s) = g_{mj}/(sC_j + g_{mj})$, $j = 1, 2$, is the transfer function for each signal path. Ideally, $H_1(s) = H_2(s)$ and the CMRR is infinite. However, due to fabrication tolerances, $H_1(s)$ and $H_2(s)$ are different. As a result, the CMRR is finite.

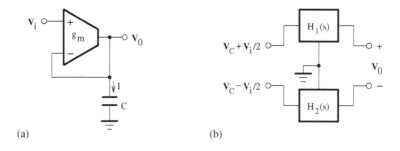

(a)   (b)

FIGURE 7.64
First-order lowpass filter: (a) Single-ended and (b) differential structures.

7.2.5 Summer circuits

A summer circuit is used to add, or subtract, analog voltage signals. It can be based on a noninverting or inverting topology.

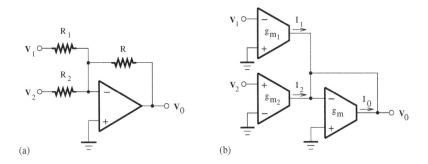

(a)　　　　　　(b)

FIGURE 7.65
Summer circuits based on (a) voltage amplifiers and (b) transconductance amplifiers.

The circuit diagram of an inverting summer based on a voltage amplifier is depicted in Figure 7.65(a). The currents due to the voltages V_1 and V_2 and flowing through the resistors R_1 and R_2, respectively, are added at the virtual ground terminal of the amplifier, and the sum is converted to a voltage with the reversed polarity by the feedback resistor, R. Ideally, $V^- = 0$ and the application of Kirchhoff's current law at the inverting node of the amplifier gives

$$\frac{V_1}{R_1} + \frac{V_2}{R_2} = -\frac{V_0}{R}.\tag{7.188}$$

Hence,

$$V_0 = -\frac{R}{R_1}V_1 - \frac{R}{R_2}V_2.\tag{7.189}$$

This circuit can be extended to more than two inputs because the input signals are isolated from each other due to the virtual ground.

A summer can also be implemented using transconductance amplifiers, as illustrated in Figure 7.65(b). The input transconductors operate as a voltage-to-current converter, while the output transconductor is configured as a current-to-voltage converter. It can be shown that

$$I_1 + I_2 = -I_0,\tag{7.190}$$

where $I_1 = -g_{m_1}V_1$, $I_2 = g_{m_2}V_2$, and $I_0 = -g_m V_0$. The output voltage can then be written as

$$V_0 = \frac{g_{m_2}}{g_m}V_2 - \frac{g_{m_1}}{g_m}V_1.\tag{7.191}$$

Note that the polarity of an input signal connected to the inverting node of the input transconductor remains unchanged, and the polarity inversion is achieved by connecting the input signal to the noninverting node of the input transconductor.

7.2.6   Gyrator

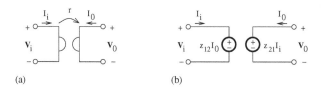

FIGURE 7.66
(a) Symbol and (b) small-signal equivalent model of a gyrator.

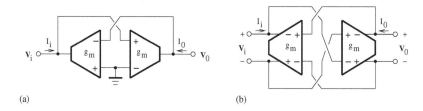

FIGURE 7.67
(a) Single-ended and (b) differential implementations of a gyrator.

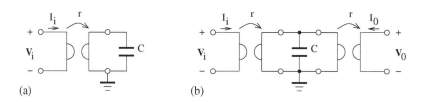

FIGURE 7.68
Equivalent circuit models of gyrator-based implementations of (a) grounded and (b) floating inductors.

In the design of integrated filters, gyrators can be used to eliminate the inductors of the LC prototype [50]. The symbol and small-signal equivalent model of a gyrator are shown in Figure 7.66. For an ideal gyrator, $V_i = z_{12}I_0$ and $V_0 = z_{21}I_i$. Provided that $z_{12} = -z_{21} = r$, we get

$$\begin{bmatrix} V_0 \\ V_i \end{bmatrix} = \mathbf{Z} \begin{bmatrix} I_0 \\ I_i \end{bmatrix}, \tag{7.192}$$

where the open-circuit impedance of a gyrator is of the form

$$\mathbf{Z} = \begin{bmatrix} 0 & -r \\ r & 0 \end{bmatrix} \tag{7.193}$$

and r is the gyrator resistance. A gyrator is then a linear two-port network, the input impedance of which is given by

$$Z_i(s) = \frac{r^2}{Z_L(s)}, \qquad (7.194)$$

where Z_L is the load impedance. Let $Z_L(s) = 1/sC$, that is, the gyrator is loaded by a capacitor C. We can obtain

$$Z_i(s) = r^2 C s = L s, \qquad (7.195)$$

where $L = r^2 C$ is the value of the simulated inductor. Single-ended and differential implementations of a gyrator are shown in Figures 7.67(a) and (b), respectively. They are based on inverting and noninverting transconductors designed to exhibit the same value $g_m = 1/r$. Here, the accuracy of the gyrator can be limited by the parasitic input and output capacitances, and finite output conductance of transconductors.

The equivalent circuit models of grounded and floating inductors are shown in Figures 7.68 (a) and (b), respectively. In the first configuration, an inductor with the impedance Z is realized, while the second network can be characterized by the following short-circuit admittance matrix:

$$\mathbf{Y} = \frac{1}{Z} \begin{bmatrix} 1 & -1 \\ -1 & 1 \end{bmatrix}. \qquad (7.196)$$

7.3 Filter characterization

A filter can be characterized by its transfer function, which is the ratio of two s-domain polynomials with real coefficients and is given by

$$H(s) = \frac{Y(s)}{U(s)} = \frac{\sum_{j=0}^{M} b_j s^j}{1 + \sum_{i=1}^{N} a_i s^i}, \qquad (7.197)$$

where $U(s)$ and $Y(s)$ are the Laplace transforms of the input and the output variables. The zeros of the denominator are called poles of the filter, while the zeros of the numerator are referred to as transmission zeros. The filter will be realizable if $N \geq M$ and will not oscillate provided $a_i > 0$. For filter stability, the real part of all poles should be lower than zero.

On the $j\omega$-axis, the filter transfer function can expressed as

$$H(j\omega) = |H(j\omega)| e^{j\phi(\omega)}. \qquad (7.198)$$

The magnitude of the transfer function can be written either in the form of a gain

$$G(\omega) = 20 \log |H(j\omega)| \qquad (7.199)$$

or an attenuation

$$A(\omega) = -20 \log |H(j\omega)|, \qquad (7.200)$$

using decibel (dB) units. The phase response, $\phi(\omega)$, is expressed in radians. The group delay of the filter is defined as

$$\tau(\omega) = -\frac{d\phi(\omega)}{d\omega}. \qquad (7.201)$$

It is often used as a specification in applications where the behavior in the time domain is of importance, for example, allpass equalizer. Note that, the transfer function of an allpass filter is of the form $H(s) = k \cdot D(-s)/D(s)$, where k is the dc gain and $D(s)$ is the transfer function denominator.

7.4 Filter design methods

Continuous-time filters process high-speed signals in applications where the linearity and accuracy specifications are relaxed, such as disk drives, communication devices, and video systems. In CMOS technology, the integration of active RC-filters, which are realized using voltage amplifiers, resistors and capacitors, may be impractical due the high resistance values required. This problem is solved in implementation approaches, such as g_m-C filters, which are only based on transconductance amplifiers and capacitors.

A filter can be categorized according to the type of function used to approximate its gain or attenuation specifications. Butterworth, Tchebychev, or Cauer approximation functions are often used in the synthesis of a stable and realizable filter prototype. To simplify the computation of filter coefficients, the transfer function is scaled to have a maximum magnitude of unity, and the frequency, ω, is normalized with respect to the passband edge frequency, ω_p, such that $\Omega = \omega/\omega_p$. The order of the transfer function is determined from the prescribed magnitude specifications in the passband,

$$1/\sqrt{1 + \delta_p^2} \leq |H(\Omega)| \leq 1, \quad \text{if} \quad 0 \leq |\Omega| \leq 1, \qquad (7.202)$$

and in the stopband,

$$|H(\Omega)| \leq 1/\sqrt{1 + \delta_s^2}, \quad \text{if} \quad |\Omega| \geq \Omega_s, \qquad (7.203)$$

where Ω_s is the normalized stopband edge frequency. The passband ripple, δ_p, and stopband ripple, δ_s, can be computed as

$$\delta_p^2 = 10^{A_p/10} - 1 \qquad (7.204)$$

and

$$\delta_s^2 = 10^{A_s/10} - 1, \qquad (7.205)$$

where the passband and stopband attenuations in dB are respectively defined as $A_p = -\min(20\log_{10}|H(\Omega)|)$ for Ω in the filter passband, and $A_s = -\max(20\log_{10}|H(\Omega)|)$ for Ω in the filter stopband.

The determination of the filter transfer function focusses primarily on low-pass prototype filters, as highpass, bandpass, or bandstop prototypes can be derived from lowpass prototypes through appropriate spectral transformations. Note that, in the case of Bessel filters, the design is often performed based on the delay or phase specifications, that is, the maximally acceptable delay errors in the frequency band of interest. The synthesis of the transfer function, which is also referred to as the approximation problem, is performed using tables of classical filter functions or software tools such as MATLAB.

There are several types of analog filters with various important characteristics. The most often used filter types are based on Butterworth, Chebyshev, Cauer (or elliptic), and Bessel (or Thomson) approximation functions. The choice of a given filter is generally made depending on the application.

Butterworth filters: They are characterized by a magnitude response that is maximally flat in the passband and is monotonically decreasing in both the passband and stopband. However, the transition band of Butterworth filters is somewhat wide.

Chebyshev filters: They are known to have a steeper roll-off than Butterworth filters. Their magnitude responses either have equal ripples in the passband and decrease monotonically in the stopband (type I) or decrease monotonically in the passband and exhibit equal ripples in the stopband (type II or inverse Chebyshev).

Cauer filters: Their magnitude responses have equal ripples in both the passband and stopband. In comparison with other filter types, Cauer filters provide the sharpest transition band and are then the most efficient from the viewpoint of requiring the smallest order to realize a given magnitude specification. However, the sharp transition band is obtained at the price of a more nonlinear phase response in the passband, especially near the passband edge.

Bessel filters: They exhibit a maximally linear phase response or maximally flat group delay in the passband. However, Bessel filters can be plagued by their largest transition band.

Butterworth, Chebyshev, Cauer, and Bessel filters are synthesized with the focus on one primordial characteristic. In practice, they are often used directly, but their zero and pole placement can also be further optimized to simultaneously satisfy several characteristic specifications.

High-order filters that simulate the behavior of doubly resistively terminated LC networks are known to exhibit the minimum sensitivity to component variations, in contrast to cascade realizations based on first-order and second-order (or biquadratic) filter structures. The frequency normalization, which corresponds to a frequency scale, should be taken into account in the determination of the nominal component values for the filter implementation. Because it is performed by dividing the variable, s, by the passband edge frequency, ω_p, the normalized variable is then of the form, $s' = s/\omega_p$, and the normalized element values can be related to actual values (R, C, and L) of the filter components as follows

$$R' = \frac{R}{\tilde{R}}, \qquad C' = \tilde{R}C\omega_p, \qquad L' = L\frac{\omega_p}{\tilde{R}}, \qquad (7.206)$$

where the normalizing resistance, $\tilde{R}$, is generally chosen to obtain well-suited and practical element values. The filter synthesis results in a circuit that should be analyzed to take into account the effect of real component imperfections and dynamic range on the overall performance.

7.4.1 First-order filter design

In general, the transfer function of a first-order filter section can be written as

$$H(s) = \frac{k_1 s + k_0}{s + \omega_c}. \qquad (7.207)$$

Table 7.1 presents the different types of first-order filters, which can be implemented using g_m-C circuits. The transfer function coefficients are then determined by the ratios of component values.

TABLE 7.1
First-Order Filter Section Classification

First-Order Filter Type	Transfer Function Coefficients
Lowpass filter	$k_1 = 0$
Highpass filter	$k_0 = 0$
Allpass filter	$k_1 = k,\ k_0 = -k\,\omega_c$

Applying Kirchhoff's current law to the first-order filter depicted in Figure 7.69, it can be shown that

$$I_1 + I_2 = I'. \qquad (7.208)$$

Since $I_1 = g_{m1}V_i$ and $I_2 = -g_{m2}V_0$, we arrive at

$$g_{m1}V_i - g_{m2}V_0 = sCV_0. \qquad (7.209)$$

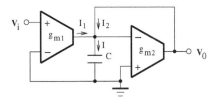

FIGURE 7.69

g_m-C implementation of a first-order lowpass filter.

The transfer function can then be computed as

$$H(s) = \frac{V_0(s)}{V_i(s)} = \frac{g_{m1}}{sC + g_{m2}}. \tag{7.210}$$

We see that $H(s)$ has a dc gain with the value, g_{m1}/g_{m2}, and a single pole at $-g_{m2}/C$. Equation (7.210) therefore characterizes a first-order lowpass filter.

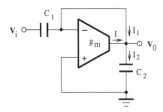

FIGURE 7.70

g_m-C implementation of a first-order highpass filter.

Proceeding with the circuit of Figure 7.70, we find an output node equation of the form

$$I_1 + I = I_2. \tag{7.211}$$

In terms of the device parameters, this equation can be rewritten as

$$sC_1(V_i - V_0) - g_m V_0 = sC_2 V_0. \tag{7.212}$$

The transfer function is then derived as

$$H(s) = \frac{V_0(s)}{V_i(s)} = \frac{sC_1}{s(C_1 + C_2) + g_m}. \tag{7.213}$$

In this case, a zero and a pole are located at 0 and $-g_m/(C_1 + C_2)$, respectively. Because the dc gain is zero, and the gain at high frequencies is of the form, $C_1/(C_1 + C_2)$, the circuit realizing $H(s)$ is referred to as a first-order highpass filter.

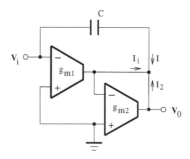

FIGURE 7.71

g_m-C implementation of a first-order allpass filter.

The circuit shown in Figure 7.71 includes two transconductance amplifiers and a capacitor. The output node equation can be written as

$$I + I_1 + I_2 = 0 \tag{7.214}$$

or equivalently,

$$sC(V_i - V_0) + g_{m1}V_i - g_{m2}V_0 = 0. \tag{7.215}$$

The transfer function is then given by

$$H(s) = \frac{V_0(s)}{V_i(s)} = \frac{sC - g_{m1}}{sC + g_{m2}}. \tag{7.216}$$

Assuming that $g_{m1} = g_{m2} = g_m$, the magnitude of $H(s)$ is equal to unity for all frequencies, and a first-order allpass filter is realized. The function $H(s)$ exhibits a pole and a zero located symmetrically on either side of the $j\omega$ axis. Hence, the phase response is readily found to be

$$\theta(\omega) = \arctan\left(-\frac{\omega}{\omega_c}\right) - \arctan\left(\frac{\omega}{\omega_c}\right) = \pi - 2\arctan\left(\frac{\omega}{\omega_c}\right), \tag{7.217}$$

where $\omega_c = g_m/C$. Note that the phase is π at $\omega = 0$, $\pi/2$ at $\omega = \omega_c$, and zero at high frequencies.

7.4.2   Biquadratic filter design methods

The design of a biquadratic or second-order filter can be based on signal-flow graphs (SFGs) or gyrators. In some cases, the resulting filter can retain the low sensitivity properties of a doubly terminated lossless network.

7.4.2.1   Signal-flow graph-based design

The SFG block diagram realizing a general biquadratic transfer function is shown in Figure 7.72. It consists of integrators and summing stages. Its transfer function is given by

$$H(s) = \frac{V_0(s)}{V_i(s)} = \frac{k_2 s^2 + k_1 s + k_0}{s^2 + \left(\dfrac{\omega_p}{Q}\right)s + \omega_p^2}, \qquad (7.218)$$

where ω_p and Q denote the pole frequency and quality factor, respectively. For the filter stability, the parameters ω_p and Q should assume only positive values.

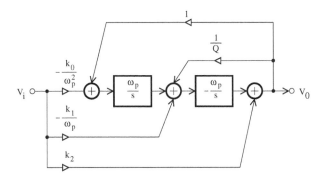

FIGURE 7.72
Signal-flow graph representation of a general biquadratic filter.

The transfer function of the g_m-C biquad circuit shown in Figure 7.73 [51] can be computed as

$$H(s) = \frac{V_0(s)}{V_i(s)} = \frac{k_2 s^2 + k_1 s + k_0}{s^2 + \left(\dfrac{\omega_p}{Q}\right)s + \omega_p^2}, \qquad (7.219)$$

where

$$k_0 = \frac{g_{m1} g_{m2} g_{m3}}{g_{m6} C_1 C_2}, \qquad (7.220)$$

$$k_1 = \frac{g_{m2} g_{m4}}{g_{m6} C_2}, \qquad (7.221)$$

$$k_2 = \frac{g_{m5}}{g_{m6}}, \qquad (7.222)$$

$$\omega_p = \sqrt{\frac{g_{m1} g_{m8}}{g_{m7} C_1}}, \qquad (7.223)$$

and

$$Q = \frac{g_{m6} C_2}{g_{m2} g_{m7}} \sqrt{\frac{g_{m1} g_{m8}}{g_{m7} C_1}}. \qquad (7.224)$$

Different biquad types can be realized depending on the choice of the transfer

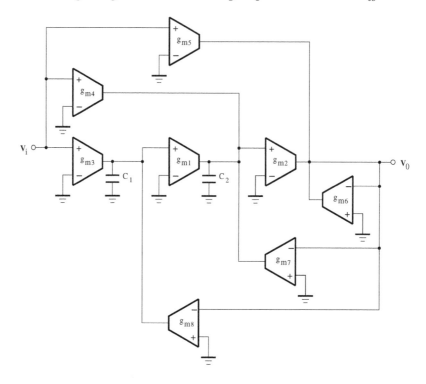

FIGURE 7.73
g_m-C implementation of a general biquadratic filter.

TABLE 7.2
Biquad Classification

Biquad Type	Transfer Function Coefficients
Lowpass filter	$k_0 = k\omega_p^2$, $k_1 = k_2 = 0$
Highpass filter	$k_0 = k_1 = 0$, $k_2 = k$
Bandpass filter	$k_0 = k_2 = 0$, $k_1 = k(\omega_p/Q_p)$
Lowpass notch filter	$k_0 = k\omega_z^2$, $k_1 = 0$, $k_2 = k$, and $\omega_z > \omega_p$
Highpass notch filter	$k_0 = k\omega_z^2$, $k_1 = 0$, $k_2 = k$, and $\omega_z < \omega_p$
Symmetrical notch filter	$k_0 = k\omega_z^2$, $k_1 = 0$, $k_2 = k$ and $\omega_z = \omega_p$
Allpass filter	$k_0 = k\omega_p^2$, $k_1 = -k(\omega_p/Q)$, $k_2 = k$

function coefficients (see Table 7.2).

- In the case of the second-order lowpass filter, the magnitude response is maximum at the frequency where $d|H(\omega)|/d\omega = 0$. This frequency is given by

$$\omega_0 = \omega_p\sqrt{1 - 1/(2Q^2)}, \tag{7.225}$$

and the maximum gain is of the form

$$H_{max} = |H(\omega_0)| = \frac{kQ}{\sqrt{1 - 1/(4Q^2)}}. \tag{7.226}$$

The dc gain is k and the gain decreases as $1/\omega^2$ at high frequencies.

• Considering the highpass filter, the maximum gain is still given by Equation (7.226), but occurs at the frequency

$$\omega_0 = \frac{\omega_p}{\sqrt{1 - 1/(2Q^2)}}. \tag{7.227}$$

The magnitude response increases as ω^2 for low frequencies and the high-frequency gain is k.

• For the bandpass filter, the -3 dB bandwidth is computed as the difference between the passband edge frequencies, that is,

$$\mathrm{BW} = \omega_2 - \omega_1 = \frac{\omega_p}{Q}, \tag{7.228}$$

where $\omega_p^2 = \omega_2\omega_1$. The maximum value of the magnitude response, which is equal to k, occurs at the frequency ω_p. The increase and decrease in the magnitude response both follow a $1/\omega$ function.

• The notch filter exhibits a magnitude response with transmission zero at $\omega = \omega_z$. For the lowpass and highpass notch filters, the maximum gain can be computed as $H_{max} = |H(\omega_0)|$, where

$$\omega_0 = \omega_p\sqrt{\frac{\dfrac{\omega_z^2}{\omega_p^2}\left(1 - \dfrac{1}{2Q^2}\right) - 1}{\dfrac{\omega_z^2}{\omega_p^2} + \dfrac{1}{2Q^2} - 1}}. \tag{7.229}$$

The dc gain and high-frequency gain are equal to $k(\omega_z^2/\omega_p^2)$ and k, respectively. In the case of the symmetrical notch filter, the -3 dB bandwidth is ω_p/Q_p, and the dc gain and high-frequency gain are equal to the same value, k.

• The second-order allpass filter is characterized by a constant gain equal to k, but its phase response changes from $0°$ to $-360°$ and takes the value $-180°$ at ω_p.

7.4.2.2 Gyrator-based design

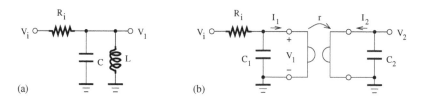

(a) (b)

FIGURE 7.74
(a) Second-order bandpass RLC prototype; (b) circuit diagram of a second-order bandpass filter based on a gyrator.

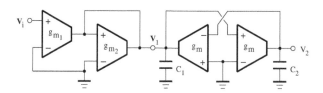

FIGURE 7.75
Single-ended g_m-C implementation of the second-order bandpass filter.

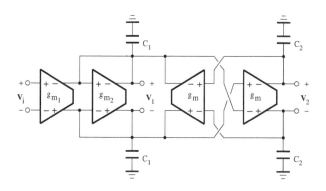

FIGURE 7.76
Differential g_m-C implementation of the second-order bandpass filter.

To achieve a low sensitivity to the variations of component values, g_m-C filters can be designed by simulating passive RLC prototypes. A second-order bandpass RLC prototype filter is shown in Figure 7.74(a). By replacing the inductor with the corresponding gyrator network, the circuit diagram of Figure 7.74(b) is derived. Using Kirchhoff's current law, we can obtain

$$\frac{V_i - V_1}{R_i} = V_1 s C_1 + I_1, \tag{7.230}$$

$$I_2 + V_2 s C_2 = 0, \tag{7.231}$$

where $V_1 = rI_2$ and $V_2 = -rI_1$. The system of Equations (7.230) and (7.231) can be solved either for a bandpass transfer function of the form

$$H_1(s) = \frac{V_1(s)}{V_i(s)} = \frac{s(r^2/R_i)C_2}{s^2r^2C_1C_2 + s(r^2/R_i)C_2 + 1} \qquad (7.232)$$

or a lowpass transfer function given by

$$H_2(s) = \frac{V_2(s)}{V_i(s)} = -\frac{r/R_i}{s^2r^2C_1C_2 + s(r^2/R_i)C_2 + 1}. \qquad (7.233)$$

The single-ended and differential g_m-C implementations of the filter are shown in Figures 7.75 and 7.76, respectively. They realize the transfer functions H_1 and H_2 with $R_i = 1/g_{m_1} = 1/g_{m_2}$ and $r = 1/g_m$. In practice, it may be preferable to use only one filter output at a time because a resistive load at the node V_1 can affect the transfer function associated with the node V_2, and vice versa.

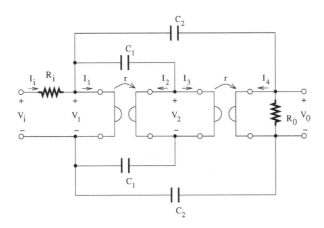

FIGURE 7.77
Circuit diagram of a second-order allpass network based on gyrators.

The circuit diagram of a second-order allpass network based on gyrators is shown in Figure 7.77. It is based on a nonreciprocal[1] network, which cannot be directly derived from LC filter prototypes. The nodal equations can be written as

$$\frac{V_i - V_1}{R_i} = I_1 + (V_1 - V_2)sC_1 + (V_1 - V_0)sC_2 \qquad (7.234)$$

$$(V_1 - V_2)sC_1 = I_2 + I_3 \qquad (7.235)$$

$$(V_1 - V_0)sC_2 = I_4 + \frac{V_0}{R_0}, \qquad (7.236)$$

---

[1]A two-pair network described by the chain matrix $\mathbf{M}$ is nonreciprocal if $\det(\mathbf{M}) \neq \pm 1$. In the special case of LC networks, $\det(\mathbf{M}) = 1$.

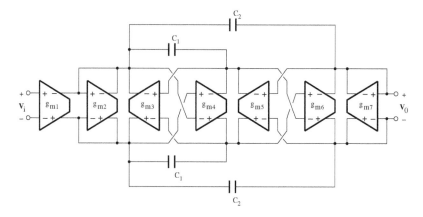

FIGURE 7.78
Differential implementation of the allpass filter based on gyrators.

where $I_1 = -V_2/r = -I_4$, $I_2 = V_1/r$, and $I_3 = -V_0/r$. Solving the system of Equations (7.234) through (7.236) gives

$$H(s) = \frac{V_0(s)}{V_i(s)} = \frac{s^2 r^2 C_1 C_2 - sr C_1 + 1}{s^2 r^2 C_1 C_2 \left(1 + \dfrac{R_i}{R_0}\right) + s C_1 \left(R_i + \dfrac{r^2}{R_0}\right) + 1 + \dfrac{R_i}{R_0}}. \quad (7.237)$$

To realize a second-order allpass transfer function, it is required to set either $r = R_i$ or $r = R_0$. Hence,

$$H(s) = \frac{V_0(s)}{V_i(s)} = k \frac{\dfrac{s^2}{\omega_0^2} - \dfrac{s}{\omega_0 Q} + 1}{\dfrac{s^2}{\omega_0^2} + \dfrac{s}{\omega_0 Q} + 1}, \quad (7.238)$$

where

$$\frac{1}{k} = 1 + \frac{R_i}{R_0}, \quad \frac{1}{\omega_0^2} = r^2 C_1 C_2, \quad \text{and} \quad \frac{1}{\omega_0 Q} = r C_1. \quad (7.239)$$

The differential implementation of the allpass filter is depicted in Figure 7.78. The gyrators and resistors are replaced by the corresponding g_m-C structures.

The design of high-order filters is generally achieved after the pole-zero pairing and gain assignment related to each low-order section obtained from the decomposition of the target transfer function. Once a suitable topology is found and the filter coefficient mapping is carried out, the scaling of the component values can be performed to obtain the desired dynamic range [53] and chip area. Remark that the distortion level at the filter output can be set under a prescribed value (e.g., 1% of the input signal magnitude) by biasing adequately the amplifiers.

7.4.3 Ladder filter design methods

High-order active filters can be derived either from LC ladder networks or using signal-flow graph techniques. The first approach consists of using various component-simulation methods, which can be based on gyrator elements or generalized immittance converters, to replace inductors (and sometimes resistors), while the second approach relies on simulating equations characterizing the operation of a lossless network prototype.

7.4.3.1 LC ladder network-based design

Active filters derived from LC ladder networks should feature a lower sensitivity to component variations in the passband in comparison to the cascade of low-order structures. The LC ladder shown in Figure 7.79 realizes a third-order lowpass elliptic filter. It can be described by

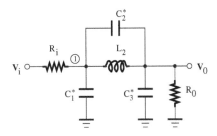

FIGURE 7.79
Third-order elliptic lowpass filter.

$$V_1 = \frac{1}{s(C_1^* + C_2^*)}\left(sC_2^*V_0 - I_{L_2} + \frac{V_i - V_1}{R_i}\right) \tag{7.240}$$

$$I_{L_2} = \frac{V_1 - V_0}{sL_2} \tag{7.241}$$

$$V_0 = \frac{1}{s(C_2^* + C_3^*)}\left(sC_2^*V_1 + I_{L_2} - \frac{V_0}{R_0}\right). \tag{7.242}$$

Let $R_i = R_0 = R$, $C_1^* = C_3^* = C^*$, and $\hat{s} = s/\omega_c$, where ω_c is the cutoff frequency. The normalized transfer function of the LC filter of Figure 7.79 can be written as

$$H(\hat{s}) = \frac{V_0(\hat{s})}{V_i(\hat{s})} = k\frac{\hat{s}^2 + \alpha}{(\hat{s} + \beta)(\hat{s}^2 + \beta_1\hat{s} + \beta_2)}, \tag{7.243}$$

where

$$k = \frac{C_2^*}{RC^*(C^* + 2C_2^*)} \tag{7.244}$$

$$\alpha = \frac{1}{L_2 C_2^*} \tag{7.245}$$

$$\beta = \frac{1}{RC^*} \tag{7.246}$$

$$\beta_1 = \frac{1}{R(C^* + 2C_2^*)} \tag{7.247}$$

$$\beta_2 = \frac{2}{L_2(C^* + 2C_2^*)} \tag{7.248}$$

and the resistor R can be chosen equal to one due to the impedance-level normalization.

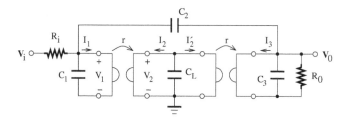

FIGURE 7.80
Single-ended circuit of the third-order elliptic lowpass filter based on gyrators.

Starting with the normalized LC network prototype derived from a given filter specification, the inductors can be eliminated with the help of gyrators. Figures 7.80 and 7.81 show the single-ended and differential circuit diagrams of the third-order elliptic lowpass filter of Figure 7.79 after the inductor substitution.

Using Kirchhoff's current laws, the analysis of the circuit depicted in Figure 7.80 results in

$$\frac{V_i - V_1}{R_i} = V_1 s C_1 + (V_1 - V_0)s C_2 + I_1 , \tag{7.249}$$

$$(V_1 - V - 0)s C_2 = V_0 s C_3 + \frac{V_0}{R_0} + I_3 , \tag{7.250}$$

$$V_2 s C_L + I_2 + I_2' = 0, \tag{7.251}$$

where $V_1 = rI_2$, $V_2 = -rI_1 = rI_3$, and $V_2 = -rI_2'$. To proceed further, it can be assumed that $C_1 = C_3 = C$ and $R_i = R_0 = R$. Solving the system of Equations (7.249) through (7.251) for the output-input voltage transfer function gives

$$H(s) = \frac{V_0(s)}{V_i(s)}$$

$$= \frac{C_2}{RC(C + 2C_L)} \frac{s^2 + \dfrac{1}{r^2 C_2 C_L}}{\left(s + \dfrac{1}{RC}\right)\left[s^2 + \dfrac{s}{R(C + 2C_L)} + \dfrac{2}{r^2 C_L(C + 2C_L)}\right]} \cdot$$

$$(7.252)$$

Comparing Equations (7.243) and (7.252), the component values for a single-ended implementation are of the form $C = C^*/R\omega_c$, $C_2 = C_2^*/r\omega_c$, and $C_L = L_2/r\omega_c$.

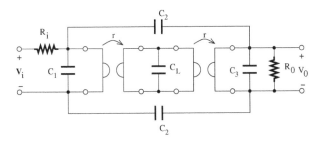

FIGURE 7.81
Differential circuit of the third-order elliptic lowpass filter based on gyrators.

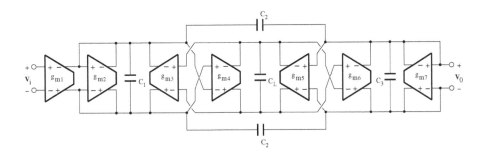

FIGURE 7.82
Differential implementation of the third-order elliptic lowpass filter based on gyrators.

The fully differential g_m-C realization of the filter is shown in Figure 7.82, where $C_1 = C_3 = C$, $g_{m1} = g_{m3} = g_{m6} = 2g_m$, and $g_{m2} = g_{m4} = g_{m5} = g_{m7} = g_m$. The resistor R of LC network and the resistor r are implemented by transconductors. Their values are of the form, $1/g_m$. Due to the inherent two-times increase of the dynamic range in the differential configuration, the sizes of the circuit components can be scaled by a factor of 1/2 and are then given by $C_j = g_m C_j^*/2\omega_c$, $j = 1, 3$, $C_2 = g_m C_2^*/2\omega_c$, and $C_L = g_m L_2^*/2\omega_c$. The resulting transfer function of the g_m-C filter reads

$$H(\hat{s}) = \frac{V_0(\hat{s})}{V_i(\hat{s})} = K \frac{\hat{s}^2 + a}{(\hat{s} + b)(\hat{s}^2 + b_1\hat{s} + b_2)}, \tag{7.253}$$

where

$$K = \frac{2g_m C_2}{C(C + 2C_2)}, \tag{7.254}$$

$$a = \frac{g_m^2}{C_L C_2}, \tag{7.255}$$

$$b = \frac{g_m}{C}, \tag{7.256}$$

$$b_1 = \frac{g_m}{C + 2C_2}, \tag{7.257}$$

and

$$b_2 = \frac{2g_m^2}{C_L(C + 2C_2)}. \tag{7.258}$$

The filter gain and cutoff frequency are determined by the values of transconductors and capacitors. However, an automatic tuning is generally required to control the transconductance fluctuations due to fabrication tolerance and temperature variations, and the filter parameters are preferably made programmable using capacitor arrays.

7.4.3.2 Signal-flow graph-based design

Another approach to design continuous-time ladder filters is based on the signal-flow graph (SFG) representation of nodal equations of an LC network. The passive filter structure is decomposed into integrator and gain stages, which can then be realized using circuits of a given type (active RC, MOSFET-C, g_m-C, etc.). The number of amplifiers is expected to be at least equal to the order of the filter to be designed.

The SFG, as shown in Figure 7.83, is derived from the LC network of Figure 7.79. The circuit diagram shown in Figure 7.84, where $C_X = C_1 + C_2$ and $C_Z = C_2 + C_3$, is then obtained using RC circuits. It consists essentially of inverting and noninverting integrators. The filter gain of active RC filter can be actually set to k times the one of the LC network provided $R_X = R_i/k$, where k is a real number. The values of the elements connected to the noninverting integrator are multiplied by a factor R, and the value of the feedback capacitor is actually $C_Y = L_2/R^2$. The circuit diagram of the g_m-C filter is depicted in Figure 7.85. Due to the absence of a summing node in g_m-C integrators, the number of active components is reduced using amplifiers with

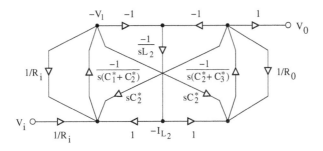

FIGURE 7.83
Signal-flow graph of a third-order elliptic lowpass filter.

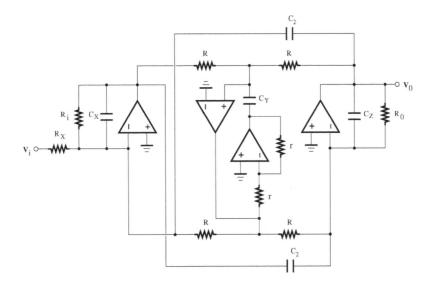

FIGURE 7.84
Active RC realization of a third-order elliptic lowpass filter.

two differential inputs. Here, the resistors simulated by transconductors assume the value $1/g_m$, where g_m is the amplifier transconductance, and the value of the capacitor related to the inductor, L_2, of the passive network is $C_3 = L_2 g_m^2$. The compensation of the 6-dB loss, which is proper to LC networks, is achieved by sizing the input transconductor to exhibit the value $2g_m$, while the transconductance of the remaining amplifiers is g_m.

Filter structures can also be derived from SFGs, which are not directly related to LC prototypes. This approach is suitable for the design of high-order allpass filters [48,54]. Let us consider the transfer function of an allpass filter given by

$$T(s) = \pm k \frac{D(-s)}{D(s)}, \tag{7.259}$$

where k is the desired constant gain and $D(s)$ is the transfer function denom-

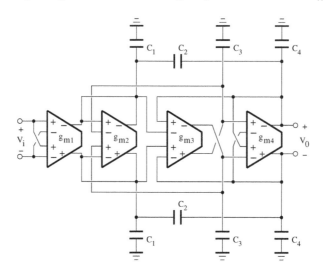

FIGURE 7.85

g_m-C implementation of the third-order elliptic lowpass filter based on the SFG technique.

inator. Assuming that $D(s) = 1 + Y(s)$ and $D(-s) = 1 - Y(s)$, we obtain

$$T(s) = \mp \left[k - \frac{2k}{1 + Y(s)} \right], \tag{7.260}$$

where $Y(s)$ is the admittance function. It can then be found that

$$Y(s) = \begin{cases} \dfrac{D(s) - D(-s)}{D(s) + D(-s)}, & \text{if N is odd} \\ \dfrac{D(s) + D(-s)}{D(s) - D(-s)}, & \text{if N is even,} \end{cases} \tag{7.261}$$

where $D(s)$ is a Hurwitz polynomial. To proceed further, the function $Y(s)$ is expanded as continued fractions. That is,

$$Y(s) = \alpha_1 s + \cfrac{1}{\alpha_2 s + \cfrac{1}{\alpha_3 s + \cdots + \cfrac{1}{\alpha_N s}}}. \tag{7.262}$$

The SFG that realizes Equations (7.260) and (7.262) is depicted in Figure 7.86(a). Its g_m-C implementation is shown Figure 7.86(b), where the capacitor values are of the form

$$C_k = \frac{1}{\alpha_k} \qquad k = 1, 2, 3, \cdots, N. \tag{7.263}$$

Table 7.3 gives the capacitor values provided by the synthesis of allpass filters

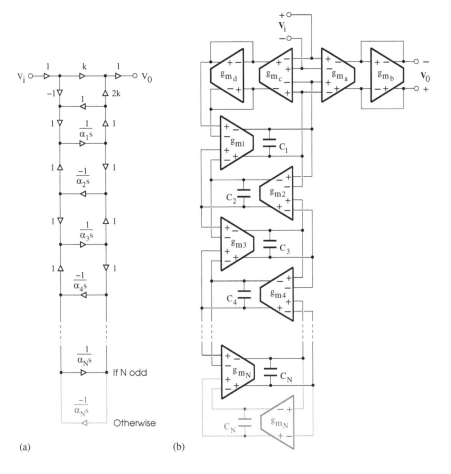

(a) (b)

FIGURE 7.86
(a) Signal-flow graph-based allpass filter; (b) g_m-C implementation of the allpass filter.

TABLE 7.3
Allpass Filter Design Equations for $N = 2$ and 3 $(a > c/b)$

Transfer Function	Capacitor Values
$T(s) = -\dfrac{1 - as + bs^2}{1 + as + bs^2}$	$C_1 = a/b;\ C_2 = 1/a.$
$T(s) = -\dfrac{1 - as + bs^2 - cs^3}{1 + as + bs^2 + cs^3}$	$C_1 = b/c;\ C_2 = \dfrac{a}{b} - \dfrac{c}{b^2};\ C_3 = \dfrac{1}{a - c/b}.$

with $N = 2$ and 3. Note that another circuit topology can be derived by decomposing $Y(s)$ into partial fractions. In any case, the resulting allpass filter can be realized such that its amplitude response remains insensitive to variations of some component values. This low-sensitivity property is especially useful in high-Q applications.

By scaling the component values of continuous-time filters conceived by one of the above design methods, the dynamic performance can be improved and the component values are set to practical IC sizes. The signal dynamic range is limited by the noise and saturation level of the amplifiers. The scaling relies on the fact that the transfer function from the filter input to the output of an amplifier will remain unchanged if the components in its forward and feedback paths are appropriately multiplied or divided by a given factor. The scaling generally ensures that the output signals of the amplifiers saturate at the same level of the input signal. The scaling factor can be made equal to either a constant or $V_{0j,max}/V_{0,max}$, where $V_{0j,max}$ and $V_{0,max}$ denote the maximum output voltage of the j-th amplifier and the filter, respectively.

7.5 Design considerations for continuous-time filters

The responses of continuous-time (CT) filters are determined by the values of the capacitors, C, resistors, R, and transconductors, g_m. Due to factors such as fabrication tolerances, temperature variation, and aging, time constants (RC products or g_m/C ratios) can drift to about 10 to 50% from their nominal values. As a consequence, the electrical characteristics of the filters do not meet the design specifications.

7.5.1 Automatic on-chip tuning of continuous-time filters

The tasks of signal processing generally encountered in the front-end section of a signal processor, namely, amplification of the desired channel to the full scale of the analog-to-digital converter, automatic gain control, and filtering to remove the interference of the adjacent channels are preferably implemented using CT circuits. An analog equalizer for read channel in magnetic recording, for instance, operates at the given data rate using lower area and power than that required by an equivalent digital structure.

In a cellular phone system, the channels can be spaced at 25-kHz intervals, and each channel needs a bandwidth of 21 kHz. The desired channel can be selected by a bandpass filter whose passband is one channel wide, only if the center frequency accuracy is strictly less than 4 kHz. For a center frequency fixed at 450 kHz, this corresponds to a precision requirement better than 0.8% (that is, 4 kHz/450 kHz). The specifications of a bandpass filter required for channel selection in communication systems are illustrated in Figure 7.87, where the parameters can be specified as follows: $f_{02} = 450$ kHz, $2f_p = 21$ kHz, and $\triangle f = 25$ kHz.

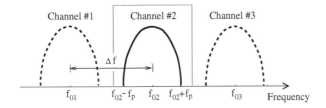

FIGURE 7.87

Bandpass filter specifications required for channel selection in communication systems.

Sigma-delta ($\Sigma\Delta$) modulators using a CT filter have the advantage of sampling the signals at higher frequencies than the ones based on discrete-time circuits. They are then suitable for the design of wireless receivers operating at the intermediate frequency. However, the noise transfer function of CT modulators can effectively suppress the quantization noise at the desired frequency only if the pole locations of the loop filter are accurately defined. Digitizing a 200-kHz band at 70 MHz, for instance, would require a center frequency precision better than 200 kHz/70 MHz, or equivalently about 0.3%.

The solution that is generally adopted is to design CT filters with an associated automatic tuning scheme, as shown in Figure 7.88 [55, 56]. Note that the reference signal (REF) provided by a crystal oscillator is connected to the frequency-tuning master only when this latter is a voltage-controlled filter. Here, the tuning scheme is based on the master-slave technique. It con-

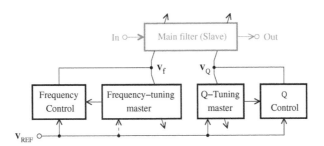

FIGURE 7.88

Block diagram of the master-slave tuning scheme.

sists of four major blocks: the main filter or slave, master, frequency control, and Q control circuits. The input signal is processed and filtered by the slave. The frequency-tuning master can be either a voltage-controlled oscillator (VCO) [31, 57] or a voltage-controlled filter (VCF) [58–60] and it models the slave's behavior that is essential for the tuning. The control circuits sense the master output signal and compare it to the one supplied by the reference source. The master is tuned until these two quantities are similar. The correction signals generated by the control circuits and used for this operation are simultaneously applied to the master and to the slave, which is then indirectly tuned.

The tuning circuit can principally be divided into two parts: the frequency and Q-tuning system. Generally, the automatic tuning of CT filters consists of locking the filter's response to an external and accurate reference. Next, various implementations of on-chip tuning approaches will be reviewed and the errors and performance limitations due to nonideal effects will also be discussed.

7.5.2 Nonideal integrator

Integrators are the principal building block of fully integrated filters. It can then be expected that the integrator nonidealities will affect the overall filter performance. Figure 7.89 shows the ideal and real frequency responses of an integrator. Ideally, an integrator has a pole at the origin and a phase shift

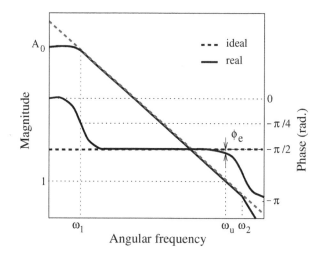

FIGURE 7.89
Frequency response of an ideal and real integrator.

of $-\pi/2$ at all frequencies. It is described by the transfer function $T_i(s) = -\omega_u/s$, where ω_u is the unity-gain frequency. However, due to the parasitic effects in real amplifiers, the frequency response exhibits further deviations for frequencies greater than ω_u. Therefore, an integrator model, which takes into account the second-order effects of additional poles, is characterized by the following transfer function,

$$T_r(s) = \frac{A_0}{\left(1 + \dfrac{s}{\omega_1}\right)\left(1 + \dfrac{s}{\omega_2}\right)}, \qquad (7.264)$$

where A_0 is the finite dc gain, and $\omega_1 = \omega_u/A_0$ and ω_2 denote the dominant and nondominant poles, respectively. It is assumed that $\omega_1 \ll \omega_u \ll \omega_2$. As depicted in Figure 7.89, the phase response decreases to values below $-\pi/2$ for higher frequencies, and ϕ_e denotes the phase error at the unity-gain frequency. This can also be confirmed by analyzing the phase variations of the function T_r given by[2]

$$\phi(\omega) = -\frac{\pi}{2} + \triangle\phi(\omega), \qquad (7.265)$$

where $\triangle\phi(\omega) = \arctan(\omega_1/\omega) - \arctan(\omega/\omega_2)$. Low dc gains and parasitic poles result respectively in leading and lagging phase errors. The phase tolerance in worst cases depends on the specifications (quality factor, pole frequency, etc.) of the filter to be designed.

7.6 Frequency-control systems

Frequency-control systems can be implemented using the phase-locked-loop-based technique or charge comparison-based technique.

7.6.1 Phase-locked-loop-based technique

7.6.1.1 Operation principle

The frequency-control system, as depicted in Figure 7.90, where V_f is the frequency tuning signal, includes a master and a phase detector. Here, the variable of interest is the phase of the filter transfer function.

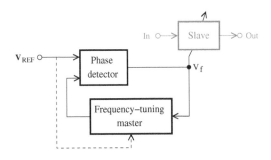

FIGURE 7.90
Frequency tuning scheme based on the phase-locked loop principle.

The entire system will operate like a phase-locked loop when the master is a VCO. This latter attempts to produce a signal that tracks the phase of the reference signal. The phase detector can consist of an analog multiplier or an exclusive OR (XOR) gate driven by a pair of comparators, followed by a loop filter (or an integrator). It measures the phase difference between the

---

[2]Note that $\arctan(1/x) = -\arctan(x) + [\pi \operatorname{sgn}(x)]/2$, provided $x \neq 0$ and $\arctan(-x) = -\arctan(x)$.

above two waveforms and produces the control signal that drives the VCO and changes its oscillation frequency in order to reduce the phase error. Because the integrators in the slave are identical to those of the VCO and are tuned by the same voltage, the frequency response of the slave is also made accurate and stable.

Another alternative consists of using a master with a VCF structure. In this case, the reference signal is also applied to the master (dashed line). Because the filter phase characteristic is monotonic, it is appropriate for the determination of the resonance frequency location with respect to the reference frequency. The phase difference between the VCF output and an accurate external clock signal is used to generate a tuning signal that alters the resonance frequency of the VCF until it equals the reference frequency.

7.6.1.2 Architecture of the master: VCO or VCF

For efficient tuning, the master should be able to model the pertinent filter characteristic. It can be realized by either a VCO or a VCF including the basic building blocks of the filter.

• VCO

A tuning scheme based on a VCO operates as a phase-locked loop. The roll-off frequency of the filter is then controlled by locking the frequency of the VCO output to the one of a reference signal. Specifically, in the case where the slave is obtained by an interconnection of biquads, the VCO can have the structure of a second-order section, the poles of which are always on the imaginary axis. The Q factor of the VCO is then infinite and harmonic oscillations can occur. In order to control the amplitude of the oscillations, nonlinear circuit components must be included in the VCO structure. This introduces extra parasitic components at the nodes where these components are connected and can significantly deteriorate the required matching, especially at high frequencies.

• VCF

A biquadratic filter with a center frequency that can be controlled by a voltage is commonly used as the VCF. Its topology must be similar to the one of the slave. The tuning operation essentially exploits the output phase characteristics of the VCF. Because the control voltage supplied by the phase detector changes the filter's pole frequency rather than the phase angle, as is in the case of a VCO master, any offset in the detection stage will result in a frequency tuning error. In order to give a relevant relative phase estimation between the input and output signals, the VCF requires a reference signal with low harmonic content, for example, a sinusoid. Therefore, a square wave cannot be used for this purpose. It should be noted that the Q-factor of the VCF must be high enough to provide a sufficient tuning sensitivity. For a master center frequency of a few megahertz, a Q-value on the order of 10 can meet the requirement. Furthermore, accurate tuning will be achieved only if the dc offsets in the loop and the nonidealities of the phase detector are reduced.

7.6.1.3 Phase detector

A phase detector can be constructed around an analog multiplier or an XOR gate.

• In the first case, let $v_{REF}(t) = V_r \sin(\omega_i t)$ and $v_m(t) = V_m \cos(\omega_i t - \phi)$ be the reference and the master output signals, respectively. The voltage at the output of the analog multiplier shown in Figure 7.91 can be written as

$$v_x(t) = \frac{1}{2} k_x V_r V_m [\sin(\phi) + \sin(2\omega_i t - \phi)], \qquad (7.266)$$

where k_x is the multiplier gain. Assuming that the high-frequency part of this

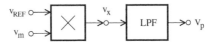

FIGURE 7.91
Block diagram of an analog multiplier-based phase detector.

signal is suppressed by the lowpass filter (LPF), the phase detector output signal is given by

$$V_p = k_p \sin(\phi), \qquad (7.267)$$

where $k_p = k_x H(0) V_r V_m / 2$ is a constant and $H(0)$ is the LPF dc gain.

• In the second case, two comparators are needed to transform the reference and master output signals (v_{REF} and v_m) into square waveforms with a 50% duty cycle. The output voltage, $\overline{v}_x$, of the XOR gate, as shown in Figure 7.92, is high whenever both input signals are different. It is processed by the LPF,

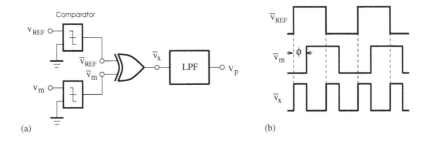

FIGURE 7.92
(a) Block diagram of a phase detector based on XOR gate; (b) input and output waveforms of the XOR gate.

and the average voltage at the output of the phase detector can be expressed as

$$V_p = k_\phi \phi, \tag{7.268}$$

where $0 < \phi < \pi$, $k_\phi = H(0)V_{DD}/\pi$ is the gain of the phase detector, and V_{DD} is the supply voltage.

Note that the value of the phase detector gain, $k_\phi = \partial V_p/\partial \phi$, must be large enough to make the capture and lock ranges insensitive to the amplitudes of the reference and master output signals.

7.6.1.4 Implementation issues

The initial lock of the PLL can be obtained over a frequency range, $\triangle f_c$, which is the capture range of the loop. Once the PLL is locked to the reference signal, the frequency range in which the frequency variations of the reference signal due to change in the operating conditions can still be tracked is called the lock range, $\triangle f_l$. Generally, $\triangle f_c$ is smaller than $\triangle f_l$ and is determined by the cutoff frequency of the LPF. Because the PLL can lock to any harmonic of the reference frequency that can pass through the loop, the cutoff frequency of the LPF must be set just below the second harmonic of the lowest frequency to be tracked. The lock range is related to the PLL dynamic behavior.

The capture and lock ranges of the PLL can be investigated using circuit simulators. This method is advantageous particularly when the macromodels used for the simulation include the different parasitic components and bias-dependent parameters [61], which can affect the tuning and the high-frequency behavior of the filter. Figure 7.93 shows the plots of the tuning voltage versus the frequency. As evidenced by the arrows in these illustrations, the simulation

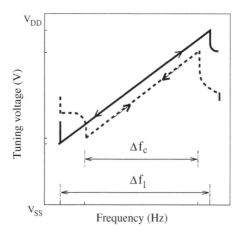

FIGURE 7.93
Lock range, $\triangle f_l$, and capture range, $\triangle f_c$, of the PLL tuning scheme.

of the capture range is done with the assumption that the PLL is initially in its locked state, in contrast to the one of the lock range.

7.6.2 Charge comparison-based technique

The effect of process variations on the resonant frequency of a g_m-C filter can be compensated for using the tuning scheme shown in Figure 7.94. It is based on the charge balancing principle as proposed in [32, 62] and consists of a charge comparator (CC) and an LPF. The dc value of the CC output voltage is extracted by the LPF and used to control the g_m value of transconductors. The tuning performance is primarily determined by the characteristics of the CC, which can be implemented by the following two circuit architectures.

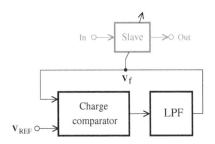

FIGURE 7.94
Frequency tuning loop based on charge comparison.

Consider the circuit diagram shown in Figure 7.95. During the first clock phase, the capacitor C is charged to the voltage V_I, which is dependent on the input dc current I_{REF} ($V_I = I_{REF}/g_m$). The difference between the charges produced by the emptying of C and the dc current NI_{REF} is transferred onto the feedback capacitor C_F during the second clock phase. Consequently,

$$\triangle q_{C_F} = TNI_{REF} - CV_I, \qquad (7.269)$$

where T is the period of the clock signal and N is the ratio of *dc* current sources. At the steady state, the average output voltage of the CC is constant. This means that there is no charge variation from the actual clock phase to the next one, and

$$\frac{g_m}{C} = \frac{f_c}{N}, \qquad (7.270)$$

where $f_c = 1/T$ is the clock frequency.

An alternative circuit diagram for the comparator is shown in Figure 7.96 [64]. Here, a reference voltage is connected to a transconductor and a switched-capacitor branch with an indirect path. The total charge transferred onto C_F can be expressed as

$$\triangle q_{C_F} = Tg_mV_{REF} - CV_{REF}. \qquad (7.271)$$

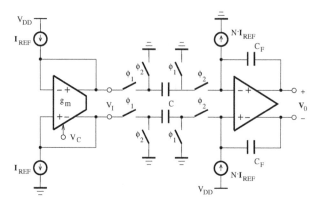

FIGURE 7.95
Circuit diagram of the charge comparator I.

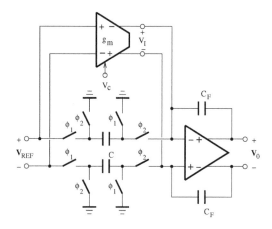

FIGURE 7.96
Circuit diagram of the charge comparator II.

With the same assumption, as previously at the steady state, the next tuning condition is met, that is,

$$\frac{g_m}{C} = f_c. \tag{7.272}$$

The tuning accuracy of the frequency is influenced by the level of matching of the g_m/C ratios achievable between the slave and the tuning circuits and the precision of the tuning circuit itself. It may be necessary to model the parasitic capacitors of the main filter and include their effects in the charge comparator capacitor C. In order to reduce the error due to the tuning loop, the offset voltages of the transconductor and the amplifier of the charge comparator must be low, and a reference signal with a sufficiently high magnitude is advisable.

7.7 Quality-factor and bandwidth control systems

Quality-factor and bandwidth control systems can be realized using the magnitude-locked-loop-based technique or envelope detection based technique.

7.7.1 Magnitude-locked-loop-based technique

The errors of the pole quality factor, Q, which is sensitive to parasitic components at high frequencies, can be reduced by the tuning scheme shown in Figure 7.97. This tuning scheme is based on a magnitude-locked loop and is a

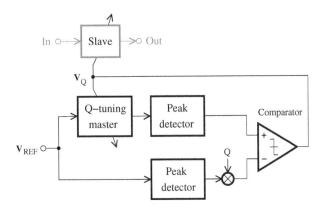

FIGURE 7.97
Magnitude-locked loop for Q tuning.

suitable method to automatically control the shape of the transfer function. Generally, the master has a biquadratic transfer function, whose magnitude is proportional to Q at the oscillation (or center) frequency. The amplitude of the reference signal is estimated by the peak detector. It is amplified by a gain factor Q and then compared to the detected amplitude of the master output voltage. This results in a signal that is used to tune the master and slave until these two quantities are equal.

An implementation of the peak detector is shown in Figure 7.98. The input voltage is first fully rectified and then lowpass filtered to generate a signal, which is related to the amplitude of the input signal. Note that the switches can be implemented using CMOS analog gates.

The VCF, the output signal of which exhibits a reduced harmonic distortion in comparison to the one of the VCO, is preferred for the master implementation. Furthermore, the Q tuning seems to work appropriately for biquadratic sections and filter with cascade topology, but it may not be useful for high-order ladder filters.

In order to reduce the errors that can be introduced by the poor high-

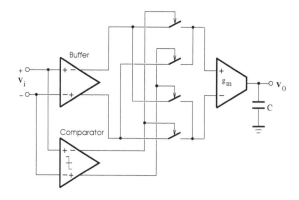

FIGURE 7.98
Circuit diagram of a peak detector.

frequency behavior of the peak detector and the offset voltage of the comparator, the alternative scheme [65] shown in Figure 7.99 can be adopted. Its

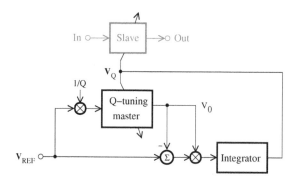

FIGURE 7.99
Magnitude-locked loop for Q tuning based on an adaptive technique.

operation is based on an adaptive algorithm of the least-mean square type. The Q-control signal, V_Q, is updated according to the following equation

$$\frac{dV_Q(t)}{dt} = \mu[V_{REF}(t) - V_0(t)]V_0(t), \qquad (7.273)$$

where V_{REF} is the reference signal that has a frequency equal to the center frequency of the filter, and V_0 is the output voltage of the master (bandpass filter). The least-mean square (LMS) algorithm will try to minimize the error signal, $V_{REF} - V_0$. Ideally, V_{REF} is equal to V_0 after the tuning because the phase shift of V_0 is zero at the center frequency.

7.7.2 Envelope detection-based technique

The bandwidth of a CT filter can be controlled using the tuning architecture shown in Figure 7.101. The tuning circuit, which is based on the envelope detection technique, is composed of a first-order LPF, a tunable biquadratic filter, two envelope detectors, and a one-phase integrator followed by a sample-and-hold (S/H) circuit [62]. The circuit diagram of the envelope detector is shown in Figure 7.102(a). The input amplifiers followed by current mirrors T_1-T_4 can be modeled by a transconductance g_m, as shown in Figure 7.102(b). During the signal detection, the capacitor C_d is charged according to the following equation,

$$C_d \frac{dV_0}{dt} = I_d - g_m V, \tag{7.274}$$

where $V = V_0 - V_i$. The range of detection is limited to the negative transitions of the signal due to the unidirectional characteristic of the current mirror. For positive transitions, g_m is reduced to zero and the output voltage increases linearly with a constant slope of value I_d/C_d.

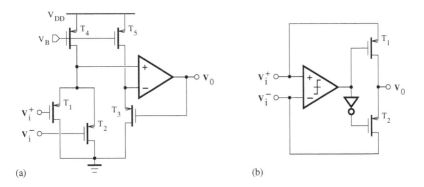

(a)   (b)

FIGURE 7.100
Circuit diagram of rectifiers ($V_i^- = -V_i^+$).

An envelope detector can also be implemented using a full-wave rectifier followed by a lowpass RC filter. Figure 7.100(a) shows the circuit diagram of a rectifier [60]. Due to the high gain of the amplifier, the voltage at the inverting and noninverting nodes should be made equal. This is achieved when a rectified version of the input signal is reproduced at the gate of T_3 or output. The rectifier shown in Figure 7.100(b) [63] is based on a comparator. Each of the transistors T_1 and T_2, which operate as a switch, is closed or open according to the input signal polarity.

An implementation of the integrator section is shown in Figure 7.103. The transconductor is connected either to the ground or to the amplifier inputs during the first and second clock phases, respectively. The charge stored on the amplifier feedback capacitor is transferred onto the hold capacitor, C_h, when there is no signal injected into the amplifier input nodes.

The detection of signal envelopes is carried out in the time domain. Assuming that the output voltage of the filter is initially set to zero, the step

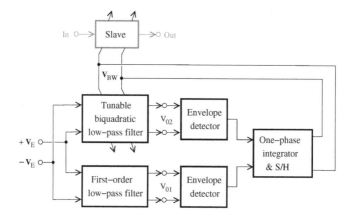

FIGURE 7.101
Bandwidth tuning loop based on envelope detection.

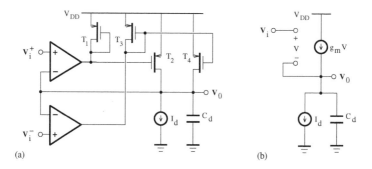

FIGURE 7.102
(a) Circuit diagram and (b) equivalent model of an envelope detector.

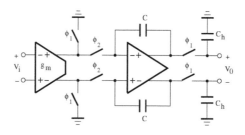

FIGURE 7.103
Circuit diagram of a one-phase integrator including the sample-and-hold function.

response can be computed as

$$v_0(t) = \mathcal{L}^{-1}\left\{\frac{H(s)}{s}\right\} \tag{7.275}$$

$$= \int_0^t h(\tau)\,d\tau, \tag{7.276}$$

where $\mathcal{L}^{-1}$ denotes the inverse Laplace transform, and $H(s)$ and $h(t)$ are the transfer function and the impulse response of the filter, respectively. For a first-order LPF with a unity dc gain, this results in

$$v_{01}(t) = E[1 - \exp(-\omega_c t)], \qquad (7.277)$$

where ω_c and E are the -3 dB cutoff frequency and the amplitude of the input step voltage, respectively. In the case of the second-order LPF, the next output voltage can be written as

$$v_{02}(t) = E[1 - h(t)]. \qquad (7.278)$$

Note that the filter dc gain and the initial conditions on the voltage are one and zero, respectively. The filter impulse response, h, is given by

$$h(t) = \frac{1}{\sqrt{1 - \dfrac{1}{4Q^2}}} \exp\left(-\frac{BW}{2}t\right) \cos\left(\sqrt{1 - \frac{1}{4Q^2}}\,\omega_0 t - \phi\right) \qquad (7.279)$$

and

$$\phi = \arctan \frac{1/2Q}{\sqrt{1 - \dfrac{1}{4Q^2}}}, \qquad (7.280)$$

where $BW = \omega_0/Q$ is an approximation of the LPF bandwidth, which is related to the center frequency, ω_0, and the Q-pole factor. This latter must be larger than $1/2$. For high Q, the only difference between the above step responses is the harmonic term that appears in the expression of h. Therefore, the envelopes measured by both detectors will have similar shapes if ω_c takes the value $BW/2$. The input signal V_E is a train of pulse with the amplitude E. In its positive transition, the detected signal at the output of the first-order and second-order LPFs correspond to an exponential charging response, whose final value is E and an exponential decay from $2E$ to E, respectively. For this reason, and because V_E is a pulse train, the second clock phase of the integrator used in the tuning loop is synchronized with the negative transition of V_E, which is the beginning of the time period where the signal envelopes can be successfully compared (see Figure 7.104). In this approach, some practical problems can limit the tuning accuracy: the level of the residual offset voltage in the transconductor required for the comparison of the filter envelopes and the precision of the envelope detectors.

7.8 Practical design considerations

Let us assume that the main filter in the master-slave tuning scheme (see Figure 7.88) is based on a biquadratic filter section with the following bandpass

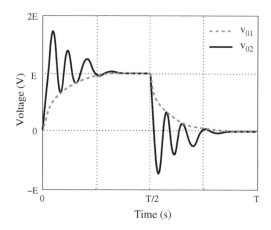

FIGURE 7.104
Plot of the transient responses (outputs $v_{01}(t)$ and $v_{02}(t)$) for $\omega_c = \text{BW}/2$.

transfer function,

$$T(s) = k\frac{\dfrac{\omega_o}{Q}s}{s^2 + \left(\dfrac{\omega_o}{Q}\right)s + \omega_o^2}, \qquad (7.281)$$

where k is a gain factor, ω_0 denotes the center frequency, and Q is the quality factor. The phase difference between the reference and the filter output signal, ϕ, is required in the frequency-tuning loop while the magnitude of the filter output signal, M, is needed in the Q-tuning circuit. Assuming that the input signal of the filter is at the same frequency than the reference signal, that is, $s = j\omega_r$, we have

$$\phi(\omega_o, Q) = \arg[T(s)] = \frac{\pi}{2} - \arctan\left(\frac{\dfrac{1}{Q}\dfrac{\omega_r}{\omega_o}}{1 - \dfrac{\omega_r^2}{\omega_o^2}}\right), \qquad (7.282)$$

$$M(\omega_o, Q) = |T(s)| = k\frac{Q\dfrac{\omega_r}{\omega_o}}{\sqrt{\left(1 - \dfrac{\omega_r^2}{\omega_o^2}\right)^2 + \left(\dfrac{1}{Q}\dfrac{\omega_r}{\omega_o}\right)^2}}, \qquad (7.283)$$

where ω_r is the frequency of the reference signal. The coupled nature of the above parameters appears in Figures 7.105 and 7.106, where the phase and magnitude surfaces are drawn. For a high Q factor, a small error in the pole frequency can result in a low output voltage, which directly translates to an inappropriate increase in the filter quality factor by the Q-tuning loop.

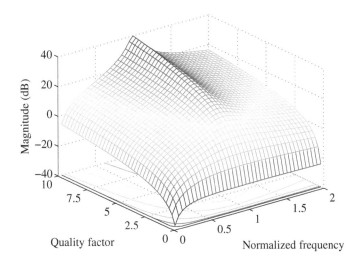

FIGURE 7.105
Magnitude surfaces.

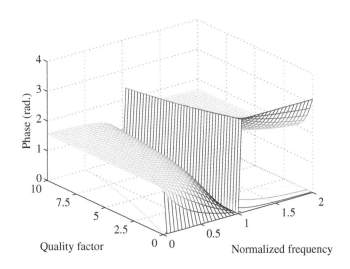

FIGURE 7.106
Phase surfaces.

Ideally, the loops can be considered independent if

$$\frac{\partial \phi(\omega_0, Q)}{\partial Q} = 0, \tag{7.284}$$

$$\frac{\partial M(\omega_0, Q)}{\partial \omega_0} = 0. \tag{7.285}$$

The next approaches that can be used to this end consist basically of making quasi-independent the two tuning loops. In the first solution, the Q-loop is designed to be much slower than the frequency loop. As a result, the loop interaction, which may lead to instability, is reduced. The second one consists of using oscillator [66] or filter structures [67] with a reduced coupling of magnitude and frequency parameters.

For the choice of the reference signal frequency, a trade-off must be made between the achievable level of the master-slave matching and the amount of reference signal feedthrough that can still be coupled to the output signal of the slave. When the reference signal is at the passband edge, that is, very close to the unity-gain frequency of the integrators in the slave, a best matching will be observed. But, this selection also results in a worst immunity to the reference signal feedthrough. On the contrary, the feedthrough will be minimum for a reference signal frequency located in the stop band due to the high attenuation of the filter, and the matching will be very poor.

7.9   Other tuning strategies

In applications where the limitations of a master-slave architecture become critical, alternative tuning strategies (tuning scheme using an external resistor, self-tuned filters) can be adopted. They exhibit the advantage of providing a tuning performance independent of the filter topology.

7.9.1   Tuning scheme using an external resistor

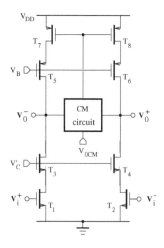

FIGURE 7.107
Circuit diagram of a differential transconductor with the common-mode circuit.

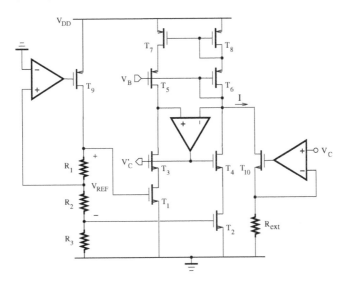

FIGURE 7.108
Tuning circuit for the transconductor using an external resistor.

The transconductance can be set in g_m-C filters by locking the filter amplifiers to an external reference resistor [70,71]. An example CMOS transconductor is shown in Figure 7.107; V_B is the bias voltage and V_C' is the internal control voltage used for the automatic tuning. This latter is supplied by the circuit shown in Figure 7.108. Here, a known voltage, V_{REF}, and a known current, I, are applied to a replica input stage of the transconductor ($T_1 - T_4$) in a feedback loop and the transconductance is then defined as the ratio of the above current and voltage. The output current, I, of the transconductor is related to the external control voltage, V_C, and the value of the transconductance is given by

$$g_m = \frac{I}{V_{REF}} = \frac{1}{R_{ext}} \times \frac{V_C}{V_{REF}}. \tag{7.286}$$

This tuning scheme can then be used to improve the accuracy of the transconductance provided that the temperature coefficients of the resistor, R_{ext}, and capacitors are sufficiently small.

7.9.2 Self-tuned filter

Generally, the matching between the master and slave is limited by the variations of the device characteristics. As a result, some percentage of uncertainty (1 to 2%) with respect to the filter parameters remains after the tuning. An improvement can be observed by increasing the device size. But this leads to higher power consumption.

A tuning scheme [68,69], the performance of which is not related to the matching accuracy, is shown in Figure 7.109. The filter is first tuned and then connected to the signal path. During the tuning process, the filter is coupled to the step-signal generator and is unavailable for the processing of the signal of interest. By comparing the square wave version of the filter output provided the inverter buffer and a reference signal, the phase and frequency detector (PFD) can generate either an Up signal or a Down signal used to drive the counter. The control circuit receives the adjustment signal and initiates the filter tuning, which is achieved via the N-bit digital-to-analog converter (DAC).

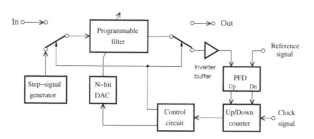

FIGURE 7.109
Block diagram of a self-tuned filter.

Here, an input step response is applied to the filter; its ringing frequency is measured and tuned to the design value. For a second-order bandpass filter with the transfer function $T(s)$ (see Equation (7.281)), the step response is given by

$$s(t) = \mathcal{L}^{-1} \left\{ \frac{H(s)}{s} \right\}$$
$$= \frac{k/Q}{\sqrt{1 - \frac{1}{4Q^2}}} \exp\left(-\frac{\omega_0 t}{2Q}\right) \sin\left(\omega_0 \sqrt{1 - \frac{1}{4Q^2}} t\right) u(t), \qquad (7.287)$$

where k is the filter gain, $u(t)$ is the unit step response, and the initial conditions were assumed to be zero. This signal crosses zero when

$$\sin\left(\omega_0 \sqrt{1 - \frac{1}{4Q^2}} t\right) = 0 \qquad (7.288)$$

and we can write

$$\frac{f_r}{f_0} = 2\sqrt{1 - \frac{1}{4Q^2}}. \qquad (7.289)$$

Note that Q is the quality factor of the filter, f_r is the ringing frequency, and $\omega_0 = 2\pi f_0$, where f_0 is the center frequency. The circuit section consisting of the PFD and Up/Down counter estimates the number of cycles N_i of the input waveform in a period, which corresponds to a number of counts, M, of the clock signal with the frequency f_c. In this way, the actual ringing frequency, f_{ra}, can be computed as

$$f_{ra} = \frac{M f_c}{N_i}. \qquad (7.290)$$

The target ringing frequency, f_{rt}, is related to the count value N_t and can be written as

$$f_{rt} = \frac{M f_c}{N_t} . \tag{7.291}$$

The objective of the tuning is to adjust the ringing frequency of the filter so that the next condition is fulfilled, that is,

$$|N_i - N_t| \leq \epsilon, \tag{7.292}$$

where ϵ denotes the residual tuning error. In this way, the overall accuracy of the resulting filter characteristics depends on the level of ϵ and frequency measurement errors.

A filter with the above tuning scheme, or self-tuned filter, can find applications in personal digital cellular systems with spaced channel. However, in situations where a CT operation of the circuit is required, two filters must be associated in a parallel configuration as shown in Figure 7.110. One filter will

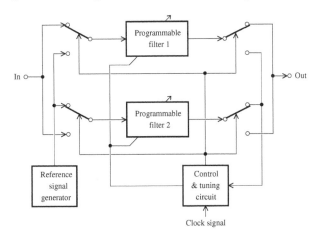

FIGURE 7.110
Self-tuned filter for the continuous-time operation.

be tuned when the other processes the input signal, and vice versa.

7.9.3 Tuning scheme based on adaptive filter technique

Here, the tuning objective [72] must be formulated in terms of a function to be minimized. This goal can be accomplished using an adaptive algorithm to update the filter coefficients (see Figure 7.111).

Although a white noise source is commonly used as the input signal in the system identification application, a sum of sinusoids, whose frequencies are chosen within the passband of the desired transfer function, H, seems to be more appropriate. That is,

$$x(t) = \sum_{l=1}^{L} \sin(2\pi f_l t). \tag{7.293}$$

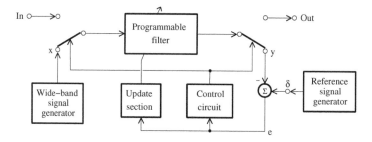

FIGURE 7.111
Tuning scheme based on adaptive filter technique.

The reference signal, δ, can then be obtained as the ideal filter response to the excitation, x. Hence,

$$\delta(t) = \sum_{l=1}^{L} a_l \sin(2\pi f_l t + \phi_l), \qquad (7.294)$$

where $a_l = |H(2\pi j f_l)|$ and $\phi_l = \arg\{H(2\pi j f_l)\}$. The tuning of the filter to meet the passband specifications can be achieved by updating the variable filter coefficients according to the least-mean-square (LMS) algorithm. During the adaptation process, the error signal, e, is given by

$$e(t) = k[\delta(t) - y(t)], \qquad (7.295)$$

where k is the error amplifier gain, and δ and y are the reference signal and filter output, respectively. Each variable coefficient of the programmable filter, labeled W, are changed to minimize the mean-squared error signal denoted as $E[e^2]$. Thus,

$$\frac{\mathrm{d}}{\mathrm{d}t} W(t) = \mu e(t)\phi(t), \qquad (7.296)$$

where μ is a small positive step size that determines the trade-off between the speed of the algorithm and the residual convergence error, and ϕ denotes the gradient signals defined as

$$\phi(t) = \left. \frac{\partial y(t)}{\partial W} \right|_{W=W(t)}. \qquad (7.297)$$

The generation of ϕ can require additional structures, which are driven by signals associated with filter states or nodes. It is worth noting that filter architectures with orthogonal states provide the advantage of improving the LMS algorithm performance.

In practice, a $\Delta\Sigma$ oscillator may be utilized to generate the sinusoids used as input and reference signals [73,74]. It combines a digital resonator structure having poles on the unit circle and a $\Delta\Sigma$ modulator. The oscillation frequency is set by the loop gain and the external clock frequency. The amplitude and phase of the signal are determined by the initial conditions. This technique results in high-quality sinusoids with spurious-free dynamic range larger than 90 dB.

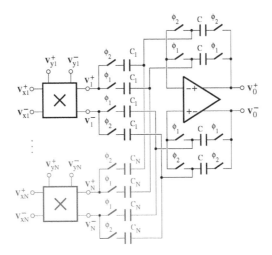

FIGURE 7.112
Multiplier with offset voltage cancelation.

The level of the tuning error can be limited by the different nonidealities of components, such as the offset voltages of the amplifier and multiplier, which can be on the order of ± 10 mV. A solution for the cancelation of the multiplier offset voltage is provided by cascading multiplier circuits and a gain stage, as shown in Figure 7.112 [75], where ϕ_1 and ϕ_2 are two complementary clock phases. The offset voltages stored on the capacitors C_k ($k = 1, 2, \cdots, N$) during the previous clock phase ϕ_1 (or ϕ_2) are used during the actual clock phase ϕ_1 (or ϕ_2) to cancel the current offset voltage contributions. The output voltage of the multi-input offset-free tunable gain stage is proportional to the ratio C_k/C and the multiplier gain. Auto-zero or chopper schemes can also be adopted to reduce the amplifier and multiplier sensitivity to offset voltages [76, 77]. With the use of these compensation techniques, the residual misadjustment can be reduced to about 60 dB, guaranteeing a reasonable accuracy for the filter characteristics.

By analyzing the filter structures, the tunability range of the characteristics can be determined. Note that the evaluation of the closed-form relations between the parameters of high-order filters is more tedious than the one of first- and second-order structures and may require the use of circuit analysis and computation programs like SPICE, MATLAB®, and Hardware Description Language (HDL).

7.10 Summary

An overview of high-performance CT circuits was provided. In advanced applications, CT circuits should be designed to exhibit a high dynamic range, a programmable bandwidth, precise tuning, high speed, low power and a small chip area. The choice of the architecture and design techniques is generally determined by the cost and performance.

Because CT filters are prone to fluctuations in their electrical parameters, an insight into the principles of operation of on-chip tuning loops is provided. A comparison of the performance is carried out in order to analyze the suitability of each tuning scheme to the high-frequency and high-Q filter applications and to choose the more convenient topology for a given design purpose. Generally, the tuning maintains a low drift of the filter parameters over power supply, temperature, and IC process variations. As a result, CT filters with less than $\pm 1\%$ pole frequency and quality factor accuracies become realizable.

7.11 Circuit design assessment

1. **Single-stage phase shift network**

   Consider each of the phase shift networks shown in Figure 7.113. Use the voltage divider principle to find the transfer functions

   $$T(s) = [V_0^+(s) - V_0^-(s)]/V_i(s).$$

   Determine and plot the magnitude and phase of T versus the frequency.

   Deduce the phase of T at the 3-dB cutoff frequency.

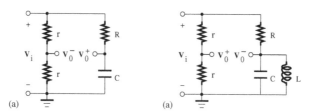

FIGURE 7.113
Circuit diagram of phase shift networks.

2. **Two-stage RC-CR phase shifters**

   For the two-stage RC-CR phase shifters of Figure 7.114, assume that the coupling and decoupling capacitors, C_C and C_D, act as a short-circuit for the entire range of operating frequencies.

Show that, for the circuit of Figure 7.114(a),

$$\frac{V_Q(s)}{V_I(s)} = \frac{s(R_1C_1 + R_2C_2)}{1 - R_1R_2C_1C_2s^2};$$

(7.298)

and for the circuit of Figure 7.114(b),

$$\frac{V_Q(s)}{V_I(s)} = \frac{1 + (R_1C_1 + R_2C_2)s - R_1R_2C_1C_2s^2}{1 - (R_1C_1 + R_2C_2)s - R_1R_2C_1C_2s^2},$$

(7.299)

where $V_Q = V_Q^+ - V_Q^-$ and $V_I = V_I^+ - V_I^-$.

Plot the magnitude and phase of the function $V_Q(s)/V_I(s)$ as a function of the frequency.

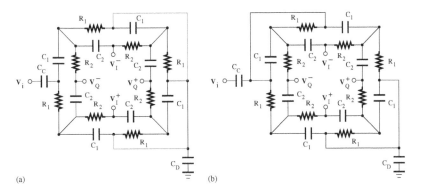

(a)   (b)

FIGURE 7.114

Circuit diagrams of two-stage RC-CR phase shifters (a) without phase imbalance and (b) without gain imbalance.

3. **Common-source amplifier stages**

Determine the transfer function of the circuit shown in Figure 7.115 using the small-signal equivalent transistor model of Figure 7.116.

Provided that the transistor operates as an ideal transconductor with the value g_m, show that the transfer function of the first and second amplifiers can be respectively reduced to

$$\frac{V_0(s)}{V_i(s)} = \frac{A_0}{1 + \tau s},$$

(7.300)

where $A_0 = -g_m R$ and $\tau = RC$, and

$$\frac{V_0(s)}{V_i(s)} = A_0 \omega_0 Q \frac{s + \dfrac{\omega_0}{Q}}{s^2 + \left(\dfrac{\omega_0}{Q}\right)s + \omega_0^2},$$

(7.301)

where $A_0 = -g_m R$, $Q = \omega_0 L/R$ and $\omega_0^2 = 1/LC$.

Verify that the zero and poles introduced by the shunt inductor result in an increase in the amplifier bandwidth.

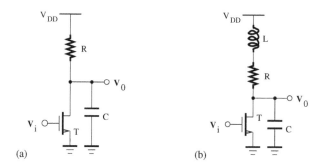

(a)   (b)

FIGURE 7.115
Circuit diagrams of (a) a simple common-source amplifier and (b) a common-source amplifier with shunt peaking.

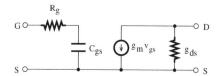

FIGURE 7.116
Small-signal equivalent transistor model.

4. **Single-stage tuned amplifier**

   The circuit diagram of a tuned amplifier is depicted in Figure 7.117. Assuming that the transistor is equivalent to transconductor with the value g_m, show that

   $$A(s) = \frac{V_0(s)}{V_i(s)} = -\frac{g_m}{C} \frac{s}{s^2 + \left(\dfrac{1}{RC}\right)s + \dfrac{1}{LC}}. \qquad (7.302)$$

   Verify that

   $$A(s) = A_0 \frac{\left(\dfrac{\omega_0}{Q}\right)s}{s^2 + \left(\dfrac{\omega_0}{Q}\right)s + \omega_0^2} \qquad (7.303)$$

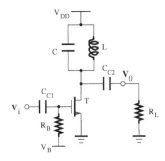

FIGURE 7.117
Circuit diagram of a tuned amplifier.

and

$$A(\omega) = A_0 \frac{1}{1 + jQ\left(\dfrac{\omega}{\omega_0} - \dfrac{\omega_0}{\omega}\right)}, \qquad (7.304)$$

where $A_0 = -g_m R$ and $Q = RC\omega_0 = R/(L\omega_0)$.

5. **SPICE analysis of low-noise amplifiers**
   Analyze the input impedance and noise of the amplifiers shown in Figure 7.118 using the SPICE program.

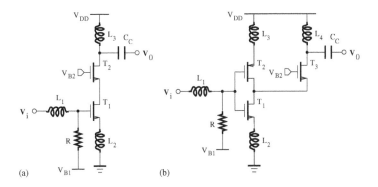

FIGURE 7.118
Circuit diagrams of single-ended low-noise amplifiers with (a) nMOS, (b) pMOS-nMOS, and (c) inductively coupled input stages.

6. **Low-noise amplifiers with improved noise factor**
   • In the low-noise amplifier shown in Figure 7.119(a) [14], the RLC input network is merged with the resistive feedback to increase the gain and reduce the noise factor.

   Assuming that the coupling capacitors, C_{C_1} and C_{C_2}, are equivalent to a short-circuit, determine the input impedance and noise factor.

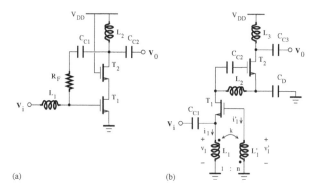

FIGURE 7.119
(a) Circuit diagram of a low-noise amplifier with RC feedback; (b) circuit diagram of a g_m-boosted low-noise amplifier using current reuse.

• Consider the low-noise amplifier circuit of Figure 7.119(a) [21]. This architecture, which is realized by cascading the g_m-boosted input stage with a common-source output stage such that there is a current reuse, has the advantage of providing a high gain even with a low power consumption. The coupling capacitors C_{C_1}, C_{C_2}, and C_{C_3}, and the decoupling capacitor C_D can be considered as a short-circuit. The i-v laws for the coupled coils, $L_1 - L_1'$, with zero initial conditions are as follows:

$$i_1 = \frac{M i_1'}{L_1} + \frac{v_1}{s L_1} \tag{7.305}$$

and

$$i_1' = \frac{v_1'}{s L_1'} + \frac{M i_1}{L_1'}, \tag{7.306}$$

where the coupling coefficient is given by

$$k = \frac{M}{\sqrt{L_1 L_1'}} \tag{7.307}$$

and M is the mutual inductance between the primary and secondary windings.

Assuming that $v_{gs_1} = -(1 + A)v_i$, where $A = kn$, show that the frequency response of the low-noise amplifier is the product of two second-order bandpass transfer functions due to the parallel resonant circuits at the drains of both transistors.

Determine the total noise figure of the amplifier using Friis' formula of the form

$$F = F_1 + \frac{F_2 - 1}{A_1}, \tag{7.308}$$

where F_1 and F_2 denote the noise figures of the first and second stages, respectively, and A_1 is the voltage gain of the first stage.

7. **Single-stage amplifier with parasitic coupling capacitors**

Let us consider the single-stage amplifier of Figure 7.120(a). Determine the voltage transfer function and input impedance using the transistor equivalent model shown in Figure 7.120(b).

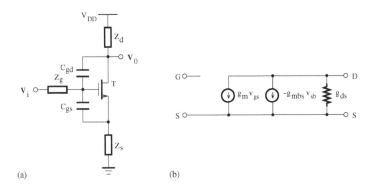

(a)   (b)

FIGURE 7.120

(a) Circuit diagram of a single-stage amplifier; (b) MOS transistor small-signal equivalent model.

8. **Pseudo-differential low-noise amplifier**

Consider the circuit diagram of a pseudo-differential low-noise amplifier with the capacitor cross-coupled input stage shown in Figure 7.121(a), where C_d represents a decoupling capacitor. The input transistors should be biased to operate in the saturation region.

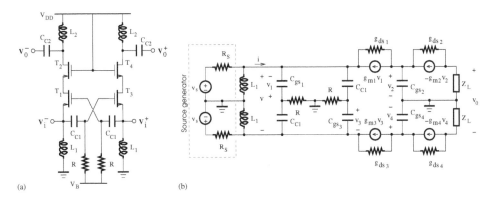

(a)   (b)

FIGURE 7.121

(a) Circuit diagram of a pseudo-differential low-noise amplifier with a capacitor cross-coupled input stage; (b) small-signal equivalent model.

Use the small-signal equivalent model depicted in Figure 7.121(b), where the input source generator consists of the voltage source v_s in series with the resistor R_s, and Z_L is the impedance of the output load, to determine the voltage gains, $G = V_0/V$ and $G_s = V_0/V_s$, and the input impedance, $Z_i = V/I$.

Compare the magnitude and phase of Z_i obtained by hand calculation to SPICE simulation results.

9. **Dynamic range of integrator**
The dynamic range can be defined as

$$\text{DR} = 20 \log_{10} \frac{V_{max}^2}{\overline{v_{in}^2}}, \tag{7.309}$$

where V_{max} is the maximum undistorted rms value of the input voltage, for which the total harmonic distortion (THD) equals 1%, and $\overline{v_{in}^2}$ is the rms value of the input referred noise integrated over the desired signal bandwidth, $\triangle f$.

Compare the dynamic range of the different continuous-time integrators using SPICE simulations

10. **Low-frequency lowpass filter**

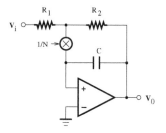

FIGURE 7.122
Circuit diagram of a low-frequency lowpass filter.

Consider the low-frequency lowpass filter of Figure 7.122. Determine the transfer function $V_0(s)/V_i(s)$.

Show that the input referred noise can be written as

$$\overline{v_{in}^2} = 4kT(R_1 + R_2) + \overline{v_{in,1/N}^2}, \tag{7.310}$$

where $\overline{v_{in,1/N}^2}$ denotes the noise contribution due to the $1/N$ scaling block, and compare $\overline{v_{in}^2}$ to the input referred noise, which can be generally observed in a similar filter structure without the current down-scaling.

Verify that the current conveyors (CCs) shown in Figure 7.123 can

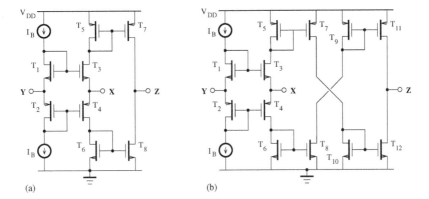

FIGURE 7.123
Circuit diagrams of (a) positive and (b) negative current conveyors.

be ideally characterized by the transfer matrix; the plus and minus signs apply to the positive and negative CCs, respectively.

$$\begin{bmatrix} i_Y \\ v_X \\ i_Z \end{bmatrix} = \begin{bmatrix} 0 & 0 & 0 \\ 1 & 0 & 0 \\ 0 & \pm 1 & 0 \end{bmatrix} \begin{bmatrix} v_Y \\ i_X \\ v_Z \end{bmatrix} \qquad (7.311)$$

Verify that the 1/N scaling can be realized by the negative cur-

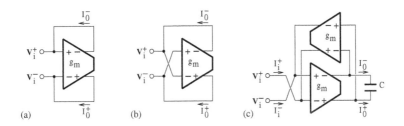

FIGURE 7.124
Current conveyor-based implementation of the scaling by 1/N.

rent conveyor with two N:1 output current mirrors (i.e., $T_9 - T_{10}$: N(W/L), $T_{11} - T_{12}$: W/L) (see Figure 7.124).

11. **Grounded resistors and inductor based on transconductors**

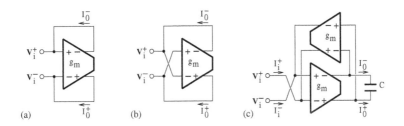

FIGURE 7.125
Circuit diagrams of transconductor-based (a) (b) resistors and (c) inductor.

In integrated filter design, active networks based on transconductance amplifiers can be used to simulate the behavior of various passive elements.

Let g_m be the amplifier transconductance. Verify that the circuits of Figures 7.125(a) and (b) can be used to realize positive and negative resistors with the value $1/g_m$, respectively.

Show that the g_m-C circuit of Figure 7.125(c) is equivalent to a grounded inductor with the value $L = C/g_m^2$.

Analyze the effect of transconductor parasitic elements (input capacitor, output capacitor and resistor) on the simulated resistors and inductor.

12. **Lowpass and highpass filter transformations**
Let

$$H(s) = \frac{1}{s^2 + \sqrt{2}s + 1} \tag{7.312}$$

be the transfer function of a second-order Butterworth lowpass filter with the normalized passband edge frequency $\Omega_p = 1$.

Use the following spectral transformation,

$$s \to \frac{\Omega_p}{\Omega'_p}s, \tag{7.313}$$

to derive the transfer function of a lowpass filter with the normalized passband edge frequency $\Omega'_p = 3\Omega_p/2$.

Determine the transfer function of the highpass filter obtained using the transformation given by

$$s \to \frac{\Omega_p \Omega'_p}{s}, \tag{7.314}$$

where $\Omega'_p = 3\Omega_p/2$ is the normalized passband edge frequency of the highpass filter.

13. **Bandpass and bandstop filter transformations**
Consider a first-order lowpass filter with a transfer function of the form

$$H(s) = k\frac{\omega_c}{s + \omega_c}, \tag{7.315}$$

where k is the gain factor and ω_c is the cutoff frequency. It is assumed that the passband edge frequency is $\Omega_p = \omega_c$.

Show that the transfer function of the bandpass filter derived using the following spectral transformation,

$$s \to \Omega_p \frac{s^2 + \Omega_l \Omega_u}{s(\Omega_u - \Omega_l)}, \tag{7.316}$$

where Ω_l and Ω_u denote the lower and upper passband edge frequencies, respectively, takes the form

$$H_{BP}(s) = k\frac{\dfrac{\omega_0}{Q}s}{s^2 + \dfrac{\omega_0}{Q}s + \omega_0^2}, \qquad (7.317)$$

where ω_0 and Q are parameters to be determined.

Let $H_{AP}(s)$ be the transfer function of an allpass filter. Verify that

$$H_{AP}(s) = 1 - 2H_{BP}(s). \qquad (7.318)$$

The transfer function of a second-order bandstop filter can be written as

$$H_{BS}(s) = k\frac{s^2 + \omega_0^2}{s^2 + \dfrac{\omega_0}{Q}s + \omega_0^2}. \qquad (7.319)$$

Find the relation between the bandstop filter parameters (ω_0 and Q) and ω_c using the next transformation

$$s \rightarrow \Omega_p\frac{s(\Omega_u - \Omega_l)}{s^2 + \Omega_l\Omega_u}, \qquad (7.320)$$

where Ω_l and Ω_u denote the lower and upper passband edge frequencies, respectively.

Verify that

$$-H_{AP}(s) = 1 - 2H_{BS}(s), \qquad (7.321)$$

where $H_{AP}(s)$ is the transfer function of an allpass filter.

14. **Analysis of second-order bandpass filter**

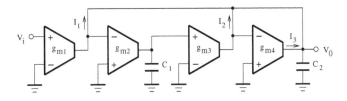

FIGURE 7.126
Circuit diagram of a second-order bandpass filter.

Find the transfer function, $H = V_0/V_i$, of the second-order bandpass filter shown in Fig 7.126 and put it into the form

$$H(s) = \frac{V_0(s)}{V_i(s)} = k \frac{\frac{\omega_0}{Q}s}{s^2 + \frac{\omega_0}{Q}s + \omega_0^2}, \qquad (7.322)$$

where

$$k = \frac{g_{m1}}{g_{m4}}, \qquad (7.323)$$

$$\omega_0 = \sqrt{\frac{1}{g_{m2}g_{m3}C_1C_2}}, \qquad (7.324)$$

$$Q = \frac{1}{g_{m4}}\sqrt{g_{m2}g_{m3}\frac{C_1}{C_2}}. \qquad (7.325)$$

Hint: At the output node, we have $V_0 = sC_2(I_1 + I_2 + I_3)$.

15. **Analysis of a general biquad**

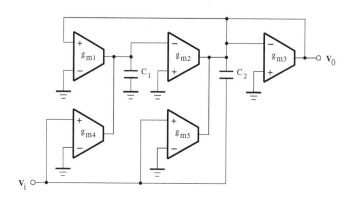

FIGURE 7.127
Circuit diagram of a biquadratic filter section.

For the biquad structure of Figure 7.127, verify that the transfer function is given by

$$H(s) = \frac{V_0(s)}{V_i(s)} = \frac{s^2 + \frac{g_{m5}}{C_2}s - \frac{g_{m2}g_{m4}}{C_1C_2}}{s^2 + \frac{g_{m3}}{C_2}s + \frac{g_{m2}g_{m4}}{C_1C_2}} \qquad (7.326)$$

16. **Transfer function synthesis**

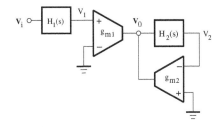

FIGURE 7.128

Circuit diagram of a network for transfer function synthesis.

Consider the network shown in Figure 7.128. Verify that

$$V_1 = H_1(s)V_i(s), \tag{7.327}$$

$$V_2 = H_2(s)V_0(s), \tag{7.328}$$

and

$$g_{m1}V_1(s) - g_{m2}V_2(s) = 0. \tag{7.329}$$

Deduce that

$$H(s) = \frac{V_0(s)}{V_i(s)} = \frac{H_1(s)}{H_2(s)}, \tag{7.330}$$

provided that $g_{m1} = g_{m2}$.

Let

$$H_2(s) = \frac{1}{s^2 + \dfrac{\omega_p}{Q_p}s + \omega_p^2} \tag{7.331}$$

and

$$H_2(s) = \frac{1}{k(s^2 + \omega_z^2)}. \tag{7.332}$$

Propose two circuits realizing the above transfer functions, and use them to build a band-reject filter.

17. **Bump equalizer**

Show that the bump equalizer depicted in Figure 7.129 realizes a transfer function of the form

$$T(s) = \frac{V_0(s)}{V_i(s)} = \frac{1 \mp kH(s)}{1 \pm kH(s)}, \tag{7.333}$$

where k is a variable gain.

In general, $|H(s)| \le 1$ and $|k| \le 1$. Assuming that

$$H(s) = \frac{s^2 - \dfrac{\omega_0}{Q}s + \omega_0^2}{s^2 + \dfrac{\omega_0}{Q}s + \omega_0^2}, \tag{7.334}$$

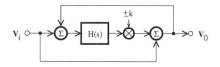

FIGURE 7.129
Block diagram of a bump equalizer.

where $Q = 1/\sqrt{3}$ and $\omega_0 = 1$ rad/s, plot the magnitude of $T(s)$ for various values of k.

18. **Active synthesis of all-pole filter**

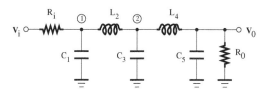

FIGURE 7.130
Circuit diagram of an all-pole LC filter.

The circuit diagram of a fifth-order all-pole LC filter is shown in Figure 7.130. Verify that the node equations can be written as

$$V_1 = \frac{1}{sC_1}\left(\frac{V_i - V_1}{R_i} - I_{L2}\right) \tag{7.335}$$

$$I_{L2} = \frac{1}{sL_2}(V_1 - V_2) \tag{7.336}$$

$$V_2 = \frac{1}{sC_3}(I_{L2} - I_{L4}) \tag{7.337}$$

$$I_{L4} = \frac{1}{sL_4}(V_2 - V_0) \tag{7.338}$$

$$V_0 = \frac{1}{sC_5}\left(I_{L4} - \frac{V_0}{R_0}\right). \tag{7.339}$$

Draw a signal-flow graph representation of the filter.

Design RC and g_m-C realizations of the all-pole filter prototype.

19. **Analysis of a fourth-order Chebyshev highpass filter**
The circuit shown in Figure 7.131 [53] is used to design a Chebyshev highpass filter with the transfer function,

$$H(s) = k\frac{s^4}{s^4 + p_3\omega_p s^3 + p_2\omega_p^2 s^2 + p_1\omega_p^3 s + p_0\omega_p^4}, \tag{7.340}$$

where ω_p is the passband edge frequency, k, p_0, p_1, p_2, and p_3 are real coefficients. Based on its signal-flow graph depicted in Figure 7.132, determine the transfer function $V_0(s)/V_i(s)$.

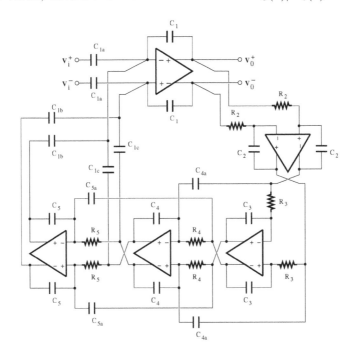

FIGURE 7.131
Circuit diagram of a fourth-order highpass RC filter. (From [53], ©2009 IEEE.)

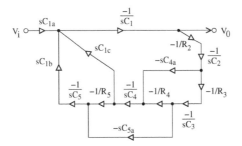

FIGURE 7.132
Signal-flow diagram of a fourth-order highpass filter.

Assuming that

$$R_2 = \frac{1}{q_1 C_2 \omega_p}, \qquad R_3 = \frac{q_1}{q_2 C_3 \omega_p}, \qquad R_4 = \frac{q_2}{q_3 C_4 \omega_p} \frac{p_1 - p_3}{p_1},$$

$$\text{and} \quad R_5 = \frac{q_3}{q_4 C_4 \omega_p} \frac{p_1(p_2 - p_0)}{p_0(p_1 - p_3)},$$

$$(7.341)$$

where q_j $(j = 1, 2, 3, 4)$ are scaling coefficients, find the transfer functions, $V_j(s)/V_i(s)$, from the filter input to the output of each amplifier.

20. **Third-order elliptic filter design**

Realize the third-order elliptic lowpass filter with the transfer function

$$H(s) = \frac{1.53210(s^2 + 1.69962)}{(s + 1.84049)(s^2 + 0.308389s + 1.41484)} \tag{7.342}$$

as a cascade connection of first-order and second-order g_m-C circuit sections, and as a g_m-C ladder circuit.

Analyze the effect of transconductor imperfections on the frequency response of each realized filter circuit using SPICE simulations.

21. **Element simulation-based filter design**

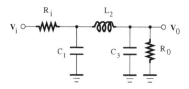

FIGURE 7.133
Lowpass LC filter prototype.

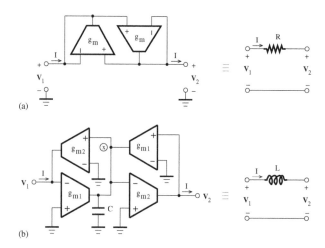

FIGURE 7.134
A floating inductor realized by a g_m-C circuit.

The design of a lowpass filter with a 3-dB frequency equal to 1 rad/s results in the LC circuit prototype shown in Figure 7.133.

Determine the transfer function $H(s) = V_0(s)/V_i(s)$.

For $R_i = R_0 = 1\ \Omega$, $C_1 = C_3 = 1$ F, and $L_2 = 2$ H, verify that

$$H(s) = \frac{V_0(s)}{V_i(s)} = \frac{1}{2}\frac{1}{s^3 + 2s^2 + 2s + 1}. \qquad (7.343)$$

Let $\phi(\omega)$ be the phase of the transfer function, $H(s)$. Show that the group delay can be written as

$$\tau(\omega) = -\frac{d\phi(\omega)}{d\omega} = \frac{2 + \omega^2 + 2\omega^4}{1 + \omega^6}. \qquad (7.344)$$

For the design of integrated filters for low-frequency applications, it is often necessary to substitute resistors and inductors, which tend to be large and sensitive to process variations, by an equivalent active network.

For ideal transconductors, verify that the circuit shown in Figure 7.134(a) realizes a floating resistor of the form $R = 1/g_m$.

Show that the circuit depicted in Figure 7.134(b) is equivalent to a floating inductor with the value $L = C/(g_{m1}g_{m2})$ using the current equation at node x.

Use SPICE to plot the frequency responses of the filter circuit obtained by substituting the resistors and inductor with their respective equivalent active network. The largest realizable capacitor is 50 pF. Each transconductor can be characterized by a single-pole model with the output capacitance $C_0 = 0.2$ pF, the output resistance $R_0 = 12$ MΩ, the input capacitance $C_i = 0.6$ pF, and a very high input resistance ($R_i = \infty$).

Bibliography

[1] J. Mitola, "The software radio architecture," pp. 26–38, *IEEE Communications Magazine*, May 1995.

[2] A. Rofougaran, G. Chang, J. J. Rael, J. Y.-C. Chang, M. Rofougaran, P. J. Chang, M. Djafari, M.-K. Ku, E. W. Roth, A. A. Abidi, and H. Samueli, "A single-chip 900-MHz spread-spectrum wireless transceiver in 1-µm CMOS–Part I: Architecture and transmitter design," *IEEE J. of Solid-State Circuits*, vol. 33, pp. 513–534, April 1998.

[3] —, "A single-chip 900-MHz spread-spectrum wireless transceiver in 1-µm CMOS–Part II: Receiver design," *IEEE J. of Solid-State Circuits*, vol. 33, pp. 535–547, April 1998.

[4] L. Levi, "Electrical transmission of energy," U.S. Patent 1,734,038, filed Aug. 12, 1918; issued Nov. 5, 1929.

[5] E. H. Armstrong, "The superheterodyne — Its origin, development, and some recent improvements," *Proc. IRE.*, vol. 12, pp. 539–552, Oct. 1924.

[6] F. M. Colebrook, "Homodyne," *Wireless World and Radio Review*, vol. 13, pp. 645–648, 1924.

[7] R. Hartley, "Single-sideband modulator," U.S. Patent 1,666,206, filed Jan. 15, 1925; issued April 17, 1928.

[8] D. K. Weaver, "A third method of generation and detection of single-sideband signals," *Proc. IRE*, vol. 44, pp. 1703-1705, 1956.

[9] J. Crols and M. Steyaert, "A single-chip 900 MHz CMOS receiver front-end with a high performance low-IF topology," *IEEE J. of Solid-State Circuits*, vol. 30, pp. 1483–1492, Dec. 1995.

[10] J. C. Rudell, J.-J. Ou, T. B. Cho, G. Chien, F. Brianti, J. A. Weldon, and P. Gray, "A 1.9 GHz wide-band IF double conversion CMOS receiver for cordless telephone applications," *IEEE J. of Solid-State Circuits*, vol. 32, pp. 2071–2088, Dec. 1997.

[11] J. A. Weldon, R. S. Narayanaswani, J. C. Rudell, L. Li, M. Otsuka, S. Dedieu, L. Tee, K.-C. Tsai, C.-W. Lee, and P. R. Gray, "A 1.75-GHz highly integrated narrow-band CMOS transmitter with harmonic-rejection mixers," *IEEE J. of Solid-State Circuits*, vol. 36, pp. 2003–2015, Dec. 2001.

[12] T. Sowlati, C. A. T. Salama, J. Sitch, G. Rabjohn, and D. Smith, "Low voltage, high efficiency GaAs class E power amplifiers for wireless transmitters," *IEEE J. of Solid-State Circuits*, vol. 30, pp. 1074–1080, Oct. 1995.

[13] K.-C. Tsai and P. R. Gray, "A 1.9-GHz, 1-W CMOS class-E power amplifier for wireless communications," *IEEE J. of Solid-State Circuits*, vol. 34, pp. 962–970, July 1999.

[14] S. Joo, T.-Y. Choi, and B. Jung, "A 2.4-GHz resistive feedback LNA in 0.13 μm CMOS," *IEEE J. of Solid-State Circuits*, vol. 44, pp. 3019–3029, Nov. 2009.

[15] W. S. Kim, X. Li, and M. Ismail, "A 2.4 GHz CMOS low noise amplifier using an inter-stage matching inductor," in *Proc. 42nd Midwest Symp. Circuits and Systems*, Aug. 1999, vol. 2, pp. 1040-1043.

[16] A. Parsa and B. Razavi, "A new transceiver architecture for the 60-GHz band," *IEEE J. of Solid-State Circuits*, vol. 44, pp. 751–762, March 2009.

[17] B. Razavi, "A 60-GHz CMOS receiver front-end," *IEEE J. of Solid-State Circuits*, vol. 41, pp. 17–22, Jan. 2006.

[18] F. Bruccoleri, E. A. M. Klumperink, and B. Nauta, "Wide-band CMOS low-noise amplifier exploiting thermal noise canceling," *IEEE J. of Solid-State Circuits*, vol. 39, pp. 275–282, Feb. 2004.

[19] D. K. Shaeffer and T. H. Lee, "A 1.5-V, 1.5-GHz CMOS low noise amplifier," *IEEE J. of Solid-State Circuits*, vol. 32, pp. 745–759, May 1997.

[20] D. K. Shaeffer and T. H. Lee, "Corrections to "A 1.5-V, 1.5-GHz CMOS low noise amplifier"," *IEEE J. of Solid-State Circuits*, vol. 40, pp. 1397-1398, June 2005.

[21] S. Shekhar, J. S. Walling, S. Aniruddhan, and D. J. Allstot, "CMOS VCO and LNA using tuned-input tuned-output circuits," *IEEE J. of Solid-State Circuits*, vol. 43, pp. 1177–1186, May 2008.

[22] F. Stubbe, S. V. Kishore, C. Hull, and V. Dellatorre, "A CMOS RF-receiver front-end for 1-GHz applications," *1998 Symposium on VLSl Circuits Digest of Technical Papers*, Honolulu, HI, pp. 80–83, June 1998.

[23] F. Behbahani, J. C. Leete, Y. Kishigami, A. Roithmeier, K. Hoshino, and A. A. Abidi, "A 2.4 GHz low-IF receiver for wideband WLAN in 0.6 μm CMOS – Architecture and front-end," *IEEE J. Solid-State Circuits*, vol. 35, pp. 1908–1916, Dec 2000.

[24] D. K. Shaeffer, A. R. Shahani, S. S. Mohan, H. Samavati, H. R. Rategh, M. d. M. Hershenson, M. Xu, C. P. Yue, D. J. Eddleman, and T. H.

Lee, "A 115-mW, 0.5-µm CMOS GPS receiver with wide dynamic-range active filters," *IEEE J. of Solid-State Circuits*, vol. 33, pp. 2219–2231, Dec. 1998.

[25] W. Zhuo, X. Li, S. Shekhar, S. H. K. Embabi, J. Pineda de Gyvez, D. J. Allstot, and E. Sanchez-Sinencio, "A capacitor cross-coupled common-gate low-noise amplifier," *IEEE Trans. on Circuits and Systems–II*, vol. 52, pp. 875–879, Dec. 2005.

[26] B. Razavi, "A millimeter-wave CMOS heterodyne receiver with on-chip LO and divider," *IEEE J. of Solid-State Circuits*, vol. 43, pp. 477–485, Feb. 2008.

[27] T.-P. Liu and E. Westerwick, "5-GHz CMOS radio transceiver front-end chipset," *IEEE J. of Solid-State Circuits*, vol. 35, pp. 1927–1933, Dec. 2000.

[28] L. S. Cutler and C. L. Searle, "Some aspects of the theory and measurement of frequency fluctuations in frequency standards," *Proc. of the IEEE*, vol. 54, pp. 136–154, Feb. 1966.

[29] D. B. Leeson, "A simple model of feedback oscillator noises spectrum," *Proc. of the IEEE*, vol. 54, pp. 329–330, Feb. 1966.

[30] A. Hajimiri and T. H. Lee, "A general theory of phase noise in electrical oscillators," *IEEE J. of Solid-State Circuits*, vol. 33, pp. 179–194, Feb. 1998.

[31] K.-S. Tan and P. R. Gray, "Fully integrated analog filters using bipolar-JFET technology," *IEEE J. of Solid-State Circuits*, vol. 13, pp. 814–821, Dec. 1978.

[32] J. Silva-Martinez, M. S. J. Steyaert, and W. Sansen, "A 10.7-MHz 68-dB SNR CMOS continuous-time filter with on-chip automatic tuning," *IEEE J. of Solid-State Circuits*, vol. 27, pp. 1843–1853, Dec. 1992.

[33] E. Hegazi, H. Sjöland, and A. A. Abidi, "A filtering technique to lower LC oscillator phase noise," *IEEE J. of Solid-State Circuits*, vol. 36, pp. 1921–1930, Dec. 2001.

[34] R. G. Meyer, "Low-power monolithic RF peak detector analysis," *IEEE J. of Solid-State Circuits*, vol. 30, pp. 65–67, Jan. 1995.

[35] S. H. Galal, H. F. Ragaie, and M. S. Tawfik, "RC sequence asymmetric polyphase networks for RF integrated transceivers," *IEEE Trans. on Circuits and Systems-II*, vol. 47, pp. 18–27, Jan. 2000.

[36] J. Kaukovuori, K. Stadius, J. Ryynänen, and K. A. I. Halonen, "Analysis and design of passive polyphase filters," *IEEE Trans. on Circuits and Systems-I*, vol. 55, pp. 3023–3037, Nov. 2008.

[37] A. Rofougaran, J. Rael, M. Rofougaran, and A. Abidi, "A 900 MHz CMOS LC-oscillator with quadrature outputs," *1996 IEEE ISSCC Digest of Technical Papers*, pp. 392–393, Feb. 1996.

[38] P. Vancorenland and M. Steyaert, "A 1.57-GHz fully integrated very low-phase noise quadrature VCO," *IEEE J. Solid-State Circuits*, vol. 37, pp. 653–656, May 2002.

[39] J. van der Tang, P. van de Ven, D. Kasperkovitz, and A. van Roermund, "Analysis and design of an optimally coupled 5-GHz quadrature LC oscillator," *IEEE J. Solid-State Circuits*, vol. 37, pp. 657–661, May 2002.

[40] A. Mirzaei, M. E. Heidari, R. Bagheri, S. Chehrazi, and A. A. Abidi, "The quadrature LC oscillator: A complete portrait based on injection locking," *IEEE J. of Solid-State Circuits*, vol. 42, pp. 1916–1932, Sept. 2007.

[41] K. Kimura, "Variable gain amplifier circuit," U.S. Patent 6,867,650, filed Dec. 9, 2002; issued March 15, 2005.

[42] R. Unbehauen and A. Cichocki, *MOS switched-capacitor and continuous-time integrated circuits and systems*, Communications and Control Engineering Series, Berlin, Germany: Springer-Verlag, 1989.

[43] Y. Tsividis, "Continuous-time filters," In Y. Tsividis and P. Antognetti, Editors, *Design of MOS VLSI circuits for telecommunications*, Chap. 11, pp. 334–371, Upper Saddle River, New Jersey: Prentice Hall, 1985.

[44] B.-S. Song, "CMOS RF circuits for data communication applications," *IEEE J. of Solid-State Circuits*, vol. 21, pp. 310–317, April 1987.

[45] Z. Czarnul, "Modification of Banu-Tsividis continuous-time integrator structure," *IEEE Trans. on Circuits and Systems*, vol. 33, pp. 714–716, July 1986.

[46] T. Ndjountche, "Linear voltage-controlled impedance architecture," *Electronics Letters*, vol. 32, pp. 1528–1529, Aug. 1996.

[47] R. L. Geiger and E. Sánchez-Sinencio, "Active filters using operational transconductance amplifiers: A tutorial," *IEEE Circuits and Dev. Mag.*, vol. 1, pp. 20–32, March 1985.

[48] T. Ndjountche and A. Zibi, "On the design of OTA-C structurally allpass filters," *Int. J. of Circuit Theory and Applications*, vol. 23, pp. 525–529, Sept.-Oct. 1995.

[49] P. O. Brackett and A. S. Sedra, "Active compensation for high-frequency effects in op-amp circuits with applications to active RC filters," *IEEE Trans. on Circuits and Systems*, vol. 23, no. 2, pp. 68–72, Feb. 1976.

[50] T. Ndjountche, R. Unbehauen, and F.-L. Luo, "Electronically tunable generalized impedance converter structures," *Int. J. of Circuit Theory and Applications*, vol. 27, pp. 517–522, 1999.

[51] R. Nawrocki and U. Klein, "New OTA-capacitor realisation of a universal biquad," *Electronics Letters*, vol. 22, pp. 50–51, Jan. 1986.

[52] A. Wyszyński, R. Schaumann, S. Szczepański, and P. V. Halen, "Design of 2.7-GHz linear OTA and a 250-MHz elliptic filter in bipolar transistor-array technology," *IEEE Trans. on Circuits and Systems*, vol. 40, pp. 19–31, Jan. 1993.

[53] F. Lin, X. Yu, S. Ranganathan, and T. Kwan, "A 70 dB MPTR integrated programmable gain/bandwidth fourth-order Chebyshev highpass filter for ADSL/VDSL receivers in 65 nm CMOS," *IEEE J. of Solid-State Circuits*, vol. 44, pp. 1290–1297, April 2009.

[54] L. Ping and J. I. Sewell, "Active and digital ladder-based allpass filters," *IEE Proceedings*, vol. 137, Pt. G., no. 6, pp. 439–445, Dec. 1990.

[55] K. R. Rao, V. Sethuraman, and P. K. Neelakantan, "Novel "follow the master" filter," *Proceedings of the IEEE*, vol. 65, pp. 1725–1726, Dec. 1977.

[56] R. Schaumann and M. A. Tan, "The problem of on-chip automatic tuning in continuous-time integrated filters," in *IEEE Proc. of the ISCAS*, 1989, pp. 106–109

[57] M. Banu and Y. Tsividis, "An elliptic continuous-time CMOS filter with on-chip automatic tuning," *IEEE J. of Solid-State Circuits*, vol. 20, pp. 1114–1121, Dec. 1985.

[58] H. Khorramabadi and P. R. Gray, "High-frequency CMOS continuous-time filters," *IEEE J. of Solid-State Circuits*, vol. 19, pp. 939–948, Dec. 1984.

[59] V. Gopinathan, Y. P. Tsividis, K.-S. Tan, and R. K. Hester, "Design considerations for high-frequency continuous-time filters and implementation of an antialiasing filter for digital video," *IEEE J. of Solid–State Circuits*, vol. 25, pp. 1368–1378, Dec. 1990.

[60] C. Yoo, S.-W. Lee, and W. Kim, "A ±1.5-V, 4 MHz CMOS continuous-time filter with a single-integrator based tuning," *IEEE J. of Solid-State Circuits*, vol. 33, pp. 18–27, Jan. 1998.

[61] C. Plett and M. A. Copeland, "A study of tuning for continuous-time filters using macromodels," *IEEE Trans. on Circuits and Systems–II*, vol. 39, pp. 524–531, Aug. 1992.

[62] T. R. Viswanathan, S. Murtuza, V. H. Syed, J. Berry, and M. Staszel, "Switched-capacitor frequency control loop," *IEEE J. of Solid-State Circuits*, vol. 17, pp. 775–778, Aug. 1982.

[63] Y. Kuraishi, K. Nakayama, K. Miyadera, and T. Okamura, "A single-chip 20-channel speech spectrum analyzer using a multiplexed switched-capacitor filter bank," *IEEE J. of Solid-State Circuits*, vol. 19, pp. 964–970, Dec. 1984.

[64] Z. Y. Chang, D. Haspeslagh, and J. Verfaillie, "A highly linear CMOS g_m-C bandpass filter with on-chip frequency tuning," *IEEE J. of Solid-State Circuits*, vol. 27, pp. 388–397, March 1997.

[65] J.-M. Stevenson and E. Sànchez-Sinencio, "An accurate quality factor tuning scheme for IF and high-Q continuous-time filters," *IEEE J. of Solid-State Circuits*, vol. 33, pp. 1970–1978, Dec. 1998.

[66] S. Pavan and Y. P. Tsvidis, "An analytical solution for a class of oscillators, and its application to filter tuning," *IEEE Trans. on Circuits and Systems–I*, vol. 45, pp. 547–556, May 1998.

[67] O. Shana'a and R. Schaumann, "Low-voltage high-speed current-mode continuous-time IC filters with orthogonal ω_0-Q tuning," *IEEE Trans. on Circuits and Systems–II*, vol. 46, pp. 390–400, April 1999.

[68] Y. Tsividis, "Self-tuned filters," *Electronics Letters*, vol. 11, pp. 406–407, June 1981.

[69] H. Yamazaki, K. Oishi, and K. Gotoh, "An accurate center frequency tuning scheme for 450-kHz CMOS g_m-C bandpass filters," *IEEE J. of Solid-State Circuits*, vol. 34, pp. 1691–1697, Dec. 1999.

[70] C. A. Laber and P. R. Gray, "A 20-MHz sixth-order BiCMOS parasitic-insensitive continuous-time filter and second-order equalizer optimized for disk-drive read channels," *IEEE J. of Solid-State Circuits*, vol. 28, pp. 462–470, April 1993.

[71] C.-S. Kim, G.-O. Cho, Y.-H. Kim, and B.-S. Song, "A CMOS 4× speed DVD read channel IC," *IEEE J. of Solid-State Circuits*, vol. 33, pp. 1168–1178, Aug. 1998.

[72] P. M. Vanpeteghem and R. Song, "Tuning strategies in high-frequency integrated continuous-time filters," *IEEE Trans. on Circuits and Systems*, vol. 36, pp. 136–139, Jan. 1989.

[73] K. A. Kozma, D. A. Johns, and A. S. Sedra, "An approach for tuning high-Q continuous-time bandpass filter," in the *IEEE Int. Symp. for Circuits and Systems*, pp. 1037–1040, 1995.

[74] M. F. Toner and G. W. Roberts, "A BIST scheme for a SNR, gain tracking and frequency response test of a sigma-delta ADC," *IEEE Trans. on Circuits and Systems*, vol. 42, pp. 1–15, Jan. 1995.

[75] T. Ndjountche and R. Unbehauen, "Improved structure for programmable filters: Application in a switched-capacitor adaptive filter design," *IEEE Trans. on Circuits and Systems*, vol. 46, pp. 1137–1147, Sept. 1999.

[76] M. C. Coln, "Chopper stabilization of MOS operational amplifiers using feed-forward techniques," *IEEE J. of Solid-State Circuits*, vol. 16, pp. 745–748, Dec. 1981.

[77] T. Ndjountche, *Dynamic analog circuit techniques for real-time adaptive networks*, Aachen, Germany: Shaker Verlag, 2000.

8

Switched-Capacitor Circuits

CONTENTS

The hardware implementation of switched-capacitor (SC) circuits [1] must have the following characteristics:

- Low power consumption

- Low chip area

However, real components are subject to several nonidealities that affect the circuit performance. Because SC structures can be configured to reduce these limitations, they appear to be suitable for interfacing and implementation of signal processing operations.

Basically, capacitors, switches, and amplifiers are necessary for the realization of SC circuits. The design is often modular, consisting of a combination of small-sized blocks (sample-and-hold, integrator, gain stage,...). Building blocks must first be optimized with respect to their transfer function sensitivities to nonideal effects before they can be connected together into the overall system.

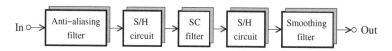

FIGURE 8.1
Block diagram of a signal processor based on a switched-capacitor filter.

The block diagram of a SC filter based processor, which can be used with analog signals, is shown in Figure 8.1. It includes additional building blocks such as sample-and-hold (S/H) or track-and-hold (T/H) circuits, anti-aliasing and smoothing filters, whose purpose is not to deliver the specified frequency sharping but rather to overcome the limitations related to the sampled-data processing. The overall filter response is primarily determined by the SC filter, provided the ratio of the sampling frequency to the cutoff frequency is much greater than one.

The anti-aliasing filter is first described in the context of sampled-data systems. In addition to describing the various properties of the basic elements, we also analyze the different compensation techniques (dummy switch, bootstrapped switch, correlated double sampling) that result in high-performance circuits. The trend toward lower supply voltages while maintaining a high dynamic range and the need for enhanced power-supply noise rejection make the use of fully differential structures mandatory.

8.1 Anti-aliasing filter

According to the Nyquist sampling theorem, a continuous-time signal can only be recovered from its samples provided the maximum frequency component of the input signal of interest is less than or equal to half of the sampling frequency. In practice, the frequency content of the signal to be sampled is limited by an anti-aliasing filter, which is a suitable continuous-time lowpass or bandpass filter [2, 3] placed before the sample-and-hold circuit as shown in Figure 8.2. In this way, aliasing is avoided by filtering out unwanted high-frequency signal components, which can be folded back into the baseband. Ideally, the anti-aliasing filter should exhibit unity gain in the passband from dc to $f_s/2$, where f_s denotes the sampling frequency, and zero gain in the stopband. But, this type of filter is difficult to implement, and it is necessary to sample the signal at a rate higher than twice the highest frequency component to relax the anti-aliasing filter specifications in practical cases.

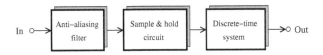

FIGURE 8.2
Building blocks of a discrete-time system.

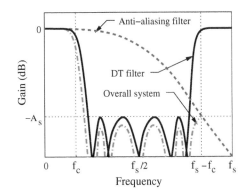

FIGURE 8.3
Discrete-time (DT) filter, anti-aliasing filter, and overall system frequency responses.

The frequency responses of a discrete-time filter, an anti-aliasing filter and the overall system are shown in Figure 8.3. The discrete-time filter has a low-pass characteristic with a cutoff frequency f_c and a stopband attenuation A_s. Note that the spectrum of every discrete-time system is replicated at multiples of the sampling frequency. The transition from the passband to the stopband of the anti-aliasing filter consists of the frequency region located between $f_s/2$ and $f_s - f_c$. A lowpass filter prototype is characterized by a cutoff (or pass-band) frequency, a stopband frequency, a maximum attenuation (or ripple) in the passband, and a minimum attenuation in the stopband. Depending on the type of application, the filter transfer function can be approximated by functions known as the Butterworth, Bessel, Chebyshev, or elliptic response, each of which has its own advantages or disadvantages. The Butterworth filter exhibits the flattest passband and lowest attenuation in the stopband. The Bessel filter has a more gradual roll-off and features a linear phase response, resulting in a constant time delay over a wide range of frequencies through the passband. The Chebyshev filter has a steeper roll-off near the cutoff frequency and ripples in the passband. The elliptic filter has the steepest roll-off and equal ripples in both passband and stopband.

As the transition band becomes smaller, the complexity of the filter architecture is increased due to the requirement for a quality factor with a high value. In this case, a multi-stage filter can then be required to meet the anti-aliasing specifications.

8.2   Capacitors

MOS capacitors are usually formed between two layers of polycrystalline silicon or metal, or between polycrystalline silicon and a heavily doped crystalline silicon [4]. Silicon dioxide (SiO_2) which is one of the most stable dielectrics, is usually used as insulator.

Ideally, the value of a MOS capacitor is given by

$$C = \frac{\varepsilon_{ox} A_p}{t_{ox}} = \frac{\varepsilon_o \varepsilon_{rox} W L}{t_{ox}} \tag{8.1}$$

where $A_p = WL$ is the area of each capacitor plate, t_{ox} is the thickness of the SiO_2 layer and $\varepsilon_{ox} = \varepsilon_o \varepsilon_{rox} \simeq 35$ pF/m is the permittivity of the SiO_2. Typical values of MOS capacitances range from 0.25 to 0.5 fF/µm² (1 fF= 10^{-15} F). The size of such capacitors is dependent on the accuracy requirements and the signal frequencies used. Capacitors are rarely made smaller than 0.1 pF. The stability of MOS capacitors with respect to the temperature and voltage difference between the input and output nodes is characterized by the temperature and voltage coefficients. Both coefficients are on the order of 100 to 110 ppm (remember that ppm means parts per million, that is 10^{-4}%) and will therefore have a negligible effect on the overall distortion of the circuits.

The smallest capacitor is generally realized in the form of a unit capacitor with a square shape ($W = L$). Larger capacitor values are formed by parallel connection of unit capacitors and possibly one fractional-valued capacitor with the same area-to-perimeter ratio as the unit capacitor. Such a design style eliminates the effect of undercutting the capacitor plates during etching on capacitance ratios and improves the matching accuracy. It also allows for regular and area efficient layout for all network capacitors.

There are unavoidable parasitic capacitances associated with MOS capacitors. The parasitic capacitance of the upper (top) plate and the lower (bottom) plate are typically 0.1 to 2% and 5 to 30% of the nominal capacitance, respectively. The effect of these parasitic capacitances can be eliminated if clever design techniques are used. To this end, in order to minimize the injection of the substrate noise into circuit nodes, the bottom plates of the capacitors should be connected only to low impedance nodes (i.e., to ground voltage sources or to the output of the amplifier) and not to virtual inputs of the amplifiers.

8.3 Switches

8.3.1 Switch description

Switches can consist of MOS transistors as shown in Figure 8.4. CMOS switches can be realized with complementary (i.e., *p*- and *n*-channel) transistors connected in parallel. The *p*- and *n*-channel transistors are controlled by appropriate positive and negative supply voltages, respectively.

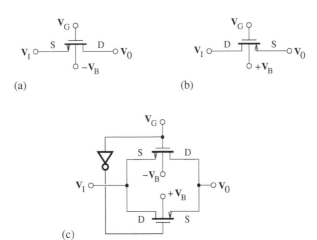

FIGURE 8.4
MOS switches: (a) nMOS transistor switch, $V_I = V_S$ and $V_0 = V_D$, (b) pMOS transistor switch, $V_I = V_D$ and $V_0 = V_S$, (c) CMOS transistor switch.

The terminal at the lower potential is considered the source for the nMOS transistor and the drain for the pMOS transistor. Therefore, the location of the drain and the source depends on the direction of the current through the channel. To avoid current dissipation through the substrate, the bulk potential must be kept lower for the nMOS device and higher for the pMOS device than the one of the source and drain. The gate voltage is used to determine the switch state. The transistor of nMOS type is in the on-state when $V_{GS} > V_T$ and the pMOS type when $V_{GS} < V_T$, where V_{GS} and V_T are the gate-source and threshold voltages, respectively. If V_{DS} is the drain-source voltage, the drain current of a transistor operating in the triode region, will be given by [5]

$$I_D = \pm K' \frac{W}{L}\left[(V_{GS} - V_T)V_{DS} - \frac{1}{2}V_{DS}^2\right], \tag{8.2}$$

where $K' = \mu C_{ox}$ is the gain for a square device, and W and L are the width and length of the channel, respectively. Note that $V_{DS} = -V_{SD}$, $V_{GS} = -V_{SG}$, $V_T > 0$ for an nMOS transistor and $V_T < 0$ for a pMOS transistor. The operating conditions of nMOS and pMOS transistors are $0 < V_{DS} < V_{GS} - V_T$ and $V_{GS} - V_T < V_{DS} < 0$, respectively. The threshold voltage, V_T, is determined according to the following equation

$$V_T = V_{T0} \pm \gamma \left(\sqrt{\pm V_{SB} + |\phi_B|} - \sqrt{|\phi_B|} \right), \tag{8.3}$$

where V_{SB} is the substrate-bulk voltage, V_{T0} is the threshold voltage for $V_{SB} = 0$, γ is the body effect factor, and ϕ_B is the approximate surface potential in strong inversion for zero back-gate bias.

For a fixed gate voltage, V_G, and for a fixed bulk voltage, V_B, the common-mode voltage, V_{CM}, and the differential-mode voltage, V_{DM}, of the switch are defined as

$$V_{CM} \hat{=} \frac{V_I + V_0}{2}, \qquad V_{DM} \hat{=} V_I - V_0, \tag{8.4}$$

where V_I and V_0 represent the voltages at the input and output terminals, respectively. The voltage V_{GS} can take the form

$$V_{GS} = \pm V_G \mp V_{CM} \pm \frac{1}{2} V_{DM}$$

and the current I_D can be rewritten as

$$I_D = \pm K' \frac{W}{L} \left(\pm V_G \mp V_{CM} - V_T \right) V_{DM}. \tag{8.5}$$

Note that, in each case, the upper and lower signs have to be applied for nMOS and pMOS devices, respectively. The above current is linearly dependent on the differential-mode voltage, V_{DM}. The current flowing through the CMOS switch can be obtained by summing the current I_D of the nMOS transistor and the one of the pMOS transistor, that is,

$$I_{CMOS} = I_{D,nMOS} + I_{D,pMOS}. \tag{8.6}$$

The on-conductances can then be defined as

$$G_{on,nMOS} = \frac{I_{D,nMOS}}{V_{DM}}, \tag{8.7}$$

$$G_{on,pMOS} = \frac{I_{D,pMOS}}{V_{DM}}, \tag{8.8}$$

and

$$G_{on,CMOS} = \frac{I_{CMOS}}{V_{DM}} \tag{8.9}$$

for the nMOS, pMOS, and CMOS switches, respectively. They are represented

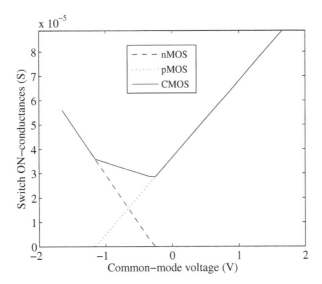

FIGURE 8.5
Switch on-conductances.

in Figure 8.5 with respect to V_{CM} based on 0.5-µm process parameters. It can be observed that nMOS and pMOS switches conduct, respectively, for lower and higher voltages and the conductivity of CMOS switches is ensured over the whole signal dynamic range. However, in low-voltage applications where transistors can be critically biased, there is no improvement in the characteristics of CMOS switches in comparison to that of nMOS and pMOS switches.

8.3.2 Switch error sources

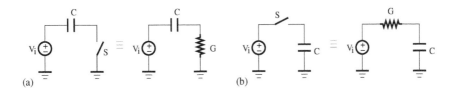

FIGURE 8.6
Simple equivalent circuit models of a (a) grounded and (b) floating switch in the on-state.

Simple equivalent circuit models of a grounded and floating switch are shown in Figures 8.6(a) and (b), respectively. The input signal is assumed to be a unit step, that is, $V_i(t) = V$ for $t \geq 0$ and zero, otherwise. The loop equation is

$$C\frac{dV_C(t)}{dt} - G(V_{CM})[V_i - V_c(t)] = 0, \qquad (8.10)$$

where $G(V_{CM})$ denotes the switch on-conductance given by [6]

$$G(V_{CM}) = G_{on} + k_G|V_{CM}|, \tag{8.11}$$

where k_G is an empirical constant and V_{CM} denotes the switch common-mode voltage. Taking into account that $V_{CM} = [V - V_C(t)]/2$ in the grounded switch configuration, while $V_{CM} = [V + V_C(t)]/2$ in the floating one, we obtain the next Riccati differential equation

$$\frac{dV_C(t)}{dt} + \{\alpha + \beta[V_c(t) - V]\}\,[V_c(t) - V] = 0. \tag{8.12}$$

The coefficients α and β are expressed for the grounded switch by

$$\alpha = \frac{G_{on}}{C} \tag{8.13}$$

and

$$\beta = \mp\frac{k_G}{2C}, \tag{8.14}$$

where the minus sign is for $V_C(0) \leq V$ and the plus sign corresponds to $V_C(0) > V$. For a floating switch, we have

$$\alpha = \frac{G_{on}}{C}\left(1 \pm \frac{K_G V}{G_{on}}\right) \tag{8.15}$$

and

$$\beta = \pm\frac{k_G}{2C}, \tag{8.16}$$

where the minus sign is valid for $V < 0$ and $V_C(0) < -V$, and the plus sign is to be used for $V \geq 0$ and $V_C(0) \geq -V$. Given the particular integral, $V_C(t) = V$, the transformation $V_C(t) \mapsto V + 1/V_C(t)$ yields a linear form of the above differential equation with the general solution

$$V_C(t) = \left[1 + \frac{(\alpha\gamma/V)\exp(-\alpha t)}{1 - \beta\gamma\exp(-\alpha t)}\right]V, \tag{8.17}$$

where γ is the integrating constant. With the assumption that $V_C(0) = 0$, the capacitor voltage may be written in the form

$$V_C(t) = [1 - \epsilon(t)]V, \tag{8.18}$$

where the relative charge transfer error, $\epsilon(t)$, with respect to the ideal case (i.e., $G(V_{CM}) = 0$) is given by

$$\epsilon = \frac{(1 - \rho)\exp(-\alpha t)}{1 - \rho\exp(-\alpha t)} \tag{8.19}$$

and

$$\rho = \frac{\beta V}{\beta V - \alpha} \, . \tag{8.20}$$

Note that $0 < |\rho| < 1$. The initial value of V_{CM} is $V/2$. For a grounded switch, the common-mode voltage can change toward 0 and

$$\exp\left[-\frac{G(V/2)}{C}t\right] \leq \epsilon(t) \leq \exp\left[-\frac{G(0)}{C}t\right] \tag{8.21}$$

while for the floating one, it can move toward V and

$$\exp\left[-\frac{G(V)}{C}t\right] \leq \epsilon(t) \leq \exp\left[-\frac{G(V/2)}{C}t\right] . \tag{8.22}$$

Thus, the charge transfer in the grounded switch structure is affected by a lower error than in the floating switch configuration. However, the floating switch may be more sensitive to the variation of the input signal.

For the circuit design, the minimum worst-case on-conductance over the signal swings should be used in order to determine the sizes of the switches. To this end, the output signal is required to approximate its final value to within an error of 0.1%.

In addition to the on-conductance, the behavior of a switch is affected by the following effects:

- Clock feedthrough

- Charge injection

- Leakage current

- Noise

Let us consider the lumped model of a single switch shown in Figure 8.7. It includes the gate-source and gate-drain overlap capacitors C_{ov}. This model can be used for the single-transistor switch, which is loaded by the capacitors C_i and C_0 and controlled by a clock phase ϕ_k ($k = 1$ or 2), as shown in the circuit of Figure 8.8.

A clock signal with a voltage swing $V_{DD} - V_{SS}$ is applied to the gate of the transistor. Because there is a voltage divider between the overlap and load capacitors during the off-state, a voltage error proportional to $C_{ov}/(C_{ov}+C_0)$ can be observed between the circuit output nodes. This error due to clock feedthrough is given by

$$\triangle V_{CF} = (V_{DD} - V_{SS})\frac{C_{ov}}{C_{ov} + C_0} \, . \tag{8.23}$$

Ideally, a switch is used to connect and disconnect two circuit nodes. However, during the turn-on phase of the MOS transistor switch, a finite amount

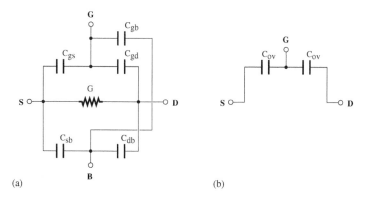

(a)   (b)

FIGURE 8.7
Equivalent circuit model for the switches: (a) on-state, (b) off-state.

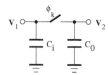

FIGURE 8.8
Single switch environment in a switched-capacitor circuit.

of mobile charges is trapped in the channel. When the switch is turned off, these charges exit through the transistor terminals. A voltage change, $\triangle V_{CI}$, can then be observed due to the fraction of channel charge, q_{inj}, injected onto C_0 [7, 8],

$$\triangle V_{CI} = \frac{q_{inj}}{C_0}, \tag{8.24}$$

where $q_{inj} = Q_a + Q_b + Q_c$ and is dependent on the input signal and the clock signal falling rate. The component Q_a is due to the charges in the strong inversion region, Q_b represents the channel charges in the weak inversion region, and Q_c represents the charges coupled through the gate-to-diffusion overlap capacitance of the transistor.

At high temperatures, the switch operation can be affected by the leakage current, I_{leak}, associated with the drain-bulk junction of the MOS transistors. For a hold time T_h, the voltage stored in the capacitor C_0 is perturbed by an amount

$$\triangle V_{leak} = \frac{I_{leak} T_h}{C_0}. \tag{8.25}$$

This voltage can be reduced if the value of C_0 is chosen in the picofarad range. Note that the leakage current has no effect on circuit operation at room temperature.

The noise due to the switch is also stored on the output capacitor. It is reduced to a voltage of $\sqrt{kT/C_0}$ in the simplified case, where the switch is equivalent in the on-state to a resistor. A sufficiently large capacitor is then required to meet the low-noise design requirement.

8.3.3 Switch compensation techniques

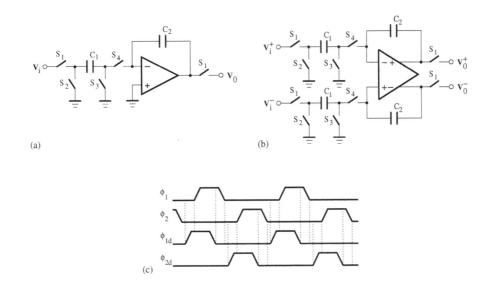

(a)

(b)

(c)

FIGURE 8.9
(a) Circuit diagram of a single-ended SC integrator; (b) circuit diagram of a differential SC integrator; (c) clocking scheme with delayed clock phases.

Many circuit techniques have been proposed for the reduction of errors in the switch operation.

Consider the circuit diagram of single-ended and fully differential SC integrators shown in Figures 8.9(a) and (b), where switches are used to connect and disconnect capacitors to circuit nodes.

Basically, the operation of SC circuits requires a clock signal with two nonoverlapping phases, ϕ_1 and ϕ_2. This is realized for the aforementioned SC integrators using the clock phase ϕ_1 for the control of switches S_1 and S_3, and the clock phase ϕ_2 for the control of switches S_2 and S_4. In practice, due to the nonideal behavior, or say charge errors, of switches, the use of minimum-sized switches controlled by a combination of two-phase clock signals with a trapezoidal shape and their delayed versions, as shown in Figure 8.9(c), may be necessary for a proper circuit operation [9]. The switch timing for the implementation of inverting and noninverting SC integrators is now as follows:
- Inverting configuration: $S_1(\phi_1)$, $S_2(\phi_2)$, $S_3(\phi_{2d})$, and $S_4(\phi_{1d})$;
- Noninverting configuration: $S_1(\phi_1)$, $S_2(\phi_2)$, $S_3(\phi_{1d})$, and $S_4(\phi_{2d})$.

Here, the switch compensation objective is to maintain almost constant the signal-independent total charge that is trapped and released by each switch during its operation. Note that the use of high-gain amplifiers with current-source output stages prevents the switch charge errors in an SC circuit to be signal dependent. In this case, because the value of the amplifier output impedance is high, the impedance load seen by switches is dominated by one of the surrounding capacitors, which are generally linear.

Switch errors can also be compensated by associating the following strategies.

One is to adopt clock signals with a very short transition time for switches in such a way that the transistor channel charges are divided equally between the source and drain and the charge injections are canceled by half-sized dummy switches [10,11] as shown in Figure 8.10. Note that the input dummy switch can be omitted for moderate value of the ratio C_i/C_0 and $\phi_2 = \overline{\phi}_1$.

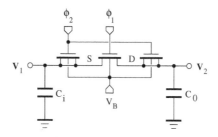

FIGURE 8.10
Charge injection cancelation using dummy switches.

The other relies on the use of differential circuit structures. For a good matching of capacitors, the voltage errors due to nonidealities are injected into the common-mode signal path and the differential signal remains unaffected in the first order.

It is interesting to note that the charge error cancelation that can be achieved in CMOS switches is not complete because the matching between the channel charges of the nMOS and pMOS transistors is poor and signal dependent.

8.4 Programmable capacitor arrays

One approach to achieve the tunability of a circuit characteristic is to use programmable capacitor arrays (PCAs) [15], which are designed to provide capacitance values selectable through a digital interface and ranging from C_{min} to C_{max} in $\triangle C$ increments. PCAs can be composed of binary-weighted capacitors and an un-switched capacitor that defines either C_{min} or C_{max}.

Let the digital word used to program each array be $b_1 b_2 \cdots b_N$, where N is the number of bits. The total capacitance of the parallel capacitor array

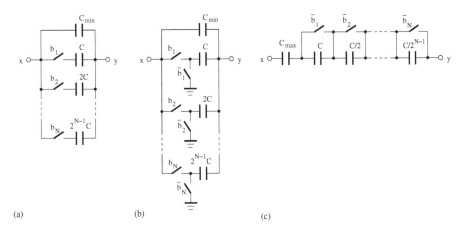

FIGURE 8.11
Programmable capacitor array: (a) Parallel configuration, (b) configuration with a constant node loading, (c) series configuration.

shown in Figure 8.11(a) is given by

$$C_T = C_{min} + \sum_{j=1}^{N} 2^{j-1} b_j C. \tag{8.26}$$

The capacitive load at the node y depends on the digital control word, because a capacitor C_j will either be connected between the nodes x and y if b_j is at the logic high or be floating if b_j is at the logic low. In the PCA of Figure 8.11(b), the capacitive load is maintained constant by switching the capacitor C_j between the node x and ground such that a capacitor never floats. For the series capacitor array depicted in Figure 8.11(c), the total capacitance can be obtained as

$$C_T = \left(\frac{1}{C_{max}} + \sum_{j=1}^{N} \frac{2^{j-1} b_j}{C} \right)^{-1}. \tag{8.27}$$

In this case, a given switch will be either open if b_j is at the logic high or closed if b_j is at the logic low.

In practice, the node y is preferably connected to a virtual ground or ground in order to reduce the PCA sensitivity to the digital switching noise. Note also that the high-frequency performance of PCAs can be affected by the parasitic components related to switches.

8.5 Operational amplifiers

An operational amplifier is one of the important active components required in the implementation of SC circuits. In the ideal case, the virtual ground induced on the inverting input node by connecting the noninverting input node to the ground is exploited to cancel the effect of parasitic capacitance on the circuit operation. Two amplifier structures can generally be used in the implementation of SC circuits:

- Operational amplifier (OA), which is equivalent to a voltage-controlled voltage source

- Operational transconductance amplifier (OTA), which is represented as a voltage-controlled current source

Because the load of the amplifiers in this kind of circuit is purely capacitive, both types of devices perform the same operation. However, it appears that designs based on OTAs are more efficient with respect to area and power consumption. The performance parameters of SC circuits, such as speed, dynamic range, and accuracy, are closely linked to the amplifier characteristics. As such, it is essential that the amplifier design reflects a number of considerations, the most important of which are summarized as follows.

Finite gain and bandwidth: The signal gain is frequency dependent. It is approximately constant at low frequencies and is typically in the range of 30 to 90 dB. The frequency at which the gain (which usually decreases in the frequency range of interest with a slope of -20 dB/decade) reaches unity or 0 dB is called unity-gain bandwidth. Generally, unity-gain bandwidths of 1 to 10 MHz are easily achieved in a simple CMOS stage, while it can range up to 100 MHz for amplifiers with a cascode structure.

Transient response time: Because an amplifier in a SC circuit is used in a clocked mode, its response time is of considerable importance. If the amplifier output signal is sampled prior to reaching the steady state, the network response will deviate from its specified value and may become distorted. The slew rate and the settling time can be used to determine the amplifier transient behavior.

The *slew rate* can be defined as the maximum rate of change of the amplifier output voltage for a step applied to the input. Its value is about 1 to 15 V/μs for classical amplifier structures and can be increased by one or even two orders of magnitude using slew-enhancement circuit techniques.

The *settling time* is the time required for the amplifier output signal to approximate its final value to within a specified error (usually 0.1 or 1%) when the input signal changes. Typically, it varies from 0.05 to 5 μs.

Linear output range: The transfer characteristic of an amplifier is linear only for a limited range of the input voltage. The maximum output signal is restricted by the supply voltages. If the amplitude of the incoming signal exceeds the maximum amplitude value, the output signal of the network will be clipped to either the minimum level or the maximum level associated with the output signal excursion of the amplifier. This can result in excessive harmonic distortions. In SC circuits, the signal amplitude for which undistorted operation is possible can be maximized by capacitance scaling.

Offset voltage: A practical amplifier can produce an output voltage even if both inputs are grounded. The dc offset voltage is random and can drift, for instance, with temperature. It can be represented by a dc voltage source which is connected in series with the noninverting input of the offset-free amplifier model and can be considered a very low-frequency noise source. The dc offset voltage may occur because of design mask errors and transistor mismatches, and is usually in the 1 to 15 mV range. It can then be minimized using good layout generations or appropriate circuit structures (e.g., correlated double sampling circuits).

For circuits with the amplifier noninverting node connected to the ground, the effect of the dc gain, A_0, and offset voltage, V_{off}, can be estimated by exploiting the following equation

$$V_0 = A_0(V^+ - V^-), \tag{8.28}$$

where $V^+ = V_{off}$. The voltage at the amplifier inverting node can be obtained as

$$V^- = -\mu V_0 + V_{off}, \tag{8.29}$$

where $\mu = 1/A_0$.

Other nonideal effects that are present in real amplifiers include, among others, noise, nonzero output resistance, imperfect common-mode signal, and power-supply rejection.

8.6 Track-and-hold (T/H) and sample-and-hold (S/H) circuits

A sample-and-hold circuit is commonly used at the interface between analog and digital systems to hold a sample of the time-varying signal for a period of time so that the high-frequency operation is facilitated. It is commonly realized by combining switching devices, such as MOS transistors, and capacitors.

Ideally, a sample-and-hold circuit takes a sample of the input signal in zero time and holds the sample value during a period T. The output signal obtained in response to a continuous-time unit impulse signal, $\delta(t)$, is known as the impulse response, which can be represented in terms of shifted unit step functions. Thus,

$$h(t) = u_s(t) - u_s(t - T), \tag{8.30}$$

where $u_s(t)$ is the unit step signal. By computing the s-transform of the impulse response, the transfer function, H_{id}, of an ideal sample-and-hold circuit can be obtained as

$$
\begin{aligned}
H_{id}(s) &= \mathcal{L}[h(t)] \\
&= \mathcal{L}[u_s(t)] - \mathcal{L}[u_s(t - T)] \\
&= (1 - e^{-Ts})\mathcal{L}[u_s(t)] \\
&= \frac{1 - e^{-Ts}}{s},
\end{aligned}
\tag{8.31}
$$

where $\mathcal{L}$ denotes the s-transform. With $s = j\omega$ and $\sin(\omega T/2) = (e^{j\omega T/2} - e^{-j\omega T/2})/2j$, it can be shown that

$$
\begin{aligned}
H_{id}(j\omega) &= \frac{1 - e^{-j\omega T}}{j\omega} \\
&= T\frac{\sin(\omega T/2)}{\omega T/2} e^{-j\omega T/2}.
\end{aligned}
\tag{8.32}
$$

Assuming that $T = 2\pi/\omega_s$, we have

$$H_{id}(j\omega) = \frac{2\pi}{\omega_s} \frac{\sin(\pi\omega/\omega_s)}{\pi\omega/\omega_s} e^{-j\pi\omega/\omega_s}. \tag{8.33}$$

Hence, the magnitude and phase of the transfer function, H_{id}, are given by

$$|H_{id}(j\omega)| = \frac{2\pi}{\omega_s} \left| \frac{\sin(\pi\omega/\omega_s)}{\pi\omega/\omega_s} \right| \tag{8.34}$$

and

$$
\angle H_{id}(j\omega) =
\begin{cases}
-\dfrac{\pi\omega}{\omega_s} & \text{if } \sin\left(\dfrac{\pi\omega}{\omega_s}\right) > 0 \\
-\dfrac{\pi\omega}{\omega_s} + \pi & \text{if } \sin\left(\dfrac{\pi\omega}{\omega_s}\right) < 0.
\end{cases}
\tag{8.35}
$$

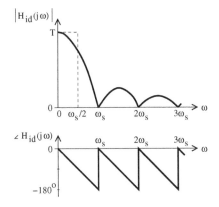

FIGURE 8.12
Frequency responses of an ideal sample-and-hold circuit.

Figure 8.12 illustrates the frequency responses of an ideal sample-and-hold circuit. In addition to a gain droop, a phase delay is introduced in the passband. By not exhibiting a sharp cutoff frequency response characteristic, the sample-and-hold circuit also transfers the aliased frequency components above one-half of the sampling rate to the output. In general, these image frequencies, whose amplitudes are not sufficiently attenuated, must be removed or attenuated by an appropriate filter.

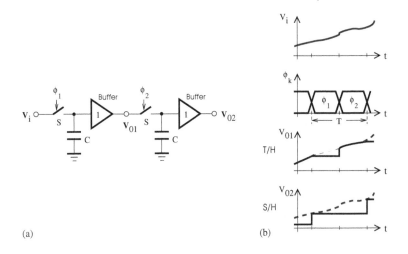

FIGURE 8.13
(a) Circuit diagram of a sampling circuit; (b) ideal representation of the sample-and-hold and track-and-hold output signals.

In practice, the sampling of a signal can be achieved using either a track-and-hold (T/H) circuit or sample-and-hold (S/H) circuit. Because the T/H circuit generally introduces a half-period delay or realizes the transfer function, $z^{-1/2}$, on the signal path, the direct implementation of the S/H or the unit-period delay operator, z^{-1}, requires a cascade of two T/H structures driven by opposite phases of a periodic clock signal, as shown in Figure 8.13(a). The T/H and S/H output signals are depicted in Figure 8.13(b) based on ideal components. The sampling instants are determined by the phases (ϕ_k, $k = 1, 2,$) of the clock signal assumed to exhibit a period T.

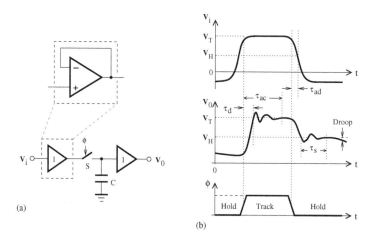

(a)

(b)

FIGURE 8.14
(a) Circuit diagram of a T/H circuit; (b) transient response of a T/H circuit.

A classical T/H circuit, as shown in Figure 8.14(a), consists of a switch, a capacitor, and unity buffer amplifiers, which are necessary to isolate the holding capacitor from the input and output load impedances. An attempt to optimize this T/H architecture using a simple structure for the active buffer and keeping the capacitors involved as small as possible tends to be limited by the exponential relationship existing between the required capacitance and resolution in number of bits. In order for the second buffer amplifier to accurately transmit the sample voltage level to the output, it should exhibit a low input bias current so as not to cause a significant variation in the capacitor charge during the hold mode and a high slew rate to rapidly react to large signal changes.

As the frequency of the input signal increases, the operation of the T/H circuit is affected by various error sources as illustrated in the transient response of Figure 8.14(b). The effects of these errors can be characterized using the following specification parameters.

Hold-to-track delay, τ_d, is the time elapsed from the initiation of the signal sample acquisition to the instant when the output starts to change in response to the input signal.

Acquisition time, τ_{ac}, is the maximum time required to acquire the input signal sample to within a specified error band (e.g., $\pm 0.1\%$ or $1/2$ least-significant bit for data converter applications) around the final value of the output.

Effective aperture delay, τ_{ad}, can be defined as the track-to-hold switching delay. It is the time difference between the propagation delays of the input signal and control signal up to the switching instant.

Track-to-hold settling time, τ_s, is the time necessary for the track-to-hold switching transient to settle to within a given error band around its final value.

Droop rate is the variation of the output as a function of the time due to leakage from the hold capacitor. It is generally specified in the hold mode with the input held at a constant dc value.

Feedthrough attenuation ratio is the fraction of the input signal that can appear at the output during the hold mode. It is a measure of the achieved isolation to prevent undesirable coupling of the input signal to the output.

In addition to drift errors, the accuracy of the T/H circuit can be affected by amplifier offset voltages and various noise sources.

In the case where amplifiers are assumed to be ideal, the equivalent model of the T/H circuit in the track mode can be derived as shown in Figure 8.15(a). The switch is modeled by a thermal noise source with the spectral density, $\overline{v_n^2}$, associated with the on-resistance, R_{on}.

- Based on the assumption that the noise signal is filtered by the transfer function, $H(j2\pi f)$, the noise spectral density at the T/H output is given by

$$\overline{v_0^2} = \int_0^{+\infty} |H(j2\pi f)|^2 \overline{v_n^2} \, \mathrm{d}f, \tag{8.36}$$

where $\overline{v_n^2} = 4kTR_{on}$, k is Boltzmann's constant, and T is the absolute temperature in Kelvin. The filter transfer function can be expressed as

$$H(j2\pi f) = \frac{1}{1 + j2\pi f R_{on} C}. \tag{8.37}$$

Hence,

$$
\overline{v_0^2} = 4kTR_{on} \int_0^{+\infty} \frac{1}{1 + (2\pi f R_{on} C)^2} \, df
$$
$$
= \frac{2kT}{\pi C} \arctan(2\pi f R_{on} C) \Big|_0^{+\infty} = \frac{kT}{C}.
$$

(8.38)

Note that $\arctan(0) = 0$ and $\lim_{x \to +\infty} \arctan(x) = \pi/2$. By making the output noise spectral density equal to the quantization noise, the capacitor value can be related to a given resolution. That is,

$$
\frac{kT}{C} = \frac{\triangle^2}{12},
$$

(8.39)

where the least-significant bit (LSB) value is represented by $\triangle = FSR/2^N$, FSR is the full-scale range, and N is the number of bits. Thus,

$$
C = 12kT \left(\frac{2^N}{FSR} \right)^2.
$$

(8.40)

For applications requiring a high resolution, the capacitor can become impractical to integrate due to the exponential increase of its value with the number of bits.

It should also be noted that the charge droop rate is reduced while the acquisition time tends to become greater as the hold capacitor value is increased. Consequently, the use of larger capacitors is not always adequate.

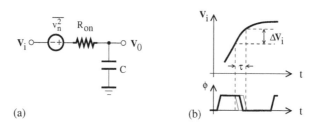

(a)   (b)

FIGURE 8.15
(a) Equivalent model of the T/H circuit; (b) illustration of the aperture uncertainty.

- Applying Kirchhoff's voltage law around the loop of the T/H equivalent model with the noise source short-circuited yields

$$
v_i = R_{on} i(t) + v_0(t),
$$

(8.41)

where $i(t) = C(dv_0(t)/dt)$. Hence,

$$
R_{on} C \frac{dv_0(t)}{dt} + v_0 = v_i.
$$

(8.42)

With the assumption that the capacitor is initially discharged, that is, $v_0 = 0$ at $t = 0$, it can be shown that

$$v_0(t) = v_i[1 - \exp(-t/R_{on}C)]. \tag{8.43}$$

For data converter applications, where the T/H circuit should settle to within the error band of ± 0.5 LSB in $t_s = \epsilon T$, it is required that

$$v_i - v_0(t_s) \ll \triangle/2, \tag{8.44}$$

where ϵ denotes a percentage (usually 50%) of the clock signal period, T. Provided the worst-case condition occurs when the input signal is set to full scale, $v_i = FSR = 2^N \triangle$, and we can obtain

$$2^N \exp(-\epsilon T/R_{on}C) \ll 1/2. \tag{8.45}$$

The switch on-resistance should be sized such that

$$R_{on} \ll \frac{\epsilon T}{2C \ln(2^{N+1})}, \tag{8.46}$$

where $T = 1/f_c$ and f_c is the frequency of the clock signal.

• The aperture uncertainty, which is also known as aperture jitter, is the random variations of the occurrence instants of the clock edge (see Figure 8.15). The amplitude of the output error due to this timing deviation can be approximated by

$$\triangle V_i \simeq \frac{dv_i}{dt}\tau. \tag{8.47}$$

Assuming the random variables are independent, the variance or power associated with the aperture uncertainty is given by

$$P_{\eta_\tau} = \mathrm{E}\left[\left(\frac{dv_i}{dt}\right)^2 \tau^2\right] = \mathrm{E}\left[\left(\frac{dv_i}{dt}\right)^2\right]\sigma_\tau^2, \tag{8.48}$$

where $\sigma_\tau^2 = \mathrm{E}[\tau^2]$.

Let us consider a sine wave input signal

$$v_i(t) = A\sin(2\pi f t), \tag{8.49}$$

where A is the amplitude. The rms value of the rate-of-change of this signal is obtained as

$$\begin{aligned}
\mathrm{E}\left[\left(\frac{dv_i}{dt}\right)^2\right] &= \frac{1}{T}\int_0^T (2\pi f A)^2 \cos^2(2\pi f t)dt \\
&= \frac{(2\pi f A)^2}{T}\int_0^T \frac{1 + \cos(4\pi f t)}{2}dt = \frac{(2\pi f A)^2}{2},
\end{aligned} \tag{8.50}$$

and the power of the aperture noise becomes

$$P_{\eta_\tau} = 2(\pi f A)^2 \, \sigma_\tau^2 . \tag{8.51}$$

The signal-to-noise ratio (SNR) due to the aperture uncertainty can be computed as

$$\text{SNR}_a = 10 \log_{10} \left(\frac{P_{v_i}}{P_{\eta_\tau}} \right), \quad \text{in dB}, \tag{8.52}$$

where $P_{v_i} = \sigma_{v_i}^2 = \text{E}[v_i^2(t)] = A^2/2$ is the input signal power or variance. Hence,

$$\text{SNR}_a = 10 \log_{10} \left(\frac{A^2/2}{2(\pi f A)^2 \, \sigma_\tau^2} \right) = -10 \log[(2\pi f)^2 \sigma_\tau^2]. \tag{8.53}$$

The effect of the aperture uncertainty is considered negligible provided the associated noise contribution remains a few decibels below the thermal noise power.

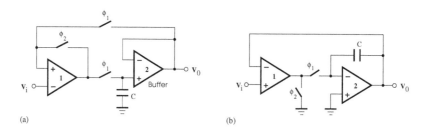

(a)   (b)

FIGURE 8.16
Circuit diagram of a closed-loop T/H circuit.

The T/H circuit accuracy can be improved using closed-loop configurations [16], which are generally known to exhibit a lower speed than open-loop structures due to amplifier stability constraints.

In the closed-loop T/H circuit of Figure 8.16(a), the signal is sampled during the clock phase ϕ_1. During the hold phase, ϕ_2, the feedback path between the amplifiers is open, the input amplifier is configured as a buffer to prevent any instability due to an abrupt change in the output level, and the charge stored on the capacitor C is maintained. One disadvantage of this T/H circuit is the undesirable variation in the charge stored on the capacitor with time due to the capacitor leakage current and amplifier input bias current.

A reduction in the output voltage droop can be achieved using the alternative T/H structure depicted in Figure 8.16(b). The capacitor used for storage of the input signal sample is connected between the virtual node and output of the second amplifier. In this way, its charge remains insensitive to the effect of grounded parasitic capacitances. The connection of the first amplifier output to the ground during the hold mode helps reduce the signal feedthrough.

8.7 Switched-capacitor (SC) circuit principle

Switched-capacitor (SC) circuits provide a high performance solution for the resistor implementations in the analog discrete-time domain. The idea is to use a capacitor C periodically switched between two circuit nodes as shown in Figure 8.17 [12]. Each switch[1] is controlled by one of the clock waveforms,

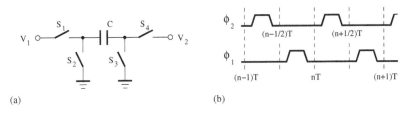

(a)   (b)

FIGURE 8.17
SC implementation of a resistor: (a) SC branch equivalent to a resistor, (b) switch clock waveforms.

ϕ_1 (or simply 1) or ϕ_2 (or 2), which are the two phases of a nonoverlapping clock signal with the frequency f_c and period T. The circuit operation is as follows: When S_2 and S_3 are closed, C is discharged; when afterwards S_1 and S_4 close, C charges to the voltage $v_1(nT) - v_2(nT)$ and the charge $\triangle q(nT) = C[v_1(nT) - v_2(nT)]$ flows from the input to the output node of the branch. With the assumption that the voltage signals are sampled at a sufficiently high rate, v_1 and v_2 can be considered constant over the sampling period. Thus, the average current transferred during one period is given by

$$\bar{i}(nT) \simeq \frac{\triangle q(nT)}{T} = \frac{v_1(nT) - v_2(nT)}{T/C} . \tag{8.54}$$

By analogy to Ohm's law for resistors, it can be concluded that a resistor of value $R \simeq T/C$ has been simulated.

The circuit of Figure 8.17, as described above, operates with a direct path between the input and output nodes. But, if the pair of switches controlled by one clock phase is now S_1 and S_3 or S_2 and S_4, the charge transfer follows an indirect path. In this way, a negative resistor can be implemented.

---

[1]The simplest realization of a switch in MOS technology is a single transistor. Its drain and source are connected to the nodes to be periodically opened and closed, and the clock signal is applied to the gate.

The direct implementation of a 10-MΩ resistor using an integrated circuit (IC) process with a sheet resistivity of polysilicon lines about 50 Ω/square needs 10^6 μm^2 chip area. In an SC design, only an area of approximately 2.10^3 μm^2 associated with a 1 pF capacitor is necessary. It has been assumed that the switches are minimum-size devices and operate with a clock frequency of 100 kHz. However, the equivalence between a switched-capacitor branch and a resistor relies on an approximation. The errors due to the use of this principle in the design can become dominant for some circuits [13,14]. For this reason, the accurate analysis of SC circuits should be done in the discrete-time domain.

The circuit diagram of an inverting SC integrator is shown in Figure 8.18, while its continuous-time equivalent models during the phases ϕ_1 and ϕ_2 are respectively depicted in Figures 8.19(a) and (b). The input voltage is piecewise

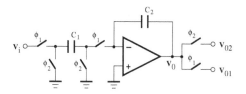

FIGURE 8.18
Circuit diagram of an SC inverting integrator.

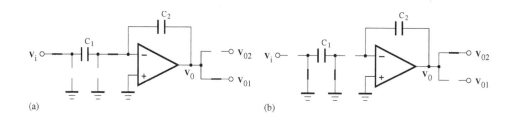

FIGURE 8.19
Continuous-time equivalent models of the SC integrator during phase (a) ϕ_1 and (b) ϕ_2.

constant and is allowed to change at the beginning of every clock phase. The amplifier is assumed to be ideal.

During the second clock phase, that is, for $(n - 1)T \le t < (n - 1/2)T$, the capacitor C_1 is discharged to the ground and $V_i[(n - 1/2)T] = 0$. The amplifier and the capacitor C_2 have been isolated since the time $(n - 1)T$; thus the voltage V_0 has maintained its value,

$$C_2 V_0[(n - 1/2)T] = C_2 V_0[(n - 1)T]. \tag{8.55}$$

During the clock phase 1, that is, for $(n - 1/2)T \le t < nT$, the application of the charge conservation principle at the inverting node of the amplifier results in

$$C_1\{V_i(nT) - V_i[(n - 1/2)T]\} + C_2\{V_0(nT) - V_0[(n - 1/2)T]\} = 0. \quad (8.56)$$

By substituting Equation (8.55) into (8.56), the following difference equation can be obtained:

$$-C_1 V_i(nT) = C_2\{V_0(nT) - V_0[(n - 1)T]\}. \quad (8.57)$$

Assuming that the first clock phase is used for the output sampling, we have $V_{01}(nT) = V_0(nT)$, and the transfer function in the z-domain reads

$$H_1(z) = \frac{V_{01}(z)}{V_i(z)} = -\frac{C_1}{C_2}\frac{1}{1 - z^{-1}}. \quad (8.58)$$

On the other hand, if the output signal is sampled using the second clock phase, that is, $V_{02}(nT) = V_0[(n - 1/2)T]$, the integrator transfer function will then become

$$H_2(z) = \frac{V_{02}(z)}{V_i(z)} = -\frac{C_1}{C_2}\frac{z^{-1/2}}{1 - z^{-1}}. \quad (8.59)$$

The $z^{-1/2}$ term in the transfer function numerator corresponds to the half-period delay introduced between the sampling instants of the input signal and the observation instants of the output signal.

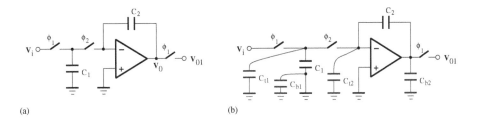

(a)　　　　　　　　　　　(b)

FIGURE 8.20
(a) Stray-sensitive SC integrator; (b) stray-sensitive SC integrator with parasitic capacitances.

An SC integrator can also be implemented as shown in Figure 8.20(a) [21]. Ideally, the capacitor C_1 is first charged to the input signal and then connected to the inverting node of the amplifier. Its switching rate is fixed by a clock signal with two nonoverlapping phases. The charge conservation equation at the end of the second clock phase, ϕ_2, can be written as

$$C_2\{V_0(nT) - V_0[(n - 1/2)T]\} = -C_1 V_i[(n - 1/2)T]. \quad (8.60)$$

During the first clock phase, ϕ_1, we have

$$C_2 V_0[(n - 1/2)T] = C_2 V_0[(n - 1)T]. \tag{8.61}$$

Combining Equations (8.60) and (8.61) yields

$$C_2\{V_0(nT) - V_0[(n - 1)T]\} = -C_1 V_i[(n - 1/2)T]. \tag{8.62}$$

Assuming that the output signal is sampled at the beginning of the phase ϕ_1, $V_0(nT) = V_{01}[(n+1/2)T]$ or equivalently, $V_0(z) = z^{1/2}V_{01}(z)$, the z-domain transfer function is obtained as

$$H(z) = \frac{V_{01}(z)}{V_i(z)} = -\frac{C_1}{C_2}\frac{z^{-1}}{1 - z^{-1}}. \tag{8.63}$$

In practice, a parasitic capacitance on the order of 0.1 to 1% of the desired capacitor value exists between the top plate and the substrate, while the one associated with the bottom plate and the substrate may be as high as 20%. Theses parasitic capacitances are illustrated in the SC integrator circuit diagram of Figure 8.20(b). The effect of parasitic capacitors, which are connected between the ground and either the virtual ground or the amplifier output, can be neglected. The accuracy of the SC integrator is then only dependent on parasitic capacitors, which are in parallel with C_1. With the equivalent input capacitor being $C_1 + C_p$, where the equivalent top plate parasitic capacitor is reduced to a single, lumped capacitor $C_p = C_{t1}$, the transfer function can take the form

$$H(z) = \frac{V_{01}(z)}{V_i(z)} = -\frac{C_1}{C_2}\left(1 + \frac{C_p}{C_1}\right)\frac{z^{-1}}{1 - z^{-1}}. \tag{8.64}$$

Thus, the capacitance ratio or integrator time constant is affected by a relative error on the order of a few percent.

Now consider the double-sampling SC integrator of Figure 8.21. In contrast to the previous one, the sampling of the input signal along with the integration is realized during both clock phases. The charge conservation law can be written as

$$-C_1' v_i[(n - 1/2)T] = C_2\{v_0[(n - 1/2)T] - v_0[(n - 1)T]\} \tag{8.65}$$

during the clock phase 2 and

$$-C_1 v_i(nT) = C_2\{v_0(nT) - v_0[(n - 1/2)T]\} \tag{8.66}$$

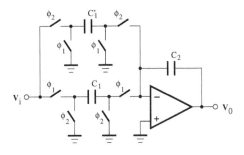

FIGURE 8.21
Double-sampling inverting SC integrator.

during the clock phase 1. From the substitution of Equation (8.65) into (8.66), we have

$$-C_1 v_i(nT) - C_1' v_i[(n-1/2)T] = C_2\{v_0(nT) - v_0[(n-1)T]\} \qquad (8.67)$$

and therefore, the z-domain transfer function is given by

$$H(z) = \frac{V_0(z)}{V_i(z)} = -\frac{C_1 + C_1' z^{-1/2}}{C_2(1 - z^{-1})}. \qquad (8.68)$$

Note that if $C_1 = C_1'$, the function $H(z)$ can be deduced from $H_1(z)$ using the transformation $z \mapsto z^{-1/2}$.

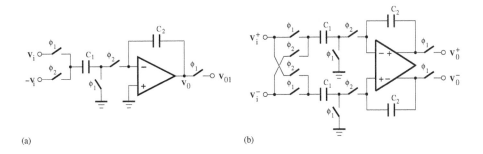

(a)   (b)

FIGURE 8.22
SC integrator: (a) Single-ended and (b) differential architectures.

Consider the SC integrator structure shown in Figure 8.22(a). During the first clock phase, ϕ_1, that is, for $(n-1)T \le t < (n-1/2)T$, the capacitor C_1 is connected between the input voltage and ground, while the feedback capacitor C_2 maintains the output voltage constant provided no other capacitor is connected to the amplifier. Hence,

$$C_2 v_0[(n-1/2)T] = C_2 v_0[(n-1)T]. \qquad (8.69)$$

During the second clock phase, ϕ_2, that is, for $(n-1/2)T < t \leq nT$, the charges due to $-v_i$ in addition to the one previously stored on C_1 are transferred to C_2. By applying the charge conservation rule, we obtain

$$C_1\{-v_i(nT) - v_i[(n-1/2)T]\}$$
$$= -C_2\{v_0(nT) - v_0[(n-1/2)T]\}. \tag{8.70}$$

Combining Equations (8.69) and (8.70) gives

$$C_1\{v_i(nT) + v_i[(n-1/2)T]\} = C_2\{v_0(nT) - v_0[(n-1)T]\}. \tag{8.71}$$

With the assumption that $v_{01}[(n+1/2)T] = v_0(nT)$, the transfer function in the z-domain is then given by

$$H(z) = \frac{V_{01}(z)}{V_i(z)} = \frac{C_1}{C_2} \frac{z^{-1/2} + z^{-1}}{1 - z^{-1}}. \tag{8.72}$$

For slow-moving signals, it can be assumed that the value of the input sample is updated only at the beginning of the first clock phase and remains constant during the second clock phase. In this case, $v_i[(n-1/2)T] = v_i[(n-1)T]$ and the integrator transfer function becomes

$$H(z) = \frac{V_{01}(z)}{V_i(z)} = \frac{C_1}{C_2} z^{-1/2} \frac{1 + z^{-1}}{1 - z^{-1}} \tag{8.73}$$

Except for the $z^{-1/2}$ term, this last transfer function can directly be transformed to the one of the continuous-time counterpart using the bilinear transform. Figure 8.22(a) shows the differential implementation of the SC integrator.

A variation up to $\pm 20\%$ can be observed in absolute capacitance values. But, with appropriate layout techniques (e.g., common-centroid layout), the matching accuracy achievable in a capacitor ratio can be as low as 0.1%. Generally, the clock signal is generated by a quartz-crystal oscillator, which is known to have an excellent stability and accuracy. SC techniques result in high-precision circuits because their transfer function coefficients are ideally determined by the clock frequency and capacitance ratios.

8.8 SC filter design

The design of SC filters is generally based on the stray-insensitive building blocks shown in Figure 8.23(a), and the transfer functions from the different inputs to the output are equivalent to various types of discrete-time integration and summation, as depicted in Figure 8.23(b).

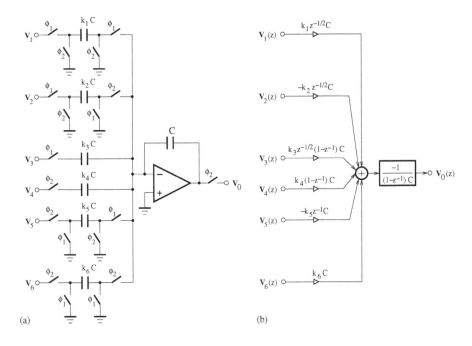

FIGURE 8.23

(a) Circuit diagram of SC building blocks and (b) the corresponding block diagrams.

The filter synthesis should be achieved in terms of discrete-time transfer functions to avoid the approximations related to the equivalence of SC circuits to continuous-time networks. Given a filter specification in the s-domain, the corresponding transfer function, $H(s)$, is mapped into the z-domain, resulting in $H(z)$, which is generally implemented using two basically different structures, the cascade of low-order sections (for biquad design, see [22–26]) and LC ladder. The s-z mapping can be achieved using the bilinear transform, which preserves the stability and is defined as

$$s = \frac{2}{T}\frac{1 - z^{-1}}{1 + z^{-1}},\tag{8.74}$$

where T is the period of the clock signal. With $s = j\omega$ and $z = e^{j\theta}$, we have

$$\omega = \frac{2}{T}\tan\frac{\theta}{2}.\tag{8.75}$$

The frequency of the analog prototype is compressed into an interval of the form $-\pi/T \leq \omega \leq +\pi/T$. The frequency mapping provided by the bilinear transform is nonlinear and the resulting frequency can be distorted, especially at high frequencies.

In general, the SC filter design should consist of the synthesis and capacitance assignment steps. The synthesis results in a signal-flow graph that realizes the filter transfer function, and the objective of the capacitance assignment, which can be carried out by an optimization scheme, is to improve the signal swing and reduce the sensitivity to the component nonidealities.

8.8.1 First-order filter

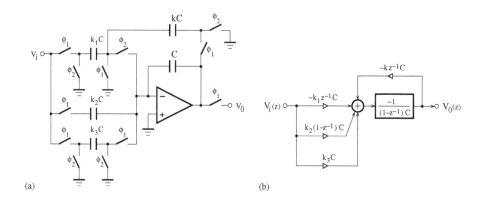

FIGURE 8.24
(a) Circuit diagram of a first-order SC filter; (b) first-order SC filter SFG.

First-order filter sections are required in the synthesis of transfer functions relying on the cascade design method. A generic structure for the first-order SC filter section is depicted in Fig 8.24(a). Based on the associated signal-flow graph (SFG) shown in Fig 8.24(b), its transfer function is given by

$$H(z) = \frac{V_0(z)}{V_i(z)} = \frac{k_1 z^{-1} - k_2(1 - z^{-1}) - k_3}{1 - (1 + k)z^{-1}}. \tag{8.76}$$

The type of the resulting filter response is determined by the choice of capacitor values as follows:
- Lowpass filter: $k_2 = k_3 = 0$

$$H_{LP}(z) = \frac{V_0(z)}{V_i(z)} = \frac{k_1 z^{-1}}{1 - (1 + k)z^{-1}} \tag{8.77}$$

- Highpass filter: $k_1 = k_3 = 0$

$$H_{HP}(z) = \frac{V_0(z)}{V_i(z)} = \frac{-k_2(1 - z^{-1})}{1 - (1 + k)z^{-1}} \tag{8.78}$$

- Allpass filter: $k_1 = 0$, $k_2 = 1$ and $k_3 = k$

$$H_{AP}(z) = \frac{V_0(z)}{V_i(z)} = \frac{z^{-1} - (1 + k)}{1 - (1 + k)z^{-1}} \tag{8.79}$$

Thus, the aforementioned circuit is capable of realizing any first-order filter function.

8.8.2 Biquad filter

Let Q and ω_0 be the pole frequency and quality factor, respectively. The s-domain transfer function of a second-order filter is given by

$$H(s) = \frac{N(s)}{D(s)} = \frac{N(s)}{1 + \dfrac{1}{Q}\dfrac{s}{\omega_0} + \left(\dfrac{s}{\omega_0}\right)^2}. \tag{8.80}$$

The type of the frequency response depends on the numerator polynomial, $N(s)$. Using the bilinear transform, the equivalent discrete-time transfer function can be written as

$$H(z) = \frac{N(z)}{D(z)} = \frac{N(z)}{1 - (2 - a - b)z^{-1} + (1 - b)z^{-2}}, \tag{8.81}$$

where

$$a = \frac{4(\omega_0 T)^2}{4 + 2(\omega_0 T/Q) + (\omega_0 T)^2} \tag{8.82}$$

and

$$b = \frac{4(\omega_0 T/Q)}{4 + 2(\omega_0 T/Q) + (\omega_0 T)^2}. \tag{8.83}$$

The poles in the z-domain can be expressed in polar form as

$$z_{p1,p2} = re^{\pm j\theta}, \tag{8.84}$$

where $0 \leq \theta \leq \pi$. Note that

$$a = 1 - 2r\cos\theta + r^2 \tag{8.85}$$

and

$$b = 1 - r^2, \tag{8.86}$$

and the stability requirement is met for $0 \leq r < 1$. With

$$\omega_z = \frac{2}{T} \tag{8.87}$$

and

$$a' = \frac{4(\omega_1 T)^2}{1 + (\omega_1 T)^2}, \tag{8.88}$$

TABLE 8.1

Numerators of Common Second-Order Filter Sections

Filter Type	$N(s)$	$N(z)$
Lowpass	$\left(1 + \dfrac{s}{\omega_z}\right)^2$	a
	$\left(1 - \dfrac{s}{\omega_z}\right)\left(1 + \dfrac{s}{\omega_z}\right)$	az^{-1}
	$\left(1 - \dfrac{s}{\omega_z}\right)^2$	az^{-2}
	$1 + \dfrac{s}{\omega_z}$	$\dfrac{a}{2}(1 + z^{-1})$
	$1 - \dfrac{s}{\omega_z}$	$\dfrac{a}{2}z^{-1}(1 + z^{-1})$
	1	$\dfrac{a}{4}(1 + z^{-1})^2$
Bandpass	$\dfrac{s}{Q\omega_0}\left(1 + \dfrac{s}{\omega_z}\right)$	$b(1 - z^{-1})$
	$\dfrac{s}{Q\omega_0}\left(1 - \dfrac{s}{\omega_z}\right)$	$bz^{-1}(1 - z^{-1})$
	$\dfrac{s}{Q\omega_0}$	$\dfrac{b}{2}(1 - z^{-1})(1 + z^{-1})$
Highpass	$\left(\dfrac{s}{\omega_0}\right)^2$	$\dfrac{a}{\omega_0^2 T^2}(1 - z^{-1})^2$
Notch	$\left(\dfrac{\omega_1}{\omega_0}\right)^2\left[1 + \left(\dfrac{s}{\omega_1}\right)^2\right]$	$\left(\dfrac{\omega_1}{\omega_0}\right)^2\dfrac{a}{a'}[1 - (2 - a')z^{-1} + z^{-2}]$
Allpass	$1 - \dfrac{s}{Q\omega_0} + \left(\dfrac{s}{\omega_0}\right)^2$	$(1 - b) - (2 - a - b)z^{-1} + z^{-2}$

where T is the clock period, Table 8.1 gives a summary of the different filter types and the corresponding numerators. Three types of notch filters can be distinguished: lowpass notch for $\omega_1 > \omega_0$, symmetrical notch for $\omega_1 = \omega_0$, and highpass notch for $\omega_1 < \omega_0$.

The general circuit diagram of the SC biquad is depicted in Figure 8.25. With reference to the corresponding SFG shown in Figure 8.26, the SC biquad

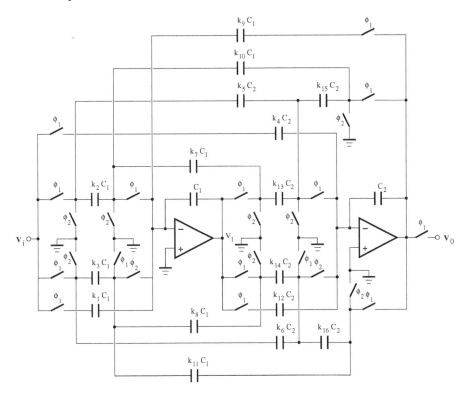

FIGURE 8.25
Circuit diagram of an SC biquad.

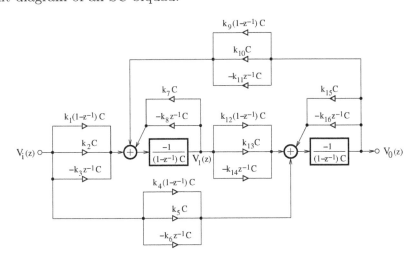

FIGURE 8.26
SFG of the SC biquad.

can be described by the following set of equations:

$$V_0(z) = -\frac{[k_5 + k_4 - (k_4 + k_6)z^{-1}]V_i(z)}{D_2(z)}$$

$$+ \frac{[k_1 + k_2 - (k_1 + k_3)z^{-1}][k_{13} + k_{12} - (k_{12} + k_{14})z^{-1}]V_i(z)}{D_1(z)D_2(z)}$$

$$+ \frac{[k_9 + k_{10} - (k_9 + k_{11})z^{-1}][k_{13} + k_{12} - (k_{12} + k_{14})z^{-1}]V_0(z)}{D_1(z)D_2(z)}$$

and

$$V_1(z) = -\frac{[k_1 + k_2 - (k_1 + k_3)z^{-1}]V_i(z)}{D_1(z)}$$
$$+ \frac{[k_5 + k_4 - (k_4 + k_6)z^{-1}][k_9 + k_{10} - (k_9 + k_{11})z^{-1}]V_i(z)}{D_1(z)D_2(z)}$$
$$+ \frac{[k_{13} + k_{12} - (k_{12} + k_{14})z^{-1}][k_9 + k_{10} - (k_9 + k_{11})z^{-1}]V_1(z)}{D_1(z)D_2(z)},$$
(8.90)

where

$$D_1(z) = 1 + k_7 - (1 + k_8)z^{-1} \tag{8.91}$$

and

$$D_2(z) = 1 + k_{15} - (1 + k_{16})z^{-1}. \tag{8.92}$$

We can then find the next transfer functions

$$H_0(z) = \frac{V_0(z)}{V_i(z)} = \frac{\alpha_0 + \alpha_1 z^{-1} + \alpha_2 z^{-2}}{\gamma_0 + \gamma_1 z^{-1} + \gamma_2 z^{-2}} \tag{8.93}$$

and

$$H_1(z) = \frac{V_1(z)}{V_i(z)} = \frac{\beta_0 + \beta_1 z^{-1} + \beta_2 z^{-2}}{\gamma_0 + \gamma_1 z^{-1} + \gamma_2 z^{-2}}, \tag{8.94}$$

where

$$\alpha_0 = (k_1 + k_2)(k_{12} + k_{13}) - (1 + k_7)(k_4 + k_5), \tag{8.95}$$

$$\begin{aligned}\alpha_1 = (1 + k_8)(k_4 + k_5) + (1 + k_7)(k_4 + k_6) \\ - (k_1 + k_2)(k_{12} + k_{14}) - (k_1 + k_3)(k_{12} + k_{13}),\end{aligned} \tag{8.96}$$

$$\alpha_2 = (k_1 + k_3)(k_{12} + k_{14}) - (1 + k_8)(k_4 + k_6), \tag{8.97}$$

$$\beta_0 = (k_4 + k_5)(k_9 + k_{10}) - (1 + k_{15})(k_1 + k_2), \tag{8.98}$$

$$\begin{aligned}\beta_1 = (1 + k_{16})(k_1 + k_2) + (1 + k_{15})(k_1 + k_3) \\ - (k_4 + k_5)(k_9 + k_{11}) - (k_4 + k_6)(k_9 + k_{10}),\end{aligned} \tag{8.99}$$

$$\beta_2 = (k_4 + k_6)(k_9 + k_{11}) - (1 + k_{16})(k_1 + k_3), \tag{8.100}$$

$$\gamma_0 = (1 + k_7)(1 + k_{15}) - (k_9 + k_{10})(k_{12} + k_{13}), \tag{8.101}$$

$$\begin{aligned}\gamma_1 = (k_9 + k_{11})(k_{12} + k_{13}) + (k_9 + k_{10})(k_{12} + k_{14}) \\ - (1 + k_8)(1 + k_{15}) - (1 + k_7)(1 + k_{16}),\end{aligned} \tag{8.102}$$

and

$$\gamma_2 = (1 + k_8)(1 + k_{16}) - (k_9 + k_{11})(k_{12} + k_{14}). \tag{8.103}$$

The denominator coefficients of the z-domain transfer functions can be related to the pole frequency, ω_0, and Q factor of an analog filter according to

$$\omega_0 = \frac{2}{T}\sqrt{\frac{\gamma_0 + \gamma_1 + \gamma_2}{\gamma_0 - \gamma_1 + \gamma_2}} \tag{8.104}$$

and

$$Q = \frac{\sqrt{(\gamma_0 + \gamma_1 + \gamma_2)(\gamma_0 - \gamma_1 + \gamma_2)}}{2(\gamma_0 - \gamma_2)} . \tag{8.105}$$

A specific filter type can be realized by selecting the appropriate capacitors to define the numerator polynomial. Note that, with reference to the SFG of Figure 8.23(b), other transfer functions can be obtained by interchanging the clock phases of some switches. Furthermore, to reduce the transfer function sensitivity to some capacitance ratios, it may be necessary to realize a biquad with more than two amplifiers [27].

For a given design, the signal levels at integrator outputs must be set by scaling the capacitances such that the amplifiers can operate with an optimal dynamic range. The maximum output signal is limited by the saturation voltage of the amplifier, while the minimum useful voltage is bounded by the noise level.

Principle of the capacitance scaling for the optimal dynamic range

Given a second-order transfer function,

$$H(z) = \frac{\delta_0 + \delta_1 z^{-1} + \delta_2 z^{-2}}{\gamma_0 + \gamma_1 z^{-1} + \gamma_2 z^{-2}}, \tag{8.106}$$

where $z = e^{j\omega T}$, the maximum value of $H(z)$ is required for the capacitance assignment that maximizes the dynamic range. That is,

$$\left. \frac{\mathrm{d}|H(e^{\omega T})|}{\mathrm{d}(\omega T)} \right|_{\omega = \omega_m} = 0. \tag{8.107}$$

The resulting equation is given by

$$\cos^2(\omega_m T) + \frac{\kappa}{\eta} \cos(\omega_m T) + \frac{\lambda}{2\eta} = 0, \tag{8.108}$$

where

$$\begin{aligned} \lambda &= 2\delta_0 \delta_2 \gamma_1 (\gamma_0 + \gamma_2) - 2\delta_1 \gamma_0 \gamma_2 (\delta_0 + \delta_2) \\ &\quad + \delta_1 (\delta_0 + \delta_2)(\gamma_0^2 + \gamma_1^2 + \gamma_2^2) - \gamma_1 (\gamma_0 + \gamma_2)(\delta_0^2 + \delta_1^2 + \delta_2^2), \end{aligned} \tag{8.109}$$

$$\eta = 2\delta_0 \delta_2 \gamma_1 (\gamma_0 + \gamma_2) - 2\delta_1 \gamma_0 \gamma_2 (\delta_0 + \delta_2), \tag{8.110}$$

and

$$\kappa = 2\delta_0 \delta_2 (\gamma_0^2 + \gamma_1^2 + \gamma_2^2) - 2\gamma_0 \gamma_2 (\delta_0^2 + \delta_1^2 + \delta_2^2). \tag{8.111}$$

Provided $(\kappa/2\eta)^2 - \lambda/2\eta \geq 0$, Equation (8.108) can be solved to

$$\omega_m = \frac{1}{T} \arccos \left(-\frac{\kappa}{2\eta} \pm \sqrt{\left(\frac{\kappa}{2\eta}\right)^2 - \frac{\lambda}{2\eta}} \right), \tag{8.112}$$

and the maximum value of transfer function, $|H(\omega_m)|$, is obtained.

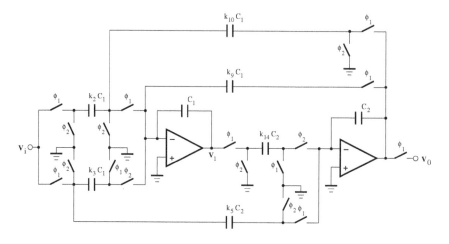

FIGURE 8.27
Circuit diagram of a bandpass SC biquad.

Application to a bandpass SC biquad

Let us consider the bandpass SC biquad shown in Figure 8.27, with the assumption that $k_5 = k_2 k_{14}$. The output and internal transfer functions are respectively given by

$$H_0(z) = \frac{V_0(z)}{V_i(z)} = \frac{\alpha_0 + \alpha_1 z^{-1} + \alpha_2 z^{-2}}{\gamma_0 + \gamma_1 z^{-1} + \gamma_2 z^{-2}} \qquad (8.113)$$

and

$$H_1(z) = \frac{V_1(z)}{V_i(z)} = \frac{\beta_0 + \beta_1 z^{-1} + \beta_2 z^{-2}}{\gamma_0 + \gamma_1 z^{-1} + \gamma_2 z^{-2}}, \qquad (8.114)$$

where

$$\alpha_0 = -k_5, \qquad (8.115)$$
$$\alpha_1 = k_5 - k_2 k_{14} = 0, \qquad (8.116)$$
$$\alpha_2 = k_2 k_{14} = -\alpha_0, \qquad (8.117)$$
$$\beta_0 = -k_2 + k_5(k_9 + k_{10}), \qquad (8.118)$$
$$\beta_1 = k_2 + k_3 - k_5 k_9, \qquad (8.119)$$
$$\beta_2 = -k_3, \qquad (8.120)$$
$$\gamma_0 = 1, \qquad (8.121)$$
$$\gamma_1 = k_9 k_{14} + k_{10} k_{14} - 2, \qquad (8.122)$$

and

$$\gamma_2 = 1 - k_9 k_{14}. \qquad (8.123)$$

Setting $k_2 = k_3 = k_5 = k$ and $k_{14} = 1$, we obtain

$$\beta_0 = k(1 + \gamma_1), \tag{8.124}$$
$$\beta_1 = k(1 + \gamma_2), \tag{8.125}$$

and

$$\beta_2 = -k. \tag{8.126}$$

It then follows that

$$\frac{\lambda}{2\eta} = 1 \tag{8.127}$$

and

$$\frac{\kappa}{2\eta} = \frac{\gamma_1^2 + (1 + \gamma_2)^2}{2\gamma_1(1 + \gamma_2)}, \tag{8.128}$$

where

$$\lambda = \begin{cases} -4\alpha_0^2\gamma_1(1 + \gamma_2) & \text{for} \quad H_0(z) \\ -4k^2\gamma_1(1 + \gamma_1 + \gamma_2)(1 + \gamma_2) & \text{for} \quad H_1(z), \end{cases} \tag{8.129}$$

$$\eta = \begin{cases} -2\alpha_0^2\gamma_1(1 + \gamma_2) & \text{for} \quad H_0(z) \\ -2k^2\gamma_1(1 + \gamma_1 + \gamma_2)(1 + \gamma_2) & \text{for} \quad H_1(z), \end{cases} \tag{8.130}$$

and

$$\kappa = \begin{cases} -2\alpha_0^2[\gamma_1^2 + (1 + \gamma_2)^2] & \text{for} \quad H_0(z) \\ -2k^2(1 + \gamma_1 + \gamma_2)[\gamma_1^2 + (1 + \gamma_2)^2] & \text{for} \quad H_1(z). \end{cases} \tag{8.131}$$

The relation between the transfer function magnitudes can be expressed as

$$|H_1(z)| = \frac{1}{k}|H_0(z)| \left| \frac{\beta_0 + \beta_1 z^{-1} + \beta_2 z^{-2}}{1 - z^{-2}} \right|. \tag{8.132}$$

Recalling the definition $z = \cos(\omega T) + j\sin(\omega T)$, it can be shown that for the frequency, ω_m, at which the transfer functions attain their maximum values, we have[2]

$$|H_1(z)| = \frac{1}{k}|H_0(z)| \sqrt{\frac{P}{4[1 - \cos^2(\omega_m T)]}}, \tag{8.133}$$

---

[2]Note that for the derivation of Equation (8.133), it was necessary to use the following trigonometric identities: $\cos^2(x) + \sin^2(x) = 1$, $\cos(2x) = 2\cos^2(x) - 1$, and $\cos(x - y) = \cos(x)\cos(y) + \sin(x)\sin(y)$.

where

$$P = \beta_0^2 + \beta_1^2 + \beta_2^2 + 2\beta_1(\beta_0 + \beta_2)\cos(\omega_m T) + 2\beta_0\beta_2[2\cos^2(\omega_m T) - 1]. \quad (8.134)$$

Because

$$\cos(\omega_m T) = -\frac{\gamma_1}{1 + \gamma_2}, \quad (8.135)$$

we arrive successively at

$$
\begin{aligned}
|H_1(z)| &= \frac{|H_0(z)|}{2}\sqrt{\frac{Q}{(1 + \gamma_2)^2 - \gamma_1^2}} \\
&= \frac{|H_0(z)|}{2}\sqrt{(1 + \gamma_2)^2 + 4(1 + \gamma_1)}, \quad (8.136)
\end{aligned}
$$

where

$$Q = (1 + \gamma_2)^2[(1 + \gamma_1)^2 + (1 + \gamma_2)^2 + 1 - 2\gamma_1^2] - 2(1 + \gamma_1)[2\gamma_1^2 - (1 + \gamma_2)^2]. \quad (8.137)$$

Hence, the maximum values of the output and internal transfer functions are equal provided,

$$\frac{1}{2}\sqrt{(1 + \gamma_2)^2 + 4(1 + \gamma_1)} = 1, \quad (8.138)$$

where $\gamma_1 = 1 - k_9$ and $\gamma_2 = k_9 + k_{10} - 2$.

In SC circuits, the high Q factor can result in a large spread of the capacitance ratios. To reduce the sensitivity of the capacitance ratio to the IC process, identical unit capacitors are generally connected in parallel to implement larger ones [28]. The implementation of large capacitance ratios would then require very small unit capacitors, whose reproduction can be affected by significant matching errors due to area variations. Furthermore, an increase in the chip area may be related to the spacing required between the units.

In practice, SC circuits exhibit a frequency response that is limited by the nonideal characteristics of components. They only operate well for input signals at much lower frequencies than the clock frequency. With a biquad consisting of a combination of an inverting and a noninverting integrator, the signals will experience one sampling period delay around the loop. As a result, this topology is less affected by the amplifier gain-bandwidth product.

The determination of capacitances should be achieved to simultaneously maximize the dynamic range and minimize the total chip area required by capacitors and the capacitance spread. The smallest capacitance can be determined based on the output noise requirement and the other capacitors are then scaled appropriately under various design trade-offs (total capacitance, sensitivity to component imperfections, ...). Capacitance scaling techniques based on analytical models can be applied to only a limited number of biquad structures. To explore all possible solutions without any restrictions, a systematic scaling approach must be numerical and rely on constrained optimization tools.

8.8.3 Ladder filter

In the cascade design of high-order filters, some of the pole Q-factors can become too high, resulting in an increase in the component-value sensitivity and a large chip area. Due to the inherent low-sensitivity in the passband, SC ladders [29–33] can then be preferred over the cascade of low-order sections.

SC ladder filters are designed using either the lossless discrete-time integrator (LDI) transform or the bilinear transform. The prototype networks or SFGs, which are used in the exact LDI design procedure adopted here, include half-unit delay terminations in order to allow the realization of any filter type. The resulting SC ladder filters are canonical, that is, the required number of amplifiers is equal to the filter order. In the case of design techniques based on the bilinear transform, the filter prototype is exactly transformed from the continuous-time domain to the discrete-time domain without any assumption on the network terminations, leading to the use of an extra T/H amplifier in the input stage.

8.9 SC ladder filter based on the LDI transform

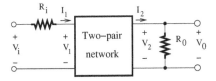

FIGURE 8.28
A doubly terminated discrete-time network.

A doubly terminated discrete-time network is depicted in Figure 8.28 [34]. The input and output terminations[3], which are assumed to be frequency dependent, are characterized by $R_i(z) = z^{-l_i/2}R_i$, $l_i = \pm 1$, and $R_0(z) = z^{l_0/2}R_0$, $l_0 = \pm 1$, respectively. The network consists of discrete-time impedances $L(z) = (z^{1/2} - z^{-1/2})L$ and $C(z) = 1/(z^{1/2} - z^{-1/2})C$. The parameters R_i, R_0, L, and C are constants. The voltages and currents are related by

$$\begin{bmatrix} V_1(z) \\ I_1(z) \end{bmatrix} = \mathbf{T} \begin{bmatrix} V_2(z) \\ I_2(z) \end{bmatrix}, \tag{8.139}$$

where $\mathbf{T}$ denotes the chain matrix of the two-pair network given by

$$\mathbf{T} = \begin{bmatrix} A(z) & B(z) \\ C(z) & D(z) \end{bmatrix}. \tag{8.140}$$

---

[3]Other forms of the terminations, but which seem to be unsuitable for the design of highpass filters, are $R_i(z) = (z^{1/2} + z^{-1/2})R_i$ and $R_0(z) = (z^{1/2} + z^{-1/2})R_0$.

With $I_1(z) = [V_i(z) - V_1(z)]/R_i(z)$ and $I_2(z) = V_2(z)/R_0(z)$, the overall transfer function is of the form

$$H(z) = \frac{V_0(z)}{V_i(z)} = \frac{1}{A(z) + \dfrac{B(z)}{R_0(z)} + R_i(z)\left[C(z) + \dfrac{D(z)}{R_0(z)}\right]} \,. \tag{8.141}$$

The reciprocal of the transfer function is $G(z) = 1/H(z)$ and the auxiliary function, $K(z)$, is defined by

$$K(z) = A(z) + \frac{B(z)}{R_0(z)} - R_i(z^{-1})\left[C(z) + \frac{D(z)}{R_0(z)}\right]. \tag{8.142}$$

The function $G(z)$ is related to the filter specifications. Provided $K(z)$ is known, the elements of the chain matrix $\mathbf{T}$ are to be determined from the following relations:

$$A(z) + \frac{B(z)}{R_0(z)} = \frac{R_i(z)K(z) + R_i(z^{-1})H(z)}{R_i(z) + R_i(z^{-1})} \tag{8.143}$$

$$C(z) + \frac{D(z)}{R_0(z)} = \frac{K(z) - H(z)}{R_i(z) + R_i(z^{-1})} \tag{8.144}$$

This can only be achieved by assuming that the following requirements are met:

- Either (i) the polynomials $A(z)$ and $D(z)$ are symmetric, $B(z)$ and $D(z)$ are anti-symmetric rational polynomials, or (ii) $A(z)$ and $D(z)$ are anti-symmetric, $B(z)$ and $D(z)$ are symmetric rational polynomials.

- The determinant of the chain matrix takes only the values ± 1.

It follows that

$$K(z)K(z^{-1}) - G(z)G(z^{-1}) = \pm[R_i(z) + R_i(z^{-1})]\left[\frac{1}{R_0(z)} + \frac{1}{R_0(z^{-1})}\right]. \tag{8.145}$$

This last equation can then be solved numerically for the function $K(z)$. The roots located in the unit circle are to be assigned to $K(z)$ and those outside to $K(z^{-1})$.

With reference to a polynomial defined by

$$P(z) = p_0 z^n + p_1 z^{n-1} + \cdots + p_n \,, \tag{8.146}$$

where p_i $(i = 0, 1, \cdots, n)$ denotes the coefficients, a balanced polynomial can be written as

$$\hat{P}(z) = z^{-n/2}P(z) = p_0 z^{n/2} + p_1 z^{n/2-1} + \cdots + p_n z^{-n/2} \,. \tag{8.147}$$

The polynomial $\hat{P}(z)$ can equivalently be represented by the equation

$$\hat{P}(z) = P^+(z) + P^-(z), \tag{8.148}$$

where the symmetric polynomial, $P^+(z)$, and the anti-symmetric polynomial, $P^-(z)$, are respectively given by

$$P^+(z) = \frac{\hat{P}(z) + \hat{P}(z^{-1})}{2} = \sum_{i=0}^{(n-1)/2} \hat{p}_i \gamma^{n-2i} \tag{8.149}$$

and

$$P^-(z) = \frac{\hat{P}(z) - \hat{P}(z^{-1})}{2} = \sum_{i=0}^{n/2} \hat{p}_i \gamma^{n-2i}. \tag{8.150}$$

The LDI variable, γ, is given by $\gamma = z^{1/2} - z^{-1/2}$, and the coefficients $\hat{p}_i$ can be computed as

$$\hat{p}_i = \begin{cases} p_0 & \text{for } i = 0, \\ p_i - \displaystyle\sum_{j=0}^{i-1} (-1)^{i-j} \binom{n-j}{i-j} \hat{p}_j & \text{otherwise.} \end{cases} \tag{8.151}$$

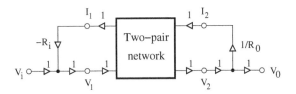

FIGURE 8.29
Signal-flow graph of a doubly terminated discrete-time network.

In the specific case of LC networks, we have $A(z) = A(z^{-1})$, $B(z) = -B(z^{-1})$, $C(z) = -C(z^{-1})$, $D(z) = D(z^{-1})$, and $\det(\mathbf{T}) = 1$. The chain matrix elements can then be written in terms of the variable γ. The rational polynomials $B(\gamma)/A(\gamma)$, $C(\gamma)/A(\gamma)$, $B(\gamma)/D(\gamma)$, and $C(\gamma)/D(\gamma)$ are reactive driving-point impedance functions. The SFG of a doubly terminated discrete-time network is illustrated in Figure 8.29. The discrete-time network, whose elements are to be determined, is decomposed into a product of low-order sections with the chain matrices $\mathbf{T}_i$, $i = 1, 2, \cdots, N$, such that

$$\mathbf{T} = \prod_{i=1}^{N} \mathbf{T}_i. \tag{8.152}$$

The decomposition of the matrix $\mathbf{T}$ from nodes 1 can be achieved recursively using the input function defined as

$$Z_1(\gamma) = \left. \frac{V_1(\gamma)}{I_1(\gamma)} \right|_{I_2=0} = \frac{C(\gamma)}{A(\gamma)}. \qquad (8.153)$$

After the extraction of the elements $A_i(\gamma)$, $B_i(\gamma)$, $C_i(\gamma)$, and $D_i(\gamma)$, corresponding to $\mathbf{T}_i$, the input function of the remaining network is given by

$$Z_1^r(\gamma) = \frac{C_i(\gamma) - A_i(\gamma)Z_1(\gamma)}{B_i(\gamma)Z_1(\gamma) - D_i(\gamma)}, \qquad (8.154)$$

which should be zero at the end of the procedure. Carrying on the extraction operation from nodes 2, the above relations become

$$Z_2(\gamma) = \left. -\frac{V_2(\gamma)}{I_2(\gamma)} \right|_{I_1=0} = \frac{B(\gamma)}{A(\gamma)} \qquad (8.155)$$

and

$$Z_2^i(\gamma) = \frac{B_i(\gamma) - A_i(\gamma)Z_2(\gamma)}{C_i(\gamma)Z_2(\gamma) - D_i(\gamma)}. \qquad (8.156)$$

Based on the input functions, the decomposition of the initial chain matrix can be achieved up to a constant reciprocal two-pair characterized by

$$\mathbf{T}_0 = \begin{bmatrix} \dfrac{1}{k} & 0 \\ 0 & k \end{bmatrix}, \qquad (8.157)$$

where k is a constant. The possibility of obtaining a network, which is incomplete, should be avoided by using a full-order input function for the chain-matrix extraction. Note that a filter of degree N is to be decomposed into N first-order network sections. The partial fraction expansion of an input function includes terms of the form k_0/γ, $2k_i\gamma/(\gamma^2 + \omega_i^2)$ and $k_\infty\gamma$, where $k_0 \geq 0$, $k_i > 0$, $k_\infty \geq 0$, and ω_i is related to the finite poles.

The signal-flow graphs of ladder filter network sections are summarized in the table illustrated in Figure 8.30 [31, 33]. Let $\mathbf{T}$ be the chain matrix of the original discrete-time filter prototype. Provided that the chain matrix $\mathbf{T}_i$ of the network section to be extracted is known, the characteristic $\mathbf{T}^r$ of the remaining network can be obtained by the following factorization,

$$\mathbf{T}^r = \mathbf{T}_i^{-1}\mathbf{T}, \qquad (8.158)$$

for the extraction from input nodes, or

$$\mathbf{T}^r = \mathbf{T}\mathbf{T}_i^{-1} \qquad (8.159)$$

for the extraction from output nodes. The structures of type I and III realize transmission zeros at infinity, as required for lowpass filters, while the ones of

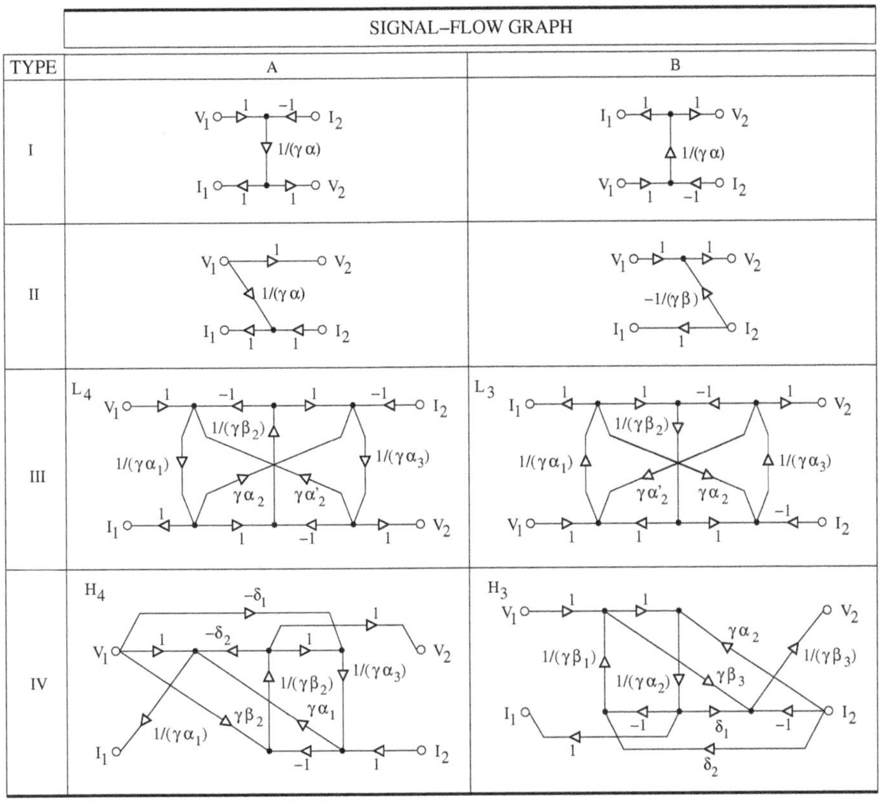

FIGURE 8.30

Signal-flow graphs of ladder filter network sections.

type II and IV implement zeros at the origin, as in the case of highpass filters. The type-I network sections correspond to the next chain matrix[4].

$$\begin{bmatrix} \alpha\gamma & 1 \\ 1 & 0 \end{bmatrix},$$ (8.160)

where $\alpha = \lim_{\gamma\to\infty}(1/\gamma Z_1(\gamma))$. We have

$$\begin{bmatrix} 1 & 0 \\ \dfrac{1}{\alpha\gamma} & 1 \end{bmatrix} \quad \text{for the type IIA}$$ (8.161)

and

$$\begin{bmatrix} 1 & \dfrac{1}{\beta\gamma} \\ 0 & 1 \end{bmatrix} \quad \text{for the type IIB,}$$ (8.162)

---

[4]With $1/Z_1(\gamma) = \tilde{k}_0/\gamma + \sum_{i=1}^{I} 2\tilde{k}_i\gamma/(\gamma^2 + \tilde{\omega}_i^2) + \tilde{k}_\infty\gamma$, the parameter α can be obtained as $\alpha = \tilde{k}_\infty = \lim_{\gamma\to\infty}(1/\gamma Z_1(\gamma))$

where $1/\alpha = \lim_{\gamma \to 0}(\gamma Z_1(\gamma))$ and $1/\beta = \lim_{\gamma \to 0}(\gamma Z_2(\gamma))$ are two positive constants[5].

The networks related to the above chain matrices contain only one integrator and can be used for the realization of all-pole filters. Remark that the network sections should not be extracted in arbitrary sequence. Using the input function Z_1, the first network section will be either of type IA, IIA if the order N of the original discrete-time network is odd, or type IB, IIB if N is even. The following extractions are achieved according to the diagram of Figure 8.31(a). With Z_2, the first network section will be either of type IA, IIA if the order N of the original discrete-time network is even or of the type IB, IIB if N is odd. The subsequent extractions are performed based on the diagram of Figure 8.31(b). From a given network section type, the arrows are directed toward the other type, which can later be extracted.

(a)   (b)

FIGURE 8.31
Signal-flow graph extraction sequences for network sections of type I and II.

The discrete-time prototype and SFG of a third-order all-pole lowpass filter are shown in Figures 8.32 and 8.33, respectively. It is assumed that $l_i = l_0 = 1$ for the terminations R_i and R_0. The resulting SC filter is shown in Figure 8.34.

For the type-III network sections, we have

$$
\begin{bmatrix}
\dfrac{(\alpha_1\alpha_3 - \alpha_2\alpha_2')\beta_2\gamma^2 + \alpha_1 + \alpha_3 - \alpha_2 - \alpha_2')\gamma}{\alpha_2\beta_2\gamma^2 + 1} & \dfrac{\alpha_1\beta_2\gamma^2 + 1}{\alpha_2\beta_2\gamma^2 + 1} \\[2ex]
\dfrac{\alpha_3\beta_2\gamma^2 + 1}{\alpha_2\beta_2\gamma^2 + 1} & \dfrac{\beta_2\gamma}{\alpha_2\beta_2\gamma^2 + 1}
\end{bmatrix}.
\tag{8.163}
$$

With $\alpha_2 = \alpha_2'$, a finite zero is realized at $\Omega = 1/\sqrt{\alpha_2\beta_2}$. As shown in the table of Figure 8.30, the network section of type IIIA includes two type IA structures, while the one of type IIIB contains two type IB structures.

The next chain matrices can be written

$$
\begin{bmatrix}
\dfrac{\alpha_2\beta_3\gamma^2 + 1}{\alpha_2\beta_3\gamma^2 + \delta_1} & \dfrac{[\beta_1(1 - \delta_1) + \beta_3(1 - \delta_2)]\alpha_2\gamma^2 - \delta_1\delta_2 + 1}{(\alpha_2\beta_3\gamma^2 + \delta_1)\beta_1\gamma} \\[2ex]
\dfrac{\beta_3\gamma}{\alpha_2\beta_3\gamma^2 + \delta_1} & \dfrac{\beta_3(\alpha_2\beta_1\gamma^2 + 1)}{\beta_1(\alpha_2\beta_3\gamma^2 + \delta_1)}
\end{bmatrix}
\tag{8.164}
$$

---

[5]Provided $Z_1(\gamma) = k_0/\gamma + \sum_{i=1}^{I} 2k_i\gamma/(\gamma^2 + \omega_i^2) + k_\infty\gamma$, the parameter α is given by $1/\alpha = k_0 = \lim_{\gamma \to 0}(\gamma Z_1(\gamma))$. Assuming that the partial fraction expansion of $Z_2(\gamma)$ takes on a form similar to the one of $Z_1(\gamma)$, the following relation can be written as well: $1/\beta = k_0 = \lim_{\gamma \to 0}(\gamma Z_2(\gamma))$.

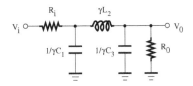

FIGURE 8.32
Circuit diagram of a third-order discrete-time lowpass ladder network.

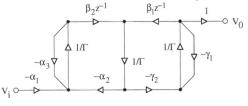

FIGURE 8.33
Signal-flow graph of a third-order lowpass ladder network ($\Gamma = 1 - z^{-1}$).

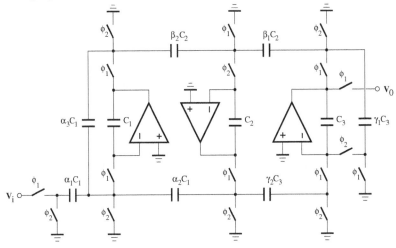

FIGURE 8.34
Circuit diagram of a third-order lowpass ladder SC filter.

for the type IVA, and

$$
\begin{bmatrix}
\dfrac{\alpha_3\beta_2\gamma^2 + 1}{\alpha_3\beta_2\gamma^2 + \delta_1} & \dfrac{\alpha_3\gamma}{\alpha_3\beta_2\gamma^2 + \delta_1} \\[2ex]
\dfrac{[\alpha_1(1 - \delta_1) + \alpha_3(1 - \delta_2)]\beta_2\gamma^2 - \delta_1\delta_2 + 1}{\alpha_1(\alpha_3\beta_2\gamma^2 + \delta_1)\gamma} & \dfrac{(1 + \alpha_1\beta_2)\alpha_3\gamma}{\alpha_1(\alpha_3\beta_2\gamma^2 + \delta_1)}
\end{bmatrix}
\tag{8.165}
$$

for the type IVB. Here, a finite zero is realized at $\Omega = \sqrt{\delta_1/\alpha_3\beta_2} = \sqrt{\delta_2/\alpha_1\beta_2}$. Remark that two type IIA (type IIB) network sections can be extracted from the structure of type IVA (type IVB).

The network sections of type III and IV can be used for the implementation of transfer functions with finite transmission zeros and at least one zero at the

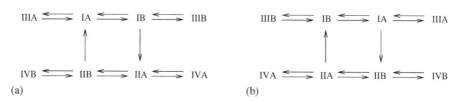

FIGURE 8.35
Signal-flow graph extraction sequences.

origin or at infinity. They should be inserted between the network sections of the type I and II, which are also taken out at the first and last steps. Depending on the choice of the input function, Z_1 or Z_2, the extraction of the remaining network sections will be performed according to the diagram shown in Figure 8.35(a) or (b), respectively.

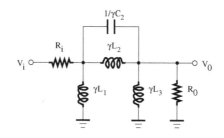

FIGURE 8.36
Circuit diagram of a third-order discrete-time highpass ladder network.

The circuit diagram of a third-order, discrete-time highpass ladder network is shown in Figure 8.36. The terminations R_i and R_0 of the discrete-time network must be chosen such that $l_i = -l_0 = 1$. The resulting SFG is depicted in Figure 8.37 and the corresponding SC filter is shown in Figure 8.38.

Given a discrete-time transfer function obtained from filter specifications, the chain parameters of the network are computed. The filter coefficients are obtained through the factorization of the two-port matrix. The SFG can then be derived and transformed into a form, which is suitable for the SC realization using parasitic-insensitive building blocks. The basic SFG transformations are included in the table of Figure 8.39.

The delays of the input and output terminations must be chosen so that a ladder filter of the order N can be realized with N amplifiers. Note that the last step of the design consists of scaling the amplifier voltage levels for the dynamic range maximization. It should be noted that depending on the SFG transformations, different filter structures realizing the same specification can be derived.

Although SC filters obtained using the design method based on the LDI

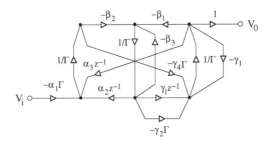

FIGURE 8.37
Signal-flow graph of a third-order highpass ladder network.

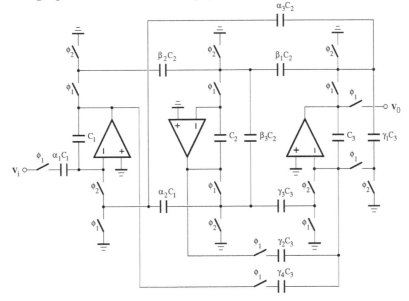

FIGURE 8.38
Circuit diagram of a third-order highpass ladder SC filter.

transform have the advantage of being simple, they can be limited by the achievable attenuation level in the stop-band. The design of high-frequency SC filters then preferably relies on the bilinear transform and building blocks that embody LDI integrators.

8.10 SC ladder filter based on the bilinear transform

8.10.1 RLC filter prototype-based design

In order to preserve a low sensitivity to fluctuations of component values in the passband, the basic design method of high-order SC filters is to model a

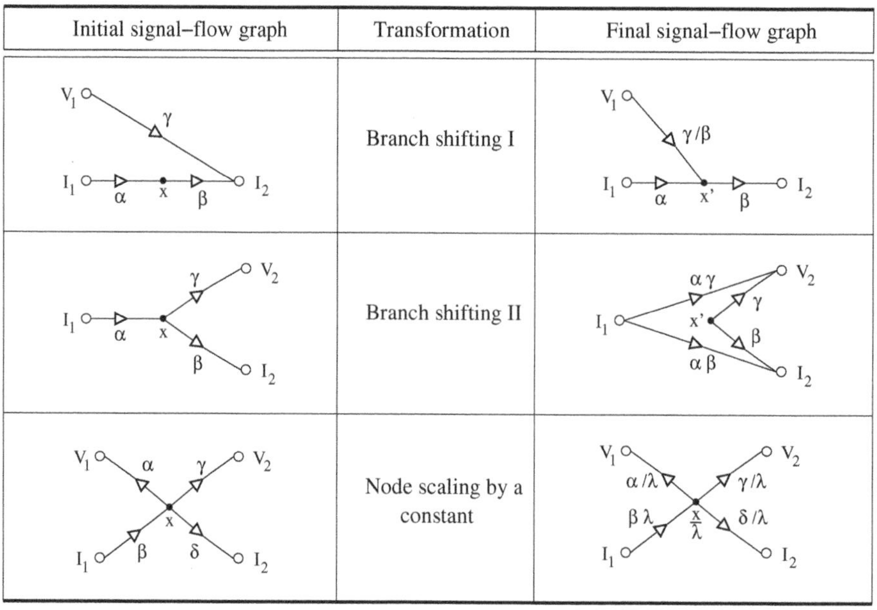

Initial signal–flow graph	Transformation	Final signal–flow graph
V_1 — γ — $I_1 \xrightarrow{\alpha} x \xrightarrow{\beta} I_2$	Branch shifting I	V_1 — γ/β — $I_1 \xrightarrow{\alpha} x' \xrightarrow{\beta} I_2$
$I_1 \xrightarrow{\alpha} x \nearrow^{\gamma} V_2 \searrow_{\beta} I_2$	Branch shifting II	$I_1 \nearrow^{\alpha\gamma} x' \nearrow^{\gamma} V_2 \searrow_{\alpha\beta} \searrow_{\beta} I_2$
$V_1 \searrow^{\alpha} \nearrow^{\gamma} V_2$ $\quad x \quad$ $I_1 \nearrow^{\beta} \searrow^{\delta} I_2$	Node scaling by a constant	$V_1 \searrow^{\alpha/\lambda} \nearrow^{\gamma/\lambda} V_2$ $\quad \frac{x}{\lambda} \quad$ $I_1 \nearrow^{\beta\lambda} \searrow^{\delta/\lambda} I_2$

FIGURE 8.39
Signal-flow graph transformations.

doubly terminated lossless passive network using SC building blocks [35]. The bilinear-transformed transfer function is exactly realized by properly arranging the phasing between adjacent SC integrators.

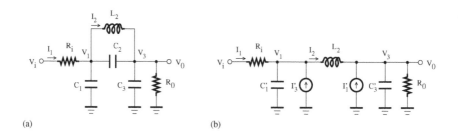

(a)   (b)

FIGURE 8.40
(a) Third-order RLC lowpass filter; (b) RLC lowpass filter after circuit transformation.

Starting from the prototype of a third-order RLC lowpass ladder filter

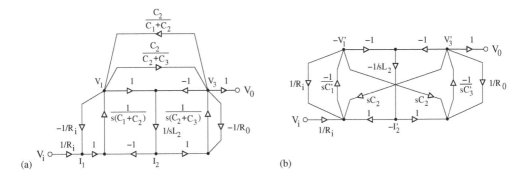

FIGURE 8.41
(a) Third-order LC lowpass filter SFG; (b) third-order LC lowpass filter SFG with the minimum number of integrator stages.

shown in Figure 8.40(a), the nodal and loop equations can be written as

$$I_1 = \frac{V_i - V_1}{R_i} \tag{8.166}$$

$$V_1 = \frac{I_1 - [I_2 + sC_2(V_1 - V_3)]}{sC_1} \tag{8.167}$$

$$I_2 = \frac{V_1 - V_3}{sL_2} \tag{8.168}$$

$$V_3 = \frac{I_2 + sC_2(V_1 - V_3)}{sC_3 + 1/R_0} \tag{8.169}$$

or, equivalently,

$$I_1 = \frac{V_i - V_1}{R_i} \tag{8.170}$$

$$V_1 = \frac{I_1 - I_2}{sC_1'} + \frac{C_2}{C_1'}V_3 \tag{8.171}$$

$$I_2 = \frac{V_1 - V_3}{sL_2} \tag{8.172}$$

$$V_3 = \frac{I_2 - \dfrac{V_3}{R_0}}{sC_3'} + \frac{C_2}{C_3'}V_1 , \tag{8.173}$$

where $C_1' = C_1 + C_2$ and $C_3' = C_2 + C_3$. The set of Equations (8.170)-(8.173) can directly be derived from the equivalent circuit of Figure 8.40(b), which is the outcome of the transformation of the initial network using Norton's theorem and has the advantage of requiring the minimum number of reactance elements. It can be represented by the SFG depicted in Figure 8.41(a), where each node represents either a voltage drop across a component or a current flowing through a component. To meet the implementation constraint of using a minimum number of inverting integrator stages, the terms of the previous equations are rearranged such that each state variable is represented by the negative integral of a weighted sum of the state variables and the input voltage. This is achieved by eliminating the current variable I_1 from the equations and scaling V_1 and I_2 by -1 , resulting in

$$-V_1 = \frac{-1}{sC_1'} \left[\frac{V_i - V_1}{R_i} - I_2 + sC_2V_3 \right] \tag{8.174}$$

$$-I_2 = -\frac{V_1 - V_3}{sL_2} \tag{8.175}$$

$$V_3 = \frac{-1}{sC_3'} \left[-sC_2V_1 - I_2 + \frac{V_3}{R_0} \right]. \tag{8.176}$$

The corresponding SFG is shown in Figure 8.41(b). It can also be derived by exploiting the proprieties related to SFG operations such as the gain scaling and node elimination.

The bilinear s-to-z transformation is given by

$$s = \frac{2}{T} \frac{1 - z^{-1}}{1 + z^{-1}} = \frac{2}{T} \frac{z^{1/2} - z^{-1/2}}{z^{1/2} + z^{-1/2}}, \tag{8.177}$$

where T denotes the clock signal period. By introducing the variable λ as follows

$$\lambda = sT/2, \tag{8.178}$$

the next expression can be derived

$$\lambda = \frac{z^{1/2} - z^{-1/2}}{z^{1/2} + z^{-1/2}} = \frac{(e^{sT/2} - e^{-sT/2})/2}{(e^{sT/2} + e^{-sT/2})/2} = \frac{\sinh(sT/2)}{\cosh(sT/2)} = \tanh(sT/2). \tag{8.179}$$

With the assumption that

$$\rho = \frac{1}{2}(z^{1/2} - z^{-1/2}) = \frac{1}{2}(e^{sT/2} - e^{-sT/2}) = \sinh(sT/2), \tag{8.180}$$

$$\nu = \frac{1}{2}(z^{1/2} + z^{-1/2}) = \frac{1}{2}(e^{sT/2} + e^{-sT/2}) = \cosh(sT/2), \tag{8.181}$$

we have

$$\lambda = \rho/\nu, \qquad \nu^2 = 1 + \rho^2, \qquad \text{and} \qquad \nu + \rho = z^{1/2}. \tag{8.182}$$

Using the s-to-z substitution, we derive the filter SFG in the λ-domain, where the capacitor C_k and inductor L_k exhibit the impedances $1/\lambda C_k$ and λL_k, respectively. The direct SC implementation should require bilinear integrators, which are more complicated than the commonly used LDI integrator. To realize the filter SFG using LDI integrators [36], it is required to divide all impedances by the complex variable ν.

Let Z_C, Z_L, and Z_R be the impedances of the capacitor C_k, inductor L_k, and resistor R_k, respectively, in the initial filter SFG. The division of all impedances by ν results in $\hat{Z}_C$, $\hat{Z}_L$, and $\hat{Z}_R$ given by

$$\hat{Z}_C = \frac{Z_C}{\nu} = \frac{1}{\lambda\nu C_k} = \frac{1}{\rho C_k}, \tag{8.183}$$

$$\hat{Z}_L = \frac{Z_L}{\nu} = \frac{\lambda L_k}{\nu} = \frac{\rho L_k}{1+\rho^2} = \left(\frac{1}{\rho L_k} + \frac{\rho}{L_k}\right)^{-1}, \tag{8.184}$$

and

$$\hat{Z}_R = \frac{Z_R}{\nu} = \frac{R_k}{\nu}, \tag{8.185}$$

respectively. After the scaling, the value C_k of the capacitor is maintained, while the inductor is transformed to a parallel combination of an inductor L_k and a capacitor $1/L_k$, and the resistor impedance becomes frequency dependent. Because the capacitor introduced by the transformation of the inductor should be taken into account in the derivation of the equivalent minimal-reactance network, the bilinear transformation and impedance scaling should first be performed.

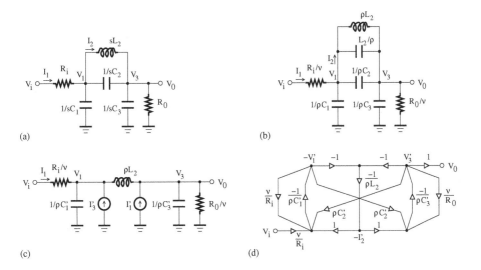

(a) (b) (c) (d)

FIGURE 8.42
Design steps of the SC ladder filter based on the third-order RLC lowpass filter prototype: (a) Third-order doubly terminated RLC filter, (b) bilinear-transformed discrete-time equivalent network, (c) modified network derived using Norton's theorem, (d) filter SFG.

The steps required for the derivation of the SFG from the third-order RLC

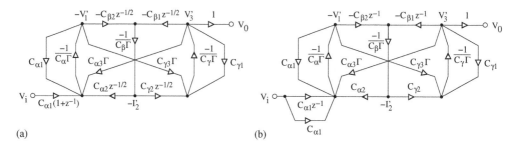

(a)   (b)

FIGURE 8.43
SFG for the SC implementation of the third-order RLC lowpass filter.

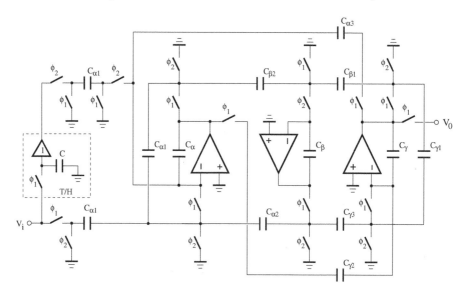

FIGURE 8.44
SC implementation of the third-order LC lowpass filter.

ladder filter are illustrated in Figure 8.42, where $C_1' = C_1 + C_2'$, $C_3' = C_2' + C_3$, $C_2' = C_2 + C_{L_2}$, and $C_{L_2} = 1/L_2$. The set of equations characterizing the filter SFG of Figure 8.42(d) can be expressed as

$$-V_1' = \frac{-1}{\rho C_1'} \left[\frac{V_i - V_1'}{R_i/\nu} - I_2' + \rho C_2 V_3' \right], \qquad (8.186)$$

$$-I_2' = -\frac{V_1' - V_3'}{\rho L_2}, \qquad (8.187)$$

$$V_3' = \frac{-1}{\rho C_3'} \left[-\rho C_2 V_1' - I_2' + \frac{V_3'}{R_0/\nu} \right]. \qquad (8.188)$$

Setting

$$\rho = \frac{z^{1/2}}{2} \Gamma \qquad \text{and} \qquad \nu = \frac{z^{1/2}}{2} (1 + z^{-1}), \qquad (8.189)$$

where $\Gamma = 1 - z^{-1}$, we can obtain

$$\frac{1}{2}[(C_1' - 1/R_i)\Gamma + 2/R_i]V_1' = \frac{1}{2}(1 + z^{-1})V_i - z^{-1/2}I_2' + \frac{1}{2}C_2'\Gamma V_3', \quad (8.190)$$

$$I_2' = z^{-1/2}\frac{2}{L_2}\frac{V_1' - V_3'}{\Gamma}, \quad (8.191)$$

$$\frac{1}{2}[(C_3' - 1/R_0)\Gamma + 2/R_0]V_3' = z^{-1/2}I_2' + \frac{1}{2}C_2'\Gamma V_1'. \quad (8.192)$$

On the other hand, a filter SFG based on SC building blocks is illustrated in Figure 8.43. It can be described by the following set of equations:

$$-V_1' = \frac{-1}{C_\alpha\Gamma}\left[C_{\alpha 1}(1 + z^{-1})V_i - C_{\alpha 1}V_1' - C_{\alpha 2}z^{-1/2}I_2' + C_{\alpha 3}\Gamma V_3'\right], \quad (8.193)$$

$$-I_2' = \frac{-1}{C_\beta\Gamma}\left[C_{\beta 2}z^{-1/2}V_1' - C_{\beta 1}z^{-1/2}V_3'\right], \quad (8.194)$$

$$V_3' = \frac{-1}{C_\gamma\Gamma}\left[-C_{\gamma 3}\Gamma V_1' - C_{\gamma 2}z^{-1/2}I_2' + C_{\gamma 1}V_3'\right]. \quad (8.195)$$

Rewriting these equations gives

$$(C_\alpha\Gamma + C_{\alpha 1})V_1' = C_{\alpha 1}(1 + z^{-1})V_i - C_{\alpha 2}z^{-1/2}I_2' + C_{\alpha 3}\Gamma V_3', \quad (8.196)$$

$$I_2' = z^{-1/2}\frac{C_{\beta 2}V_1' - C_{\beta 1}V_3'}{C_\beta\Gamma}, \quad (8.197)$$

$$(C_\gamma\Gamma + C_{\gamma 1})V_3' = C_{\gamma 2}z^{-1/2}I_2' + C_{\gamma 3}\Gamma V_1'. \quad (8.198)$$

Comparing the last sets of equations obtained from the SFGs of Figures 8.42(d) and 8.43, the initial values for the capacitors of the SC ladder filter are derived as

$$C_\alpha = (C_1' - C_{R_i})/2, \quad C_{\alpha 1} = C_{R_i}, \quad C_{\alpha 2} = 1, \quad C_{\alpha 3} = C_2'/2,$$
$$C_{\beta_1}/C_\beta = C_{\beta_2}/C_\beta = 2C_{L_2},$$
$$C_\gamma = (C_3' - C_{R_0})/2, \quad C_{\gamma 1} = C_{R_0}, \quad C_{\gamma 2} = 1, \quad C_{\gamma 3} = C_2'/2, \quad (8.199)$$

where $C_{R_i} = 1/R_i$, $C_{R_0} = 1/R_0$, and $C_{L_2} = 1/L_2$. Taking into account the normalization factor $T/2 = \lambda/s$ of the bilinear transform, these last values take the form $C_{R_i} = T/(2R_i)$, $C_{R_0} = T/(2R_0)$, and $C_{L_2} = T^2/(4L_2)$. The circuit diagram of the SC filter is shown in Figure 8.44, where a T/H circuit is required for the realization of the input branch.

Note that the capacitor values can be further scaled for the optimal dynamic range of amplifiers and the minimum total capacitance. Because the transfer function is only determined by the capacitor ratios, the capacitor values for each integrator can be normalized relative to the corresponding unswitched feedback capacitor.

In general, starting from the RLC filter prototype, the SC ladder filter can be designed by following the following steps:

- Perform the bilinear transformation;

- Scale all impedances;

- Construct the minimal-reactance network;

- Derive the filter SFG; and

- Convert the filter SFG into an SC implementation.

8.10.2 Transfer function-based design of allpass filters

In the continuous-time domain, the transfer function of an N-th-order allpass filter can generally be written as

$$H(s) = \pm \frac{D(-s)}{D(s)}, \tag{8.200}$$

where

$$D(s) = \sum_{k=0}^{N} a_k s^k$$

and $a_N = 1$. To meet the stability requirements, the roots of $D(s)$ must be in the left-half s-plane, or equivalently, all of the coefficients a_k should be positive, as is the case for Hurwitz polynomials. By performing the bilinear transformation, the equivalent transfer function in the z-domain is given by

$$H(z) = \pm \frac{z^{-N} D(z^{-1})}{D(z)}, \tag{8.201}$$

where

$$D(z) = \sum_{l=0}^{N} a_l z^{-l}$$

and $a_0 = 1$. With the numerator being the mirror-image polynomial of the denominator, it can be shown that the magnitude of the transfer function, $|H(e^{j\omega})|$, is equal to unity for any frequency, and the phase response is

$$\theta(\omega) = \arg[H(e^{j\omega})] = \begin{cases} -N\omega - 2\mathrm{Arg}[D(e^{j\omega})], & \text{if } H(j\omega) \geq 0 \\ \pi - N\omega - 2\mathrm{Arg}[D(e^{j\omega})], & \text{if } H(j\omega) < 0, \end{cases} \tag{8.202}$$

where it is assumed that the phase of a positive real number is zero while the one of a negative real number is π.

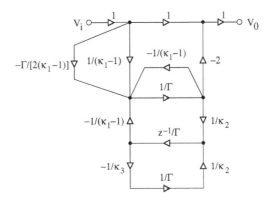

FIGURE 8.45
Signal-flow graph of a third-order allpass ladder network.

Let us consider a third-order allpass filter with the transfer function $H(z)$ given by

$$H(z) = -\frac{z^3 D(z^{-1})}{D(z)}, \qquad (8.203)$$

where $D(z)$ is a third-order polynomial with real coefficients. The design of the corresponding SC circuit is based on the next decomposition [37]

$$H(z) = 1 - 2\frac{z+1}{2}\frac{1}{1 + \tilde{Y}(z)}, \qquad (8.204)$$

where

$$\tilde{Y}(z) = \frac{1}{2}(z-1) + \frac{1}{2}(z+1)Y(z) \qquad (8.205)$$

and the admittance function[6], $Y(z)$, is obtained as

$$Y(z) = \frac{D(z) - z^{-3}D(z^{-1})}{D(z) + z^{-3}D(z^{-1})}. \qquad (8.206)$$

To proceed further, $Y(z)$ can be expressed in a continued fraction expansion [38] of the form

$$Y(z) = z^{1/2}Y(\gamma), \qquad (8.207)$$

where

$$Y(\gamma) = \kappa_1\gamma + \cfrac{1}{\kappa_2\gamma + \cfrac{1}{\kappa_3\gamma}} \qquad (8.208)$$

and $\gamma = z^{1/2} - z^{-1/2}$. Note that κ_1 is assumed to be greater than one, and κ_2 and κ_3 are positive constants. The third-order allpass transfer function can

---

[6]Note that, depending on the sign of the transfer function, the ladder network synthesis can be achieved either with $Y(z)$ or $1/Y(z)$.

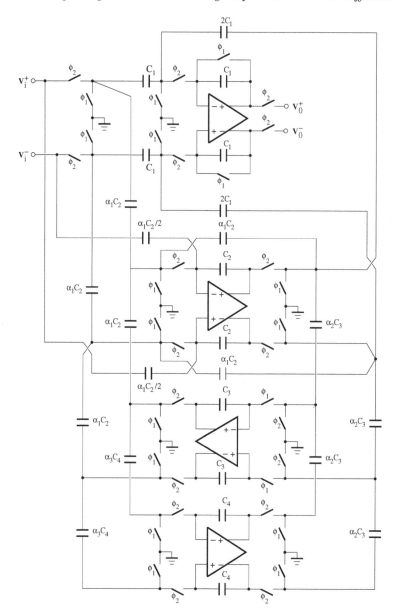

FIGURE 8.46
Circuit diagram of a third-order allpass ladder SC filter.

be realized by the SFG depicted in Figure 8.45. The resulting SC circuit is shown in Figure 8.46, where the coefficients α_i $(i = 1, 2, 3)$ can be obtained as

$$
\alpha_i = \begin{cases} \dfrac{1}{\kappa_1 - 1}, & \text{for } i = 1 \\ \dfrac{1}{\kappa_i}, & \text{otherwise.} \end{cases} \tag{8.209}
$$

Here, the SC implementation exploits the availability of signals with both polarities in fully differential structures.

In general for $N > 3$, the admittance function, $Y(z)$, can be decomposed into either a continued fraction expansion or a partial fraction expansion, and the design of an N-th order allpass filter results in two different SC implementations, which require $N + 1$ amplifiers.

8.11 Effects of the amplifier finite gain and bandwidth

Integrators are commonly used in filter design. In this analysis, the stray-insensitive circuit of Figure 8.47 is considered. It can operate as an inverting SC integrator when the switch timings are $S_1(\phi_2)$ and $S_2(\phi_1)$, or a noninverting SC integrator when switch timings are $S_1(\phi_1)$ and $S_2(\phi_2)$. The amplifier

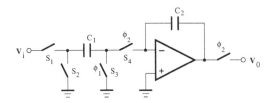

FIGURE 8.47
Lossless discrete-time stray-insensitive integrator.

will have, as in practice, an open loop gain, A, with a finite dc gain[7], A_0, and a finite transition frequency (or unity-gain frequency), f_t. Thus, the inverting input terminal, v^-, of the amplifier will not be a true virtual ground but will be dependent on the output signal, as $v_0 = A(v^+ - v^-)$.

The equivalent models of the amplifiers, which can be used in the implementation of SC circuits, are shown in Figure 8.48. With the amplifier including an ideal output buffer (see Figure 8.48(a)), the charge transfer is independent of the capacitive load, but the additional pole, which is related to the output voltage follower, can affect the stability of the overall circuit. Generally, the speed performance is then improved for IC designs based on operational transconductance amplifiers (see Figure 8.48(b)).

The amplifier is first described as an ideal voltage-controlled voltage source with a finite dc gain and infinite bandwidth. This model is useful when the distortions due to the finite dc gain must be evaluated. However, in the context of high-frequency applications, the influence of the finite bandwidth on the settling speed becomes dominant. In this case, the analysis is done by modeling

---

[7]An OTA is generally characterized by a finite transconductance gain, g_m. For a finite output impedance, Z_0, the OTA can be modeled as a voltage-controlled voltage source with the finite dc gain $A_0 = g_m Z_0$.

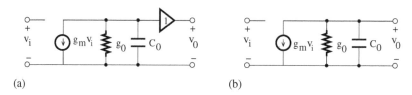

(a) (b)

FIGURE 8.48
Equivalent models of (a) an operational amplifier with an ideal output buffer and (b) an operational transconductance amplifier.

the amplifier as an ideal voltage-controlled voltage source with a finite unity-gain frequency and an infinite dc gain [39–41].

For simplicity, the switches are assumed to have a negligible on-resistance and the input signal is considered constant during the charge transfer between the capacitors. The values of all necessary voltages are evaluated at the end of each clock phase, resulting in a set of difference equations that can be readily transformed into the z-domain. The error function, $E(z)$, is then extracted from the circuit transfer function according to the next equations:

$$H(z) = \frac{V_0(z)}{V_i(z)} = H_{id}(z)E(z) \tag{8.210}$$

and

$$E(z)|_{z=\exp(j\omega T)} = [1 + m(\omega)]\exp[j\theta(\omega)], \tag{8.211}$$

where T is the clock period, and $m(\omega)$ and $\theta(\omega)$ are the error magnitude and phase, respectively. For small error values, that is, $m(\omega)$ and $\theta(\omega) \ll 1$, the next approximation can be made

$$E(z)|_{z=\exp(j\omega T)} \simeq 1 + m(\omega) + j\theta(\omega), \tag{8.212}$$

which can be conveniently used for estimating the error magnitude and phase on a complete network. In the independent analysis of the errors due to A_0 and f_t, the error function can be expressed as

$$E(z) = E_{A_0}(z)E_{f_t}(z), \tag{8.213}$$

corresponding to the following expressions,

$$m(\omega) = m_{A_0}(\omega) + m_{f_t}(\omega) \tag{8.214}$$

and

$$\theta(\omega) = \theta_{A_0}(\omega) + \theta_{f_t}(\omega), \tag{8.215}$$

which give the magnitude and phase errors, respectively.

8.11.1 Amplifier dc gain

The gain error is independent of the integrator switch phasing and a dual analysis of the inverting and noninverting structure is not useful.

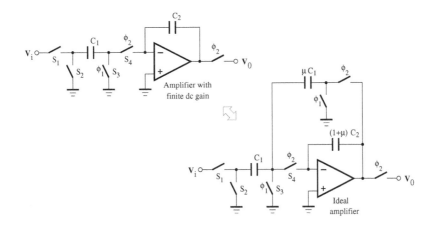

FIGURE 8.49
Equivalent circuit model for the analysis of the finite dc gain effect on the lossless discrete-time stray-insensitive integrator ($\mu = 1/A_0$).

The application of the charge conservation law at the inverting node of the amplifier leads to the following difference equations:

$$C_2 \left\{ v^-[(n - 1/2)T] - v_0[(n - 1/2)T] \right\} = C_2 \left\{ v^-[(n - 1)T] - v_0[(n - 1)T] \right\} \tag{8.216}$$

at the end of the second clock phase, ϕ_2, or say for $(n-1)T \leq t < (n-1/2)T$, and

$$C_1[v_i(nT) - v^-(nT)]$$
$$= C_2 \left(v^-(nT) - v_0(nT) - \left\{ v^-[(n - 1/2)T] - v_0[(n - 1/2)T] \right\} \right) \tag{8.217}$$

at the end of the first clock phase, ϕ_1, that is, for $(n - 1/2)T \leq t < nT$, where T is the clock signal period. Taking into account the fact that $v^- = -\mu v_0 + V_{off}$, and combining Equations (8.216) and (8.217) to eliminate the term $v_0[(n - 1/2)T]$, we obtain

$$C_1[v_i(nT) + \mu v_0(nT)] + C_2(1 + \mu) \left\{ v_0(nT) - v_0[(n - 1)T] \right\} = 0. \tag{8.218}$$

From the above equation, the z-domain transfer function can be found as

$$H(z) = \frac{V_0(z)}{V_i(z)} = \frac{-C_1}{C_2(1 - z^{-1}) \left[1 + \mu + \mu \dfrac{C_1}{C_2} \dfrac{1}{1 - z^{-1}} \right]}, \tag{8.219}$$

where $\mu = 1/A_0$. As illustrated in Figure 8.49, this transfer function is equivalent to the one of an ideal integrator (i.e., designed with the ideal amplifier) but with a feedback capacitor of the form $(1 + \mu)C_2$ and an extra feedback path providing a damping term equal to μC_1.

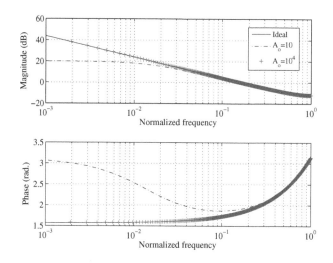

FIGURE 8.50

Finite dc gain effects on the integrator frequency responses.

In the frequency domain, we have

$$H(j\omega) = H_{id}(j\omega) \left[1 + \mu \left(1 + \frac{C_1}{2C_2}\right) - j\mu \frac{C_1}{2C_2 \tan(\omega T/2)}\right]^{-1}, \qquad (8.220)$$

where

$$H_{id}(j\omega) = -\frac{C_1}{C_2} \frac{e^{j(\omega T/2)}}{j \sin(\omega T/2)}. \qquad (8.221)$$

The term in brackets is the same for the two types of integrator. Then, one can identify the next gain and phase errors

$$m_{A_0}(\omega) = m = -\mu \left(1 + \frac{C_1}{2C_2}\right) \qquad (8.222)$$

and

$$\theta_{A_0}(\omega) = \mu \frac{C_1}{2C_2 \tan(\omega T/2)}. \qquad (8.223)$$

The gain error is found to be frequency independent and can therefore be modeled as an element variation of the integration capacitor. This suggests the use of pre-distortion methods in order to reduce this error. When the frequency increases, the phase error will approach zero.

As confirmed by the test results of Figure 8.50 (see also [39]), the dc gain should be greater than 60 dB in order to reduce the integrator deviation to an acceptable level (i.e., approximately 0.1%).

8.11.2 Amplifier finite bandwidth

In order to analyze the influence of the finite bandwidth on the transfer function of the OTA-based integrator, the amplifier is modeled as an ideal voltage-controlled current source with the short-circuit transconductance, g_m, and an infinite output impedance. Its unity-gain frequency, f_t, is given by $f_t = g_m/(2\pi C_T)$, where C_T is the total load capacitance. Thus, the bandwidth of the OTA depends on the capacitive loading of the circuit; as such, it varies from phase to phase.

The inverting integrator has no delay between the sample instants of the input and output signals. The noninverting one, in contrast, is realized with a half-clock period delay between these sample instants for a 50% duty clock pattern. This delay influences the phase error due to the amplifier finite bandwidth.

8.11.2.1 Inverting integrator

No charge will be transferred by the input branch during the clock phase 1. The charge transfer from the input signal will take place during the clock phase 2. Thus, the amplifier input voltage $v_x(t)$ will be discontinuous at the end of the clock phase 1.

During the clock phase 1, that is, $(n-1)T \le t < (n-1/2)T$, the amplifier is disconnected from the input signal and we have

$$v_x(t) - v_0(t) = v_x[(n-1)T] - v_0[(n-1)T] \tag{8.224}$$

or in a differential form

$$\frac{dv_0(t)}{dt} = \frac{dv_x(t)}{dt}. \tag{8.225}$$

In the time domain, the OTA can be described as

$$g_m v_x(t) + C_L \frac{dv_0(t)}{dt} = 0 \tag{8.226}$$

Substituting Equation (8.225) into (8.226) and solving for the value of $v_x(t)$ at $t = (n-1/2)T$ as

$$v_x[(n-1/2)T]^- = v_x[(n-1)T]e^{-k_1 T/2}, \tag{8.227}$$

where the superscript $(-)$ denotes "the time instant just before $t = (n-1/2)T$", and $k_1 = g_m/C_L$. At the end of the clock phase 1, $v_x[(n-1/2)T]^-$ will almost be zero. For $t = (n-1/2)T$, Equation (8.224) can then be written as

$$v_0[(n-1/2)T] = v_0[(n-1)T] - v_x[(n-1)T]. \tag{8.228}$$

During the clock phase 2, that is, for $(n-1/2)T \le t < nT$, the charge conservation at the inverting terminal reduces to

$$v_0(t) - v_0[(n-1/2)T] = \frac{C_1 + C_2}{C_2} v_x(t) - \frac{C_1}{C_2} v_i(t). \tag{8.229}$$

Assuming that the input voltage remains constant during the clock phase, the next differential equation is obtained:

$$\frac{dv_0(t)}{dt} = \left(1 + \frac{C_1}{C_2}\right)\frac{dv_x(t)}{dt}. \tag{8.230}$$

In this phase, the OTA is described by

$$g_m v_x(t) + C_1\frac{dv_x(t)}{dt} + C_L\frac{dv_0(t)}{dt} = 0. \tag{8.231}$$

Solving the differential equation obtained by combining Equations (8.230) and (8.231), the value of $v_x(t)$ at $t = nT$ can be written as

$$v_x(nT) = v_x[(n-1/2)T]^+ e^{-k_2T/2}, \tag{8.232}$$

where

$$k_2 = \frac{g_m}{C_1 + C_L\left(1 + \dfrac{C_1}{C_2}\right)}, \tag{8.233}$$

and

$$v_0(nT) = v_0[(n-1/2)T] + \left(1 + \frac{C_1}{C_2}\right)v_x[(n-1/2)T]^+ e^{-k_2T/2} - \frac{C_1}{C_2}v_i(nT). \tag{8.234}$$

Note that $v_x[(n-1/2)T]^+$ can be determined from the initial conditions as

$$v_x[(n-1/2)T]^+ = k_0 v_i(nT), \tag{8.235}$$

where $k_0 = C_1(C_2+C_L)/[C_1C_2+C_L(C_1+C_2)]$. To proceed further, Equations (8.228) and (8.235) are substituted into Equation (8.234) and the result is subsequently transformed into the z-domain. Then,

$$H(z) = \frac{V_0(z)}{V_i(z)} = H_{id}(z)E(z), \tag{8.236}$$

where

$$H_{id}(z) = -\frac{C_1}{C_2}\frac{1}{1 - z^{-1}} \tag{8.237}$$

and

$$E(z) = 1 - (1 + \frac{C_2}{C_1})k_0 e^{-k_2T/2} + z^{-1}\frac{C_2}{C_1}k_0 e^{-k_2T/2}. \tag{8.238}$$

From Equation (8.238), the gain and phase errors can be computed according to Equation (8.211) as

$$m(\omega) \simeq -\left(1 + \frac{C_2}{C_1}\right)\left(1 - \frac{1}{1 + \dfrac{C_1}{C_2}}\cos(\omega T)\right)k_0 e^{-k_2T/2} \tag{8.239}$$

and

$$\theta(\omega) \simeq -\frac{C_2}{C_1} k_0 \sin(\omega T) e^{-k_2 T/2} \omega T. \tag{8.240}$$

Figure 8.51 shows the frequency responses of the integrator for different values

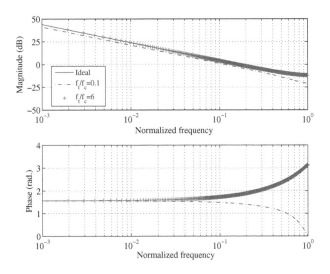

FIGURE 8.51

Finite bandwidth effects on the integrator frequency responses.

of the ratio f_t/f_c.

8.11.2.2 Noninverting integrator

For the noninverting integrator, the charge transfer between the capacitors will take place at the end of the clock phase 2. The analysis of the circuit during the clock phase 1 is very similar to the situation of the inverting integrator.

From Equations (8.224) and (8.225), we have

$$v_x[(n-1/2)T] = v_x[(n-1)T]e^{-k_1 T/2} \simeq 0 \tag{8.241}$$

and

$$v_0[(n-1/2)T] = v_0[(n-1)T]. \tag{8.242}$$

During the clock phase 2, that is, $(n-1/2)T \leq t < nT$, the charge conservation law at the inverting input node of the OTA reads

$$v_0(t) - v_0[(n-1/2)T] = \frac{C_1 + C_2}{C_2} v_x(t) + \frac{C_1}{C_2} v_i[(n-1/2)T]. \tag{8.243}$$

In the differential form, the above equation can be reduced to Equation (8.230). Because Equation (8.231) is still valid for the description of the OTA, we can obtain

$$v_x(nT) = v_x[(n-1/2)T]^+ e^{-k_2 T/2} \tag{8.244}$$

and

$$v_0(nT) = v_0[(n-1/2)T] + \frac{C_1 + C_2}{C_2} v_x[(n-1/2)T]^+ e^{-k_2 T/2} + \frac{C_1}{C_2} v_i[(n-1/2)T].$$

$$(8.245)$$

Initially,

$$v_x[(n-1/2)T]^+ = -k_0 v_i[(n-1/2)T]. \tag{8.246}$$

The analysis results in the next transfer function given by

$$H(z) = \frac{V_0(z)}{V_i(z)} = H_{id}(z)E(z), \tag{8.247}$$

where

$$H_{id}(z) = \frac{C_1}{C_2} \frac{z^{-1/2}}{1 - z^{-1}} \tag{8.248}$$

and

$$E(z) = 1 - \left(1 + \frac{C_2}{C_1}\right) k_0 e^{-k_2 T/2}. \tag{8.249}$$

The associated magnitude and phase errors are

$$m(\omega) = -\left(1 + \frac{C_1}{C_2}\right) k_0 e^{-k_2 T/2} \tag{8.250}$$

and

$$\theta(\omega) = 0. \tag{8.251}$$

In this case, the phase error is eliminated. This is due to an increase in the settling time of the OTA resulting from the additional half-clock period available between the sampling time of the input signal and the beginning of the charge transfer between the capacitors.

In practice, the errors due to the finite bandwidth become negligible by choosing the amplifier unity-gain frequency to be about five times as large as the clock frequency.

8.12 Settling time in the integrator

The circuit diagram of an amplification stage with its capacitive load is depicted in Figure 8.52(a). The amplifier is assumed to have the transfer characteristic shown in Figure 8.52(b). The output current is given by

$$i_0 = \begin{cases} g_m v, & \text{for } |v| < V_m \\ \text{sign}(v)I_m, & \text{otherwise}, \end{cases} \tag{8.252}$$

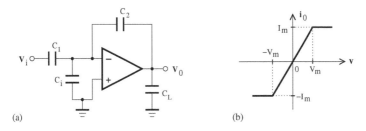

FIGURE 8.52

(a) Amplifier with capacitive loads; (b) voltage-to-current characteristic of the amplifier.

where v denotes the input voltage of the amplifier. The amplifier operates in the linear region with the transconductance g_m provided $|v| < V_m$; otherwise, the output current will be limited by $\pm I_m$ due to the nonlinear effect of the slew rate.

Let us consider the case of the noninverting integrator connected to a step input voltage. Initially,

$$-C_1 v_i = (C_1 + C_i + C_2 + C_L')v(0) \tag{8.253}$$

and

$$v(0) = -\frac{C_1 v_i}{C_1 + C_i + C_2 + C_L'}, \tag{8.254}$$

where $C_L' = C_0 \| C_L$ and C_i denotes the amplifier input capacitance. The output voltage can be written as

$$v_0(0) = \frac{C_2}{C_2 + C_L'} v(0). \tag{8.255}$$

From the next charge conservation equation,

$$-(C_1 + C_i)[v - v(0)] = C_2\{[v - v(0)] - [v_0 - v_0(0)]\}, \tag{8.256}$$

we have

$$v = \beta v_0 - \beta \alpha v_i, \tag{8.257}$$

where

$$\alpha = \frac{C_1}{C_2}, \tag{8.258}$$

and

$$\beta = \frac{C_2}{C_1 + C_i + C_2}. \tag{8.259}$$

With a single-pole model of the amplifier, the differential equation of the circuit is

$$i_0 + \frac{v_0}{r_0} + C_2 \frac{d}{dt}(v - v_0) + C_L' \frac{dv_0}{dt} = 0, \qquad (8.260)$$

where $r_0 = 1/g_0$. It can be reduced to

$$\tau \frac{dv_0}{dt} + v_0 = -r_0 i_0, \qquad (8.261)$$

where

$$\tau = r_0[(1 - \beta)C_2 + C_L']. \qquad (8.262)$$

For $v \geq V_m$, the output current takes the value I_m and

$$v_0(t) = r_0 I_m \left[1 - \exp\left(-\frac{t}{\tau}\right)\right] + v_0(0) \exp\left(-\frac{t}{\tau}\right). \qquad (8.263)$$

At the end of the slewing period, $t = T_{SR}$ and $v(t) = V_m$. Assuming that $\tau \gg T_{SR}$, we can obtain

$$T_{SR} = \left[1 - \frac{V_m + \beta(\alpha v_i - r_0 I_m)}{k(v_0(0) - r_0 I_m)}\right] \tau. \qquad (8.264)$$

For $v < V_m$, the amplifier operates linearly. With the initial condition at

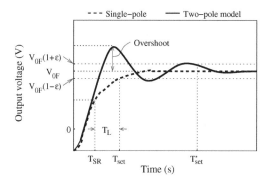

FIGURE 8.53
Amplifier step responses.

$t = T_{SR}$, the output voltage is given by

$$v_0(t) = \frac{\alpha \delta}{1 + \delta} v_i \left[1 - \exp\left(-\frac{1 + \delta}{\tau}(t - T_{SR})\right)\right]$$
$$+ v_0(T_{SR}) \exp\left(-\frac{1 + \delta}{\tau}(t - T_{SR})\right), \qquad (8.265)$$

where $\delta = \beta g_m r_0$. The settling period, T_{set}, is defined as the time required by the output response to remain within ϵ (generally, about 0.1%) of its final value, V_{0F} (see Figure 8.53). It can be written as

$$T_{set} = T_{SR} + T_L, \tag{8.266}$$

where

$$T_L = \frac{\tau}{1+\delta} \ln\left(\frac{v_0(T_{SR}) - V_{0F}}{\epsilon V_{0F}}\right) \tag{8.267}$$

and

$$V_{0F} = \frac{\alpha\delta}{1+\delta} v_i. \tag{8.268}$$

The slew rate is approximately given by $SR \simeq (V_{0F} + v_0(0))/T_{SR}$. The amplifier sometimes exhibits a second pole, which is nondominant. In this case, an overshoot can appear in the output transient response, as shown in Figure 8.53.

8.13 Amplifier dc offset voltage limitations

The dc offset voltage is one of the limiting factors in the context of low-voltage applications. It consists of a deterministic and a random component. Because the former is caused by improper bias conditions, it can be reduced to a negligible value by careful circuit design. The latter contributes substantially to the offset voltage and is determined by the level of matching achievable between identical transistors.

To analyze the effect of the offset voltage on the integrator, a voltage source, V_{off}, of unknown polarity is connected to one of the inputs of the amplifier supposed to be ideal. The integrator output voltage can be expressed as

$$V_0(z) = H_{id}(z)V_i(z) + H_0(z)V_{off}, \tag{8.269}$$

where $H_{id}(z)$ is the transfer function of the integrator based on the ideal amplifier characteristics and $H_0(z) = C_1/[C_2(1 - z^{-1})]$.

Hence, the offset voltage has essentially a scaling effect on the output voltage. This effect will be practically negligible if an amplifier with low offset voltage (that is, in the range of microvolt) is used.

8.14 Computer-aided analysis of SC circuits

Generally, large-scale integrated circuits are partitioned into small building blocks that can be more easily analyzed. However, because the interactions between the different blocks and parasitic effects may play a significant role, they must be estimated and included in the analysis procedure. This requires the use of an accurate and efficient computer-aided design (CAD) program [42, 43] to simulate the entire circuit. Moreover, even for small-sized networks, the analysis of nonideal effects requires a CAD tool.

The most widely used circuit simulation program is SPICE (Simulation Program with Integrated Circuit Emphasis). It offers a more advanced circuit analysis at the transistor level. However, the SPICE analysis of switched-capacitor circuits, particularly in the frequency domain, is limited by the required amount of computer memory. Hence, we first review the spectral analysis of SC circuits.

Frequency response of a double-sampling lossless discrete-time integrator

Let us find the continuous-time frequency response of the circuit of Figure 8.54. Because this circuit does not have a continuous feedthrough path,

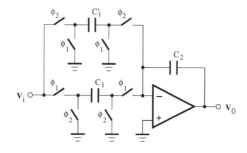

FIGURE 8.54
Double-sampling lossless discrete-time integrator.

its output is always piecewise-constant. Applying Kirchhoff's charge conservation law, for the clock phase 1 and 2, we obtain

$$-C_1 z^{-1/2} V_i^2(z) + C_2 V_0^1(z) - C_2 z^{-1} V_0^2(z) = 0 \qquad (8.270)$$

and

$$-C_1 z^{-1/2} V_i^1(z) + C_2 V_0^2(z) - C_2 z^{-1} V_0^1(z) = 0, \qquad (8.271)$$

respectively. Solving the above equations with respect to the output voltages, we get

$$V_0^1(z) = \frac{C_1}{C_2} \frac{z^{-1}}{1 - z^{-1}} V_i^1(z) + \frac{C_1}{C_2} \frac{z^{-1/2}}{1 - z^{-1}} V_i^2(z), \qquad (8.272)$$

$$V_0^2(z) = \frac{C_1}{C_2} \frac{z^{-1/2}}{1 - z^{-1}} V_i^1(z) + \frac{C_1}{C_2} \frac{z^{-1}}{1 - z^{-1}} V_i^2(z), \qquad (8.273)$$

or in matrix form

$$\begin{bmatrix} V_0^1(z) \\ V_0^2(z) \end{bmatrix} = H(z) \begin{bmatrix} 1 & z^{1/2} \\ z^{1/2} & 1 \end{bmatrix} \begin{bmatrix} V_i^1(z) \\ V_i^2(z) \end{bmatrix}, \qquad (8.274)$$

where

$$H(z) = \frac{C_1}{C_2} \frac{z^{-1}}{1 - z^{-1}}. \tag{8.275}$$

The output voltage is observed during both subintervals. The uniform sam-

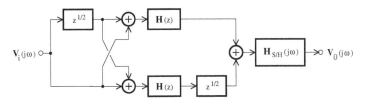

FIGURE 8.55
Block diagram representation of the double-sampling lossless discrete-time integrator.

pling can be modeled mathematically using the following discrete-time input signals, $v_i^1(t)$ and $v_i^2(t)$:

$$v_i^1(t) = v_i(t) \sum_{n=-\infty}^{+\infty} \delta(t - nT), \tag{8.276}$$

$$v_i^2(t) = v_i(t) \sum_{n=-\infty}^{+\infty} \delta(t - nT - T/2). \tag{8.277}$$

The Fourier transform of the sampled input signals is expressed as

$$V_i^1(\omega) = \frac{1}{T} \sum_{n=-\infty}^{+\infty} V_i(\omega - n\omega_s), \tag{8.278}$$

$$V_i^2(\omega) = \frac{1}{T} \sum_{n=-\infty}^{+\infty} V_i(\omega - n\omega_s)e^{j(\omega - n\omega_s)T/2}, \tag{8.279}$$

where $\omega_s = 2\pi/T$ and $V_i(\omega)$ is the Fourier transform of the input signal, $v_i(t)$. From the equivalent continuous-time system shown in Figure 8.55, the S/H output spectra of the integrator can be determined as

$$V_0(\omega) = H_{SH}(j\omega) \left[(1 + z^{1/2})H(z)\right]_{z=e^{j\omega T}} \sum_{n=-\infty}^{+\infty} V_i(\omega - n\omega_s)e^{j(\omega - n\omega_s)T/2}$$

$$+ H_{SH}(j\omega) \left[(1 + z^{1/2})H(z)\right]_{z=e^{j\omega T}} \sum_{n=-\infty}^{+\infty} V_i(\omega - n\omega_s), \tag{8.280}$$

where

$$H_{SH}(j\omega) = \frac{1 - e^{-j\omega T/2}}{j\omega} \tag{8.281}$$

After some simplification, the next expression can be deduced.

$$V_0(\omega) = \frac{C_1}{C_2} \frac{1}{j\omega T} \sum_{n=-\infty}^{+\infty} V_i(\omega - n\omega_s) \left(1 - e^{-jn\omega_s T/2}\right) \qquad (8.282)$$

For the baseband, that is, $n = 0$, we obtain

$$V_0(\omega) = \frac{C_1}{C_2} \frac{2}{j\omega T} V_i(\omega). \qquad (8.283)$$

This equation indicates that if the clock frequency is greater than the maximum signal frequency, the double sampling integrator will behave exactly like an analog integrator.

Frequency response of SC circuits

The analytical spectral analysis of SC circuits can be complicated for large structures. The use of the circuit model of Figure 8.56 allows us to obtain the frequency response directly from the z-domain transfer function based on the direct substitution $z = e^{j\omega T}$. The impulse sampling stage transforms

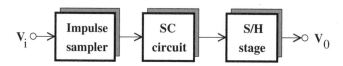

FIGURE 8.56
Frequency domain representation of an SC circuit.

the analog signal into an analog discrete-time signal. It should be emphasized here that, in accordance with the sampling theorem[8], the maximum frequency of the incoming signal is limited to less than the Nyquist rate, that is, $\omega_s/2 = \pi/T$. The SC circuit can then be represented as a discrete-time signal processor. Its output signal is received and transformed into a staircase signal by the S/H stage.

In general, the frequency domain transfer function can be written as

$$H(j\omega) = \sum_{l=1}^{M} \left[\sum_{k=1}^{M} H_{SH}^l(j\omega)\, H^{(k,l)}(z) \Big|_{z \,=\, e^{j\omega T}} \right], \qquad (8.284)$$

where

$$H_{SH}^l(j\omega) = e^{-j\omega(\tau_l/2)} \frac{\tau_l}{T} \frac{\sin(\omega\tau_l/2)}{\omega\tau_l/2}. \qquad (8.285)$$

Here, $H^{(k,l)}$ is the transfer function of the analog discrete-time circuit in the

---

[8] The sampling theorem can be stated as follows: A band-limited lowpass signal, $x(t)$, with spectrum $X(j\omega) = 0$ for all $|\omega| > \omega_m$ is uniquely and completely described by a set of samples values $x(nT)$ taken at uniformly spaced time instants separated by $T = \pi/\omega_m$ seconds or less.

z-domain, the superscripts k and l mean that the input voltage is sampled at the beginning of the k-th clock phase and the output voltage is observed during the l-th clock phase. It should be emphasized that the input signal is sampled M times per period of the clock signal and that the output signal is maintained constant during the subinterval of observation, τ_l.

The simulation of SC circuits can be computationally intensive due to their time-varying nature. A program such as SPICE or HSPICE [44] can only provide the transient analysis of SC circuits. In order to obtain the frequency response, it is necessary to use specialized programs [45, 46] supporting only behavioral-level descriptions of the circuit, or SpectreRF [47,48]. In contrast to HSPICE, SpectreRF relies on performing analysis about a periodic operating point. It can then be applied to predict SC circuit characteristics in the time and frequency domains using nonideal components at the transistor level.

8.15   T/H and S/H circuits based on SC circuit principle

T/H and S/H circuits based on a clock waveform with more than two clock phases can be realized using SC circuit techniques. From the analysis of their associated continuous-time subcircuits, it appears that the circuit settling time is limited by the clock period. As a consequence, the number of clock phases must be limited to the minimum (i.e., two) for high-frequency applications.

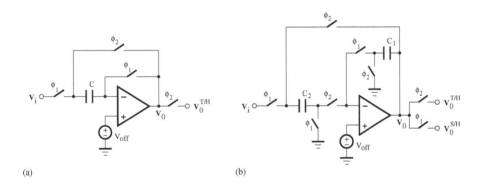

(a)   (b)

FIGURE 8.57
(a) Offset-compensated T/H circuit; (b) S/H circuit with an improved speed performance.

The T/H circuit depicted in Figure 8.57(a) [17] operates with a two-phase clock signal. Initially, the switches controlled by the first clock phase, ϕ_1, are closed and the capacitor C is connected between the input voltage and the inverting node of the amplifier, which is configured to operate as unity-gain buffer. During the second clock phase, ϕ_2, the capacitor C is included in the amplifier feedback path, and the amplifier output is equal to a delayed version of the input signal. Because the capacitor C always remains connected to the inverting node of the amplifier, the offset voltage contribution stored on C during ϕ_1, that is, for $(n-1)T \leq t < (n-1/2)T$, is canceled by the one produced during ϕ_2, that is, for $(n-1/2)T \leq t < nT$. Taking into account the amplifier finite gain, the application of the charge conservation rule at the amplifier inverting node during ϕ_2 yields

$$V^-(nT) - V_0(nT) = V^-[(n-1/2)T] - V_i[(n-1/2)T]. \tag{8.286}$$

In general, we have

$$V^- = -\mu V_0 + V_{off}, \tag{8.287}$$

where $\mu = 1/A_0$ and A_0 is the amplifier dc gain. We can then combine (8.287) and the expression, $V^-[(n-1/2)T] = V_0[(n-1/2)T]$, obtained at the end of ϕ_1, to write

$$V_0[(n-1/2)T] = V_{off}/(1+\mu). \tag{8.288}$$

Finally, by substituting Equations (8.287) and (8.288) into Equation (8.286), the output voltage is expressed as

$$V_0(nT) = \frac{V_i[(n-1/2)T]}{1+\mu} + \mu \frac{V_{off}}{(1+\mu)^2}. \tag{8.289}$$

Ideally, the dc gain is infinite and the offset voltage is negligible, resulting in $V_0(nT) = V_i[(n-1/2)T]$. Due to the fact that the output is reset during the first clock phase, the resulting speed appears to be critically limited by the required value of the amplifier slew rate and settling time.

An S/H circuit, as shown in Figure 8.57(b), that does not require a high-speed amplifier [18]. During the clock phase ϕ_1, the capacitor C_2 is charged up to the input voltage, while C_1 is connected between the inverting node and output of the amplifier. During the clock phase ϕ_2, that is, for $(n-1/2)T \leq t < nT$, the capacitor C_2 is placed in the amplifier feedback loop and C_1 is connected as an output load. Hence,

$$C_2[V^-(nT) - V_0(nT)] = -C_2 V_i[(n-1/2)T]. \tag{8.290}$$

At the end of the next clock phase ϕ_1 occurring for $nT \leq t < (n+1/2)T$, the charge conservation equation can be written as

$$C_1\left\{V^-[(n+1/2)T] - V_0[(n+1/2)T]\right\} = -C_1 V_0[(n)T]. \tag{8.291}$$

Because the voltage at the amplifier inverting node is given by $V^- = -\mu V_0 + V_{off}$, where $\mu = 1/A_0$ and A_0 is the amplifier dc gain, we can obtain

$$V_0(nT) = \frac{V_i[(n-1/2)T] + V_{off}}{1+\mu} \tag{8.292}$$

and

$$V_0[(n+1/2)T] = \frac{V_0(nT) + V_{off}}{1 + \mu}. \tag{8.293}$$

In the case where the amplifier is assumed to be ideal, it can be observed that, once per clock period, a sample of the input signal is transferred to the output, whose level is maintained constant up to the next update.

The S/H circuit shown in Figure 8.58(a) consists of capacitors, an amplifier, an inverting voltage buffer, and switches, which are controlled by a clock signal with the two nonoverlapping phases, ϕ_1 and ϕ_2. During the

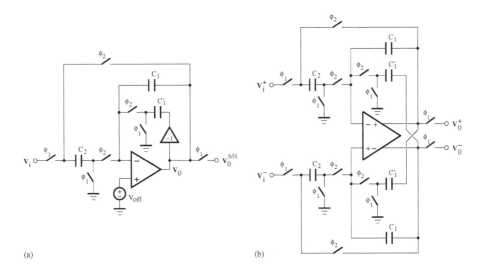

(a) (b)

FIGURE 8.58
Circuit diagram of the unity gain S/H circuit ($C_1' = C_1$): (a) single-ended and (b) differential implementations.

first half of the clock signal period (or the first clock phase, ϕ_1), that is, $(n-1)T \leq t < (n-1/2)T$, the capacitor C_2 is charged to V_i, while the capacitors C_1 and C_1' are charged to V_0 and $-V_0$, respectively. Next, the amplifier and the feedback capacitors C_1 are isolated; thus the amplifier output voltage of the previous phase is maintained. From the charge conservation equations, we can obtain

$$V_i[(n-1/2)T] = V_i[(n-1)T] \tag{8.294}$$

and

$$V_0[(n-1/2)T] = V_0[(n-1)T]. \tag{8.295}$$

During the second clock phase, ϕ_2, that is, $(n-1/2)T \leq t < nT$, all capacitors are connected to the amplifier. Because the charges due to the output voltages and stored on the capacitors C_1 and C_1' have opposite signs, they ideally cancel each other and the new charge redistribution is only determined by the charges stored on the capacitors C_2. The application of the charge conservation rule at the amplifier inverting node gives

$$
\begin{aligned}
C_2 & \{V_0(nT) - V^-(nT) - V_i[(n-1/2)T]\} \\
& + C_1\{-V_0(nT) - V^-(nT) - (-V_0[(n-1/2)T])\} \\
& = C_1\{V^-(nT) - V_0(nT) - (V^-[(n-1/2)T] - V_0[(n-1/2)T])\}.
\end{aligned}
\tag{8.296}
$$

Taking into account the fact that $V^- = -\mu V_0 + V_{off}$, and substituting Equations (8.294) and (8.295) into Equation (8.296), we arrive at

$$
[(1+\mu)C_2 + 2\mu C_1]V_0(nT) - \mu C_1 V_0[(n-1)T] = C_2 V_i[(n-1)T] + (C_1+C_2)V_{off}.
\tag{8.297}
$$

Using the z-domain transform, it can be shown that

$$
V_0(z) = H_i(z)V_i(z) + H_0(z)V_{off},
\tag{8.298}
$$

where

$$
H_i(z) = \pm \frac{z^{-1}}{1 + \mu + \mu \dfrac{C_1}{C_2}\left(2 - z^{-1}\right)}
\tag{8.299}
$$

and

$$
H_0(z) = \frac{1 + \dfrac{C_1}{C_2}}{1 + \mu + \mu \dfrac{C_1}{C_2}\left(2 - z^{-1}\right)}.
\tag{8.300}
$$

Here, V_{off} represents the amplifier offset voltage and $A_0 = 1/\mu$ is the amplifier dc gain. The sign of the transfer function, H_i, is determined by the signal polarity. Ideally, V_{off} is negligible and $\mu \ll 1$, such that

$$
V_0(z) = \pm z^{-1} V_i(z).
\tag{8.301}
$$

Hence, the input signal is sampled and held for a full-clock period. Figure 8.58(b) [19] shows the differential implementation of the S/H circuit.

The aforementioned S/H structure has the advantage of preventing large signal variations at the amplifier output during the sampling phase, thus reducing the effect of delays caused by the transient response. However, it can only be used to implement a delay with unity gain. Generally, an additional gain stage is required to perform the amplification or attenuation function.

An alternative architecture is shown in Figure 8.59(a) [20]. It features a variable gain and operates as follows. During the first clock phase, ϕ_1, that is, $(n-1)T \leq t < (n-1/2)T$, the capacitor C_1 is connected and charged to the voltages V_i. A similar situation takes place between the capacitor C_2' and the

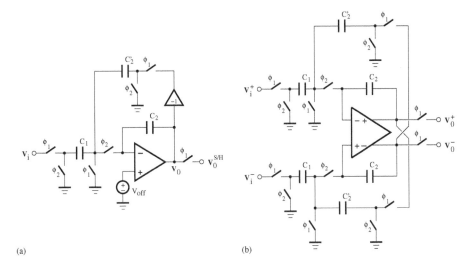

FIGURE 8.59
Circuit diagram of the S/H circuit with a variable gain ($C_2' = C_2$): (a) Single-ended and (b) differential implementations.

voltage V_0. The feedback capacitor C_2 remains connected between the output and inverting input of the amplifier. Hence,

$$V_i[(n-1/2)T] = V_i[(n-1)T] \tag{8.302}$$

and

$$V_0[(n-1/2)T] = V_0[(n-1)T]. \tag{8.303}$$

During the second clock phase, ϕ_2, that is, $(n-1/2)T \le t < nT$, all capacitors are connected to the amplifier inverting input and a cancelation between the same charges stored respectively on C_2 and C_2' takes place. The charge conservation equation can be written as

$$-C_1\{V^-(nT) + V_i[(n-1/2)T]\} + C_2\{-V^-(nT) + V_0[(n-1/2)T]\}$$
$$= C_2\{V^-(nT) - V_0(nT) - (V^-[(n-1/2)T] - V_0[(n-1/2)T])\}. \tag{8.304}$$

Using the relationship, $V^- = -\mu V_0 + V_{off}$, in addition to Equations (8.302) and (8.303), we obtain

$$[(1+2\mu)C_2 + \mu C_1]V_0(nT) - \mu C_2 V_0[(n-1)T] = C_1 V_i[(n-1)T] + (C_1+C_2)V_{off}. \tag{8.305}$$

The output-input relationship can then be described in the z-domain by

$$V_0(z) = H_i(z)V_i(z) + H_0(z)V_{off}, \tag{8.306}$$

where

$$H_i(z) = \pm\frac{C_1}{C_2}\frac{z^{-1}}{1 + \mu\left(2 + \dfrac{C_1}{C_2}\right) - \mu z^{-1}} \tag{8.307}$$

and

$$H_0(z) = \frac{1 + \dfrac{C_1}{C_2}}{1 + \mu\left(2 + \dfrac{C_1}{C_2}\right) - \mu z^{-1}}. \tag{8.308}$$

Here, V_{off} represents the amplifier offset voltage and $A_0 = 1/\mu$ is the amplifier dc gain. In the ideal case, that is, when V_{off} is negligible and $\mu \ll 1$, we obtain

$$V_0(z) = \pm\frac{C_1}{C_2}z^{-1}V_i(z). \tag{8.309}$$

In addition to being sampled and held for a full clock period, the input signal is scaled by a factor set by the capacitor ratio. The differential implementation of the S/H circuit shown in Figure 8.59(b) offers the advantage of generating the inverting version of the output voltage without the need of an inverting gain stage.

8.16 Circuit structures with low sensitivity to nonidealities

The operation of SC circuits is strongly influenced by the amplifier characteristics. The purpose of the circuit techniques discussed in this section is to relax the constraints that must be imposed on the amplifier structure.

Two techniques are described in this section in order to improve the performance of SC circuits. They are respectively based on the use of additional active components or the correlated double sampling (CDS) scheme. In a CDS-based structure, the error signal is sampled and stored on a capacitor. It is then subtracted from the samples of the input signal in order to achieve the compensation. The dc offset voltage, which is constant over the time, can be cancelled in this way. This can also be the case of any noise source that varies very slowly.

In the following analysis, the circuit output voltage, V_0, is written as

$$V_0(z) = H_i(z)V_i(z) + H_0(z)V_{off}. \tag{8.310}$$

Note that V_i is the input voltage; V_{off} and H_0 denote the offset voltage and its associated transfer function, respectively; and $H_i = H_{id} \cdot E$, where

$$H_{id}(z) = \begin{cases} -\dfrac{C_1}{C_2}\dfrac{1}{1 - z^{-1}} & \text{for the integrator} \\[2ex] -\dfrac{C_1}{C_2} & \text{for the gain stage} \end{cases} \tag{8.311}$$

and the error function, E, is due to the amplifier dc gain. For convenience, a 50% duty cycle clock pattern is assumed and the amplifier dc gain, A_0, will be indicated by $\mu = 1/A_0$.

8.16.1   Integrators

• The circuit technique presented in [49] for the realization of an integrator with a low sensitivity to the amplifier dc gain uses an unity-gain buffer. If this latter is ideal, a phase-error free integrator will be obtained. Figure 8.60 shows the compensated integrator circuit. By inspection of the circuit, it can

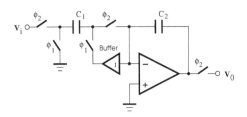

FIGURE 8.60
Phase-error free integrator.

be found that during the clock phase 1, the output voltage will not change and therefore the error signal will be stored on the input capacitor C_1. The charge transfer takes place during the clock phase 2. Then, we have

$$v_0[(n - 1/2)T] = v_0[(n - 1)T] \qquad (8.312)$$

during the clock phase 1 and

$$C_1 \{v_i(nT) + \mu v_0(nT) - (1 - \alpha)\mu v_0[(n - 1/2)T] - \alpha V_{off}\} \\ + (1 + \mu)C_2 \{v_0(nT) - v_0[(n - 1)T]\} = 0 \qquad (8.313)$$

during the clock phase 2. The buffer gain has the value $1 - \alpha$, where α is the gain error. The functions E and H_0 are given by

$$E(z) = \cfrac{1}{1 + \mu + \mu\dfrac{C_1}{C_2} + \alpha\mu\dfrac{C_1}{C_2}\dfrac{z^{-1}}{1 - z^{-1}}} \qquad (8.314)$$

and

$$H_0(z) = \alpha\frac{C_1}{C_2(1 - z^{-1})}E(z), \qquad (8.315)$$

respectively. This technique will be efficient if α is on the order of a few percent and A_0 is greater than 100.

• The schematic diagram of an integrator based on the CDS scheme [50] is depicted in Figure 8.61. The charge conservation principle can be used to

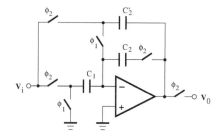

FIGURE 8.61
CDS compensated integrator I.

derive the following equations

$$
\begin{aligned}
C_1 \{-v_i[(n-1)T] + \mu v_0[(n-1/2)T] - \mu v_0[(n-1)T]\} \\
+ C_2' \{-v_i[(n-1)T] - (1+\mu)v_0[(n-1/2)T] + v_0[(n-1)T] + V_{off}\} = 0
\end{aligned}
$$
(8.316)

during the clock phase 1 and

$$
\begin{aligned}
C_1 \{v_i(nT) + \mu v_0(nT) - \mu v_0[(n-1/2)T]\} \\
+ (1+\mu)C_2 \{v_0(nT) - v_0[(n-1)T]\} = 0
\end{aligned}
$$
(8.317)

during the clock phase 2. The capacitor C_2' is included in the feedback path during the clock phase 1 and enables C_1 to discharge. The charge previously stored on C_2' is then used for the compensation purpose. The resulting error function can be written as

$$
E(z) = \frac{1 - \dfrac{\mu C_1}{\mu C_1 + (1+\mu)C_2'}\left(1 - \dfrac{C_2'}{C_1}\right)z^{-1}}{1 + \dfrac{\mu C_1}{(1+\mu)C_2}\dfrac{1}{1-z^{-1}}\left(1 - \dfrac{C_2' + \mu C_1}{\mu C_1 + (1+\mu)C_2'}z^{-1}\right)}.
$$
(8.318)

The contribution of the dc offset voltage to the output signal can be deduced from the transfer function

$$
H_0(z) = \frac{\dfrac{-\mu C_2'}{\mu C_1 + (1+\mu)C_2'}}{1 + \dfrac{\mu C_1}{(1+\mu)C_2}\dfrac{1}{1-z^{-1}}\left(1 - \dfrac{C_2' + \mu C_1}{\mu C_1 + (1+\mu)C_2'}z^{-1}\right)} H_{id}(z).
$$
(8.319)

The compensation strategy is effective when $C_2' = C_1$. Furthermore, the compensated integrator in a configuration with N inputs will require N capacitors C_2'.

- The circuit of Figure 8.62 [51, 52] also relies on CDS switching for the

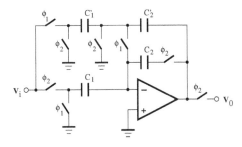

FIGURE 8.62
CDS compensated integrator II.

amplifier nonideality reduction. If $v_0[(n-1/2)T]$ can be made equal to $v_0[(n-1)T]$, the transfer function, H_i, will be free of a phase error and the gain error will be $-\mu(1+C_1/C_2)$. This ideal situation is only approximated by the next difference equations valid at the end of the clock phase 1:

$$
\begin{aligned}
C_1 \{-v_i[(n-1)T] + \mu v_0[(n-1/2)T] - \mu v_0[(n-1)T]\} \\
+ C_1' \{v_i[(n-1/2)T] + \mu v_0[(n-1/2)T] - V_{off}\} \\
+ C_2' \{(1+\mu)v_0[(n-1/2)T] - v_0[(n-1)T] - V_{off}\} = 0.
\end{aligned}
\tag{8.320}
$$

By using the charge conservation laws at the end of the clock phase 2, we arrive at

$$
\begin{aligned}
C_1 \{v_i(nT) + \mu v_0(nT) - \mu v_0[(n-1/2)T]\} \\
+ C_2(1+\mu) \{v_0(nT) - v_0[(n-1)T]\} = 0.
\end{aligned}
\tag{8.321}
$$

The functions E and H_0 can then be derived from the z-domain output-input relation as

$$
E(z) = \frac{1 + \dfrac{\mu C_1}{\mu(C_1+C_1')+(1+\mu)C_2}\left(C_1' z^{-1/2} - C_1 z^{-1}\right)}{1 + \dfrac{\mu C_1}{(1+\mu)C_2}\dfrac{1}{1-z^{-1}}\left(1 - \dfrac{\mu C_1 + C_2'}{\mu(C_1+C_1')+(1+\mu)C_2'}z^{-1}\right)}
\tag{8.322}
$$

and

$$
H_0(z) = \frac{\dfrac{-\mu(C_1'+C_2')}{\mu(C_1+C_1')+(1+\mu)C_2}}{1 + \dfrac{\mu C_1}{(1+\mu)C_2}\dfrac{1}{1-z^{-1}}\left(1 - \dfrac{\mu C_1 + C_2'}{\mu(C_1+C_1')+(1+\mu)C_2'}z^{-1}\right)} H_{id}(z),
\tag{8.323}
$$

respectively. The choice of capacitors C_1' and C_2' is dictated by the minimization of the transfer function deviations. If the input voltage varies very slowly, that is, $v_i[(n-1/2)T] = v_i[(n-1)T]$, $C_1' = 2C_1$ and $C_2' = C_2$, the integrator dc gain will be reduced to

$$H_i(1) = H_{id}(z)E(z)|_{z=1} = \mu^{-2} \frac{1 + \mu\left(1 + 4\dfrac{C_1}{C_2}\right)}{1 + 2\dfrac{C_1}{C_2}}. \qquad (8.324)$$

Thus, the effective gain is now a function of μ^{-2} (or A_0^2) rather than A_0, as in the case of the uncompensated structure.

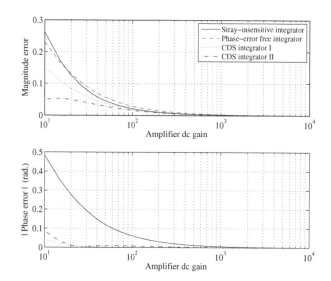

FIGURE 8.63
Plots of the integrator gain and phase errors ($C_1/C_2 = 2$ and $f/f_c = 0.05$).

The effects of the amplifier dc gain on the different integrator structures are shown in Figure 8.63. The CDS compensated integrators seem to offer better performance than other structures, but they also require extra switches and capacitors.

• A goal in designing an SC integrator that is insensitive to amplifier non-idealities is to maintain the input and feedback capacitors connected at the summing nodes during both clock phases, so that the charge transfer between the capacitors is made without an error. Based on this design principle, an implementation of the CDS technique was proposed in [52], but the output of the resulting circuit (see Figure 8.62) is valid only during one clock phase. Furthermore, this circuit works well only under the assumption that the input voltage is approximately constant over two consecutive half-clock cycles.

An integrator circuit that operates without constraints on the input voltage is depicted in Figure 8.64 [58]. Note that the second input path capacitor of

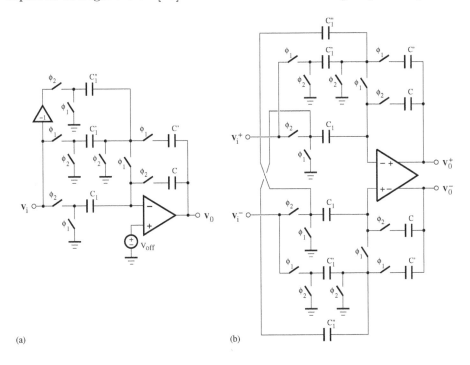

(a)   (b)

FIGURE 8.64
Circuit diagram of a double-sampling, low-sensitivity integrator ($C_1 = C_1' = C_1''$ and $C = C'$): (a) Single-ended and (b) differential versions.

the integrator of [52], which is two times the first input path capacitor value, is divided into two equal capacitors: one of these is connected to the input voltage and the other is connected to a negative version of the input voltage. Ideally, the application of the charge conservation law at the amplifier input nodes leads to the following difference equation,

$$Cv_0(nT) = Cv_0[(n-1)T] - C_1v_i(nT), \qquad (8.325)$$

for the clock phase 2. In this modified SC integrator, the capacitor C_1'' is charged during the clock phase 1 by a negative version of the input voltage. In this way, its charge is used in the next clock phase (clock phase 2) to cancel the charge generated by the capacitor C_1, which is always connected at the summing node. Then, the next relation can be written as

$$Cv_0[(n-1/2)T] = Cv_0[(n-3/2)T] - C_1v_i[(n-1/2)T], \qquad (8.326)$$

for the clock phase 1. It appears that the output $v_0[(n-1/2)T]$ in the clock phase 1 and the output $v_0(nT)$ in the clock phase 2 represent a discrete-time integration of a given input signal sample.

Taking into account the effect of the finite dc gain, $A_0 = 1/\mu$, and offset voltage, V_{off}, of the operational amplifier, the application of the charge conservation law at the end of the clock phase ϕ_1, which is characterized by $(n-1)T \leq t < (n-1/2)T$, gives

$$
\begin{aligned}
C_1\{\mu v_0[(n-1/2)T] - v_i[(n-1)T] - \mu\{v_0[(n-1)T]\} \\
+ C_1'\{v_i[(n-1/2)T] + \mu\{v_0[(n-1/2)T] - V_{off}\} \\
+ C''_1\{\mu v_0[(n-1/2)T] + v_i[(n-1)T] - V_{off}\} \\
+ C_2(1+\mu)\{v_0[(n-1/2)T] - v_0[(n-3/2)T]\} = 0.
\end{aligned}
\tag{8.327}
$$

For the clock phase ϕ_2, or the time instants defined by $(n-1/2)T \leq t < nT$, the charge conservation law can be written as

$$
C_1\{v_i(nT) + \mu v_0(nT) - \mu v_0[(n-1/2)T]\} + C_2(1+\mu)\{v_0(nT) - v_0[(n-1)T]\} = 0
\tag{8.328}
$$

Combining Equations (8.327) and (8.327) and taking the z-transform of the resulting difference equation, we can obtain

$$
V_0(z) = H_i(z)V_i(z) + H_0(z)V_{off},
\tag{8.329}
$$

where $H_i = H_{id} \cdot E$. Here,

$$
H_{id}(z) = \pm \frac{C_1}{C} \frac{1}{1 - z^{-1}},
\tag{8.330}
$$

$$
E(z) = \frac{1 + \mu \dfrac{C_1}{C} x^{-1} z^{-1/2}}{1 + \mu + \dfrac{\mu \dfrac{C_1}{C}}{1 - z^{-1}}\left[1 - \mu \dfrac{C_1}{C} x^{-1} z^{-1}\left(1 + \dfrac{1 + \mu}{\mu \dfrac{C_1}{C}} z^{-1/2}\right)\right]},
\tag{8.331}
$$

$$
H_0(z) = H_{id}(z) \cdot \frac{2\mu \dfrac{C_1}{C} x^{-1}}{1 + \mu + \dfrac{\mu \dfrac{C_1}{C}}{1 - z^{-1}}\left[1 - \mu \dfrac{C_1}{C} x^{-1} z^{-1}\left(1 + \dfrac{1 + \mu}{\mu \dfrac{C_1}{C}} z^{-1/2}\right)\right]},
\tag{8.332}
$$

and

$$
x = 1 + \mu + 3\mu \frac{C_1}{C},
\tag{8.333}
$$

where H_{id} represents the integrator ideal transfer function and E gives an estimation of the transfer function deviation caused by the finite amplifier gain. At dc, that is, for $z = 1$, the transfer function, H_i, is reduced to

$$
|H_i(1)| = \mu^{-2} \frac{1 + \mu + 4\mu \dfrac{C_1}{C}}{2 \dfrac{C_1}{C}}.
\tag{8.334}
$$

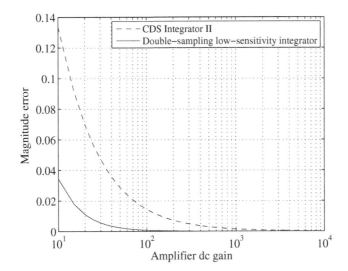

FIGURE 8.65
Gain errors of low-sensitivity integrators ($C_1 = C$, $f = 100$ kHz and $f_c = 2$ MHz).

Therefore, the effective gain of the amplifier is squared ($\mu^{-2} = A_0^2$).

The error function magnitudes were computed as $1 - |E|$, for the structures of Figures 8.62 and 8.64. As Figure 8.65 demonstrates, the double-sampling circuit topology is far superior in magnitude accuracy. The action of the SC integrator on the offset voltage is characterized by the transfer function H_0. Due to the CDS switching, the offset voltage contribution is primarily determined by the size of only two capacitors (here, C_1' and C_1''), as is the case for the integrator shown in Figure 8.62.

8.16.2 Gain stages

A conventional gain stage structure is shown in Figure 8.66 [53]. Ideally, its output voltage in the z-domain can be written as

$$V_0(z) = -(C_1/C_2)V_i(z), \qquad (8.335)$$

Taking into account the finite dc gain, we obtain

$$V_0(z) = -\frac{C_1}{C_2}\frac{1}{1 + \mu\left(1 + \dfrac{C_1}{C_2}\right)}V_i(z) \qquad (8.336)$$

The resulting gain is then affected by an error term dependent on μ. For typical component values, a gain error on the order of 1% is to be expected.

In order to attenuate the above-mentioned deviation, the CDS technique

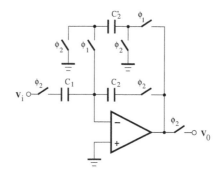

FIGURE 8.66
Uncompensated gain stage.

can be used for the design of gain stage circuits.

- The circuit shown in Figure 8.67 was proposed in [51]. During the clock

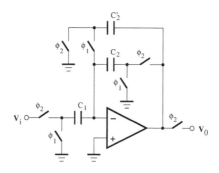

FIGURE 8.67
CDS compensated gain stage I.

phase 1, C_2' plays the role of the feedback capacitor and the others capacitors are connected between the amplifier inputs and the ground. The charge conservation equation is then

$$
\begin{aligned}
C_1 \left\{ -v_i[(n-1)T] + \mu v_0[(n-1/2)T] - \mu v_0[(n-1)T] \right\} & \\
+ C_2 \left\{ \mu v_0[(n-1/2)T] - (1+\mu)v_0[(n-1)T] \right\} & \\
+ C_2' \left\{ (1+\mu)v_0[(n-1/2)T] - v_0[(n-1)T] - V_{off} \right\} = 0. &
\end{aligned}
$$
(8.337)

During the clock phase 2, the capacitor C_1 is connected to the input voltage and a charge transfer can take place between C_1 and C_2. This results in the following equation,

$$
\begin{aligned}
C_1 \left\{ v_i(nT) + \mu v_0(nT) - \mu v_0[(n-1/2)T] \right\} & \\
+ C_2 \left\{ (1+\mu)v_0(nT) - \mu v_0[(n-1/2)T] \right\} = 0. &
\end{aligned}
$$
(8.338)

From the above equations, we can obtain

$$E(z) = \frac{1 - \dfrac{\mu(C_1 + C_2)}{C_2' + \mu(C_1 + C_2 + C_2')} z^{-1}}{1 + \mu\left(1 + \dfrac{C_1}{C_2}\right) - \dfrac{\mu(C_1 + C_2)}{C_2' + \mu(C_1 + C_2 + C_2')}\left[1 + \dfrac{C_2'}{C_2} + \mu\left(1 + \dfrac{C_1}{C_2}\right)\right] z^{-1}}$$

(8.339)

and

$$H_0(z) = \frac{C_2'}{C_2}$$

$$\times \frac{\dfrac{\mu(C_1 + C_2)}{C_2' + \mu(C_1 + C_2 + C_2')}}{1 + \mu\left(1 + \dfrac{C_1}{C_2}\right) - \dfrac{\mu(C_1 + C_2)}{C_2' + \mu(C_1 + C_2 + C_2')}\left[1 + \dfrac{C_2'}{C_2} + \mu\left(1 + \dfrac{C_1}{C_2}\right)\right] z^{-1}} .$$

(8.340)

It can be observed that the gain is frequency dependent due to the CDS high-pass filter effect.

• The circuit diagram of another gain stage that relaxes the amplifier specifications is shown in Figure 8.68 [51]. Here, the input and output branches are

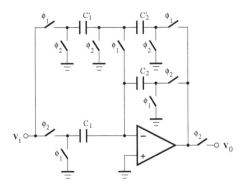

FIGURE 8.68
CDS compensated gain stage II.

duplicated in order to provide an anticipatory amplification during the clock phase 1. The circuit operation can be described by

$$C_1 \left\{ - v_i[(n-1)T] + \mu v_0[(n-1)T] - \mu v_0[(n-1)T] \right\}$$
$$+ C_2 \left\{ \mu v_0[(n-1/2)T] - (1+\mu)v_0[(n-1)T] \right\}$$
$$+ C_1' \left\{ v_i[(n-1/2)T] + \mu v_0[(n-1/2)T] - V_{off} \right\}$$
$$+ C_2' \left\{ (1+\mu)v_0[(n-1/2)T] + V_{off} \right\} = 0.$$

(8.341)

During the clock phase 2, C_1 and C_2 form the signal path around the amplifier and the appropriate output signal is generated according to the following relation:

$$C_1\left\{v_i(nT) + \mu v_0(nT) - \mu v_0[(n-1/2)T]\right\}$$
$$+ C_2\left\{(1+\mu)v_0(nT) - \mu v_0[(n-1/2)T]\right\} = 0. \tag{8.342}$$

In this case, we have

$$E(z) = \frac{1 + \dfrac{\mu(C_1+C_2)}{C_2' + \mu(C_1+C_1'+C_2+C_2')}\left(\dfrac{C_1'}{C_1}z^{-1/2} - z^{-1}\right)}{\left(1+\mu+\mu\dfrac{C_1}{C_2}\right)\left(1 - \dfrac{\mu(C_1+C_2)}{C_2' + \mu(C_1+C_1'+C_2+C_2')}z^{-1}\right)} \tag{8.343}$$

and

$$H_0(z) = \frac{C_2'}{C_2}\left(1+\frac{C_1'}{C_2'}\right)\frac{\dfrac{\mu(C_1+C_2)}{C_2' + \mu(C_1+C_1'+C_2+C_2')}}{\left(1+\mu+\mu\dfrac{C_1}{C_2}\right)\left(1 - \dfrac{\mu(C_1+C_2)}{C_2' + \mu(C_1+C_1'+C_2+C_2')}z^{-1}\right)}. \tag{8.344}$$

The capacitors C_1' and C_2' can be chosen to satisfy the relations $C_1' = C_1$ and $C_2' = C_2$.

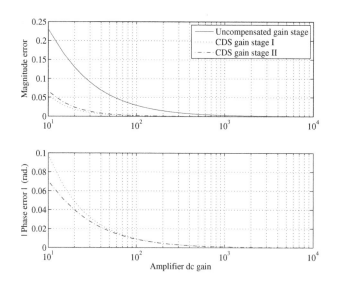

FIGURE 8.69
Plots of the gain stage errors due to the amplifier finite dc gain ($C_1/C_2 = 2$ and $f/f_c = 0.05$).

The error functions of the aforementioned gain stages are plotted in Figure 8.69. The performance provided by CDS-compensated gain stages appears to be superior to the ones of the other circuit structures.

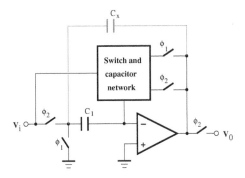

FIGURE 8.70

SC circuit with a transient spike compensation based on the use of small feedback capacitors.

In high-frequency applications, the amplifier is designed to have a moderate dc gain and a slew rate of about hundreds of millivolts per nanosecond (mV/ns). Because some improved building blocks operate with a nonoverlapping switching scheme, the amplifier may saturate during the brief intervals, when the two clock phases take their low level, and in which the amplifier does not have a negative feedback path. One solution to reduce the effect of the resulting transient spike is to use a small capacitor C_x [54] as shown in Figure 8.70 in order to maintain a closed loop around the amplifier.

Other circuit configurations for low-sensitivity structures were proposed in [55–57] (see Circuit design assessment 7 at the chapter end). The principle was to perform a preliminary charge transfer before the desired one in order to obtain a close approximation of the amplifier error signal. This latter was then stored on an auxiliary capacitor and subsequently used for compensation. However, the presence of a stray-sensitive node can substantially limit the precision in practical realizations.

8.17   Low-voltage SC circuits

Due to the increasing importance of portable systems for data processing in instrumentation and multimedia communication applications, the analog circuitry of modern mixed-signal integrated circuits has to operate with low supply voltage. The use of switched capacitor techniques results in circuits having a high accuracy and a good dynamic range. However, the low supply voltage does not allow a suitable control of the switches whose overdrive is signal dependent (e.g., input switch and switches at the amplifier output). Figure 8.71 shows the on-conductances of a CMOS switch operating with the

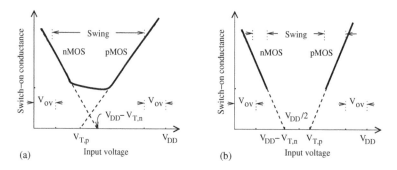

FIGURE 8.71
On-conductances of a CMOS switch operating (a) normally and (b) with a low supply voltage.

appropriate and low supply voltages, respectively. By decreasing the supply voltage to about $V_{T,n} + |V_{T,p}|$, there is a range of the input voltage around $V_{DD}/2$ for which the nMOS and pMOS transistors will not conduct (see Figure 8.71(b)). In the low-voltage circuit design, it is preferable to use either the nMOS or pMOS transistor as a switch. A supply voltage of at least, $V_T + V_{ov}$, where V_T is the transistor threshold voltage and V_{ov} is the highest voltage level of the signal to be switched, is then required [59].

Three approaches have been proposed in the literature for the design of low-voltage switched-capacitor circuits:

1. The first approach is to use lower threshold voltage transistors [60]. But this technique suffers from the high cost associated with the required process technology. Furthermore, the switch-off leakage is much higher than in the classical case.

2. The second one consists of using voltage multipliers to generate the clock voltages that can drive the switches [61].

3. The third alternative is the switched-amplifier (SA) technique [62, 63]. In this case, the critical switches are eliminated and their functions are realized by the amplifier, which can be turned off and on by a clock signal. This can realized, for instance, by a switch introduced between the amplifier core and the power supply line.

The circuit diagram of a bootstrapped switch [64] is shown in Figure 8.72. It is equivalent to the simplified structure of Figure 8.73. The states of the switch are determined by the phases, ϕ_1 and ϕ_2, of the clock signal. Transistors T_3 and T_5 are used to allow a design within the reliability limits of the IC process. The voltages V_{DS} and V_{GD} sustained by T_2 during its on state are decreased by T_3. Due to the regulation achieved by T_5, the voltage V_{GS} of T_4 should not exceed the supply voltage, V_{DD}.

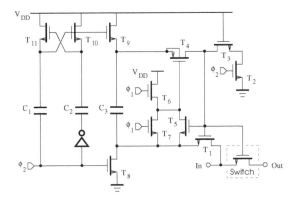

FIGURE 8.72

Circuit diagram of a bootstrapped switch.

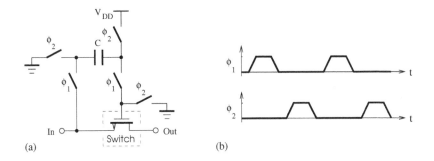

FIGURE 8.73

(a) Principle of a bootstrapped switch; (b) timing diagram.

The switch is off when ϕ_1 is low and ϕ_2 is high. The capacitor C_3, which is isolated from the switch by T_1 and T_4, is charged by V_{DD}. It should be sufficiently large to mitigate the effect of parasitic capacitances on the boosted clock signal. To reduce the error due to the subthreshold charge leakage, the switch implemented by the nMOS transistor T_9 is controlled by a level-shifted signal with the levels V_{DD} and $2V_{DD}$. The basic idea for the signal generation is to use the charge pump circuit consisting of capacitors C_1 and C_2, which are connected to V_{DD} via the cross-coupled transistors T_{10} and T_{11}, and an inverter.

During the on state of the switch, that is, when ϕ_1 is high and ϕ_2 is low, a connection is established by T_7 between the capacitor C_3 and the gate of T_4, and the gate of the switch is then bootstrapped to $V_{DD} + V_i$, where V_i is the input voltage. The device T_1 maintains constant the voltage V_{GS} across the switch.

The on-resistance of the switch is given by

$$R_{on} = \frac{1}{K(V_{DD} - V_T)}, \tag{8.345}$$

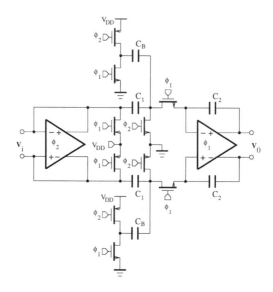

FIGURE 8.74
Circuit diagram of an integrator based on the switched-amplifier technique.

where $K = \mu_n C_{ox}(W/L)$ is the transconductance parameter and V_T is the threshold voltage. Due to the V_T dependence, it is sensitive to the body effect of the transistor.

A fully differential SA integrator is depicted in Figure 8.74. The nMOS switches are connected to ground, while the pMOS switches are related to the supply voltage, V_{DD}. The branch including the capacitor C_B is used to optimize the dynamic range. In the steady state, it can be shown that

$$C_1 V_{0,dc} + C_B V_{DD} = C_1 V_{DD} \tag{8.346}$$

where $V_{0,dc} = V_{DD}/2$ provided that $C_B = C_1/2$. It was assumed that the input dc level is set to the ground. Note that the amplifier architecture should be chosen with the objective of minimizing the turn-on time, which can limit the achievable sampling frequency. Furthermore, the set of transfer functions that can be realized with the SA approach is limited due to the fact that the amplifier output is defined and can be used only during one clock phase.

8.18 Summary

SC circuits used for the implementation of S/H and T/H circuits, gain stages, and integrators should be designed to be less sensitive to component nonidealities. This can be achieved by optimizing the circuit performance based on the analysis of the limitations due to the practical characteristics of components (amplifiers, switches and capacitors), and by introducing some refinements at the circuit level. Furthermore, in the specific case of filter design, the circuit accuracy can also depend on the synthesis method or filter architecture. Hence, SC ladder filters are less sensitive to capacitance mismatches or fluctuations in the passband than cascaded biquad structures.

8.19 Circuit design assessment

1. **Offset-free tunable gain stage**

   Consider the offset-free tunable gain stage shown in Figure 8.75, where ϕ is 50% duty cycle clock signal.

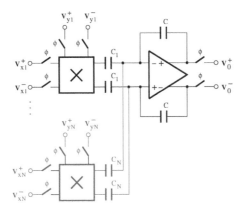

FIGURE 8.75
Offset-free tunable gain stage.

Analyze the charge transfer taking place between the capacitors C_j and C to prove that the output voltage is not affected by the offset voltage contributions due to the multipliers and amplifier.

Assuming that the amplifier gain is $A = 1/\mu$, show that the output voltage can be expressed as

$$V_0(z) = \pm \frac{1}{C(1+\mu)} \sum_{j=1}^{N} C_j \cdot \triangle V_j(z), \qquad (8.347)$$

where $v_0 = v_0^+ - v_0^-$, and $\triangle v_j$ is the voltage across each capacitor C_j.

2. **Integrator with input parasitic capacitors**

   The circuit diagram of an SC inverting integrator is depicted in

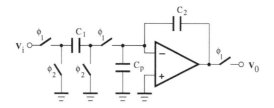

FIGURE 8.76
SC inverting integrator.

Figure 8.76. Taking into account the effect of the finite dc gain, $A_0 = 1/\mu$, and parasitic capacitor, C_p, show that the transfer function can be derived as

$$H(j\omega) = \frac{V_0(j\omega)}{V_i(j\omega)}$$

$$= H_{id}(j\omega)\left[1 + \mu\left(1 + \frac{C_p + C_1/2}{C_2}\right) - j\mu\frac{C_1}{2C_2\tan(\omega T/2)}\right]^{-1},$$

$$(8.348)$$

where

$$H_{id}(j\omega) = -\frac{C_1}{C_2}\frac{e^{j(\omega T/2)}}{j\sin(\omega T/2)}.$$

$$(8.349)$$

3. Track-and-hold circuit

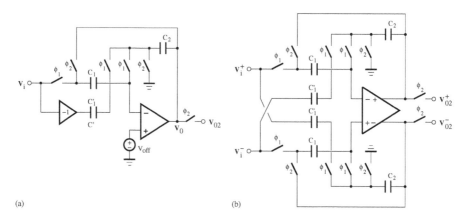

(a)   (b)

FIGURE 8.77
Circuit diagram of a low-sensitivity T/H ($C_1 = C_2 = C$ and $C_1' = 2C$): (a) Single-ended and (b) differential versions.

The circuit of Figure 8.77 is designed with $C_1 = C_2 = C$ and $C_1' = 2C$.

Let v^- be the voltage at the inverting node of the amplifier. Verify that

$$C\big(v^-[(n-1/2)T] - v_0[(n-1/2)T] - \{v^-[(n-1)T] - v_i[(n-1)T]\}\big) = 0 \tag{8.350}$$

during the clock phase, ϕ_2, that is, for $(n-1)T \le t < (n-1/2)T$, and

$$\begin{aligned}
C\big(v_i(nT) - v^-(nT) - \{v_0[(n-1/2)T] - v^-[(n-1/2)T]\}\big) \\
+ 2C\big(-v_i(nT) - v^-(nT) - \{-v_i[(n-1)T] - v^-[(n-1)T]\}\big) \\
= C\{v^-(nT) - v_0(nT) + v_0[(n-1/2)T]\}
\end{aligned} \tag{8.351}$$

during the clock phase, ϕ_1, that is, for $(n-1/2)T \le t < nT$.

With the assumption that $v_{02}(nT) = v_0[(n-1/2)T]$, show that

$$V_{02}(z) = z^{-1/2}\frac{(1+\mu-\mu z^{-1})V_i(z) + (1+\mu)V_{off}}{(1+\mu)(1+4\mu) - \mu(4+3\mu)z^{-1}}, \tag{8.352}$$

where $\mu = 1/A_0$, A_0 and V_{off} are the dc gain and offset voltage of the amplifier, respectively.

In the case of an amplifier with ideal characteristics (dc gain, bandwidth, offset voltage), deduce that $V_0(z) = z^{-1/2}V_i(z)$.

4. **Analysis of a sample-and-hold circuit**
   Consider the circuit structure shown in Figure 8.78, where $C_1 = C_1'$. During the first clock phase, the capacitor C_1 charges up to V_i, and the capacitor C_1' charges up to $-V_i$. The charges acquired by C_1 and due to the input voltages cancel those acquired by C_1'. During the second clock phase, the capacitor C_1 is included in the amplifier feedback path while the capacitors C_1' and C_2 are discharged.

   Assuming that v^- is the voltage at the inverting node of the amplifier, verify that

$$v^-[(n-1/2)T] - v_0[(n-1/2)T] - \{v^-[(n-1)T] - v_i[(n-1)T]\} = 0 \tag{8.353}$$

during the clock phase, ϕ_2, that is, for $(n-1)T \le t < (n-1/2)T$, and

$$\begin{aligned}
C_1\big(v_i(nT) - v^-(nT) - \{v_0[(n-1/2)T] - v^-[(n-1/2)T]\}\big) \\
+ C_1'\{-v_i(nT) - v^-[(n-1/2)T]\} = C_2[v^-(nT) - v_0(nT)]
\end{aligned} \tag{8.354}$$

during the clock phase, ϕ_1, that is, for $(n-1/2)T \le t < nT$.

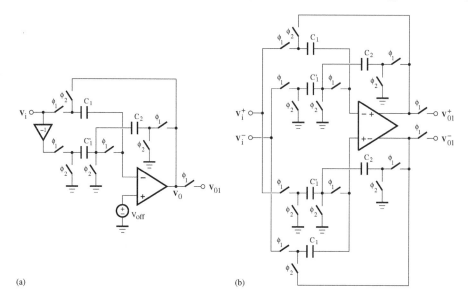

(a) (b)

FIGURE 8.78
Circuit diagram of an S/H structure ($C_1 = C_1'$): (a) Single-ended and (b) differential versions.

Taking into account the dc gain $A_0 = 1/\mu$ and offset voltage V_{off} of the amplifier, determine the circuit transfer function, $H(z)$.

In the ideal case, deduce that $v_0(nT) = v_i[(n-1)T]$ or $V_0(z) = z^{-1}V_i(z)$.

5. **Improved bootstrapped switch technique**
   The bootstrapped switch of Figure 8.79 [65] was designed to be less sensitive to the body effect of the transistor. During the on state, the voltage $V_{DD} + V_i + V_T$ is applied to the gate of the switch. This latter is grounded during the off state.

   Show that the on-resistance of the switch is given by

   $$R_{on} = \frac{1}{K(V_{DD} - V_B)}, \qquad (8.355)$$

   where K is the transconductance parameter of the switch, $V_B = \sqrt{2I_B/K_1}$, and $K_1 = \mu_n C_{ox}(W/L)_1$.

   What is the effect of the mismatch between the switch and T_1 on the circuit operation?

6. **Analysis of first-order filter sections**
   With the assumption that the input signal is slow moving, that is, $V_i(n-1/2) = V_i(n)$, show that the transfer function of the filter circuit of Figure 8.80 can be written as

   $$H(z) = \frac{V_0(z)}{V_i(z)} = \frac{\alpha - \beta + \beta z^{-1}}{1 + \gamma - z^{-1}}. \qquad (8.356)$$

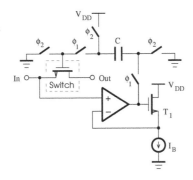

FIGURE 8.79
Circuit diagram of a bootstrapped switch.

Determine the type of filter corresponding to $\alpha = 1$, $\beta = 2$ and $\gamma = 1$.

Consider the circuit diagram of a first-order allpass filter shown in Figure 8.81.
Show that the filter transfer function is given by

$$H(z) = \frac{V_0(z)}{V_i(z)} = \frac{\alpha z^{-1} - \beta}{\alpha - \beta z^{-1}} \qquad (8.357)$$

What is the advantage of this circuit structure with respect to the mismatch between the capacitor ratios?

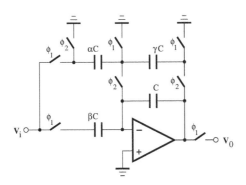

FIGURE 8.80
Circuit diagram of a first-order allpass filter.

7. **Low-sensitivity single-sampling integrator**
Given $\mu = 1/A_0$, where A_0 is the amplifier dc gain, and $C_3 = C_2$, verify the following relations for the single-ended version of the low-sensitivity integrator shown in Figure 8.82 [56].

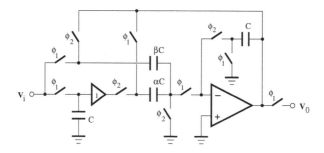

FIGURE 8.81
Circuit diagram of a first-order structurally allpass filter.

During the clock phase 1, that is, for $(n-1)T \le t < (n-1/2)T$,

$$C_3\{\mu v_0[(n-1/2)T] - \mu v_0[(n-1)T] - v_x[(n-1)T]\}$$
$$+ C_2\{(1+\mu)v_0[(n-1/2)T] - v_0[(n-1)T] + v_x[(n-1)T] - V_{off}\} = 0$$
$$(8.358)$$

During the clock phase 2, that is, for $(n-1/2)T \le t < nT$,

$$C_1[v_i(nT) - v_x(nT)] + C_2\{v_0(nT) - (1+\mu)v_0[(n-1/2)T] - v_x(nT) + V_{off}\}$$
$$(8.359)$$

and

$$v_x(nT) - \mu\{v_0[(n-1/2)T] - v_0(nT)\} = 0 \qquad (8.360)$$

because there is no current flowing through C_3.
Show that the output voltage can be written as

$$V_0(z) = H_i(z)V_i(z) + H_0(z)V_{off}, \qquad (8.361)$$

where

$$H_0(z) = -\frac{\mu}{1+2\mu}H_i(z), \qquad (8.362)$$

$$H_i(z) = H_{id}(z)E(z), \qquad (8.363)$$

$$H_{id}(z) = -\frac{C_1}{C_2}\frac{1}{1-z^{-1}}, \qquad (8.364)$$

and

$$E(z) = \frac{1}{1+\mu+\mu\dfrac{C_1}{C_2}\dfrac{1}{1-z^{-1}} - \mu\dfrac{1+\mu}{1+2\mu}\dfrac{C_1}{C_2}\dfrac{z^{-1}}{1-z^{-1}}}. \qquad (8.365)$$

Repeat the above question with the assumption that $v_0[(n-1/2)T] = v_0(nT)$.
What is the effect of a parasitic capacitor, C_p, connected to the stray-sensitive node x, on the error function $E(z)$?

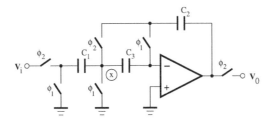

FIGURE 8.82

Circuit diagram of a single-sampling integrator.

8. **Design of an anti-aliasing filter**

   In DSP applications, aliasing can occur whenever the input signal contains spectral components at frequencies greater than one half of the sampling frequency.

FIGURE 8.83

Block diagram of an anti-aliasing filter.

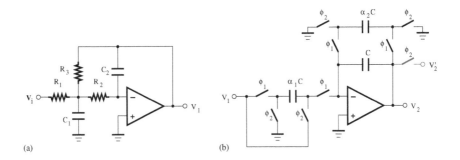

FIGURE 8.84

(a) Circuit diagram of a second-order RC active lowpass filter (LPF); (b) circuit diagram of a first-order cos SC filter.

   Suppose that we are required to design an anti-aliasing filter to restrict the bandwidth of a 200-kHz input signal to be sampled at $f_s = 800$ kHz. Because the sampling frequency is only four times the signal passband, it may be difficult to attenuate high-frequency components that can be aliased into the passband using conventional pre-filter based only on RC circuits. Figure 8.83 shows the block diagram of the anti-aliasing filter. The switched-capacitor section is assumed to operate with a sampling frequency of 1.6 MHz.

   Show that the transfer function of the RC biquad depicted in Fig-

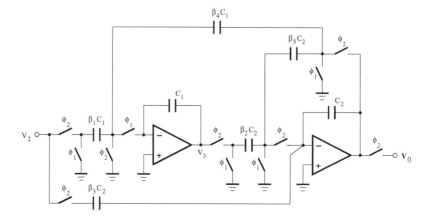

FIGURE 8.85
Circuit diagram of a second-order SC lowpass filter.

ure 8.84(a) is of the form

$$H_1(s) = \frac{V_1(s)}{V_i(s)} = \frac{-k\omega_0^2}{s^2 + \left(\dfrac{\omega_0}{Q}\right)s + \omega_0^2}, \tag{8.366}$$

where

$$k = \frac{R_3}{R_1}, \tag{8.367}$$

$$\omega_0 = \sqrt{\frac{1}{R_2 R_3 C_1 C_2}}, \tag{8.368}$$

and

$$Q = \frac{\sqrt{\dfrac{C_1}{R_2 R_3 C_2}}}{\dfrac{1}{R_1} + \dfrac{1}{R_2} + \dfrac{1}{R_3}}. \tag{8.369}$$

The RC filter section is designed to provide a 40-dB attenuation for frequency components at 1.4 MHz. Using the Butterworth approximation, the normalized transfer function can be obtained as

$$H_1(s') = \frac{-k'\omega_0'^2}{s'^2 + \left(\dfrac{\omega_0'}{Q'}\right)s' + \omega_0'^2} = \frac{-1}{s'^2 + \sqrt{2}s' + 1}, \tag{8.370}$$

where

$$s' = \frac{s}{\omega_p}, \qquad (8.371)$$

$$k' = \frac{R_3'}{R_1'}, \qquad (8.372)$$

$$\omega_0' = \sqrt{\frac{1}{R_2' R_3' C_1' C_2'}}, \qquad (8.373)$$

and

$$Q' = \frac{\sqrt{\dfrac{C_1'}{R_2' R_3' C_2'}}}{\dfrac{1}{R_1'} + \dfrac{1}{R_2'} + \dfrac{1}{R_3'}}. \qquad (8.374)$$

Assuming that $R_1' = R_2' = R_3'$ and $C_1' = C_2'$, determine k', ω_0', Q', and the normalized values, R_j' $(j = 1, 2, 3)$ and C_j' $(j = 1, 2)$, of the filter components.

Find the values of the filter components using the relations, $C_j = C_j'/\omega_p \tilde{R}$ and $R_j = R_j' \tilde{R}$, where $\tilde{R} = 1$ kΩ.

Consider the first-order cos SC filter depicted in Figure 8.84(b). Assuming that the amplifier is ideal, write the charge conservation equations during each of both clock phases.

Deduce that the transfer function can be put into the form

$$H_2(z) = \frac{V_2'(z)}{V_1(z)} = -\frac{\alpha_1 z^{-1/2}(1 + z^{-1/2})}{1 + \alpha_2 - z^{-1}}. \qquad (8.375)$$

Determine the ratio α_1/α_2 by setting the dc gain, $|H_2(z)|_{z=1}$, equal to unity.

For the second-order SC lowpass filter illustrated in Figure 8.85, use the signal-flow graph to show that the transfer function can be written as

$$H_3(z) = \frac{V_0(z)}{V_2(z)} = -\frac{\beta_3 + (\beta_1\beta_2 - 2\beta_3)z^{-1} + \beta_3 z^{-2}}{1 + \beta_5 + (\beta_2\beta_4 - \beta_5 - 2)z^{-1} + z^{-2}}. \qquad (8.376)$$

Verify that $\beta_1/\beta_4 = 1$ if the dc gain, $|H_3(z)|_{z=1}$, is equal to unity.

Using the bilinear transform and the Butterworth approximation function, determine the coefficients β_j $(j = 1, 2, 3, 4, 5)$ based on the requirement that the overall anti-aliasing filter should provide more than 40 dB attenuation for frequencies above 600 kHz.

9. Design of a bandpass SC ladder filter

After prewarping the passband and stopband specifications from the discrete-time domain to the continuous-time domain, the bandpass RLC ladder filter prototype of Figure 8.86(a) was found to satisfy the target filtering requirements.

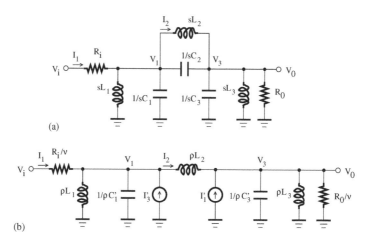

(a)

(b)

FIGURE 8.86

(a) Continuous-time and (b) discrete-time bandpass filter prototypes.

Given $\rho = (z^{1/2} - z^{-1/2})/2$, verify that the next design step should result in a discrete-time filter prototype, as shown in Figure 8.86(b), where

$$C_1' = C_1 + \frac{1}{L_1} + C_2 + \frac{1}{L_2}, \tag{8.377}$$

$$C_2' = C_2 + \frac{1}{L_2}, \tag{8.378}$$

$$C_3' = C_3 + \frac{1}{L_3} + C_2 + \frac{1}{L_2}, \tag{8.379}$$

$$I_1' = \rho C_2' V_1, \tag{8.380}$$

and

$$I_3' = \rho C_2' V_3. \tag{8.381}$$

Derive the corresponding z-domain SFG and convert it into an SC filter.

Bibliography

[1] T. Ndjountche, *Dynamic analog circuit techniques for real-time adaptive networks,* Aachen, Germany: Shaker Verlag, 2000.

[2] L. Larson and G. C. Temes, "Signal conditioning and interface circuits," In S. K. Mitra and J. F. Kaiser, Editors, *Handbook for digital signal processing,* Chap. 10, pp. 677–720, New York, NY: John Wiley & Sons, 1993.

[3] S. Hirano and E. Hayara, "Pre- and postfiltering effects on SCF stopband attenuation," *IEEE Trans. on Circuits and Systems,* vol. 38, pp. 547–551, May. 1991.

[4] D. J. Allstot and W. C. Black Jr., "Technological design considerations for monolithic MOS switched-capacitor filtering systems," *Proc. of the IEEE,* vol. 71, pp. 967–986, Aug. 1983.

[5] R. Unbehauen and A. Cichocki, *MOS switched-capacitor and continuous-time integrated circuits and systems,* Berlin, Germany: Springer Verlag, 1989.

[6] A. Robertini, J. Goette, and W. Guggenbühl, "Nonlinear distortions in SC-integrators due to nonideal switches and amplifiers," *Proc. of 1989 IEEE Int. Symp. on Circuits and Systems,* pp. 1692–1695, May 1989.

[7] B. Sheu and C. Hu, "Switch-induced error voltage on switched capacitor," *IEEE J. of Solid-State Circuits,* vol. 19, pp. 519–525, Aug. 1984.

[8] G. Wegmann, E. A. Vittoz, and F. Rahali, "Charge injection in analog MOS switches," *IEEE J. of Solid-State Circuits,* vol. 22, pp. 1091–1097, Dec. 1987.

[9] D. G. Haigh and B. Singh, "A switching scheme for SC filters which reduces the effect of parasitic capacitances," in *Proc. 1983 IEEE Int. Symp. on Circuits and Systems,* pp. 586–589, April 1983.

[10] E. Suarez, P. R. Gray, and D. A. Hodges, "All–MOS charge redistribution analog-to-digital conversion techniques–Part II," *IEEE J. of Solid-State Circuits,* vol. 10, pp. 379–385, Dec. 1975.

[11] C. Eichenberger and W. Guggenbühl, "On charge injection in analog MOS switches and dummy switch compensation techniques," *IEEE Trans. on Circuits and Systems*, vol. 37, pp. 256–264, Feb. 1990.

[12] K. Martin and A. S. Sedra, "Strays-insensitive switched-capacitor filters based on the bilinear z-transform," *Electronics Letters*, vol. 19, pp. 365–366, June 1979.

[13] Y. Tsividis, "Analytical and experimental evaluation of the switched-capacitor filter and remarks on the resistor/switched capacitor correspondence," *IEEE Trans. on Circuits and Systems*, vol. 26, pp. 140–144, Feb. 1979.

[14] K. Martin and A. S. Sedra, "Transfer function deviations due to resistor-SC equivalence assumption in switched-capacitor simulation of LC ladders," *Electronics Letters*, vol. 16, pp. 387–389, May 1980.

[15] A. M. Durham, J. B. Hughes, and W. Redman-White, "Circuit architectures for high linearity monolithic continuous-time filtering," *IEEE Trans. on Circuits and Systems–II*, vol. 39, pp. 651–657, Sept. 1992.

[16] D. A. Johns and K. Martin, *Analog integrated circuit design*, New York, NY: John Wiley & Sons, 1997.

[17] Y. A. Haque, R. Gregorian, R. W. Blasco, R. A. Mao, and W. E. Nicholson, "A two chip PCM voice CODEC with filters," *IEEE J. of Solid-State Circuits*, vol. 14, pp. 961–969, Dec. 1979.

[18] J. J. F. Rijns and H. Wallinga, "Stray-insensitive switched-capacitor sample-delay-hold buffers for video frequency applications," *Electronics Letters*, vol. 27, no. 8, pp. 639–640, April 1991.

[19] G. Nicollini, P. Confalonieri and D. Senderowicz, "A fully differential sample-and-hold circuit for high-speed applications," *IEEE J. of Solid-State Circuits*, vol. 24, pp. 1461–1465, Oct. 1989.

[20] T. Ndjountche and R. Unbehauen, "Analog discrete-time basic structures for adaptive IIR filters," *IEE Proc. -Circuits Devices and Systems*, vol. 147, pp. 250–256, Aug. 2000.

[21] B. J. Hosticka, R. W. Brodersen, and P. R. Gray, "MOS sampled data recursive filters using switched capacitor integrators," *IEEE J. of Solid-State Circuits*, vol. 12, pp. 600–608, Dec. 1977.

[22] P. E. Fleischer and K. R. Laker, "A family of active switched-capacitor biquad building blocks," *Bell Systems Tech. J.*, vol. 58, pp. 2235–2269, Dec. 1979.

[23] E. I. El-Masry, "Strays-insensitive state-space switched-capacitor filters," *IEEE Trans. on Circuits and Systems*, vol. 30, pp. 474–488, July 1983.

[24] S. Signell, "On selectivity properties of discrete-time linear networks," *IEEE Trans. on Circuits and Systems*, vol. 31, pp. 275–280, March 1984.

[25] U. Weder and A. Moeschwitzer, "Comments on "Design techniques for improved capacitor area efficiency in switched-capacitor biquads"," *IEEE Trans. on Circuits and Systems*, vol. 37, pp. 666–668, May 1990.

[26] W.-H. Ki and G. C. Temes, "Optimal capacitance assignment of switched-capacitor biquads," *IEEE Trans. on Circuits and Systems*, vol. 42, pp. 334–342, June 1995.

[27] A. Petraglia and S. K. Mitra, "Switched-capacitor equalizers with digitally programmable tuning characteristics," *IEEE Trans. on Circuits and Systems*, vol. 38, pp. 1322–1331, Nov. 1991.

[28] J. L. McCreary, "Matching properties, and voltage and temperature dependence of MOS capacitors," *IEEE J. of Solid-State Circuits*, vol. 16, pp. 608–616, Dec. 1981.

[29] D. A. Vaughan-Pope and L. T. Burton, "Transfer function synthesis using generalized doubly terminated two-pair networks," *IEEE Trans. on Circuits and Systems*, vol. 24, pp. 79–88, Feb. 1977.

[30] M. Kaneko and M. Onoda, "Z-domain exact design for LDI leapfrog switched-capacitor filters," *Proc. of 1984 IEEE Int. Symp. on Circuits and Systems*, pp. 304–307, May 1984.

[31] A. Kaelin and G. S. Moschytz, "Exact design of arbitrary parasitic-insensitive elliptic SC-ladder filter in the z-domain," *Proc. of 1988 IEEE Int. Symp. on Circuits and Systems*, pp. 2485–2488, June 1988.

[32] S. O. Scanlan, "Analysis and synthesis of switched-capacitor state-variable filters," *IEEE Trans. on Circuits and Systems*, vol. 28, pp. 85–93, Feb. 1981.

[33] A. Muralt, *The design of switched-capacitor filters based on doubly-terminated two-pair signal-flow graphs*, PhD thesis, Swiss Federal Institute of Technology, ETH-Center, Zurich, Switzerland, 1993.

[34] E. S. K. Liu, L. E. Turner, and L. T. Bruton, "Exact synthesis of LDI and LDD ladder filters," *IEEE Trans. on Circuits and Systems*, vol. 31, pp. 369–381, April 1984.

[35] R. B. Datar and A. S. Sedra, "Exact design of strays-insensitive switched-capacitor ladder filters," *IEEE Trans. on Circuits and Systems*, vol. 30, pp. 888–898, Dec. 1983.

[36] M. S. Lee and C. Chang, "Switched-capacitor filters using the LDI and bilinear transformations," *IEEE Trans. on Circuits and Systems*, vol. 28, pp. 265–270, April 1981.

[37] B. Nowrouzian, "A new synthesis technique for the exact design of switched-capacitor LDI allpass networks," *Proc. of the 1990 IEEE Int. Symp. on Circuits and Systems*, pp. 2185–2188, 1990.

[38] A. M. Davis, "A new *z*-domain continued fraction expansion," *IEEE Trans. on Circuits and Systems*, vol. 29, pp. 658–662, Oct. 1982.

[39] K. Martin and A. S. Sedra, "Effects of the op amp finite gain and bandwidth on the performance of switched-capacitor filters," *IEEE Trans. on Circuits and Systems*, vol. 28, pp. 822–829, Aug. 1981.

[40] D. B. Ribner and M. A. Copeland, "Biquad alternatives for high-frequency switched-capacitor filters," *IEEE J. of Solid-State Circuits*, vol. 20, pp. 1085–1095, Dec. 1985.

[41] W. M. C. Sansen, H. Qiuting, and K. A. I. Halonen, "Transient analysis of charge transfer in SC filters-gain error and distortion," *IEEE J. of Solid-State Circuits*, vol. 22, pp. 268–276, April 1987.

[42] J. Vlach, K. Singhal, and M. Vlach, "Computer oriented formulation of equations and analysis of switched-capacitor networks," *IEEE Trans. on Circuits and Systems*, vol. 31, pp. 753–765, Sept. 1984.

[43] N. Fröhlich, B. M. Riess, U. A. Wever, and Q. Zheng, "A new approach for parallel simulation of VLSI circuit on a transistor level," *IEEE Trans. on Circuits and Systems*, vol. 45, pp. 601–613, June 1998.

[44] *HSPICE simulation and analysis user guide*, Release U-2003.03-PA, Synopsys, March 2003.

[45] S. C. Fang, Y. Tsividis, and O. Wing, "SWITCAP: A switched-capacitor network analysis program, part I: Basic features," *IEEE circuits and Dev. Mag.*, vol. 5, pp. 4–10, Dec. 1983.

[46] ———, "SWITCAP: A switched-capacitor network analysis program, part II: Advanced applications," *IEEE circuits and Dev. Mag.*, vol. 5, pp. 41–46, Dec. 1983.

[47] *Affirma RF Simulator (SpectreRF) Theory (v446)*, Cadence Design Systems, June 2000.

[48] *Affirma RF Simulator (SpectreRF) User Guide (v446)*, Cadence Design Systems, April 2001.

[49] G. Fischer and G. S. Moschytz, "SC Filters for high frequencies with compensation for finite-gain amplifiers," *IEEE Trans. on Circuits and Systems*, vol. 32, pp. 1050–1056, Oct. 1985.

[50] G. C. Temes and K. Haug, "Improved offset-compensation schemes for the switched-capacitor circuits," *Electronics Letters*, vol. 20, pp. 508–509, June 1984.

[51] L. E. Larson, K. W. Martin, and G. C. Temes, "GaAs switched-capacitor circuits for high-speed signal processing," *IEEE J. of Solid-State Circuits*, vol. 22, pp. 971–981, Dec. 1987.

[52] L. E. Larson and G. C. Temes, "Switched-capacitor building blocks with reduced sensitivity to finite amplifier gain, bandwidth, and offset voltage," *Proc. of 1987 IEEE Int. Symp. on Circuits and Systems*, pp 334–338, May 1987.

[53] R. Gregorian, K. W. Martin, and G. C. Temes, "Switched-capacitor circuit design," *Proc. of the IEEE*, vol. 71, pp. 941–966, Aug. 1983.

[54] H. Matsumoto and K. Watanabe, "Spike-free SC circuits," *Electronics Letters*, vol. 8, pp. 428–429, 1987.

[55] K. Nagaraj, K. Singhal, T. R. Viswanathan, and J. Vlach, "Reduction of finite-gain effect in switched-capacitor filters," *Electronics Letters*, vol. 21, pp. 664–665, June 1985.

[56] K. Nagaraj, J. Vlach, T. R. Viswanathan, and K. Singhal, "Switched-capacitor integrator with reduced sensitivity to amplifier gain," *Electronics Letters*, vol. 22, pp. 1103–1105, Oct. 1986.

[57] K. Nagaraj, T. R. Viswanathan, K. Singhal, and J. Vlach, "Switched-capacitor circuits with reduced sensitivity to amplifier gain," *IEEE Trans. on Circuits and Systems*, vol. 34, pp. 571–574, May 1987.

[58] T. Ndjountche and R. Unbehauen, "Improved structures for programmable filters: Application in a switched-capacitor adaptive filter design," *IEEE Trans. on Circuits and Systems*, vol. 46 , pp. 1137–1147, Sept. 1999.

[59] E. A. Vittoz, "The design of high-performance analog circuits on digital CMOS chips," *IEEE J. of Solid-State Circuits*, vol. 20, pp. 657–665, June 1985.

[60] Y. Matsuya and J. Tamada, "1 V power supply low-power consumption A/D conversion technique with swing suppression noise shapping," *IEEE J. of Solid-State Circuits*, vol. 29, pp. 1524–1530, Dec. 1994.

[61] G. Nicollini, A. Nagari, P. Confalonieri, and C. Crippa, "A −80 dB THD 4 V_{pp} switched-capacitor filter for 1.5V battery-operated systems," *IEEE J. of Solid-State Circuits*, vol. 31, pp. 1214–1219, Aug. 1996.

[62] J. Crols and M. Steyaert, "Switched-opamp: An approach to realize full CMOS switched-capacitor circuits at very low power supply voltages," *IEEE J. of Solid-State Circuits*, vol. 29, pp. 936–942, Aug. 1994.

[63] A. Baschirotto and R. Castello, "A 1-V 1.8 MHz CMOS switched-opamp SC filter with rail-to-rail output swing," *IEEE J. of Solid-State Circuits*, vol. 32, pp. 1979–1987, Dec. 1997.

[64] A. M. Abo and P. Gray, "A 1.5-V, 10-bit, 14.3-Ms/s CMOS pipeline analo-to-digital converter," *IEEE J. of Solid-State Circuits*, vol. 34, pp. 599–606, May 1999.

[65] A. K. Ong, V. I. Prodanov, and M. Tarsia, "A method for reducing the variation in on resistance of a MOS sampling switch," *Proc. of 2000 IEEE Int. Symp. on Circuits and Systems*, vol. V, pp. 437–440, May 2000.

9

Data Converter Principles

CONTENTS

Data converters, or specifically analog-to-digital converters (ADCs) and digital-to-analog converters (DACs), play an important role in the design of data acquisition units in communication and microprocessor-based instrumentation systems. They include analog and digital building blocks and form the main interface component in mixed-signal processing systems (for some typical applications, see Table 9.1).

TABLE 9.1

Data Converter Specifications for Some Applications

Applications	Resolution (bits)	Sampling Frequency
Audio device	14–24	< 200 kHz
Digital oscilloscope	8	150 MHz
Magnetic read channel equalizer	6–8	(50–200) MHz
Wireless local area network	6–10	(1–50) MHz
Digital video camera	8–12	20 MHz
TV baseband processor	8–10	20 MHz
Modem	8–10	(10–20) MHz

• The process of converting an analog signal into a digital sequence, as illustrated in Figure 9.1, involves three operations: sampling, quantization, and coding. After the filtering operation, a sample-and-hold circuit first picks up the signal representative, which is maintained constant for the duration required by the converter to provide a digital word. Note that a continuous-time

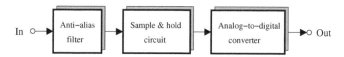

FIGURE 9.1
Typical ADC system.

signal with the maximum frequency, f_m, can adequately be represented by its samples acquired at the rate $f_s \geq f_N = 2f_m$, where f_N is termed the Nyquist rate or frequency. To maintain the frequency content of the input signal within the Nyquist bandwidth, an analog lowpass filter, also known as an anti-aliasing filter, is placed before the sample-and-hold circuit. Ideally, this filter should attenuate signal components with a frequency above $f_s/2$. The sampled signal is then transformed into one of a finite set of prescribed values by a quantizer, the levels of which can be uniformly or nonuniformly spaced. The transfer characteristic and error, e, of uniform quantizer, whose implementation is the simplest of both, is shown in Figure 9.2. The error caused by the quantization

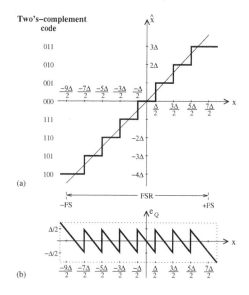

FIGURE 9.2
(a) Transfer characteristic and (b) quantization error of a 3-bit ADC.

is defined as the difference between the discrete output level and the actual analog input, $e_Q = \hat{x} - x$. It is in the range $\pm\Delta/2$ as long as the quantizer does not saturate. The transfer characteristic of Figure 9.2 belongs to a quantizer of the mid-tread type because it follows the input axis about zero. As a result, the output is insensitive to small input variation in the absence of signal in contrast to the mid-riser characteristic, which is supported by the output axis about zero. The coding then consists of assigning a unique binary number to each quantization level. Assuming that the number of bits, N, is 3, the 2^N

quantization levels can be coded in an N-bit number representation. The step size of the converter, Δ, represents the least significant bit (LSB) of the digital number and is given by $\Delta = FSR/2^N$, where FSR is the quantizer range or full-scale range (FSR). By using the two's complement code, the sign of the input sample is determined by the most significant bit (MSB). A real number, X, which can be represented as

$$b_1 b_2 b_3 \cdots b_N, \tag{9.1}$$

corresponds to the value

$$-b_1 2^0 + b_2 2^{-1} + b_3 2^{-2} + \cdots + b_N 2^{-(N-1)}, \tag{9.2}$$

where b_1 is the MSB. Other codes can also be used depending on signal characteristics and the desired application; however, the two's complement representation is a convenient way of representing signed numbers and the most suitable for addition and subtraction operations.

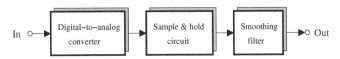

FIGURE 9.3
Typical DAC system.

• A digital-to-analog conversion stage, as shown in Figure 9.3, generally contains a DAC, a sample-and-hold (S/H) circuit, and a lowpass filter (LPF). The DAC is used to transform a finite number of digital codes into the corresponding analog discrete-time signal. Its transfer characteristic is shown in Figure 9.4 for a bipolar input code. In contrast to the ADC, which exhibits a quantization error due to the fact that any voltage within a given step size is mapped to the same output code, the DAC uniquely assigns each input code to an output level without an inherent error. Therefore, DACs do not directly realize the inverse function of ADCs.

In general, the DAC output signal, X_0, can ideally be put in the form

$$X_0 = G \cdot X_{REF} \left(K_1 \frac{D}{2^N} + K_2 \right), \tag{9.3}$$

where G is the gain, X_{REF} is the reference signal, D is the decimal equivalent of the binary input code, and K_1 and K_2 are the gain and offset constants, respectively. In the case of unipolar conversion, $K_1 = 1$ and $K_2 = 0$, and the output range is from 0 to $G \cdot X_{REF}$. For bipolar DACs based on the offset binary input coding, the constants are chosen as $K_1 = 2$ and $K_2 = -1$ to produce an output swing between $-G \cdot X_{REF}$ and $G \cdot X_{REF}$. Note that the two's complement representation is converted into offset binary code only by inverting the MSB.

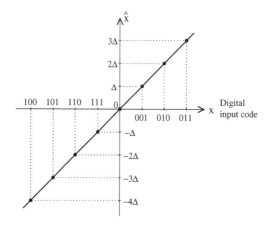

FIGURE 9.4
Transfer characteristic of a 3-bit DAC.

The digital-to-analog conversion process should be realized with the highest fidelity and minimal lag in time. The S/H introduces a delay in the output of the DAC to allow the current sample at the DAC output to reach the steady state. An LPF, often referred to as a smoothing (or reconstruction) filter, is used to remove the frequency components above $f_s/2$ from the converter output. It also smoothes the signal provided by the S/H by removing all sharp discontinuities.

Various architectures are known for the implementation of data converters. They can be divided into two main groups: Nyquist and oversampling converters. Nyquist converters operate at a sampling rate close to the Nyquist frequency or slightly higher than twice the bandwidth of the input signal; therefore their output data rate can be very high. On the other hand, the operation of oversampling converters, which can achieve a higher resolution even with low-precision components, requires a sampling rate that is several times higher than the Nyquist frequency. By relying on the averaging of multiple samples performed by a digital filter for each conversion, oversampling converters feature a longer acquisition time than the one of Nyquist converters, which process each input sample independently.

9.1 Binary codes

One feature of data converters such as DAC and ADC is that either the input or output is in digital form and can be represented using binary codes.

Digital systems generally use a binary number coding rather than the most familiar decimal number representation. For binary codes, the base or radix is 2 and the digits are called bits and take the values 0 or 1. Positive integers with the value in the range 0 to $2^N - 1$ can be encoded using N bits. Starting with the LSB, which has a weight of 2^0 or equivalently, 1, the weight is increased by a factor of 2 from one bit to the next and up to the value of the MSB weight, 2^{N-1}. The binary representation is said to be positional because the location of a bit in the resulting sequence determines its weight. The value of a binary number corresponds to the sum of the weights of all nonzero bits.

In practice, the operating range of a data converter is bounded by the full scale (FS). Scaling down to the full-scale range requires representing all of the numbers as fractions [1]. Any binary integer can be set into a fractional format by dividing its value by 2^N. With the binary point assumed to be at the left of the MSB, it is possible to encode the numbers in the range from 0 to $1 - 2^{-N}$ of the converter full-scale. The bit weight, which is inversely proportional to a power of 2, varies from $1/2$ for the MSB down to $1/2^N$ for the LSB.

9.1.1 Unipolar codes

TABLE 9.2
Common Unipolar Codes for 4-Bit Converters

Decimal Number	Fraction of the FS	Natural Binary	Gray Code	8421 BCD
15	15/16	1111	1000	1001 0011
14	14/16	1110	1001	1000 0111
13	13/16	1101	1011	1000 0101
12	12/16	1100	1010	0111 0010
11	11/16	1011	1110	0110 1000
10	10/16	1010	1111	0110 0010
9	9/16	1001	1101	0101 0110
8	8/16	1000	1100	0101 0000
7	7/16	0111	0100	0100 0011
6	6/16	0110	0101	0011 0111
5	5/16	0101	0111	0011 0001
4	4/16	0100	0110	0010 0101
3	3/16	0011	0010	0001 1000
2	2/16	0010	0011	0001 0010
1	1/16	0001	0001	0000 0110
0	0	0000	0000	0000 0000

Unipolar codes are used to represent signals with a predetermined sign. The most common unipolar codes, as shown in Table 9.2 for 4-bit converters, are natural binary, Gray, and binary-coded decimal (BCD) representations.

The natural binary representation is a positional system of numeration that uses the digits 0 and 1 and a radix of 2. Each successive digit of the binary code, which is generally adopted for a data converter, represents an inverse power of 2. The all-zero code corresponds to the zero-scale signal, while the all-one code is associated to the signal value, which is one LSB below the full scale.

Gray codes are used in the design of encoders for data converters in order to prevent errors due to the fact that all the required bit transitions between any two neighboring numbers may not occur at the same time. In general, the Gray code can be obtained by rearranging a binary sequence such that there is only one bit change between two adjacent numbers. The possibility to obtain multiple Gray code representations for a given number then increases with the resolution. The 4-bit example shown in Table 9.2 is known as binary-reflected Gray codes. Starting with a binary sequence where all bits are set to zero; binary-reflected Gray codes are formed by successively changing the state of the right-most bit so as to produce a new sequence. Let b_k and g_k be the bits of the natural binary code and Gray code, respectively. The conversion from the binary to Gray code is based on the next algorithm:

$$g_N = b_N \tag{9.4}$$
$$g_k = b_{k+1} \oplus b_k \quad \text{for} \quad k = N - 1 \quad \text{down to} \quad 1, \tag{9.5}$$

while the Gray-to-binary conversion is obtained as follows:

$$b_N = g_N \tag{9.6}$$
$$b_k = b_{k+1} \oplus g_k \quad \text{for} \quad k = N - 1 \quad \text{down to} \quad 1, \tag{9.7}$$

where $\oplus$ denotes the exclusive or (XOR) logic operation and N is the number of bits of the code.

In the BCD code, each decimal digit, D_k, with a value of 0 through 9 is replaced by its 4-bit binary equivalent, $b_{k,j}$, where $j = 1, 2, 3, 4$. Table 9.2 shows 8421 BCD codes of the first two fractional digits of FS fractions. The designation 8421 indicates the binary weights of the four bits representing each digit. Note that different versions of BCD codes can be obtained using other weight combinations. A further digit can be appended to a BCD code just by adding another 4-bit sequence. One advantage of BCD over binary representations is that the range of numbers that can be represented is not limited. The BCD code is particularly useful for interfacing to printing or display devices, which can process individual decimal digits (e.g., digital multimeter, digital instruments for physical measurements). On the other hand, the natural binary code for a given number requires fewer bits than the corresponding BCD code. The BCD representation is a relatively inefficient coding due to the fact that only 10 of the 2^4 or 16 combinations allowed by any 4-bit sequence are exploited.

Note that additional bit combinations are sometimes used in the BCD code in order to take into account the data sign or other meaningful indications.

9.1.2 Bipolar codes

TABLE 9.3

Common Bipolar Codes for 4-Bit Converters

Decimal Number	Fraction of the FS	Sign Magnitude	One's Complement	Offset Binary	Two's Complement
7	7/8	0111	0111	1111	0111
6	6/8	0110	0110	1110	0110
5	5/8	0101	0101	1101	0101
4	4/8	0100	0100	1100	0100
3	3/8	0011	0011	1011	0011
2	2/8	0010	0010	1010	0010
1	1/8	0001	0001	1001	0001
0	0	0000 1000	0000 1111	1000	0000
−1	−1/8	1001	1110	0111	1111
−2	−2/8	1010	1101	0110	1110
−3	−3/8	1011	1100	0101	1101
−4	−4/8	1100	1011	0100	1100
−5	−5/8	1101	1010	0011	1011
−6	−6/8	1110	1001	0010	1010
−7	−7/8	1111	1000	0001	1001
−8	−8/8	-	-	0000	1000

Bipolar codes are used to represent signals that can be either positive or negative. The most common bipolar codes, which are sign magnitude, one's complement, offset binary, and two's complement representations, are shown in Table 9.3 for a resolution of 4 bits.

In the sign-magnitude representation, the bit in the MSB position is reserved for the number sign and the remaining bits indicate the number magnitude. The sign bit can be either 0 for positive numbers or 1 for negative numbers. The sign-magnitude representation has the drawback of having two different codes for zero and requiring a rather complex hardware for the realization of arithmetic operations.

The one's complement representation is formed by inverting each bit of the natural binary code for the number to be converted. It can also be obtained by subtracting each bit of the natural binary code from one. The bit in the MSB position can be set either to 0 for positive numbers, or 1 for negative numbers. Even if the one's complement representation can help reduce the algorithm complexity for some arithmetic operations, it still leads to ambiguity because there are two different codes for zero.

The offset binary coding is a binary representation that is shifted so that a signal with the zero value corresponds to the mid-scale code, that is, the code consisting of a one at the MSB position followed by zeros at all the remaining bit positions. The all-zero code is then used for the negative full scale and the all-one code is assigned to the signal value that is one LSB below the positive full scale.

In the two's complement representation, zero and positive signal values have the same code as in the natural binary format while the negative signal values are represented by forming the two's complement of the corresponding positive number. The two's complement is formed by complementing each bit of the binary code and then adding one LSB without taking into account any carry-out. The MSB is either 0 for positive numbers, or 1 in the case of negative numbers. The two's-complement representation is an efficient coding approach for bipolar signal values in microprocessors because it allows the use of only an adder to implement both the addition and subtraction. Note that the conversion from the offset binary format to two's complement code only requires the inversion of the MSB logic state.

9.1.3   Remarks

The choice of a number representation system has repercussions on the complexity of algorithm implementations for arithmetic operations and the input or output interface of the data converter with other circuits. The two's complement representation is used in most digital systems, and hence is commonly considered for data converter implementations. Sign-magnitude and BCD codes are mainly used for instrumentation applications.

Number expressions and dynamic ranges in common binary representations are summarized in Table 9.4, where X_0 denotes the encoded signal value and X_{FS} designates the data converter full-scale. Note that the variable X can stand for a voltage or current signal.

In some data converter configurations, it is required to use the aforementioned codes with all bits inverted, also known as complementary codes. In differential structures, the required inversion is carried out simply by permuting the input or output nodes.

9.2   Data converter characterization

In addition to the errors introduced by the quantization process [2], the performance of data converters can be affected by device nonlinearities.

9.2.1   Quantization errors

Let x and $\hat{x}$ be the input and output samples of the quantizer, respectively. According to the rounding quantizer model depicted in Figure 9.5, the quan-

TABLE 9.4

Number Expressions and Dynamic Range in Common Binary Representations

Representation	Range

Natural binary

$$\frac{X_0}{X_{FS}} = \sum_{k=1}^{N} b_k 2^{-k} \qquad\qquad 0 \leq \frac{X_0}{X_{FS}} \leq 1 - 2^{-N}$$

8421 BCD

$$\frac{X_0}{X_{FS}} = \sum_{k=1}^{N} D_k 10^{-k}$$

$$\text{where } D_k = \sum_{j=1}^{4} b_{k,j} 2^{-j+1} \qquad 0 \leq \frac{X_0}{X_{FS}} \leq 1 - 10^{-N}$$

Offset binary

$$\frac{X_0}{X_{FS}} = -1 + \sum_{k=1}^{N} b_k 2^{-k+1} \qquad -1 \leq \frac{X_0}{X_{FS}} \leq 1 - 2^{-N+1}$$

Sign-magnitude

$$\frac{X_0}{X_{FS}} = (-1)^{b_1} \sum_{k=2}^{N} b_k 2^{-k+1} \qquad 2^{-N+1} - 1 \leq \frac{X_0}{X_{FS}} \leq 1 - 2^{-N+1}$$

One's complement

$$\frac{X_0}{X_{FS}} = \left(2^{-N+1} - 1\right) b_1 + \sum_{k=2}^{N} b_k 2^{-k+1} \qquad 2^{-N+1} - 1 \leq \frac{X_0}{X_{FS}} \leq 1 - 2^{-N+1}$$

Two's complement

$$\frac{X_0}{X_{FS}} = -b_1 + \sum_{k=2}^{N} b_k 2^{-k+1} \qquad -1 \leq \frac{X_0}{X_{FS}} \leq 1 - 2^{-N+1}$$

tization error is defined as

$$e_Q(k) = \hat{x}(k) - x(k) \tag{9.8}$$

$$\hat{x}(k) = Q(x(k)), \tag{9.9}$$

where Q denotes the quantizer operation. Its value should not exceed half of the quantization level,

$$-\frac{\Delta}{2} < e_Q(n) \leq \frac{\Delta}{2}, \tag{9.10}$$

where Δ is the quantizer step size. However, for input signals with a high dynamic range, the samples that go over the quantizer limit are clipped and e_Q can be greater than $\Delta/2$. Note that, for the rounding quantizer, the signal values that are below an integer multiple of Δ located between two adjacent transitions are quantized to the lower level; otherwise they should be mapped to a higher level.

The converter performance can be described by the signal-to-noise ratio (SNR) given by

$$\text{SNR} = 10\log_{10}\left(\frac{P_x}{P_Q}\right) \quad \text{in dB}, \tag{9.11}$$

where $P_x = \sigma_x^2 = \text{E}[x^2(k)]$ is the input signal power or variance, and $P_Q = \sigma_Q^2 = \text{E}[e_Q^2(k)]$ is the variance or power of the quantization noise.

The SNR can also be expressed in terms of root-mean square (rms) amplitudes. By exploiting the fact that the average power is proportional to the square of the signal rms amplitude, we can write that

$$\text{SNR} = 10\log_{10}\left(\frac{A_x^2}{A_Q^2}\right) = 20\log_{10}\left(\frac{A_x}{A_Q}\right) \quad \text{in dB}, \tag{9.12}$$

where A is the rms amplitude. The logarithmic decibel scale helps describe signal level ratios, that span many orders of magnitude, with numbers of modest size without losing information.

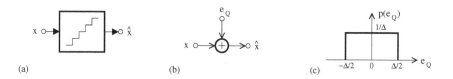

(a) (b) (c)

FIGURE 9.5
(a) Quantizer; (b) linear quantizer model; (c) probability density function of the quantization error.

It is convenient to deal with an input signal, x, that is zero mean, stationary, and uncorrelated with e_Q. Furthermore, by assuming that e_Q is a uniformly distributed white noise sequence over the interval $-\Delta/2$ to $\Delta/2$, the probability function p (see Figure 9.5(c)) is given by

$$p(e_Q) = \begin{cases} \dfrac{1}{\Delta}, & \text{for} \quad |e_Q| \leq \dfrac{\Delta}{2} \\ 0, & \text{otherwise.} \end{cases} \tag{9.13}$$

The power of the quantization noise can be obtained as

$$P_Q = \sigma_Q^2 = \int_{-\Delta/2}^{\Delta/2} e_Q^2 p(e_Q) de_Q = \frac{1}{\Delta}\int_{-\Delta/2}^{\Delta/2} e_Q^2 de_Q , = \frac{\Delta^2}{12} \tag{9.14}$$

where Δ is the quantizer step size.

Compute the power of the discrete-time sinusoidal signal given by

$$x(k) = A\sin(\omega k), \tag{9.15}$$

where $\omega = 2\pi(f/f_s)$, f is the frequency of the sinusoid, f_s represents the sampling frequency, and A is the amplitude.

The power is defined as

$$P_x = \sigma_x^2 = \mathrm{E}[x^2(k)] = \frac{1}{N}\sum_{k=0}^{N-1} x^2(k). \tag{9.16}$$

With the substitution of $x(k)$ and using the fact that the cosine has a zero mean, we obtain

$$P_x = \frac{A^2}{N}\sum_{k=0}^{N-1}\sin^2(\omega k) = \frac{A^2}{N}\sum_{k=0}^{N-1}\frac{1}{2}[1 - \cos(2\omega k)] = \frac{A^2}{2}. \tag{9.17}$$

The power is proportional to the square of the sinusoidal signal amplitude.

Assuming a sinusoidal input signal with an amplitude equal to half of the quantizer full-scale range, the power can be written as

$$P_x = \frac{(FSR/2)^2}{2}, \tag{9.18}$$

where FSR denotes the full-scale range. With the full-scale range of the N-bit quantizer given by $FSR = 2^N\Delta$, the SNR can take the next form

$$\mathrm{SNR} = 10\log_{10}\left(\frac{3}{2}2^{2N}\right)$$

$$= 6.02N + 1.76 \quad (\mathrm{dB}). \tag{9.19}$$

The SNR increases by about 6 dB for each additional bit of the quantizer. However, practical implementations rarely achieve the theoretical SNR due to various imperfections associated to circuit components.

An approach to improve the SNR can consist of using the oversampling technique, which distributes the power of the quantization noise over a wider frequency band. A digital filter is then required to reduce the quantization noise to a great extent without affecting the signal of interest, thereby increasing the number of bits.

By increasing the value of the sampling rate, the initial power of the quantization noise, which remains unchanged, can be expressed as a function of the quantization noise spectral density, E_Q. Hence,

$$P_Q = \int_{-f_s/2}^{f_s/2} |E_Q(f)|^2 df = |E_Q(f)|^2 \int_{-f_s/2}^{f_s/2} df = |E_Q(f)|^2 f_s .$$ (9.20)

Consequently, we have

$$|E_Q(f)|^2 = \frac{\Delta^2}{12} \frac{1}{f_s} .$$ (9.21)

The quantized signal is then processed by an ideal lowpass filter with the frequency response

$$H(f) = \begin{cases} 1, & |f| \leq f_B \\ 0, & \text{otherwise,} \end{cases}$$ (9.22)

where f_B represents the cutoff frequency. The resulting quantization noise power is now due to the spectral contributions of E_Q, which are confined between $-f_B$ and f_B. That is,

$$\begin{aligned} P_Q' &= \int_{-f_s/2}^{f_s/2} [E_Q(f)H(f)]^2 df = \int_{-f_B}^{f_B} [E_Q(f)]^2 df \\ &= \frac{\Delta^2}{12} \frac{2f_B}{f_s} = \frac{\Delta^2}{12} \frac{1}{OSR}, \end{aligned}$$ (9.23)

where $OSR = f_s/2f_B$ is the oversampling ratio. Recalling the signal-to-noise ratio definition, we have

$$\text{SNR'} = 10\log_{10}\left(\frac{P_x}{P_Q'}\right) = 10\log_{10}\left(\frac{3}{2}2^{2N}\right) + 10\log_{10}(OSR).$$ (9.24)

Thus

$$\text{SNR'} = 6.02N + 1.76 + 10\log_{10}(OSR) \quad (\text{dB}).$$ (9.25)

It should be emphasized that for every doubling of the OSR, the signal-to-noise ratio is increased by 3 dB or equivalently 0.5 bit of resolution. With the oversampling technique, the sampling frequency must be multiplied by a factor of 2^{2N} to yield an increase of N bits. It should be noted that oversampling has the advantage of relaxing the requirements of the anti-aliasing or smoothing analog filter.

An efficient architecture used to improve the resolution without requiring an excessive high oversampling is the delta-sigma ADC [3], which modulates the quantization noise so that its magnitude is attenuated in the signal band and increased for out-of-band frequencies. The quantization noise outside the signal bandwidth is reduced using a digital filter, which can also adjust the rate of the output data, if necessary. The delta-sigma converter can achieve a

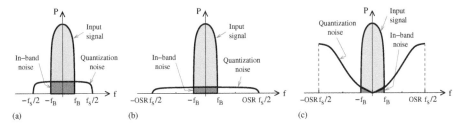

FIGURE 9.6
Power spectrum of the quantization noise: (a) Nyquist ADC, (b) oversampling ADC, (c) delta-sigma ADC.

resolution up to 24 bits without the need for high-precision components. For its implementation, the choice of an architecture depends on the characteristics required for the quantization noise shaping.

The frequency spectrum of the digitized signal, which includes the contributions of the input signal and the quantization noise, is shown in Figure 9.6 for Nyquist, oversampling, and delta-sigma ADCs. Due to the oversampling, the quantization noise inherent in the conversion process is spread over a large band of frequencies, including the signal bandwidth. For delta-sigma converters, only a small fraction of the quantization noise falls in the frequency range of interest.

9.2.2 Errors related to circuit components

In addition to the quantization errors, the overall noise available at the converter output is related to the component noise and the nonuniform allocation of the sampling instants or clock jitter. In the particular case of ADCs, the contribution due to the comparator ambiguity should also be taken into account. Furthermore, the dynamic performance of data converters is limited by the frequency characteristics of passive and active components.

- The noise caused by the different components of a converter can be analyzed using equivalent models. It is dominated by the thermal and flicker noise contributions.

- For the jitter analysis, let us consider an input sinusoid

$$x(t) = A\sin(2\pi ft + \phi), \qquad (9.26)$$

where A is the amplitude, and f and ϕ are the frequency and phase of the signal, respectively. Ideally, the sampling instants should be given by

$$t_k = kT, \quad k = 0, 1, 2, \cdots, K - 1. \qquad (9.27)$$

However, they are affected by errors, δ_k, due to the clock jitter and can be written as

$$t_k = kT + \delta_k. \tag{9.28}$$

The samples of the input signal can be obtained using

$$\bar{x}(k) = A\sin(\Omega k + J_k + \phi), \tag{9.29}$$

where the digital angular frequency, Ω, and the phase jitter, J_k, are given by $\Omega = 2\pi fT$ and $J_k = 2\pi f\delta_k$, respectively. The error due to the jitter can be computed as

$$e_J(k) = \bar{x}(k) - x(k), \tag{9.30}$$

where $x(k)$ is the uniformly sampled version of the input signal, that is, $x(k) = A\sin(\Omega k + \phi)$. The SNR contribution of the jitter is

$$\text{SNR} = 10\log_{10}\left(\frac{\sigma_x^2}{\sigma_Q^2}\right) \quad \text{in dB}, \tag{9.31}$$

where $\sigma_x^2 = \text{E}[x^2(k)]$ is the input signal variance and $\sigma_Q^2 = \text{E}[e_J^2(k)]$ is the jitter noise variance. The value of σ_Q^2 is dependent on the jitter statistical model and ADCs generally exhibit a clock jitter in the range of 0.5 to 2 ps. In practice, high-precision oscillators are used in conjunction with phase-locked loop or delay-locked loop circuits to minimize the effects of clock and timing errors.

• Input signals around the decision level of the comparator, which often consists of an amplifier stage followed by a latch, may result in ambiguous output codes. Let v_Q and $v_{\overline{Q}}$ be the voltages related to the positive and negative outputs of the comparator. The difference , $v_d = v_Q - v_{\overline{Q}}$, can be obtained by solving a differential equation of the form

$$\frac{dv_d(t)}{dt} - \frac{1}{\tau}v_d(t) = 0, \tag{9.32}$$

where the time constant, τ, characterizes the latch ability to resolve intermediate voltage levels and depends on the loading conditions, transistor parameters, and the IC process. That is,

$$v_d(t) = v_d(0)\exp(t/\tau), \tag{9.33}$$

with $v_d(0)$ being the initial condition at the beginning of the metastable region. Ideally, the latch requires an extra time, T, to generate a valid output, which can be processed by the next circuit section. This requires that v_d reaches a given value, say $V_{FS}/2$, within the time period, T. Otherwise, an inaccurate decision will be made if $v_d(T) < V_{FS}/2$. Taking into account the fact that v_d is equal to the amplifier output at $t = 0$, we have

$$v_d(0) = A(|v_i(0) - V_{th}|) = A\Delta v_i(0), \tag{9.34}$$

where v_i is the input voltage, and V_{th} and A denote the threshold level and gain of the comparator, respectively. Hence, the failure condition reads

$$\Delta v_i(0) < \frac{V_{FS}}{2A} \exp(-T/\tau) = \Delta V_i \,. \tag{9.35}$$

Assuming uniformly distributed input sample over the comparator range, the probability, P_e, to produce an uncertain output voltage is given by

$$P_e = P[\Delta v_i(0) < \Delta V_i] = \frac{2\Delta V_i}{V_L} \,, \tag{9.36}$$

where V_L is the effective LSB voltage. The error likelihood P_e can be considered an additive contribution to the quantization noise, that is,

$$\sigma_Q^2 = \frac{\Delta^2}{12}(1 + P_e). \tag{9.37}$$

Note that P_e should include the contribution of all comparators used in the ADC.

Data converters are limited by various error sources. Static errors affect the accuracy of converters during the conversion of dc signals, whereas dynamic errors essentially degrade the high-speed performance. Offset, gain, differential nonlinearity, and integral nonlinearity errors are generally associated with static performance of data converters. Their impact on the signal level can also be characterized in the frequency domain by estimating dynamic characteristics such as the signal-to-noise ratio, total harmonic distortion, and spurious-free dynamic range. On the other hand, dynamic errors can also be related to the limitations (acquisition time, settling time, glitches) of the transient response.

9.2.3 Static errors

The converter can be affected by the following static errors [2, 4] (see Figures. 9.7 through 9.14) when it transforms a signal. These errors are most commonly expressed in LSB units, or as a percentage of the converter FSR.

- *Offset error* — The offset error corresponds to the converter output deviation obtained by applying an input signal with a zero-scale to the converter. It can be either positive or negative, and affects all the output data in the same way. (see Figures 9.7 and 9.11).

- *Gain error* — The gain of the transfer characteristic is given by the slope of the straight line joining the two endpoints. The gain error results in a slope difference between the ideal and real converters. All the codes exhibit the same percentage of deviation.

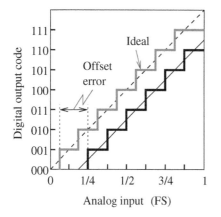

FIGURE 9.7
ADC offset error.

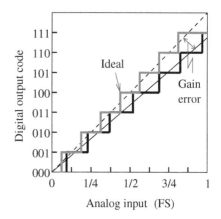

FIGURE 9.8
ADC gain error.

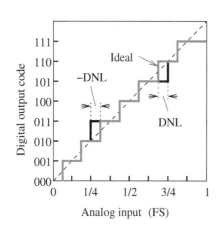

FIGURE 9.9
ADC differential nonlinearity.

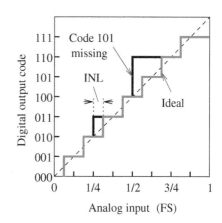

FIGURE 9.10
ADC integral nonlinearity.

- *Differential nonlinearity (DNL) error* — The DNL is the deviation of either the step width for the ADC (see Figure 9.15) or the step height for the DAC from the ideal value of 1 LSB. In the specific case of a data converter, where a quantization step size can be associated with each code k, it is defined as

$$\mathrm{DNL}_k = \frac{\triangle_k}{V_{LSB}} - 1, \tag{9.38}$$

where $\triangle_k$ represents the actual quantization step size for the code k and V_{LSB} is the ideal quantization step size. Generally, the highest value of $|\mathrm{DNL}_k|$ is considered the DNL of the data converter. The missing code is the result of a DNL equal to or less than -1 LSB. In this case, the corresponding step does not appear in the transfer characteristic. It is possible that the

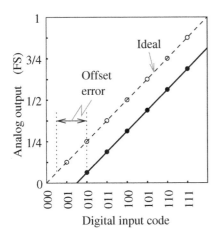

FIGURE 9.11
DAC offset error.

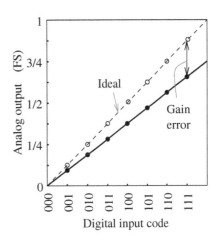

FIGURE 9.12
DAC gain error.

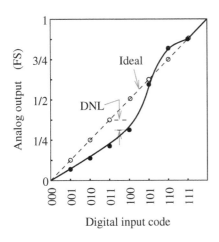

FIGURE 9.13
DAC differential nonlinearity.

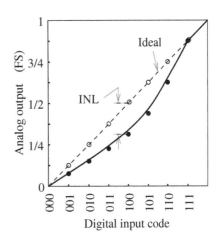

FIGURE 9.14
DAC integral nonlinearity.

converter becomes nonmonotonic for DNL values greater than 1 LSB. The magnitude of the converter output diminishes as the input data increases.

- *Integral nonlinearity (INL) error* — The INL denotes the deviation at any point of the transfer characteristic of output data from a reference straight line drawn through the zero and full scale. Its value can depend on the definition of the two endpoints. Assuming that the computation of quantization levels is possible, we can obtain

$$\mathrm{INL}_k = \left| \frac{V_k - V_0}{V_{LSB}} - k \right|, \qquad (9.39)$$

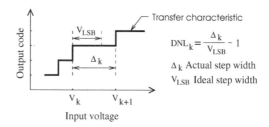

FIGURE 9.15
Illustration of the DNL error.

where V_k is threshold level associated with the code k, V_0 is the threshold level corresponding to the lowest transition code, and V_{LSB} is the ideal quantization step size. It can also be shown that

$$\text{INL}_k = \sum_{j=0}^{k} DNL_j \,, \tag{9.40}$$

where the first output code index is supposed to be zero.

Linearity errors are more important than the ones due to offset and gain deviations, which can be adjusted using a suitable calibration technique.

9.2.4 Dynamic errors

In addition to static errors, the converter performance is also affected by errors whose root is related to the time-varying nature of the input signals. The signal-to-noise ratio (SNR), total harmonic distortion (THD), signal-to-noise and distortion (SINAD) ratio, effective number of bits (ENOB), and spurious-free dynamic range (SFDR), together with specifications in time domain such as the settling time, slew rate, and glitch impulse, are generally used to specify the converter dynamic performance. Parameters that are normally specified in dB, can also be given in units of dBc (decibels to carrier) when the absolute power of the fundamental is used as the reference, or dBFS (decibels to full-scale) when the power of the fundamental is extrapolated to the converter full-scale range.

In general, all the spectral components available at the converter output and that are different from the one of the input signal are considered to be noise. However, a better insight into the conversion process is provided by separately estimating the noise floor and harmonic distortion levels. Hence, the harmonic distortion components are excluded in the measurement of the SNR, but are taken into account in the determination of the SINAD.

• The SNR is the ratio of the power of the fundamental input signal to the noise power, which is caused by all spectral components from dc to half of the sampling frequency, excluding noise at dc and harmonic distortion components. By considering a full-scale input signal and only the noise due to quantization errors, the SNR in decibels (dB) of a Nyquist converter is given by

$$SNR = 10 \log_{10} \left(\frac{P_1}{P_Q} \right) = 6.02N + 1.76, \qquad (9.41)$$

where P_1 is the power of the first harmonic component or fundamental signal, P_Q denotes the noise power, and N is the number of bits. In the case of oversampling converters, the SNR includes additional terms depending on the OSR and modulator feedback structure.

- The dynamic range (DR) is the ratio of the power of a full-scale sinusoidal input signal to the power of the noise delivered by the converter with inputs shorted together.

- The THD is the ratio of the rms sum of the powers of the harmonic components above the fundamental frequency to the power of the fundamental signal at the converter output. The THD can be derived as

$$THD = 10 \log_{10} \left(\frac{P_D}{P_1} \right), \qquad (9.42)$$

where P_D is the sum of the powers, P_j $(j \geq 2)$, of all distortion spectral components. The distortion measurement is realized with an input signal whose amplitude is generally 0.5 dB to 1 dB below the full scale to avoid clipping, and only takes into account harmonic components within the Nyquist bandwidth. In practice, the distortion effects are directly observable on time-domain waveforms for THD value about -30 dB.

- The SINAD is the ratio of the power of the fundamental signal to the power of all the remaining spectral (except dc) components below half of the sampling frequency. It can be expressed as

$$SINAD = 10 \log_{10} \left(\frac{P_1}{P_Q + P_D} \right), \qquad (9.43)$$

where P_D is the sum of all distortion spectral component powers. The SINAD is also known as the signal-to-noise and distortion ratio (SNDR) and is measured in dB at a specified input frequency and sampling rate. In the case of an ideal converter, the quantization error is the only source of noise and the SINAD is reduced to the SNR.

- The ENOB specifies the resolution that an ideal converter would realize in order to exhibit the same SINAD as the one measured on the real converter.

It can be derived for Nyquist converters as

$$ENOB = (SINAD - 1.76)/6.02. \tag{9.44}$$

The difference between the ENOB and the nominal number of bits indicates the impact of circuit imperfections on the conversion process.

Assuming that an input signal with the same amplitude and frequency is used for all measurements, determine the relationship between the SINAD, SNR, and THD.

From the definitions of the SNR, THD, and SINAD, we obtain

$$\frac{P_Q}{P_1} = 10^{-SNR/10}, \tag{9.45}$$

$$\frac{P_D}{P_1} = 10^{THD/10}, \tag{9.46}$$

and

$$\frac{P_Q + P_D}{P_1} = 10^{-SINAD/10}. \tag{9.47}$$

It can then be shown that

$$\frac{P_1}{P_Q + P_D} = \left(\frac{P_Q}{P_1} + \frac{P_D}{P_1}\right)^{-1} = \left(10^{-SNR/10} + 10^{THD/10}\right)^{-1}. \tag{9.48}$$

Hence,

$$SINAD = 10\log_{10}\left(\frac{P_1}{P_Q + P_D}\right)$$
$$= -10\log_{10}\left(10^{-SNR/10} + 10^{THD/10}\right). \tag{9.49}$$

The degradation of the SINAD for high frequencies is primarily due to the fact that the importance of distortion effects increases with the input signal frequency.

• The SFDR can be obtained as the ratio of power of the fundamental signal to the power of the highest spurious component in the converter spectrum (excluding dc). It is generally plotted as a function of the test signal amplitude. The SFDR is commonly used in communication applications, as an indication of the usable dynamic range.

• The two-tone intermodulation distortion (IMD) is measured by connecting the converter input to the sum of two sinusoidal signals with the same magnitude, but having slightly different frequencies. It is computed as the ratio of the power of the worst third-order intermodulation product to the power of either input tone. The IMD is a key specification used in the selection of building blocks for multi-carrier communication systems. With the use of test signals at frequencies f_1 and f_2, the determination of the third-order IMD can rely only on the distortion products occurring at $2f_2 - f_1$ and $2f_1 - f_2$, or relatively near the fundamental signals, as the distortion components at the higher frequencies, or say at $2f_2 + f_1$ and $2f_1 + f_2$, usually fall outside the passband, where they can be filtered out easily.

It should be noted that the specifications of different converters can be fairly compared only if the measurements are realized with the same fundamental input frequency or are valid over the same bandwidth, which is defined as the frequency range over which the input signal can be converted with an amplitude attenuation less than or equal to 3 dB.

• The glitch impulse represents the undesired signal transients that can appear at the DAC output. It is characterized by measuring its area at the mid-scale code transition where the logic states of the maximum number of bits are changed. The unit of measurement can be chosen as $nV \cdot s$, for instance.

• The settling time is the time elapsed from the application of the input signal until the converter output reaches and remains within a given error range about the final value. It can include components such as a short delay time related to the propagation delay, the slew time required for the output to reach its highest value, and the ring time needed by the output to recover from the overload condition associated with the slewing and to settle to within the specified error band, which can be a certain percent of the final value or $\pm 1/4$ LSB. The settling time is usually specified for a full-scale transition.

• The latency time denotes the time elapsed from the initiation of one input conversion to next. It includes the conversion time, which is defined as the time required for a single conversion, and data retrieval time. For ADCs, the latency time is measured by assuming that the signal is already sampled.

9.3   Summary

Data converters are generally designed to meet the requirements of various applications such as imaging, video, instrumentation, control, and communication systems. For most commonly used architectures, the specifications for the resolution and sampling frequency generally extend beyond the range from 8 bits, 250 Msps (digital oscilloscope) to 24 bits, 2.5 Msps (seismic monitoring systems) for ADCs, and from 8 bits, 330 Msps (digital radio modulation) to 24 bits, 200 ksps (DVD systems) for DACs. The distribution of resolution versus sampling frequency can furnish insight into the performance limitations of data converters.

The resolution appears to be limited at low frequencies by the component mismatches and thermal noise. It tends to be reduced on the order of one bit when the sampling frequency is increased by two times. This is due to the enhanced effect related to the comparator ambiguity and clock jitter at high frequencies. Furthermore, the maximum sampling frequency of the converter cannot exceed the transition frequency of the transistor, which is determined by the IC process.

Data converters used in portable equipment should meet the requirement of low supply voltage and low power consumption. In this case, the important limitation, which is introduced on the dynamic range and speed, depends on the architecture and IC technology.

Bibliography

[1] B. M. Gordon, "Linear electronic analog/digital conversion architectures, theirs origins, parameters, limitations, and applications," *IEEE Trans. on Circuits and Systems*, vol. 25, no. 7, pp. 391–418, July 1978.

[2] Understanding data converters, Application notes, Texas Instruments, 1999.

[3] S. R. Norsworthy, R. Schreier, and G. C. Temes, Eds., *Delta-sigma data converters: Theory, design, and simulation*, New York, NY: Wiley-IEEE Press, 1996.

[4] J. R. Naylor, "Testing digital/analog and analog/digital converters," *IEEE Trans. on Circuits and Systems*, vol. 25, no. 7, pp. 526–538, July 1978.

10

Nyquist Digital-to-Analog Converters

CONTENTS

Digital-to-analog converters (DACs) enable the interfacing of digital systems with the real world. They are used to transform binary code into an analog output signal, either in the form of a voltage or current.

Ideally, signals from dc up to the Nyquist frequency, which is defined as the half of the DAC sampling frequency, can be generated by a converter. In practice, in addition to the increased difficulty to meet the specifications of the reconstruction filter, the converter performance tends to degrade significantly when approaching the Nyquist frequency. Nyquist DACs are based on various architectures (binary-weighted, thermometer-coded, and segmented structures) and conversion techniques such as voltage scaling, current scaling, charge scaling or redistribution, and hybrid methods. In each of these types of DACs, the conversion is achieved by summing all the output signal contributions associated with the different bits of the input digital code. The resolution of high-speed DACs based on each of the aforementioned techniques is generally limited by the circuit size and complexity. One approach adopted to reduce the number of elements (transistors, resistors, and capacitors) is to use a segmented DAC. In a segmented converter, a first stage decodes the most significant bit (MSB) part of the input digital code and a second stage is driven by the remaining least significant bits (LSBs). A further issue with high-resolution DACs is that the linearity and monotonicity of the conversion characteristic is limited by component matching. The effect of mismatches, which are generally caused by IC process gradients, can be alleviated either by adding a calibration stage to the DAC or by laser trimming components to adjust their values after wafer fabrication.

10.1 Digital-to-analog converter (DAC) architectures

The transformation of a digital code by an N-bit digital-to-analog (DAC) can rely on the use of switches to select the appropriate reference voltages, which are then summed to provide a discrete time signal. Generally, a lowpass filter is used to smooth the DAC output signal. The performance of the converter is related to the choice of the IC technology and switching scheme.

10.1.1 Binary-weighted structure

The simplest and area-efficient way to control the switches consists of using the digital input directly. However, the resulting structure has some drawbacks. The components, which determine the reference voltages, are required to be binary weighted. They should be matched to within $\pm 1/2$ LSB to meet the specifications of a high-precision DAC. This is equivalent to achieving a relative accuracy of less than $1/2^N$ LSB for the MSB. The statistical spread appears to be a limiting factor for the realization of this objective, and the monotonicity of the DAC is not guaranteed. Furthermore, the output data can exhibit a large spike or glitch at the mid-code transition, such as from 0111 to 1000 in a 4-bit example, due to the required asymmetrical switching. Hence, the state of all bits is modified. The LSBs all can switch faster than the MSB and the DAC output will first attempt to change toward zero before returning to the right state. A glitch can be characterized by measuring its energy, which is expressed in units of picovolt-seconds for high-performance DACs. A sample-and-hold circuit can be used to maintain the DAC output constant during the code transition, reducing in this way the glitch effect.

10.1.2 Thermometer-coded structure

An N-bit thermometer-coded DAC consists of 2^N unit elements, which are connected to switches, whose control signals are generated by a binary-to-thermometer decoder. The switching of only one element is needed when the input code changes by 1 LSB. That is, the relative accuracy to be realized is $1/2$, corresponding to a relaxed matching constraint. Furthermore, the glitch energy is considerably reduced. The monotonicity is guaranteed because the output remains constant or follows the variation of the input code. The primary inconvenience of the thermometer-coded DAC is the need for a large chip area.

10.1.3 Segmented architecture

A segmented DAC is based on an array of binary-weighted elements directly controlled by the L least significant bits and an array of unit elements steered by the remaining $(N - L)$ bits, which are thermometer encoded. It is then a compromise between both aforementioned structures and can combine the high accuracy and conversion rate.

Note that the glitch problem at the mid-code can also be solved by never turning off elements as the digital code is increased. This results in the interpolated architecture. However, the drawback in this case is related to the requirement of a complicated switching scheme.

10.2 Voltage-scaling DACs

In response to a digital input code, voltage-scaling DACs produce an output voltage by exploiting the voltage-divider principle. As simple in operation as a digitally controlled potentiometer, voltage-scaling DACs can easily be designed to meet the high-speed and low-power specifications without affecting their inherent monotonicity. While such DAC architectures are generally widely used for applications requiring a resolution not exceeding 8 bits, they are not suitable in cases where the resolution is high. This is due to the fact that the required number of resistors and switches increases exponentially with the resolution, making the resulting chip area very large.

10.2.1 Basic resistor-string DAC

The simplest architecture used to design a DAC based on the voltage divider principle consists of a series of resistors connected between two supply voltages, switches controlled by the digital input code, and a buffer to drive low impedance loads. Each digital code is converted by selecting a node between two resistors so that the appropriate voltage level is transferred to the converter output.

A 3-bit resistor-string DAC based on a binary-tree structure is shown in Figure 10.1. For an N-bit resolution, such a converter requires 2^N resistors and $2(2^N - 1)$ switches. Using the voltage divider rule, the voltage at node j of the resistor string can be expressed as

$$V_j = \frac{j-1}{2^N} V_{REF} \quad j = 1, 2, \cdots, 2^N , \tag{10.1}$$

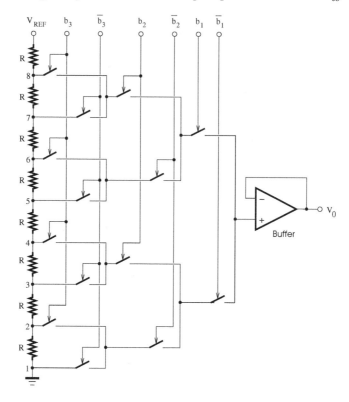

FIGURE 10.1
Block diagram of a 3-bit binary-decoded resistor-string DAC.

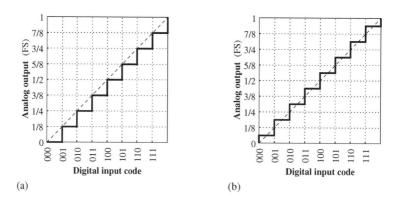

FIGURE 10.2
Transfer characteristics of a 3-bit DAC.

where V_{REF} is the reference voltage. The input and output characteristics of the unipolar DAC of Figure 10.1 are shown in Figure 10.2(a). Ideally, the DAC establishes a unique correspondence between an input digital code and a reference voltage. Because the reference levels are separated by an LSB or equivalently $V_{REF}/2^N$, the characteristic can be shifted upward by LSB/2, as shown in Figure 10.2(b), where the output level corresponding to the zero

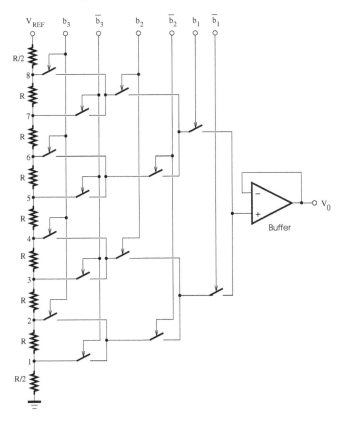

FIGURE 10.3
Block diagram of a 3-bit binary-decoded resistor-string DAC with code transitions at multiples of LSB/2.

digital code is now LSB/2. The resistor-string DAC based on the output characteristic of Figure 10.2(b) is depicted in Figure 10.3. In comparison with the previous structure, the overall number of resistors increases by one. Using the topmost and bottommost resistors with a value of R/2, the voltage at the node j of the resistor string is given by

$$V_j = \frac{j - 1/2}{2^N} V_{REF} \quad j = 1, 2, \cdots, 2^N. \quad (10.2)$$

In addition to its simple structure, the resistor-string DAC exhibits an inherent monotonicity [1]. For increasing input values, a DAC with a monotonic characteristic generates strictly increasing output values. The linearity of the resistor-string DAC then appears to be less affected by resistor mismatches. However, the resistor-string DAC architecture is limited by the number of resistors and switches, which grows exponentially as the resolution increases.

The number of switches can be reduced using a N-to-2^N decoder, as shown in Figure 10.4 for a resolution of 3 bits. Because only one switch is included in the signal path for a given digital code conversion, the effects of switch

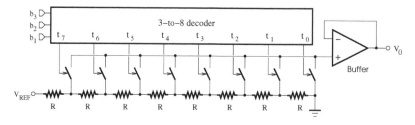

FIGURE 10.4
Block diagram of a 3-bit resistor-string DAC with a 3-to-8 decoder.

resistances and parasitic capacitance are reduced, yielding an improved speed performance.

TABLE 10.1
Truth Table and Logic Equations of the 3-to-8 Decoder

	Binary Code			1-out-of-8 Code								
	b_1	b_2	b_3	t_0	t_1	t_2	t_3	t_4	t_5	t_6	t_7	
0	0	0	0	1	0	0	0	0	0	0	0	$t_0 = \bar{b}_1 \cdot \bar{b}_2 \cdot \bar{b}_3$
1	0	0	1	0	1	0	0	0	0	0	0	$t_1 = \bar{b}_1 \cdot \bar{b}_2 \cdot b_3$
2	0	1	0	0	0	1	0	0	0	0	0	$t_2 = \bar{b}_1 \cdot b_2 \cdot \bar{b}_3$
3	0	1	1	0	0	0	1	0	0	0	0	$t_3 = \bar{b}_1 \cdot b_2 \cdot b_3$
4	1	0	0	0	0	0	0	1	0	0	0	$t_4 = b_1 \cdot \bar{b}_2 \cdot \bar{b}_3$
5	1	0	1	0	0	0	0	0	1	0	0	$t_5 = b_1 \cdot \bar{b}_2 \cdot b_3$
6	1	1	0	0	0	0	0	0	0	1	0	$t_6 = b_1 \cdot b_2 \cdot \bar{b}_3$
7	1	1	1	0	0	0	0	0	0	0	1	$t_7 = b_1 \cdot b_2 \cdot b_3$

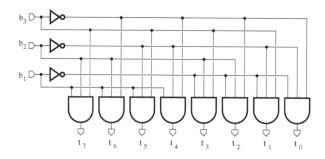

FIGURE 10.5
Circuit diagram of a 3-to-8 decoder.

The digital input code is translated into a format suitable for the switch control by the 3-to-8 decoder, which can be implemented using various architectures.

TABLE 10.2

Truth Table of the 3-Bit Binary-to-Thermometer Decoder and 1-out-of-8 Encoder

	Binary Code			Thermometer Code							1-out-of-8 Code							
	b_1	b_2	b_3	T_0	T_1	T_2	T_3	T_4	T_5	T_6	t_0	t_1	t_2	t_3	t_4	t_5	t_6	t_7
0	0	0	0	0	0	0	0	0	0	0	1	0	0	0	0	0	0	0
1	0	0	1	1	0	0	0	0	0	0	0	1	0	0	0	0	0	0
2	0	1	0	1	1	0	0	0	0	0	0	0	1	0	0	0	0	0
3	0	1	1	1	1	1	0	0	0	0	0	0	0	1	0	0	0	0
4	1	0	0	1	1	1	1	0	0	0	0	0	0	0	1	0	0	0
5	1	0	1	1	1	1	1	1	0	0	0	0	0	0	0	1	0	0
6	1	1	0	1	1	1	1	1	1	0	0	0	0	0	0	0	1	0
7	1	1	1	1	1	1	1	1	1	1	0	0	0	0	0	0	0	1

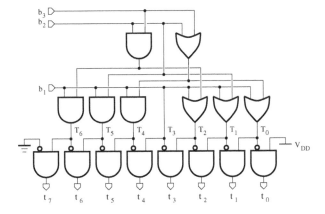

FIGURE 10.6

Circuit diagram of a modular 3-to-8 decoder.

The 3-to-8 decoder detects the occurrence of a specific bit combination in the input code, as illustrated by the truth table in Table 10.2. The decoding leads to the activation of only the output with the subscript that is the same as the decimal equivalent of the input code. Figure 10.5 shows the circuit diagram of a 3-to-8 decoder using inverters and AND gates. This implementation is directly related to the decoder logic equations and is simple. However, the maximum number of digital inputs that a single AND gate can support can be quickly reached as the input code resolution is increased.

For high resolutions, an implementation solution of the N-to-2^N decoder is to use an N-bit binary-to-thermometer decoder, followed by a 1-out-of-2^N encoder. This results in a structure that is modular and realizable with two-input gates. In the specific case of $N = 3$, let T_p, $p = 0, 1, 2 \cdots, 6$, and t_q,

$q = 0, 1, 2 \cdots , 7$, denote respectively the thermometer code and the 1-out-of-8 code for the binary code, b_k, $k = 1, 2, 3$. Table 10.2 shows the correspondence between the logic states of b_k and T_p, and all possible combinations of T_p and t_q. The logic equations derived from the truth table are given by

$$
\begin{aligned}
&T_0 = b_1 + b_2 + b_3 , \quad T_1 = b_1 + b_2 , \quad T_2 = b_1 + b_2 \cdot b_3 , \quad T_3 = b_1 , \\
&T_4 = b_1 \cdot b_2 + b_1 \cdot b_3 , \quad T_5 = b_1 \cdot b_2 , \quad T_6 = b_1 \cdot b_2 \cdot b_3 ,
\end{aligned}
\tag{10.3}
$$

and

$$
t_j =
\begin{cases}
\overline{T}_0 & \text{for} \quad j = 0 \\
T_{j-1} \cdot \overline{T}_j & \text{for} \quad j = 1, 2, \cdots , 6 \\
T_6 & \text{for} \quad j = 7.
\end{cases}
\tag{10.4}
$$

The circuit realization of the 3-to-8 decoder is shown in Figure 10.6. It should be noted that the required number of gates can be high as the converter resolution increases.

Design a 4-bit binary-to-thermometer decoder based on a 3-bit binary-to-thermometer decoder.

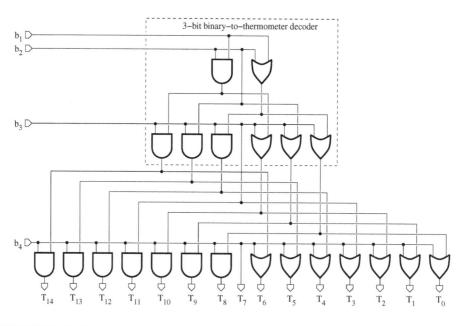

FIGURE 10.7

Circuit diagram of a 4-bit binary-to-thermometer decoder.

The circuit diagram of a 4-bit binary-to-thermometer decoder is shown in Figure 10.7. The encoding of a 4-bit binary code produces a thermometer code with j consecutive zeros followed by

$15 - j$ consecutive ones. The number of ones in the thermometer code is equal to the decimal equivalent of the binary input code. The logic equations of the 4-bit binary-to-thermometer decoder are given by

$$
\begin{aligned}
T_0 &= b_1 + b_2 + b_3 + b_4 \,, & T_1 &= b_2 + b_3 + b_4 \,, \\
T_2 &= b_1 \cdot b_2 + b_3 + b_4 \,, & T_3 &= b_3 + b_4 \,, \\
T_4 &= b_1 \cdot b_3 + b_2 \cdot b_3 + b_4 \,, & T_5 &= b_2 \cdot b_3 + b_4 \,, \\
T_6 &= b_1 \cdot b_2 \cdot b_3 + b_4 \,, & T_7 &= b_4 \,, \\
T_8 &= (b_1 + b_2 + b_3) \cdot b_4 \,, & T_9 &= (b_2 + b_3) \cdot b_4 \,, \\
T_{10} &= (b_1 \cdot b_2 + b_3) \cdot b_4 \,, & T_{11} &= b_3 \cdot b_4 \,, \\
T_{12} &= (b_1 \cdot b_3 + b_2 \cdot b_3) \cdot b_4 \,, & T_{13} &= b_2 \cdot b_3 \cdot b_4 \,, \\
T_{14} &= b_1 \cdot b_2 \cdot b_3 \cdot b_4 \,.
\end{aligned}
\tag{10.5}
$$

In general, the required number of AND and OR gates for an N-bit binary-to-thermometer decoder is equal to $2^N - (N+1)$.

Switches are used to connect the different reference voltages provided by the resistor string to the output. Fully decoded architectures have the advantage of providing a monotonic output, even if the resistances drift from their nominal values. The output resistance of the DAC is dependent on the digital code. This results in a code-dependent settling time when charging a capacitive output load. The switches must be appropriately sized or compensated to reduce the effect of this nonideality on the converter performance.

10.2.2 Intermeshed resistor-string DAC

The complexity of the digital decoder can be reduced by using a DAC with intermeshed resistor strings [2, 3]. The reference voltage is subdivided into 2^P coarse voltage segments using a P-bit coarse resistor string with a low impedance. Each coarse resistor is related in parallel to a Q-bit fine resistor string with a relatively high impedance. The conversion of an input digital code then consists of selecting the nodes of the fine resistor string connected to the appropriate coarse voltage segments. With an overall resolution of N bits, where $N = P + Q$, the resulting intermeshed resistor-string DAC required a P-to-2^P decoder and a Q-to-2^Q decoder. Because the integers P and Q are less than N, the complexity of the required decoders remains low in comparison with a single N-to-2^N decoder. Furthermore, by connecting in parallel each coarse resistor with a fine resistor string, the converter output impedance value and its variations with the input code are reduced, resulting in an improved settling time.

The circuit diagram of a 5-bit intermeshed resistor-string DAC is shown in Figure 10.8. The proper switches are selected by splitting the 5-bit input code in the three MSBs and the two LSBs, that are applied to a 3-to-8 decoder and

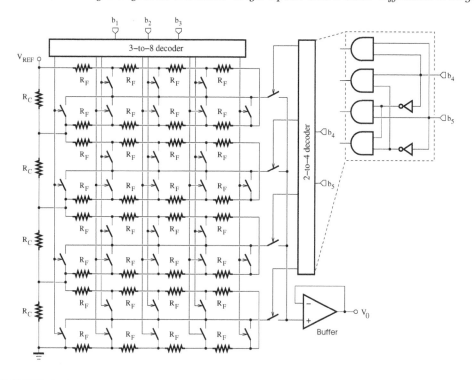

FIGURE 10.8

Circuit diagram of a 5-bit intermeshed resistor-string DAC.

a 2-to-4 decoder, respectively. Latches can be inserted between the decoders and switches, if necessary, to reduce glitch errors during a code transition. In general, an N-bit intermeshed resistor-string DAC requires 2^P resistors R_C in addition to 2^N resistors R_F. For resolutions greater than 8 bits, the overall number of resistors can be too high to be practical.

10.2.3 Two-stage resistor-string DAC

The number of passive elements in the DAC can be further reduced by adopting a design technique based on a segmented architecture [4], that consists of a cascade of P-bit and Q-bit resistor-string DACs through buffer amplifiers to realize a resolution of N bits, where $N = P + Q$. In a segmented DAC, the input code is partitioned so that the P MSBs are processed by the first stage of the converter, while the remaining Q LSBs are converted by the second stage. The switch control signals are then obtained using P-to-2^P and Q-to-2^Q decoders. Figure 10.9 shows a segmented converter based on two 3-bit resistor-string DACs. Buffer amplifiers are required to prevent the second-stage resistor string from loading the first-stage resistor string.

With the switch configuration defined by the P MSBs of a given input code, the first resistor string can generate two reference voltage levels of the

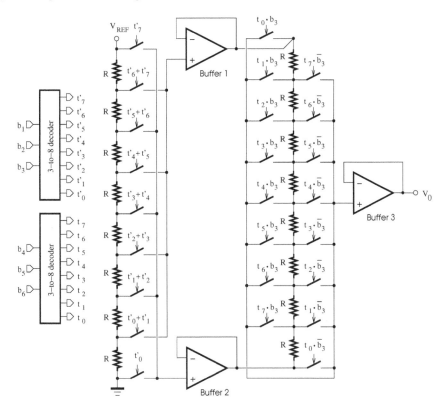

FIGURE 10.9

Circuit diagram of a 6-bit two-stage resistor-string DAC.

form

$$V_p = p\frac{V_{REF}}{2^P} \tag{10.6}$$

and

$$V_{p+1} = (p+1)\frac{V_{REF}}{2^P}, \tag{10.7}$$

where p is the decimal equivalent of the P MSBs and $p = 0, 1, 2, \cdots, 2^P - 1$. These voltages are applied at the inputs of the buffer amplifiers, whose outputs are connected to the top and bottom of the second resistor string. Using the superposition theorem, the converter output voltage generated in accordance with the remaining Q LSBs of the input code can be expressed as

$$V_0 = q\frac{V_{01}}{2^Q} + (2^Q - q)\frac{V_{02}}{2^Q} = V_{02} + q\frac{V_{01} - V_{02}}{2^Q}, \tag{10.8}$$

where V_{01} and V_{02} denote the output voltages of the first and second buffer amplifiers, respectively; q is the decimal equivalent of the Q LSBs; and $q = 0, 1, 2, \cdots, 2^Q - 1$. The output voltage is then generated by linearly interpolating between V_{01} and V_{02}. The step size by which the output voltage of the second resistor-string DAC can change has a value of $(V_{01} - V_{02})/2^Q$. If the buffer amplifiers are assumed to be ideal, $V_{01} = V_{p+1}$ and $V_{02} = V_p$, the voltage V_0 will be given by

$$V_0 = \left(p \cdot 2^Q + q \right) \frac{V_{REF}}{2^N}. \tag{10.9}$$

With k being the decimal representation of the input code, the converter output voltage is reduced to

$$V_0 = k \frac{V_{REF}}{2^N}, \tag{10.10}$$

where $k = p \cdot 2^Q + q$. For an N-bit unipolar DAC, the value of k can vary from 0 to $2^N - 1$.

A resistor-string DAC is inherently monotonic. However, the monotonicity in a segmented string DAC can be affected by the offset voltage of practical buffer amplifiers. By taking into account the offset voltages V_{off1} and V_{off2} of the first and second buffer amplifiers, we have $V_{01} = V_{p+1} + V_{off1}$ and $V_{02} = V_p + V_{off2}$. The converter output voltage is then given by

$$V_0 = k \frac{V_{REF}}{2^N} + V_{off}, \tag{10.11}$$

where

$$V_{off} = V_{off2} + q \frac{V_{off1} - V_{off2}}{2^Q}. \tag{10.12}$$

Assuming that the buffer offset voltages are identical, the output voltage for each input code is shifted by the same amount and the monotonicity of the converter characteristic is preserved. However, this last requirement is rarely met in practice.

When q reaches its highest value, $2^Q - 1$, the voltage V_0 becomes

$$V_0 = V_{02} + (2^Q - 1) \frac{V_{01} - V_{02}}{2^Q} = V_{01} - \frac{V_{01} - V_{02}}{2^Q}, \tag{10.13}$$

where $V_{01} = V_{p+1} + V_{off1}$ and $V_{02} = V_p + V_{off2}$. Hence, the output voltage is one step size below $V_{p+1} + V_{off1}$. The next increase of the converter output occurs when the input codes of the second and first resistor-string DACs are respectively reset and incremented by one.

With a switching scheme designed such that only the value of V_{02} is changed to $V_{p+2} + V_{off2}$ as a result of the digital code decoding, the digital-to-analog conversion can preserve the numerical order of the input codes. Because V_{p+1} and V_{p+2} are two adjacent voltage levels of the first resistor-string DAC, we have

$$V_{p+2} - V_{p+1} = V_{REF}/2^P. \tag{10.14}$$

In the worst case, $V_{off2} = -V_{off1}$, and a monotonic conversion is guaranteed provided the magnitude of the buffer offset voltage remains lower than

$V_{REF}/2^{P+1}$, or equivalently, half of the LSB of the first resistor-string DAC. The polarity of the voltage supplied to the second resistor string is reversed whenever the input code of the first resistor-string DAC is incremented by one. This polarity inversion should be compensated by appropriately changing the order used in the selection of the tap voltages of the second resistor-string DAC. The switches responsive to the Q-bit LSBs are then configured such that the tap voltages can be selected in either a top-down or a bottom-up fashion.

The circuit diagram of a 6-bit two-stage resistor-string DAC using the foregoing reversal switching scheme is depicted in Figure 10.9. The decoder responsive to the three LSBs selects one of the eight tap voltages of the second resistor string, starting with the node connected to either V_{01} or V_{02}, depending on the state of b_3.

The two-stage architecture can be used to implement DAC with a resolution on the order of 16 bits. The required number of resistors and switches is reduced to $2^P + 2^{N-P}$ for a resolution of N bits. However, the main disadvantages are the offset voltage and settling-time requirements placed on the design of buffer amplifiers to meet the monotonicity and conversion speed specifications.

In general, even with the use of improved IC process technology, resistor-string DACs can feature nonlinearity errors only in the 4 LSB range, and are then essentially used in closed-loop applications such as motor control or process control, where the INL and DNL specifications are undemanding.

10.3   Current-scaling DACs

Current-scaling DACs generally consist of a number of current sources that are selectively switched into a summing node in response to a digital input code. Various structures can be used in the implementation of a switchable current source, while the sum of currents is generally formed at the inverting node of an amplifier.

10.3.1   Binary-weighted resistor DAC

A binary-weighted resistor DAC is realized by summing the current contributions associated to the different bits of a digital input code. It then consists of weighted current sources, which can simply be a set of resistors with power-of-two values and connected in parallel to either the reference voltage or the ground, depending on each bit state. Figure 10.10 shows the circuit diagram of a binary-weighted resistor DAC. This architecture allows only unipolar

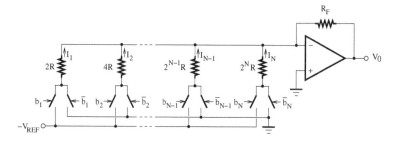

FIGURE 10.10
Circuit diagram of a binary-weighted resistor DAC.

digital-to-analog conversion. The output voltage can be computed as,

$$V_0 = -R_F(I_1 + I_2 + I_3 + \cdots + I_N), \tag{10.15}$$

where N represents the DAC resolution. Because various binary-weighted resistors are used to convert the reference voltage V_{REF} into the currents I_k, the voltage V_0 is found to be

$$V_0 = V_{REF}\frac{R_F}{R}\left(\frac{b_1}{2} + \frac{b_2}{2^2} + \frac{b_3}{2^3} + \cdots + \frac{b_N}{2^N}\right) = V_{REF}\frac{R_F}{R}\sum_{k=1}^{N}\frac{b_k}{2^k}. \tag{10.16}$$

The converter full scale, which is the difference between the output voltages for the highest and smallest input codes, is obtained as

$$FS = \frac{(2^N - 1)R_F}{2^N R}V_{REF}. \tag{10.17}$$

The resolution of the binary-weighted DAC depends on the precision achieved for the resistor values. The ratio between the resistor values associated with the MSB, b_1, and the LSB, b_N, is given by

$$\frac{R_1}{R_N} = \frac{2R}{2^N R} = \frac{1}{2^{N-1}}. \tag{10.18}$$

In the specific case of an 8-bit binary-weighted resistor DAC, the range of resistor values can vary by a factor of 128. The spread of resistances can then be very large for high resolutions, resulting in a difficulty to design resistors with accurate values using classical integrated-circuit fabrication methods. Furthermore, by increasing the DAC resolution, the current related to the MSB can be slightly smaller than the sum of all currents due to the other remaining bits, leading to a nonmonotonic conversion characteristic. In general, the binary-weighted resistor architecture is used with a resolution not exceeding 4 bits as a building block in the design of large systems.

10.3.2 R-2R ladder DAC

One of the most common DAC structures is based on the R-2R resistor ladder network, which is symmetric and uses only two values of resistors, thus greatly simplifying the matching requirements. The block diagram of DAC using the R-2R ladder network is shown in Figure 10.11. The switches are controlled by digital logic. A 2R resistor can be connected either to the amplifier inverting node or to the ground. In the first case, the corresponding bit, b_k, assumes the high state, while it is in the low state in the latter. Ideally, the load of

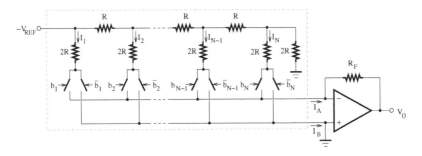

FIGURE 10.11
Block diagram of R-2R ladder DAC.

each 2R resistor has a resistance of 2R. That is, the currents can be obtained according to

$$I_1 = 2I_2 = 4I_3 = \cdots = 2^{N-1}I_N, \qquad (10.19)$$

where $I_1 = -V_{REF}/(2R)$. The DAC output is given by

$$V_0 = -R_F \sum_{k=1}^{N} b_k I_k \qquad (10.20)$$

$$= \frac{R_F V_{REF}}{R} \sum_{k=1}^{N} \frac{b_k}{2^k}, \qquad (10.21)$$

where $I_k = -V_{REF}/(2^k R)$ and b_k is either 1 or 0. The voltage V_0 is proportional to the switched-on bits. The R-2R DAC can exhibit a lower noise and nonlinearity errors of only ± 1 LSB. However, without involving some amount of resistor trimming to reduce the matching errors, the resolution of an R-2R DAC is limited to 12 bits. Furthermore, the glitch caused by switch timing differences has a more significant effect on R-2R structures than on resistor-string architectures. Hence, the R-2R DAC is less attractive for glitch-sensitive applications such as waveform generation.

10.3.3 Switched-current DAC

In general, switched-current DACs or current-steering DACs are preferred for applications requiring a resolution on the order of 10 bits and several megahertz of signal bandwidth. By driving a resistive load directly without the need for a voltage buffer, it can exhibit a very high power efficiency because almost all power is directed to the output load. Furthermore, the switched-current DAC has the advantage of featuring a low power consumption and requiring a small chip area for low to medium resolutions. Switched-current DACs can be designed using binary-weighted, or thermometer coded current array, as shown in Figure 10.12, where R_L is the load resistor. The number of current sources required to achieve N bits of resolution is, respectively, N and $2^N - 1$ for the binary-weighted and thermometer coded architectures because the converter state without any switched-current contribution represents the zero input code.

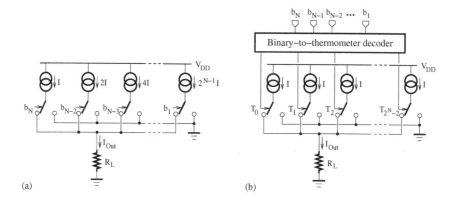

FIGURE 10.12
Block diagram of the (a) binary-weighted and (b) thermometer-coded switched-current DACs.

A binary-weighted DAC consists of current sources that are sized to have power-of-two values and are associated with switches directly controlled by the input code. Even though it has a simple structure, its operation can be limited by mismatches between the current sources. The worst-case error affecting the conversion transfer characteristic can be observed at mid-code transitions, or when the MSB current source is enabled, and all other current sources are disconnected from the output. As it is difficult to match the value of the MSB current source and the sum of remaining current sources to within 0.5 LSB, the DAC monotonicity cannot be guaranteed. Furthermore, glitches can arise in the DAC output as a result of the effects of asynchronous current switching and parasitic capacitive coupling.

For a thermometer coded DAC, $2^N - 1$ current sources of equal weight are required to achieve a resolution of N bits. Each unit current source is connected to a switch controlled by a signal generated by the binary-to-thermometer decoder. For each increase in the input digital code by one, only one additional current source is switched to the converter output, and the connection configuration of the other current sources is maintained unchanged. This greatly reduces the effect of glitches and the monotonicity of the DAC characteristic is guaranteed. However, the main drawback of the thermometer-coded DAC is the large chip area required as the resolution is increased.

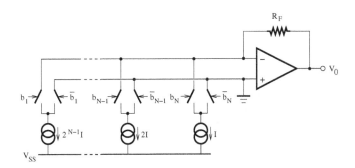

FIGURE 10.13
Block diagram of a binary-weighted switched-current DAC.

Switched-current DACs can also be designed to generate an output voltage, as shown in Figure 10.13, for a converter of the binary-weighted type. The DAC operation principle relies on summing the digitally selected outputs of a set of current sources. Each bit is assigned to a switch, which directs the current flow either to the common summing node or to the ground, depending on the bit state. The current direction adopted here allows the inverting amplifier of the DAC to generate a positive output voltage of the form

$$V_0 = R_F I (b_N + 2b_{N-1} + \cdots + 2^{N-1} b_1)$$

$$= 2^N R_F I \sum_{k=1}^{N} \frac{b_k}{2^k}, \tag{10.22}$$

where I denotes the LSB current and N is the converter resolution. Note that the operation of the switched-current DAC with an output current can be affected by the finite output impedance.

The circuit diagram of a binary-weighted switched-current DAC is depicted in Figure 10.14. Each current source is formed by superposing a power-of-two multiple of unit currents obtained by duplicating the current I. This is realized by a current mirror with the number of output transistors associated with a current source equal to the weight of the corresponding bit.

The major drawbacks of the thermometer-coded switched-current DAC are the complexity of the decoder and the high number of the required current sources. On the other hand, the binary-weighted switched-current DAC architecture is limited by its large element spread. A common practice is to divide a high-resolution DAC into subconverters with a small number of bits, and then either combine their output currents with a weighting network or scaling the weights of the current sources used in the subconverters [8]. The MSBs, which require a high level of accuracy, are thermometer coded and the LSBs

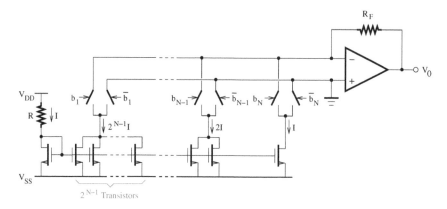

FIGURE 10.14
Circuit diagram of a binary-weighted switched-current DAC.

can either be implemented using the binary-weighted or thermometer-coded design technique.

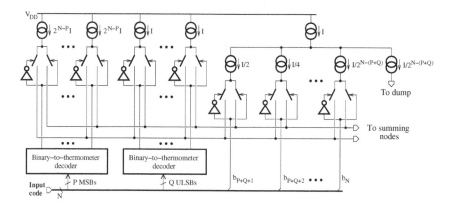

FIGURE 10.15
Block diagram of a segmented switched-current DAC.

The block diagram of a segmented switched-current DAC is shown in Figure 10.15, and its transistor implementation is depicted in Figure 10.16. Current sources with weight values determined according to the subconverter resolutions are used to ensure the monotonicity of the conversion characteristic. They are implemented using p-channel MOS transistors, which can allow an output swing from the ground level. The segmented DAC combines thermometer-decoded current cells, which are associated with the P most significant bits (MSBs) and the Q upper least significant bits (ULLSBs), and binary-decoded current sources, which determine the $N - (P + Q)$ lower least significant bits (LLSBs). With the LLSB transistor array being driven by an ULLSB transistor, an extra LLSB transistor connected to the ground is required to make the sum of the LSBs equal to 1 ULLSB.

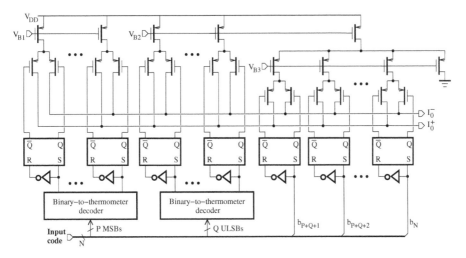

FIGURE 10.16
Transistor implementation of a segmented switched-current DAC.

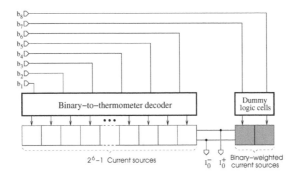

FIGURE 10.17
Block diagram of a $(6+2)$-bit segmented DAC.

Note that, by eliminating the intermediate ULSB segment, the segmented DAC can be reduced to an upper array of thermometer-decoded MSB current sources and an additional MSB current source loaded by a lower array of binary-decoded LSB current sources. In this way, the use of multiple reference currents, which can be difficult to match in practice to ensure the linearity between the DAC segments, is no longer necessary. Fig 10.17 shows a DAC architecture including a thermometer-encoded segment controlled by the 6 MSBs and a binary-weighted segment connected to the 2 LSBs.

With the number of components required in the thermometer-coded segment of the DAC remaining large for high resolutions, the current sources and switches can be arranged in a two-dimensional pattern to enable efficient routing of switch control signals.

In the structure of Figure 10.18 [9, 10], the output signal is generated by summing the contributions of the thermometer-coded and binary-weighted

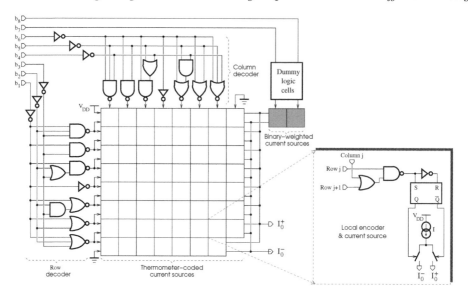

FIGURE 10.18

Block diagram of a $(2 \times 3 + 2)$-bit segmented DAC.

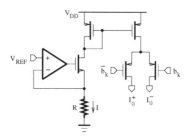

FIGURE 10.19

Circuit diagram of a single switched-current cell with a bias circuit.

DAC segments, which are respectively controlled by the 6 MSBs and 2 LSBs. The number of thermometer-coded current sources, which are sized to have a value of 4 LSBs and arranged in a matrix form, is 63 since no current source is enabled for the input code 0. The values of the two binary-weighted current sources are 2 LSBs and 1 LSB, respectively. The first 3 MSBs steer the column decoder, the next 3 bits are applied to the row decoder, and the remaining 2 bits are equalized to have equal delay with the MSBs and used to select the binary-weighted current sources.

By using a two-stage decoding process for the 6 MSBs, the decoder can be implemented with a minimum gate delay. The first stage is carried out by the row and column decoders, and the second stage is completed by a local decoder provided for each current cell. A given current source is enabled according to the output state of the local decoder in response to the selection signals generated by the row and column decoders. The output signal of each local decoder can be derived simply by combining two of the row selection signals *Row* j and *Row* $(j+1)$ and one of the column selection signals *Column* j $(j = 1, 2, \cdots, 7)$. This is due to the fact that, for the conversion of a given input code, the matrix of current sources is configured to exhibit, at most, three types of rows. The first rows in which all the current sources are enabled are followed in succession by a row in which only the first current sources are enabled, and the remaining last rows in which all the current sources are disabled. As the digital input code is increased, current cells are enabled along the rows.

The decoders are implemented with inverters, NAND and NOR gates. The skew between the signals of the row and column decoders is eliminated by the latch connected to the switching transistors of each current cell. Figure 10.19 shows the detailed circuit diagram of a switched-current source, which is necessary to build the DAC. The current source is implemented by applying the reference voltage, V_{REF}, to the noninverting input of an amplifier with a high input impedance, which helps maintain the voltage at the inverting input also equal to V_{REF}. The current flowing through the resistor R as a result of the conduction of the MOSFET included in the negative feedback of the amplifier is given by

$$I = V_{REF}/R. \tag{10.23}$$

The duplication or scaling of the current I can then be achieved using a current mirror.

The matching errors at the edge of the array can be eliminated by surrounding the active current cells with layers of dummy cells. Because the smallest difference between the values of the current sources is 1 LSB, a DNL error of 0.5 LSB can still be achieved with a relative current mismatch as large as $\pm 12.5\%$. The segmented DAC architecture has the advantage of reducing the linearity error caused by random errors in the current source array. However, the INL characteristic is especially sensitive to graded and symmetrical errors caused, respectively, by a voltage drop along the power supply lines and thermal distribution inside the DAC chip. In practice, for converters with resolutions greater than 8 bits, the linearity error due to nonuniform current sources can be reduced using improved switching schemes or calibration circuits [11, 12].

The performance of switched-current DACs can also be affected by the limited output impedance of current sources. A different number of current cells is connected to the output, depending on the digital input code. This results in a nonlinearity caused by the variation of the output impedance, which can be reduced by using current sources with a cascode configuration as shown in Figure 10.20, where each switch is implemented using a pMOS transistor with the gate connected to the inverted input code.

A cascode current source has the advantage of featuring an output impedance greater than the one of a single transistor. To provide the highest possible voltage swing at the output terminal, the current-source bias circuit

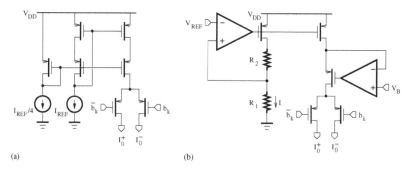

FIGURE 10.20
(a) Circuit diagram of a cascode current source; (b) circuit diagram of an active cascode current source.

is designed to maintain the transistors in the cascode configuration in the saturation region.

For the normal operation of the structure illustrated in Figure 10.20(a), two reference currents are required to establish the dc bias levels. The output current is approximately equal to the reference current I_{REF}, provided the sizes of transistors are identical.

In the current source of Figure 10.20(b), a pair of active cascode transistors is used to further improve the output impedance. The output of the operational amplifier having the noninverting input connected to the reference voltage, V_{REF}, and the inverting input connected to the voltage at the node between resistors R_1 and R_2, drives the transistor gate to set the delivered current at the value $I = V_{REF}/R_1$. The voltage divider including R_1 and R_2 helps maintain the transistor drain at the voltage $V_{REF}(1 + R_2/R_1)$. The duplication of the current I is realized by applying the amplifier output to the gate of other transistors. An additional amplifier uses the bias voltage V_B applied to its noninverting input to regulate the voltage at its inverting node, which is connected between the drain and source of transistor. In this way, the output impedance of the current source is enhanced by the gain of the amplifier in the negative feedback loop. For high values of the output current, the performance of the active gain enhancement technique may be limited by stability requirements, which can become critically dependent on the parasitic capacitances.

10.4 Charge-scaling DAC

A typical charge-scaling DAC exploits the principle of charge transfer between capacitors for the conversion of the digital input code. Charge-scaling DACs [5] that are capable of achieving a high linearity can be fabricated using CMOS technology.

The circuit diagram of a binary-weighted capacitor-array DAC is shown in Figure 10.21, where b_1 and b_N are the MSB and LSB, respectively. With the assumption that all capacitors are initially discharged, the converter operation requires a two-phase, nonoverlapping clock signal. The converter is based on the charge redistribution principle. The amplifier is assumed to have an offset voltage V_{off} and an infinite dc gain.

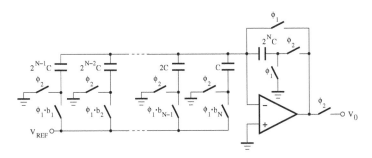

FIGURE 10.21
Circuit diagram of a binary-weighted capacitor-array DAC.

During the first clock phase, the amplifier operates as a unity-gain voltage follower. Each input capacitor $2^{N-k}C$ is connected between the amplifier inverting node and either the reference voltage or the ground, depending on whether the corresponding bit is high or low, while the capacitor $2^N C$ is connected between the amplifier inverting node and the ground. The charge stored on the input capacitors is of the form

$$\sum_{k=1}^{N} \left(2^{N-k}b_k CV_{REF} - 2^{N-k}CV_{off}\right) = CV_{REF}\sum_{k=1}^{N} b_k 2^{n-k} - (2^N - 1)CV_{off}$$

(10.24)

and the capacitor $2^N C$ is charged to V_{off}. Note that the capacitor array has a total capacitance of $(2^N - 1)C$. During the second clock phase, the input capacitors are connected between the ground and the amplifier inverting node and are then charged to V_{off}, while the voltage $V_{off} - V_0$ is applied across the capacitor $2^N C$, which is now included in the amplifier feedback path. Applying the charge conservation rule at the amplifier inverting node, we obtain

$$2^N C(V_{off} - V_0) - 2^N CV_{off}$$
$$= -(2^N - 1)CV_{off} - CV_{REF}\sum_{k=1}^{N} b_k 2^{N-k} + (2^N - 1)CV_{off}.$$

(10.25)

Hence, the resulting output voltage can be expressed as

$$V_0 = V_{REF}\sum_{k=1}^{N} \frac{b_k}{2^k},$$

(10.26)

where V_{REF} is the reference voltage. For resolutions greater than 8 bits, the capacitance spread, which increases exponentially with the number of bits, can become too high, thereby greatly limiting the achievable matching accuracy.

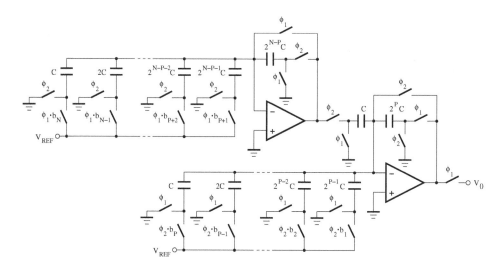

FIGURE 10.22
Circuit diagram of an N-bit cascaded binary-weighted capacitor-array DAC.

One technique to reduce the capacitance spread can consist of using an architecture with two conversion stages as shown in Figure 10.22. The capacitor array of the first conversion stage is responsive to the $(N - P)$ LSBs of the N-bit input code, while the P MSBs are applied to the capacitor array of the second conversion stage. The final result is produced by exploiting the inverting node of the amplifier used in the second conversion stage to combine the output of the first stage with the signal due to the input code MSBs. The output voltage, V_0, can then be computed as

$$V_0 = V_{REF} \sum_{k=1}^{P} \frac{b_k}{2^k} + \frac{1}{2^P} V_{REF} \sum_{k=P+1}^{N} \frac{b_k}{2^{k-P}} = V_{REF} \sum_{k=1}^{N} \frac{b_k}{2^k}. \quad (10.27)$$

Note that the reference signal sampling and the charge transfer phase take place during opposite clock phases in the first and second stage of the DAC. Using the above two-stage architecture, the capacitance spread for an N-bit DAC can be reduced from $1/2^N$ to $\max(1/2^P, 1/2^{N-P})$.

Because the charge is simply redistributed between the capacitors of a charge-scaling DAC, the power consumption is significantly reduced compared with other DAC structures. However, precise clock signals are indispensable for the control of the charge transfer; otherwise glitches may appear at the converter output. In addition, because of the charge slowly leaking from the capacitors over time, the accuracy of charge-scaling DACs starts to decrease within a few milliseconds after the beginning of the conversion. Charge-sharing DACs then appear to be unsuitable for general-purpose DAC applications, and are preferably used in successive-approximation ADCs, whose conversion cycle, by lasting only about a few microseconds, is generally ended well before the effect of the leakage current can become significant.

10.5   Hybrid DAC

An effective method to reduce the component spread and still achieve a high resolution is to use hybrid DAC architectures. By dividing the input code into two subwords to be processed by different sections of the DAC, it is possible to realize the conversion using a reduced number of components.

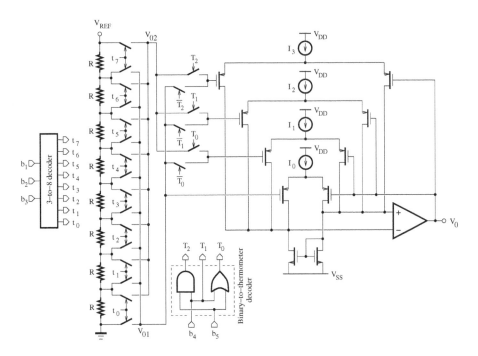

FIGURE 10.23
Circuit diagram of a 5-bit unipolar hybrid DAC including a resistor-string network and an interpolation amplifier stage ($P = 3, Q = 2$).

The unipolar hybrid DAC, as shown in Figure 10.23, includes a 3-bit resistor string and a 2-bit interpolation amplifier stage. This architecture has the advantages of improving the conversion speed and output settling characteristics [6].

With an N-bit input code assumed to be of the form $N = P + Q$, the resistor string is composed of 2^P resistors and the interpolation amplifier requires 2^Q differential transistor pairs, each having their source nodes connected to the corresponding current source. Two consecutive voltages, V_{01} and V_{02}, of the resistor string are selected according to the decoding of the P MSBs and connected to the interpolation amplifier by switches controlled by the Q LSBs. All the inputs of the interpolation amplifier, except the first one which is driven by V_{01}, can be switched to either V_{01} or V_{02}.

In general, the output voltage representing the DAC input code can be obtained as

$$V_0 = V_{01} + q \frac{V_{02} - V_{01}}{2^Q}, \tag{10.28}$$

where V_{02} is greater than V_{01} and q is the decimal equivalent of the Q LSBs.

Both the resistor string and the interpolation sections are inherently monotonic, and the monotonicity of the entire DAC is related to the operation principle consisting of adding the value interpolated from V_{01} and V_{02} to the voltage V_{01}.

Let us assume that all the inputs of the interpolation amplifier are initially connected to V_{01}. The negative feedback from the amplifier output forces V_0 to be equal to V_{01} and all the differential transistor pairs become balanced. Furthermore, all differential transistor pairs are actively loaded with the same current mirror, whose output current should be the same as the input current.

By incrementing the Q LSBs by one, one of the interpolation amplifier inputs is switched from V_{01} to V_{02}. The drain current flowing through the corresponding input transistor is then decreased by $\triangle I_D$. The negative feedback provided by the amplifier responds to this imbalance by inducing an increase in $\triangle V_0$ in the output voltage, where $\triangle V_0$ is equal to $(V_{02} - V_{01})/2^Q$. The drain current through each of the transistors with the gates connected to V_0, except the transistor forming a differential stage with the transistor connected to V_{02} and whose drain current is increased by $\triangle I_D$, is augmented by $\triangle I_D/3$ due to the variations in their gate-source voltages. Because the input and output currents of the current mirror are to be maintained equal, the drain current through each of the input transistors connected to V_{01} is increased by $\triangle I_D/3$.

In general, for a given value of the P MSBs, the interpolation controlled by the Q LSBs is realized to allow variation in the output voltage from V_{01} to $V_{01} + (2^Q - 1)(V_{02} - V_{01})/2^Q$ in a step of $(V_{02} - V_{01})/2^Q$.

Due to the transistor sizes required to determine the transconductances in the differential stages of the interpolation amplifier, the effect of parasitic capacitances becomes critical as the number P of LSBs exceeds 8. Taking into account the fact that the converter linearity is also limited by mismatches in the resistor string, the maximum resolution of a DAC including a resistor-string network followed by an interpolating amplifier stage should be on the order of 16 bits.

The aforementioned hybrid DAC uses an amplifier in the interpolation stage. To achieve a high resolution, the amplifier should have a large common-mode rejection ratio to maintain the accuracy over the entire input range and a low offset voltage. An alternative design solution can consist of combining a resistor-string network with a capacitor array DAC. The resulting DACs can exploit the operation principles of switched-capacitor circuits to reduce the effect of the amplifier offset voltage and to achieve a bipolar conversion with a single reference voltage. In addition, the operation of the required amplifier is not affected by common-mode input signals.

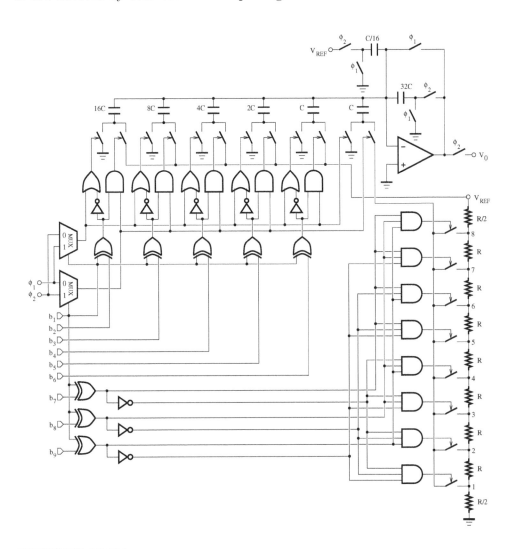

FIGURE 10.24
Circuit diagram of a bipolar hybrid DAC with a 3-bit resistor-string LSB network and a 5-bit binary-weighted capacitor MSB array.

The circuit diagram of a bipolar hybrid DAC is depicted in Figure 10.24. It is based on the voltage-scaling and charge-scaling principles [7]. This DAC structure features a resolution of 8 bits plus the sign bit and has the advantage of performing a bipolar conversion with only a single reference voltage. The input code is supposed to be in two's complement format, which is commonly found in DSP applications.

The absolute value of the two's complement representation equivalent to a negative value is obtained by first inverting all the bits, and then adding one to the result. For the realization of the first step, it is usual to perform the exclusive OR operation on the sign bit, b_1, and each of the remaining bits, $b_2 - b_9$. To implement the second step, the DAC is designed such that the output swings associated with positive and negative input codes can appear to be respectively shifted upward and downward by an offset voltage corresponding to $1/2$ LSB. The MSBs of the input code are applied to a 5-bit binary-weighted capacitor array, while the remaining bits are used to control a 3-bit resistor-string network with the topmost and bottommost resistors chosen to have the value of $R/2$ so that the aforementioned offset voltages can be produced. The operation of adding a value of $-1/2$ LSB to the output voltage of the DAC is required to set the converter output for the zero input code to 0 V.

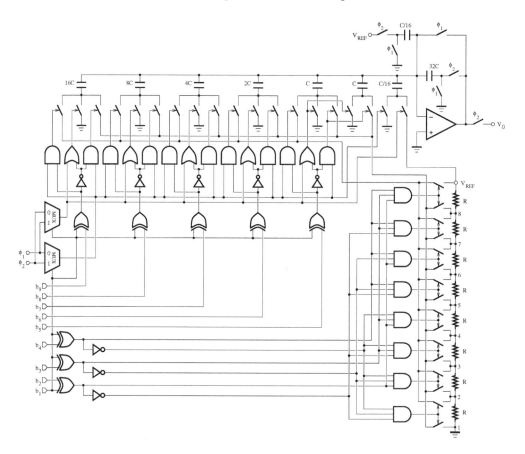

FIGURE 10.25
Circuit diagram of a bipolar hybrid DAC with a 3-bit resistor-string MSB network and a 5-bit binary-weighted capacitor LSB array.

The DAC operation requires two nonoverlapping clock phases. In the case of a positive input code, the multiplexer configuration defined by the state of the sign bit, b_1, which is a logic low, allows the reference voltage sampling and charge transfer to occur respectively during the clock phases ϕ_1 and ϕ_2, and the DAC operates as a noninverting gain stage. For a negative input code, the state of b_1 is a logic high, leading to the role interchange between ϕ_1 and ϕ_2 in the input branch, and the DAC is now equivalent to an inverting gain stage.

The output voltage of the DAC will be expressed as

$$V_0 = V_{REF} \left[\frac{b_2}{2} + \frac{b_3}{2^2} + \cdots + \frac{b_6}{2^5} + \frac{1}{2^5} \left(\frac{b_7}{2} + \frac{b_8}{2^2} + \frac{b_9}{2^3} \right) + \frac{1}{2} \frac{1}{2^8} \right] - \frac{1}{2} \frac{V_{REF}}{2^8}$$

(10.29)

if $b_1 = 0$, and

$$V_0 = -V_{REF} \left[\frac{\overline{b}_2}{2} + \frac{\overline{b}_3}{2^2} + \cdots + \frac{\overline{b}_6}{2^5} + \frac{1}{2^5} \left(\frac{\overline{b}_7}{2} + \frac{\overline{b}_8}{2^2} + \frac{\overline{b}_9}{2^3} \right) + \frac{1}{2} \frac{1}{2^8} \right] - \frac{1}{2} \frac{V_{REF}}{2^8}$$

(10.30)

if $b_1 = 1$. Hence,

$$V_0 = \begin{cases} \dfrac{V_{REF}}{2^8} \left(2^7 b_2 + 2^6 b_3 + \cdots + 2^1 b_8 + 2^0 b_9 \right) & \text{if } b_1 = 0, \\[2mm] -\dfrac{V_{REF}}{2^8} \left(2^7 \overline{b}_2 + 2^6 \overline{b}_3 + \cdots + 2^1 \overline{b}_8 + 2^0 \overline{b}_9 + 1 \right) & \text{if } b_1 = 1. \end{cases}$$

(10.31)

Note that the conversion is not affected by the dc offset voltage of the amplifier. The DAC accuracy is mainly determined by the achievable component matching.

Another approach for the hybrid-DAC design consists of using the MSBs to control a resistor-string network and the LSBs to drive a binary-weighted capacitor array. Figure 10.25 shows the circuit diagram of a bipolar hybrid DAC with a 3-bit resistor-string MSB network and a 5-bit binary-weighted capacitor LSB array. The input digital code is supposed to be in two's complement representation.

A set of node voltages is defined on the resistor string as the result of the reference voltage division by resistors with equal values. After the MSB decoding, two adjacent nodes of the resistor string are selected and connected to the binary-weighted capacitors, depending on the decoded state of the remaining LSBs. If the decoded state of an LSB is a high logic, the corresponding capacitor will be switched between the node with the highest voltage and ground; and if it is a low logic, the capacitor switching will take place between the lowest voltage and ground. In both cases, the initial position of the switches and the polarity of the output voltage are determined by the sign bit, b_1.

10.6 Configuring a unipolar DAC for the bipolar conversion

In general, DACs can be designed to operate in either a unipolar or bipolar mode. But, unipolar DACs can be associated with a differential amplifier stage to achieve conversions with a bipolar output-voltage range.

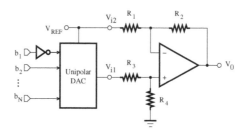

FIGURE 10.26
Circuit diagram of a unipolar DAC with a bipolar output-voltage range.

The circuit diagram of a unipolar DAC with a bipolar output-voltage range is shown in Figure 10.26. Here, the DAC uses an input code in the offset binary representation and an incoming two's complement code is converted to the offset binary format by inverting the MSB, b_1. Using the voltage divider principle, we have

$$V^+ = \frac{R_4}{R_3 + R_4} V_{i1} \tag{10.32}$$

and

$$V^- = \frac{R_2}{R_1 + R_2} V_{i2} + \frac{R_1}{R_1 + R_2} V_0, \tag{10.33}$$

where $V_{i1} = \sum_{k=1}^{N}(b_k/2^k)$ and $V_{i2} = V_{REF}$. For the usual case of an amplifier with a high dc gain, the relation $V^+ = V^-$ is exploited to express the output voltage in the form

$$V_0 = \frac{R_4}{R_3 + R_4}\left(1 + \frac{R_2}{R_1}\right)\sum_{k=1}^{N}\frac{b_k}{2^k} - \frac{R_2}{R_1}V_{REF}. \tag{10.34}$$

When $R_2/R_1 = R_4/R_3$, the DAC output stage operates as a differential amplifier, and

$$V_0 = \frac{R_2}{R_1}\left(\sum_{k=1}^{N}\frac{b_k}{2^k} - V_{REF}\right) \tag{10.35}$$

The DAC output voltage depends on the resistor ratios. However, due to mismatches in the resistor ratios, the effect of common-mode signals on the converter characteristics can become critical. Furthermore, the amplifier should be designed to exhibit a low offset voltage.

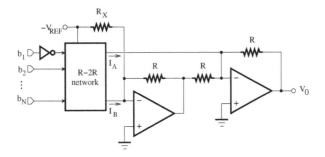

FIGURE 10.27
Circuit diagram of a bipolar R-2R ladder DAC.

An alternative design solution consists of using a differential stage with more than one amplifier to improve the common mode rejection. In the special case of DACs based on R-2R resistor network, a converter with a bipolar output range can be realized as shown in Figure 10.27. From the analysis of an R-2R network, the currents I_A and I_B can be expressed as

$$I_A = -\frac{V_{REF}}{2R} \sum_{k=1}^{N} \frac{b_k}{2^k} \qquad (10.36)$$

and

$$I_B = -\frac{V_{REF}}{2R} \sum_{k=1}^{N} \frac{\overline{b_k}}{2^k}. \qquad (10.37)$$

The output voltage of the DAC is given by

$$V_0 = -R(I_A - I_B) - \frac{R}{R_X} V_{REF}. \qquad (10.38)$$

Exploiting the sum formula for geometric series[1], it can be deduced that

$$I_A + I_B = -\frac{V_{REF}}{2R} \sum_{k=1}^{N} \frac{1}{2^k} = -\frac{V_{REF}}{R} \left(1 - \frac{1}{2^N}\right) \qquad (10.39)$$

and

$$I_A - I_B = 2I_A - (I_A + I_B) = -\frac{V_{REF}}{R} \left(-1 + \sum_{k=1}^{N} \frac{b_k}{2^k}\right) - \frac{V_{REF}}{R} \frac{1}{2^N}. \quad (10.40)$$

---

[1]For the geometric series with the common ratio r, the sum S is given by

$$S = \sum_{k=0}^{n} r^k = 1 + r + r^2 + \cdots + r^n = \frac{1 - r^{n+1}}{1 - r}$$

and if the sum is taken starting at $k = 1$, we have

$$\sum_{k=1}^{n} r^k = \frac{r(1 - r^n)}{1 - r}.$$

This last expression without the term $V_{REF}/(2^N R)$ is proportional to the offset binary representation of the DAC input code. By choosing the value of the resistor R_X such that

$$R_X = 2^N R, \qquad (10.41)$$

the DAC output voltage is reduced to

$$V_0 = V_{REF} \left(-1 + \sum_{k=1}^{N} \frac{b_k}{2^k} \right). \qquad (10.42)$$

The performance of this DAC architecture depends on the achievable matching between the resistors.

When an ac source is used as the reference voltage, it is amplified by a factor determined by the digital input code. In this case, the converter is then referred to as a multiplying DAC. Two-quadrant or four-quadrant multiplications are performed, depending on whether the multiplying DAC is designed for unipolar or bipolar conversion.

10.7 Algorithmic DAC

The algorithmic DAC operates according to a selection tree structure. It can be efficiently implemented with less power dissipation and fewer circuit components than other conversion architectures for a given resolution and bandwidth [13].

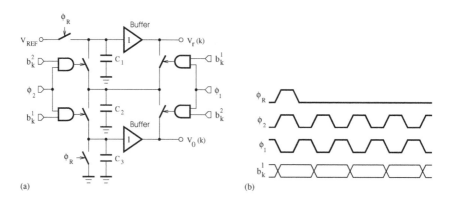

FIGURE 10.28
(a) Circuit diagram and (b) clock timing of an algorithmic DAC.

Starting from the MSB, a reference voltage, V_{REF}, is added or not to the previous output, depending on the state (low or high) of the present bit, b_k^l

$(l = 1, 2)$. During the k-th cycle, the output voltage, V_0, is computed as

$$V_0(k) = V_0(k-1) + b_k^l V_r(k), \tag{10.43}$$

where

$$V_0(0) = 0 \tag{10.44}$$

$$V_r(k) = \begin{cases} V_{REF}, & \text{if } k = 0, \\ \dfrac{V_r(k-1) - V_0(k-1)}{2}, & \text{otherwise.} \end{cases} \tag{10.45}$$

At the end of N cycles, the analog output is obtained as the sum of the voltages associated to each bit of the digital input code. An implementation of the algorithmic DAC with its clock timing is shown in Figure 10.28. The converter consists of three identical capacitors C_1, C_2, and C_3 and two unity buffers. Initially, the pulse signal, ϕ_R, is used to allow the capacitors C_1 and C_3 to be charged by the initial voltages $V_r(0) = V_{REF}$ and $V_0(0) = 0$. The charge transfer is controlled by two nonoverlapping clock signals, ϕ_1 and ϕ_2, and the state of the bits, b_k^l. For the high state of the bits, $V_r(k-1)$ is held by C_1 and $V_0(k-1)$ is updated because the charge stored on C_2 during the clock phase ϕ_1 is redistributed between C_2 and C_3 during the clock phase ϕ_2. In the case of low-state bits, the charge produced by $V_0(k-1)$ on C_3 remains unchanged and the update of $V_r(k-1)$ is achieved as the result of the charge sharing between C_2, which was first connected to $V_0(k-1)$, and C_1.

However, it should be mentioned that component nonidealities (mismatch, charge injection, clock feed-through) limit the resolution of the algorithmic DAC to no more than 10 bits.

10.8 Summary

Depending on the trade-offs between the power consumption, resolution, conversion speed, and latency, Nyquist rate DACs can be designed using parallel, pipeline, or serial architecture.

Parallel DAC structures can require a high power consumption to meet the speed and settling requirements of the amplifier, which is used to perform the charge transfer or current summing. Pipeline DACs can help reduce the power consumption and the spread of component values. However, this is achieved at the cost of an increase in latency time. Serial DAC structures often require a low circuit area, but they can be very slow. Hybrid or segmented DACs can be adopted to reduce the spread of component values, and thereby the effect of the amplifier loading on the converter performance.

10.9   Circuit design assessment

1.   Analysis of R-2R DACs

A multiplying DAC can be implemented as shown in Figure 10.29.

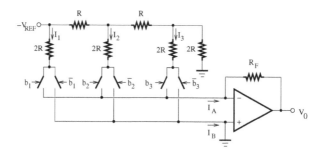

FIGURE 10.29
Multiplying DAC based on current-mode R-2R network.

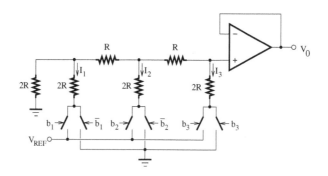

FIGURE 10.30
DAC based on voltage-mode R-2R network.

The R-2R network generates a current that is proportional to the input digital code and is converted by the feedback structure consisting of the amplifier and resistor R_F to the required voltage level.

Find the output voltage V_0 as a function of the input digital code.

Show that the output resistance of the R-2R network is given by

$$R_0 = \frac{3R}{(3/2)b_1 + (9/2^3)b_2 + (33/2^5)b_3 + (3/2^3)b_2b_3}. \qquad (10.46)$$

To reduce the effects of nonlinear errors on the converter performance, the DAC can be realized using the voltage-mode R-2R network, as illustrated in Figure 10.29.

Verify that the output resistance of the R-2R network is constant and equal to R.

Determine the output voltage V_0.

2. Implementation of the R-2R DAC using MOS transistors

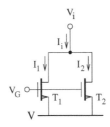

FIGURE 10.31
Current division principle based on two transistors.

The two-transistor circuit of Figure 10.31 can be used to divide an input current I_i into two components, I_1 and I_2. Assuming that V and V_G are chosen so that the transistors remain in the on-state, show that

$$\frac{I_1}{I_2} = \frac{W_1/L_1}{W_2/L_2}, \tag{10.47}$$

where W_i and L_i $(i = 1, 2)$ are the width and length of the corresponding transistors, respectively.

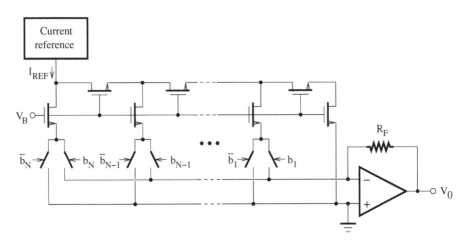

FIGURE 10.32
Transistor implementation of a R-2R DAC.

Using the current division principle [14], it is possible to implement an R-2R ladder using MOS transistors as shown in Figure 10.32. To improve the linearity, the transistors are assumed to operate in the triode region. The switches are realized using transistors driven either by b_k or $\bar{b}_k$.

Use SPICE simulations to verify that the small-signal equivalent resistance seen between the drain and source terminals of the MOS-FETs is not identical throughout the transistor network.

Ideally, each stage of the transistor network splits its input current into two equal parts, that is,

$$I_{0,k} = I_{i,k}/2 = b_k 2^{-k} I_{REF} . \tag{10.48}$$

Due to transistor mismatches, the current division is affected by an error of the form

$$\Delta I_{0,k} = \epsilon_k I_{REF} . \tag{10.49}$$

Assuming that the requirements INL < 0.5 LSB and DNL < 1 LSB are met, provided that $\max(\sum_k |b_k \epsilon_k|) < 1/2^{N+1}$, determine the worst-case error for a converter resolution, $N = 8$ bits.

3. **Segmented DAC**

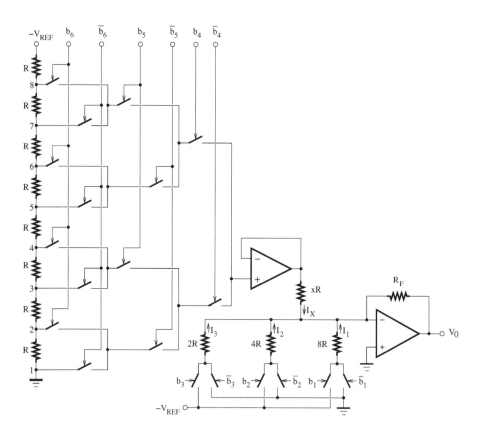

FIGURE 10.33
Segmented DAC.

Consider the segmented DAC depicted in Figure 10.33, which is based on resistor string and binary weighted resistors.

Assuming ideal amplifiers, determine x to realize a 6-bit segmented DAC.

Express the output voltage, V_0, as a function of the input digital code, b_k, $k = 1, 2, 3, 4, 5, 6$.

Put the output voltage into the form

$$V_0 = V_{FS} \sum_{k=1}^{6} \frac{b_k}{2^k},$$

where V_{FS} is the full-scale output voltage to be determined.

Assuming that the DAC is designed to feature a slew rate, $SR = 0.75$ V/μs, use the equation, $f_{max} = 1/t_{max}$, where $t_{max} = \triangle V_0/SR$ and $\triangle V_0 = V_{FS}$, to estimate the maximum data rate, f_{max}, for $V_{FS} = 5$ V.

4. **Binary-weighted charge-scaling DAC**

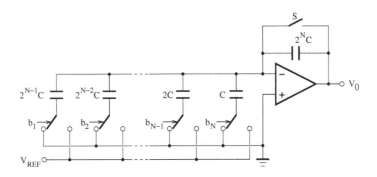

FIGURE 10.34
Binary-weighted charge-scaling DAC.

The circuit diagram of a N-bit binary-weighted charge-scaling DAC is shown in Figure 10.34. A digital-to-analog conversion starts with the discharge of all capacitors, and each switched capacitor is then connected either to the ground or the reference voltage, depending on the state of its corresponding bit.

With the assumption that the operational amplifier is ideal, determine the output voltage, V_0, as a function of the input digital code, b_k, $k = 1, 2, 3, \cdots, N - 1$.

In practical DAC implementations, the capacitors exhibit a variation of $\pm \triangle C$. Compute the worst-case integral nonlinearity, INL, and differential nonlinearity, DNL, using the formulas [15]

$$INL(k) = V_0(code\ k)|_{C \pm \triangle C} - V_0(code\ k)|_C$$

$$INL = INL(1)$$

and

$$DNL = \frac{V_0(1000\cdots0)|_{C+\triangle C} - V_0(0111\cdots1)|_{C-\triangle C}}{V_{LSB}} - 1,$$

where $V_{LSB} = V_{REF}/2^N$ and, in the case of the DNL, it is assumed that the MSB capacitor takes its maximum value, while the remaining capacitors exhibit their minimum values.

Determine the number of bits, N, if the capacitor tolerance, $\triangle C/C$, is equal to $\pm 0.5\%$ and $INL = \pm 0.5$ LSB.

Bibliography

[1] A. R. Hamadé, "A single chip all-MOS 8 bit A/D converter," *IEEE J. of Solid-State Circuits*, vol. 13, no. 6, pp. 785–791, Dec. 1978.

[2] A. Abrial, J. Bouvier, J.-M. Fournier, P. Senn, and M. Veillard, "A 27-MHz digital-to-analog video processor," *IEEE J. of Solid-State Circuits*, vol. 23, no. 6, pp. 1358–1369, Dec. 1988.

[3] M. J. M. Pelgrom, "A 10-b 50-MHz CMOS D/A converter with 75-Ω buffer," *IEEE J. of Solid-State Circuits*, vol. 25, no. 6, pp. 1347–1352, Dec. 1990.

[4] P. Holloway, "A Trimless 16b digital potentiometer," *1984 IEEE ISSCC Digest of Technical Papers*, p. 66–67, 320–321, Feb. 1984.

[5] R. Gregorian and G. Amir, "A single chip speech synthesizer using a switched-capacitor multiplier," *IEEE J. of Solid-State Circuits*, vol. 18, no. 1, pp. 65–75, Feb. 1983.

[6] A. Yilmaz, "LSB interpolation circuit and method for segmented digital-to-analog converter," U.S. Patent 6,246,351, filed October 7, 1999; issued June 12, 2001.

[7] M. Kokubo, S. Nishita, and K. Yamakido, "Interpolative D/A converter," U.S. Patent 4,652,858, filed April 16, 1986; issued March 24, 1987.

[8] A. R. Bugeja, B.-S. Song, P. L. Rakers, and S. F. Gillig, "A 14-b, 100-MS/s CMOS DAC designed for spectral performance," *IEEE J. of Solid-State Circuits*, vol. 34, pp. 1719–1732, Dec. 1999.

[9] T. Miki, Y. Nakamura, M. Nakaya, S. Asai, Y. Akasaka, and Y. Horiba, "An 80-MHz 8-bit CMOS D/A converter," *IEEE J. of Solid-State Circuits*, vol. 21, pp. 983–988, Dec. 1986.

[10] J. M. Fournier and P. Senn, "A 130-MHz 8-b CMOS video DAC for HDTV applications," *IEEE J. of Solid-State Circuits*, vol. 26, pp. 1073–1077, July 1991.

[11] Y. Nakamura, T. Miki, A. Maeda, H. Kondoh, and N. Yazawa, "A 10-b 70-MS/s CMOS D/A converter," *IEEE J. of Solid-State Circuits*, vol. 26, pp. 637–642, April 1991.

[12] G. A. M. Van der Plas, J. Vandenbussche, W. Sansen, M. S. J. Steyaert, and G. G. E. Gielen, "A 14-bit intrinsic accuracy random walk CMOS DAC," *IEEE J. of Solid-State Circuits*, vol. 34, pp. 1708–1718, Dec. 1999.

[13] K. Watanabe, G. C. Temes, and T. Tagami, "A new algorithm for cyclic and pipeline data conversion," *IEEE Trans. on Circuits and Systems*, vol. 37, no. 2, pp. 249–252, Feb. 1990.

[14] C. M. Hammerschmied and Q. Huang, "Design and implementation of an untrimmed MOSFET-only 10-bit A/D converter with −79-dB THD," *IEEE J. of Solid-State Circuits*, vol. 33, pp. 1148–1157, Aug. 1998.

[15] P. E. Allen and D. R. Holberg, *CMOS analog circuit design*, 2nd ed., New York, NY: Oxford University Press, 2002.

11

Nyquist Analog-to-Digital Converters

CONTENTS

In general, analog-to-digital converters (ADCs) are required for any application where an analog or continuous-time signal must be processed by digital systems. A variety of architectures is available for the design of ADCs with different characteristics, such as resolution, bandwidth, sampling frequency, power consumption, latency, and chip area.

For the special case of Nyquist ADCs, the sampling frequency can be at least two times the maximum frequency of the input signal. This converter group includes, but is not limited to, successive approximation register (SAR) ADC, integrating ADC, flash ADC, pipelined ADCs, and cyclic ADC. All techniques for analog-to-digital conversion rely on at least one operation of comparison between the input signal and a reference level. The flash ADC, which uses one comparator for each comparison, exhibits a higher speed than the SAR ADC, whose operation involves various comparisons realized by the same comparator. Because of the parallelism of the flash ADC, the number of comparators grows exponentially with the resolution, leading to an increase in the power consumption and chip area, and a decrease in the bandwidth as a result of an augmentation in the input capacitance. Some variations of flash architecture, such as the folding and interpolating ADC, and pipelined ADC, have been proposed in order to reduce the effect of some of these limitations.

In general, the trade-offs between the converter characteristics (speed, resolution, power consumption, size, linearity) play an important role in determining the better ADC architecture for a given application. For instance, a significantly faster conversion is usually achieved at the price of a reduction in the initial resolution. Due to the component matching requirement, the resolution of flash ADC is limited to 9 bits. SAR ADCs most commonly exhibit a resolution in the range from 8 to 16 bits and provide a low power consumption as well as a small IC size. They are ideal for many real-time applications. On the other hand, resolutions up to 18 bits can be achieved by integrating ADCs, which can work well also with low-level signals. However, this architecture generally has a low conversion speed and is only suitable for applications such as portable instruments, where the signal bandwidth remains low.

ADC architectures offer different compromises between the performance metrics. They are then chosen to be consistent with the specifications of the target application.

11.1 Analog-to-digital converter (ADC) architectures

Generally, trade-offs between resolution, power consumption, chip area size, conversion time, static performance, and dynamic performance play an important role in the choice of the proper ADC architecture for a given application (data acquisition, measurement, voiceband audio, image and video, wireless communication systems).

11.1.1 Successive approximation register ADC

• The block diagram of a successive approximation register (SAR) ADC is shown in Figure 11.1. It consists of a digital control logic, a clock generator, a comparator, a successive approximation register (SAR), a digital-to-analog converter (DAC), and an output buffer based on latches. The operations of resetting the SAR, enabling the clock signal, cycling each bit, and halting the clock signal at the end are conducted by the control logic.

For positive input signals, the operation principle of an SAR ADC can be based on the algorithm illustrated in Table 11.1. The SAR ADC exploits the concept of the binary search algorithm to find the nearest digital code to an input voltage.

The principle is to compare the analog version of various DAC digital codes with the input analog signal. Starting with the DAC most-significant bit (MSB), each bit is initially set to the logic high state. If the signal to be converted is higher than the one generated by the DAC and which actually represents one-half of the full scale (FS), the initial value of the MSB is maintained and the comparison procedure continues with the next DAC output signal of (3/4) FS. Otherwise, the MSB is reset to the logic low state. In this case, the output signal of the DAC required for the next comparison step corresponds to (1/4) FS. This successive conversion (see Figure 11.2) continues until the least-significant bit (LSB) is reached, that is, the DAC output is

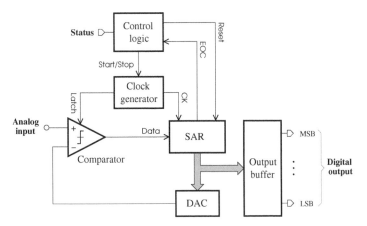

FIGURE 11.1
Block diagram of an SAR ADC.

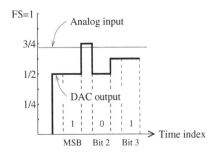

FIGURE 11.2
Three-bit successive approximation of the input signal.

within $\pm(1/2)$ LSB of the input voltage. For an N-bit ADC with 2^N levels of resolution, the conversion is carried out in N clock periods.

The SAR of Figure 11.3 is built with two sets of N-bit registers (control and data registers), the function of which is to store and guess the conversion result, respectively. Each flip-flop in the data register is sequentially set by the control register to a state such that on the next rising edge of the clock pulse, the current value of the input data is transferred to the output. This approach has the advantage of simplifying the layout design, which consists of reproducing each bit cell containing two D flip-flops. However, this approach can require a large chip area (19 D flip-flops for an 8-bit SAR). Furthermore, while the flip-flops of the control register use the same clock signal, the clock input of a given flip-flop of the data register is obtained from the next flip-flop output. As a result, the SAR output can be affected by a three flip-flop propagation delay in the worst case. If this delay is comparable to the DAC settling time or the comparator response time, its effect can become significant [1, 2].

The design of Figure 11.4 makes use of a single register with $N+1$ JK flip-

TABLE 11.1
The Computation Scheme of an SAR ADC

Begin

1. Initialization:
Acquire a sample of V_i
Specify the initial time index, $k \leftarrow 1$
Set the most significant bit, b_1, to the logic high and clear the remaining bits
Assign $V_r^1 = V_{FS}/2$

2. Repeat

(a) Compare V_i with V_r^k
    If $V_i \geq V_r^k$ then
        return the actual logic state of the bit under test
    else
        invert the logic state of the bit under test
    End If
(b) Adjust the time index, $k \leftarrow k + 1$
(c) Set the bit b_k to the logic high
(d) Update V_r^k
    If $V_i \geq V_r^{k-1}$ then
    $$V_r^k \leftarrow V_r^{k-1} + \frac{V_{FS}}{2^k}$$
    else
    $$V_r^k \leftarrow V_r^{k-1} - \frac{V_{FS}}{2^k}$$
    End If

until the DAC output is within $\pm(1/2)$ LSB of the input voltage, V_i

End

flops for the generation of the digital code and the end of conversion signal. The k-th cell consists of a JK flip-flop and two OR gates. The input terminals are labeled **Reset** and **Data**, and the N-bit code and **End** represent the output variables. The JK flip-flops are cleared by **Reset** = 1, and then **Reset** = 0 is maintained. A 1 at the **Data** input shows that the analog version of the SAR code provided by the DAC is greater than the input signal and the current approximation must be reduced.

A given flip-flop is initialized to 0 and held in this state as long as $X_k = 1$. On the other hand, its behavior will be described by referring to the truth table of a JK flip-flop if $X_k = 0$ (see Appendix A). The output of the flip-flop

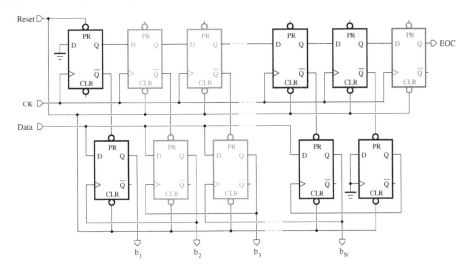

FIGURE 11.3

Circuit diagram of an SAR using two sets of registers (b_1 is the MSB and b_N is the LSB).

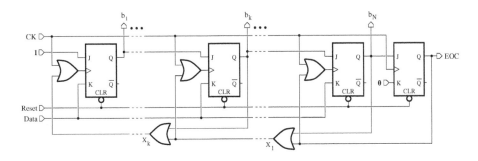

FIGURE 11.4

Circuit diagram of an SAR with a reduced number of registers.

is changed to 1 for $J = 1$. This state is hold during the next clock pulse if **Data** = 0, otherwise the current output bit is modified to 0. As the next bit is set to 1, X_k will take the value 1 and the flip-flop output state will be maintained. At this step, a further change can only be initiated by the **Reset** signal.

Because capacitors are easily fabricated in CMOS technologies, the charge redistribution technique is generally adopted in the SAR ADC implementation [3]. Figure 11.5 shows the block diagram of an SAR ADC exploiting the charge redistribution on weighted capacitors. Note that the extra LSB capacitor, C', of value C is required to make the total value of the capacitor array equal to $2^N C$, where N is the number of bits, so that a binary voltage division can be performed by switching any weighted capacitor of the DAC. The timing diagram of the SAR ADC is illustrated in Figure 11.6. A start-of-conversion (SOC) signal is used to initiate the conversion process, while the

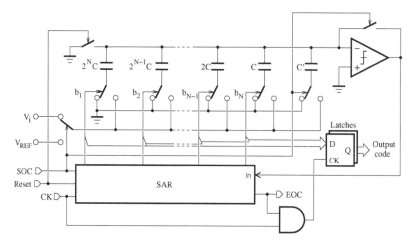

FIGURE 11.5
Block diagram of a charge-redistribution unipolar SAR ADC ($C' = C$).

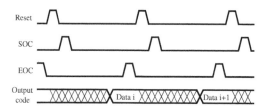

FIGURE 11.6
Timing diagram of the SAR ADC.

end-of-conversion (EOC) signal is asserted by the SAR to indicate the conversion completion. Before each conversion cycle, the **Reset** signal is enabled to initiate the discharge of capacitors through the ground connection and the reset of all the bits in the SAR to the logic low. Note that a clock signal is generally necessary for proper operation of the SAR ADC, even if it does not have to be synchronized with other control signals. Its frequency depends on the conversion resolution and speed. The conversion process consists of a sequence of three operations: the sample phase, the hold phase, and the redistribution phase (or bit testing mode).

In the *sample phase*, all capacitors are connected to the input voltage, V_i, and the comparator feedback switch is closed. The voltage V_C across the capacitor array at the end of the sampling period is actually

$$V_C = V_i - V_{off},\qquad(11.1)$$

where V_i is assumed to be positive, and the offset voltage, V_{off}, plays the role of the comparator threshold voltage.

During the *hold phase*, the comparator feedback switch is open and the bottom plates of capacitors are connected to ground. Because the charge on

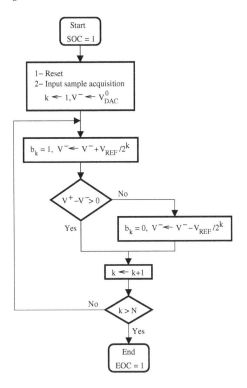

FIGURE 11.7
Binary search algorithm of the SAR ADC.

the top plate is conserved, the top plate potential goes to V_C and the voltage applied to the negative terminal of the comparator can be expressed as

$$V^- = -V_C. \tag{11.2}$$

The input node is now connected to the reference voltage, V_{REF}, instead of the input voltage.

The *redistribution phase* begins with the determination of b_1 or MSB. The largest capacitor is then connected to the reference voltage and the equivalent circuit of the capacitor array seen from the input node is a voltage divider consisting of two equal capacitors $2^N C$. Hence, the voltages at the negative and positive terminals of the comparator are given by

$$V^- = V_{REF}/2 - V_C = -V_i + V_{off} + V_{REF}/2, \tag{11.3}$$
$$V^+ = V_{off}. \tag{11.4}$$

The logic state of the MSB depends on the sign of the voltage difference, $V^+ - V^-$, provided by the comparator. It will remain unchanged, that is, $b_1 = 1$, if $V^+ - V^- > 0$, or equivalently $V_i > V_{REF}/2$. But, if $V^+ - V^- < 0$, then $V_i < V_{REF}/2$, and the MSB will be set back to the logic low, $b_1 = 0$, by switching back the largest capacitor to the ground. The determination of the bit from b_2 to b_N also proceeds in the same manner. The bottom plate of the capacitor associated with the bit b_k is connected to the reference voltage, and the previous value of V^- is increased by $V_{REF}/2^k$ as a result of the voltage division carried out by the capacitor array. The final logic state is assigned to the bit under test, and the SAR either maintains the capacitor connected to V_{REF} or connects it to the ground, depending on the polarity of $V^+ - V^-$ detected by the comparator. At the end of the conversion, we have

$$V^- = -V_i + V_{REF}\left(\frac{b_1}{2} + \frac{b_2}{2^2} + \cdots + \frac{b_{N-1}}{2^{N-1}} + \frac{b_N}{2^N}\right) \qquad (11.5)$$

and the value of V^- should be as close to zero as possible. The valid output code can then be stored in the output buffer consisting of latches.

The conversion process of the charge-redistribution unipolar SAR ADC is based on the algorithm shown in Figure 11.7, where $V_{DAC}^0 = -V_i + V_{off}$. Furthermore, the code transitions of the transfer characteristic actually occur at k LSB, where k is an integer. However, the quantization error will be reduced if the code transitions can arise at $k/2$ LSB. This can be achieved by switching the lower plate of the capacitor C' to $V_{REF}/2$ rather than the ground after the sample phase.

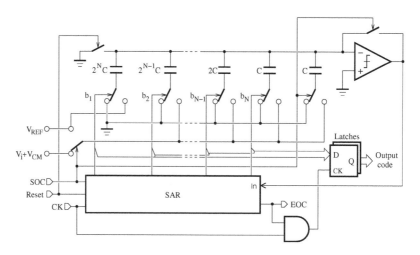

FIGURE 11.8
Block diagram of a charge-redistribution bipolar SAR ADC.

To allow the conversion of positive and negative input voltages, the capacitor switching scheme of the charge redistribution SAR ADC must be somewhat modified. Figure 11.8 shows the block diagram of a charge-redistribution bipolar SAR ADC, whose operation involves the next steps.

In the *sample phase*, the largest capacitor is connected to the reference voltage, V_{REF}, while the remaining capacitors are connected to the input voltage, V_i, and the comparator feedback switch is closed. The charge Q_C

stored on capacitors is given by

$$Q_C = 2^{N-1}C(V_i + V_{CM} - V_{off}).\qquad(11.6)$$

During the *hold phase*, the comparator feedback switch is open and the bottom plates of capacitors are now switched to the ground. Because the capacitor array is equivalent to a voltage divider consisting of two equal capacitors, the voltage applied to the negative terminal of the comparator now becomes

$$V^- = -(V_i + V_{CM})/2 + V_{off}.\qquad(11.7)$$

The input node is actually connected to the reference voltage, V_{REF}, instead of the input voltage.

The *redistribution phase* begins with the determination of b_{N-1} or MSB. The SAR is initialized so that the MSB is set to the logic high. With the assumption that the comparator threshold is modeled as an offset voltage, V_{off}, in series with the positive input, we have

$$V^+ = V_{off}.\qquad(11.8)$$

In the case where $V^+ - V^- > 0$, that is, if $V_i > 0$, the comparator output will be a logic high and the state of the MSB will remain unchanged. However, if $V^+ - V^- < 0$, then $V_i < 0$, and the comparator output will be a logic low. The SAR is then configured to set the first bit of the output code to a logic low and switch the MSB capacitor from the ground to V_{REF}. Here, the MSB is used as a sign bit. For the determination of each of the remaining bits, the corresponding capacitor is switched from the ground to V_{REF} and the comparison trial proceeds as in the unipolar case. At the end of the conversion, the voltage V^- can be written as

$$V^- = -\frac{V_i + V_{CM}}{2} + \frac{V_{REF}}{2}\left(-b_{N-1} + \frac{b_{N-2}}{2} + \cdots + \frac{b_1}{2^{N-2}} + \frac{b_0}{2^{N-1}}\right),\qquad(11.9)$$

where V_{CM} denotes the common-mode voltage. An appropriate choice of V_{CM} is useful for handling bipolar input signals. It can be assumed that the converter input signal is biased about

$$V_{CM} = V_{REF} + V_{LSB} = V_{REF}(1 + 1/2^N),\qquad(11.10)$$

where $V_{LSB} = V_{REF}/2^N$. Hence, the converter features a two's complement output coding with the zero code, which is actually 1/2 LSB above the mid-scale. It should be noted that the converter full-scale still exhibits a dynamic range of V_{REF} as in the case of the charge redistribution unipolar SAR ADC.

An alternative design approach can consist of modifying the switching scheme of the SAR ADC to control the connection between the reference voltage terminal and either V_{REF}^+ for a positive input or V_{REF}^- for a negative input. The *sample phase* and *hold phase* remain similar to the ones of the unipolar structure, except that the sign bit should be detected by the comparator just before the *redistribution phase* and used to select the suitable reference voltage. This latter is required for the determination of the remaining bits of the output code.

The accuracy of the charge redistribution SAR ADC is mainly limited by the achievable capacitor matching. Because the settling time of the comparator increases as the minimum overdrive signal becomes smaller for high resolutions, the time delay of the comparator appears to be a critical limiting factor in the achievable speed of the data conversion. Furthermore, the stability requirements of the comparator should be specified, taking into account the fact that the offset voltage estimation is performed in the unity-gain configuration.

• Due to mismatches of passive components, resolutions higher than 12 bits can only be realized by means of self-calibrating converters [4]. This technique is illustrated with the ADC shown in Figure 11.9. The data conversion,

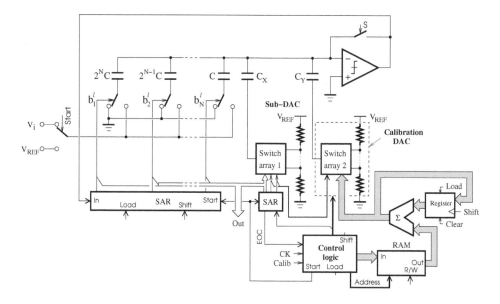

FIGURE 11.9
Block diagram of a self-calibrating SAR ADC. (Adapted from [4], ©1983 IEEE.)

which is achieved in three steps, is based on the charge redistribution. First, the reset switch is closed and the comparator is configured as a unity gain buffer. The top and bottom plates of capacitors are then connected to the virtual ground and input voltage, respectively. During the next phase, the reset switch is open while the bottom plates are switched to the ground. Due to the charge conservation, the potential at the top plates is changed from V_i to $-V_i$. The last operation consists of finding the digital version of the analog input signal. The conversion process starts with the determination of the MSB. The bottom plate of the largest capacitor is connected to V_{REF}. The connection will be maintained if the comparator output is high. Otherwise, the capacitor is switched to the ground. The following MSB and the remaining bits are determined in the same way. The MSBs are resolved by the binary weighted capacitor DAC. The resistor-string DAC, which is connected to the coupling capacitor C_X, provides the LSBs. Basically, it is not affected by the differential nonlinearity.

Ideally, the voltage contribution of the binary-weighted capacitor DAC at the comparator input is given by

$$V = \frac{V_{REF}}{2^N} \sum_{j=1}^{N} 2^{j-1} b_j^l , \qquad (11.11)$$

where V_{REF} is the reference voltage and b_j^l $(l = 0, 1)$ is the logic value of the j-th bit. Due to the fluctuations of the IC fabrication process, the value of a weighted capacitor can be written as

$$C_j = 2^{j-1} C(1 + \epsilon_j), \quad j = 1, 2, \cdots, N, \qquad (11.12)$$

where ϵ_j denotes the matching error and C is the unit capacitor, which is the ratio of the total capacitance C_T to 2^N, that is,

$$C = \frac{C_T}{2^N} . \qquad (11.13)$$

The voltage V then becomes

$$\hat{V} = \frac{V_{REF}}{C_T} \sum_{j=1}^{N} C_j b_j^l \qquad (11.14)$$

Substituting Equations (11.12) and (11.13) into Equation (11.14), we obtain

$$\hat{V} = \frac{V_{REF}}{2^N} \sum_{j=1}^{N} 2^{j-1}(1 + \epsilon_j) b_j^l . \qquad (11.15)$$

The error voltage can be computed as

$$\triangle V = \hat{V} - V = \sum_{j=1}^{N} V_{\epsilon j} b_j^l , \qquad (11.16)$$

where

$$V_{\epsilon j} = \frac{V_{REF}}{2^N} 2^{j-1} \epsilon_j . \qquad (11.17)$$

The calibration stage is necessary for the reduction of nonlinearities introduced by the capacitor mismatches. Starting from the largest capacitor to the smallest one, the different error contributions are estimated. For a given capacitor C_k, this is achieved first by connecting V_{REF} to all capacitors except

C_k. Then, the charge is redistributed by switching all capacitors except C_k to the ground. It follows that the charge stored at the top plate of the capacitors is

$$\triangle Q = V_{REF}(2C_k - C_T). \tag{11.18}$$

Substituting Equations (11.12) and (11.13) into Equation (11.18), it can be shown that

$$\triangle Q = CV_{REF}2^k \epsilon_k \tag{11.19}$$

and the corresponding residual voltage is given by

$$V_{xk} = \frac{\triangle Q}{C_T} = 2V_{\epsilon k}. \tag{11.20}$$

Starting from the MSB capacitor, the residual and error voltages can be computed as

$$V_{\epsilon k} = \begin{cases} \dfrac{V_{xN}}{2}, & \text{if } k = N, \\ \dfrac{1}{2}\left(V_{xk} - \displaystyle\sum_{j=k+1}^{N} V_{\epsilon j}\right), & \text{otherwise.} \end{cases} \tag{11.21}$$

The correction signals are obtained from the digital version of the residual voltages. A random access memory (RAM) is used to store the error voltages $V_{\epsilon j}$, which are estimated during the calibration cycle carried out just after the converter power-up. If the k-th bit assumes the high level, the error $V_{\epsilon k}$ will be added to the ones accumulated from the MSB through the $(k-1)$-th bit and the result is stored in the accumulator. Otherwise, $V_{\epsilon k}$ is discarded and the previous error voltage contained in the accumulator remains unchanged. The accumulator output is then converted by the calibration DAC into an analog signal, which is summed with the appropriate sign to the output voltage of the binary-weighted capacitor DAC. The calibration is equivalent to the cancelation of the error voltages due to capacitor mismatches from the corresponding ADC output signal during the normal conversion cycle. It works well provided the linearity error on the coupling capacitor $C_X = C$ is less than $1/2^M$ for an M-bit sub-DAC. In the layout, C_X should preferably be located near the center of the array to counteract the effect of fabrication errors.

11.1.2 Integrating ADC

By using time to quantize a signal, which represents the integral or average of an input voltage over a fixed period of time, an integrating ADC can exhibit good linearity performance and high-frequency noise rejection. In comparison with a successive approximation converter, the integrating ADC is simple, low cost, and slow. Generally, its speed is approximately 500 times smaller than the one of a typical successive approximation converter. Integrating ADCs can be implemented using single-slope, dual-slope, or multiple-slope architectures. The primary advantage of a dual-slope ADC over a single-slope architecture is that the final conversion result is insensitive to errors in the component values. A multiple slope ADC still features the advantages of a dual-slope converter, but its conversion speed can be greatly increased at the cost of extra hardware complexity.

The dual-slope architectures are suitable for digitizing low bandwidth signals in instrumentation devices such as the digital multimeter, which require a high resolution (10 to 20 bits or $3\frac{1}{2}$ to $5\frac{1}{2}$ digits) and a low conversion speed in the range of 100 sps (samples per second).

The dual-slope ADC, as shown in Figure 11.10, uses an analog integrator with switched inputs, a comparator, a control logic, and a counter. Fig 11.11 shows the conversion timing of the converter. The input voltage is assumed to be positive.

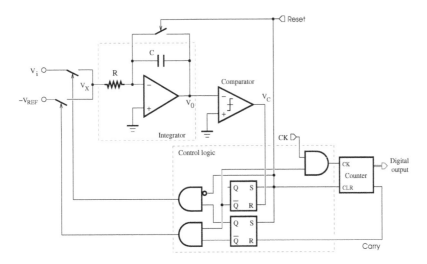

FIGURE 11.10
Circuit diagram of a dual-slope ADC.

The operation of the dual-slope ADC starts with a reset phase, which consists of shorting the capacitor C to drive the integrator output to zero and clearing the counter, followed by the input signal integration phase and the reference signal integration phase.

– Input signal integration phase
After the reset phase, the unknown input signal V_i is applied to the integrator, while at the same time the process of counting clock pulses is started. The integrator output voltage is given by

$$V_0(t) = -\frac{1}{\tau}\int_0^t V_i(t)\mathrm{d}t, \qquad (11.22)$$

where $\tau = RC$ is the time constant of the integrator. The voltage V_i is then integrated for a duration T_c, which is generally known as the charging (or

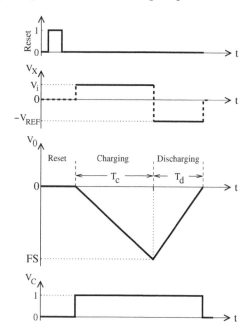

FIGURE 11.11
Conversion timing of the dual-slope ADC.

ramp-up) time. Assuming that V_i is time invariant, the output of the integrator at the end of the integrating phase can be written as

$$V_0(T_c) = -\frac{V_i}{\tau} \cdot T_c. \qquad (11.23)$$

Here, the integration of the input signal takes place until the N-bit counter overflows; this corresponds to a fixed time of $T_c = 2^N T$, where T is the clock period. Note that the integrator output voltage is directly proportional to the input signal.

– **Reference signal integration phase**
When the input signal integration phase is completed, the counter is reset. A known reference voltage, V_{REF}, with a polarity opposite to that of V_i is connected to the integrator input and the counting of clock pulses resumes. The integrator output voltage is now ramping down at a constant slope according to the expression

$$V_0(t) = -\int_{T_c}^{t} \frac{-V_{REF}}{\tau} dt + V_0(T_c) = \frac{V_{REF}}{\tau}(t - T_c) - \frac{V_i}{\tau} \cdot T_c. \qquad (11.24)$$

The counter is stopped after a discharging (or ramp-down) duration of T_d when the comparator output takes the logic low level because the integrator output reaches zero, that is, $V_0(T_d) = 0$. Hence,

$$\frac{V_{REF}}{\tau} \cdot T_d - \frac{V_i}{\tau} \cdot T_c = 0. \tag{11.25}$$

The time T_d is then given by

$$T_d = T_c \cdot \frac{V_i}{V_{REF}}. \tag{11.26}$$

On the other hand, T_d can be expressed in terms of the clock period, T, as

$$T_d = N_r \cdot T, \tag{11.27}$$

where N_r is the number of clock periods recorded during the connection of V_{REF} to the integrator input. Thus, it can be found that

$$N_r = 2^N \frac{V_i}{V_{REF}}. \tag{11.28}$$

The digital output of the counter is proportional to the magnitude of the input signal, and is independent of the integrator time constant, which is supposed not to change during an individual conversion cycle. When the input voltage is negative, the dual-slope ADC operates according to the same basic principle as described above for a positive input voltage, except that all the signal polarities are reversed.

The accuracy of the dual-slope ADC seems to be only affected by the fluctuations of the reference voltage and clock timing. But, the behavior of a practical circuit can also be plagued by the nonideal characteristics of the amplifier, MOS switches and capacitors, and the response time or switching delay of the comparator. Figure 11.12 shows the effects of the offset voltage and input over-range on the integrator output waveform. Additional conversion phases are required in order to improve the accuracy of the dual-slope ADC.

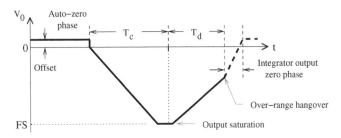

FIGURE 11.12
Effects of the offset voltage and input over-range on the integrator output waveform.

The dual-slope ADC can be designed to allow bipolar operation and to incorporate a phase for the converter offset compensation. Figure 11.13 shows

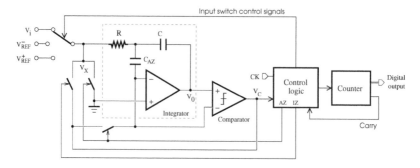

FIGURE 11.13

Circuit diagram of the dual-slope ADC with an auto-zero capacitor.

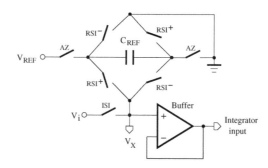

FIGURE 11.14

Circuit diagram of the input section of a bipolar dual-slope ADC with a single reference voltage.

the circuit diagram of a dual-slope ADC with an auto-zero (AZ) capacitor [5]. The conversion cycle includes an auto-zero phase, an input signal integration (ISI) phase, a reference signal integration (RSI) phase, and an integrator-output zero (IZ) phase.

During the *auto-zero phase*, the converter input is connected to the ground and a feedback loop is closed around the system such that the error voltage can appear across the auto-zero capacitor, C_{AZ}, whose charge is used for offset voltage correction during the subsequent phases. By including all active components in the loop, the accuracy of the auto-zero calibration is limited only by the noise of the system. Note that a larger value of C_{AZ} should be used to minimize the noise sensibility for a converter with a small resolution. Typically, the capacitor C_{AZ} is at least two times greater than C.

In the *input signal integration phase*, the auto-zero switch is open and the voltage V_i is connected to the converter input. The input signal polarity is determined at the end of this phase.

During the *reference signal integration phase*, the control circuit connects the reference signal with the polarity such that the integrator output can be driven with a fixed slope to the zero level established in the auto-zero phase.

In the *integrator output zero phase*, the switch controlled by the IZ control signal is closed and the negative feedback from the output of the comparator to the converter input drives the integrator output to zero. This phase is used to completely discharge the feedback capacitor of the integrator following the occurrence of an over-range condition (see Figure 11.12), which is characterized by the integrator output remaining far from the initial level after the maximum time allowed to the reference signal integration phase.

Due to the need to accurately estimate the occurrence time of the zero-crossing at the end of the reference signal integration phase, the operation speed of the dual-slope ADC is generally slow. Although the frequency of the clock signal may be increased, the use of a high clock rate is limited by the delay of the comparator in detecting the zero-crossing. For a typical converter, the comparator delay should not be greater than the duration of one half clock pulse, yielding a clock frequency on the order of a few hundred kilohertz based on a comparator with a delay of a few microseconds. An improvement in the performance may be achieved by modifying the structure of the conventional dual-slope ADC such that the discharging time of the integrator can be measured with a precision greater than the pulse width of the clock signal.

In the approach relying on the use of two reference voltages to design a dual-slope ADC with a bipolar input range, the symmetry of the converter transfer characteristic may be affected by the independent fluctuation of each reference source. A solution can consist of using the input section of the dual-slope ADC shown in Figure 11.14, where the charge stored by a capacitor can induce balanced reference voltages [6, 7]. Switches controlled by the signal RSI^+ and RSI^- are closed for the positive and negative input signals, respectively.

During the AZ phase or an idle phase, the reference capacitor, C_{REF}, is charged by the reference voltage. For a stable storage of the charge due to the reference voltage, the capacitor C_{REF} must exhibit a low leakage current and the effect of stray capacitances appearing on the reference capacitor nodes must also be minimized. A capacitor in the microfarad range is required to prevent rollover error, which arises as a consequence of the difference in the reference voltage value for positive and negative input signals. The polarity of the input signal determined by the comparator at the end of the signal integration phase is used by the control logic to connect the capacitor C_{REF} to the input buffer such that the integrator output can return to the zero level. Hence, depending on the polarity of the analog input, the reference capacitor is used to generate either a positive or negative voltage during the reference signal integration phase.

The block diagram of a dual-slope ADC including an amplification and feedback circuitry to improve the conversion accuracy [8] is depicted in Figure 11.15. The conversion phases are illustrated on the integrator output shown in Figure 11.16.

Initially, a conventional dual-slope integration consisting of an *input signal integration phase* and a *reference signal integration phase* is performed.

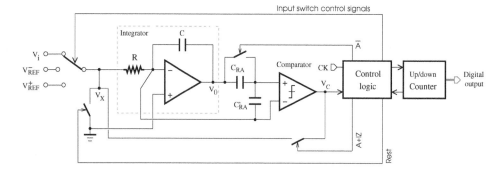

FIGURE 11.15

Circuit diagram of the dual-slope ADC with extra amplification and feedback circuitry.

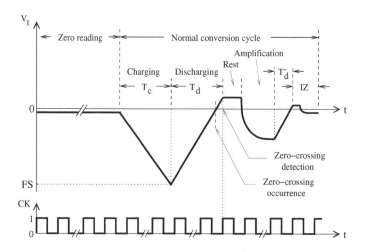

FIGURE 11.16

Integrator output during each conversion phase.

The converter operates by integrating an unknown input analog signal for a predetermined time period, T_c, and then integrating a known reference signal until the integrator output reaches a predetermined zero level. Due to the comparator delay, the zero-crossing is actually detected on the first clock pulse after its occurrence, even if the integrator output reached the zero level within the clock period. The duration, T_d, of the reference signal integration phase measured by counting clock pulses is then larger than its true value. Because the integrator output continues to ramp down below the zero level until the end of the clock period, a residual voltage appears across capacitors C and C'_{RA}.

A *rest phase* is required to maintain constant the residual integrator charge, which is then scaled up by a given negative factor, $-k$, and fed back to the input node of the integrator amplifier in order to account for the measurement error on the phase duration during the *amplification phase*.

In general, the absolute value of the multiplication factor depends on the converter number system and is of the form, $k = 2^p$ for a binary system, where p is the resolution increase in number of bits and $k = 10^q$ for a decimal system, where q represents the resolution increase in number of digits. Typically, the residual voltage is multiplied by -8 in a binary converter or -10 in a decimal converter. In practice, the multiplication of the residual integrator charge is achieved by sizing capacitors C_{RA} and C'_{RA} such that

$$C'_{RA} = kC_{RA}. \tag{11.29}$$

At the start of the amplification phase, the comparator output is fed back to the integrator input, the voltage across C_{RA} is zero, and the charge stored on C'_{RA} is

$$Q = C'_{RA}V_{res} = kC_{RA}V_{res}, \tag{11.30}$$

where V_{res} is the residual voltage. To establish a virtual ground at the positive input of the comparator, a charge transfer is initiated between C'_{XR} and C_{XR}, ending with the induction of a voltage across the capacitor C_{XR} given by

$$V_{C_{RA}} = -Q/C_{RA} = -kV_{res}. \tag{11.31}$$

The voltage across the capacitor C is also $-kV_{res}$ because $V_{C'_{RA}}$ is actually almost equal to zero.

The converter enters a second *reference signal integration phase*, where the capacitor C_{RA} is short-circuited again, and the feedback connection between the comparator output and integrator input is open. To keep the effect of the charge redistribution between the capacitors negligible, the capacitor C is designed to be much larger than the capacitor C'_{RA}. Thus, for this phase, the initial value of the voltage at the positive input of the comparator remains close to $-kV_{res}$.

A second *reference signal integration phase* is initiated. The time, T'_d, which is measured as the number of clock pulses required by the integrator output to cross the zero level again, is proportional to the residual error. The net count resulting from the subtraction in the same scale of T'_d from T_d represents the value of the time effectively needed by the integrator output to cross the zero level. This calibration scheme can be implemented using up-down counters incremented and decremented by the control logic.

The accuracy of the actual time measurement can be further improved by resorting to subsequent *rest, amplification,* and *reference signal integration phases.*

After the occurrence of an over-range condition, the comparator output and the input analog signal are of opposite polarity. The feedback connection realized between the comparator output and the integrator input during the *integrator output zero phase* forces the integrator output voltage to change in a direction such that its magnitude is decreased toward zero, thereby causing the discharge of the capacitor C.

In this approach, the zero reading, that is, the result of the conversion with the input voltage shorted to the ground, should be subtracted from each measurement. This helps minimize the effect of offset voltages on the conversion process.

For proper operation of a dual-slope ADC, the period of each clock pulse should be greater than the comparator delay. Hence, the conversion speed is primarily limited by the comparator delay, which is dependent on the converter overdrive, defined as the maximum integrator output voltage swing divided by the maximum number of clock pulses during the reference signal integration phase. By augmenting the number of times, the rest, amplification, and reference signal integration phases are repeated, a decrease in the duration of each phase becomes feasible, thereby yielding an overall conversion with a faster speed.

11.1.3   Flash ADC

The most straightforward way to perform the N-bit analog-to-digital conversion is to compare the sampled-and-held version of an analog signal with 2^N reference voltages. The flash ADC, which generally consists of a resistive divider, comparators, and a binary encoder, is based on this principle. The foregoing track-and-hold (T/H) circuit samples the analog input and holds it for half a clock cycle. This operation can be performed by exploiting the inherent sampling properties of latched comparators. However, the use of a T/H circuit is preferred because the operation of the comparator array is often affected by clock skews, which can degrade converter linearity.

TABLE 11.2
Thermometer Code for a 3-Bit Unipolar Flash ADC (with $\triangle = V_{REF}/8$)

	Input Range	Thermometer Code							Binary Code		
		T_6	T_5	T_4	T_3	T_2	T_1	T_0	b_1	b_2	b_3
7	$13\triangle/2 < V_i < 8\triangle$	1	1	1	1	1	1	1	1	1	1
6	$11\triangle/2 < V_i < 13\triangle/2$	0	1	1	1	1	1	1	1	1	0
5	$9\triangle/2 < V_i < 11\triangle/2$	0	0	1	1	1	1	1	1	0	1
4	$7\triangle/2 < V_i < 9\triangle/2$	0	0	0	1	1	1	1	1	0	0
3	$5\triangle/2 < V_i < 7\triangle/2$	0	0	0	0	1	1	1	0	1	1
2	$3\triangle/2 < V_i < 5\triangle/2$	0	0	0	0	0	1	1	0	1	0
1	$\triangle/2 < V_i < 3\triangle/2$	0	0	0	0	0	0	1	0	0	1
0	$0 < V_i < \triangle/2$	0	0	0	0	0	0	0	0	0	0

The block diagram of a 3-bit flash ADC is shown in Figure 11.17. In this case, $N = 3$ and $2^3 - 1 = 8$ comparators are used. Furthermore, the input signal is supposed to be positive. The reference voltage at the k-th node can

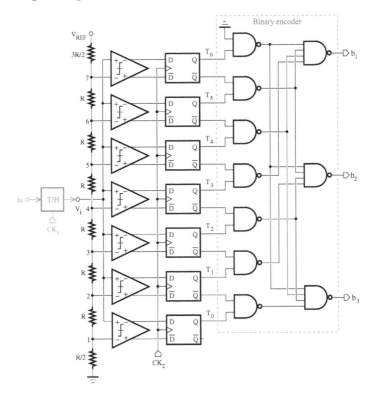

FIGURE 11.17
Block diagram of a 3-bit unipolar flash ADC.

be expressed as

$$V_k^- = V_{REF} \frac{R/2 + (k-1)R}{R/2 + (2^N - 2)R + 3R/2} = k\frac{V_{REF}}{2^N} - \frac{V_{REF}}{2^{N+1}}. \tag{11.32}$$

The difference between the reference voltages of two adjacent comparators is $V_{REF}/2^N$, that is,

$$\triangle = V_{k+1}^- - V_k^- = \frac{V_{REF}}{2^N}, \tag{11.33}$$

where $\triangle$ denotes the voltage level of an LSB. The output of a comparator will be at the high state if its input voltage is higher than the reference voltage. Otherwise, the comparator output is at the low state. The outputs of all comparators form a thermometer code that is then converted into a binary code by an encoder. Table 11.2 lists the thermometer and binary coding schemes for a 3-bit word. Note that the number of consecutive 1s in the thermometer code corresponds to the count of reference voltages less than the input signal. By choosing the bottom resistor in the comparator resistor string to be $R/2$, the transitions of the ADC transfer characteristic are at multiples of $\triangle/2$.

In practice, due to mismatches and imperfections in the reference resistor string and in the latched comparators, as well as high speed limitations, errors can be introduced in the thermometer code produced at the comparator

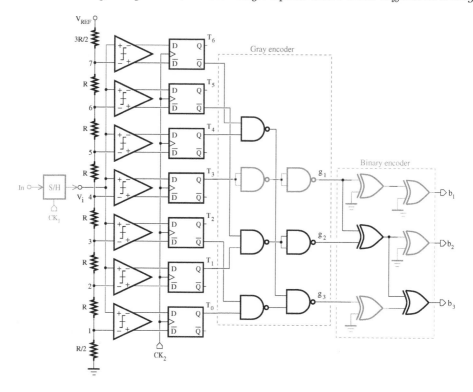

FIGURE 11.18

Block diagram of a flash ADC with improved encoder.

TABLE 11.3

Encoder Truth Table ($\triangle = V_{REF}/8$)

	Thermometer Code							Gray Code			Binary Code		
	T_6	T_5	T_4	T_3	T_2	T_1	T_0	g_1	g_2	g_3	b_1	b_2	b_3
7	1	1	1	1	1	1	1	1	0	0	1	1	1
6	0	1	1	1	1	1	1	1	0	1	1	1	0
5	0	0	1	1	1	1	1	1	1	1	1	0	1
4	0	0	0	1	1	1	1	1	1	0	1	0	0
3	0	0	0	0	1	1	1	0	1	0	0	1	1
2	0	0	0	0	0	1	1	0	1	1	0	1	0
1	0	0	0	0	0	0	1	0	0	1	0	0	1
0	0	0	0	0	0	0	0	0	0	0	0	0	0

outputs. They show up as bubbles[1], which are zeros surrounded by ones, or vice versa. This may happen in the specific case where a comparator switches

---

[1] The use of the term "bubble" is justified by the fact that such a code error is analogous to a bubble occurring in the mercury of a thermometer.

more slowly than expected, so that the output is latched before reaching the final state. When the bubble error cannot be detected by the digital encoder, the ADC will produce an output code not representative of the input signal value.

TABLE 11.4

Thermometer-to-Gray and Gray-to-Binary Encodings

$N = 3$	$g_1 = T_3$	$b_1 = g_1$
	$g_2 = T_1 \cdot \overline{T_5}$	$b_2 = g_2 \oplus b_1$
	$g_3 = T_0 \cdot \overline{T_2} + T_4 \cdot \overline{T_6}$	$b_3 = g_3 \oplus b_2$
$N = 4$	$g_1 = T_7$	$b_1 = g_1$
	$g_2 = T_3 \cdot \overline{T_{11}}$	$b_2 = g_2 \oplus b_1$
	$g_3 = T_1 \cdot \overline{T_5} + T_9 \cdot \overline{T_{13}}$	$b_3 = g_3 \oplus b_2$
	$g_4 = T_0 \cdot \overline{T_2} + T_4 \cdot \overline{T_6} + T_8 \cdot \overline{T_{10}} + T_{12} \cdot \overline{T_{14}}$	$b_4 = g_4 \oplus b_3$

The effect of bubble errors can be reduced by first converting the thermometer code into the Gray code, which is then used for the computation of the binary code. Gray encoding itself has no correction ability. But its tolerance to bubbles is due to the fact that only one bit changes between adjacent codes, leading to a small difference between the ideal and incorrect codes. The logic equations for the thermometer-to-Gray and Gray-to-binary encodings are summarized in Table 11.4 for the 3-bit and 4-bit flash ADCs. The circuit diagram of a 3-bit flash ADC including a Gray-code-based encoder is shown in Figure 11.18. In order to equalize the propagation delay of the different signal paths, additional NAND and XOR gates were inserted in the encoder according to the following Boolean algebraic identities:

$$A = \overline{\overline{A}} \tag{11.34}$$

$$A = A \oplus 0, \tag{11.35}$$

where A is a logic variable. Note that long wiring structures can be required to logically combined signals as the number of bit is increased, resulting in an irregular circuit layout. Furthermore, latches can be introduced between the logic stages to improve the operation speed and to reduce the metastability error probability.

Due to their regular structure, ROM-based encoders are preferred over gate-based encoders for converter implementations with a resolution greater than 5 bits. A ROM can consist of bit lines, word lines, and MOS-type memory cells. The storage of a logic 0 is carried out using an n-channel pull-down transistor, while no connection is required for the storage of a logic 1. A precharged logic is generally used to eliminate the static power dissipation and to help keep the pull-up and pull-down transistors as close as possible to the minimum size. However, the converter speed can be limited by the ROM pre-charge time. Ideally, the 1-out-of-2^N code obtained by detecting the location of the 1-to-0 transition in the thermometer code is commonly used to enable a single word line of the ROM. Hence, only the bit representing the location of the detected transition can take the logic 1 and all the remaining bits are at the logic 0. However, if bubbles exist in the thermometer code, there will be multiple 1-to-0 transition points and the 1-out-of-2^N encoder will select more than one ROM line. As a result, errors are introduced in the binary output, which is actually the representation of the bitwise logical OR between the ROM lines.

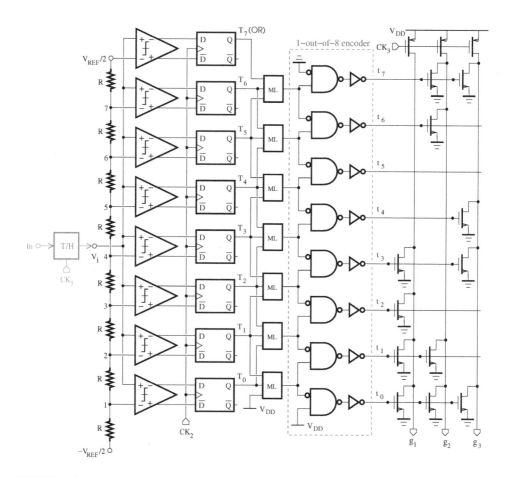

FIGURE 11.19
Block diagram of a 3-bit flash ADC with the bubble error correction based on a majority logic (OR: over-range).

An approach for the bubble correction consists of using a majority logic function [9] whose output takes the same logic state as the greater number of inputs. The block diagram of the 3-bit flash ADC with the bubble error correction based on a majority logic (ML) is illustrated in Figure 11.19. With the assumption that $T_{-1} = 1$ and T_7 represents the over-range signal, the corrected thermometer code, T_i^*, is given by the Boolean equation,

$$T_i^* = T_{i-1} \cdot T_i + T_i \cdot T_{i+1} + T_{i-1} \cdot T_{i+1} \,, \tag{11.36}$$

where $i = 0, 1, \cdots, 6$. The state of a given bit in the thermometer code is then determined by the one of its two neighbors. The ML encoder can efficiently suppress the bubbles, which affect the 1-0 transition in the thermometer code. However, the power consumption and chip area can be increased because the required number of gates tends to be high. The ROM line selection can be carried out using a thermometer to 1-out-of-2^N encoder based on two-input AND gates. This simple encoder will operate correctly only if there is no bubble error in the thermometer code.

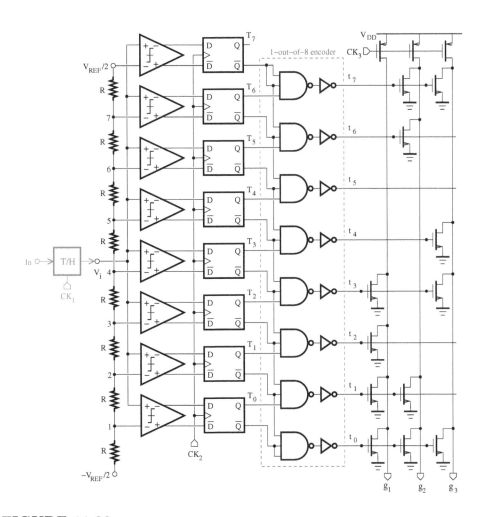

FIGURE 11.20
Block diagram of a 3-bit flash ADC with the bubble error correction based on three-input AND gates.

In general, the occurrence likelihood of bubbles becomes less important as their number becomes greater. Because the bubbles are mainly introduced near the one-to-zero transition point in the thermometer code, the encoder can be designed to correct only certain single bubbles. Figure 11.20 shows the block diagram of the 3-bit flash ADC with the bubble error correction based on three-input AND gates [10]. The correction is limited to bubble errors, which do not affect the 1-to-00 transition in the thermometer code. The possible states of the different signals in response to a given input are shown in Table 11.5 for the 3-bit flash ADC. The appropriate word line in the Gray ROM encoder is activated by the signals t_i, which are given by the following Boolean equation,

$$
t_i = \begin{cases} T_6 \cdot \overline{T_7}, & \text{for} \quad i = 7 \\ T_{i-1} \cdot \overline{T_i} \cdot \overline{T_{i+1}}, & \text{for} \quad i = 1, 2, \cdots, 6 \\ \overline{T_0} \cdot \overline{T_1}, & \text{for} \quad i = 0, \end{cases} \tag{11.37}
$$

where T_i denotes a bit of the thermometer code. Note that T_7 represents the over-range signal generated as a result of the comparison between the input signal and $V_{REF}/2$.

TABLE 11.5
Truth Table of the 3-Bit Bipolar Flash ADC with a Three-Input AND Encoder

Input Range	Thermometer Code $T_6\ T_5\ T_4\ T_3\ T_2\ T_1\ T_0$							1-out-of-8 Code $t_7\ t_6\ t_5\ t_4\ t_3\ t_2\ t_1\ t_0$								Gray Code $g_1\ g_2\ \ \ g_3$		
$3\triangle < V_i < 4\triangle$	1	1	1	1	1	1	1	1	0	0	0	0	0	0	0	1	0	0
$2\triangle < V_i < 3\triangle$	0	1	1	1	1	1	1	0	1	0	0	0	0	0	0	1	0	1
$\triangle < V_i < 2\triangle$	0	0	1	1	1	1	1	0	0	1	0	0	0	0	0	1	1	1
$0 < V_i < \triangle$	0	0	0	1	1	1	1	0	0	0	1	0	0	0	0	1	1	0
$-\triangle < V_i < 0$	0	0	0	0	1	1	1	0	0	0	0	1	0	0	0	0	1	0
$-2\triangle < V_i < -\triangle$	0	0	0	0	0	1	1	0	0	0	0	0	1	0	0	0	1	1
$-3\triangle < V_i < -2\triangle$	0	0	0	0	0	0	1	0	0	0	0	0	0	1	0	0	0	1
$-4\triangle < V_i < -3\triangle$	0	0	0	0	0	0	0	0	0	0	0	0	0	0	1	0	0	0

The fundamental difference between the ML and three-input AND encoders can be illustrated by comparing the ideal thermometer code and error-correction results of the examples included in Table 11.6, where the first T_i column represents the ideal case and bold characters are used to show the bits affected by bubble errors. The ML encoder relies on the best-expectation principle and the detection of the 1-to-0 transition, while the three-input AND encoder is based on the detection of the 1-to-00 transition. In both cases, the error correction cannot be ensured for all bubble types.

An N-bit flash ADC generally requires at least $2^N - 1$ comparators, whose reference voltages are set by a resistor string. The value of each comparator

TABLE 11.6

Illustration of Three Correction Examples of Bubbles in the ML and Three-Input AND Encoders

| | | ML Encoder | | | | | | | | | Three-input AND Encoder | | | | | |
| | | Case 1 | | | Case 2 | | | Case 3 | | | Case 1 | | Case 2 | | Case 3 | |
i	T_i	T_i	T_i^*	t_i	T_i	T_i^*	t_i	T_i	T_i^*	t_i	T_i	t_i	T_i	t_i	T_i	t_i
7	0	0	0	0	0	0	0	0	0	0	0	0	0	0	0	0
6	0	0	0	0	0	0	0	0	0	0	0	0	0	0	0	0
5	0	0	0	0	0	0	0	0	0	0	0	1	0	0	0	0
4	0	**1**	0	1	0	0	0	0	0	0	**1**	0	0	1	0	1
3	1	**0**	1	0	1	0	1	1	0	0	**0**	0	1	0	1	0
2	1	1	1	0	**0**	1	0	**0**	0	0	1	0	**0**	0	**0**	0
1	1	1	1	0	1	1	0	**0**	0	1	1	0	1	0	**0**	1
0	1	1	1	0	1	1	0	1	1	0	1	0	1	0	1	0

output depends on whether or not the input voltage exceeds the corresponding reference voltage. A set of all comparator outputs, which form the so-called thermometer code, is first scaled to the appropriate logic levels by latches in order to mitigate the metastability problem and then converted into a form of binary data. The resulting data with N-bit resolution is applied to the output buffer, which can be implemented using D latches. Note that an additional comparator is often used to indicate the presence of a signal over-range. Because the metastability, which is characterized by an output state between the logic level high and logic level low, is due to the violation of setup and hold time, it can be reduced by allowing more time for comparator regeneration. The speed of the overall structure depends on the one of the comparator and the propagation delay of the digital section. But with the use of pipeline latches, the comparison and code conversion can ideally be achieved in a single clock phase allowing the flash ADC to operate at higher frequencies.

The block diagram of a flash ADC based on the above principle is shown in Figure 11.21. For a given analog input voltage, the thermometer code should ideally exhibit only a single 1-to-0 transition. But in practice, the converter components may be subject to nonidealities such as finite bandwidth, noise, and mismatches, leading to the introduction of bubbles in the thermometer code. This problem can be alleviated by using an intermediate Gray encoding stage instead of converting directly from the thermometer code to the binary code. Furthermore, the robustness to bubbles can be enhanced by inserting an adequate error correction stage before the digital encoder.

Note that the output generated by a thermometer-to-binary encoder is a binary representation of the number of 1s in the thermometer code. Hence, a

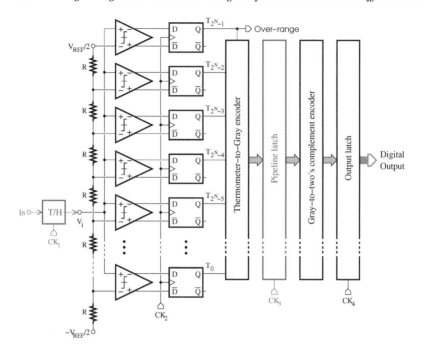

FIGURE 11.21
Block diagram of a flash ADC.

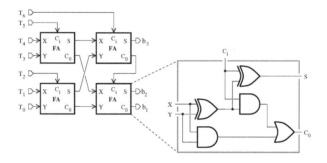

FIGURE 11.22
Tree encoder based on ones-counter for a 3-bit flash ADC.

ones-counter based on full adders can also be used as a digital encoder. Figure 11.22 shows the block diagram of a tree encoder for a 3-bit flash ADC. This approach is insensitive to bubbles, which maintain constant the overall number of 1s in the thermometer code. However, the propagation delay of the tree encoder increases linearly with the converter resolution. Bit-level pipelining, which consists of inserting a register between logic blocks, can then be used to reduce the critical path. This will incur an increase in the hardware overhead and power consumption.

High-resolution flash ADCs generally require a large number of compara-

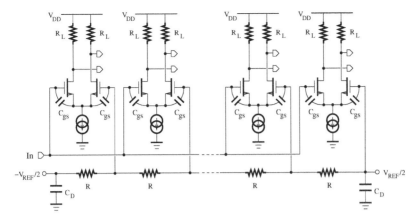

FIGURE 11.23
Equivalent circuit model of the flash ADC input stage.

tors. This can result in a large input capacitor, considerable chip area and very high power consumption. Figure 11.23 shows the equivalent circuit model of the flash ADC input stage, where C_D is used to decouple the reference voltages. The gate-source capacitors, C_{gs}, of the transistors in the comparator input stage form a parasitic capacitor between the comparator inputs. The input signal is then capacitively coupled to the resistive reference network, thereby leading to a variation in the reference voltages. The total resistance of the resistive reference network should be kept low enough to maintain the maximum feedthrough error lower than 1 LSB. This requirement can be relaxed by dividing the resistive reference network into subsections with an adequate size using decoupling capacitors, especially in cases where there is an important increase in the power consumption.

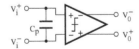

FIGURE 11.24
Comparator with a parasitic capacitor between the inputs.

At high frequencies, the input parasitic capacitor of the comparator, as illustrated in Figure 11.24, couples the input signal with the resistive network used for the generation of the reference voltages. To avoid a variation in reference voltages, which can affect the comparator threshold levels, the maximum resistance of the resistive network must be estimated using the equivalent circuit model shown in Figure 11.25, where it is assumed that the high and low reference voltages are fully decoupled. For a flash ADC with a resolution of N bits, we have

$$R = R_T/2^N \qquad (11.38)$$

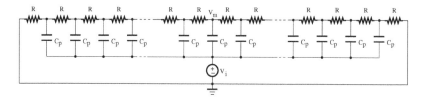

FIGURE 11.25
Equivalent circuit model for the maximum resistance derivation in the flash ADC.

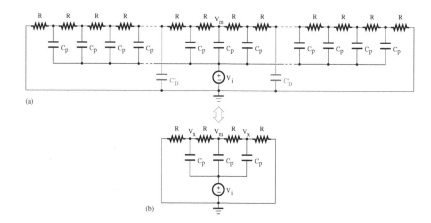

FIGURE 11.26
(a) Equivalent circuit model with decoupling capacitors; (b) equivalent circuit model of a subsection of the resistive reference network.

and

$$C_p = C_T/2^N \,, \tag{11.39}$$

where R_T denotes the total resistance of the resistive network and C_T is the total parasitic capacitance. Generally, the resulting resistance is low for high-speed ADCs, and the power consumption of the reference network, which is of the form V_{REF}^2/R_T, is increased. With the use of decoupling capacitors as shown in Figure 11.26(a), where C_D' is the decoupling capacitor, the total resistance of the resistive network can be increased without deteriorating the level of the feedthrough error. The effect of the decoupling is to split the equivalent circuit model required for the worst-case estimation of R_T into q almost identical subcircuits. Simulations of the equivalent subcircuit depicted in Figure 11.26(b) show that the voltage at the middle node, V_m, is most affected by the input signal feedthrough error.

Using Kirchhoff's current law, the next node equations can be obtained,

$$(V_i - V_m)sC_p = 2\frac{V_m - V_x}{R} \tag{11.40}$$

$$(V_i - V_x)sC_p = \frac{V_x}{R} + \frac{V_x - V_m}{R}. \tag{11.41}$$

Combining Equations (11.40) and (11.41), we get

$$\frac{V_m}{V_i} = \frac{(sRC_p)^2 + 4sRC_p}{(sRC_p)^2 + 4sRC_p + 2} = \frac{(sR_TC_T)^2 + 2^{2N+2}sR_TC_T}{(sR_TC_T)^2 + 2^{2N+2}sR_TC_T + 2^{4N+1}}. \tag{11.42}$$

Let f be the frequency of the input voltage V_i. With the assumption that $s = j2\pi f$ and $\pi f R_T C_T \ll 1$, we can find the following expression:

$$\left|\frac{V_m}{V_i}\right| \simeq \frac{\pi}{4}fR_TC_T. \tag{11.43}$$

In the worst-case, the voltages V_m and V_i can be expressed in LSB units, leading to

$$\left|\frac{V_m}{V_i}\right| = q\frac{k}{2^N}, \tag{11.44}$$

where k represents the feedthrough level and q is the decoupling period. The maximum resistance is then defined by

$$R_T \leq q\frac{4k}{\pi f C_T 2^N}. \tag{11.45}$$

Because q is greater than 1, the decoupling capacitors will increase the total resistance by a factor of q.

Furthermore, the attainable resolution is limited by the accuracy of the reference voltages and comparator offset voltage. Typically, for a 10-bit converter with 2 V_{p-p} input signal the comparator should resolve less than about 2 mV. This requirement is difficult to meet in CMOS technology, and various techniques for reducing the offset-voltage effects such as chopper stabilization and auto-zeroing result in an increase in the power consumption and a reduction in the conversion speed. Therefore, a good compromise may be to design a flash ADC for applications requiring less than 8 bits for the data representation.

Note that flash ADCs can also be implemented without the input T/H circuit, which is generally based on open-loop structures in high-speed applications. In this case, a higher dynamic performance is required for comparators.

11.1.4 Averaging ADC

Device mismatches often limit the differential and integral nonlinearity characteristics and the resolution of classical flash ADC architectures. Averaging techniques can be used to minimize the mismatch effect.

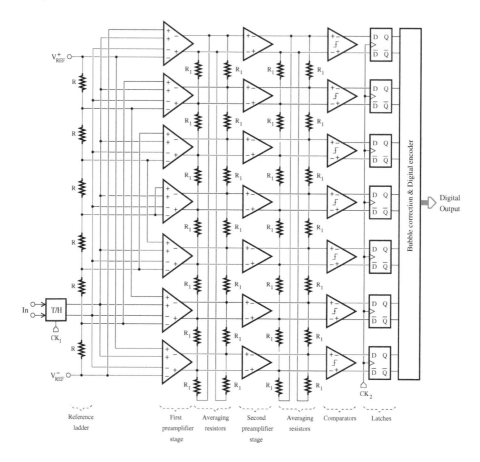

FIGURE 11.27
Circuit diagram of a flash ADC with averaging stages.

Figure 11.27 shows the circuit diagram of a flash ADC with two averaging stages. The inherent symmetry of differential circuits is exploited to use the same number of reference nodes as in single-ended structures. The mirror-image preamplifiers with respect to the reference midpoint can then be connected oppositely to reference voltage nodes. The requirement of maintaining the gain high enough in order to reduce the effect of the input-referred offset on the subsequent stages can be met by cascading two stages of preamplification and averaging. By inserting lateral resistors between the outputs of neighboring preamplifier stages, the random component of currents is reduced and the effect of offsets on the ADC performance is attenuated [11]. The offset reduction is dependent on the value of the preamplifier output resistance, R_0, and the averaging resistors, R_1. Using the superposition principle with the consideration that the signal and reference voltage sources are disconnected, the inputs of the first stage of preamplifiers are reduced to the offset voltages.

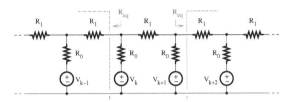

FIGURE 11.28
Equivalent circuit of the preamplifier array with an averaging resistor network.

Figure 11.28 shows the equivalent circuit of a preamplifier array with an averaging resistor network supposed to be infinite. The equivalent resistance, R_{eq}, seen to the right and to the left of each output node is given by

$$R_{eq} = R_1 + R_0 \parallel R_{eq} \Rightarrow R_{eq} = \frac{1}{2}R_1 + \sqrt{\frac{1}{4}R_1^2 + R_1 R_0}. \qquad (11.46)$$

In practice, the length of the averaging resistive network is limited and the equivalent resistance, especially at the termination nodes, may differ from the above value.

The preamplifier transfer characteristics are illustrated in Figure 11.29(a). The zero-crossing points of the preamplifiers at the array edges are shifted inward. This leads to undesirable variations in the INL mean value near the two edges of the preamplifier array, as illustrated in Figure 11.29(b). Furthermore, the standard deviation of the offset voltage is larger for preamplifiers at the boundaries than the one at the center.

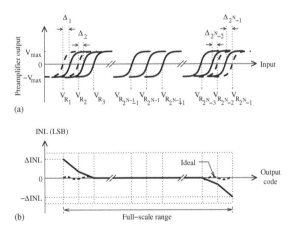

FIGURE 11.29
(a) Preamplifier transfer characteristics; (b) effect of the asymmetrical preamplifier boundaries on the converter linearity.

To avoid any discontinuity at the edges of the averaging network, over-range dummy preamplifiers, whose outputs remain unused, are often added at each edge of the array. The complementary outputs of the first and last over-range preamplifiers are cross-connected through a pair of resistors to ensure that every preamplifier in the array sees the same effective load resistance and has a balanced number of preamplifiers contributing to its output [12]. A 6-bit ADC, for instance, includes an array of 63 preamplifiers, and about 9 dummy preamplifiers are required at each array end to create a linear behavior at the edges of the input full scale. However, the addition of dummy preamplifiers results in a reduction in the available headroom for the input voltage, and an increase in the input capacitance and power dissipation.

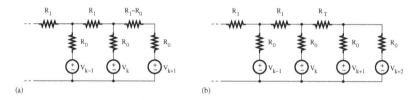

FIGURE 11.30
Equivalent circuits of one edge of the averaging network with a resized termination resistor: (a) Case $R_1 > R_0$, (b) case $R_1 < R_0$.

In the specific case of resistor-loaded preamplifiers, the number of dummy preamplifiers can be reduced by resizing the resistors at either edge of the averaging network, as shown in Figure 11.30 [13], where the index k is used for the lowermost or uppermost in-range preamplifier. Provided $R_1 > R_0$, the electrical load symmetry at the averaging network edges can be restored without resorting to the use of dummy preamplifiers by changing the first and last resistances to $R_1 - R_0$. In the cases where $R_1 < R_0$, the first and last averaging resistors will be short-circuited and the value of the next resistors will be modify as follows:

$$R_T = \begin{cases} \dfrac{3R_1 - R_0}{2}, & \text{if } R_1 < 3R_0, \\ 2R_1 - \dfrac{R_0}{3}, & \text{if } R_1 < 6R_0. \end{cases} \quad (11.47)$$

Otherwise, additional preamplifiers with short-circuited averaging resistors will be connected at the network edges to reduce the equivalent load resistance so that a positive value can be found for R_T.

For a given preamplifier, the magnitude of offset reduction provided by the averaging network depends on the ratio R_0/R_1 and the number of neighboring preamplifiers operating in the nonsaturated region. By increasing the ratio R_0/R_1 to improve the offset reduction, the overall output resistance seen by the preamplifiers can be reduced, leading to a decrease in the gain and an increase in the input-referred offset for the subsequent stages (amplifiers, or comparators). On the other hand, an augmentation in the number of nonsaturated preamplifiers can also improve the offset reduction and can contribute to a substantial increase in the preamplifier gain. Because this can be achieved by stepping up the overdrive voltage, there is a great repercussion on the power consumption. In practice, it is necessary to use a high number of nonsaturated preamplifiers and a low value of the ratio R_0/R_1 to maintain the effects of resistor mismatches to an acceptable level. At least a two times reduction in the offset voltage of a flash converter can be achieved by choosing the ratio R_0/R_1 between 0.5 and 1. The use of edge termination resistors does not significantly increase the power consumption and area of the converter. However, it can effectively mitigate the effect of preamplifier offset voltages only when the averaging window is narrow and the specification of matching termination resistors is less stringent.

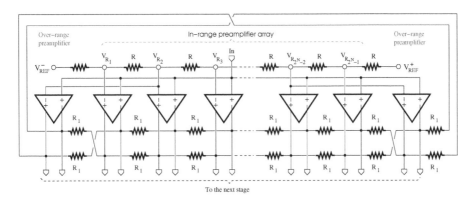

FIGURE 11.31
Preamplifier array with the averaging resistor network based on the triple cross-connection.

An alternative low-power and low-area design solution is to use an averaging resistor network based on the triple cross-connection [14], as shown in Figure 11.31. The effect of zero-crossing shifts can be compensated by introducing a cross-connection and an over-range preamplifier at each boundary. A proper termination can be realized due to the symmetry provided by the cross-connection at the center. To counteract the reduction in the effective transconductance at the boundary due to the negative transconductance of the over-range preamplifier, the over-range preamplifier should be designed such that its input linear region extends along that of the adjacent in-range preamplifier. Furthermore, it is desirable that the over-range preamplifier operate with the same reference voltage as the penultimate preamplifier.

11.1.5 Folding and interpolating ADC

An analog preprocessing structure, which performs the folding and interpolation operations, can be used to reduce respectively the number of comparators and preamplifiers needed in the flash ADC [15–18]. Figure 11.32 shows a sec-

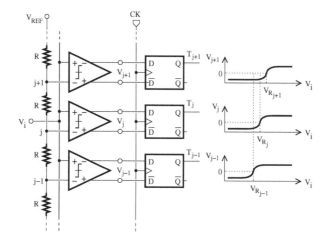

FIGURE 11.32

A section of the processing path in a flash ADC.

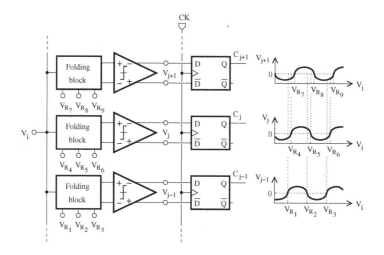

FIGURE 11.33

A section of the processing path in a folding ADC.

tion of the processing path in a flash ADC. For N-bit conversion, the input signal should be compared with at least $2^N - 1$ voltage reference levels and a different comparator is used for each possible digital code. By introducing folding stages, as shown in Figure 11.33 for three reference voltages, the locations of the zero-crossings in the signals now determine the cyclic code transitions to be identified by comparators. The folding factor F (here $F = 3$) indicates the number of zero-crossings included in the transfer characteristic. One comparator detects F reference voltages rather than one, as is the case in the full flash architecture, and the reference voltages of the folding amplifier should be sufficiently far apart to ensure a proper operation of comparators.

Let us consider a folding ADC including four folding amplifiers, whose

TABLE 11.7
Thermometer and Circular Codes of Numbers from 0 to 3

Circular Code				Thermometer Code			
C_3	C_2	C_1	C_0	T_2	T_1	T_0	
0	0	0	0	0	0	0	0
0	0	0	1	0	0	1	1
0	0	1	1	0	1	1	2
0	1	1	1	1	1	1	3
1	1	1	1	0	0	0	0
1	1	1	0	0	0	1	1
1	1	0	0	0	1	1	2
1	0	0	0	1	1	1	3

reference voltages are chosen as follows:

$$
\begin{aligned}
F_1 &: \quad (1/16)V_{REF} \quad (5/16)V_{REF} \quad (9/16)V_{REF} \quad (13/16)V_{REF} \\
F_2 &: \quad (2/16)V_{REF} \quad (6/16)V_{REF} \quad (10/16)V_{REF} \quad (14/16)V_{REF} \\
F_3 &: \quad (3/16)V_{REF} \quad (7/16)V_{REF} \quad (11/16)V_{REF} \quad (15/16)V_{REF} \\
F_4 &: \quad (4/16)V_{REF} \quad (8/16)V_{REF} \quad (12/16)V_{REF} \quad (16/16)V_{REF}
\end{aligned}
$$

The first cycle of the circular code, which is produced at the comparator outputs, is shown in Table 11.7. The equivalent thermometer code can be derived using a circular-to-thermometer encoder based on the following Boolean equations:

$$T_2 = C_2 \oplus C_3 \quad T_1 = C_1 \oplus C_3 \quad \text{and} \quad T_0 = C_0 \oplus C_3 \tag{11.48}$$

and implemented using XOR gates. There are two folds in one cycle. With a folding factor of 4 as in this example, the input full range is divided into two cycles or four folds. Because the folding characteristic is redundant, a cycle pointer is generally necessary to resolve the ambiguity in the output code.

A high degree of folding can result in a decrease in the signal bandwidth, which is caused by the presence of parasitic capacitors at the folder output nodes. In practice, F is then limited to eight and more quantization levels can be obtained using a parallel configuration of folding blocks. The necessary number of folding amplifiers or folders can still be high. The amplifier number is reduced using an interpolator with the factor I between two consecutive folding blocks. The interpolation is based on the signal division and can be realized by a resistor ladder, as shown in Figure 11.34, where $V_{R4} = (V_{R1} + V_{R7})/2$, $V_{R4} = (V_{R2} + V_{R8})/2$, and $V_{R6} = (V_{R3} + V_{R9})/2$.

The folding operation relies on mapping the input waveform into a repetitive signal, whose frequency is multiplied by the folding factor. Here, the

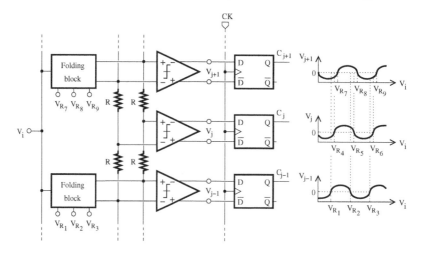

FIGURE 11.34
A section of the processing path in a folding and interpolating ADC.

amplitude quantization is transformed into the detection of zero-crossings of the folding signal. Figure 11.35 shows the circuit diagram of the j-th folding stage, where F is the folding factor. The output of the odd- and even-numbered differential transistor pairs are cross-coupled. The input reference voltages are defined by a resistive network. The polarity of the signal changes each time the input voltage attains a reference level.

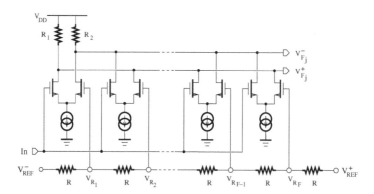

FIGURE 11.35
Implementation of a folding amplifier.

The interpolation can be realized using passive elements, and active elements and components. The circuit diagram of an interpolation circuit with a differential structure is shown in Figure 11.36(a). The voltage division is realized by a resistor ladder driven by nMOS source followers [18]. The interpolated signals are denoted by V_{i_k}, where $k = 1, 2, \cdots, I+1$, and I is the interpolation factor. The output signals of two neighboring folding stages can also be interpolated as shown in Figure 11.36(b). Due to the fact that only the zero-crossings contain the useful information, an accurate gain is not required for any amplifier of the interpolation circuit. The interpolation amplifiers can simply consist of differential transistor pairs, which can operate with a low supply voltage. Here, the interpolation by a factor of the form $I = 2^p$ is achieved by cascading p amplifier sections and $2^q + 1$ amplifiers are required in the q-th section.

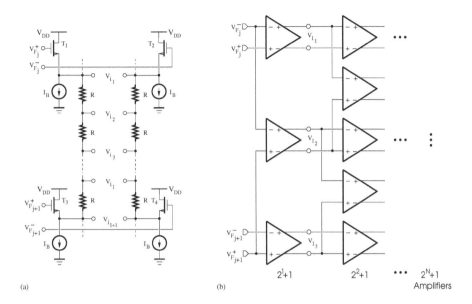

(a)　　　　(b)

FIGURE 11.36
Implementations of interpolating circuits using (a) resistors and (b) differential amplifiers.

Note that current-division techniques can also be exploited in order to implement the interpolation stage. This approach relies on the use of current comparators or I-to-V converters and can have the drawback of reducing the bandwidth of the folder circuit due to the extra node, which can be introduced in the signal path.

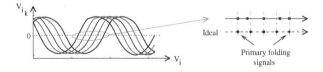

FIGURE 11.37
Illustration of the input stage nonlinearity effects on interpolated signals.

In the case of interpolations by a factor higher than 2, a systematic error is introduced in the position of zero-crossings due to the effect of the non-linear transfer characteristic of folding amplifiers on the signal values near the boundaries of the input range. Figure 11.37 shows signals obtained as a result of the interpolation by a factor of 4. Because of the symmetry of the transfer characteristic, the zero-crossing location of the interpolated signal in the middle remains unaffected. In comparison with the ideal case, where the zero-crossings should be uniformly distributed, the other two interpolated zero-crossings tend to move outward. This can be attributed to the amplitude mismatch between the folding signals and the interpolated signals. One way to restore the ideal zero-crossing points can be to extend the operation range of the folding amplifier.

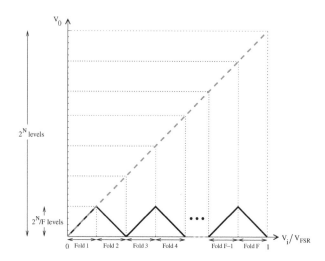

FIGURE 11.38
Transfer characteristic of an ideal folding ADC.

Ideally, the input-output characteristic of the folding amplifier is periodic and consists of piece-wise linear segments, or folds, whose number is related to the folding factor. As this leads to a repetitive code at the comparator output, a cycle pointer or coarse converter is needed in addition to a fine converter. The coarse converter ascertains in which period of the folding amplifier transfer characteristic the input signal lies. The transfer characteristic of a converter based on the folding principle is illustrated in Figure 11.38, where the input range is assumed to be divided into F regions or folds. Note that two folds constitute a folding cycle. An N-bit full flash ADC requires 2^N quantization levels, while a folding ADC needs $2^N/F$ levels, allowing the use of the folding signal signs for the determination of the LSBs and F levels for the generation of the MSBs of the output code.

The folding and interpolating converter, as shown in Figure 11.39, consists of two ADCs operating in parallel. The coarse ADC quantizes the input signal and provides the MSBs, while the fine ADC, which includes an analog preprocessing stage, is used for the generation of the LSBs. The outputs at the boundaries of the resistor interpolation structure are generally cross-connected to alleviate the effects caused by the asymmetrical nature of the network edges. As a result, the translational symmetry of the impulse response

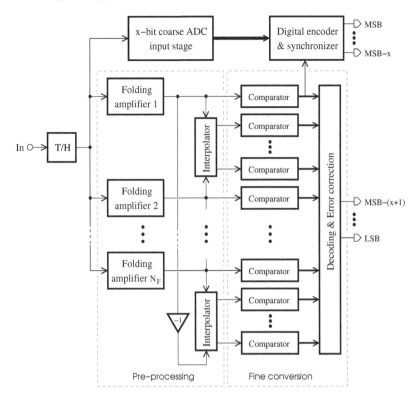

FIGURE 11.39
Block diagram of a folding and interpolating ADC.

of the resistor network is preserved. The mitigation of the delay effect introduced by the analog preprocessing between the MSBs and the other bits is achieved by a bit synchronization block. Otherwise, a misalignment of the different bits can be observed for high-frequency inputs. The decoder transforms the comparator output signals into a binary code.

The total resolution of the folding ADC is given by

$$N = N_{MSB} + N_{LSB}, \tag{11.49}$$

where N_{MSB} and N_{LSB} are the numbers of bits resolved in the coarse and fine converters, respectively. Let F be the folding factor, I denote the interpolation factor, N_F represent the number of primary folding signals, and N_I be the total number of interpolated signals [19]. We have

$$F = 2^{N_{MSB}} \tag{11.50}$$

and

$$N_I = 2^{N_{LSB}}, \tag{11.51}$$

while the total number of folding amplifier is computed as the ratio between the total number of primary folds and the interpolation factor,

$$N_F = N_I/I. \tag{11.52}$$

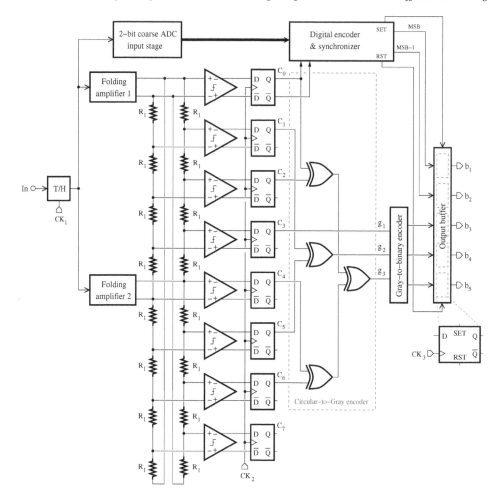

FIGURE 11.40
Block diagram of a 5-bit folding and interpolating ADC.

It can be shown that $N_F F = 2^N / I$. In practice, F and I are assumed to be a power of two. For a given resolution, the choice of F, I, and N_F is determined by the trade-off to be made between the bandwidth, speed, and power consumption of the converter.

The numbers of comparators in flash and folding ADCs are summarized in Table 11.8. For the same resolution, the flash ADC requires more comparators than the folding ADC. The use of the folding technique then results in an important reduction in the comparator number as the converter resolution is increased.

A 5-bit full flash ADC includes at less thirty-one comparators. Using a folding and interpolating architecture based on a 2-bit coarse ADC and a 3-bit fine ADC, as shown in Figure 11.40, the number of comparators can be reduced, leading to a decrease in chip area and power consumption [20]. In this case, the reference voltages for the folding amplifier F_1 and F_2 are,

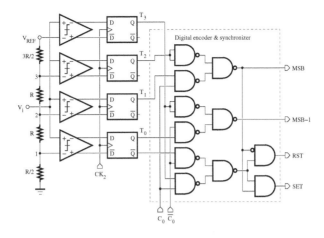

FIGURE 11.41
Block diagram of the coarse ADC.

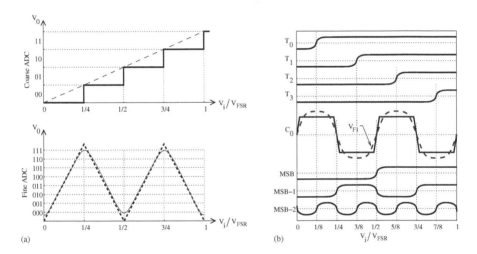

FIGURE 11.42
(a) Transfer characteristic of a 5-bit folding and interpolating ADC; (b) illustration of the synchronization between the coarse ADC bits and MSB-2.

respectively, as follows:

$$F_1: \quad (1/32)V_{REF} \quad (9/32)V_{REF} \quad (17/32)V_{REF} \quad (25/32)V_{REF}$$
$$F_2: \quad (5/32)V_{REF} \quad (13/32)V_{REF} \quad (21/32)V_{REF} \quad (29/32)V_{REF}$$

Only two primary folded signals are generated and the remaining ones are recovered by interpolation. By using two interpolators with a factor of 4, the fine ADC required eight comparators. The output code provided by these comparators is repeated for every sixteen quantization levels and the overall input range of the converter involves two folding cycles. Taking into account the four comparators required in the coarse ADC, which is based on the flash

TABLE 11.8

Number of Comparators in Flash ADC and Folding ADC versus the Converter Resolution

	Number of Comparators			
	5 bits	6 bits	7 bits	8 bits
Flash ADC	31	63	127	255
Folding ADC with a 2-bit coarse ADC	11	19	35	67
Folding ADC with a 3-bit coarse ADC	11	15	23	39

architecture, as shown in Figure 11.41, the total comparator count for the folding and interpolating converter is twelve.

The transfer characteristic of the 5-bit folding and interpolating ADC is shown in Figure 11.42(a). Due to the finite slew rate of the folding amplifier, the triangular characteristic is approximated by a sinusoidal-like waveform. The comparator outputs of the fine ADC form a circular code, which can be converted into Gray representation using XOR gates. The circular-to-Gray encoder is based on the next Boolean expressions:

$$g_1 = C_3 \tag{11.53}$$

$$g_2 = C_5 \oplus C_1 \tag{11.54}$$

$$g_3 = (C_6 \oplus C_4) \oplus (C_2 \oplus C_0). \tag{11.55}$$

Note that the bit C_7, which is not actually connected to the encoder, can be useful in the case where the circular code must be represented as a thermometer code.

Note that the transition between a group of 1s and a group of 0s in the circular code can also be detected using two-input XOR gates, whose Boolean equations are expressed as follows:

$$c_j = \begin{cases} \overline{C_0 \oplus C_7}, & \text{if } j = 0 \\ C_j \oplus C_{j-1}, & \text{if } j = 1, 2, \cdots, 7. \end{cases} \tag{11.56}$$

The XOR gate outputs, c_j, are then to be applied to a ROM structure for the conversion to Gray or binary representation.

For the coarse ADC, which is based on the flash architecture, the comparator outputs constitute a thermometer code to be converted into the binary representation. In addition to the minimum of three comparators required by a 3-bit flash converter, the coarse ADC also includes a comparator for the generation of the over-range signal. Ideally, the bit transitions of the coarse and fine ADCs should be exactly synchronized, as illustrated in Figure 11.42(b) for the 3 MSBs. However, this is not the case in practice because the coarse and fine ADCs operate independently and exhibit different delays. A bit-synchronization section is associated with the digital encoder of the coarse ADC to prevent errors from occurring in the output code, as shown in Figure 11.43. It should be noted that the critical regions are located near the MSB transitions. The coarse ADC bits can then be expressed as

$$MSB = C_0 \cdot T_1 + T_2 \tag{11.57}$$

$$MSB - 1 = \overline{C_0} \cdot T_0 + T_3, \tag{11.58}$$

where C_0 is obtained from the fine ADC and is used as a bit-synchronization signal.

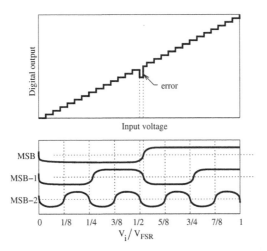

FIGURE 11.43
Effect of the misalignment between the coarse ADC bits and MSB-2.

For proper operation of the folding and interpolating ADC, the output code should be set to the maximum or minimum value, respectively, provided the input voltage is greater than the highest reference voltage or lower than the lowest reference voltage. However, the MSBs saturate at the end of the input range while the LSBs wrap around due to the circular nature of the folding signals. For instance, the code 00000 is changed to 00111 as the input voltage decreases; and with the input voltage increasing, the code 00111 becomes 00000. An over-range and under-range detection mechanism is required to set or reset the output latches. The signals to detect the over-range and under-range conditions are given by

$$RST = \overline{MSB} \cdot \overline{IR} \tag{11.59}$$

$$SET = MSB \cdot \overline{IR}, \tag{11.60}$$

where

$$IR = (C_0 + T_0) \cdot (\overline{C_0} + \overline{T_3}). \tag{11.61}$$

The logic state of IR can be used to flag out-of-range signals. Table 11.9

TABLE 11.9

Circular and Binary Codes for the 5-Bit Folding and Interpolating ADC

Circular Code								Binary Code				
C_7	C_6	C_5	C_4	C_3	C_2	C_1	C_0	MSB	MSB-1	MSB-2	MSB-3	LSB
0	0	0	0	0	0	0	0	0	0	0	0	0
0	0	0	0	0	0	0	1	0	0	0	0	1
0	0	0	0	0	0	1	1	0	0	0	1	0
0	0	0	0	0	1	1	1	0	0	0	1	1
0	0	0	0	1	1	1	1	0	0	1	0	0
0	0	0	1	1	1	1	1	0	0	1	0	1
0	0	1	1	1	1	1	1	0	0	1	1	0
0	1	1	1	1	1	1	1	0	0	1	1	1
1	1	1	1	1	1	1	1	0	1	0	0	0
1	1	1	1	1	1	1	0	0	1	0	0	1
1	1	1	1	1	1	0	0	0	1	0	1	0
1	1	1	1	1	0	0	0	0	1	0	1	1
1	1	1	1	0	0	0	0	0	1	1	0	0
1	1	1	0	0	0	0	0	0	1	1	0	1
1	1	0	0	0	0	0	0	0	1	1	1	0
1	0	0	0	0	0	0	0	0	1	1	1	1
0	0	0	0	0	0	0	0	1	0	0	0	0
0	0	0	0	0	0	0	1	1	0	0	0	1
0	0	0	0	0	0	1	1	1	0	0	1	0
0	0	0	0	0	1	1	1	1	0	0	1	1
0	0	0	0	1	1	1	1	1	0	1	0	0
0	0	0	1	1	1	1	1	1	0	1	0	1
0	0	1	1	1	1	1	1	1	0	1	1	0
0	1	1	1	1	1	1	1	1	0	1	1	1
1	1	1	1	1	1	1	1	1	1	0	0	0
1	1	1	1	1	1	1	0	1	1	0	0	1
1	1	1	1	1	1	0	0	1	1	0	1	0
1	1	1	1	1	0	0	0	1	1	0	1	1
1	1	1	1	0	0	0	0	1	1	1	0	0
1	1	1	0	0	0	0	0	1	1	1	0	1
1	1	0	0	0	0	0	0	1	1	1	1	0
1	0	0	0	0	0	0	0	1	1	1	1	1

summarizes the output codes of the 5-bit folding and interpolating ADC. The MSB and MSB-1 delivered by the coarse ADC are necessary for the identification of each fold, and the MSB-2, MSB-3, and LSB are derived from the circular code available at the comparator outputs of the fine ADC.

CMOS folding and interpolating converters can achieve resolutions of 8 to

10 bits at sampling frequencies comparable to that of a flash ADC and are suitable for low-power applications. However, the converter operation can be affected by nonideal effects, such as offsets in the comparators required for the zero-crossing detection. Furthermore, without a front-end track-and-hold circuit, the converter performance can be limited by distortions due to the nonlinear transfer characteristic of folders.

11.1.6 Sub-ranging ADC

A solution for the reduction of the hardware growth with the number of bits resolved can consist of achieving the analog-to-digital conversion in two steps as shown in Figure 11.44 [21]. The input is first tracked-and-held and then digitized by the L-bit ADC to produce the MSBs, which are applied to the DAC and the result is subtracted from the T/H output signal. This difference denotes the residue of the first step and is extended to the full scale by a 2^L amplification before being processed by the (N-L)-bit ADC, which determines the LSBs. A summer can then combine a delayed version of the MSBs and the LSBs to yield the final N-bit word.

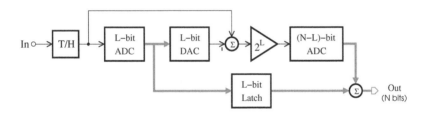

FIGURE 11.44
Block diagram of a sub-ranging ADC.

Sub-ranging ADCs can reach an accuracy of 10 to 12 bits and possess the latency of two clock cycles. This latter is the delay between the instant of the conversion initiation of an input sample and the instant at which the corresponding digital data are being made available. With a speed comparable to the one of a flash ADC, a sub-ranging ADC uses far less hardware for the same resolution. For example, an N-bit flash converter requires $2^N - 1$ comparators, while a two-stage sub-ranging structure needs only $2(2^{N/2} - 1)$ comparators. However, the design of high-resolution sub-ranging ADCs remains limited by practical circuit nonidealities such as component mismatches, charge injection, offset, noise, and finite amplifier gain and bandwidth.

11.1.7 Pipelined ADC

A pipelined ADC is another type of sub-ranging ADC, derived by breaking a high-resolution conversion into multiple steps. Pipelined converters are attractive for applications such as image and video processing, digital communication, and instrumentation that require a resolution from 10 to 16 bits and data throughput greater than 5 Ms/s.

The pipelined ADC architecture shown in Figure 11.45 includes a T/H circuit, a cascade of $M - 1$ coarse conversion stages, followed by a fine converter. The cascaded stages are structurally similar, consisting of a sub-ADC (SADC), a DAC, and an amplifier with a gain factor of 2^{n_k}. Each of these stages, for instance, performs a coarse conversion of the incoming full-scale ramp signal and generates a residue signal that corresponds to an amplified version of the quantization error. The parameter n_k $(k = 1, 2, \cdots, M - 1)$ represents the number of bits resolved by the k-th stage of the converter, and the fine converter exhibits a resolution of n_M bits.

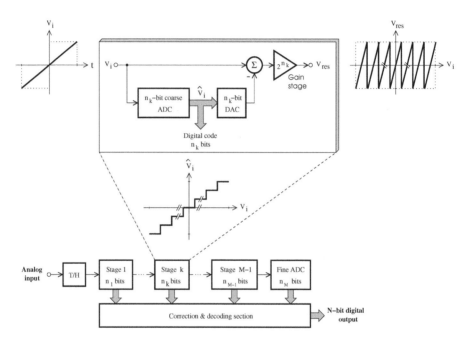

FIGURE 11.45
Block diagram of a pipelined ADC.

The first processing step is performed by the T/H circuit and amounts to acquiring a sample of the input signal. The resulting output is digitized by the SADC to provide the first n_1 MSBs, which are also transformed into the corresponding analog voltage by the n_1-bit DAC. The residue obtained by subtracting the n_1-bit DAC output from the sampled-and-held input is amplified by a factor of 2^{n_1} so that its maximum swing is readjusted to the ADC full scale. This amplified residue is passed to the subsequent stages, where identical operations are performed. Finally, the different digital codes are appropriately synchronized to eliminate the clock delay and combined to produce the required N-bit full ADC resolution. Thus,

$$N = \sum_{k=1}^{M-1} n_k + n_M .$$ (11.62)

Note that the amplification of the residue signal helps keep the signal level constant, allowing the use of the same reference source for all sub-ADCs and yielding a reduced sensibility of the last stages to circuit imperfections such as noise, offset voltages, switch charge injection, and parasitic-loading capacitors.

Furthermore, the insertion of a track-and-hold circuit between the pipelined stages to allow a concurrent operation of all stages can result in a high conversion throughput. However, this is achieved at the cost of increased latency and power consumption.

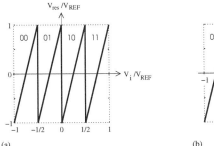
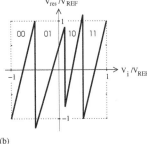

(a)   (b)

FIGURE 11.46
(a) Ideal and (b) practical residue-input signal transfer characteristics.

Let us consider a 2-bit pipelined stage for the analysis of the circuit imperfection effects on the residue waveform for a full-scale ramp input signal [22, 23]. The ideal relationship between the amplified residue signal, V_{res}, and the input signal is shown in Figure 11.46(a). The reference levels $(0, \pm V_{REF}/2)$ of the flash SADC set the position of transitions, whose magnitude at the code boundary is determined by the DAC and the gain of the interstage amplifier. The different output voltage contributions of the DAC are $\pm V_{REF}/4$ and $\pm 3V_{REF}/4$. The signal V_{res} is a linearly increasing function of the input, the magnitude of which is between two adjacent thresholds of the SADC comparator. Once the input signal reaches a threshold level, the signal V_{res} abruptly changes in the opposite direction while remaining within the full scale. Note that the residue is multiplied by the interstage gain of 4 to exactly fit the full-scale of the next stage.

Practical converters can exhibit some linearity errors (see Figure 11.46(b)) such as missing codes. This can be the result of a missing decision level or the loss of a digital code at a DAC input caused by a residual voltage, which exceeds the actual conversion range due to the gain error of the interstage amplifier and offset voltages.

The over-range errors of the residual voltage can be corrected by reducing the interstage gain to 2 (see Figure 11.47(a)) [22]. For a full-scale input, the signal V_{res} of the pipelined stage k should remain between $-V_{REF}/2$ and $V_{REF}/2$. Any excursion of V_{res} outside this range is considered an error, which can be detected by adding extra quantization levels to the stage $k+1$. The SADC outputs representing these quantization levels are used by the digital correction to either increment or decrement the output code of the stage k for a residue signal greater than $V_{REF}/2$ or less than $-V_{REF}/2$, respectively. This correction approach works successfully for decision errors as large as $\pm V_{REF}/4$ or $\pm 1/2$ LSB.

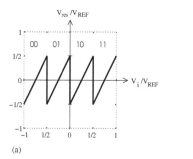
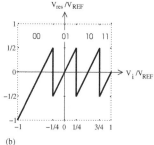

(a)   (b)

FIGURE 11.47
Residue-input signal transfer characteristics (a) with a gain of 2 and (b) with $V_{REF}/4$ offset voltage.

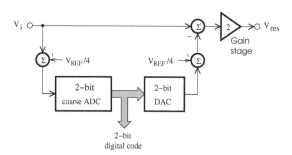

FIGURE 11.48
An equivalent model of the 2-bit stage with offset adjustments.

The arithmetic of the correction section can be reduced to an addition operation using the stage architecture of Figure 11.48. By appending a $-V_{REF}/4$ offset voltage to the SADC and DAC, the signal V_{res} now varies from $-V_{REF}$ to $V_{REF}/2$. Because the locations of the SADC decision levels and the DAC reference voltages are uniformly shifted by the offset value, a negative over-range condition is prevented for errors up to $\pm V_{REF}/4$. This eliminates the requirement for the subtraction function in the correction logic.

As shown in Figure 11.47(b), the top decision level can be fixed at $V_{REF}/4$ for each stage except the last one. The comparator with the threshold at

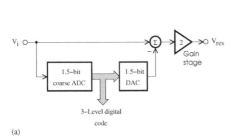 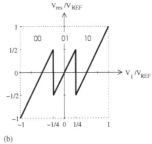

(a) (b)

FIGURE 11.49

(a) A 1.5-bit stage of the pipelined ADC; (b) residue-input signal transfer characteristic.

$3V_{REF}/4$ is not required because the amplified residue signal falling above the expected range can be detected by the next stage. The MSB of each pipelined stage can then be resolved by an SADC with only 1.5-bit resolution. In this case, the comparator reference voltages of the SADC are at $\pm V_{REF}/4$ and the output levels of the DAC include $\pm V_{REF}/2$ and 0. Table 11.10 summarizes the characteristics of the 2-bit and 1.5-bit pipelined stages.

TABLE 11.10

Characteristics of 2-Bit and 1.5-Bit Pipelined Stages

	2-Bit/stage	1.5-Bit/stage
Input range	From $-V_{REF}$ to V_{REF}	From $-V_{REF}$ to V_{REF}
SADC threshold levels	$-V_{REF}/2,\ 0,\ V_{REF}/2$	$-V_{REF}/4,\ V_{REF}/4$
Number of comparators	3	2
DAC reference levels	$-3V_{REF}/4,\ -V_{REF}/4,$ $V_{REF}/4,\ 3V_{REF}/4$	$-V_{REF}/2,\ 0,\ V_{REF}/2$
Output digital code	00, 01, 10, 11	00, 01, 10
Inter-stage gain	4	2

Here, the nonideal effect of the component is reduced by introducing a redundancy in the pipelined ADC. That is, the resolution of the converter should be less than the sum of the ones provided by single stages. The extra bits are eliminated at the output by the digital correction.

For the switched-capacitor (SC) implementation, the block diagram of the 1.5-bit pipelined stage can be scaled as shown in Figure 11.50. This allows the use of an adequate clocking scheme to maintain a synchronization between the signal paths. Let V_i be the input voltage and V_{REF} denote the reference voltage. Ideally, the amplified residue signal generated by the k-th stage can then be written as

$$V_{res(k+1)} = 2V_{res(k)} - D_k V_{REF}, \qquad k = 1, 2, \cdots, M-1, \qquad (11.63)$$

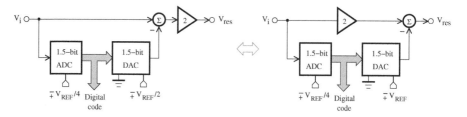

FIGURE 11.50

Scaling of the 1.5-bit pipelined stage for the switched-capacitor implementation.

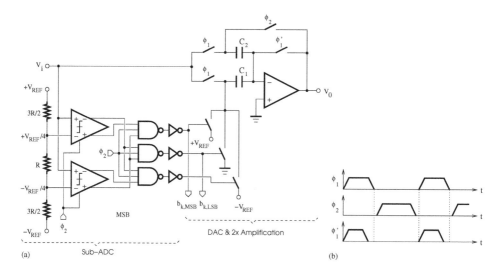

FIGURE 11.51

(a) A switched-capacitor implementation of the 1.5-bit stage circuit; (b) timing diagram.

where

$$D_k = \begin{cases} -1, & \text{if } V_{res(k)} < -V_{REF}/4 \\ 0, & \text{if } -V_{REF}/4 < V_{res(k)} < V_{REF}/4 \\ 1, & \text{if } V_{res(k)} > V_{REF}/4 \end{cases} \quad (11.64)$$

and $V_{res(1)} = V_i$. A single-ended SC implementation of the 1.5-bit pipelined stage is depicted in Figure 11.51. This circuit includes a flash SADC and a multiplying DAC (MDAC), which performs the multiplication by two followed by the digital-to-analog conversion of the SADC output, the track-and-hold function and the two times amplification of the input signal, and the subtraction. It operates with a two-phase nonoverlapping clock signals. To reduce the effect of the difference in propagation delay between the signal paths, the charges due to the held and reconverted versions of the input signal are transferred toward the output during the same clock phase, ϕ_2. The comparators used in the SADC can be designed using an input-sensing preamplifier

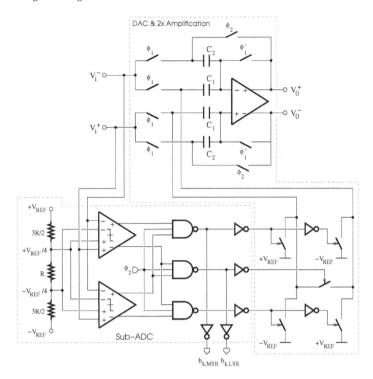

FIGURE 11.52

Differential version of the switched-capacitor implementation of the 1.5-bit stage circuit.

followed by a regenerative latch, and the control signals of the MDAC are derived from the outputs of a thermometer-to-binary encoder. The amplifier should be designed with a dc gain and bandwidth such that it can settle within $\pm 1/2$ LSB accuracy in one half of the clock signal period. The output signal, V_0, is given by

$$
V_0 = \begin{cases}
\left(1 + \dfrac{C_1}{C_2}\right) V_i - V_{REF}, & \text{if } V_i > V_{REF}/4, \\[2mm]
\left(1 + \dfrac{C_1}{C_2}\right) V_i, & \text{if } -V_{REF}/4 \le V_i \le V_{REF}/4, \\[2mm]
\left(1 + \dfrac{C_1}{C_2}\right) V_i + V_{REF}, & \text{if } V_i < -V_{REF}/4.
\end{cases} \qquad (11.65)
$$

The stage gain of 2 is realized for $C_1 = C_2$. By requiring only two capacitors, the amplifier loading requirement of the resulting 1.5-bit pipelined stage is relaxed, yielding a reduced sensibility to noise and an improved operation speed.

A differential version of the 1.5-bit pipelined stage is shown in Figure 11.52 [24]. The zero reference is implemented by shorting the DAC output together. It should be noted that differential circuits have the advantage of increasing the signal dynamic range.

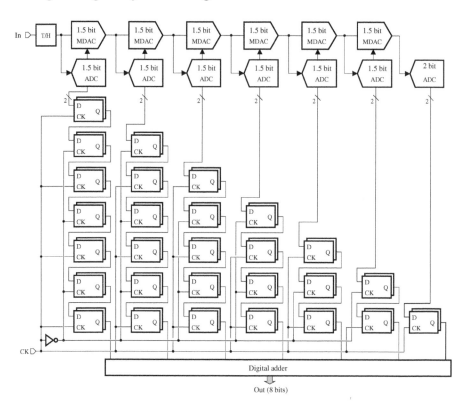

FIGURE 11.53
Block diagram of an 8-bit pipelined ADC including a digital correction stage.

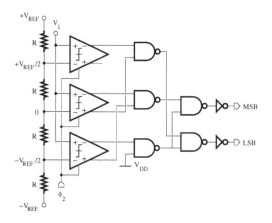

FIGURE 11.54
Circuit diagram of a 2-bit flash ADC.

In a pipelined ADC, the first stage generates the MSB of the resulting digital code, while subsequent stages increase the resolution of the input signal conversion by delivering additional bits in less significant positions of the output code. Figure 11.53 shows the block diagram of an 8-bit pipelined ADC, which is based on 1.5-bit stage with an inter-stage amplifier gain of 2. The amplified residue of each pipelined stage, except the last one, is quantized by the following stage. Because the last stage does not need to generate a residue, it can be implemented as the 2-bit flash ADC shown in Figure 11.54. A resistor network is used to set the reference levels of the three latched comparators to $-V_{REF}/2$, 0, and $V_{REF}/2$, respectively, and the thermometer-to-binary encoder is based on few logic gates.

Adjacent stages of the pipelined ADC should be driven by the clock signals with opposite phases to ensure a concurrent operation. There is a delay of a half clock cycle between the instants at which the outputs of two consecutive stages are available. One input signal sample then requires several clock cycles to be processed by the overall pipelined ADC. The output codes ($b_{k,MSB}$ and $b_{k,LSB}$, $k = 1, 2, \cdots, M$) of the different pipelined stages are synchronized using an array of D latches, and then summed to produce the converter output code with B_1 being the MSB.

By exploiting the 0.5-bit redundancy on each stage to correct any decision error in the previous adjacent stage, the concept of the error correction circuit can be illustrated as follows:

$$
\begin{array}{cccccccc}
b_{1,MSB} & b_{1,LSB} & & & & & & \\
& b_{2,MSB} & b_{2,LSB} & & & & & \\
& & \cdots & \cdots & & & & \\
& & & \cdots & \cdots & & & \\
& & & & b_{M-2,MSB} & b_{M-2,LSB} & & \\
& & & & & b_{M-1,MSB} & b_{M-1,LSB} & \\
& & & & & & b_{M,MSB} & b_{M,LSB} \\
+ & & & & & & & \\
\hline
B_1 & B_2 & \cdots & \cdots\;\cdots & & B_{N-2} & B_{N-1} & B_N
\end{array}
$$

This correction scheme works well as long as the comparator offset magnitudes are not so high (i.e., less than $V_{REF}/4$) as to cause missing codes. In this example, the digital correction should remove the redundancy and deliver an 8-bit output data. It can simply be based on addition. Thus, the correction consists of summing the MSB of a given stage with the LSB of the previous stage. Here, the digital correction cannot be extended to the LSB of the last stage. This implies that either the last stage should be implemented as a 2-bit full flash ADC or using two 1.5-bit stages with the LSB of the final stage being excluded from the converter output code. Figure 11.55 shows the block diagram of a digital adder that may be used in the 1.5-bit/stage pipelined ADC. The propagation of the carry bit is achieved in the direction of the output code MSB. Each 1.5-bit pipelined stage effectively provides 1 bit to the overall resolution. The resulting output data is coded in offset binary format, which is identical with the two's complement representation except that the MSB must be inverted.

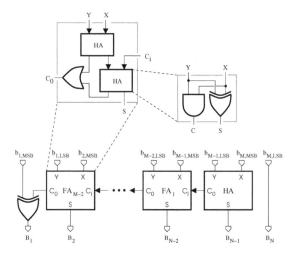

FIGURE 11.55

Block diagram for the digital correction stage (HA: half adder; FA: full adder).

For the 1.5-bit/stage pipelined ADC, the sequence of output digits does not form a conventional binary number representation because the bits are signed. The 1.5-bit stage generates three digits and exhibits a radix of 2. Because the number of digits is greater than the radix, there is a redundancy in the sequence of output digits. Hence, more than one combination of digits can be used to represent the same magnitude of a signal sample. The 1.5-bit/stage pipelined ADC is then equivalent to a redundant signed digit system.

It should be noted that, from the input to the output of the converter, the input-referred noise contribution of the pipelined stages is reduced after each stage due to the cumulative scaling effect of the interstage gain on the signal level. Hence, capacitors can be scaled down toward the later stages to reduce the power consumption [25].

Consider both circuit diagrams of Figure 11.56, which represent the different versions of the MDAC generally used in the implementation of the 1.5 bit per stage pipelined ADC architecture. During the clock phase ϕ_1, $(n - 1 < t \le n - 1/2)$, according to the MDAC circuit depicted in Figure 11.56(a), the input voltage is sampled onto capacitors C_1 and C_2, and the amplifier is configured as a unity-gain follower. Assuming that A_0 is the amplifier dc gain, it can be shown that

$$V_0(n - 1/2) = A_0[V^+(n - 1/2) - V^-(n - 1/2)], \qquad (11.66)$$

where $V^+(n - 1/2) = V_{off}$ and $V^-(n - 1/2) = V_0(n - 1/2)$. During the clock phase ϕ_2, $(n - 1/2 < t \le n)$, the capacitor C_1 is connected to $D_k V_{REF}$, while C_2 acts as a feedback capacitor. The application of the charge conservation law at the negative input terminal of the amplifier

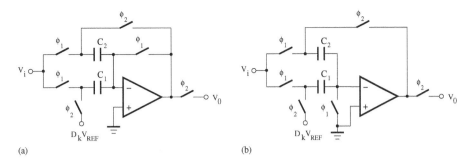

FIGURE 11.56
Two configurations of the SC MDAC for the 1.5-bit stage.

yields

$$C_1 \left\{ [D_k V_{REF} - V^-(n)] - [V_i(n - 1/2) - V^-(n - 1/2)] \right\}$$
$$= C_2 \left\{ [V^-(n) - V_0(n)] - [V^-(n - 1/2)) - V_i(n - 1/2)] \right\} \tag{11.67}$$

and

$$V_0(n) = A_0[V^+(n) - V^-(n)], \tag{11.68}$$

where $V^+(n) = V_{off}$. Because

$$V^-(n - 1/2) = V_0(n - 1/2) = \frac{\mu}{1 + \mu} V_{off} \tag{11.69}$$

and

$$V^-(n) = V_{off} - \mu V_0(n), \tag{11.70}$$

where $\mu = 1/A_0$, it can be shown that

$$\left[1 + \mu \left(1 + \frac{C_1}{C_2}\right)\right] V_0(n)$$
$$= \left(1 + \frac{C_1}{C_2}\right) V_i(n - 1/2) - \frac{C_1}{C_2} D_k V_{REF} + \frac{\mu}{1 + \mu} \left(1 + \frac{C_1}{C_2}\right) V_{off}. \tag{11.71}$$

Finally, the output-input relationship in the z-domain can be derived as

$$V_0(z) = \frac{\left(1 + \dfrac{C_1}{C_2}\right) z^{-1/2} V_i(z) - \dfrac{C_1}{C_2} D_k V_{REF} + \dfrac{\mu}{1 + \mu} \left(1 + \dfrac{C_1}{C_2}\right) V_{off}}{1 + \mu \left(1 + \dfrac{C_1}{C_2}\right)}. \tag{11.72}$$

An alternative MDAC structure is shown in Figure 11.56(b). Its operation is similar to the first one, except during the phase ϕ_2, where the amplifier is now used in the open-loop configuration and $V^-(n - 1/2) = 0$. The output-input relationship in the z-domain is then of the form

$$V_0(z) = \frac{\left(1 + \dfrac{C_1}{C_2}\right) z^{-1/2} V_i(z) - \dfrac{C_1}{C_2} D_k V_{REF} + \left(1 + \dfrac{C_1}{C_2}\right) V_{off}}{1 + \mu \left(1 + \dfrac{C_1}{C_2}\right)}. \quad (11.73)$$

For a sufficiently high dc gain, μ tends to zero and it appears that the second MDAC exhibits an increased sensibility to the effect of the offset voltage. Ideally, the input signal is multiplied by a factor of 2 for $C_1 = C_2$.

Assuming that the effects of the amplifier imperfections are negligible, the linearity of the converter can still be limited by the capacitor mismatch. With α_k being the ratio mismatch between C_1 and C_2, that is, $C_1 = (1 + \alpha_k)C_2$, the output voltage of the MDAC can be expressed as

$$V_0(z) = 2(1 + \alpha_k/2)z^{-1/2} V_i(z) - D_k(1 + \alpha_k)V_{REF}. \quad (11.74)$$

In order to fulfill the N-bit resolution requirement, any deviation error due to a component imperfection should be no larger than LSB/2, or $1/2^{N-k-1}$ for the k-th pipelined stage.

The correction logic in the 1.5-bit/stage pipelined ADC is implemented as an addition with the carry propagation of the overlapping correction bits. Although this technique is very simple, it can only correct the offset voltage effect of SADC comparators. The resulting resolution is then limited to 10 bits due to the effect of errors introduced by the finite gain and bandwidth of the amplifier, capacitor mismatches, and noise on the accuracy of the digital-to-analog conversion and interstage amplifier gain.

The use of laser wafer trimming to adjust the values of circuit components during the IC test results in the enhancement of the converter resolution. But this approach can be limited by aging, and temperature variations, which can affect the matching accuracy, and the extra production cost

To enable a high resolution, various other circuit techniques (digital calibration, capacitor averaging, dithering, gain-boosting method) can be used [26–29], generally at the cost of an increased circuit complexity and power consumption, and a reduced conversion speed.

With a uniform per-stage resolution, the design is modular, but the use of a multi-bit stage at the converter input can greatly relax the matching and noise requirements for the following stages. In [30], a converter resolution of 14 bits is achieved using a 4-bit pipelined stage followed by eight 1.5-bit stages and 3-bit flash ADC. Note that the stage resolution is generally not greater than 4 bits to maintain the converter power consumption at an acceptable level.

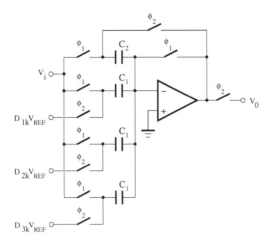

FIGURE 11.57
Circuit diagram of the 2.5-bit SC MDAC.

The principle of the redundant sign digit coding can also be exploited in the design of pipelined stages with a resolution greater than 1.5 bits.

In the specific case of a 2.5-bit stage, the SADC has six reference levels of the forms, $\pm V_{REF}/8$, $\pm 3V_{REF}/8$, and $\pm 5V_{REF}/8$. In the ideal case, the amplified residue signal generated by the k-th stage can be obtained as

$$V_{res(k+1)} = 4V_{res(k)} - D_k V_{REF}, \qquad k = 1, 2, \cdots, M-1, \qquad (11.75)$$

where

$$D_k = \begin{cases} -3, & \text{if} \quad V_{res(k)} < -5V_{REF}/8 \\ -2, & \text{if} \quad -5V_{REF}/8 < V_{res(k)} < -3V_{REF}/8 \\ -1, & \text{if} \quad -3V_{REF}/8 < V_{res(k)} < -V_{REF}/8 \\ 0, & \text{if} \quad -V_{REF}/8 < V_{res(k)} < V_{REF}/8 \\ 1, & \text{if} \quad V_{REF}/8 < V_{res(k)} < 3V_{REF}/8 \\ 2, & \text{if} \quad 3V_{REF}/8 < V_{res(k)} < 5V_{REF}/8 \\ 3, & \text{if} \quad V_{res(k)} > 5V_{REF}/8 \end{cases} \qquad (11.76)$$

and $V_{res(1)} = V_i$. To maintain the residue output voltage within $\pm V_{REF}$, the value of the comparator offset voltage should not exceed $\pm V_{REF}/8$. The representation in the binary format of the SADC comparator outputs can result in seven codes, which are 000, 001, 010, 011, 100, 101, and 110.

The circuit diagram of a 2.5-bit MDAC is shown in Figure 11.57, where C_1 and C_2 should have the same value. During the first phase of the clock signal, the capacitors C_1 are connected to the input voltage. In the second clock phase, the capacitor C_2 is included in the amplifier feedback loop, while the capacitors C_1 are set to the appropriate values of the reference voltage. Ideally, the output voltage is given by

$$V_0(z) = \left(1 + \frac{3C_1}{C_2}\right)z^{-1/2}V_i(z) - \frac{C_1}{C_2}(D_{1k} + D_{2k} + D_{3k})V_{REF}, \quad (11.77)$$

where D_{1k}, D_{2k}, and D_{3k} are equal to -1, 0, or 1, depending on the output of the SADC. The MDAC output signal corresponding to a given SADC digital code is generated by appropriately switching each of the three capacitors C_1 to one of the reference levels, $-V_{REF}$, 0, or V_{REF}.

11.1.8 Algorithmic ADC

Algorithmic or cyclic analog-to-digital converters (ADCs) [31–34], which are based on the binary division principle, are useful in applications requiring low power consumption and chip area. The core of an algorithmic ADC, as shown in Figure 11.58, consists of a T/H circuit, precision multiply-by-two amplifier, comparator, and summer.

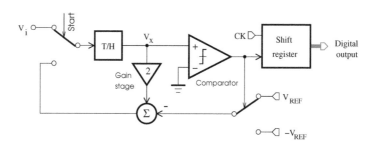

FIGURE 11.58
Block diagram of an algorithmic ADC.

Let k be the number of conversion cycles and b_k the bit to be determined. The residue signal, V_x, which is considered the partial remainder of a division, can be expressed as

$$V_{x(k)} = \begin{cases} 2V_i, & \text{for} \quad k = 1 \\ 2V_{x(k-1)} + (-1)^{b_{k-1}}V_{REF}, & \text{otherwise,} \end{cases} \quad (11.78)$$

where

$$b_k = \begin{cases} 1, & \text{if} \quad V_{x(k-1)} \geq V_{REF} \\ 0, & \text{if} \quad V_{x(k-1)} < V_{REF} \end{cases} \quad (11.79)$$

and $k = 1, 2, 3, \cdots, N$. The iterative execution of this algorithm produces a digital output code in the offset binary representation. At each conversion step, the operation of the algorithmic ADC involves a magnitude comparison, a bit selection, and a decision-dependent summation.

The input signal is first selected using the input switch before being sampled by the T/H circuit. The resulting signal, V_x, is applied at the input of the comparator. If V_x is greater than zero, the selected bit will be set to the high logic state and V_{REF} will be subtracted from $2V_x$; otherwise, this bit will take the low logic state and V_{REF} will be added to $2V_x$. By closing the feedback loop through the operation of the input switch, the residue is sent back to the input of the T/H circuit and the determination of the remaining LSB bits proceeds in the same manner. It should be emphasized that each bit of the digital output code is kept in the shift register after its determination.

Starting with the MSB, each bit b_k is determined sequentially, depending on the polarity of $V_{x(k)}$. By iterating up to $k = N$, we can obtain

$$V_{x(N)} = 2^{N-1} \left[V_i + \left(\sum_{k=1}^{N-1} (-1)^{b_k} 2^{-k} \right) V_{REF} \right]. \qquad (11.80)$$

The signal V_x can be considered the residue generated after the determination of each bit. Hence, the set of bits b_k $(k = 1, 2, \cdots N)$ is a binary representation of a fractional number equal to V_i/V_{REF}, where V_i is a bipolar signal with a range from $-V_{REF}$ to V_{REF}.

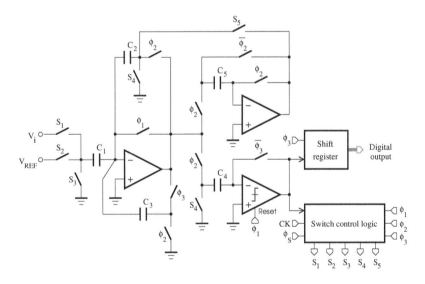

FIGURE 11.59
Circuit diagram of an algorithmic ADC.

To achieve a high resolution, the algorithmic ADC circuit should be designed such that its sensitivity to component imperfections is minimized [35, 36]. For a given resolution, the deviation of the converter characteristic due to these nonidealities should be maintained well below 1/2 LSB. Figure 11.59 shows the circuit diagram of a ratio-independent algorithmic ADC, which requires a three-phase clock signal to overcome the effects of capacitor mismatches and offset voltages. To perform the voltage multiplication by a factor

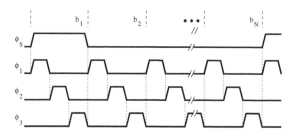

FIGURE 11.60
Clock signals for the algorithmic ADC.

of 2 independently of the capacitor ratio, the charge of the input capacitor C_3 due to the residue voltage, V_x, is transferred onto the feedback capacitor, C_2, initially charged to V_x. The operation of the ratio-independent algorithmic ADC is better understood by analyzing the charge transfer between capacitors during the three phases of the clock signal.

During the first phase, a sample of the input signal is connected to the converter and the charge due to the input voltage is stored on the capacitor C_1.

In the second phase, C_2 is connected to the amplifier output, a charge transfer takes place between C_1 and C_2, and the output of the input amplifier is set to the voltage V_x, which is used to charge the capacitors C_4 and C_5 of the next stages.

During the third phase, the charge on C_2 is transferred onto C_3, while the charge on C_5 is maintained and the MSB is resolved by the comparator. Note that the charge transfer is not affected by C_1 because one of its input node is floating.

For the first phase of the determination of the next bit, the capacitor C_1 is charged to V_{REF} and the charge produced by V_x on C_5 is transferred onto C_2.

In the second phase, C_2 acts as a feedback capacitor and its initial charge is combined with the ones on C_1 and C_3 to update the value of the residue signal, thereby resulting in the summation of the two charge contributions associated with the input voltage and due to the reference voltage. The derived residue is used to charge the capacitors C_4 and C_5 connected to the output of the input amplifier.

During the third phase, the state of the output bit is determined by the comparator and the charge on C_2 is transferred onto C_3.

These last three phases are then repeated for each of remaining bits to be resolved. The ratio-independent algorithmic ADC requires three clock phases for the determination of each bit; therefore an N-bit conversion is performed in $3N$ clock phases.

The clock signals for a bipolar conversion can be generated according to the Boolean logic equations of Table 11.11. In the case of a unipolar conversion

TABLE 11.11
Switch Control Signals for a Bipolar Conversion

	MSB	Bit b_k $(k > 1)$	Bit b_k
S_1	$\phi_S \cdot \phi_1$		$\phi_S \cdot \phi_1$
S_2		$\overline{\phi}_S(\overline{b}_{k-1}\phi_1 + b_{k-1} \cdot \phi_2)$	$\overline{\phi}_S(\overline{b}_{k-1}\phi_1 + b_{k-1} \cdot \phi_2)$
S_3	$\phi_S \cdot \phi_2$	$\overline{\phi}_S(\overline{b}_{k-1}\phi_2 + b_{k-1} \cdot \phi_1)$	$\phi_S \cdot \phi_2 + \overline{\phi}_S(\overline{b}_{k-1}\phi_2 + b_{k-1} \cdot \phi_1)$
S_4	$\phi_S(\phi_1 + \phi_3)$	$\overline{\phi}_S \cdot \phi_3$	$\phi_S(\phi_1 + \phi_3) + \overline{\phi}_S \cdot \phi_3$
S_5		$\overline{\phi}_S \cdot \phi_1$	$\overline{\phi}_S \cdot \phi_1$

with the input signal varying between 0 and V_{REF}, the converter switching scheme should be slightly modified. It is necessary to exchange the control signals for switches S_2 and S_3 during the determination of the MSB. As a result, the first residue will be of the form $V_{x(1)} = V_i - V_{REF}$.

The accuracy of the algorithmic ADC is affected by the nonlinearities due to charge injections of switches, and offset voltages of the amplifier and comparator. The amplifier can be designed to have enough gain and speed such that the deviations due to the finite gain and settling time are greatly reduced. But, the remaining uncompensated component imperfections limit the achievable resolution to about 10 bits.

11.1.9   Time-interleaved ADC

A solution for the design of data converters operating with a speed beyond the fundamental technological limit can consist of interleaving in time more than one ADC.

One approach to control the sampling operation in time-interleaved structures is to use several clock signals, as shown in Figure 11.61, where M is the number of parallel channels. Ideally, the effective sampling rate of the resulting structure is increased by M times in comparison to the one of a single ADC. However, due to the difference in the delays between successive sampling instants, dynamic distortions can be observed in the spectrum of the output signal. This results in a degradation of the overall ADC performance. A way to reduce the timing mismatches is to use the structure of Figure 11.62, where the input signal is sampled at a high rate and then multiplexed over M parallel channels to be processed by ADCs. The drawback of this approach is the limited input bandwidth due to the increased capacitive loading of the input sampling stage.

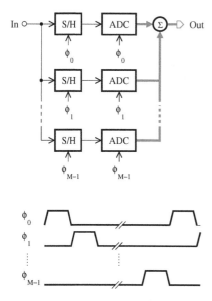

FIGURE 11.61
Time-interleaved ADC with identical parallel channels.

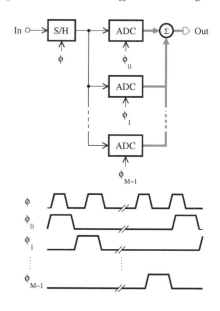

FIGURE 11.62
Time-interleaved ADC with a single-input S/H circuit.

By relying on the definition

$$\widehat{\mathbf{V}}_i(\omega) = \sum_{n=-\infty}^{+\infty} v_i(t_n)e^{-j\omega nT} \tag{11.81}$$

to compute the signal spectrum as if the signal has been sampled uniformly, the digital output spectrum of the converter processing a sine wave will exhibit line spectra whose magnitudes are frequency dependent.

Let $v_i(t_n)$ denote the sequences obtained by sampling the band-limited input signal at the instants t_n given by

$$t_n = (n + \alpha_m)T, \tag{11.82}$$

where $T = 1/f_s$ is the sampling period, α_m ($m = 0, 1, \cdots, M-1$) is the timing offset measured in percentage of the sampling period T, and $n = kM + m$. The spectrum of the nonuniformly sampled signal [37] can be computed as

$$\widehat{\mathbf{V}}_i(\omega) = \sum_{n=-\infty}^{+\infty} v_i(t_n)e^{-j\omega t_n}. \tag{11.83}$$

That is[2],

$$\widehat{\mathbf{V}}_i(\omega) = \frac{1}{T} \sum_{k=-\infty}^{+\infty} A(k)\mathbf{V}_i\left(\omega - k\frac{2\pi}{MT}\right), \qquad (11.93)$$

where

$$A(k) = \frac{1}{M} \sum_{m=0}^{M-1} e^{-jk\alpha_m(2\pi/M)} e^{-jkm(2\pi/M)} \qquad (11.94)$$

and $\mathbf{V}_i(\omega)$ is the Fourier transform of $v_i(t)$. For $\alpha_m = 0$, the signal spectrum is reduced to

$$\widehat{\mathbf{V}}_i(\omega) = \frac{1}{T} \sum_{k=-\infty}^{+\infty} \mathbf{V}_i\left(\omega - k\frac{2\pi}{MT}\right), \qquad (11.95)$$

which corresponds to a uniformly sampled signal. Classical time-interleaved

---

[2]The signal spectrum, $\widehat{\mathbf{V}}_i(\omega)$, is derived as follows:

$$\widehat{\mathbf{V}}_i(\omega) = \sum_{n=-\infty}^{+\infty} v_i(t_n)e^{-j\omega t_n} \qquad (11.84)$$

$$= \sum_{k=-\infty}^{+\infty} \sum_{m=0}^{M-1} v_i((kM + m + \alpha_m)T)e^{-j\omega(kM+m+\alpha_m)T} \qquad (11.85)$$

Using

$$v_i((kM + m + \alpha_m)T) = \frac{1}{2\pi} \int_{-\infty}^{+\infty} V_i(\Omega)e^{-j\Omega(kM+m+\alpha_m)T}d\Omega \qquad (11.86)$$

and permuting the order of the first summation and integration, we obtain

$$\widehat{\mathbf{V}}_i(\omega) = \sum_{m=0}^{M-1} \frac{1}{2\pi} \int_{-\infty}^{+\infty} V_i(\Omega)\left[\sum_{k=-\infty}^{+\infty} e^{-j(\Omega-\omega)kMT}\right] e^{-j(\Omega-\omega)(m+\alpha_m)T}d\Omega. \qquad (11.87)$$

Since

$$\sum_{k=-\infty}^{+\infty} e^{-j(\Omega-\omega)kMT} = \sum_{k=-\infty}^{+\infty} \frac{2\pi}{MT}\delta\left(\Omega - \omega + k\frac{2\pi}{MT}\right), \qquad (11.88)$$

it can be shown that

$$\widehat{\mathbf{V}}_i(\omega) = \sum_{k=-\infty}^{+\infty} \sum_{m=0}^{M-1} \frac{1}{MT} \int_{-\infty}^{+\infty} V_i(\Omega)\delta\left(\Omega - \omega + k\frac{2\pi}{MT}\right)e^{-j(\Omega-\omega)(m+\alpha_m)T}d\Omega. \qquad (11.89)$$

Hence,

$$\widehat{\mathbf{V}}_i(\omega) = \frac{1}{MT} \sum_{k=-\infty}^{+\infty} \sum_{m=0}^{M-1} V_i\left(\omega - k\frac{2\pi}{MT}\right)e^{-jk(2\pi/MT)(m+\alpha_m)T} \qquad (11.90)$$

and finally,

$$\widehat{\mathbf{V}}_i(\omega) = \frac{1}{T} \sum_{k=-\infty}^{+\infty} A(k)V_i\left(\omega - k\frac{2\pi}{MT}\right), \qquad (11.91)$$

where

$$A(k) = \frac{1}{M} \sum_{m=0}^{M-1} e^{-jk\alpha_m(2\pi/M)} e^{-jkm(2\pi/M)}. \qquad (11.92)$$

ADC structures can be affected by the following sources of error.

- Timing skew errors

Let us consider a sinusoidal, $v_i(t) = \sin(\omega_i t)$, whose Fourier's transform is given by

$$\mathbf{V}_i(\omega) = j\pi[\delta(\omega + \omega_i) - \delta(\omega - \omega_i)]. \tag{11.96}$$

With the assumption that $\omega_i = 2\pi f_i$, where f_i is the frequency, the substitution of Equation (11.96) into (11.93) leads to the next equation,

$$\widehat{\mathbf{V}}_i(\omega) = \frac{2\pi}{T} \sum_{k=-\infty}^{+\infty} \left[A(k)\delta\left(\omega + \omega_i - k\frac{2\pi}{MT}\right) + B(k)\delta\left(\omega - \omega_i - k\frac{2\pi}{MT}\right) \right], \tag{11.97}$$

where

$$A(k) = -\frac{1}{2jM} \sum_{m=0}^{M-1} e^{j\alpha_m(2\pi f_i/f_s)} e^{-jkm(2\pi/M)}, \tag{11.98}$$

and

$$B(k) = \frac{1}{2jM} \sum_{m=0}^{M-1} e^{-j\alpha_m(2\pi f_i/f_s)} e^{-jkm(2\pi/M)}. \tag{11.99}$$

Note that $A(k) = B^*(M - k)$, where $*$ denotes the notation for the complex conjugate. Due to timing errors between the ADC clock signals, pairs of line spectra centered at $\pm f_i + mf_s/M$ ($m = 1, 2, \cdots, M - 1$) appear in the output spectrum. The corresponding magnitudes are given by $|A(k)|$ and $|B(k)|$, respectively.

Practical time-interleaved ADCs exhibit a clock skew error of a few picoseconds. That is, the value of α_m computed from the discrete Fourier transform of the converter output signal is used to control programmable delays with picosecond resolution or clock signal generators. However, the major drawback of this approach is the high complexity of the digital hardware needed for the algorithm implementation [38]. In order to address the problem of timing skew mismatches, the generation of clock signals can be controlled by a delay-locked loop. Another alternative can consist of using the structure of Figure 11.62. By not resetting between samples, the full-speed single sample and hold (S/H) at the front-end provides subsequent circuit sections the whole clock period to operate on the held signal and eliminates in this way the timing skew errors.

- Gain and offset dispersions

The distortions due to the gain and offset dispersions can be modeled by assuming an input sinusoidal input signal of the form

$$v_i(t) = A_m \sin(\omega_i t) + V_m, \tag{11.100}$$

where $m = 0, 1, \cdots, M - 1$. The corresponding Fourier transform is given by

$$\mathbf{V}_i(\omega) = j\pi A_m[\delta(\omega + \omega_i) - \delta(\omega - \omega_i)] + 2\pi V_m \delta(\omega). \qquad (11.101)$$

Substituting Equation (11.101) into (11.93), the output spectrum can be written as

$$\widehat{\mathbf{V}}_i(\omega) = \frac{2\pi}{T} \sum_{k=-\infty}^{+\infty} A(k) \left[\delta\left(\omega + \omega_i - k\frac{2\pi}{MT}\right) + \delta\left(\omega - \omega_i - k\frac{2\pi}{MT}\right) \right]$$

$$+ \frac{2\pi}{T} \sum_{k=-\infty}^{+\infty} V(k)\delta\left(\omega - k\frac{2\pi}{MT}\right),$$

$$(11.102)$$

where

$$A(k) = -\frac{1}{2jM} \sum_{m=0}^{M-1} A_m e^{-jkm(2\pi/M)}, \qquad (11.103)$$

and

$$V(k) = \frac{1}{M} \sum_{m=0}^{M-1} V_m e^{-jkm(2\pi/M)}. \qquad (11.104)$$

The gain error results in sidebands centered at $\pm f_i + m f_s/M$. However, the components in each pair of line spectra have the same magnitude, $|A(k)|$, in contrast to distortions caused by clock skew mismatches.

The dispersion of the offset among the channels gives rise to distortions which can be observed in the frequency domain as tones at each path sampling frequency, f_s/M, and its integer multiples. The magnitude of these spectral lines is determined by $|V(k)|$.

Because the distortion power of gain and offset errors is not frequency dependent, it can then be compensated using appropriate circuit calibration techniques [39, 40].

In high-resolution ADCs, the thermal noise appears to be the most important nonideality. Specifically, the increase in the number of bits implies a reduction in the noise level, which can be achieved by augmenting the component sizes and equivalently the power consumption.

11.2 Summary

In practice, ADCs operating at Nyquist rate are difficult to implement and may require a high power consumption, especially for high resolutions. Nyquist

ADCs exhibit a quantization noise, which is uniformly spread from 0 to approximately half the sampling rate or clock signal frequency, and individually convert each input signal sample into a digital output code. They can be categorized into SAR, integrating, flash, sub-ranging, and pipelined architectures. In general, the achievable resolution and speed are limited by various noise contributions, and component and timing mismatches. For a given application, the ADC design is determined by the trade-off that can be achieved between the resolution, speed, chip area, and power consumption.

Flash ADCs can provide a resolution ranging from 5 to 9 bits at the sampling rate, while counting ADCs can offer a resolution of 10 to 20 bits at a speed 2^N times smaller than the sampling rate and with a latency equal to the product of 2^N and the clock period, where N is the number of bits. On the other hand, pipelined ADCs can achieve a resolution of 10 to 14 bits by operating at the sampling rate, while SAR ADCs can provide a resolution of 8 to 16 bits at a frequency equal to the sampling rate divided by the resulting number of bits. SAR and pipelined ADCs exhibit an identical latency equal to the product of the resolution in bits and the clock period, but they can be designed to meet the requirement of low power consumption.

11.3   Circuit design assessment

1. **Buffer amplifier for data converter interfacing**

   – Consider a buffer amplifier with the response to a step input given by

   $$v_0(t) = V_m[1 - \exp(-t/\tau)], \qquad (11.105)$$

   where V_m is the maximum amplitude of the output signal and τ is the time constant. In an application requiring a resolution of N bits, the amplifier should be designed such that $[v_0(t) - V_m]/V_m \leq 1/2^{N+1}$ for $t = t_s = 1/(2f_s)$, where f_s is the sampling frequency.

   Determine the time constant τ.

   – In the case, where the input signal is a sinusoid of the form

   $$v_i(t) = V_{FS} \sin 2\pi f_B t, \qquad (11.106)$$

   where V_{FS} and f_B represent the signal full-scale amplitude and bandwidth frequency, respectively, find the maximum aperture timing error $t_a = \Delta V/[(dv_i(t)/dt)|_{t=0}]$, assuming that the resulting amplitude error $|\Delta V|$ is to be less than a half of the least-significant bit (LSB) and $LSB = V_{FS}/2^N$.

2. **Quantizer model**

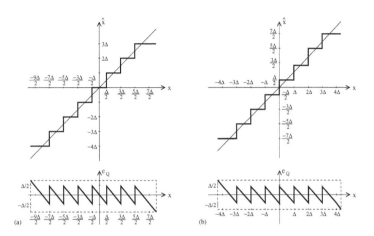

FIGURE 11.63

Characteristics and errors of (a) midtread and (b) midrise quantizers.

In analog-to-digital converters, each sample of the input signal is quantized to fit a finite resolution. This quantization process can be modeled using a characteristic and error function, as illustrated in Figures 11.63(a) and (b) in the case of midtread (a) and midrise (b) quantizers, respectively. Assuming that the quantization error, $e_Q = \hat{x} - x$, is a stationary process and uncorrelated with the input signal, x, it can be seen that the probability density of the quantization error is uniformly distributed between $-\triangle/2$ and $\triangle/2$.

For both quantizers, show that

$$E(e_Q) = \int_{-\Delta/2}^{\Delta/2} e_Q p(e_Q) de_Q = 0 \qquad (11.107)$$

and

$$\sigma_Q^2 = E(e_Q^2) = \int_{-\Delta/2}^{\Delta/2} e_Q^2 p(e_Q) de_Q = \frac{\triangle^2}{12}, \qquad (11.108)$$

where $p(e_Q)$ is the probability density of the quantization error and $\triangle$ is the quantizer step size.

3. **A model for the sampling jitter estimation**
   The output of a S/H circuit, y_k, can be computed as

$$y_k = x(kT + \delta_k) + n_k, \quad k \in \mathbb{Z}, \qquad (11.109)$$

where $x(t)$ is the input signal, T is the sampling period, n_k is the additive sampling noise, and δ_k denotes the error due to the deviation in the sampling instant.

Let

$$x(t) = A\cos(\omega_0 t + \phi) \tag{11.110}$$

be a sinusoid signal with the amplitude A, the initial phase ϕ, and the angular frequency ω_0. Using the assumption $x(kT + \delta_k) \simeq x(kT) + \delta_k x'(kT)$, where x' represents the first derivative of x, verify that

$$y_k \simeq A\cos(\omega_0 kT + \phi) + \epsilon_k , \tag{11.111}$$

where $\epsilon_k = -A\omega_0 \delta_k \sin(\omega_0 kT + \phi) + n_k$.

Show that the variance of the error term ϵ_k can be written as

$$E[\epsilon_k^2] = E[A^2 \omega_0^2 \delta_k^2 \sin^2(\omega_0 kT + \phi) + n_k^2] \tag{11.112}$$

$$= \frac{A^2 \omega_0^2}{2}\sigma_\delta^2 - \frac{A^2 \omega_0^2}{2}\sigma_\delta^2 \cos(2\omega_0 kT + 2\phi) + \sigma_n^2 , \tag{11.113}$$

where $\sigma_\delta^2 = E[\delta_k^2]$ and $\sigma_n^2 = E[n_k^2]$.

Propose a procedure based on the Fourier transform for the computation of the variances σ_δ^2 and σ_n^2.

4. **Switched-capacitor SAR ADC**

   The circuit architecture shown in Fig 11.64 [41] achieves the conversion of bipolar signals into a digital code, the MSB of which indicates the polarity. The control signals S_j $(j = 1, 2, \cdots, 14)$ are defined in Table 11.12, $C_1 = C_2$, and $C_5 = C_6$. The clock signal ϕ_S determines the sampling and conversion phases. The sign of the input signal is detected during the on-state of ϕ_P. The bit b_0 will be set either to 1 if $V_i \geq 0$ or to 0 if the input signal is negative. Its value is maintained at the output of one latch during the next conversion period fixed by ϕ_H.

   Let k denote the conversion cycle. The sampled analog input signal, $V_i(k)$, which is stored on C_1, is compared to the DAC output, $V(k)$, available on C_2. The sign of the threshold voltage generated by the DAC is similar to the one of the input signal. Show that the voltage $V_c(k)$ at the inverting input node of the comparator is given by

   $$V_c(k) = \frac{-C_1 V_i(k) + C_2 V(k)}{C_1 + C_2 + C_p} + V_{off} , \tag{11.114}$$

   where C_p is the parasitic capacitance at the comparator input node and V_{off} is the comparator *dc* offset voltage.

   Propose a gate-level implementation of the control circuit.

   The DAC is realized using an S/H and an amplifier circuit with the gain of $1/2$. Its output signal can be written as

   $$V(k) = -(-1)^{b_0} V_{REF}\left(2^{-k} + \sum_{j=0}^{k-1} b_j 2^{-j}\right) . \tag{11.115}$$

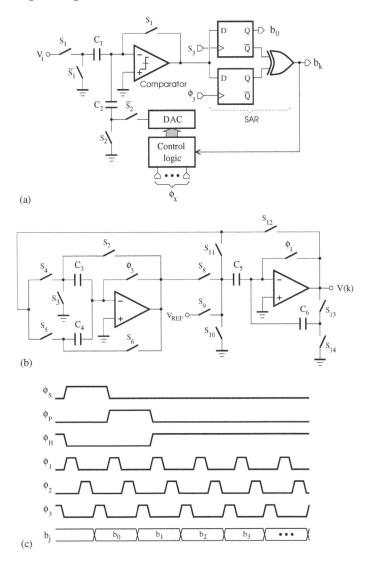

(a)

(b)

(c)

FIGURE 11.64
(a) Block diagram of a SAR ADC, $x = S, P, H, 1, 2, 3$; (b) SC DAC; (c) clock and digital signal waveforms.

Depending on the value of b_0, the positive or negative charge due to V_{REF} and stored on C_5 is transferred onto C_4. Compare the results of the theoretical analysis of the DAC circuit to SPICE simulations.

Analyze the dependence of the converter resolution to the amplifier finite gain and mismatching between C_5 and C_6.

5. **SNR degradation due to clock skew errors**
The analysis (see Subsection 11.1.9) of timing skew errors between the ADC clock signals of time-interleaved converters shows that pairs of line spectra centered around $\pm f_i + m f_s / M$ ($m = $

TABLE 11.12

Digital Signals for the SAR ADC Switch Control

$S_1 : \phi_S \phi_1$	$S_2 : \phi_S + \phi_1 \phi_P$	$S_3 : \phi_S \phi_3$
$S_4 : b_k \overline{\phi_S} \phi_3$	$S_5 : \phi_S \phi_3 + \overline{b_k}\,\overline{\phi_S} \phi_3$	$S_6 : b_k \phi_1 \phi_H$
$S_7 : \overline{b_k} \phi_1 \phi_H$	$S_8 : \phi_1 \phi_H$	$S_9 : \phi_S \phi_2$
$S_{10} : \phi_S \phi_1 + \overline{b_k} \phi_1 \phi_P$	$S_{11} : \overline{\overline{b_k} \phi_S \phi_3}$	$S_{12} : \phi_S \phi_3 + \overline{\phi_S}(\phi_2 + \phi 3)$
$S_{13} : \phi_S \phi_2 + \overline{b_k} \phi_S \phi_3 + \overline{\phi_S}(\phi_2 + \phi 3)$		$S_{14} : \phi_S \phi_1 + b_k \phi_1 \phi_P$

$1, 2, \cdots, M - 1)$ appear in the output spectrum. The corresponding magnitudes are given by $|A(k)|$ and $|B(k)|$, respectively, and

$$A(k) = -\frac{1}{2jM} \sum_{m=0}^{M-1} e^{j\alpha_m (2\pi f_i/f_s)} e^{-jkm(2\pi/M)} \qquad (11.116)$$

$$B(k) = \frac{1}{2jM} \sum_{m=0}^{M-1} e^{-j\alpha_m (2\pi f_i/f_s)} e^{-jkm(2\pi/M)}, \qquad (11.117)$$

where f_i is the frequency of the input sine wave, f_s is the sampling frequency, and α_m is the relative error in the sampling instants with respect to the clock signal period.

Verify that the SNR due to clock skew errors is given by

$$SNR = 10 \log_{10} \left(\frac{P_i}{P_\eta} \right), \qquad (11.118)$$

where the noise power is provided by the formula

$$P_\eta = P - P_i \qquad (11.119)$$

$$P_i = |A(0)|^2 + |B(0)|^2 \qquad (11.120)$$

is the power of the input signal, and by using the Parseval's relation[3], the output signal power is estimated as

$$P = P_A + P_B = \frac{1}{2} \qquad (11.122)$$

---

[3] Let $x(n)$ be an N-point sequence, and $X(k)$ its discrete Fourier transform. The next equation,

$$\sum_{n=0}^{N-1} |x(n)|^2 = \frac{1}{N} \sum_{k=0}^{N-1} |X(k)|^2 \qquad (11.121)$$

is known as Parseval's relation.

with

$$P_A = \sum_{k=0}^{M-1} |A(k)|^2 = \frac{1}{4}, \tag{11.123}$$

$$P_B = \sum_{k=0}^{M-1} |B(k)|^2 = \frac{1}{4}. \tag{11.124}$$

Note that $A(k)$ and $B(k)$ can be considered the discrete Fourier transform of the sequences $-(1/2jM)\mathrm{e}^{j\alpha_m(2\pi f_i/f_s)}$ and $(1/2jM)\mathrm{e}^{-j\alpha_m(2\pi f_i/f_s)}$, respectively.

Bibliography

[1] H. T. Russel, Jr., "An improved successive-approximation register design for use in A/D converters," *IEEE Trans. on Circuits and Systems*, vol. 25, pp. 550–554, July 1978.

[2] C. K. Yuen, "Another design of the successive approximation register for A/D converters," *Proc. of the IEEE*, vol. 67, pp. 873–874, May 1979.

[3] J. L. McCreary and P. R. Gray, "All-MOS charge redistribution analog-to-digital conversion techniques - Part I," *IEEE J. of Solid-State Circuits*, vol. 10, pp. 371–379, Dec. 1975.

[4] H.-S. Lee and D. A. Hodges, "Self-calibration technique for A/D converters," *IEEE Trans. on Circuits and Systems*, vol. 30, pp. 188–190, March 1983.

[5] L. L. Evans, "High speed integrating analog-to-digital converter," U.S. Patent 4,395,701, filed March 25, 1980; issued July 26, 1983.

[6] B. W. Phillips, "Bipolar dual-ramp analog-to-digital converter," U.S. Patent 3,906,486, filed July 2, 1973; issued September 16, 1975.

[7] N. H. Strong, "Instantaneous gain changing analog to digital converter," U.S. Patent 4,588,983, filed June 17, 1985; issued May 13, 1986.

[8] B. J. Rodgers and C. R. Thurber, "A monolithic $\pm 5\frac{1}{2}$-digit BiMOS A/D converter," *IEEE J. of Solid-State Circuits*, vol. 24, pp. 617–626, Jun. 1989.

[9] C. W. Mangelsdorf, "A 400-MHz input flash converter with error correction," *IEEE J. of Solid-State Circuits*, vol. 25, pp. 184–191, Feb. 1990.

[10] C. L. Portmann and T. H. Y. Meng, "Power-efficient metastability error reduction in CMOS flash A/D converters," *IEEE J. of Solid-State Circuits*, vol. 25, pp. 1132–1140, Aug. 1996.

[11] K. Kattmann and J. Barrow, "A technique for reducing differential nonlinearity errors in flash A/D converters," *1991 IEEE ISSCC Digest of Technical Papers*, pp. 170–171, Feb. 1991.

[12] M. Choi and A. A. Abidi, "A 6-b 1.3-Gsample/s A/D converter in 0.35-μm CMOS," *IEEE J. of Solid-State Circuits*, vol. 36, pp. 1847–1858, Dec. 2001.

[13] P. C. S. Scholtens and M. Vertregt, "A 6-b 1.6-Gsample/s flash ADC in 0.18-µm CMOS using averaging termination," *IEEE J. of Solid-State Circuits*, vol. 27, pp. 1599–1609, Dec. 2002.

[14] X. Jiang and M.-C. F. Chang, "A 1-GHz signal bandwidth 6-bit CMOS ADC with power-efficient averaging," *IEEE J. of Solid-State Circuits*, vol. 40, pp. 532–535, Feb. 2005.

[15] R. E. J. Van de Grift, I. W. J. M. Rutten, and M. Van der Veen, "An 8-bit video ADC incorporating folding and interpolation techniques," *IEEE J. of Solid-State Circuits*, vol. 22, pp. 944–953, Dec. 1987.

[16] R. J. van de Plassche and P. Baltus, "An 8-bit 100-MHz full-Nyquist analog-to-digital converter," *IEEE J. of Solid-State Circuits*, vol. 23, pp. 1334–1344, Dec. 1988.

[17] M. P. Flynn and D. J. Allstot, "CMOS folding A/D converter with current-mode interpolation," *IEEE J. of Solid-State Circuits*, vol. 31, pp. 1248–1257, Sept. 1996.

[18] A. G. W. Venes and R. J. van de Plassche, "An 80-MHz, 8-b CMOS folding A/D converter with distributed track-and-hold preprocessing," *IEEE J. of Solid-State Circuits*, vol. 31, pp. 1846–1853, Dec. 1996.

[19] S. Limotyrakis, K. Nam, and B. A. Wooley, "Analysis and simulation of distortion in folding and interpolating A/D converters," *IEEE Trans. on Circuits and Systems–II*, vol. 49, pp. 161–169, March 2002.

[20] Y. Li and E. Sánchez-Sinencio, "A wide input bandwidth 7-bit 300-Msamples/s folding and current-mode interpolating ADC," *IEEE J. of Solid-State Circuits*, vol. 38, pp. 1405–1410, Aug. 2003.

[21] T. C. Verster, "A method to increase the accuracy of fast serial-parallel analog-to-digital converters," *IEEE Trans. on Electronic Computers*, EC-13, pp. 471–473, 1964.

[22] S. H. Lewis, H. S. Fetterman, G. F. Gross, Jr., R. Ramachandran, and T. R. Viswanathan, "A 10-b 20-Msample/s analog-to-digital converter," *IEEE J. of Solid-State Circuits*, vol. 27, pp. 351–358, March 1992.

[23] T. Cho and P. R. Gray, "A 10-b, 20-Msample/s, 35-mW pipeline A/D converter," *IEEE J. of Solid-State Circuits*, vol. 30, pp. 166-172, March 1995.

[24] A. M. Abo and P. R. Gray, "A 1.5-V, 10-bit, 14.3-MS/s CMOS pipeline analog-to-digital converter," *IEEE J. of Solid-State Circuits*, vol. 34, pp. 599–606, May 1999.

[25] D. W. Cline and P. R. Gray, "A power optimized 13-b 5-Msamples/s pipelined analog-to-digital converter in 1.2-μm CMOS," *IEEE J. of Solid-State Circuits*, vol. 31, pp. 294–303, March 1996.

[26] B.-S. Song, M. F. Tompsett, and K. R. Lakshmikumar, "A 12-bit 1-Msample/s capacitor error-averaging pipeline A/D converter," *IEEE J. of Solid-State Circuits*, vol. 23, pp. 1324–1333, Dec. 1988.

[27] Y. Chiu, P. R. Gray, and B. Nikolić, "A 14-b 12-MS/s CMOS pipeline ADC with over 100-dB SFDR," *IEEE J. of Solid-State Circuits*, vol. 39, pp. 2139–2151, Dec. 2004.

[28] E. Siragusa and I. Galton, "A digitally enhanced 1.8-V 15-bit 40-MSample/s CMOS pipelined ADC," *IEEE J. of Solid-State Circuits*, vol. 39, pp. 2126–2138, Dec. 2004.

[29] M. Daito, H. Matsui, M. Ueda, and K. Iizuka, "A 14-bit 20-MS/s pipelined ADC with digital distortion calibration," *IEEE J. of Solid-State Circuits*, vol. 41, pp. 2417–2423, Nov. 2006.

[30] W. (W.) Yang, D. Kelly, I. Mehr, M. T. Sayuk, and L. Singer, "A 3-V 340-mW 14-b 75-Msample/s CMOS ADC with 85-dB SFDR at Nyquist input," *IEEE J. of Solid-State Circuits*, vol. 36, pp. 1931–1936, Dec. 2001.

[31] R. H. McCharles, V. A. Saletore, W. C. Black. Jr., and D. A. Hodges, "An algorithmic analog-to-digital converter," *1977 IEEE ISSCC Digest of Technical Papers*, section IX, pp. 96–97, Feb. 1977.

[32] C.-C. Lee, "Switched-capacitor circuit analog-to-digital converter," U.S. Patent 4,529,965, filed May 3, 1983; issued July 16, 1985.

[33] P. W. Li, M. Chin, P. R. Gray, and R. Castello, "A ratio-independent algorithmic analog-to-digital conversion technique," *IEEE J. of Solid-State Circuits*, vol. 19, pp. 828–836, Dec. 1984.

[34] K. Nagaraj, "Efficient circuit configurations for algorithmic analog to digital converters," *IEEE Trans. on Circuits and Systems–II*, vol. 40, pp. 777–785, Dec. 1993.

[35] H. Onodera, T. Tateishi, and K. Tamaru, "A cyclic A/D converter that does not require ratio-matched components," *IEEE J. of Solid-State Circuits*, vol. 23, pp. 152–158, Feb. 1988.

[36] H. Matsumoto and K. Watanabe, "Improved switched-capacitor algorithmic analogue-to-digital convertor," *Electronic Letters*, vol. 21, pp. 430–431, March 1985.

[37] Y.-C. Jenq, "Perfect reconstruction of digital spectrum from nonuniformly sampled signals," *IEEE Trans. on Instrum. Meas.*, vol. 46, pp. 649–652, June 1997.

[38] Y.-C. Jenq, "Digital spectra of nonuniformly sampled signals: Fundamentals and high-speed waveform digitizers," *IEEE Trans. on Instrum. Meas.*, vol. 37, pp. 245–251, June 1988.

[39] T. Ndjountche and R. Unbehauen, "Adaptive calibration techniques for time-interleaved ADCs," *Electronics Letters*, vol. 37, pp. 412–414, March 2001.

[40] T. Ndjountche, F.-L. Luo, and R. Unbehauen, "A High-frequency double-sampling second-order delta-sigma modulator," *IEEE Trans. on Circuits and Systems–II*, vol. 52, pp. 841–845, Dec. 2005.

[41] S. Ogawa and K. Watanabe, "A switched-capacitor successive-approximation A/D converter," *IEEE Trans. on Instrum. and Meas.*, vol. 42, pp. 847–853, Aug. 1993.

12

Delta-Sigma Data Converters

CONTENTS

In comparison with other analog-to-digital converters (ADCs) and digital-to-analog converters (DACs), delta-sigma ($\Delta\Sigma$) data converters (or generally oversampling data converters) exhibit a reduced sensitivity to analog component matching. $\Delta\Sigma$ converters are usually the best choice for applications requiring a resolution greater than 20 bits. In these converters, the input signal can be sampled at a rate much greater than the Nyquist frequency (i.e., twice the bandwidth or highest frequency of the signal being sampled), and the quantization noise is shaped by the modulator to be low in the signal band and high in the out-band spectrum. The specifications of either the analog anti-aliasing filter for the analog-to-digital conversion or the smoothing filter for the digital-to-analog conversion are then relaxed due to the signal oversampling, and the remaining out-band noise is attenuated by a filter. The actual reduction of in-band noise power level depends on the modulator structure and the oversampling ratio (OSR) and a high resolution is achieved with a penalty in speed, as the modulator hardware has to operate at the oversampling rate, and an increased complexity of the filter hardware. Oversampling data converters then present a trade-off between speed and resolution.

$\Delta\Sigma$ modulators can be exploited in the implementation of calibration stages required to improve the linearity of Nyquist converters. Built-in self-test is another application for analog signal synthesis where $\Delta\Sigma$ modulators are ideally suited.

12.1 Delta-sigma analog-to-digital converter

$\Delta\Sigma$ ADCs can be implemented using either continuous-time (CT) or discrete-time (DT) filters, as shown in Figures 12.1 and 12.2. The signal sampling is performed at the input node in the DT case by an S/H circuit, while it is implemented after the filtering in the CT structure. In both types, the system consists of a $\Delta\Sigma$ modulator followed by a digital decimator. A filter, a quantizer, and a DAC are used as building blocks of the modulator, which is based on an output-feedback structure. Note that even modulators based on a CT filter possess an equivalent in the DT domain, where the converter design is generally achieved.

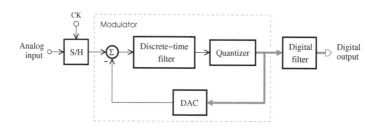

FIGURE 12.1
Block diagram of a DT $\Delta\Sigma$ ADC.

During the data conversion, the modulator feedback forces the average

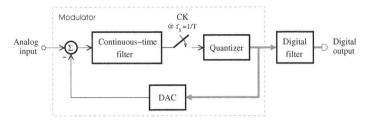

FIGURE 12.2
Block diagram of a CT $\Delta\Sigma$ ADC.

value of the quantized signal to follow one of the input signals and the quantization noise in the signal band is attenuated. A decimation filter is required to eliminate the out-of-band noise and to reduce the sampling rate of the modulator output signal.

12.1.1 Time domain behavior

At each processing step, a $\Delta\Sigma$ modulator generates a digital estimation of the signal, which is subtracted from the actual sample of the input signal and the digital conversion of the resulting sequence is achieved such that the output and input signals tend to be equal on average.

In the general case, a single-stage modulator can be described in the time domain using the following equations

$$x(n) = \sum_{j=0}^{J_S} h_S(j)s(n-j) + \sum_{j=1}^{J_Q} h_Q(j)e_Q(n-j), \tag{12.1}$$

$$e_Q(n) = y(n) - x(n), \tag{12.2}$$

$$y(n) = Q[x(n)], \tag{12.3}$$

where h_S and h_Q are the impulse responses of the signal transfer function (STF) and quantization noise transfer function (QNTF), respectively; x denotes the output state of the modulator filter, J_S and J_Q represent the length of the STF and QNTF, respectively; and Q is the equivalent piecewise-constant function of the quantizer. The modulator will be considered *stable* for a given input and initial conditions if the signal samples x are bounded and the quantizer is not overloaded.

In the special case of the first-order $\Delta\Sigma$ modulator shown in Figure 12.3, the sum ϵ of the input signal s and the output $\tilde{y}$ of a 1-bit feedback DAC is applied to an integrator, whose output x is connected to the input of a comparator delivering the digital sequence y available at the modulator output, which

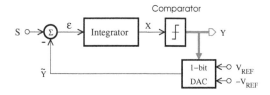

FIGURE 12.3
Block diagram of a first-order $\Delta\Sigma$ modulator.

is then used to drive the 1-bit feedback DAC. Assuming that the supply voltages of the comparator are V_{DD} and $-V_{SS}$, and the DAC reference voltages are $\pm V_{REF}$, the modulator can be described using the next time-domain equations,

$$x(n) = x(n-1) + \epsilon(n-1), \tag{12.4}$$
$$y(n) = Q[x(n)], \tag{12.5}$$
$$\epsilon(n) = s(n) - \tilde{y}(n), \tag{12.6}$$

where

$$Q[x(n)] = \begin{cases} H, & \text{if} \quad x(n) \geq 0 \\ L, & \text{if} \quad x(n) < 0 \end{cases} \tag{12.7}$$

and

$$\tilde{y}(n) = \begin{cases} V_{REF}, & \text{if} \quad y(n) = 1 \\ -V_{REF}, & \text{if} \quad y(n) = 0. \end{cases} \tag{12.8}$$

At the modulator output, a logic high state, H, corresponds to a voltage level of about V_{DD} and a logic low state, L, is represented by approximately $-V_{SS}$.

Let a dc input signal of 0.25 V be applied to the modulator and the reference voltages of the DAC be ± 1 V. The state and output sequences of the modulator are given in Table 12.1, where the initial conditions are specified in the row associated with $n = 0$. It can be observed that the state and output sequences for $n \in [2, 9]$ are periodically repeated starting from $n = 10$. For the first-order modulator, the quantization noise is correlated with the input signal and appears not to be entirely random. The allowed input range is from V_{REF} to $-V_{REF}$, resulting in a converter full-scale range (FSR) of $2V_{REF}$, or say 2 V. A 0.25-V input signal is 1.25 V above the lower -1-V limit of the FSR, that is, the input represents $(1.25/2) \times 100 = 62.5\%$ of the FSR. By averaging the first 8 samples of the output sequence, the number of bits at the high state is 5, leading to the H-state density given by $(5/8) \times 100 = 62.5\%$.

In practice, the modulator output is decoded using a digital low-pass filter that averages every given number of samples, which

TABLE 12.1

State and Output Sequences of the First-Order Modulator

n	$s(n)$	$x(n)$	$y(n)$	$\tilde{y}(n)$	$\epsilon(n)$
0	×	0.10	H	1	-1.00
1	0.25	-0.90	L	-1	1.25
2	**0.25**	**0.35**	**H**	**1**	**-0.75**
3	0.25	-0.40	L	-1	1.25
4	0.25	0.85	H	1	-0.75
5	0.25	0.10	H	1	-0.75
6	0.25	-0.65	L	-1	1.25
7	0.25	0.60	H	1	-0.75
8	0.25	-0.15	L	-1	1.25
9	0.25	1.10	H	1	-0.75
10	**0.25**	**0.35**	**H**	**1**	**-0.75**
11	0.25	-0.40	L	-1	1.25
12	0.25	0.85	H	1	-0.75
13	0.25	0.10	H	1	-0.75
14	0.25	-0.65	L	-1	1.25
15	0.25	0.60	H	1	-0.75

can be increased to improve the overall resolution. For input signals around the mid-scale, the H-state density is about 50% in the modulator output sequence. An increase in the input signal toward the higher limit of the FSR results in an augmentation of the H-state density, while a decrease in the input signal toward the lower limit of the FSR induces a reduction in the H-state density.

Due to oversampling, a $\Delta\Sigma$ modulator makes use of the available speed to exchange the resolution in time for that in amplitude.

12.1.2 Linear model of a discrete-time modulator

A linear model of a discrete-time modulator can be obtained based on the assumptions that different feed-ins to the filter are used by the input and feedback signals, and the quantization is done with an additive error. With reference to Figure 12.4, the output signal of the modulator can be computed as

$$Y(z) = H_S(z)S(z) + H_Q(z)E_Q(z) \tag{12.9}$$

where

$$H_S(z) = \frac{Y(z)}{S(z)} = \frac{qH(z)}{1 + qH(z)}, \tag{12.10}$$

$$H_Q(z) = \frac{Y(z)}{E_Q(z)} = \frac{1}{1 + qH(z)}, \tag{12.11}$$

and H represents the z-domain transfer function of the loop filter, H_S denotes the signal transfer function (STF), H_Q is the quantization noise transfer function (QNTF), E_Q represents the quantization error, and q $(q > 0)$ is the quantizer gain. To simplify the analysis, the modulator is generally modeled by replacing the quantizer with a unity-gain element followed by an additive noise source. Hence,

$$H_S(z) = 1 - H_Q(z) \tag{12.12}$$

and the QNTF determines the modulator performance and stability. It should be emphasized that a modulator realized with real components is a nonlinear system and the coupling between the signal and quantization noise is neglected in the above description.

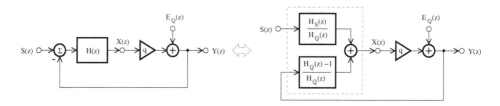

FIGURE 12.4
Block diagram of the DT $\Delta\Sigma$ modulator linear model.

$\Delta\Sigma$ modulators are generally described by the order of the loop filter, the characteristic of which determines the shape of the noise spectrum. Lowpass filters are generally used to meet the desired resolution in audio applications, while modulators based on a bandpass filter are preferred for the digitization of high-frequency band-limited signals in telecommunication systems. In the z-domain, an arbitrary L-th order lowpass modulator can be transformed into a bandpass modulator of order $2L$ using the transformation

$$z^{-1} \to -z^{-2}. \tag{12.13}$$

In this way, the zeros of the QNTF are shifted from dc to $f_s/4$, where f_s is the sampling frequency.

The resolution achievable with a single-bit $\Delta\Sigma$ ADC is limited and can generally be improved using converters based on high-order filters [4] or multi-bit quantizers [10].

12.1.3 Modulator dynamic range

Consider a $\Delta\Sigma$ modulator with order L operating with the oversampling ratio $OSR = f_s/f_N = f_s/(2f_{max})$, where f_s is the sampling frequency, f_N is the Nyquist frequency, and f_{max} is the highest spectral component present in the input signal. The quantization noise is shaped by a transfer function of the form

$$H_Q(z) = (1 - z^{-1})^L .$$ (12.14)

In the frequency domain, we have

$$H_Q(jf) = \left[1 - \exp\left(-j2\pi \frac{f}{f_s}\right)\right]^L = \left[2j\sin\left(\pi \frac{f}{f_s}\right)\exp\left(-j\pi \frac{f}{f_s}\right)\right]^L$$ (12.15)

and

$$|H_Q(jf)| = \left[2\sin\left(\pi \frac{f}{f_s}\right)\right]^L .$$ (12.16)

Figure 12.5 shows the plot of H_Q magnitudes for $L = 1, 2, 3, 4, 5$. The quan-

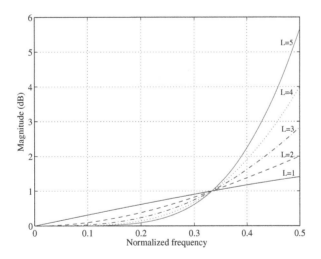

FIGURE 12.5
Plot of H_Q magnitudes for $L = 1, 2, 3, 4, 5$.

tization noise suppression over the low-frequency signal band is improved as the value of L, or equivalently the modulator order, is increased.

Let $\triangle$ be the quantizer step size. Assuming that the quantization error, e_Q, is evenly distributed between $-\triangle/2$ and $\triangle/2$, the mean value of e_Q is zero and the probability density of e_Q can expressed in the form

$$
p(e_Q) = \begin{cases} \dfrac{1}{\triangle}, & \text{if } e_Q \in [-\triangle/2, \triangle/2], \\ 0, & \text{otherwise.} \end{cases} \tag{12.17}
$$

The variance of the quantization noise, σ_Q^2, is then given by

$$
\sigma_Q^2 = \int_{-\infty}^{\infty} e_Q^2 p(e_Q) \mathrm{d}e_Q = \frac{1}{\triangle} \int_{-\triangle/2}^{\triangle/2} e_Q^2 \mathrm{d}e_Q = \frac{\triangle^2}{12}. \tag{12.18}
$$

Generally, the input signal is sampled at the frequency f_s, and the spectral density of the quantization noise to be filtered is supposed to remain constant between 0 and $f_s/2$. That is,

$$
\sigma_Q^2 = \int_0^{f_s/2} p_i \mathrm{d}f = p_i \int_0^{f_s/2} \mathrm{d}f = p_i(f_s/2) \tag{12.19}
$$

or, equivalently,

$$
p_i = \sigma_Q^2 \frac{1}{f_s/2} = \frac{\triangle^2}{12} \frac{2}{f_s}. \tag{12.20}
$$

The spectral density of the quantization noise at the modulator output can be written as

$$
p_0 = |H_Q(jf)|^2 p_i = \frac{2^{2L} \triangle^2}{6 f_s} \sin^{2L}\left(\pi \frac{f}{f_s}\right). \tag{12.21}
$$

The power of the quantization noise in the Nyquist frequency range is given by

$$
P_Q = \int_0^{f_N/2} p_0(f) \mathrm{d}f. \tag{12.22}
$$

Because the value of the oversampling ratio $OSR = f_s/f_N$ is generally high, we have $f_N \ll f_s$. As a consequence, $0 \leq f \leq f_N/2 \ll f_s$ and $sin(\pi f/f_s) \simeq \pi f/f_s$. Hence,

$$
P_Q \simeq \frac{2^{2L} \triangle^2}{6 f_s} \int_0^{f_N/2} \left(\pi \frac{f}{f_s}\right)^{2L} \mathrm{d}f. \tag{12.23}
$$

Finally, we obtain

$$
P_Q \simeq \frac{\triangle^2}{12} \frac{\pi^{2L}}{2L+1} \left(\frac{1}{OSR}\right)^{2L+1}. \tag{12.24}
$$

In the case of a rounding quantizer, each input sample is assigned to the nearest quantization level. The quantization error is then limited to the range of $-\triangle/2$ to $\triangle/2$, and

$$
\triangle = \frac{\mathrm{FSR}}{2^B - 1}, \tag{12.25}
$$

where B is the number of bits of the quantizer. In the case of a sinusoidal signal with a peak-to-peak amplitude equal to the quantizer full-scale range, FSR, the average power is

$$P_S = \sigma_S^2 = \mathrm{E}[s^2(n)] = \frac{\mathrm{FSR}^2}{8}. \tag{12.26}$$

The dynamic range (DR) of the modulator can then be expressed as

$$DR^2 = \frac{P_S}{P_Q}$$

$$= \frac{\left(\dfrac{\mathrm{FSR}}{2\sqrt{2}}\right)^2}{\dfrac{\mathrm{FSR}^2}{12(2^B-1)^2}\dfrac{\pi^{2L}}{2L+1}\left(\dfrac{1}{OSR}\right)^{2L+1}} = \frac{3}{2}\frac{2L+1}{\pi^{2L}}(2^B-1)^2\,OSR^{2L+1} \tag{12.27}$$

or in decibels,

$$DR(\text{in dB}) = 10\log_{10}\left(\frac{3}{2}\frac{2L+1}{\pi^{2L}}(2^B-1)^2\right) + 10(2L+1)\log_{10}OSR. \tag{12.28}$$

Thus, the dynamic range, DR, is a function of the filter order and the over-

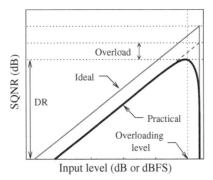

FIGURE 12.6
Curve of the signal-to-quantization noise ratio (SQNR) versus the input level.

sampling ratio, OSR. The multiplication of the OSR by a factor of 2 results in an increase in the DR on the order of $3(2L+1)$ dB, or equivalently, $(L+1/2)$ bits of resolution. Note that the dynamic range given by Equation (12.28) may be considered an upper bound because it is based on a linear model of the modulator. Furthermore, the stability requirements of modulators with an order equal to or greater than 2 are only met by using design techniques or structures that can constrain the modulator dynamic range well below this upper bound.

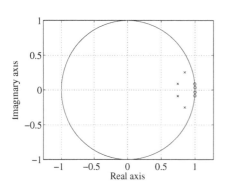

FIGURE 12.7

Pole-zero plot of the NTF.

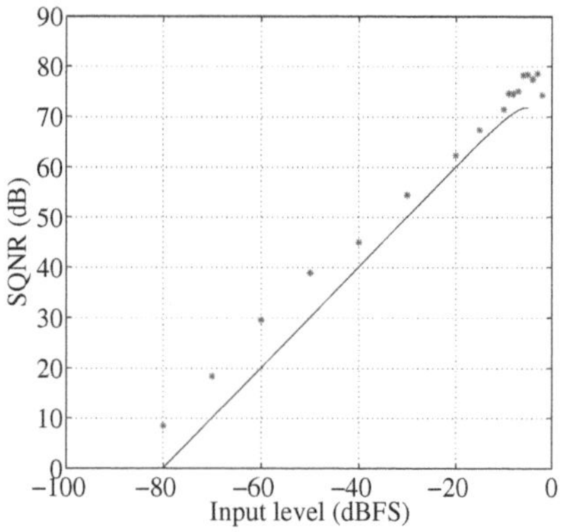

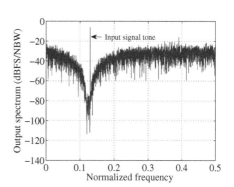

FIGURE 12.8

SQNR curve of the $\Delta\Sigma$ modulator.

The curve of the signal-to-quantization noise ratio (SQNR) versus the input level is depicted in Figure 12.6. A linear scaling effect is observed between the modulator SQNRs obtained by relying on an ideal model or a practical chip. Furthermore, a premature clipping occurs in the practical SQNR at high input levels because the slew rate of active components appears to be limited. Note that the input level can be evaluated in dB, or in dBFS (decibels relative to full scale), provided the output spectrum is normalized so that a full-scale sine wave can appear at 0 dB.

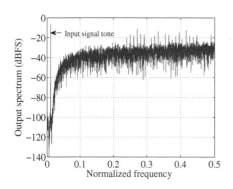

FIGURE 12.9

Output power spectrum of a fourth-order lowpass $\Delta\Sigma$ modulator.

FIGURE 12.10

Output power spectrum of a fourth-order bandpass $\Delta\Sigma$ modulator.

A typical way to synthesize a $\Delta\Sigma$ modulator is to select the integrator type and signal paths such that all zeros of the noise transfer function are located on the unit circle, or equivalently at $z = 1$. This approach has the advantage of reducing the complexity of the modulator architecture, but the achievable SNR and DR may be limited.

Given a modulator order, the above performance characteristics can be improved by selecting a noise transfer function with zeros on the unit circle and poles inside the unit circle [5, 6], as shown in Figure 12.7. The poles can be chosen to be identical to the ones of a filter approximation function, such as Butterworth or Chebyshev polynomial, while the zeros are optimally placed on the unit circle. In this case, the noise transfer function is of the form $H_Q(z) = N(z)/D(z)$, where N and D are two polynomials, and it is assumed that the realizability constraint, $\lim_{z \to \infty} H_Q(z) = 1$, and stability requirement are met. The resulting SQNR curve is depicted in Figure 12.8. The fourth-order lowpass modulator considered here exhibits a maximum signal-to-noise ratio of about 80 dB at an oversampling ratio of 32.

The DR formula is still valid in the case of $2L$-th bandpass modulators obtained by applying the dc-to-$f_s/4$ transformation to L-th lowpass modulator prototypes. For lowpass modulators, the STF is designed as a lowpass filtering function and the QNTF is chosen as a highpass function, while, for bandpass modulators, the STF is a bandpass function and the QNTF is a band-reject or notch function. The output spectra of fourth-order lowpass and bandpass modulators are depicted in Figures 12.9 and 12.10, respectively. The output spectrum of the bandpass modulator is determined relative to the noise-power bandwidth (NBW).

12.1.4 Continuous-time modulator

Continuous-time (CT) modulators are more suitable for high-frequency applications. However, they are more sensitive to clock jitter than their SC counterparts.

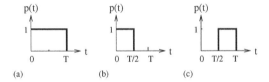

FIGURE 12.11
DAC waveforms: (a) NRZ, (b) RZ, (c) HRZ.

In CT modulators, the sampling of the signal takes place in the loop and the stability must be analyzed in the DT domain. The design of the modulator filter then results in the equivalent transfer function $H(z)$. Note that the unstable behavior will be observed when the poles are located far away from the zeros even for input signals with a low amplitude, while the performance is compromised by placing the poles very close to the zeros. Let us assume that the operation of a 1-bit quantizer can be described by the following expression,

$$p(t) = \begin{cases} 1, & \text{if} \quad p_1 T \leq t \leq p_2 T \\ 0, & \text{otherwise,} \end{cases} \tag{12.29}$$

where $p_1, p_2 \in [0, 1]$ and T is the sampling period. In the s-domain, this corresponds to the zero-order hold pulse transfer function given by

$$P(s) = (\mathrm{e}^{-sp_1 T} - \mathrm{e}^{-sp_2 T})/s. \tag{12.30}$$

The type of the DAC pulse, which is determined by the values of p_1 and p_2, is nonreturn-to-zero (NRZ) for $p_1 = 0$ and $p_2 = 1$, return-to-zero (RZ) for $p_1 = 0$ and $p_2 = 1/2$, and half-clock-period delayed return-to-zero (HRZ) for $p_1 = 1/2$ and $p_2 = 1$ (see Figure 12.11). The CT transfer function, $H(s)$, can be obtained using the impulse invariant transformation as follows,

$$\mathcal{Z}^{-1}\{H(z)\} = \mathcal{L}^{-1}\{P(s)\hat{H}(s)\}|_{t=nT}, \tag{12.31}$$

where $\mathcal{Z}^{-1}$ and $\mathcal{L}^{-1}$ denote the inverses of the z-transform and Laplace transform, respectively; $H(z)$ is the transfer function of the DT filter; and $\hat{H}(s)$ represents the transfer function of the CT filter. This last equation can equivalently be written in the time domain as

$$h(nT) = \left. [p(t) * \hat{h}(t)] \right|_{t=nT} = \left. \left(\int_{-\infty}^{\infty} p(t) * \hat{h}(t - \tau)d\tau \right) \right|_{t=nT}, \tag{12.32}$$

where $*$ represents the time convolution, $h(nT)$ is the impulse response of the DT filter, $p(t)$ is the impulse response of the DAC, and $\hat{h}(t)$ denotes the impulse response of the CT filter.

The classical approach used for the design of a CT modulator consists of first choosing the appropriate z-domain QNTF, $H_Q(z)$, that meets the required specifications and convert it to the DT loop transfer function $H(z) = (H_Q(z)-1)/H_Q(z)$. The CT filter transfer function, $\hat{H}(s)$, is then obtained by solving the equation of the impulse invariant transformation with a symbolic math program or numerical methods. Whenever possible, the DT transfer function of the modulator can be decomposed into partial fractions of the form $H_i(z)$, $i = 1, 2$, and the equivalent CT function is derived using the results of Table 12.2, where f_s is the clock signal frequency [7]. Note that l'Hopital's rule was exploited to obtain the s-domain equivalent functions in the specific case where $z_k = 1$ (i.e., the poles are located at dc). The impulse invariant method has the advantage of resulting in circuits with a low complexity in comparison with other approaches.

In practice, the settling behavior of the DAC and quantizer are affected by the excess loop delay due to the nonzero switching time of transistors in the quantizer latch and DAC, and timing jitter.

It can be observed that in the CT structure, the clock jitter, which causes a variation in the width of DAC pulses, disturbs the sum of the input signal and quantization noise, because the sampling occurs at the quantizer rather than the input. In the DT case, only the input signal is affected. As a result, the signal-to-noise ratio (SNR) of CT modulators is more severely affected by the timing jitter in the quantizer clock than the SNR of the equivalent

TABLE 12.2

Impulse-Invariant Transformation of Functions with Single and Double Poles

z-domain	s-domain
$H_1(z) = \dfrac{z^{-1}}{1 - z_k z^{-1}}$	$\hat{H}_1(s) = \begin{cases} \dfrac{s_k}{q_1(s - s_k)}, & \text{if } z_k \neq 1 \\[2ex] \dfrac{f_s}{(p_2 - p_1)s}, & \text{if } z_k = 1 \end{cases}$
$H_2(z) = \dfrac{z^{-2}}{(1 - z_k z^{-1})^2}$	$\hat{H}_2(s) = \begin{cases} \dfrac{(q_2 s_k + q_1 f_s)s - q_2 s_k^2}{z_k q_1^2 (s - s_k)^2}, \\[1ex] \qquad\qquad\qquad \text{if } z_k \neq 1 \\[2ex] \dfrac{-f_s \left(1 - \dfrac{p_1 + p_2}{2}\right)s + f_s^2}{(p_2 - p_1)s^2}, \\[1ex] \qquad\qquad\qquad \text{if } z_k = 1 \end{cases}$
$H_3(z) = \dfrac{z^{-3}}{(1 - z_k z^{-1})^3}$	$\hat{H}_3(s) = \begin{cases} \dfrac{r_2 f_s s^2 + r_1 f_s^2 s + r_0 f_s^3}{z_k^2 q_1^3 (s - s_k)^3}, \\[1ex] \qquad\qquad\qquad \text{if } z_k \neq 1 \\[2ex] \dfrac{r f_s s^2 - f_s^2 \left(\dfrac{3}{2} - \dfrac{p_1 + p_2}{2}\right)s + f_s^3}{(p_2 - p_1)s^3}, \\[1ex] \qquad\qquad\qquad \text{if } z_k = 1 \end{cases}$

where
$f_s = 1/T$
$s_k = \ln(z_k)/T$
$q_1 = z_k^{1-p_1} - z_k^{1-p_2}$
$q_2 = (1 - p_2)z_k^{1-p_2} - (1 - p_1)z_k^{1-p_1}$
$r_0 = (q_4/2)s_k^3$
$r_1 = -q_4 s_k^2 + q_3 s_k + q_1^2$
$r_2 = (q_4/2)s_k - q_3$
$r = 1 + [p_1(p_1 - 9) + p_2(p_2 - 9) + 4p_1 p_2]/12$
$q_3 = (3/2 - p_1)(z_k^{1-p_1})^2 + (3/2 - p_2)(z_k^{1-p_2})^2 + (p_1 + p_2 - 3)z_k^{1-p_1} z_k^{1-p_2}$
and
$q_4 = (1 - p_1)(2 - p_1)(z_k^{1-p_1})^2 + (1 - p_2)(2 - p_2)(z_k^{1-p_2})^2$
$\qquad\qquad + [p_1(p_1 + 3) + p_2(p_2 + 3) - 4(1 + p_1 p_2)]z_k^{1-p_1} z_k^{1-p_2}$

Adapted from [7], ©1999 IEEE.

DT versions. It should be noted that modulators with a NRZ DAC are less sensitive to the clock jitter than the one with RZ or HRZ DACs.

Furthermore, the performance of CT $\Delta\Sigma$ modulators can be affected by the so-called excess delay, which is required by the quantizer to update its output. As a result, the DAC pulse can extend beyond T (or the clock period end) and the order of the equivalent DT loop filter is now one unit higher than the CT filter order. In general, solutions at the circuit level (appropriate selection of the DAC pulse, feedback coefficient tuning, use of extra feedback paths) can be used for the compensation.

12.1.5   Lowpass delta-sigma modulator

Lowpass $\Delta\Sigma$ modulators are based on discrete-time integrators with a delay, and whose transfer function is

$$I(z) = \frac{z^{-1}}{1 - z^{-1}} \tag{12.33}$$

and a comparator. Note that the term $1/(1 - z^{-1})$ is generally realized by a switched-capacitor integrator, but the delay z^{-1} introduced in the transfer function numerator can be implemented using appropriate clock signals at the integrator input and output, or at the integrator input and to drive the quantizer.

12.1.5.1   Single-stage modulator with a 1-bit quantizer

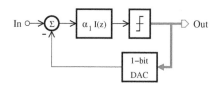

FIGURE 12.12
Block diagram of a first-order modulator.

The block diagram of a first-order modulator is shown in Figure 12.12. It consists of an integrator, a 1-bit quantizer or comparator, and a 1-bit DAC. With the assumption that $\alpha_1 = 1$, the STF and QNTF are given by

$$H_S(z) = z^{-1} \qquad \text{and} \qquad H_Q(z) = 1 - z^{-1}. \tag{12.34}$$

This structure has a large dynamic range, and is simple and less sensitive to the component nonidealities. However, the quantization noise can be signal dependent and not statistically uncorrelated with the input signal as it is usually assumed. As a consequence, single-frequency tones appear in the modulator output spectrum for slowly varying input signals. This effect can be prevented by whitening the quantization noise through *dithering*. It consists of adding a pseudo-random sequence, which is independent and uncorrelated with the input signal, at the quantizer input. The transfer function from the dither input to the modulator output must be proportional to the one of the quantization noise. Furthermore, the magnitude of the dither signal should be chosen so that the quantizer cannot overload. The increase of the number of state variables (integrator input and output) can also help to reduce tones by preventing the formation of a repeating bit pattern at the modulator output.

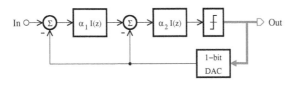

FIGURE 12.13
Block diagram of a second-order modulator.

The modulator shown in Figure 12.13 achieves a second-order shaping of the quantization noise. It uses two integrators in the filter loop. In comparison to the first-order structure, the number of internal states is increased and the occurrence likelihood of spectral tones is reduced. Here for $\alpha_2/2 = 2\alpha_1 = 1$, the STF and QNTF can be written in the form

$$H_S(z) = z^{-2} \quad \text{and} \quad H_Q(z) = (1 - z^{-1})^2. \tag{12.35}$$

Here, the QNTF exhibits two zeros at dc and can provide an improved attenuation of the noise in the baseband compared to the one of the first-order modulator.

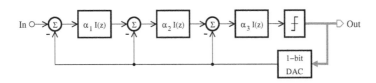

FIGURE 12.14
Block diagram of a third-order modulator.

The third-order modulator of Figure 12.14 is derived by adding an integrator stage and a feedback path to the second-order structure. The STF and

QNTF can be respectively expressed as

$$H_S(z) = \frac{\displaystyle\prod_{i=1}^{3} \alpha_i [I(z)]^3}{1 + \displaystyle\sum_{i=1}^{3} \prod_{j=i}^{3} \alpha_j [I(z)]^{3-i+1}} \tag{12.36}$$

$$= \frac{\alpha_1 \alpha_2 \alpha_3 z^{-3}}{D(z)} \tag{12.37}$$

and

$$H_Q(z) = \frac{1}{1 + \displaystyle\sum_{i=1}^{3} \prod_{j=i}^{3} \alpha_j [I(z)]^{3-i+1}} \tag{12.38}$$

$$= \frac{(1 - z^{-1})^3}{D(z)}, \tag{12.39}$$

where

$$D(z) = 1 + (\alpha_3 - 3)z^{-1} + [\alpha_3(\alpha_2 - 2) + 3]z^{-2} + [\alpha_3(\alpha_2\alpha_1 - \alpha_2 + 1) - 1]z^{-3}. \tag{12.40}$$

Due to the nonlinear nature of $\Delta\Sigma$ modulators, the stability can depend on characteristics such as the input signal level or initial conditions. Simulations show that the use of a multi-bit quantizer, which can better accommodate large signals than a single-bit quantizer, is necessary to stabilize a single-loop modulator with the QNTF of the form $H_Q(z) = (1 - z^{-1})^L$, where $L > 2$. An L-th-order modulator will then be stable if the quantizer has $B \geq L + 1$ bits of resolution. However, higher-order single-loop modulators with a single-bit quantizer can be made stable by matching the QNTF to a more general highpass or band-reject transfer function. This is generally achieved either using the Butterworth or inverse-Chebyshev filter approximations to find the QNTF and then suitable zeros are added to the numerator in order to improve the attenuation of the baseband quantization noise or by numerically finding the poles and zeros of the QNTF, which can provide a more effective shaping of the baseband quantization noise out of the band of interest, with the help of computer-aided design tools. It may then be necessary to increase the complexity of the loop filter structure, as shown in Figures 12.15 and 12.16. These third-order modulator structures are based on single-stage topologies with feedforward and feedback paths and can allow the QNTF zeros to be spread over the signal bandwidth instead of being all placed at dc.

Single-bit modulators with an order equal to or greater than three possess a dynamic range lower than the one predicted by Equation (12.28) due to the attenuation required in the signal path in order to meet the loop stability

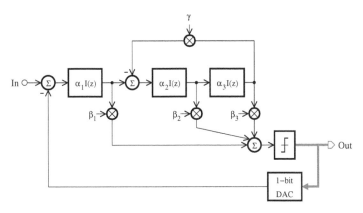

FIGURE 12.15
Block diagram of the third-order modulator with a feedforward summation and local resonator feedback.

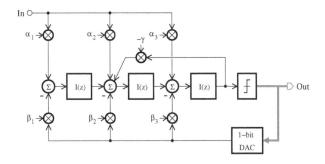

FIGURE 12.16
Block diagram of the third-order modulator with feedforward input paths, distributed feedbacks, and a local resonator feedback.

condition. The design objective is to find the coefficient combination that provides the maximal dynamic range, while maintaining the modulator stability.

Most of the methods for generating the loop coefficients from modulator specifications focus on synthesizing a QNTF based on a filtering function. Because a delay-free loop around a quantizer is not implementable, the associated QNTF should have the property that $\lim_{z \to \infty} H_Q(z) = 1$. Let the noise power gain (NPG) be defined as

$$NPG = \frac{1}{\pi} \int_0^\pi (|H_Q(e^{j\omega})|)^2 d\omega. \tag{12.41}$$

The modulator must be designed such that the NPG limitation is satisfied. That is,

$$\text{NPG}_{min} \le \text{NPG} \le \text{NPG}_{max}, \tag{12.42}$$

where NPG_{min} is determined by the acceptable level of in-band tones, and

NPG_{max} depends on the stability requirement that is affected by the modulator order and the maximum power of the input dc signal. To take into account the effect of coefficient variations due to component imperfections, the upper bound of the NPG must be selected with a safety margin from the instability border. The resulting NPG is generally a function of the in-band noise suppression, the OSR, and the modulator order.

12.1.5.2 Dithering

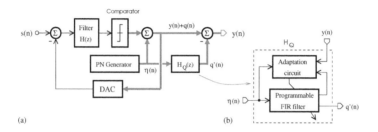

FIGURE 12.17
(a) Block diagram of a modulator with dithering; (b) adaptive filter-based realization of $H_Q(z)$.

In general, the time-domain output waveform of $\Delta\Sigma$ modulators can be affected by idle tones or pattern noise, which appears as periodic impulses whose peak levels are much greater than their rms values. This behavior, which is due to the correlation between the quantizer error and the dc level of the input signal, is undesirable, especially in audio applications. A solution consists of using dithering, which can be realized by adding a pseudo-random sequence to the quantizer input. As a result, the quantization noise is made almost independent of the modulator input signal and asymptotically white in some cases.

The block diagram of a modulator with dithering is shown in Figure 12.17(a). Here, the dithering signal, which is a digital sequence provided by a pseudo-random number generator, can be added to the comparator output [8], instead of the input. Because it is shaped by the modulator in the same way as the quantization noise, its cancelation at the modulator output is realized using a filter section with the transfer function, H_Q, and a subtractor. The noise transfer function, H_Q, is determined by the specifications of the loop filter, and is implemented using either a conventional digital filter or an adaptive filter [9], as illustrated in Figure 12.17(b). This latter approach can provide a better tracking of the modulator response.

12.1.5.3 Design examples

The block diagram of a second-order lowpass modulator is shown in Figure 12.18, where $\alpha_1 = 1/2$, $\alpha_2 = 2$. This modulator is implemented as a fully differential circuit depicted in Figure 12.19. It is based on switched-capacitor

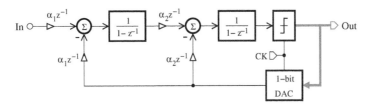

FIGURE 12.18
Block diagram of a second-order lowpass DT modulator.

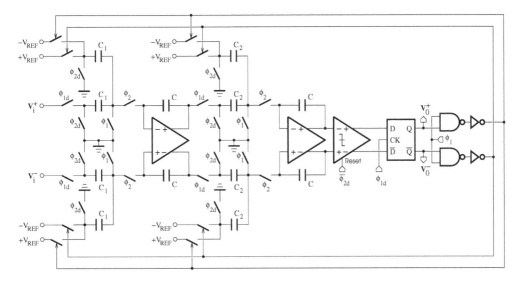

FIGURE 12.19
Circuit diagram of a second-order lowpass DT modulator.

(SC) integrator and operates with nonoverlapping two-phase clock signals. The phase 1 includes ϕ_1 and ϕ_{1d}, while ϕ_2 and ϕ_{2d} constitute the phase 2. The comparator can provide erroneous decisions due to the fact that its decision time generally increases for input signals with a low magnitude. It is then followed by a flip-flop, which can reduce the bit error due to the metastability. The switching of the reference voltages is controlled by a digital circuit, which consists of NAND gates and inverters with a buffer function. Simulation results show that the dynamic range of the modulator increases by 15 dB for every doubling of the OSR.

The block diagram of a second-order lowpass modulator, which uses a continuous-time filter, is illustrated in Figure 12.20, where $\alpha_1 = \alpha_2 = 1$, $\omega_1 = \omega_2 = 1$, $\beta_1 = 1.5$, and $\beta_2 = 1$. The implementation of this modulator shown in Figure 12.21 is based on g_m-C operational amplifier integrators. The output signal of the filter is quantized by the latched comparator and then processed by the D flip-flops, which drive the two 1-bit switched-current DACs used in the feedback path.

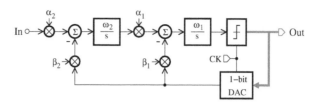

FIGURE 12.20
Block diagram of a second-order lowpass CT modulator.

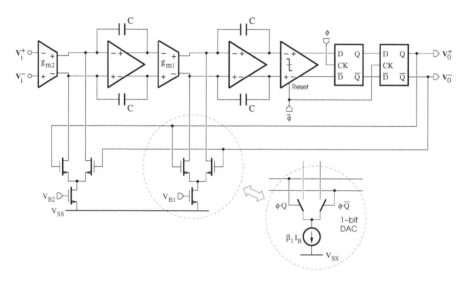

FIGURE 12.21
Circuit diagram of a second-order lowpass CT modulator.

Because $\hat{H}(s)$ cannot uniquely be determined by the impulse invariant transformation, a suitable CT filter prototype is generally used to solve Equation (12.31). For the filter of Figure 12.21, it can be assumed that the DAC pulse is of the NRZ type and the filter has the following transfer function:

$$\hat{H}(s) = \frac{\beta_1}{s} + \frac{\beta_2}{s^2}. \tag{12.43}$$

Here, the equivalent DT transfer function can be obtained as

$$H(z) = \mathcal{Z}\left\{\mathcal{L}^{-1}\left[P(s)\hat{H}(s)\right]_{t=nT}\right\}. \tag{12.44}$$

Because the s-transform of the DAC pulse is given by

$$P(s) = \frac{1 - e^{-sT}}{s}, \tag{12.45}$$

we have

$$H(z) = (1 - z^{-1})\mathcal{Z}\left\{\mathcal{L}^{-1}\left[\frac{\hat{H}(s)}{s}\right]_{t=nT}\right\}. \tag{12.46}$$

Noting that

$$\mathcal{Z}\left\{\mathcal{L}^{-1}\left[\frac{1}{s^2}\right]_{t=nT}\right\} = \frac{Tz^{-1}}{(1 - z^{-1})^2} \tag{12.47}$$

$$\mathcal{Z}\left\{\mathcal{L}^{-1}\left[\frac{1}{s^3}\right]_{t=nT}\right\} = \frac{T^2 z^{-1}(1 + z^{-1})}{(1 - z^{-1})^3} \tag{12.48}$$

and $T = 1$, the equivalent z-domain transfer function can be expressed as

$$H(z) = \frac{(2\beta_1 + \beta_2)z^{-1} + (-2\beta_1 + \beta_2)z^{-2}}{2(1 - z^{-1})^2}. \tag{12.49}$$

In the case where $\beta_1 = 1.5$ and $\beta_2 = 1$, we can obtain

$$H(z) = \frac{2z^{-1} - z^{-2}}{(1 - z^{-1})^2}. \tag{12.50}$$

The CT modulator of Figure 12.20 is then equivalent to a second-order DT modulator with the QNTF given by

$$H_Q(z) = \frac{1}{1 + H(z)} = (1 - z^{-1})^2. \tag{12.51}$$

Note that the design of a CT modulator can be based on other types of DAC pulses. In order to obtain the output at any time between two consecutive sampling instants, the expression of the equivalent DT transfer function should be rewritten as

$$H(z) = \mathcal{Z}_m\left\{\mathcal{L}^{-1}\left[P(s)\hat{H}(s)\right]_{t=nT}\right\}, \tag{12.52}$$

where $\mathcal{Z}_m$ denotes the modified z-transform.

In the analysis of the excess loop delay, the transfer function provided by the modified z-transform, which is based on the assumption that the delay occurs at the output of the CT filter, is similar but not identical to the one obtained by solving the impulse invariant transformation equation in the time domain, where the fact that the delay occurs prior to the DAC pulse, as it is the case in practical circuit, can be taken into account.

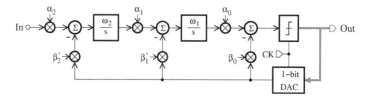

FIGURE 12.22
Block diagram of a second-order lowpass CT modulator with an extra feedback path.

$$\alpha_0 = \alpha_1 = \alpha_2 = 1, \ \omega_1 = \omega_2 = 1.$$

One approach to compensate for the excess loop delay consists of adding extra feedback paths to the modulator. In the modulator block diagram shown in Figure 12.22, the additional path connects the DAC output to the summer inserted between the last integrator and the quantizer. The transfer function of the CT filter now becomes

$$\hat{H}(s) = \beta_0 + \frac{\beta_1'}{s} + \frac{\beta_2'}{s^2}. \tag{12.53}$$

Considering an excess loop delay equal to τ, the z-domain equivalent transfer function of the filter can be obtained as

$$H_\tau(z) = \mathcal{Z}\left\{\mathcal{L}^{-1}\left[P_\tau(s)\hat{H}(s)\right]_{t=nT}\right\}, \tag{12.54}$$

where T is the clock period, and the Laplace transform of the DAC pulse, $P_\tau(s)$, is given by

$$P_\tau(s) = \int_\tau^{T+\tau} e^{-st} dt \tag{12.55}$$

$$= \frac{e^{-s\tau} - e^{-s(T+\tau)}}{s} \tag{12.56}$$

$$= \frac{e^{-s\tau} - e^{-sT}}{s} + e^{-sT}\frac{1 - e^{-s\tau}}{s}. \tag{12.57}$$

Using the following impulse invariant transformations

$$\mathcal{Z}\left\{\mathcal{L}^{-1}\left[\frac{P_\tau(s)}{s}\right]_{t=nT}\right\} = \frac{(1-\tau)z^{-1}}{1-z^{-1}} + z^{-1}\frac{\tau z^{-1}}{1-z^{-1}} \tag{12.58}$$

and

$$\mathcal{Z}\left\{\mathcal{L}^{-1}\left[\frac{P_\tau(s)}{s^2}\right]_{t=nT}\right\} = \frac{(1-\tau)^2 z^{-1} + (1-\tau^2)z^{-2}}{2(1-z^{-1})^2}$$

$$+ z^{-1}\frac{\tau(2-\tau)z^{-1} + \tau^2 z^{-2}}{2(1-z^{-1})^2}, \tag{12.59}$$

it can be shown that

$$H_\tau(z) = \frac{az^{-1} + bz^{-2} + cz^{-3}}{2(1 - z^{-1})^2}, \tag{12.60}$$

where

$$a = 2\beta_0 + 2\beta_1'(1 - \tau) + \beta_2'(1 - \tau)^2, \tag{12.61}$$

$$b = -4\beta_0 - 2\beta_1'(1 - 2\tau) + \beta_2'(1 + 2\tau - 2\tau^2), \tag{12.62}$$

and

$$c = 2\beta_0 - 2\beta_1'\tau + \beta_2'\tau^2. \tag{12.63}$$

The compensation for the excess loop delay is achieved provided the transfer functions $H_\tau(z)$ and $H_\tau(z)$ are matched. Setting Equation (12.49) equal to (12.60) yields

$$\beta_0 = \beta_1\tau + \beta_2\tau^2/2 \tag{12.64}$$

$$\beta_1' = \beta_1 + \beta_2\tau \tag{12.65}$$

and

$$\beta_2' = \beta_2. \tag{12.66}$$

The aforementioned method for the derivation of the feedback coefficients may become cumbersome and impractical due the nonideal characteristics (parasitic poles and zeros) of amplifiers used in the integrator design. As a consequence, numerical techniques and behavioral simulations are often used in practice.

Note that, by operating with symmetric signals, differential architectures have the advantage of reducing the inter-symbol interference effects caused by unequal rise and fall times of the DAC pulses.

12.1.5.4 Modulator architectures with a multi-bit quantizer

A multi-bit $\Delta\Sigma$ modulator is realized by using multi-bit quantizer (or say, a B-bit ADC) and DAC. The main advantage of multi-bit modulators is the increase in the dynamic range by about $20\log_{10}(2^B - 1)$ dB compared to that of modulators with a single-bit quantizer. This is due to the fact that the power of the quantization noise is proportional to the square of the quantizer step size, which is reduced by increasing the number of quantization levels in the converter range. The oversampling ratio required to achieve a given conversion resolution can then be reduced. In the design of high-order ($L > 2$), single-loop modulators, the use of a multi-bit quantizer helps prevent the instability due to quantizer overload and observed in most single-bit structures. In the case of low-order ($L \leq 2$) modulators, an improved attenuation of tones can be expected because the quantization noise is more randomly distributed as the number of bits of the quantizer is increased.

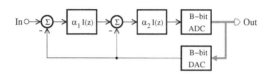

FIGURE 12.23
Block diagram of a multi-bit second-order modulator.

The block diagram of a multi-bit second-order modulator is depicted in Figure 12.23, where $\alpha_1 = 1/2$, $\alpha_2 = 2$, and $I(z) = z^{-1}/(1 - z^{-1})$. The multi-bit DAC used in the feedback loop can be implemented using either current-steering circuits, wherein transistor matching is essential to obtain a high linearity, or charge-redistribution circuits, in which the capacitor matching is required. By modeling the quantizer as an additive source with the quantization noise E_Q and the multi-bit DAC as an additive source with the nonlinearity error E_D, the modulator output, Y, is given by

$$Y(z) = z^{-2}S(z) + (1 - z^{-1})^2 E_Q(z) - E_D(z), \qquad (12.67)$$

where S denotes the input signal. Hence, the performance of the multi-bit modulator is limited by the nonlinearity of the B-bit internal DAC, which results in distortions directly added to the input signal. Modulator designs were reported using digital correction techniques [10] or dynamic element matching (DEM) methods such as data weighted averaging (DWA) to reduce the effect of the DAC mismatch errors that can be introduced in the signal baseband.

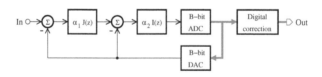

FIGURE 12.24
Block diagram of a multi-bit second-order modulator with digital calibration.

The general architecture used for the digital calibration is illustrated by the block diagram shown in Figure 12.24. During the calibration, the system is configured to allow the estimation of DAC errors that are stored in memory and subsequently used for correction.

The internal ADC is generally of the flash type and its hardware complexity increases exponentially with the bit resolution. That is, a suitable choice for the number of bits, B, is on the order of 4. The block diagram of the digital calibration required to meet the accuracy of 16 bits using a second-order modulator with an OSR of 128 is shown in Figure 12.25 [10]. It is based on the length truncation of data stored in a random access memory (RAM). In the worst case, the 4-bit internal DAC is assumed to exhibit a linearity of 9 bits. The modulator output is transferred to the address lines for the selection of the corresponding RAM word with the length of 10 bits (1 sign bit + 9 bits). This latter is reduced to 3 bits by a first-order digital $\Delta\Sigma$ modulator. The compressed RAM word, $\mathbf{e}$, is then added to $\mathbf{x}$, which is a 10-bit delayed version of the modulator output. The modulator output sequences form the four MSBs, the fifth MSB is set to 1 to assign the positive sign of $\mathbf{x}$ to the addition result for any value of $\mathbf{e}$, and the remaining bits are zero. The 10-to-4 bit truncation of $\mathbf{s}$ is then achieved using the structure shown in Figure 12.26. Note that the characteristic of the truncator is determined by the signal resolution predicted by simulations to be at least 18 bits. The scheme depicted in Figure 12.27 is used to store the conversion errors of the 4-bit DAC in the RAM. The analog equivalent of the digital input code generated by a 4-bit counter is applied to the initial multi-bit $\Delta\Sigma$ modulator operating as a single bit converter. The decimation stage is based on counters, which can provide an 18-bit word for each digital input code, which is held for 2^{18} clock periods. The error data to be stored in the RAM is computed as the difference between the converted and the original input code. The overall calibration requires $2^4 \times 2^{18}$ clock periods and is achieved off-line. However, with a DAC structure that can operate with multiple inputs and outputs, such as the resistor-string converter, the modulator can be duplicated to allow a background calibration.

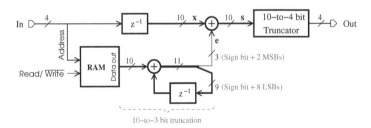

FIGURE 12.25
Block diagram of the digital calibration. (From [10], ©1993 IEEE.)

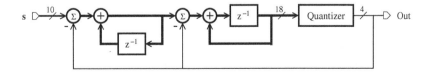

FIGURE 12.26
Block diagram of the modulator used for the 10-to-4 bit truncation.

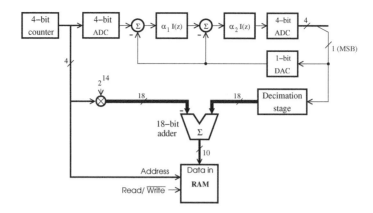

FIGURE 12.27
Scheme for the storage of the DAC error data in the RAM. (From [10], ©1993 IEEE.)

An alternative technique used to mitigate the effect of component mismatch in multi-bit DACs is dynamic element matching (DEM), which consists of using an algorithm to assign randomly the DAC unit elements to the code being converted. In this way, the linearity error, which is generated in the case where some mismatched elements are more frequently selected than others over a given time period, is modulated at frequencies outside the signal band. The DEM technique can be implemented using a shifter controlled by a suitable selection logic. However, the clock frequency of the resulting modulator can be limited by the time delay introduced in the feedback path by this additional logic.

In comparison with single-bit topologies, multi-bit modulators offer a better performance (an increase of the dynamic range by 6 dB per additional bit as a result of the reduced quantization noise, an improved attenuation of tones because the randomness assumption of the quantization noise is better satisfied as the number of quantization levels is increased) without increasing the OSR, but they can be limited by the stringent linearity requirement placed on the feedback DAC.

12.1.5.5 Cascaded modulator

The use of a high-order filter structure or a multi-bit internal quantizer can be adopted to improve the dynamic range of a modulator without increasing the oversampling ratio. However, each of these design solutions is known to be limited by potential shortcomings. High-order single-loop modulators may become prone to instability due to quantizer overload caused by large signals or the integrator initial conditions. The performance of modulators with a multi-bit quantizer is affected by the nonlinearity of the internal DAC and the increasing loading of amplifiers. A suitable design alternative can consist of performing high-order filtering through a cascade of low-order structures to ensure modulator stability and using a multi-bit quantizer only in the final stage, whose noise cancelation logic also attenuates the nonlinearity of the multi-bit DAC in the signal baseband [11]. The resulting implementation is known as a multistage or cascaded modulator. Here, cascaded modulators are realized using only first-order and second-order structures, which feature relaxed stability criteria. In this way, the dynamic range of the cascaded modulator can be larger than the one of single-loop structures, provided an adequate matching is achieved between the loop coefficients.

The input signal of the first stage is S and the subsequent stages are fed by a signal, which is either the inverted version of the quantization noise generated by the previous stage or the output of last integrator in the previous stage, and which can be computed from the following equations:

$$E_{Qi}(z) = Y_i(z) - \frac{X_i(z)}{\alpha_i} \qquad \text{or} \qquad X_i(z) = \alpha_i(Y_i(z) - E_{Qi}(z)) \qquad i = 1, 2, 3$$

(12.68)

for the first-order modulator, and

$$E_{Q1}(z) = Y_1(z) - \frac{X_2(z)}{\alpha_1 \alpha_2} \qquad \text{or} \qquad X_2(z) = \alpha_1 \alpha_2(Y_1(z) - E_{Q1}(z)) \quad (12.69)$$

in the case of the second-order modulator. Here, E_{Q1}, E_{Q2}, and E_{Q3} are the quantization noises; Y_1, Y_2, and Y_3 denote the modulator outputs; and X_1, X_2, and X_3 represent the integrator outputs; and α_1, α_2, and α_3 are scaling coefficients. The purpose of the scaling process is to maximize the overload level by using all the available signal swing at the output of each integrator without clipping. To keep the modulator output independent of the integrator coefficients, the output of the last integrator in a given stage is multiplied by a factor proportional to the inverse of the product of all the integrator coefficients of that stage before being summed at the input of the next stage.

Second-order modulator

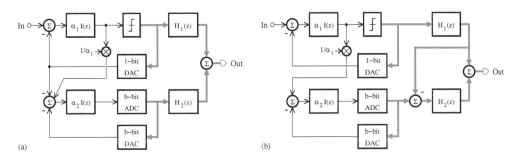

(a)   (b)

FIGURE 12.28
Block diagrams of 1-1 cascaded lowpass modulators.

The second-order architectures depicted in Figure 12.28 require two first-order modulator stages. It can be assumed that $\alpha_1 = \alpha_2 = 1$.

- Second-order, 1-1 cascaded lowpass modulator (a)

In the modulator structure of Figure 12.28(a), the quantization noise of the first stage is estimated, inverted, and applied to the next stage. The linear analysis in the z-domain of the modulator yields the following equations,

$$Y_1(z) = z^{-1}S(z) + (1 - z^{-1})E_{Q1}(z), \tag{12.70}$$

$$Y_2(z) = -z^{-1}E_{Q1}(z) + (1 - z^{-1})E_{Q2}(z), \tag{12.71}$$

and

$$Y(z) = H_1(z)Y_1(z) + H_2(z)Y_2(z), \tag{12.72}$$

where Y_1 and Y_2 denote the outputs of the first and second stages, respectively, and E_{Q1} and E_{Q2} represent the quantization noises of the first and second stages, respectively.

- Second-order, 1-1 cascaded lowpass modulator (b)

For the implementation illustrated by the block diagram of Figure 12.28(b), the output of the first integrator is connected to the input of the second modulator stage. The following expressions can be derived:

$$Y_1(z) = z^{-1}S(z) + (1 - z^{-1})E_{Q1}(z) \tag{12.73}$$

$$Y_2(z) = z^{-1}X_1(z) + (1 - z^{-1})E_{Q2}(z) \tag{12.74}$$

and

$$Y(z) = H_1(z)Y_1(z) + H_2(z)[Y_2(z) - H_1(z)Y_1(z)], \tag{12.75}$$

where Y_1 and Y_2 denote the outputs of the first and second stages, respectively, and E_{Q1} and E_{Q2} represent the quantization noises of the first and second stages, respectively. The output, X_1, of the integrator in the first stage can be obtained as

$$X_1(z) = z^{-1}(S(z) - E_{Q1}(z)), \tag{12.76}$$

and it can then be found that

$$Y_2(z) = z^{-2}(S(z) - E_{Q1}(z)) + (1 - z^{-1})E_{Q2}(z). \tag{12.77}$$

By choosing the transfer functions, H_1 and H_2, of the digital cancelation logics for the modulators of Figure 12.28 as

$$H_1(z) = z^{-1} \qquad \text{and} \qquad H_2(z) = 1 - z^{-1}, \tag{12.78}$$

the overall output should ideally exhibit only the quantization noise of the last modulator stage. Hence,

$$Y(z) = H_S(z)S(z) + H_Q(z)E_{Q2}(z), \tag{12.79}$$

where the STF and QNTF are of the form

$$H_S(z) = z^{-2} \quad \text{and} \quad H_Q(z) = (1 - z^{-1})^2, \tag{12.80}$$

respectively.

Third-order modulator

A third-order lowpass modulator can be implemented by the 1-1-1 cascaded or 2-1 cascaded structures as shown in Figure 12.29, where $\alpha_1 = \alpha_2 = \alpha_3 = 1$, or Figure 12.30, where $\alpha_1 = 1/2$, $\alpha_2 = 2$, $\alpha_3 = 1$.

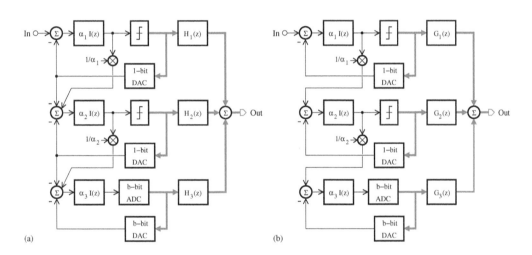

FIGURE 12.29
Block diagrams of 1-1-1 cascaded lowpass modulators.

• Third-order, 1-1-1 cascaded lowpass modulator (a)
With reference to the modulator of Figure 12.29(a), we have

$$Y_1(z) = z^{-1}S(z) + (1 - z^{-1})E_{Q1}(z), \tag{12.81}$$
$$Y_2(z) = -z^{-1}E_{Q1}(z) + (1 - z^{-1})E_{Q2}(z), \tag{12.82}$$
$$Y_3(z) = -z^{-1}E_{Q2}(z) + (1 - z^{-1})E_{Q3}(z), \tag{12.83}$$

and

$$Y(z) = H_1(z)Y_1(z) + H_2(z)Y_2(z) + H_3(z)Y_3(z), \tag{12.84}$$

where Y_1, Y_2, and Y_3 are the outputs of the first, second, and third stages, respectively; and E_{Q1}, E_{Q2}, and E_{Q3} represent the quantization noises of the first, second, and third stages, respectively. To remove the quantization noises of the two first stages from the modulator output, it is necessary to use digital circuit sections with the transfer functions

$$H_1(z) = z^{-2}, \qquad H_2(z) = z^{-1}(1 - z^{-1}), \qquad \text{and} \qquad H_3(z) = (1 - z^{-1})^2 . \tag{12.85}$$

• Third-order, 1-1-1 cascaded lowpass modulator (b)

In the case of the structure depicted in Figure 12.29(b), the following expressions can be derived:

$$Y_1(z) = z^{-1}S(z) + (1 - z^{-1})E_{Q1}(z) \tag{12.86}$$

$$Y_2(z) = z^{-1}X_1(z) + (1 - z^{-1})E_{Q2}(z) \tag{12.87}$$

$$Y_3(z) = z^{-1}X_2(z) + (1 - z^{-1})E_{Q3}(z) \tag{12.88}$$

and

$$Y(z) = G_1(z)Y_1(z) + G_2(z)Y_2(z) + G_3(z)Y_3(z), \tag{12.89}$$

where Y_1, Y_2, and Y_3 are the outputs of the first, second and third stages, respectively; and E_{Q1}, E_{Q2}, and E_{Q3} represent the quantization noises of the first, second, and third stages, respectively. The output of the first integrator can be computed as

$$X_1(z) = z^{-1}(S(z) - E_{Q1}(z)) \tag{12.90}$$

and Y_2 becomes

$$Y_2(z) = z^{-2}(S(z) - E_{Q1}(z)) + (1 - z^{-1})E_{Q2}(z), \tag{12.91}$$

while the output of the second integrator is obtained as

$$X_2(z) = z^{-2}(S(z) - E_{Q1}(z)) - z^{-1}E_{Q2} \tag{12.92}$$

and Y_3 can take the form

$$Y_3(z) = z^{-3}(S(z) - E_{Q1}(z)) - z^{-2}E_{Q2}(z) + (1 - z^{-1})E_{Q3}(z). \tag{12.93}$$

Here, the cancelation of the quantization noise of the first previous stages is achieved with the following transfer functions:

$$G_1(z) = z^{-3}, \qquad G_2(z) = z^{-2}(1 - z^{-1}), \qquad \text{and} \qquad G_3(z) = (1 - z^{-1})^2 . \tag{12.94}$$

• Third-order, 2-1 cascaded lowpass modulator (a)

The modulator structure shown in Figure 12.30(a) can be described by

$$Y_1(z) = z^{-2}S(z) + (1 - z^{-1})^2 E_{Q1}(z), \tag{12.95}$$

$$Y_2(z) = -z^{-1}E_{Q1}(z) + (1 - z^{-1})E_{Q2}(z), \tag{12.96}$$

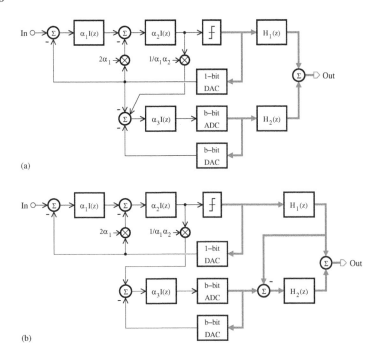

(a)

(b)

FIGURE 12.30
Block diagrams of 2-1 cascaded lowpass modulators.

and

$$Y(z) = H_1(z)Y_1(z) + H_2(z)Y_2(z), \tag{12.97}$$

where Y_1 and Y_2 are the outputs of the first and second stages, respectively; and E_{Q1} and E_{Q2} represent the quantization noises of the first and second stages, respectively.

• Third-order, 2-1 cascaded lowpass modulator (b)
In the case of the modulator structure of Figure 12.30(b), we have

$$Y_1(z) = z^{-2}S(z) + (1 - z^{-1})^2 E_{Q1}(z), \tag{12.98}$$

$$Y_2(z) = z^{-1}X_2(z) + (1 - z^{-1})E_{Q2}(z), \tag{12.99}$$

and

$$Y(z) = H_1(z)Y_1(z) + H_2(z)[Y_2(z) - H_1(z)Y_1(z)], \tag{12.100}$$

where Y_1 and Y_2 are the outputs of the first and second stages, respectively; and E_{Q1} and E_{Q2} represent the quantization noises of the first and second stages, respectively. The output of the second integrator in the first stage, X_2, is given by

$$X_2(z) = z^{-2}S(z) + z^{-1}(z^{-1} - 2)E_{Q1}(z) \tag{12.101}$$

and Y_2 takes the form

$$Y_2(z) = z^{-3}S(z) + z^{-2}(z^{-1} - 2)E_{Q1}(z) + (1 - z^{-1})E_{Q2}(z). \qquad (12.102)$$

In the cases of the 2-1 cascaded structures depicted in Figure 12.30, the transfer functions of the cancelation logic are given by

$$H_1(z) = z^{-1} \qquad \text{and} \qquad H_2(z) = (1 - z^{-1})^2, \qquad (12.103)$$

and the overall output of the modulator can be written as

$$Y(z) = H_S(z)S(z) + H_Q(z)E_{Q3}(z), \qquad (12.104)$$

where the STF and QNTF are given by

$$H_S(z) = z^{-3} \qquad \text{and} \qquad H_Q(z) = (1 - z^{-1})^3. \qquad (12.105)$$

Ideally, the 2-1 and 1-1-1 cascaded modulators can be designed to realize the same STF and QNTF. However, because the cancelation of the quantization noise generated by the first stage is achieved after the second-order shaping in the 2-1 cascaded modulator, the required component matching is more relaxed than in the 1-1-1 cascaded structure, which is based on first-order stages.

Fourth-order modulator

The block diagrams of fourth-order lowpass modulators are depicted in Figure 12.31, where $\alpha_1 = 1/2$, $\alpha_2 = 2$, and $\alpha_3 = \alpha_4 = 1$, and Figure 12.32, where $\alpha_1 = \alpha_3 = 1/2$ and $\alpha_2 = \alpha_4 = 2$.

• Fourth-order, 2-1-1 cascaded lowpass modulator (a)
With reference to the 2-1-1 cascaded modulator structure shown in Figure 12.31(a), we can obtain

$$Y_1(z) = z^{-2}S(z) + (1 - z^{-1})^2 E_{Q1}(z), \qquad (12.106)$$
$$Y_2(z) = -z^{-1}E_{Q1}(z) + (1 - z^{-1})E_{Q2}(z), \qquad (12.107)$$
$$Y_3(z) = -z^{-1}E_{Q2}(z) + (1 - z^{-1})E_{Q3}(z), \qquad (12.108)$$

and

$$Y(z) = H_3'(z)[H_1(z)Y_1(z) + H_2(z)Y_2(z)] + H_3(z)Y_3(z), \qquad (12.109)$$

where Y_1, Y_2, and Y_3 are the outputs of the first, second, and third stages, respectively; and E_{Q1}, E_{Q2}, and E_{Q3} represent the quantization noises of the first, second, and third stages, respectively.

• Fourth-order, 2-1-1 cascaded lowpass modulator (b)

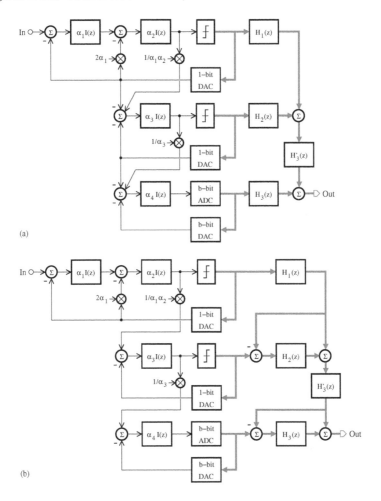

FIGURE 12.31
Block diagrams of 2-1-1 cascaded lowpass modulators.

For the 2-1-1 cascaded modulator structure depicted in Figure 12.31(b), it can be shown that

$$Y_1(z) = z^{-2}S(z) + (1 - z^{-1})^2 E_{Q1}(z), \qquad (12.110)$$

$$Y_2(z) = z^{-1}X_2(z) + (1 - z^{-1})E_{Q2}(z), \qquad (12.111)$$

$$Y_3(z) = z^{-1}X_3(z) + (1 - z^{-1})E_{Q3}(z), \qquad (12.112)$$

and

$$Y(z) = H_3(z)[Y_3(z) - Y'(z)] + Y'(z), \qquad (12.113)$$

where

$$Y'(z) = H_3'(z)\{H_1(z)Y_1(z) + H_2(z)[Y_2(z) - H_1(z)Y_1(z)]\}. \qquad (12.114)$$

Here, Y_1, Y_2 and Y_3 are the outputs of the first, second, and third stages, respectively; and E_1, E_2, and E_3 represent the quantization noises of the first, second, and third stages, respectively. The output, X_2, of the second integrator in the first stage is

$$X_2(z) = z^{-2}S(z) + z^{-1}(z^{-1} - 2)E_{Q1}(z) \tag{12.115}$$

and Y_2 can be written as

$$Y_2(z) = z^{-3}S(z) + z^{-2}(z^{-1} - 2)E_{Q1}(z) + (1 - z^{-1})E_{Q2}(z). \tag{12.116}$$

For the output, X_3, of the integrator in the second stage, the following expression is obtained

$$X_3(z) = z^{-3}S(z) + z^{-2}(z^{-1} - 2)E_{Q1}(z) - z^{-1}E_{Q2}(z), \tag{12.117}$$

and

$$Y_3(z) = z^{-4}S(z) + z^{-3}(z^{-1} - 2)E_{Q1}(z) - z^{-2}E_{Q2}(z) + (1 - z^{-1})E_{Q3}(z). \tag{12.118}$$

The transfer functions of the cancelation logic for the 2-1-1 cascaded structures shown in Figure 12.31 are derived such that the quantization noises of the first two stages should be removed from the overall output of the modulator. That is,

$$H_1(z) = H_3'(z) = z^{-1}, \quad H_2(z) = (1 - z^{-1})^2, \quad \text{and} \quad H_3(z) = (1 - z^{-1})^3. \tag{12.119}$$

• Fourth-order, 2-2 cascaded lowpass modulator (a)

The analysis of the 2-2 cascaded modulator structure shown in Figure 12.32(a) yields

$$Y_1(z) = z^{-2}S(z) + (1 - z^{-1})^2 E_{Q1}(z), \tag{12.120}$$

$$Y_2(z) = z^{-2}E_{Q1}(z) + (1 - z^{-1})^2 E_{Q2}(z), \tag{12.121}$$

and

$$Y(z) = H_1(z)Y_1(z) + H_2(z)Y_2(z), \tag{12.122}$$

where Y_1 and Y_2 are the outputs of the first and second stages, respectively; and E_{Q1} and E_{Q2} represent the quantization noises of the first and second stages, respectively.

• Fourth-order, 2-2 cascaded lowpass modulator (b)

The 2-2 cascaded modulator structure of Figure 12.32(b) can be characterized by equations of the form

$$Y_1(z) = z^{-2}S(z) + (1 - z^{-1})^2 E_{Q1}(z), \tag{12.123}$$

$$Y_2(z) = z^{-2}X_2(z) + (1 - z^{-1})^2 E_{Q2}(z), \tag{12.124}$$

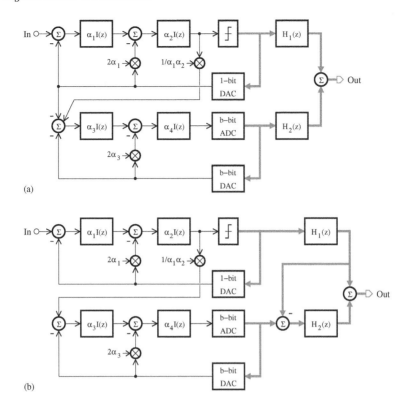

(a)

(b)

FIGURE 12.32
Block diagrams of 2-2 cascaded lowpass modulators.

and

$$Y(z) = H_2(z)[Y_2(z) - H_1(z)Y_1(z)] + H_1(z)Y_1(z), \qquad (12.125)$$

where Y_1 and Y_2 are the outputs of the first and second stages, respectively; and E_{Q1} and E_{Q2} represent the quantization noises of the first and second stages, respectively. By expressing the output, X_2, of the second integrator in the first stage as

$$X_2(z) = z^{-2}S(z) + z^{-1}(z^{-1} - 2)E_{Q1}(z), \qquad (12.126)$$

we obtain

$$Y_2(z) = z^{-4}S(z) + z^{-3}(z^{-1} - 2)E_{Q1}(z) + (1 - z^{-1})^2 E_{Q2}(z). \qquad (12.127)$$

For the 2-2 cascaded modulators of Figure 12.32, the digital cancelation logic suppresses the quantization noise of the first stage, provided that

$$H_1(z) = z^{-2} \qquad \text{and} \qquad H_2(z) = (1 - z^{-1})^2. \qquad (12.128)$$

The output of the above fourth-order modulator is of the form

$$Y(z) = H_S(z)S(z) + H_Q(z)E_{Q3}(z), \qquad (12.129)$$

where the STF and QNTF are given by

$$H_S(z) = z^{-4} \qquad \text{and} \qquad H_Q(z) = (1 - z^{-1})^4 . \tag{12.130}$$

In practice, the cancelation of the quantization noise due to the first modulator stages is limited by the matching level achievable between the loop gains.

12.1.5.6 Effect of the multi-bit DAC nonlinearity

Given an L-th order cascaded modulator with k stages, the nonlinearity of the multi-bit DAC used in the last stage can be modeled as an additive noise characterized by the z-domain function, E_D. The linear analysis yields a more general equation of the modulator output, Y, as follows,

$$Y(z) = H_S(z)S(z) + H_Q(z)E_{Qk}(z) + H_D(z)E_D(z), \tag{12.131}$$

where S denotes the input signal, E_{Qk} is the quantization noise of the last stage, and STF and QNTF are respectively given by

$$H_S(z) = z^{-L} \qquad \text{and} \qquad H_Q(z) = (1 - z^{-1})^L . \tag{12.132}$$

Here, the DAC nonlinearity is attenuated in the signal baseband by the transfer function

$$H_D(z) = H_k(z) = (1 - z^{-1})^\eta , \tag{12.133}$$

where η represents the number of integrators used in the stages 1 to $k - 1$. This can be understood by observing that the output of the multi-bit DAC is actually fed back to the input of the last stage, but it can be considered a signal shaped by an η-th order transfer function from the perspective of the overall modulator. Note that the 2-1-1 and 2-2 cascaded modulators realize the same STF and QNTF, but the errors due to the feedback multi-bit DAC are shaped by third- and second-order transfer functions, respectively. As a result, the 2-1-1 cascaded modulator features better attenuation of the DAC nonlinearities in the baseband than does the 2-2 cascaded structure.

12.1.5.7 Quantization noise shaping and inter-stage coefficient scaling

To avoid clipping at high levels of the quantization noise, it may be necessary to use additional inter-stage scaling coefficients in cascaded modulators as shown in the block diagram of Figures 12.33 and 12.34. The 2-1-1 cascaded modulator uses two inter-stage gains, κ_1 and κ_3, and the DAC feedback signal is scaled by κ_2 and κ_4; while for the 2-2 cascaded structure, κ_1 denotes the inter-stage gain and κ_2 is the scaling coefficient of the DAC feedback signal. The QNTF can be obtained as

$$H_Q(z) = \frac{(1 - z^{-1})^4}{\kappa}, \tag{12.134}$$

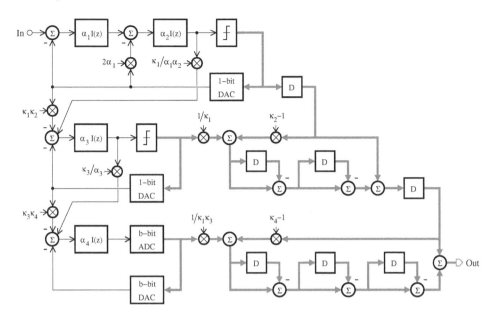

FIGURE 12.33
Block diagram of a 2-1-1 cascaded lowpass modulator with scaling coefficients.

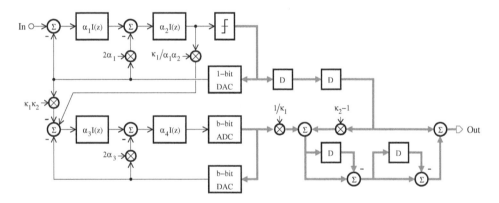

FIGURE 12.34
Block diagram of a 2-2 cascaded lowpass modulator with scaling coefficients.

where κ is respectively equal to $\kappa_1\kappa_3$ and κ_1 for the 2-1-1 and 2-2 cascaded modulators. The design objective is to find the modulator coefficients that yield the maximum dynamic range. The level of the signal transferred from one stage to the next is set to avoid a premature overload by appropriately selecting the values of the coefficients κ_1, κ_2, κ_3, and κ_4. The multiplication of the last stage output with a factor $1/\kappa$ that is greater than 1 leads to a decrease in the modulator resolution by $\log_2(1/\kappa)$ bits. Thus, a trade-off must be made between the minimization of the quantization noise in the baseband and the achievable improvement of the overload condition due to the $1/\kappa$ scaling effect. To simplify the implementation of the digital circuit for the noise cancelation, the factor $1/\kappa$ is generally chosen as a power of 2. Note that a feedback path is added to the cancelation logics for each of the coefficients κ_2 and κ_4, that is different from unity.

12.1.6 Bandpass delta-sigma modulator

For a given technology, bandpass $\Delta\Sigma$ modulators are generally dedicated for the analog-to-digital conversion of signals with a higher frequency than the one supported by lowpass modulators. This is due to the fact that the bandwidth frequency of a bandpass modulator is limited to $f_s/(2\,\text{OSR})$, where OSR is the oversampling ratio and f_s is the sampling frequency, instead of the signal frequency as it is the case for a lowpass modulator, thus making possible the conversion of signals with frequencies up to $f_s/2$. Bandpass modulators can then find applications in the digitalization of intermediate frequency signals in wireless receivers.

The key parameters in the design of $\Delta\Sigma$ modulators for a specified signal-to-noise ratio (SNR) and dynamic range (DR) are the OSR, the order or structure of the loop filter, and the quantizer resolution. A bandpass modulator can be designed to have the passband center frequency located anywhere between 0 and $f_s/2$. However, the class of bandpass modulators with the passband centered around $f_s/4$ seems to exhibit some advantages. It can easily be derived from lowpass prototypes using the z^{-1} to $-z^{-2}$ transformation. The resulting bandpass modulator can be implemented using building blocks with a reduced complexity because it is based on second-order resonators with the transfer function given by

$$R(z) = I(\hat{z})|_{\hat{z}^{-1}=-z^{-2}} = \left.\frac{\hat{z}^{-1}}{1-\hat{z}^{-1}}\right|_{\hat{z}^{-1}=-z^{-2}} = -\frac{z^{-2}}{1+z^{-2}}. \qquad (12.135)$$

The function R is characterized by a resonance occurring at the frequency $f_s/4$ due to the pair of complex poles located at $z = \pm j$.

In the general case, the transformation of the lowpass prototype into a bandpass modulator can be achieved using the following z-variable substitution,

$$z^{-1} \rightarrow -z^{-1}\frac{z^{-1}-\alpha}{1-\alpha z^{-1}}, \qquad (12.136)$$

where $\alpha = \cos 2\pi(f_0/f_s)$ and f_0 represents the desired center frequency. Assuming that $f_0 = f_s/4$, the above expression is reduced to the transformation, $z^{-1} \rightarrow -z^{-2}$. However, this general approach has the inconvenience of not preserving the dynamic properties of the lowpass prototype and may result in a circuit implementation with increased complexity.

12.1.6.1 Single-loop bandpass delta-sigma modulator

The second-order bandpass modulator shown in Figure 12.35 is derived by performing the dc-to-$f_s/4$ transformation on a first-order lowpass prototype.

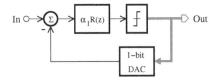

FIGURE 12.35
Block diagram of a second-order bandpass modulator.

Assuming that $\alpha_1 = 1$, the STF and QNTF of the bandpass modulator are respectively given by

$$H_S(z) = -z^{-2} \quad \text{and} \quad H_Q(z) = 1 + z^{-2}. \tag{12.137}$$

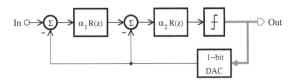

FIGURE 12.36
Block diagram of a fourth-order bandpass modulator.

The fourth-order bandpass modulator depicted in Figure 12.36 is obtained from the single-loop second-order lowpass modulator. In the case where $\alpha_1 = 1/2$ and $\alpha_2 = 2$, the STF and QNTF can be expressed as

$$H_S(z) = z^{-4} \quad \text{and} \quad H_Q(z) = (1 + z^{-2})^2. \tag{12.138}$$

12.1.6.2 Cascaded bandpass delta-sigma modulator

The SNR can be increased using a higher OSR or higher-order filter. However, the power consumption due to the settling requirements of amplifiers becomes a constraint in modulators with a high value of OSR. Because higher-order modulators based on a single loop may be prone to instability, an alternative is to use cascaded structures, provided that the mismatch between the analog and digital transfer functions is maintained at an acceptable level.

The block diagrams of 4-2 cascaded bandpass modulators, as shown in Figure 12.37, are derived from the 2-1 cascaded lowpass modulators. To proceed further, it can be assumed that $\alpha_1 = 1/2$, $\alpha_2 = 2$, and $\alpha_3 = 1$. With the transfer functions of the cancelation logic being given by

$$H_1(z) = z^{-2} \quad \text{and} \quad H_3(z) = (1 + z^{-2})^2, \tag{12.139}$$

the overall output of the bandpass modulators can be expressed as

$$Y(z) = H_S(z)S(z) + H_Q(z)E_2(z), \tag{12.140}$$

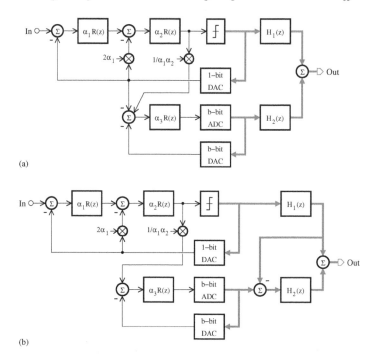

(a)

(b)

FIGURE 12.37
Block diagrams of 4-2 cascaded bandpass modulators.

where S is the input signal, E_2 is the quantization noise of the second stage, the STF and QNTF are given by

$$H_S(z) = -z^{-6} \qquad \text{and} \qquad H_Q(z) = (1 + z^{-2})^3. \qquad (12.141)$$

For the design of a bandpass modulator based on the z^{-1} to $-z^{-2}$ transformation, the zeros $(z = \pm j)$ of the QNTF are located at $f_s/4$ instead of dc (or say, $z = 1$), as is the case for lowpass modulators. In addition, this transformation has the advantage of preserving the stability property of the lowpass prototype.

12.1.6.3 Design examples

The block diagram of a DT, second-order bandpass modulator is illustrated in Figure 12.38(a). It is derived from a first-order lowpass filter prototype using the z^{-1} to $-z^{-2}$ transformation. As a result, the integrator is replaced by a resonator with the transfer function $-z^{-2}/(1 + z^{-2})$, which can be implemented using various SC topologies. The resonator implementation using a loop of two undamped SC integrators is generally plagued by transfer function errors due to the finite dc gain and bandwidth of the amplifier. The bandpass modulator is then preferably realized using a two-path structure, as depicted in Figure 12.38(b), where ϕ_1 and ϕ_2 represent both nonoverlapping clock phases, and each path is clocked in a time-interleaved fashion. The

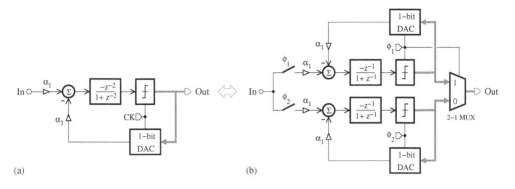

(a)   (b)

FIGURE 12.38
Block diagram of a second-order bandpass DT modulator.

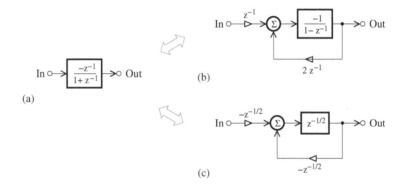

(a)   (b)   (c)

FIGURE 12.39
Block diagrams of alternative realizations of the transfer function, $H(z) = -z^{-1}/(1 + z^{-1})$.

z-domain transfer function of the filter is obtained as

$$H(z) = R(z^2) = -\frac{z^{-1}}{1 + z^{-1}}. \qquad (12.142)$$

It should be noted that the effective sampling frequency of the overall modulator is two times the one of a single path, as implied by the variable z^2. However, without a careful circuit and layout technique, the modulator dynamic range may be limited by path mismatches, which appear as spurious frequency components in the output spectrum.

Depending on the filter topologies, the accuracy of the transfer function pole location can be limited by capacitor mismatches. Figure 12.39 shows the block diagram of various structures that can be used to precisely realize the highpass transfer function, $H(z)$.

Based on the filter block diagram of Figure 12.39(b), the circuit diagram of the modulator was realized as shown in Figure 12.40. During the sampling

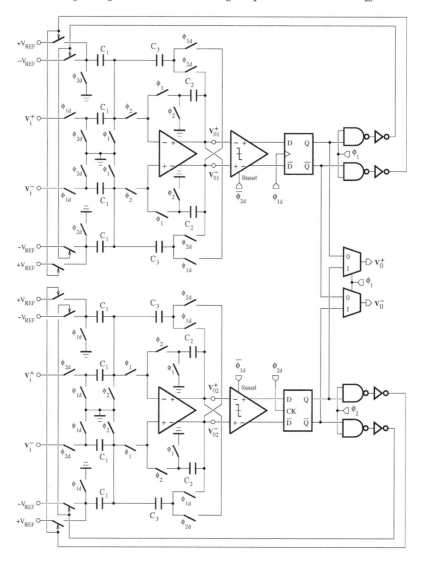

FIGURE 12.40
Circuit diagram of the second-order bandpass DT modulator based on a first-order highpass filter.

phase, the capacitors C_1 are connected to the input and reference voltages, respectively; the charge stored on capacitors C_2 and C_3 are proportional to the output voltage and the inverse of the output voltage, respectively; and the capacitors C_2 are included in the feedback path around the amplifier. During the integrating phase, the capacitors C_3 are switched to the amplifier feedback path and a charge transfer takes place between the capacitors C_1 and C_3; the capacitors C_2 become a load connected to the output voltage. That is,

$$C_3 V_{0j}(n) = C_1[V_i(n-1) + V_{REF}(n-1)] - C_3 V_{0j}(n-1) \quad j = 1, 2 \quad (12.143)$$

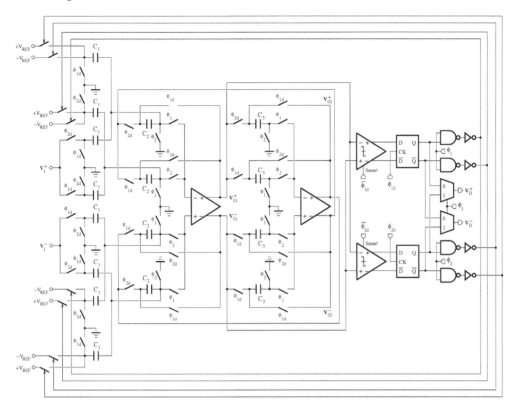

FIGURE 12.41
Circuit diagram of the second-order bandpass DT modulator based on a half-delay cell.

and

$$V_{0j}(z) = H(z)[V_i(z) + V_{REF}(z)], \tag{12.144}$$

where

$$H(z) = \pm \frac{C_1}{C_3} \frac{z^{-1}}{1 + z^{-1}}. \tag{12.145}$$

The transfer function $H(z)$ was derived with the assumption that the clock phasing at the filter input and at the level of the comparator latch is such that a delay of one clock period exists from the input to output (leading to the term z^{-1} in the numerator). Note that the sign of the transfer function can be modified by simply reversing the positive and negative input nodes of the differential implementation. Furthermore, the signal is sampled during the first clock signal and the charge transfer occurs during the second clock phase for one path; whereas for the other path, we have the signal sampling on the second phase and the charge transfer on the first phase. Because the capacitors C_3 are used in the amplifier feedback path and to invert the polarity of the output voltage, the numerator of the filter transfer function is not affected by capacitor mismatches and the pole location can remain on the unit circle. The value of C_2 is not critical for the z-domain design, but it must be chosen so that a low level of the thermal noise and an adequate settling of the amplifier during the sampling phase can be achieved.

The discrete-time filter in the double-sampling bandpass modulator can also be implemented using two half-delay cells, as depicted in Figure 12.39(c). This approach is selected for the SC implementation of the modulator shown in Figure 12.41. The sampling of the signal during one clock phase and the charge transfer during the next clock phase required for the realization of the half delay are performed by the same capacitor, or say C_2 and C_3 for the first and second stages, respectively. As a result, the pole of the filter transfer function remains insensitive to capacitor mismatches. Then, we can write

$$C_1[V_i(n-1/2) + V_{REF}(n-1/2)] = C_2[V_{01}(n) + V_{02}(n-1/2)], \quad (12.146)$$
$$C_3[V_{02}(n-1/2) - V_{01}(n-1)] = 0, \quad (12.147)$$

and

$$V_{02}(z) = H(z)[V_i(z) + V_{REF}(z)], \quad (12.148)$$

where

$$H(z) = \pm \frac{C_1}{C_2} \frac{z^{-1}}{1 + z^{-1}}. \quad (12.149)$$

The term z^{-1} included in the numerator of the $H(z)$ transfer function is due to the fact that there is a delay of one clock period between the sampling instants at the filter input and at the level of the comparator latch. The size of capacitors C_1 is determined by the dc gain of the filter, while C_2 and C_3 can be unit size capacitors.

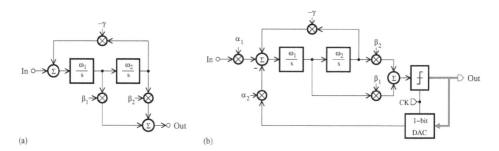

(a)　　　　　　　　　　　　　(b)

FIGURE 12.42
Block diagrams of (a) a CT resonator and (b) a second-order bandpass CT modulator.

The CT bandpass modulator to be designed is based on a second-order resonator. Let X be the input signal of the CT resonator shown in Figure 12.42(a), and X_1 and X_2 the output variables of the first and second integrators, respectively. With the resonator output signal given by

$$Y(s) = \beta_1 X_1(s) + \beta_2 X_2(s), \quad (12.150)$$

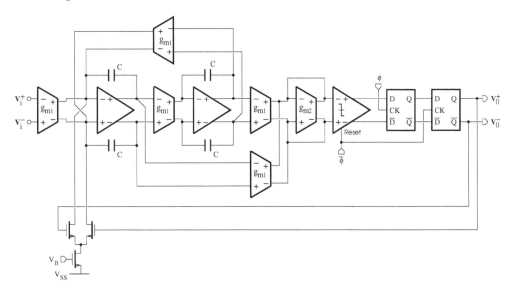

FIGURE 12.43

Circuit diagram of a second-order bandpass CT modulator.

where

$$X_1(s) = \frac{s\omega_1}{s^2 + \gamma\omega_1\omega_2}X(s), \qquad (12.151)$$

and

$$X_2(s) = \frac{\omega_1\omega_2}{s^2 + \gamma\omega_1\omega_2}X(s), \qquad (12.152)$$

the transfer function is obtained as

$$\hat{H}(s) = \frac{Y(s)}{X(s)} = \frac{\beta_1\omega_1 s + \beta_2\omega_1\omega_2}{s^2 + \gamma\omega_1\omega_2}, \qquad (12.153)$$

where β_1, β_2, ω_1, ω_2, and γ are real coefficients.

In the case of a second-order CT modulator designed with the loop coefficients $\beta_1 = -1/2$, $\beta_2 = -1/2$, $\gamma = 1$, and $\omega_1 = \omega_2 = \omega_0$, the transfer function of the resonator is given by

$$\hat{H}(s) = -\frac{(\omega_0/2)s + \omega_0^2/2}{s^2 + \omega_0^2}. \qquad (12.154)$$

For an NRZ DAC pulse, the equivalent z-domain transfer function can be derived as

$$H(z) = (1 - z^{-1})\mathcal{Z}\left\{\mathcal{L}^{-1}\left[\frac{\hat{H}(s)}{s}\right]_{t=nT}\right\}. \qquad (12.155)$$

The partial-fraction expansion of the function $\hat{H}(s)/s$ takes the form

$$\frac{\hat{H}(s)}{s} = -\frac{1/2}{s} - \frac{-(1/2)s + \omega_0/2}{s^2 + \omega_0^2}. \tag{12.156}$$

Because

$$\mathcal{Z}\left\{\mathcal{L}^{-1}\left[\frac{1}{s}\right]_{t=nT}\right\} = \frac{1}{1 - z^{-1}} \tag{12.157}$$

$$\mathcal{Z}\left\{\mathcal{L}^{-1}\left[\frac{s}{s^2 + \omega_0^2}\right]_{t=nT}\right\} = \frac{1 - z^{-1}\cos\omega_0 T}{1 - 2z^{-1}\cos\omega_0 T + z^{-2}} \tag{12.158}$$

$$\mathcal{Z}\left\{\mathcal{L}^{-1}\left[\frac{\omega_0}{s^2 + \omega_0^2}\right]_{t=nT}\right\} = \frac{z^{-1}\sin\omega_0 T}{1 - 2z^{-1}\cos\omega_0 T + z^{-2}} \tag{12.159}$$

and the center frequency of the modulator is located at $f_s/4$, that is,

$$\omega_0 T = \frac{\pi}{2}, \tag{12.160}$$

where T is the period of the clock signal, the transfer function $H(z)$ is reduced to

$$H(z) = -\frac{1 - z^{-1}}{2}\left(\frac{1}{1 - z^{-1}} - \frac{1}{1 + z^{-2}} + \frac{z^{-1}}{1 + z^{-2}}\right). \tag{12.161}$$

Finally, the DT version of the filter transfer function is obtained as

$$H(z) = -\frac{z^{-1}}{1 + z^{-2}}. \tag{12.162}$$

Note that different CT filters can be derived, depending on the numerator implementation of the DT transfer function due to the occurrence of the signal sampling inside the modulator loop. For the function $H(z)$ given by Equation (12.162), it is assumed that the digital section of the modulator exhibits a delay of one clock period, or equivalently z^{-1}. However, it is also possible to consider that the digital stage between the comparator and the DAC has no delay, and the DT transfer function of the filter is

$$H(z) = -\frac{z^{-2}}{1 + z^{-2}}. \tag{12.163}$$

That is, we need to have the set of coefficients, $\beta_1 = 1/2$, $\beta_2 = -1/2$, $\gamma = 1$, and $\omega_1 = \omega_2 = \omega_0$, yielding the CT resonator transfer function

$$\hat{H}(s) = -\frac{-(\omega_0/2)s + \omega_0^2/2}{s^2 + \omega_0^2}. \tag{12.164}$$

The resonator can be implemented with g_m-C operational amplifier integrators, a local feedback, and two feedforward gain stages. In this way, the coefficients of the transfer function can be related to the values of capacitances and transconductances.

The modulator design can start by choosing a second-order z-domain QNTF given by

$$H_Q(z) = 1 + z^{-2}. \tag{12.165}$$

The loop transfer function is derived as

$$H(z) = \frac{1 - H_Q(z)}{H_Q(z)} = -\frac{z^{-2}}{1 + z^{-2}}. \tag{12.166}$$

To remove the delay introduced by the digital section, here z^{-1}, the loop transfer function is multiplied by $1/z^{-1}$. This results in the transfer function of the DT filter, which is then converted to the equivalent CT filter of the form

$$\hat{H}(s) = -\frac{(\omega_0/2)s + \omega_0^2/2}{s^2 + \omega_0^2}, \tag{12.167}$$

where $\omega_0 = \pi/(2T)$ and T is the period of the clock signal. This conversion is achieved using the impulse invariant transform and assuming a DAC pulse of the NRZ type. Figure 12.42(b) shows the block diagram of the bandpass modulator. The same value is assigned to the coefficients α_1, α_2, and γ, which can be used to scale the signal level at the input of the first integrator, while $\beta_1 = -1/2$, $\beta_2 = -1/2$, $\gamma = 1$, and $\omega_1 = \omega_2 = \omega_0$. The modulator implementation with g_m-C operational amplifier circuits is illustrated by the circuit diagram shown in Figure 12.43, where $g_{m2} = 2g_{m1}$ and $\omega_0 = g_{m1}/C$.

Unlike DT modulators using a SC filter, whose coefficients are related to capacitor ratios, CT modulators based on a g_m-C operational amplifier filter have coefficients that are determined by the absolute values of transconductances and capacitors. A matching on the order of 1% is achievable between g_m/C values. However, the absolute g_m/C value can be subject to fluctuations of about 20% due to CMOS IC process variations. A solution may consist of using an on-chip tuning circuit or a calibration stage based on a programmable transconductor or capacitor array.

12.1.7 Decimation filter

Once the modulator has transformed the analog samples into a low-resolution code with the frequency $\text{OSR} \cdot f_s$ much greater than two times the highest spectral component of the input signal, a digital filter is used to attenuate the out-of-band quantization noise, and high-frequency interferences that can be aliased into the passband. The decimation of the modulator output bit stream at the Nyquist rate is achieved by a lowpass filter followed by a down-sampler with the factor D equal to the OSR. The aliasing in the decimation process is avoided by pre-filtering the signal samples to eliminate the components above the frequency $f_s/(2D)$.

The SNR is simply increased when the resolution changes from the low resolution to B bits as a result of the down-sampling and filtering. However, it should be noted that there is no one-to-one correspondence between the input and output samples as it is the case for some ADCs and each input sample value contributes to the whole train of output samples.

The filter architecture should be chosen to have a linear-phase frequency response[1] and minimize the hardware complexity. These requirements can be met by a moving-average filter [12], the transfer function of which is given by

$$G(z) = \sum_{i=0}^{D-1} z^{-i} = \begin{cases} D, & \text{if} \quad z = 1 \\ \dfrac{1 - z^{-D}}{1 - z^{-1}}, & \text{otherwise,} \end{cases} \tag{12.168}$$

where $D = f_s/f_D$ is the decimation ratio and f_D is the decimation frequency. In the frequency domain, that is, for $z = e^{j\omega T}$, we have

$$G(\omega) = \left(\frac{e^{j\omega DT/2} - e^{-j\omega DT/2}}{e^{j\omega T/2} - e^{-j\omega T/2}} \right) \frac{e^{-j\omega DT/2}}{e^{-j\omega T/2}} \tag{12.169}$$

and

$$G(\omega) = \frac{\sin(\omega DT/2)}{\sin(\omega T/2)} e^{-j\omega(D-1)T/2}, \tag{12.170}$$

where $T = 1/f_s$ and f_s is the sampling frequency. It was assumed that $\sin(x) \hat{=} (e^{jx} - e^{-jx})/2j$. The filter zeroes are located uniformly at multiples of the decimation frequency, f_s/D. The magnitude response is obtained as

$$|G(\omega)| = \begin{cases} D, & \text{if} \quad \omega = 0 \\ \left| \dfrac{\sin(\omega DT/2)}{\sin(\omega T/2)} \right|, & \text{otherwise;} \end{cases} \tag{12.171}$$

and the phase response is given by

$$\angle G(\omega) = \begin{cases} -(D-1)\dfrac{\omega T}{2}, & \text{if} \quad G(\omega) \geq 0 \\ -(D-1)\dfrac{\omega T}{2} \pm \pi, & \text{if} \quad G(\omega) < 0. \end{cases} \tag{12.172}$$

---

[1]A filter is assumed to have a linear phase if its transfer function can be written as $G(\omega) = \alpha e^{-j\tau\omega} G_R(\omega)$, where α and τ are complex and real constants, respectively, and $G_R(\omega)$ is a real-valued function of ω.

Note that the phase response changes linearly with the frequency. The stop-band attenuation of a single filter section with the transfer function $G(z)$ is limited. Given a modulator with the order L, the required decimation filter should have a transfer function of the form [13]

$$H(\omega) = [G(\omega)]^N, \tag{12.173}$$

where $N = L+1$, in order to reduce the effect of an aliasing of the out-of-band noise on the baseband signal. Figure 12.44 shows the frequency responses of a decimation filter with N sections. Generally, they will have $D - 1$ spectral zeros, $\lfloor D/2 \rfloor$ of which are located between 0 and $f_s/2$, where $\lfloor x \rfloor$ denotes the largest integer not greater than x. By sampling a signal at a rate of f_s/D, the baseband of interest should be restricted to the frequency range from 0 to $f_b = f_s/(2D)$. The worst-case distortion occurs at the edge frequency of the baseband, which is characterized by $\omega_b = \pi(f_s/D)$ and where the transfer function magnitude of the decimation filter is

$$|H(\omega_b)| = [\sin(\pi/D)]^{-N}. \tag{12.174}$$

Within the signal bandwidth, a decimation filter with N sections then exhibits a maximum attenuation of

$$\text{Droop} = -20N \log_{10}[\sin(\pi/D)] \quad \text{dB}. \tag{12.175}$$

Generally, the signal components in the frequency range from $k(f_s/D) - f_b$ to $k(f_s/D) + f_b$, $k = 1, 2, \ldots, \lfloor D/2 \rfloor$ can alias back into the signal baseband. In practical applications, the useful signal bandwidth is then selected such that $f_b \ll f_s/(2D)$ to improve the filter attenuation in aliasing bands centered around multiples of the decimation frequency.

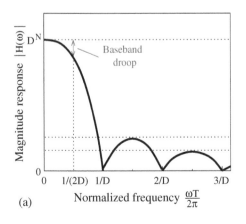

(a)

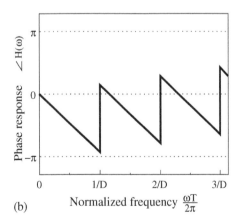

(b)

FIGURE 12.44

(a) Magnitude and (b) phase responses of a decimation filter with N sections.

By moving the numerator of the filter transfer function after the rate

(a)

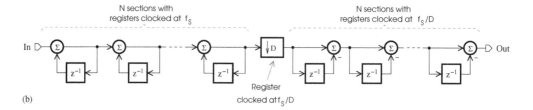

(b)

FIGURE 12.45

(a) Single-stage decimation filter and (b) its CIC filter-based implementation.

change stage where the sampling rate can be reduced to f_s/D (see Figure 12.45(a)), the power consumption of the resulting structure is minimized. Figure 12.45(b) shows an implementation of the decimation filter, which consists of N integrators using shift registers clocked at the sampling rate f_s, a rate change stage that can be realized using a shift register operating at the decimation rate, f_s/D, and N differentiators based on shift registers with a clock signal of f_s/D. Such an implementation, known as a cascaded integrator-comb (CIC) filter[2], has the advantage of requiring only registers and adders.

Data loss due to arithmetic overflow is avoided in the decimation filter by using a two's complement representation and registers with a sufficiently large length. If the input data of the decimation filter are encoded using a two's complement binary format with B_{in} bits, the magnitude of the output samples will be bounded by H_{max} satisfying the following equation,

$$H_{max} = 2^{B_{in}-1} \sum_{n=0}^{D-1} |h(n)|. \tag{12.176}$$

Next, the transfer function of the decimation filter can be expressed as

$$H(z) = \sum_{n=0}^{D-1} h(n)z^{-n} = \left[\sum_{n=0}^{D-1} z^{-n} \right]^N. \tag{12.177}$$

The decimation stage can be considered a finite-impulse response (FIR) filter

---

[2] The decimation CIC filter is also called a *sinc* filter. This is due to the fact that the frequency magnitude response can be expressed in terms of *sinc* functions, where $sinc(x)$ is defined as $\sin(x)/x$.

producing an output word, which represents a weighted average of its most recent input samples.

A simplistic analysis of a first-order modulator shows that the average value of the input signal is contained in the serial output bitstream. The decimation filter following the modulator reduces the sampling rate of the signal sequence and increases the signal resolution by averaging the input bitstream. In the case of a decimation by a factor D, this is achieved by transferring one out of every D signal samples to the output.

Let us assume that the modulator delivers a 1-bit output bitstream of the form

$$HLHLHHLH,$$

where H and L denote the high and low logic states, respectively, to be downsampled by a factor 8. The output of the decimation filter is related to the H-state density, given by $5/8 = 0.625$, that is, the signal sequence with an overall of 8 states has 5 states at the logic high level. Because $0.625 = 1 \times 2^{-1} + 0 \times 2^{-2} + 1 \times 2^{-3}$, the input signal can be interpreted as a 3-bit number with the binary representation, 101. Here, the decimation process results in a reduction of the sampling rate by a factor of 8 and the conversion of the 1-bit serial input into a 3-bit number.

Because the product of positive coefficient polynomials is a polynomial with positive coefficients, all the coefficients $h(n)$ are positive, and

$$\sum_{n=0}^{D-1} |h(n)| = \sum_{n=0}^{D-1} h(n) = H(1) = D^N . \tag{12.178}$$

Assuming that B_{out} is the number of bits of the output data, we have

$$H_{max} \leq 2^{B_{out}} \tag{12.179}$$

and

$$B_{out} = \lceil N \log_2(D) + B_{in} \rceil, \tag{12.180}$$

where $\lceil x \rceil$ is the smallest integer not less than x. By considering the least significant bit (LSB) to be the bit number zero, the most significant bit (MSB) number B_{max} in the output is given by

$$B_{max} = \lceil N \log_2(D) + B_{in} - 1 \rceil. \tag{12.181}$$

Although the overall dc gain of a CIC decimation filter with N sections is finite, individual integrators, whose gain is infinite at dc, can be subjected to numerical overflows. If two's complement arithmetic is used, and if the sum of more than two numbers is guaranteed not to overflow, then overflows in partial sums will be interpreted as either the most positive or most negative representable number, depending on its sign. The difference between any two successive samples computed by the following differentiator cancels out the overflow provided the data in all filter sections are represented with the same MSB position as the one of the output samples, as set by the word length, B_{max}.

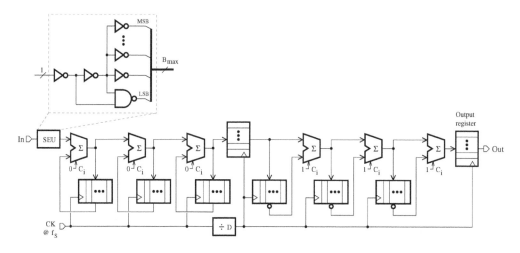

FIGURE 12.46

Circuit diagram of a CIC decimation filter with three sections.

The circuit diagram of a CIC decimation filter with three sections is illustrated in Figure 12.46. This structure can be used as a decimation stage for a single-bit second-order modulator. The sign extension unit (SEU) increases the resolution and converts the binary input data stream into the two's complement representation with B_{max} bits. This is achieved by converting a sample with a high logic level to 1 in two's complement and a sample with a low logic level to -1 in two's complement. The subtraction in the differentiator is realized using the adder carry-in node to add one to the sum of one input sequence and the complement obtained by inverting the bits of the other input sequence. A hard-wired logic realization of the CIC decimation filter only requires registers, adders, and a clock divider. However, other implementations may use a digital signal processor, which relies on multiply and accumulate, and address increment and decrement (or data shifting) operations.

The word length of registers and adders are related to the overall dc gain of the decimation filter and the input data word length. In applications with a decimation factor greater than 64, the number of bits required for the data representation can become excessively high as the number of filter sections increases, leading to a structure that is prohibitively difficult to implement. Because registers and adders are sized to have the same MSB, the word length can be scaled down by discarding the least significant bits (LSBs) to the bit position in any section that cannot grow beyond the LSB of the output word. The number of LSBs to be truncated or rounded from one filter section to another can be determined by assuming that the truncation error at the filter's output uniformly bounds the error incurred at the intermediate sections.

The use of truncation to reduce the word length in the N sections of a CIC decimation filter can result in a total of $2N + 1$ error sources. The errors associated with the truncation at integrator and differentiator inputs are labeled with j running from 1 to N and from $N+1$ to $2N$, respectively. The truncation of the data available at the output of the decimation filter produces an error source specified by $2N + 1$. If B_j represents the number of bits truncated at the j-th section, the truncation or rounding error has a uniform distribution with a width of

$$E_j = \begin{cases} 0, & \text{when using the full precision} \\ 2^{B_j}, & \text{otherwise.} \end{cases} \tag{12.182}$$

The corresponding mean is $E_j/2$ in the case of truncation; otherwise it is zero, and the variance is respectively given by

$$\sigma_j^2 = \frac{E_j^2}{12}. \tag{12.183}$$

The noise error introduced at the j-th section propagates through the filter. It can be verified that the overall mean at the filter output is a function of only the contributions of the first and last error sources. The statistical dispersion of all errors is better tracked by the variance, which is then used for the derivation of the design criteria. Assuming that the noise sources are mutually uncorrelated, the variance at the filter output due to the truncation at the j-th stage is

$$\sigma_{T_j}^2 = \sigma_j^2 F_j^2, \tag{12.184}$$

where

$$F_j^2 = \begin{cases} \displaystyle\sum_{n=0}^{(D-1)N+j-1} h_j^2(n), & j = 1, 2, \ldots, N \\ \displaystyle\sum_{n=0}^{2N+1-j} h_j^2(n), & j = N+1, N+2, \ldots, 2N \\ 1, & j = 2N+1 \end{cases} \tag{12.185}$$

and $h_j(n)$ is the impulse response from the j-th error source to the output. The overall variance at the filter output can be written as

$$\sigma_T^2 = \sum_{j=1}^{2N+1} \sigma_{T_j}^2. \tag{12.186}$$

In practice, it can be assumed that the variance from the first $2N$ error sources is at least as small as the variance of the last error source and the overall error is evenly spread out between these $2N$ sources. Hence,

$$\sigma_{T_j}^2 \leq \frac{1}{2N}\sigma_{T_{2N+1}}^2, \quad j = 1, 2, 3, \ldots, 2N, \quad (12.187)$$

where

$$\sigma_{T_j}^2 = \frac{1}{12}2^{2B_j}F_j^2. \quad (12.188)$$

For each filter section, the number of LSBs that can be thrown aside is then

$$B_j = \left\lfloor \frac{1}{2}\log_2 \frac{6}{N} + \log_2 \sigma_{T_{2N+1}} - \log_2 F_j \right\rfloor \quad (12.189)$$

for $j = 1, 2, 3, \ldots, 2N$. With the length of the output data being B_{out}, we can get the number of LSBs discarded at the output as follows,

$$B_{2N+1} = B_{max} - B_{out} + 1. \quad (12.190)$$

Note that the truncation of B_{2N+1} bits at the filter output (or rounding the filter output) will produce an output noise with variance $2^{2B_{2N+1}}/12$. Figure 12.47 shows the distribution of the register word-length in a CIC decimation filter.

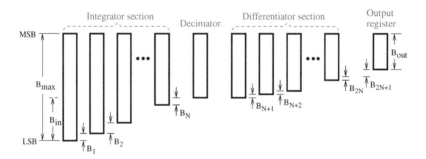

FIGURE 12.47
Register word-lengths in a CIC decimation filter.

Note that alternate decompositions of the transfer function, $H(z)$, can be exploited to arrive at other implementation structures of the CIC decimation filter.

The decimation filter can be realized using a single stage architecture as shown in Figure 12.48(a). With the FIR filter commonly used for the decimation, the required order (or length) is generally proportional to the sampling frequency and inversely proportional to the width of the transition band. Because the decimation filter should provide a narrow transition band, a hardware-efficient design then relies on a multistage architecture [14] as shown

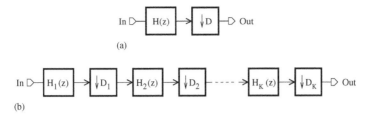

(a)

(b)

FIGURE 12.48

(a) Single-stage and (b) multistage decimation filters.

in Figure 12.48(b), where the earlier sections have a lower order than the latter ones. In this way, the input signal with a high sampling rate is processed by a filter that exhibits a large transition width, and therefore possesses a low order. For the following sections, both sampling frequency and transition width are reduced and an acceptable filter length can still be achieved.

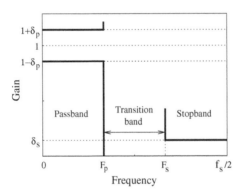

FIGURE 12.49

Lowpass filter specifications.

Let the lowpass filter required by a single stage decimation be characterized by the specifications shown in Figure 12.49, where δ_p and δ_s are the passband and stopband ripples, respectively, and F_p and F_s denote the edge frequencies of the passband and stopband, respectively. Conventional design methods such as the Parks-McClellan algorithm and linear programming can be used to find the FIR filter length and coefficients such that the magnitude frequency response, $|H(\omega)|$, meets the following requirements:

$$||H(\omega)| - 1| \leq \delta_p \qquad \text{for} \quad 0 \leq \omega \leq \Omega_p \tag{12.191}$$

and

$$|H(\omega)| \leq \delta_s \qquad \text{for} \quad \Omega_s \leq \omega \leq \omega_s/2, \tag{12.192}$$

where $\Omega_p = 2\pi F_p$ and $\Omega_s = 2\pi F_s$.

With a multistage decimation filter [16], the overall decimation ratio D is factored into the product,

$$D = \prod_{k=1}^{K} D_k \qquad (12.193)$$

where K is the number of stages, and each independent section within the structure decimates the signal by D_k, which is a positive integer. The k-th stage must have a ripple less than δ_p/K in the passband defined by

$$0 \le f \le F_p, \qquad (12.194)$$

where F_p is supposed to be the highest frequency in the original signal, and a ripple less than δ_s in the stopband specified by

$$F_k - \frac{F_K}{2} \le f \le \frac{F_{k-1}}{2}, \qquad k = 1, 2, \ldots, K, \qquad (12.195)$$

where F_k is the sampling frequency of the k-th stage. Hence,

$$F_k = \frac{F_{k-1}}{D_k} \qquad (12.196)$$

and $F_K = F_0/D$, where F_0 represents the input sampling frequency, Df_s, of the first stage. The passband ripple of the individual filters is selected with the aim of maintaining the overall passband ripple within the bounds set by δ_p. For each stage, the stopband ripple, and the edge frequencies of the passband and stopband are determined such that the effects of aliasing on the baseband spanning from 0 to $F_0/2D$ can be eliminated. Note that the transition band of the last stage is the same as the one of the single-stage architecture, but the sampling frequency is somewhat reduced.

The decimation filter should be designed to provide sufficient attenuation of unwanted high-frequency signals that can be aliased into the baseband, and to feature a hardware-efficient structure. By increasing the stopband attenuation using more CIC filter sections, the passband droop is also increased. To compensate for the limitations of the CIC structure, a typical multistage decimation filter will then consist of a CIC filter, followed by half-band FIR filters and a FIR compensation filter.

An FIR filter can be described by the transfer function

$$H(z) = \sum_{n=0}^{N-1} h(n)z^{-n}, \qquad (12.197)$$

where N is the length of the filter and $h(n)$ denotes the filter coefficients. In the case of a lowpass FIR filter constrained to be a linear phase system, the coefficients must satisfy the condition

$$h(n) = h(N - 1 - n). \qquad (12.198)$$

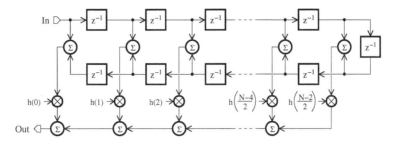

FIGURE 12.50
Direct form structure of a linear-phase FIR filter with N even.

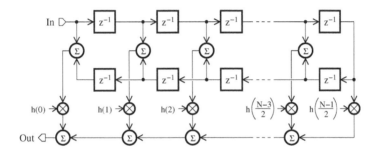

FIGURE 12.51
Direct form structure of a linear-phase FIR filter with N odd.

For N even, the filter transfer function can be decomposed as

$$H(z) = \sum_{n=0}^{\frac{N}{2}-1} h(n)z^{-n} + \sum_{n=N/2}^{N-1} h(n)z^{-n} \tag{12.199}$$

$$= \sum_{n=0}^{\frac{N-2}{2}} h(n)z^{-n} + \sum_{n=0}^{\frac{N-2}{2}} h(N-1-n)z^{-(N-1-n)} \tag{12.200}$$

$$= \sum_{n=0}^{\frac{N-2}{2}} h(n) \left[z^{-n} + z^{-(N-1-n)} \right], \tag{12.201}$$

and for N odd, we have

$$H(z) = \sum_{n=0}^{\frac{N-1}{2}-1} h(n)z^{-n} + h\left(\frac{N-1}{2}\right)z^{-(N-1)/2} + \sum_{n=\frac{N-1}{2}+1}^{N-1} h(n)z^{-n}$$

(12.202)

$$= \sum_{n=0}^{\frac{N-3}{2}} h(n)z^{-n} + h\left(\frac{N-1}{2}\right)z^{-(N-1)/2} + \sum_{n=0}^{\frac{N-3}{2}} h(N-1-n)z^{-(N-1-n)}$$

(12.203)

$$= \sum_{n=0}^{\frac{N-3}{2}} h(n)\left[z^{-n} + z^{-(N-1-n)}\right] + h\left(\frac{N-1}{2}\right)z^{-(N-1)/2}.$$

(12.204)

As a result, the filter exhibits a symmetric $h(n)$ and the number of coefficients is reduced to either $N/2$ for N even or $(N+1)/2$ for N odd. Figures 12.50 and 12.51 show the direct form structures of a linear-phase FIR filter, where N is even and odd, respectively. By reversing the signal flow-graph of the direct form FIR filters while maintaining the same transfer function, the transpose form structures of Figures 12.52 and 12.53 can be derived in the cases, where N is even and odd, respectively. The main advantage of transpose architectures is that adders are naturally pipelined without introducing additional latency. Note that pipelining consists of adding latches or flip-flops between logic sections to reduce the critical path and increase the throughput of a system.

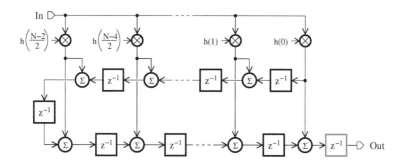

FIGURE 12.52
Transpose form structure of a linear-phase FIR filter with N even.

The block diagram of a two-fold decimation filter based on the direct form structure is depicted in Figure 12.54(a). This implementation features the computational complexity of the FIR filter and is not hardware efficient. An improvement can be obtained using a polyphase structure and then swapping the position of the filter and down-sampler, as shown in Figure 12.54(b). In general, for an FIR filter of length N, the next polyphase decomposition can

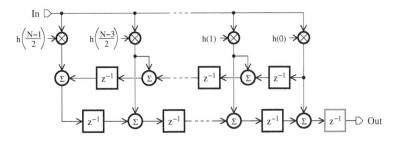

FIGURE 12.53
Transpose form structure of a linear-phase FIR filter with N odd.

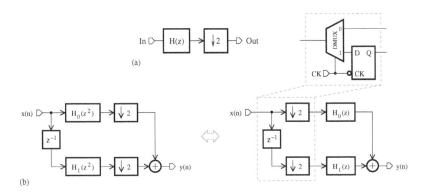

FIGURE 12.54
Block diagrams of two-fold decimation filters based on (a) direct form and (b) polyphase structures.

be obtained [17]:

$$
H(z) = \begin{cases}
\displaystyle\sum_{n=0}^{\frac{N-2}{2}} h(2n)z^{-2n} + z^{-1} \sum_{n=0}^{\frac{N-2}{2}} h(2n+1)z^{-2n}, & \text{if } N \text{ even} \\[4ex]
\displaystyle\sum_{n=0}^{\frac{N-1}{2}} h(2n)z^{-2n} + z^{-1} \sum_{n=0}^{\frac{N-3}{2}} h(2n+1)z^{-2n}, & \text{if } N \text{ odd.}
\end{cases} \tag{12.205}
$$

Let

$$
h_i(n) = h(2n+i) \qquad i = 0, 1 \tag{12.206}
$$

denote the n-th filter coefficient of the i-th polyphase component. The transfer function of the FIR filter can be written as

$$
H(z) = H_0(z^2) + z^{-1}H_1(z^2), \tag{12.207}
$$

where

$$H_0(z) = \begin{cases} \displaystyle\sum_{n=0}^{\frac{N-2}{2}} h_0(n)z^{-n}, & \text{if } N \text{ even} \\ \displaystyle\sum_{n=0}^{\frac{N-1}{2}} h_0(n)z^{-n}, & \text{if } N \text{ odd} \end{cases} \qquad (12.208)$$

and

$$H_1(z) = \begin{cases} \displaystyle\sum_{n=0}^{\frac{N-2}{2}} h_1(n)z^{-n}, & \text{if } N \text{ even} \\ \displaystyle\sum_{n=0}^{\frac{N-3}{2}} h_1(n)z^{-n}, & \text{if } N \text{ odd}. \end{cases} \qquad (12.209)$$

Basically, the filter coefficients $h(n)$ were grouped into even- and odd-numbered samples.

For N even, the number of multipliers and delay units is reduced by making use of the fact that the filter coefficients of polyphase components exist in mirror image pairs. This leads to the block diagrams of a linear-phase FIR filter with a decimation factor of 2, as illustrated in Figures 12.55 and 12.56, where $\lceil x \rceil$ denotes the smallest integer not less than x.

For N odd, the implementation of the decimation filter is made hardware efficient by exploiting the coefficient symmetry. With $\lfloor x \rfloor$ being the largest integer less than or equal to x, Figures. 12.57 and 12.58 show the resulting block diagrams of a linear-phase FIR filter with a decimation factor of 2.

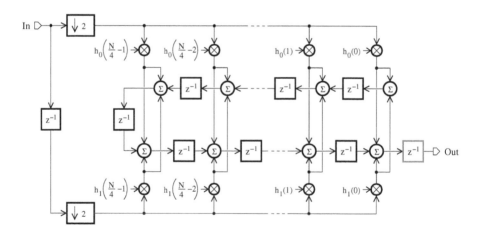

FIGURE 12.55
Block diagram of a linear-phase FIR filter with a decimation factor of 2 (N even and $N/2$ even).

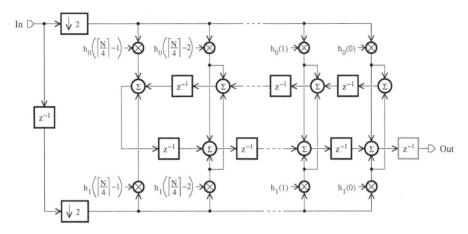

FIGURE 12.56
Block diagram of a linear-phase FIR filter with a decimation factor of 2 (N even and $N/2$ odd).

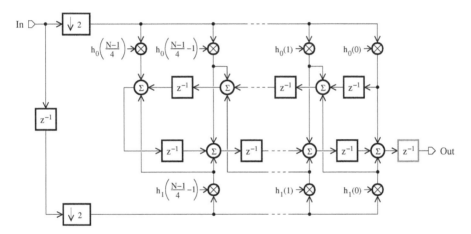

FIGURE 12.57
Block diagram of a linear-phase FIR filter with a decimation factor of 2 (N odd and $(N-1)/2$ even).

The impulse response of a half-band FIR filter [18, 19] is characterized by

$$h(n) = \begin{cases} \alpha, & \text{if} \quad n = (N-1)/2 \\ 0, & \text{if} \quad n = 2k-1, \quad k \neq (N+1)/4 \quad k = 1, 2, \ldots, (N-1)/2 \\ h(2k), & \text{if} \quad n = 2k, \quad k = 0, 1, 2, \ldots, (N-1)/2, \end{cases}$$

(12.210)

where α is usually $1/2$ and the filter length is of the form, $N = 4L - 1$, with L being an integer. All coefficients with n odd, except for $n = (N-1)/2$, are then zero. As a result, the z-domain transfer function can be expressed

$$H(z) + H(-z) = 2\alpha.$$

(12.211)

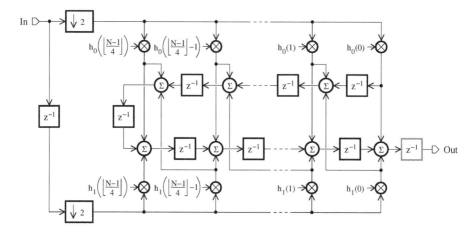

FIGURE 12.58
Block diagram of a linear-phase FIR filter with a decimation factor of 2 (N odd and $(N-1)/2$ odd).

Assuming that $z = e^{j\omega T}$, we have $H(z) \leftrightarrow H(\omega)$ and $H(-z) \leftrightarrow H(\omega - \pi)$, so that

$$H(\omega) + H(\omega - \pi) = 2\alpha. \tag{12.212}$$

Hence, the frequency response is symmetric with respect to one quarter of the sampling frequency, or $f_s/4$. That is,

$$F_p + F_s = f_s/2, \tag{12.213}$$

where F_p and F_s are the edge frequencies of the passband and stopband, respectively, and the passband and stopband ripples should be identical, namely, $\delta_p = \delta_s$.

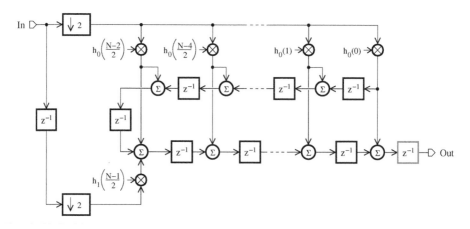

FIGURE 12.59
Block diagram of a half-band FIR filter with a decimation factor of 2.

Because a half-band filter is required to have a magnitude response with the value of $1/2$ at $f_s/4$, it is only suitable for a decimation by a factor of 2. Figure 12.59 shows the block diagram of a half-band FIR filter with a two-fold decimation. The realization of a higher decimation ratio can then be addressed by cascading a series of such filters.

In practice, the decimation filter should have a flat passband response and narrow transition band in order to avoid the signal distortion. These features are not generally offered by the CIC and half-band filters. A solution can then consist of using a compensation FIR filter to cancel the passband droop introduced by the CIC filter and to narrow the overall passband. This FIR filter is designed to have the inverted version of the CIC filter passband response and can achieve a decimation by a factor 2. To prevent the amplification of signal components near the stopband edge frequency of the CIC filter, a good rule of thumb is to limit the upper passband frequency of the compensation FIR filter to about $1/4$ the frequency of the first null in the frequency response of the CIC filter.

An approach for overcoming the threat of runtime overflow in FIR filters is to estimate the maximum value of the signal at the output of the k-th stage, $y_{max,k}$, and to find the worst-case wordlength of each adder and the size of the corresponding output register using to the following equation,

$$y_{max,k} = x_{max} \sum_{n=0}^{k-1} |h(n)| \le 2^{B_k} , \qquad (12.214)$$

where x_{max} is the maximum value of the input signal, $h(n)$ denotes the filter coefficients, and B_k is the required number of bits. To minimize the hardware complexity, rounding to lower word-length or bit truncation can be performed at each filter stage if the additional noise generated will not affect the target accuracy.

By representing filter coefficients in the canonic signed digit (CSD) form [20, 21], multiplications by a constant value can be transformed into a sequence of shift operations, additions and subtractions. An important reduction in the power consumption, area, and latency in FIR circuits can then be achieved, especially when shift operations are simply carried out at the wiring level, and dedicated shifters are not required. The radix-2 CSD expression of a fractional filter coefficient h is given by

$$h = \sum_{k=P}^{L-1} a_k 2^{p_k} , \qquad (12.215)$$

where $a_k \in \{-1, 0, 1\}$, $P \le p_k \le M - 1$ and $M - P$ denotes the number of ternary digits. Note that a CSD code contains no adjacent nonzero digits, and the -1 value is represented by $\bar{1}$.

Starting from the LSB and proceeding toward the MSB, the bits b_k of a two's complement number can be converted into CSD digits a_k for $k = 0, 1, 2, \cdots, N-1$. It is assumed that $c_0 = 0$ and $b_N = b_{N-1}$, and the conversion is realized according to the following equation,

$$c_{k+1} = \lfloor (c_k + b_k + b_{k+1})/2 \rfloor \tag{12.216}$$

$$a_k = b_k + c_k - 2c_{k+1}, \tag{12.217}$$

where c_k is the input auxiliary carry, c_{k+1} represents the output auxiliary carry, and $\lfloor x \rfloor$ denotes the largest integer less than or equal to x.

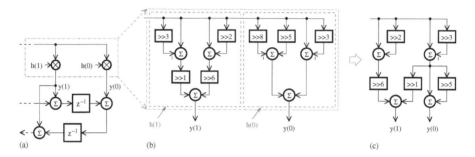

FIGURE 12.60
(a) Section of a transpose filter; filter coefficient implementations (b) without and (c) with sub-expression sharing.

The CSD representation of a number is unique and it has the advantage of containing the fewest number of nonzero digits, which represent additions or subtractions. A substantial hardware saving can then be realized in the implementation of the multiplication with a CSD coefficient.

A section of a linear-phase filter with the coefficients $h(0) = 0.90234$ and $h(1) = 0.45703$ is shown in Figure 12.60(a). The coefficient values in the 10-bit two's complement representation are given by

$$h(0) = 2^{-1} + 2^{-2} + 2^{-3} + 2^{-6} + 2^{-7} + 2^{-8} = 0.111001110 \tag{12.218}$$

$$h(1) = 2^{-2} + 2^{-3} + 2^{-4} + 2^{-6} = 0.011101001, \tag{12.219}$$

whereas in the 10-digit radix-2 CSD representation, we have

$$h(0) = 2^0 - 2^{-3} + 2^{-5} - 2^{-8} = 1.00\bar{1}0100\bar{1}0 \tag{12.220}$$

$$h(1) = 2^{-1} - 2^{-4} + 2^{-6} + 2^{-8} = 0.100\bar{1}01010. \tag{12.221}$$

By exhibiting a reduced number of nonzero ternary digits, the CSD code appears to be suitable for the implementation of the filter coefficients, as shown in Figure 12.60(b). Note that the symbol $>> i$ indicates the right shift by i bit positions due to the term 2^{-i} and the symbol $<< i$ represents the left shift by i bit positions related to the term 2^i. Provided that the transposed direct-form structure is used for the FIR filter realization, common sub-expressions can be shared between the coefficients to reduce the overall number of operations.

The filter coefficient implementation shown in Figure 12.60(b) is based on the identities

$$h(0) = (-2^{-3} + 1)(2^{-5} + 1) \quad \text{and} \quad h(1) = 2^{-1}(-2^{-3} + 1) + 2^{-6}(2^{-2} + 1).$$
(12.222)

For high-order filters, the design of hardware efficient structures is performed using algorithms [22–24] to combine sub-expressions occurring often in coefficients.

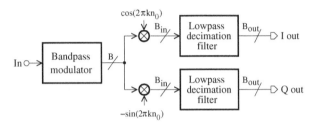

FIGURE 12.61
Architecture of a bandpass modulator with a decimation filter.

The decimation stage for a bandpass modulator can consist of bandpass filters and sample rate down-converters. Because bandpass modulators are generally used to digitize signals at an intermediate-frequency stage of a wireless receiver based on the direct-conversion scheme, where in-phase (I) and quadrature-phase (Q) signals are required to track any changes in magnitude and phase of the incoming message, the decimation filter can be realized, as shown in Figure 12.61. The two mixers, respectively, perform digital sine and cosine multiplications at the same sample rate as the one of the bandpass modulator. The center frequency of the input signal is then shifted to baseband or dc, and I/Q versions of the modulator output are generated. This allows the use of the lowpass decimation filter, which is easier to implement and more hardware efficient than the bandpass decimation filter. However, the word-length of the input signal can still be long enough to significantly increase the filter hardware resource.

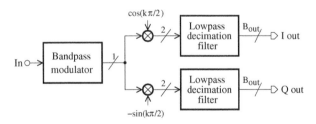

FIGURE 12.62
Architecture of a bandpass $f_s/4$-modulator with a decimation filter.

When the sampling frequency, f_s, is four times the center frequency, f_0, of the modulator, the ratio $n_0 = f_0/f_s$ is reduced to $1/4$ and the coefficients in the multiplications are given by

$$\cos(k\pi/2) = 1, 0, -1, 0, 1, 0, -1, 0, \cdots \quad \text{for} \quad k = 0, 1, 2, 3, 4, 5, 6, 7, \cdots$$
(12.223)

$$-\sin(k\pi/2) = 0, -1, 0, 1, 0, -1, 0, 1, \cdots \quad \text{for} \quad k = 0, 1, 2, 3, 4, 5, 6, 7, \cdots$$
(12.224)

Because the multiplication operations are reduced to selectively not change, nullify, or invert the input signal, they can be implemented using a few logic gates in the case of a single-bit modulator. Figure 12.62 shows a bandpass $f_s/4$-modulator with the decimation filter.

Remark

In special cases, the CIC decimation filter can be implemented using polyphase structures [14].

Consider a CIC decimation filter consisting of N cascaded sections. Its transfer function is

$$H(z) = \left(\frac{1 - z^{-D}}{1 - z^{-1}} \right)^N.$$
(12.225)

By choosing the decimation factor to be a power of 2, we have $D = 2^M$, where M is an integer. The transfer function $H(z)$ can be written as

$$H(z) = \left(\sum_{i=0}^{D-1} z^{-i} \right)^N = \left(\sum_{i=0}^{2^M-1} z^{-i} \right)^N = \prod_{i=0}^{M-1} (1 + z^{-2^i})^N.$$
(12.226)

The commutative rule for multirate systems can be used to transform the block diagram of the CIC decimation filter shown in Figure 12.63(a) into the polyphase structure of Figure 12.63(b), which is realized by cascading M identical sections with a decimation factor of 2. The transfer function of each section can be decomposed as follows:

$$(1 + z^{-1})^N = \sum_{j=0}^{N} h(j) z^{-j} = H_0(z^2) + z^{-1} H_1(z^2),$$
(12.227)

where H_0 and H_1 denote the polyphase components. This implementation is based on FIR filters. The datapath size increases by N bits for each section and the overflow is prevented by setting the minimum word length for the k-th section to $B_{in} + kN$

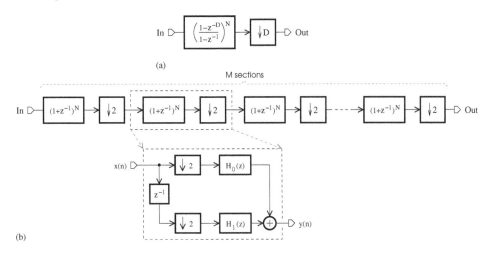

(a)

(b)

FIGURE 12.63
(a) Single-stage CIC decimation filter and (b) its polyphase FIR implementation for $D = 2^M$.

bits, where B_{in} is the number of bits at the filter input and $k = 1, 2, \cdots, M$. Because the sampling frequency is successively decreased by a factor of 2, the power consumption component, which is proportional to the sampling frequency, can be reduced in comparison with the one of decimation structures where this is not the case.

12.2 Delta-sigma digital-to-analog converter

Delta-sigma ($\Delta\Sigma$) digital-to-analog converters (DACs) are typically used in applications where a high linearity is preferred over a high bandwidth.

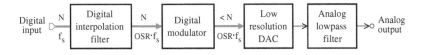

FIGURE 12.64
Block diagram of a delta-sigma DAC.

The block diagram of a delta-sigma DAC is depicted in Figure 12.64. It comprises a digital interpolation filter, a digital delta-sigma modulator, a low-resolution DAC, and an analog lowpass filter. The input signal is assumed to be a stream of digital words with N bits. It is processed by a digital interpolation filter, which raises the data rate to $OSR \cdot f_s$, where OSR is the oversampling ratio and f_s is the sampling rate, by inserting $OSR - 1$ equidistant zero-valued samples between two consecutive samples of the input sequence. The oversampled signal is then supplied to a digital modulator or noise shaper, which reduces the word-length, generally to 1 bit. An analog version of the modulator output is provided by the reconstruction stage, which includes a low-resolution DAC and a lowpass (smoothing) filter.

12.2.1 Interpolation filter

The interpolation by an integer factor of I, which results in an increase in the output sampling rate, is the process of inserting $I - 1$ zeros between successive samples of the input signal, followed by the filtering of the undesired spectral images. For large values of I, the hardware-efficient implementation of the interpolation filter is generally based on a multistage structure [16].

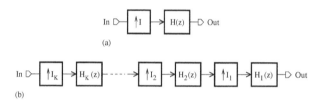

FIGURE 12.65
(a) Single-stage and (b) multistage interpolation filters.

Figure 12.65(a) shows the block diagram of a single-stage interpolation filter. The overall interpolation ratio I, which is equal to the oversampling ratio of the modulator, OSR, can also be realized by a cascade of K stages, each achieving a sampling rate increase of I_k (see Figure 12.65(b)). It can then be factored as

$$I = \prod_{k=1}^{K} I_k , \qquad (12.228)$$

where K is the number of stages. The sampling frequency, F_k, at the output of the k-th stage is given by

$$F_{k-1} = I_k F_k \qquad k = K, K - 1, \cdots, 1, \qquad (12.229)$$

where it is assumed that the output sampling frequency is $F_0 = I f_s$ with f_s being the sampling frequency of the signal to be interpolated. Here, the stages are numbered backward from K to 1 to show that, for a given rate change and number of stages, an interpolation filter is the dual of a decimation filter, and the input signal is applied to the K-th filter stage. The up-sampling process also creates images of the original spectrum centered at multiples of the original sampling frequency. For the k-th stage, lowpass filters characterized by the transfer functions $H_k(z)$ are used to remove the images of the baseband signal at frequencies above $\omega = \pi(f_s/I_k)$. This requirement can be met by a filter designed to have a passband ripple δ_p/K, a stopband ripple δ_s, a passband specified by

$$0 \leq f \leq F_p, \tag{12.230}$$

and a stopband of the form

$$F_k - \frac{F_K}{2} \leq f \leq \frac{F_{k-1}}{2}, \tag{12.231}$$

where F_K is the input sampling rate, f_s. It should be noted that the K-th filter stage has the smaller transition band.

Generally, multistage structures can exhibit reduced filter lengths and computational complexity as compared to single-stage designs. This is due to the fact that the specifications of individual interpolation filters are relaxed and low-order filters can provide a sufficient attenuation of unwanted high-frequency signals that can be aliased into the baseband. A hardware-efficient implementation of the interpolation filter will then consist of a compensation FIR filter, followed by half-band FIR filters and a CIC interpolation filter. The role of the compensation filter, which is designed to have the inverse frequency response of the CIC filter, is to pre-equalize the passband droop of the CIC structure.

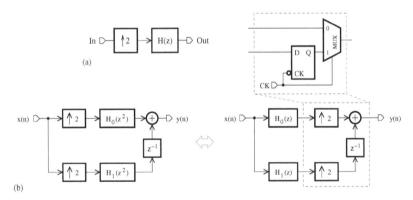

FIGURE 12.66

Block diagram of twofold interpolation filters based on (a) direct form and (b) polyphase structures.

The block diagram of a twofold interpolation filter based on the direct form structure, as shown in Figure 12.66(a), features the computational complexity of the FIR filter and is not hardware efficient. The number of operations required for the signal processing in interpolation filters can be reduced using a polyphase structure and then swapping the position of the filter and down-sampler as shown in Figure 12.66(b).

By grouping the filter coefficients $h(n)$ into even- and odd-numbered sam-

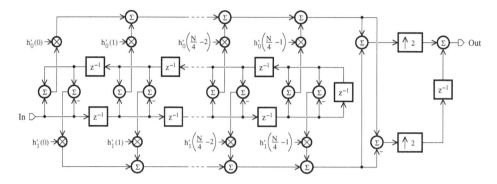

FIGURE 12.67
Block diagram of a linear-phase FIR filter with an interpolation factor of 2 (N even and $N/2$ even).

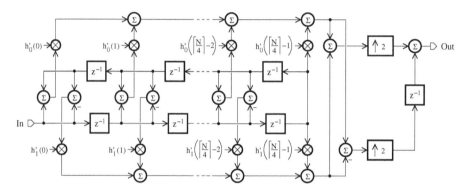

FIGURE 12.68
Block diagram of a linear-phase FIR filter with an interpolation factor of 2 (N even and $N/2$ odd).

ples, the transfer function of an FIR filter can be expressed as

$$H(z) = H_0(z^2) + z^{-1}H_1(z^2), \tag{12.232}$$

where

$$H_0(z) = \begin{cases} \displaystyle\sum_{n=0}^{\frac{N-2}{2}} h_0(n)z^{-n}, & \text{if} \quad N \quad \text{even} \\ \displaystyle\sum_{n=0}^{\frac{N-1}{2}} h_0(n)z^{-n}, & \text{if} \quad N \quad \text{odd} \end{cases} \tag{12.233}$$

and

$$H_1(z) = \begin{cases} \displaystyle\sum_{n=0}^{\frac{N-2}{2}} h_1(n)z^{-n}, & \text{if} \quad N \quad \text{even} \\ \displaystyle\sum_{n=0}^{\frac{N-3}{2}} h_1(n)z^{-n}, & \text{if} \quad N \quad \text{odd.} \end{cases} \tag{12.234}$$

In the case of linear-phase FIR filters, the coefficients are symmetric because $h(n) = h(N - 1 - n)$.

When N is even, we have $h_0(n) = h_1(N/2 - 1 - n)$. Hence, the coefficients of the polyphase components are not symmetric and exist in time-reversed pairs that can be realized using filter structures with symmetric and anti-symmetric impulse responses [26].

For $N/2$ even, a new set of filter coefficients is defined as

$$h_0'(n) = \frac{1}{2}[h_0(n) + h_0(N/2 - 1 - n)] \tag{12.235}$$

$$h_1'(n) = \frac{1}{2}[h_0(n) - h_0(N/2 - 1 - n)], \tag{12.236}$$

where $n = 0, 1, 2, \cdots, N/4 - 1$. The transfer functions of the polyphase components can then be given by

$$H_0(z) = \sum_{n=0}^{N/4-1} h_0'(n)(z^{-n} + z^{-(N/2-1-n)}) + \sum_{n=0}^{N/4-1} h_1'(n)(z^{-n} - z^{-(N/2-1-n)}), \tag{12.237}$$

$$H_1(z) = \sum_{n=0}^{N/4-1} h_0'(n)(z^{-n} + z^{-(N/2-1-n)}) - \sum_{n=0}^{N/4-1} h_0'(n)(z^{-n} - z^{-(N/2-1-n)}). \tag{12.238}$$

The block diagram of the resulting linear-phase FIR filter with an interpolation factor of 2 is depicted in Figure 12.67.

For $N/2$ odd, we have

$$h_0'(n) = \frac{1}{2}[h_0(n) + h_0(N/2 - 1 - n)] \tag{12.239}$$

$$h_1'(n) = \frac{1}{2}[h_0(n) - h_0(N/2 - 1 - n)], \tag{12.240}$$

where $n = 0, 1, 2, \cdots, \lceil N/4 \rceil - 2$, and

$$h_0'(\lceil N/4 \rceil - 1) = \frac{1}{2}h_0(\lceil N/4 \rceil - 1) \tag{12.241}$$

$$h_1'(\lceil N/4 \rceil - 1) = -\frac{1}{2}h_0(\lceil N/4 \rceil - 1), \tag{12.242}$$

where $\lceil x \rceil$ denotes the function that returns the smallest integer not less than x. The following expressions can then be derived:

$$H_0(z) = \sum_{n=0}^{\lceil N/4 \rceil-2} h_0'(n)(z^{-n} + z^{-(N/2-1-n)}) + h_0'(\lceil N/4 \rceil - 1)z^{-(\lceil N/4 \rceil-1)}$$

$$+ \sum_{n=0}^{\lceil N/4 \rceil-2} h_1'(n)(z^{-n} - z^{-(N/2-1-n)}) + h_1'(\lceil N/4 \rceil - 1)z^{-(\lceil N/4 \rceil-1)} \tag{12.243}$$

$$H_1(z) = \sum_{n=0}^{\lceil N/4 \rceil - 2} h_0'(n)(z^{-n} + z^{-(N/2-1-n)}) + h_0'(\lceil N/4 \rceil - 1)z^{-(\lceil N/4 \rceil - 1)}$$

$$- \left[\sum_{n=0}^{\lceil N/4 \rceil - 2} h_0'(n)(z^{-n} - z^{-(N/2-1-n)}) + h_1'(\lceil N/4 \rceil - 1)z^{-(\lceil N/4 \rceil - 1)} \right]$$

$$(12.244)$$

The block diagram of the resulting linear-phase FIR filter with an interpolation factor of 2 is shown in Figure 12.68.

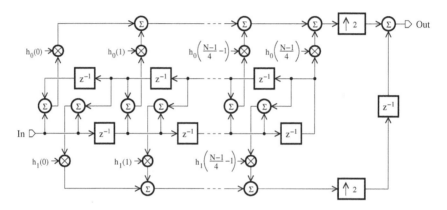

FIGURE 12.69
Block diagram of a linear-phase FIR filter with an interpolation factor of 2 (N odd and $(N-1)/2$ even).

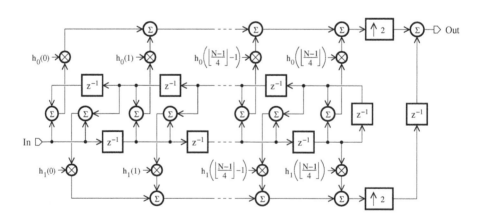

FIGURE 12.70
Block diagram of a linear-phase FIR filter with an interpolation factor of 2 (N odd and $(N-1)/2$ odd).

In the case of FIR filters with N odd and half-band FIR filter, the coefficients of the polyphase components are also symmetric, and the interpolation structures can be obtained by transposing the polyphase structures of the

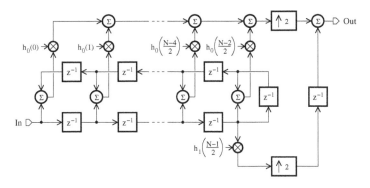

FIGURE 12.71
Block diagram of a half-band FIR filter with an interpolation factor of 2.

corresponding decimation filter. The block diagrams of linear-phase two-fold FIR filters with N odd are shown in Figures 12.69 and 12.70, for $(N-1)/2$ even and $(N-1)/2$ odd, respectively. Note that $\lfloor x \rfloor$ is the largest integer less than or equal to the number x. Figure 12.71 shows the block diagram of a half-band FIR filter with the interpolation factor of 2.

The polyphase structure offers increased efficiency in both size and speed. This is due to the fact that the filtering operation occurs at the lower sampling-rate side of the system, and the coefficient symmetry is exploited to derive an optimal form of the filter using resource sharing.

For large rate changes, a cascaded integrator-comb (CIC) filter [12] has an advantage over an FIR structure with respect to hardware efficiency in the context of a high-speed operation. The higher interpolation factor in a multistage architecture can then be achieved in the CIC filter following the polyphase FIR systems. Two equivalent block diagrams based on a noble identity for multirate structures are shown in Figure 12.72(a). A CIC interpolation filter, as illustrated in Figure 12.72(b), includes a differentiator section, an up-sampler or zero-stuff circuit, and an integrator section. The differentiator consists of a register and a subtractor. The zero-stuff circuit is realized using a 2-to-1 multiplexer controlled by a binary counter. The integrator is composed of a register and adder. The differentiator register is clocked at the input sampling frequency while the counter and integrator register are clocked at the output sampling frequency.

Because the effect of finite word-lengths may be critical in the integrator section, where overflows can cause very large errors and lead to instability due to the unity feedback, rounding is not allowed for integrators. Furthermore, the data widths in the CIC interpolation filter should be set by the worst-case gain at the output of each section to accommodate the maximum value of the signal.

Consider the realization of the interpolation by a factor I_1 using a CIC structure. The transfer function of the interpolation filter can be shown to

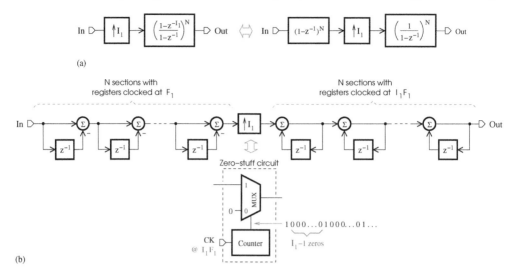

(a)

(b)

FIGURE 12.72

(a) Single-stage CIC interpolation filter and (b) its implementation.

be

$$H_1(z) = \left(\frac{1 - z^{-I_1}}{1 - z^{-1}} \right)^N, \tag{12.245}$$

where N is the number of sections. The CIC interpolation filter [12] is realized by cascading N differentiators, an up-sampler, and N integrators. Each differentiator exhibits the transfer function

$$H_D(z) = 1 - z^{-1}. \tag{12.246}$$

With $z = e^{-j\omega T}$, the frequency response is given by

$$H_D(\omega) = 2j \sin(\omega T/2) e^{-j\omega T/2}. \tag{12.247}$$

The magnitude, which is then of the form

$$|H_D(\omega)| = 2|\sin(\omega T/2)|, \tag{12.248}$$

has a maximum value of 2 because the absolute value of a sine function is bounded by one. Hence, the maximum gain from the input to the output of the k-th section of the interpolation filter can be written as

$$G_j = \begin{cases} 2^j, & j = 1, 2, \cdots, N \\ 2^{2N-j} I_1^{j-N-1}, & j = N+1, \cdots, 2N, \end{cases} \tag{12.249}$$

where, for $j > N$, we have

$$H_1(z) = (1 - z^{-I_1})^{2N-j} \left(\frac{1 - z^{-I_1}}{1 - z^{-1}} \right)^{j-N} \tag{12.250}$$

and the factor $1/I_1$ is introduced to account for the $I_1 - 1$ zeros inserted by the up-sampler between the input samples. For an input data stream with B_{in} bits, the minimum data width at the j-th section is

$$B_j = \lceil B_{in} + \log_2 G_j \rceil. \tag{12.251}$$

However, as $G_N > G_{N+1}$, the data width of the last differentiator is larger than the one of the first integrator. Consequently, either the data width of the last differentiator must be reduced to

$$B_N = B_{in} + N - 1 \tag{12.252}$$

when the two's complement arithmetic is employed, or the data width at each integrator should be increased by one to ensure filter stability. To obtain an output data with B_{out} bits, the number of LSBs discarded should be

$$B_T = B_{2N} - B_{out}. \tag{12.253}$$

Note that some LSBs will be truncated only if the effect of arithmetic errors at the filter output is maintained at an acceptable level.

12.2.2 Digital modulator

Generally, the digital modulator is based on a noise shaping loop, where either the output signal or the quantizer error signal is fed back. This leads to two possible structures, known as the output-feedback and error-feedback modulators.

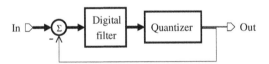

FIGURE 12.73
Block diagram of an output-feedback modulator.

The block diagram of the output-feedback modulator is shown in Figure 12.73. It consists of a digital filter and quantizer, which has a truncation function. In this case, the quantizer should provide an output data of B-bit length consisting of the input signal MSBs. Let H be the transfer function of the filter. Assuming that the quantizer can be modeled as an additive noise source, the modulator output, V_0, can be related to the input voltage, V_i, in the z-domain by

$$V_0(z) = H_S(z)V_i(z) + H_Q(z)E_Q(z), \tag{12.254}$$

where

$$H_S(z) = \frac{H(z)}{1 + H(z)}, \tag{12.255}$$

$$H_Q(z) = \frac{1}{1 + H(z)}, \tag{12.256}$$

and E_Q is the quantization error signal. Ideally, the quantization noise has to be suppressed at the frequency band allocated to the signal. The quantizer and the loop filter must be designed to ensure the stability of the modulator.

Another architecture of the digital modulator, referred to as the error-feedback scheme [25], is depicted in Figure 12.74. Here, the feedback signal involves the quantizer error. Assuming that H is the filter transfer function,

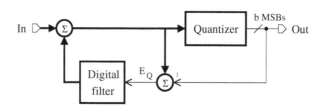

FIGURE 12.74
Block diagram of an error-feedback modulator.

the output of this structure can be expressed in the z-domain as

$$V_0(z) = V_i(z) + H_Q(z)E_Q(z), \tag{12.257}$$

where the QNTF is given by

$$H_Q(z) = 1 - H(z) \tag{12.258}$$

and E_Q is the error signal introduced by the quantizer. Note that the STF is reduced to 1. Due to the high sensitivity to component nonidealities, the error feedback scheme is only suitable for digital implementations.

One way to improve the attenuation of the quantization noise in the baseband is to increase the order of the digital modulator, which is similar to the one of the loop filter. There are various structures for the realization of digital modulators based on QNTFs with a highpass frequency response.

The block diagram of a second-order error-feedback modulator with a B-bit quantizer, which extracts the MSBs of its input signal, is shown in Figure 12.75. The remaining LSBs are accumulated in the feedback path until they overflow into MSBs and thus contribute to the output. The quantization error is shaped by a second-order transfer function of the form

$$H_Q(z) = (1 - z^{-1})^2, \tag{12.259}$$

and the transfer function of the filter is given by

$$H(z) = 2z^{-1} - z^{-2}. \tag{12.260}$$

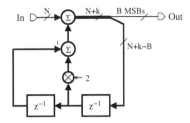

FIGURE 12.75
Block diagram of a second-order error-feedback modulator with a B-bit quantizer.

The modulator implementation is greatly simplified because all multiplier coefficients are powers of 2 and the multiplication can be reduced to shift and add operations.

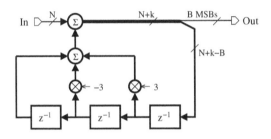

FIGURE 12.76
Block diagram of a third-order error-feedback modulator with a B-bit quantizer.

The block diagram of a third-order error-feedback modulator is depicted in Figure 12.76. By placing the zeros of the QNTF on the unit circle, we have

$$H_Q(z) = (1 - z^{-1})^3 \,, \tag{12.261}$$

and the transfer function of the filter is given by

$$H(z) = 3z^{-1} - 3z^{-2} + z^{-3} \,. \tag{12.262}$$

Hence, the filter used in the modulator has an FIR frequency response. The modulator implementation may not be hardware efficient due to the circuit complexity required for multiplications by non-power-of-2 coefficients.

A third-order error-feedback modulator can also be implemented by the 2-1 cascaded structure shown in Figure 12.77, where it assumed that $B \geq b$. The first stage is a second-order b-bit modulator, and the inverted version of its quantization noise is applied to the second stage, which is based on a first-order B-bit modulator. Let S be the modulator input signal. The linear

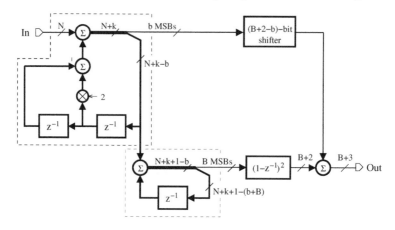

FIGURE 12.77

Block diagram of a third-order, 2-1 cascaded error-feedback modulator.

analysis yields the equations

$$Y_1(z) = S(z) + (1 - z^{-1})^2 E_{Q1}(z), \tag{12.263}$$

$$Y_2(z) = -E_{Q1}(z) + (1 - z^{-1}) E_{Q2}(z), \tag{12.264}$$

and

$$Y(z) = Y_1(z) + (1 - z^{-1})^2 Y_2(z), \tag{12.265}$$

where Y_1 and Y_2 denote the outputs of the first and second stage, respectively; E_{Q1} and E_{Q2} represent the quantization noises of the first and second stage, respectively; and the transfer function of the cancelation section is chosen in the form, $(1 - z^{-1})^2$. In order to correctly combine the output signals, Y_1 and Y_2, the output of the first stage must be shifted left to ensure the alignment of both MSBs. The output of the digital modulator can then be written as

$$Y(z) = S(z) + (1 - z^{-1})^3 E_{Q2}(z). \tag{12.266}$$

The realization of the differentiator at the output of the second stage using digital circuits has the advantage of being more accurate than the implementation in the analog domain. However, the bit growth introduced in the error-cancelation path has the effect of imposing stringent matching requirements on the unit elements used in the implementation of the output DAC.

The single-bit quantization is generally preferred, because the two-level DAC is inherently linear. However, the use of a multi-bit quantizer results in an increase of the modulator resolution due to the reduction of the noise effect both inside and outside the signal band and the elimination of the spectral tones that can be problematic in single-bit structures. Note that a B-bit quantizer can be implemented in modulators based on the two's complement arithmetic by simply truncating the multi-bit input code to its B MSBs. Aside from this advantage, multi-bit converters require calibration schemes for the cancelation of the distortion caused by element mismatches in the DAC.

A high resolution can also be obtained using high-order modulators. This approach is limited by the stability constraint set by the nonoverload requirement of the quantizer. A digital modulator with a QNTF of the form $(1-z^{-1})^L$ will remain stable if the number of bits, B, in the quantizer output is such that $B \geq L+1$ and the input signal does not exceed the midpoint of the last quantization interval [27]. This is a sufficient but not necessary criterion for the modulator stability, which depends on the input signal. To ensure stability in special cases, simulations can be carried out to determine the allowed maximum magnitude of the input signal.

The representation of the modulator output in the analog domain is achieved by a DAC. The latter produces a signal containing the replica of the digital input and the additive quantization noise. An analog lowpass filter is then designed to eliminate the noisy signal, which lies outside the bandwidth of interest.

12.3 Nyquist DAC design issues

In DAC implementations using an array of unit elements, the output signal is the sum of all contributions due to the unit elements, which are selected based on the input code. Any element mismatch may then affect DAC operation and linearity. Dynamic element-matching (DEM) calibrations can be used to convert mismatch errors into a zero-mean white noise, or to remove the noise caused by mismatches out of the frequency band of interest [1–3, 43]. The objective is to randomize the switching of DAC unit elements such that the mismatch errors are averaged out. This is achieved taking into account the fact that the bit weights are equal, and the unit element switching associated with each digital code conversion can be performed in an arbitrary way.

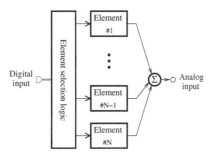

FIGURE 12.78
Block diagram of a DAC with dynamic element-matching calibration.

The general architecture of a DEM DAC is depicted in Figure 12.78.

In mismatch-scrambling DEM DACs, mismatch errors are turned into white noise, while for mismatch-shaping DEM DACs, mismatch errors are spectrally shaped by an appropriate filtering function. A straightforward implementation of the digital circuit section, which is needed in DEM DACs, may substantially increase the hardware complexity. Various approaches, such as data-weighted averaging, tree-structured, butterfly shuffler, and vector feedback methods, are often adopted to trade the hardware complexity for a lower degree of randomization.

12.3.1 Data-weighted averaging technique

Data-weighted averaging (DWA) algorithms equally select the DAC elements taking part in the data conversion, such that matching errors average to zero over a given time period. The re-selection of a given element is possible only after the choice of all the others. As a consequence, the processing of consecutive input codes should require different DAC units.

Assuming that the deviation from the nominal value of the i-th unit elements of the DAC is denoted by ϵ_i, we have

$$\sum_{i=1}^{I} \epsilon_i = 0, \tag{12.267}$$

where I is the number of elements. A given input code, which can be written as

$$X = qI + r, \quad 0 \leq r < I, \tag{12.268}$$

where q and r are two integers, is converted with the mismatch error

$$\Delta X = q \sum_{i=1}^{I} \epsilon_i + \sum_{i=1}^{J} b_i \epsilon_i \tag{12.269}$$

$$= \sum_{i=1}^{J} b_i \epsilon_i . \tag{12.270}$$

Here, $\sum_{i=1}^{J} b_i = r$ and b_i is either 0 or 1. To reduce the effect of the residual error term on the converter performance, it is advisable to choose b_i randomly.

The result of the conversion is obtained by adding the DAC codes at successive time instants, k ($k = 0, 1, \cdots, K$). That is, ΔX is written in the z-domain as the product of the error associated with the initial DAC code and the function $1 + z^{-1} + \cdots + z^{-K}$, which can be approximated by $1 - z^{-1}$ for large K. The error caused by mismatch is then first-order shaped.

The block diagram of the N-bit $\Delta\Sigma$ modulator depicted in Figure 12.79 includes a DWA circuit [31], which can reject the tones caused by mismatches of the DAC unit elements out of the baseband. The DWA implementation of Figure 12.80 consists of an adder, a shift register, binary-to-thermometer encoders, AND and exclusive-OR gates, inverters, and multiplexers. The number of unit elements actually required is $N = K + L$, where K and L are the number of elements assuming an ideal DAC and the number of additional elements due to the DWA, respectively. Note that the value of L determines the noise-shaping performance of the DWA technique. For a 4-bit modulator, N can be chosen to be $15 + 1$. The binary-to-thermometer decoder should have 16 outputs and the last one can be maintained at the low state.

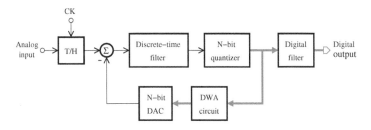

FIGURE 12.79
Block diagram of an N-bit delta-sigma modulator including a DWA circuit.

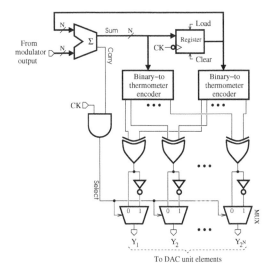

FIGURE 12.80
A digital implementation of the data-weighted averaging technique.

12.3.2 Element selection logic based on a tree structure and butterfly shuffler

Assuming a 3-bit (or eight-element) DAC, Figures 12.81 and 12.82 show the block diagrams of the element selection logic (ESL) or encoder based on a tree structure and butterfly shuffler, respectively. In both schemes, the number of unit elements is given by 2^N, and the switching section comprises $\log_2(N)$ layers, where N is the number of bits.

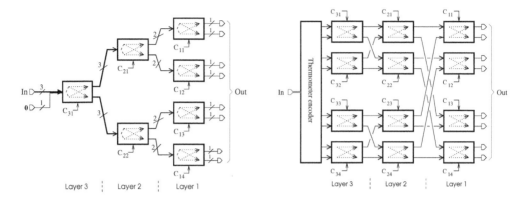

FIGURE 12.81
Tree-structured ESL.

FIGURE 12.82
Butterfly shuffler ESL.

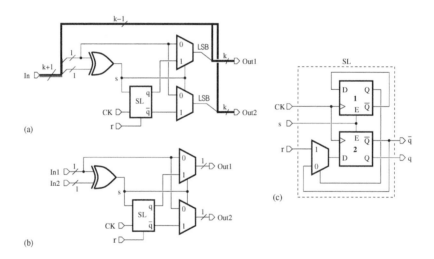

FIGURE 12.83
Circuit diagram of a switching block with (a) $(k+1)$-bit and (b) 2-bit input words; (c) implementation of the selection logic (SL).

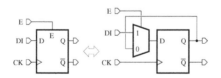

FIGURE 12.84
Circuit diagram of D flip-flops with enable.

- For the approach based on a tree structure, a combination of the N-bit binary code to be converted and a zero, which represents the first bit (LSB) of the input data, is applied to the encoder. This latter consists of $2^N - 1$ switching blocks, the operation of which is equivalent to signal processing functions of the form

$$y_1^{ij} = \frac{x^{ij} + C_{ij}}{2}, \qquad (12.271)$$

$$y_2^{ij} = \frac{x^{ij} - C_{ij}}{2}, \qquad (12.272)$$

where x^{ij} and y_k^{ij} ($k = 1, 2$) are the input and output signals, respectively, and C_{ij} denotes the difference between the top and bottom outputs of the switching block j on the layer i. To simplify the hardware implementation, other definitions, which satisfy the number conservation rule, may be adopted for switching function. It can be assumed, for instance, that

$$C_{ij} = \begin{cases} 0, & \text{if} \quad x^{ij} \quad \text{is even} \\ \pm 1, & \text{if} \quad x^{ij} \quad \text{is odd}. \end{cases} \qquad (12.273)$$

The operation of the encoder is equivalent to an N-to-2^N transformation followed by a scrambling. By processing an input with $(k + 1)$-bits, the switching block provides two k-bit outputs.

- The butterfly shuffler ESL can only operate with an even number of DAC unit elements. It can be composed of a thermometer encoder, which achieves the N-to-2^N conversion, and a selection stage consisting of $N2^{N-1}$ cells called swappers or switching blocks, which can be described as follows:

$$y_1^{ij} = \frac{x_1^{ij} + x_2^{ij}}{2} + C_{ij} \frac{x_1^{ij} - x_2^{ij}}{2} \qquad (12.274)$$

$$y_2^{ij} = \frac{x_1^{ij} + x_2^{ij}}{2} - C_{ij} \frac{x_1^{ij} - x_2^{ij}}{2}, \qquad (12.275)$$

where x_k^{ij} and y_k^{ij} ($k = 1, 2$) are the input and output signals, respectively; i denotes the layer number; and j represents the position of the swapper in the layer. Depending on the level of the signal C_{ij}, the switching block may either pass the inputs directly to the corresponding outputs or assign the inputs reversely to the outputs on each clock cycle. The butterfly shuffler ESL (see Figure 12.82) can perform efficiently even if it allows only a selected set of connections. It should be noted that the association of each of the N inputs to all possible N outputs, would require a digital encoder with N factorial paths, or a large die area.

The resulting noise of the DAC is a linear combination of the data, C_{ij}, with weighting factors, which are linearly related to the unit element errors. As long as each selection logic shapes C_{ij} by a specific transfer function, the static errors introduced by the element mismatch in the DAC will also be modeled in the same way.

For a $(k + 1)$-bit input word, the switching block can be implemented, as shown in Figure 12.83(a). The $(k + 1)$-bit input word is split into the upper $k - 1$ bits and the lower 2 bits, which are used to appropriately assign the least significant bit (LSB) of the output data. When the length of the input word becomes equal to 2 bits, the switching block can be realized using the structure of Figure 12.83(b). For a first-order mismatch shaping, the selection logic (SL) can be designed, as illustrated in Figure 12.83(c) [28–30]. The D flip-flops are enabled by the parity signal, s, provided by the XOR gate connected to the 2 LSBs of the input code, and the clock frequency is fixed at the data rate. The dither signal, r, which is delivered by a pseudo-random noise generator (or linear feedback shift register) whose output assumes the values 0 or 1 with the same likelihood and is used to eliminate spurious tones in the DAC output spectrum, should be uncorrelated with the input code. The selection sequence is determined by the output signals (Q and $\overline{Q}$) of the second D flip-flop. The circuit diagram of D flip-flops with enable is shown in Figure 12.84. A logical 1 at the **enable** node E allows the transfer of the data D to the flip-flop, and the truth table of the overall structure is then similar to the one of a conventional D-flip-flop. The previous state of the flip-flop is maintained in the cases where $E = 0$.

It should be noted that a tree-structured DEM DAC seems to offer a more efficient hardware implementation in the case of high-order mismatch shaping functions than a butterfly shuffler DEM DAC.

12.3.3 Vector feedback DEM DAC

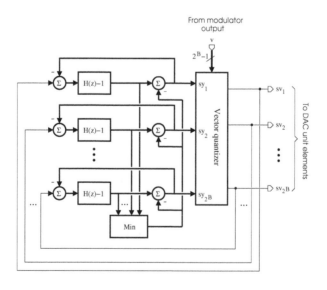

FIGURE 12.85
Vector feedback ESL.

An alternative element selection approach with the advantage of providing a high-order spectral shaping of the DAC noise can be based on the vector feedback structure [3, 44], which uses a sorting mechanism to determine the DAC unit elements to be selected for each conversion.

The block diagram of a vector feedback ESL is depicted in Figure 12.85. It consists of a vector quantizer, a min block (or a smallest-element sorter), two adders, and a filter. The vector feedback ESL processes a signal vector, whose length is equal to the number of DAC unit elements, that is, $2^B - 1$ for a resolution of B bits. The signal vector, sy, is sorted and quantized in such a way that the DAC elements associated to the v largest sy components are enabled by the corresponding sv bits, while the remaining elements are deactivated. For each conversion, the number of bits set to 1 in sv must be equal to the number of 1s in the thermometer code of v. To keep the signal values within the range, which is fixed by the finite precision arithmetic, the smallest of all filter output signals is subtracted from the signal available at each filter output.

The noise contribution due to element mismatch is shaped in the z-domain by $H(z)$, where $H(z) - 1$ is the filter transfer function. For applications with lowpass or bandpass signals, the mismatch noise is efficiently removed from the band of interest by highpass and band-reject filters, respectively.

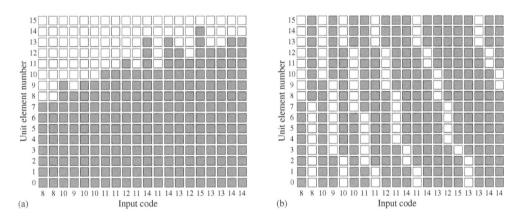

FIGURE 12.86
DAC unit-element usage: (a) Thermometer coding, (b) first-order shaping.

A conventional 4-bit thermometer-coded DAC consists of 16 unit elements. Ideally, all unit elements should be matched. However, in practice, they can be slightly different due to IC process variations. Assuming that the DAC unit elements exhibit random errors with a standard deviation of 1%, a first-order noise-shaping scheme with the transfer function of the form $H(z) = 1 - z^{-1}$ is used to improve the SNR of the DAC. Figure 12.86 shows the unit-element usage for the thermometer coding and first-order noise shaping, respectively. The number of unit elements that are in the on state (boxes filled in gray) corresponds to the decimal equivalent of the input digital code.

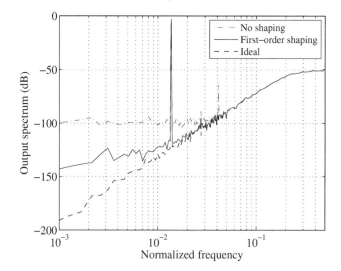

FIGURE 12.87

Output power spectrum of third-order lowpass modulators.

Figure 12.87 illustrates the output power spectrum of third-order lowpass modulators using thermometer-coded and first-order noise-shaped DACs. The correction scheme improves the SNR for low frequencies by first-order shaping the mismatch-induced noise. In order to whiten the noise generated by the selection algorithm itself, a dither signal can be included in the correction scheme.

Let us consider a vector, $\mathbf{x}$, with four different elements, x_0, x_1,

TABLE 12.3

Boolean Expressions for the Min Block Design

$\text{Min}(x_0, x_1, x_2, x_3) = x_0$ $C_{x0} = C_0 \cdot C_1 \cdot C_2 \cdot \overline{C}_3 \cdot \overline{C}_4 + C_0 \cdot C_1 \cdot C_2 \cdot C_3 \cdot \overline{C}_5 + C_0 \cdot C_1 \cdot C_2 \cdot C_4 \cdot C_5$
$\text{Min}(x_0, x_1, x_2, x_3) = x_1$ $C_{x1} = \overline{C}_0 \cdot \overline{C}_1 \cdot \overline{C}_2 \cdot C_3 \cdot C_4 + \overline{C}_0 \cdot C_1 \cdot C_3 \cdot C_4 \cdot \overline{C}_5 + \overline{C}_0 \cdot C_2 \cdot C_3 \cdot C_4 \cdot C_5$
$\text{Min}(x_0, x_1, x_2, x_3) = x_2$ $C_{x2} = \overline{C}_0 \cdot \overline{C}_1 \cdot \overline{C}_2 \cdot \overline{C}_3 \cdot C_5 + C_0 \cdot \overline{C}_1 \cdot \overline{C}_3 \cdot \overline{C}_4 \cdot C_5 + \overline{C}_1 \cdot C_2 \cdot \overline{C}_3 \cdot C_4 \cdot C_5$
$\text{Min}(x_0, x_1, x_2, x_3) = x_3$ $C_{x3} = \overline{C}_0 \cdot \overline{C}_1 \cdot \overline{C}_2 \cdot \overline{C}_4 \cdot \overline{C}_5 + C_0 \cdot \overline{C}_2 \cdot \overline{C}_3 \cdot \overline{C}_4 \cdot \overline{C}_5 + C_1 \cdot \overline{C}_2 \cdot C_3 \cdot \overline{C}_4 \cdot \overline{C}_5$

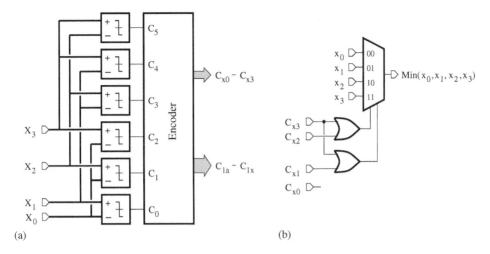

(a) (b)

FIGURE 12.88
(a) Generation of the signals C_{x0}-C_{x3} and C_{1a}-C_{1x}; (b) implementation of the Min block for an input vector with four elements.

x_2, and x_3. To design the Min block and vector quantizer, six digital multi-bit comparators are necessary. They are configured, as shown in Figure 12.88(a), to perform the following operations:

$$C_0 : x_1 > x_0 \quad C_1 : x_2 > x_0 \quad C_2 : x_3 > x_0 \quad C_3 : x_2 > x_1$$
$$C_4 : x_3 > x_1 \quad C_5 : x_3 > x_2 \quad \overline{C}_0 : x_0 > x_1 \quad \overline{C}_1 : x_0 > x_2$$
$$\overline{C}_2 : x_0 > x_3 \quad \overline{C}_3 : x_1 > x_2 \quad \overline{C}_4 : x_1 > x_3 \quad \overline{C}_5 : x_2 > x_3$$

The output signal of a comparator is set to either the logic high state or the logic low state, depending on the comparison result. A dichotomy technique can be used to sort a given set of elements. This is achieved by iteratively defining mutually exclusive subsets such that a tree hierarchy can emerge. Here, the sorting process provides six ordered combinations with a given element being the minimum. The total number of possible combinations is then 6 × 4, or 24.

The circuit diagram of the Min block is depicted in Figure 12.88(b). There is no connection between the C_{x0} input and any of the encoder gates because all inputs are in the logic low state when the 00 code is selected. It is assumed that only one of the Boolean functions, C_{x0}, C_{x1}, C_{x2}, or C_{x3}, can be set to the logic high state at a time. Following the design process of combinational logic circuits, each of the Boolean expressions, C_{x0}, C_{x1}, C_{x2}, and C_{x3}, whose logic state can be related to the fact that the corresponding element is minimum or not, can be derived as shown in Table 12.3.

To design the vector quantizer, the occurrence condition for each

TABLE 12.4

Boolean Functions Useful for the Sorting Procedure Implementation

$x_3 > x_2 > x_1 > x_0$	$x_3 > x_2 > x_0 > x_1$
$C_{1a} = C_0 \cdot C_1 \cdot C_2 \cdot C_3 \cdot C_4 \cdot C_5$	$C_{1b} = \overline{C}_0 \cdot C_1 \cdot C_2 \cdot C_3 \cdot C_4 \cdot C_5$
$x_3 > x_1 > x_2 > x_0$	$x_3 > x_1 > x_0 > x_2$
$C_{1c} = C_0 \cdot C_1 \cdot C_2 \overline{C}_3 \cdot C_4 \cdot C_5$	$C_{1d} = C_0 \cdot \overline{C}_1 \cdot C_2 \overline{C}_3 \cdot C_4 \cdot C_5$
$x_3 > x_0 > x_2 > x_1$	$x_3 > x_0 > x_1 > x_2$
$C_{1e} = \overline{C}_0 \cdot \overline{C}_1 \cdot C_2 \cdot C_3 \cdot C_4 \cdot C_5$	$C_{1f} = \overline{C}_0 \cdot \overline{C}_1 \cdot C_2 \overline{C}_3 \cdot C_4 \cdot C_5$
$x_2 > x_3 > x_1 > x_0$	$x_2 > x_3 > x_0 > x_1$
$C_{1g} = C_0 \cdot C_1 \cdot C_2 \cdot C_3 \cdot C_4 \cdot \overline{C}_5$	$C_{1h} = \overline{C}_0 \cdot C_1 \cdot C_2 \cdot C_3 \cdot C_4 \cdot \overline{C}_5$
$x_2 > x_1 > x_3 > x_0$	$x_2 > x_1 > x_0 > x_3$
$C_{1i} = C_0 \cdot C_1 \cdot C_2 \cdot C_3 \cdot \overline{C}_4 \cdot \overline{C}_5$	$C_{1j} = C_0 \cdot C_1 \cdot \overline{C}_2 \cdot C_3 \cdot \overline{C}_4 \cdot \overline{C}_5$
$x_2 > x_0 > x_3 > x_1$	$x_2 > x_0 > x_1 > x_3$
$C_{1k} = \overline{C}_0 \cdot C_1 \cdot \overline{C}_2 \cdot C_3 \cdot C_4 \cdot \overline{C}_5$	$C_{1l} = \overline{C}_0 \cdot C_1 \cdot \overline{C}_2 \cdot C_3 \cdot \overline{C}_4 \cdot \overline{C}_5$
$x_1 > x_3 > x_2 > x_0$	$x_1 > x_3 > x_0 > x_2$
$C_{1m} = C_0 \cdot C_1 \cdot C_2 \cdot \overline{C}_3 \cdot \overline{C}_4 \cdot C_5$	$C_{1n} = C_0 \cdot \overline{C}_1 \cdot C_2 \cdot \overline{C}_3 \cdot \overline{C}_4 \cdot C_5$
$x_1 > x_2 > x_3 > x_0$	$x_1 > x_2 > x_0 > x_3$
$C_{1o} = C_0 \cdot C_1 \cdot C_2 \cdot \overline{C}_3 \cdot \overline{C}_4 \cdot \overline{C}_5$	$C_{1p} = C_0 \cdot C_1 \cdot \overline{C}_2 \cdot \overline{C}_3 \cdot \overline{C}_4 \cdot \overline{C}_5$
$x_1 > x_0 > x_3 > x_2$	$x_1 > x_0 > x_2 > x_3$
$C_{1q} = C_0 \cdot \overline{C}_1 \cdot \overline{C}_2 \cdot \overline{C}_3 \cdot \overline{C}_4 \cdot C_5$	$C_{1r} = C_0 \cdot \overline{C}_1 \cdot \overline{C}_2 \cdot \overline{C}_3 \cdot \overline{C}_4 \cdot \overline{C}_5$
$x_0 > x_3 > x_2 > x_1$	$x_0 > x_3 > x_1 > x_2$
$C_{1s} = \overline{C}_0 \cdot \overline{C}_1 \cdot \overline{C}_2 \cdot C_3 \cdot C_4 \cdot C_5$	$C_{1t} = \overline{C}_0 \cdot \overline{C}_1 \cdot \overline{C}_2 \cdot \overline{C}_3 \cdot C_4 \cdot C_5$
$x_0 > x_2 > x_3 > x_1$	$x_0 > x_2 > x_1 > x_3$
$C_{1u} = \overline{C}_0 \cdot \overline{C}_1 \cdot \overline{C}_2 \cdot C_3 \cdot C_4 \cdot \overline{C}_5$	$C_{1v} = \overline{C}_0 \cdot \overline{C}_1 \cdot \overline{C}_2 \cdot C_3 \cdot \overline{C}_4 \cdot \overline{C}_5$
$x_0 > x_1 > x_3 > x_2$	$x_0 > x_1 > x_2 > x_3$
$C_{1w} = \overline{C}_0 \cdot \overline{C}_1 \cdot \overline{C}_2 \cdot \overline{C}_3 \cdot \overline{C}_4 \cdot C_5$	$C_{1x} = \overline{C}_0 \cdot \overline{C}_1 \cdot \overline{C}_2 \cdot \overline{C}_3 \cdot \overline{C}_4 \cdot \overline{C}_5$

of the possible combinations of the vector elements to be sorted is translated into a Boolean expression. The block diagram of the vector quantizer is shown in Figure 12.89. It consists of four 2-1 multiplexers and 24 decoders followed by tri-state buffers. The use of tri-state buffers allows all decoders to share the same output line. Table 12.4 summarizes the Boolean functions used as control signals in the vector quantizer.

In practice, the implementation of DACs based on the vector feedback ESL can be affected by the stability problems of high-order modulator loops and

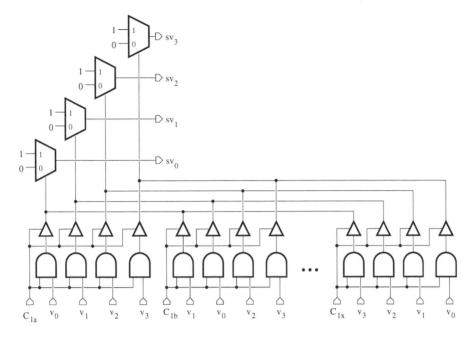

FIGURE 12.89
Implementation of the vector quantizer for a DAC with four unit elements.

is limited to resolutions less than 4 bits due to the high number of logic gates (e.g., $(B-1)!$ digital comparators and $(B-1)! \times B$ decoders for a resolution of B bits) required by the vector quantizer.

12.4 Data converter testing and characterization

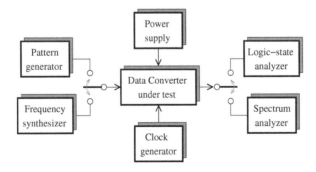

FIGURE 12.90
Block diagram of a test setup.

The performance of data converters can be characterized by static and dynamic parameters. Static linearity, which can be obtained by comparing the ideal and real transfer characteristics of the converter, is generally specified through DNL and INL errors. The analysis of the converter output samples can also provide dynamic measures such as signal-to-noise ratio (SNR), signal-to-noise and distortion (SINAD), effective number of bits (ENOB), spurious-free dynamic range (SFDR), and harmonic distortions. The data converter testing [32–34] is achieved by means of microprocessor-based instrumentation due to the complexity of the required signal processing algorithms. A typical test setup for a data converter is shown in Figure 12.90. The following analysis methods can be used for the converter characterization.

12.4.1 Histogram-based testing

As shown in Figure 12.91, the histogram of an ideal ADC processing a dc signal consists of equal-sized bins for all output codes. When the converter transfer characteristic exhibits a nonlinearity, the bins will not have the same size due to the fact that some output codes occur more frequently than others. Generally, a periodic input sequence is applied to the ADC. The output data

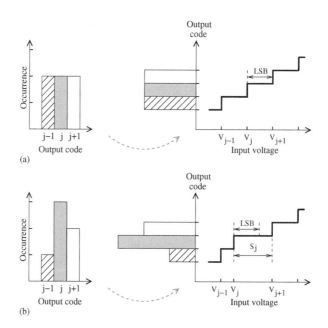

FIGURE 12.91
Correspondence between the histogram and ADC transfer characteristic: (a) Ideal case, (b) nonideal case.

are collected as a series of records, each of which contains a given number of samples. They are represented in the form of a normalized histogram or code density showing the occurrence frequency of each converter code. The converter characteristics can then be determined by comparing the code density data to the ideal distribution density function. In the absence of offset errors, for instance, the histogram should be symmetrical. Gain deviations affect the histogram width. A zero occurrence is the result of a missing code. Note that a large spike generally corresponds to a high DNL.

Let Λ_i be the number of occurrences of code i, Λ_T be the total number of samples, and $P(i)$ be the occurrence probability of code i or the bin width of the ideal converter. The next definitions can be adopted for the DNL and INL expressed in LSBs:

$$DNL(i) = \frac{\Lambda_i}{\Lambda_T P(i)} - 1 \tag{12.276}$$

$$DNL = \max|DNL(i)| \tag{12.277}$$

$$INL(i) = \sum_{j=1}^{i} \mathrm{DNL}(j) \tag{12.278}$$

$$INL = \max|INL(i)|, \tag{12.279}$$

where $i = 0, 1, \cdots, 2^{B+1} - 2$. Note that the definition of the DNL cannot be applied to the last code and Λ_i/Λ_T is the histogram bin width of the converter under test (Λ_i is the number of occurrences of each code i and Λ_T denotes the total number of acquired codes). If the signal applied to the converter is assumed to be a sinusoid of the form

$$x = A\sin(2\pi ft), \tag{12.280}$$

the probability density function is given by[3]

$$p_X(x) = \frac{1}{\pi\sqrt{A^2 - x^2}}, \tag{12.283}$$

where A denotes the amplitude. The parameter P for a given input sample is defined as

$$P(i) = \int_{V_i}^{V_{i+1}} p_X(x)\mathrm{d}x, \tag{12.284}$$

---

[3]Let $X = \Phi(\Theta) = A\sin(\Theta)$. With $\theta = \Phi^{-1}(x)$, the probability density function (pdf) is derived from the following expression,

$$p_X(x) = p_\Theta(\theta)\left|\frac{\mathrm{d}\theta}{\mathrm{d}x}\right|, \tag{12.281}$$

where the phase, Θ, of the sine wave is assumed to be a random variable uniformly distributed between $-\pi/2$ and $+\pi/2$ with the pdf given by

$$p_\Theta(\theta) = \begin{cases} \frac{1}{\pi}, & \text{if } \theta \in (-\frac{\pi}{2}, +\frac{\pi}{2}) \\ 0, & \text{otherwise.} \end{cases} \tag{12.282}$$

where V_i and V_{i+1} are the lower and higher transition levels, respectively. That is,

$$P(i) = \frac{1}{\pi} \left[\arcsin\left(\frac{V_{i+1}}{A}\right) - \arcsin\left(\frac{V_i}{A}\right) \right]. \qquad (12.285)$$

Taking the cosine of both sides of the above relation and using trigonometric relations[4], we can obtain

$$V_{i+1}^2 - [2V_i \cos(\pi P(i))] V_{i+1} + V_i^2 - A^2 \left[1 - \cos^2(\pi P(i))\right] = 0. \qquad (12.288)$$

This quadratic equation can be solved for V_{i+1}. As a result,

$$V_{i+1} = V_i \cos(\pi P(i)) + \sin(\pi P(i))\sqrt{A^2 - V_i^2}, \qquad (12.289)$$

where only the positive square root was retained so that V_{i+1} can be greater than V_i. With the assumption that the first decision level is fixed at $-A$, the other decision levels can be computed as

$$V_{i+1} = -A \cos\left(\frac{\pi \Sigma \Lambda_i}{\Lambda_T}\right), \qquad (12.290)$$

where P is replaced by the measured frequency of occurrence, $\Sigma \Lambda_i / \Lambda_T$, and $\Sigma \Lambda_i$ denotes the total number of codes included in the bins 1 through i. A missing code corresponds to a DNL equal to -1. The record length and ratio of the sampling rate to the signal frequency are chosen so that dynamic errors (in-phase distortion, information redundancy, etc.) are negligible.

12.4.2 Spectral analysis method

Fast Fourier transform data are used to characterize linearity and noise properties of the ADC in the frequency domain. The output provided by an ADC, which processes a sine-wave signal, comprises a tone at the input frequency, harmonics, spurious components, dc offset, and a broadband term characterizing the different kinds of noise. The power estimation of each narrowband component can be affected by the energy leaking from neighboring tones. A solution can then consist of using a suitable window function prior to the Fourier transform. The next parameters can be deduced from the spectrum data.

- The SNR is a measure of the broadband noise introduced by the converting and sampling process into the signal band. It is the ratio of the root-mean

---

[4]Given two numbers x and y,

$$\cos(x - y) = \cos(x) \cos(y) + \sin(x) \sin(y) \qquad (12.286)$$

and

$$\cos(\arcsin(x)) = \sqrt{1 - x^2}. \qquad (12.287)$$

square (rms) value or power of the output signal to the one of the sum of all other frequency components below the Nyquist rate, except those representing dc and harmonics of the fundamental.

- The dynamic range (DR) is the ratio of the rms value of a full-scale sinusoidal input signal to the rms noise delivered by the converter with inputs shorted together. It is limited by the Nyquist frequency.

- The total harmonic distortion (THD) is the ratio of the fundamental to the sum of the harmonics, which can be identified from the noise floor. It can also be expressed as a percentage.

- The SINAD[5] is the ratio of the power in the fundamental frequency bin to that in all other bins, including harmonics. It can also be computed as $(\text{SNR}^2 + \text{THD}^2)^{1/2}$.

- The SFDR is the difference in rms magnitudes of the fundamental and the highest spur, which is not due to dc offset.

In another approach to performing a spectral analysis of the converter, the input signal is assumed to be the sum of two sine waves with the same amplitude, and frequencies equal to f_1 and f_2, respectively. The inter-modulation distortion (IMD) provides the ratio of the rms sum of inter-modulation components at frequencies $if_1 \pm jf_2$ in the spectrum to the rms value of the input signal, where i and j are integers different from zero. The inter-modulation order is given by $i + j$. In a practical implementation, the spectral leakage is eliminated either by assuming a coherent relationship between the sampling frequency, f_s, and input frequencies, that is, $m/f_s = m_1/f_1 = m_2/f_2$, where the integers m_1 and m, and m_2 and m are respectively prime of each other, or by applying a filtering window such as the Blackmann Harris function to the data.

The noise power test, as shown in Figure 12.92, consists of analyzing the output samples delivered by a converter, which processes a limited band of white noise provided by a generator. The Fourier transform is used to evaluate the noise power ratio (NPR), which is the measure of all contributed errors in the frequency domain. However, the fundamental frequency and dc offset are discarded in the computation. The ENOB can be written as

$$\text{ENOB} = N - \frac{1}{2} \log_2 \frac{\text{NPR}}{\sigma_Q^2}, \tag{12.291}$$

where N is the number of bits of the converter and σ_Q^2 denotes the theoretical quantization noise.

---

[5]The SINAD is also known as signal-to-noise and distortion ratio (SNDR).

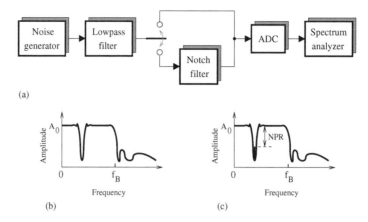

FIGURE 12.92
(a) NPR test setup; (b) signal spectrum at the output of the notch filter; (c) output signal spectrum of the ADC.

Dynamic specifications are generally expressed in decibels (dB). However, they can be referenced to the converter FSR, which is constant, before being transformed into decibels. This results in parameters, whose unit is dBFS or say decibels relative to full-scale.

12.4.3 Walsh transform-based transfer function estimation

The transfer function of an N-bit converter can be represented as the sum of a given number of Walsh functions adequately weighted by the Walsh coefficients. These latter can be obtained by reconstructing the ADC output data using the Walsh transform. To achieve a good resolution, the number of points considered for the computation should be a power of 2 multiple of 2^N. The comparison of the ideal and real transfer functions can then provide the ADC error parameters.

12.4.4 Testing using sine-fit algorithms

The ENOB characterization of an ADC, which processes a sampled sine wave, is carried out by reconstructing the input signal based on the four parameters (amplitude, frequency, phase, and dc offset) computed from the output data. The signal samples of the original input are then subtracted from the ones of the synthesized sine wave to estimate the average noise power. The ENOB at a given input frequency can then be computed as

$$\text{ENOB} = \log_2 \left(\frac{FSR}{\sqrt{12} \cdot RMSE} \right), \tag{12.292}$$

where FSR is the full-scale range of the ADC and $RMSE$ is the root-mean square of the digitized signal or the noise power provided by the test procedure. The achievable accuracy is limited by the convergence performance of the estimation algorithm, and the validity of the stochastic model is guaranteed only for a restricted range of the ratio between the number of ADC quantization levels and the one of the acquired samples.

Note that a pattern generator instead of a frequency synthesizer is required for the DAC testing. The ADC, which can be used to deliver a digital version of the analog output necessary for the different computations, can limit the speed and precision of the evaluation. To test DAC in the frequency-domain, the solution can consist of using analog spectrum analysis techniques. The level of harmonic distortions can then be related to the transfer characteristic deviations.

Generally, the power consumption of data converters increases with performance characteristics such as the dynamic range and bandwidth. The figure of merit (FOM) measures the efficiency with respect to the dissipated power. It is defined as

$$\text{FOM} = \frac{\text{DR} \times \text{BW}}{P}, \tag{12.293}$$

where DR and P are the dynamic range and the total power dissipation of the converter, respectively, and BW is the signal bandwidth.

12.5 Delta-sigma modulator-based oscillator

Generally, the on-chip generation of high-quality signals is required in built-in self-test structures for mixed-signal circuits. An approach to resolve this problem is to use $\Delta\Sigma$ modulator-based oscillators, which can deliver signals with a spurious-free dynamic range on the order of 90 dB.

A $\Delta\Sigma$ modulator-based oscillator consists of a loop including a digital resonator with poles on the unit circle and an $N \times N$-bit multiplier, which is implemented by the combination of a $\Delta\Sigma$ modulator with a multiplexer to reduce the silicon area and timing delay. The 1-bit pattern used to control the multiplexer switching is available at the output of the $\Delta\Sigma$ modulator, which should have a unity signal transfer function. It contains the sinusoidal signal generated by the resonator and the out-of-band quantization noise, which can be suppressed by a filter. An analog signal can be obtained by cascading a 1-bit DAC with the oscillator.

The block diagram of a lowpass $\Delta\Sigma$ oscillator [35] is shown in Figure 12.93. It includes two integrators, a lowpass $\Delta\Sigma$ modulator, and a 2-1 multiplexer. The delay of one clock period introduced on the signal path by the second-order lowpass $\Delta\Sigma$ modulator shown in Figure 12.94 is compensated for by using one nondelayed integrator. Let x_1 and x_2 be the state variables associated with the output of the first and second integrators, respectively. We can

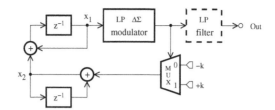

FIGURE 12.93
Lowpass $\Delta\Sigma$ modulator-based oscillator.

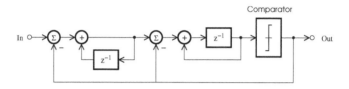

FIGURE 12.94
Block diagram of a second-order lowpass $\Delta\Sigma$ modulator.

write

$$x_1(n) = x_1(n-1) + x_2(n-1) \tag{12.294}$$
$$x_2(n) = -kx_1(n) + x_2(n-1). \tag{12.295}$$

Using the z-transform, X_1 and X_2 can be eliminated and the next characteristic equation is derived,

$$z^{-2} - (2-k)z^{-1} + 1 = 0. \tag{12.296}$$

To ensure the oscillation, the roots of the above equation should be conjugate complex and located on the unit circle. This is the case for $0 < k < 4$, and

$$z_{1,2}^{-1} = \frac{2 - k \pm j\sqrt{k(4-k)}}{2}. \tag{12.297}$$

The angular frequency of oscillation, ω_0, can then be related to the coefficient, k, and the period of the clock signal, T, according to

$$\tan(\omega_0 T) = \pm\frac{\sqrt{k(4-k)}}{2-k}. \tag{12.298}$$

A bandpass $\Delta\Sigma$ oscillator [36], as shown in Figure 12.95, offers the advantage of possessing a greater usable bandwidth while operating at a sample rate comparable to that of a lowpass structure. It uses two registers (blocks denoted by z^{-1}) included in a loop with a multiplier (coefficient l), a bandpass $\Delta\Sigma$ modulator, and a 2-1 multiplexer. Figure 12.96 shows the block diagram of a fourth-order bandpass $\Delta\Sigma$ modulator with a signal transfer function equal

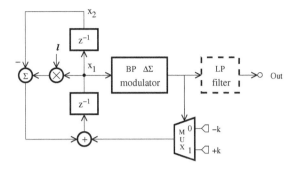

FIGURE 12.95
Bandpass $\Delta\Sigma$ modulator-based oscillator.

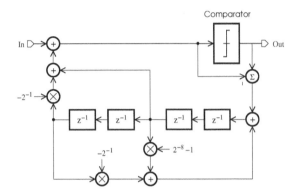

FIGURE 12.96
Block diagram of a fourth-order bandpass $\Delta\Sigma$ modulator.

to 1, such that the signal level is not modified. By inspection of the oscillator, we can derive

$$x_1(n) = -x_2(n-1) + lx_1(n-1) - kx_1(n-1), \tag{12.299}$$
$$x_2(n) = x_1(n-1), \tag{12.300}$$

where x_1 and x_2 denote the state variables of the register outputs. These last equations can be transformed to the z-domain as

$$z^{-2} + (k-l)z^{-1} + 1 = 0. \tag{12.301}$$

Solving for z^{-1}, the roots of the characteristic equation (12.301) are given by

$$z_{1,2}^{-1} = \frac{l-k \pm j\sqrt{4-(k-l)^2}}{2}, \tag{12.302}$$

where $|k-l| < 2$. Hence, the oscillation frequency can be obtained from the next expression,

$$\tan(\omega_0 T) = \pm\frac{\sqrt{4-(k-l)^2}}{l-k}. \tag{12.303}$$

The multiplication coefficient l can be chosen to be a power of 2 to reduce the hardware complexity. Further reduction is achieved for $l = 0$.

For both oscillator structures, a discrete-time sinusoidal signal of the form

$$x(n) = A\sin(\omega_0 Tn + \phi) \qquad (12.304)$$

can be obtained at the node labeled x_1. The amplitude A and the phase ϕ are dependent on the coefficient k (and l), and the initial conditions, I_1 and I_2, of the first and second registers.

Principles of the time division multiplexing [37] can be exploited for the generation of two-tone signals. This is realized by replacing the 2-1 multiplexer with a 4-1 multiplexer and each register with a pair of registers. As a result, the effective clock frequency is divided by a factor of two.

12.6 Digital signal processor interfacing with data converters

Due to the difference of processing speed and electrical characteristics existing between input/output (I/O) devices, such as data converters, and the computer processing unit (CPU) of a microprocessor (digital signal processor(DSP), micro-controller), interface chips are required to synchronize data transfer between the CPU and I/O devices. Generally, an interface chip is composed of control registers, data registers, status registers, data direction registers, and control circuit [38]. Control registers include data bits, whose states determine the parameters of the I/O operation. The data transfer direction for each I/O pin is set by the corresponding bit of the data direction registers. The data register is used as a buffer to temporarily store the data being transferred to or from the CPU. The status registers store bits providing information on the progression of the I/O operation. Because access to the data bus is allowed to only one I/O device at a time, an address decoder is used to generate chip-select or chip-enable signals for each device at the request of the microprocessor.

The data transfer between the microprocessor and I/O devices can be either parallel or serial. Parallel communications are based on the use of several wires to simultaneously transmit data. Serial communications involve transmitting digital data, sequentially, over only one wire. To achieve a high transfer speed, parallel data transmissions are preferred, while serial links are the better option when the interconnection hardware overhead should be kept minimal.

Due to the typical speed difference between the microprocessor and I/O devices, a synchronization mechanism is required for proper data transmission. Various types of synchronization can be used to interface I/O devices.

A simple synchronization technique is to design the software such that it can initiate the communication and then wait a fixed amount of time for the I/O operation to complete. This method is known as blind cycle counting because it provides no information about the outcome of the I/O operation back to the microprocessor.

In the gadfly or busy waiting approach, the software routine includes loops that can check the I/O status until the completion of data transfer. This approach is suitable only for I/O operations with a small wait time.

The periodic polling is based on the principle of continually checking the status of the I/O operation to detect whether it is complete. By continuously monitoring the status register, the microprocessor can notice the end of the data transfer. It can then retrieve data and proceed further according to the programmed instructions.

The interrupt technique requires more complex hardware and software, but has the advantage of efficiently using the microprocessor CPU. An interrupt request is generated either when the I/O device is ready or to acknowledge a successful data transfer. As a result, the CPU forces a branch-out of the current program sequence to the appropriate interrupt service routine (ISR). Prior to the transfer of the control to the ISR, the CPU state must be saved on the stack. This is necessary because the program execution should resume after returning from the ISR. However, the achievable response time may be limited due to the microprocessor latency time (the time elapsed between the generation of an interrupt request and the servicing of the corresponding I/O device).

Another I/O synchronization technique is based on direct memory access (DMA). DMA controllers can transfer data from I/O devices directly to the main memory, and vice versa, without the intervention of the CPU. They can generate an address sequence to access blocks of data and manage access priorities. Here, the load of the CPU is reduced and a higher data throughput can be achieved by manipulating data blocks.

Interfacing a DSP with data converters involves both physical connections and software routines that steer the transmission of data. Data converters used in the interface implementation should exhibit more flexibility. This is achieved using a set of on-chip registers to achieve programmability and control the data flow. The write (WR) and read (RD) operation of each register is determined by a given signal. The communication between a digital signal processor (DSP) and data converters can be done either in parallel or serially [39]. A DSP with only one type (parallel or serial) of port can still communicate with any I/O device, provided data can be converted from parallel to serial form, and vice versa. Figures 12.97(a) and (b) show the circuit diagrams, which realize the parallel-to-serial and serial-to-parallel transformation, respectively. The first structure uses time-division multiplexing for the placement of N-bit input data in a single channel, while the second structure relies on the delay, which can be introduced on a data stream by shift registers. Note that various architectures

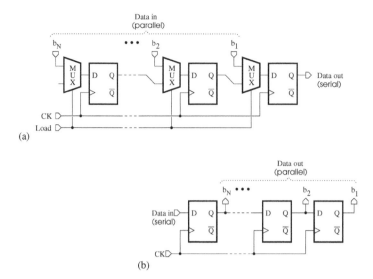

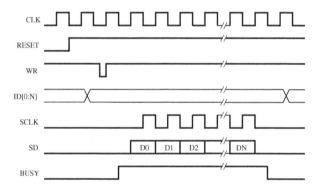

FIGURE 12.97

(a) Parallel-to-serial converter; (b) serial-to-parallel converter.

FIGURE 12.98

Timing waveforms for the parallel-to-serial converter.

are available for the same interface type, which is efficiently implemented as a combination of hardware and software.

It should be noted that the above converters include additional input and output nodes in a data acquisition environment. The timing waveforms are shown in Figure 12.98 for the specific case of the parallel-to-serial converter. After the initialization step steered by the RESET signal, the input data $ID[0:N]$ are applied to the circuit and the write (WR) pulse is enabled. The signal BUSY changes to the high level and data are transferred 1 bit after the other to the serial output (SD) under control of the clock signal (CLK). The initiation and end of the transmission are detected from the information in SCLK, which is an inverted version of CLK.

12.6.1 Parallel interfacing

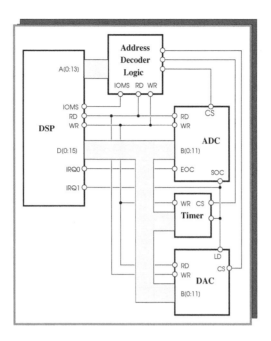

FIGURE 12.99
Parallel interfacing of a DSP.

The block diagram of a parallel interface implementation for a fixed-point DSP is depicted in Figure 12.99. Data are transferred between the DSP memory and ADC outputs (WR operation) or DAC inputs (RD operation) in one clock period. Due to the short memory access time of high-speed processors, the data transfer flow must be regulated by programming the DSP to insert wait-states in the converter access cycle. Alternatively, the DSP can include a different external input/output memory space (IOMS) for converters or other nonmemory peripherals. Each data bit requires a pin, as well as the control signals [WR, RD, and chip select (CS)]. The timer must generate an interrupt request (IRQ), which determines the start of the conversion (SOC) of the ADC or the load of data (LD) into the DAC. The end-of-conversion (EOC) goes high to indicate that the conversion is complete and ADC output data are ready to be read.

The converter resolution can be lower than the one of the DSP data bus. The appropriate connection is then determined by the number representation system. For instance, the right justification of buses is needed for binary coding, while the left justification provides an adequate transfer in the case of two's complement representations. This latter situation can be applied to the interface structure of Figure 12.99, where a 12-bit ADC and 14-bit DAC are used. The MSB (B11) of the ADC should be joined to D15 down to the LSB (B0) wired to D4. The DAC inputs must be connected to the data bus starting from B13 to D15 through B0 to D2.

Parallel interfacing has the advantage of higher transfer speed, but it results in a chip package with a high number of pins.

12.6.2   Serial interfacing

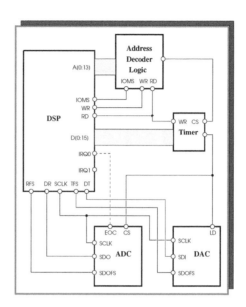

FIGURE 12.100
Serial interfacing of a DSP.

By interfacing serially a DSP, the number of pins can be reduced. This approach is constrained by the requirement that the transfer rate must be greater than the required data bandwidth. Various serial protocols based on different bit encoding and basic packet structure are available for the communication between the DSP and data converters. Serial ports (SPORTs) can be used to transmit or receive data words of length 4 to 16 bits. A DSP is able to communicate in both directions simultaneously, that is, in full duplex mode. In contrast to microcontrollers, DSPs use a frame sync (FS) signal to indicate the beginning of the data stream and can operate with a continuous serial clock (SCLK) signal together with FS pulses. For a microcontroller, the data transfer takes place with respect to the SCLK signal, which must be active. The data synchronization can be conducted either by the DSP or data converters, but it is often convenient to have the sample timing being determined by the ADC and DAC.

The block diagram of a serial interface is shown in Figure 12.100. The DSP features pins corresponding to the data receive (DR), data transmit (DT), receive frame sync (RFS), and transmit frame sync (RFS) operations. Before the start of the transfer, synchronization pulses must be generated on the corresponding pin. When the SPORT is enabled, the digital data from the ADC are sent out on SDO, and the ones from SDI are transmitted to the

DAC. The EOC flag is raised at the end of the analog-to-digital conversion and can be reset to account for the DSP interrupt signal. The timer is used to generate the chip select (CS) and DAC load (LD) inputs.

12.7 Built-in self-test structures for data converters

Due to the increase in circuit complexity, testing is becoming an integral part of the integrated circuit design. Built-in self-test (BIST) structures provide the advantage of reducing the test cost and improving the testing accuracy in high-density circuits.

BIST structures based on code density test principles can be used to determine low-frequency spectral characteristics of data converters [40, 41]. The generation of the test signal can rely on the use of pattern memory and a DAC. To reduce the required chip area, the digital version of the signal, which is available at the memory output, is transformed into an analog waveform by a 1-bit DAC, whose linearity is generally excellent. Thus, the quality of the signal is primarily determined by the number of samples, which is bounded by the memory size. By using a linear ramp as a test stimulus, the code width associated with the converter output signal can be computed. The number of occurrences in each bin should be equal in the ideal case, and any deviation can be related to the imperfection of a practical converter. The DNL for a given input sample is derived by subtracting the ideal code width from the measured code width. The sum of the DNL from the first up to the current code is equal to the INL. The accuracy of the DNL and INL determination is limited by the noise and quantization errors to about 0.05 LSB. For data converters embedded in a mixed-signal circuit, including a digital signal processor (DSP), the self-test program and test data can be stored in the read-only memory (ROM) and random access memory (RAM). However, BIST structures using logic gates can feature a low area and a high speed.

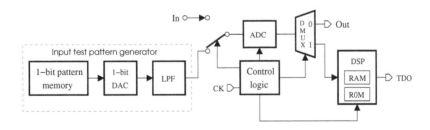

FIGURE 12.101
BIST structure for ADCs.

The block diagram of the BIST structure for the ADC is shown in Fig-

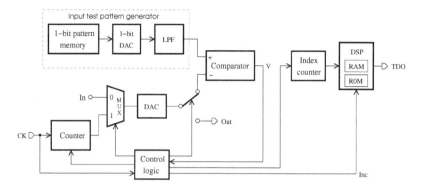

FIGURE 12.102
BIST structure for DACs.

ure 12.101, where CK is the clock signal and TDO is the test data output. It consists of an input test generator including a pattern memory, a 1-bit DAC and a lowpass filter (LPF), a control logic, a DSP, and the ADC to be tested. Let the test pattern be a linear ramp, the magnitude of which is greater than the full-scale range of the ADC. The output code of the converter can be written as b_{2^N-1}, b_{2^N-2}, $\cdots$, b_0, where N denotes the number of bits, and b_{2^N-1} and b_0 are the MSB and LSB, respectively. The DNL and INL can be estimated from the array of $2^N - 2$ elements obtained by excluding the MSB and LSB, which correspond to non-doubly-bounded input ranges, and the occurrence number, Λ_i, of each code b_i. That is,

$$\mathrm{DNL}_i = \frac{\Lambda_i}{\Lambda} - 1 \tag{12.305}$$

and

$$\mathrm{INL}_i = \begin{cases} 0, & \text{if } i = 0 \\ \mathrm{INL}_{i-1} + \dfrac{\mathrm{DNL}_i + \mathrm{DNL}_{i-1}}{2}, & \text{otherwise,} \end{cases} \tag{12.306}$$

where

$$\Lambda = \frac{\displaystyle\sum_{i=1}^{2^N-2} \Lambda_i}{2^N - 2}. \tag{12.307}$$

The above static parameters can be expressed as a fraction of the LSB.

The BIST architecture for the DAC, as shown in Figure 12.102, includes a test pattern generator, a counter for digital code generation, an analog comparator, an index counter, a DSP, a control logic, and the circuit under test, which is a DAC. An analog version of the encoded sawtooth signal stored in the memory of the pattern generator and the DAC output signal are applied respectively to the positive and negative input nodes of the analog comparator. During the test, the output signal, V, delivered by the analog comparator gives

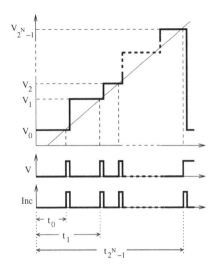

FIGURE 12.103
Operation principle of a DAC BIST structure.

an estimation of the magnitude levels associated with the different input codes of the DAC. It will assume the high or low state if the signal level at V^+ is greater or lower than the one at V^-. By detecting the rising edge of V, the control logic can increment the index counter, the content of which represents the different indexes, t_i, to be stored in the DSP memory. The DNL and INL (in fraction of the LSB) of the DAC can be derived as

$$\mathrm{DNL}_i = \frac{t_i - t_{i-1}}{\Lambda} - 1 \qquad (12.308)$$

and

$$\mathrm{INL}_i = \begin{cases} 0, & \text{if } i = 0 \\ \mathrm{INL}_{i-1} + \mathrm{DNL}_i, & \text{otherwise}, \end{cases} \qquad (12.309)$$

where

$$\Lambda = \frac{t_{2^N-1} - t_0}{2^N - 1} \qquad (12.310)$$

and N is the number of bits of the DAC. Note that $i = 1, \cdots, 2^N - 2$ and $\mathrm{INL}_{2^N-1} = 0$ because the determination of the DAC parameters relies on the use of a linearized output line, whose support points are located in the first and last levels of the transfer characteristic.

The BIST performance depends on the quality of the test signal generated on-chip. Figure 12.103 illustrates the testing principle when the DAC transfer characteristic is a monotonically increasing function and the levels of the adjacent codes are sufficiently separated to be detected by the analog comparator. In cases where these last requirements are not fulfilled for the codes i and $i + 1$, after the estimation of t_i, the determination process of the next index t_{i+1} should be restarted with the code $i + 1$ held constant at the DAC input.

12.8   Circuit design assessment

1.   Nyquist data converter analysis

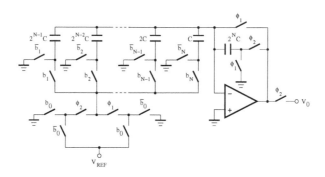

FIGURE 12.104
Block diagram of a charge redistribution DAC.

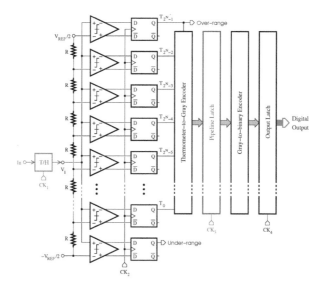

FIGURE 12.105
Block diagram of a flash ADC.

● Consider the DAC depicted in Figure 12.104, which consists of a binary weighted capacitor array.

In the ideal case, show that the output voltage is of the form

$$V_0 = (-1)^{b_0} V_{REF} \sum_{k=1}^{N} 2^{-k} b_k , \qquad (12.311)$$

where b_0 is the sign bit, b_k $(k = 1, 2, \cdots, N)$ represents the magnitude bit, and V_{REF} is the reference voltage.

Verify that the total capacitance required to achieve a resolution of N bits is $C_T = (2^{N+1} - 1)C$.

Assuming that the saturation level of the amplifier output voltage is αV_{DD}, where $\alpha = 0.8$ and $V_{DD} = 2.5$ V, solve the equation $|V_0| \leq \alpha V_{DD}$ with $C = 1$ pF to determine the maximum value of N.

• For the flash ADC of Figure 12.105, verify that the numbers of comparators and resistors required to achieve a resolution of N bits are $2^N + 1$ and 2^N, respectively.

Let the differential nonlinearity (DNL) be defined as the difference between an actual step width and the ideal value of 1 least significant bit (LSB). For each code T_k $(k = 0, 1, \cdots, 2^N - 1)$, the DNL is given by

$$\text{DNL}_k = \frac{\triangle_k}{V_{LSB}} - 1, \qquad (12.312)$$

where $V_{LSB} = V_{REF}/2^N$, and $\triangle_k$ is the actual step size associated with the code T_k.

Assuming that the differential input voltage of each comparator is of the form $V^+ - V^- + V_{off}$, where V_{off} is the offset voltage, and V^+ and V^- are the voltage levels applied to the noninverting and inverting node, respectively, show that

$$\text{DNL}_k = \begin{cases} V_{off}, & \text{if } k = 2^N - 1 \\ 0, & \text{otherwise.} \end{cases} \qquad (12.313)$$

For $V_{off} = 10$ mV and $V_{REF} = 2.5$ V, determine the maximum achievable resolution, N_{max}, by solving the equation $\text{DNL}_k \leq V_{LSB}/2$, which guarantees an effective number of bits equal to N_{max}.

2. Truncation quantizer model

Delta-sigma digital modulators rely on a truncation quantizer to reduce the number of bits of digital code. The conversion of digital code x into a truncated version $\hat{x}$ incurs a quantization error defined as $e_Q = \hat{x} - x$.

− Considering the characteristic and quantization error of the truncation quantizer shown in Figure 12.106(a) in the case of the two's complement representation, the probability density function is given

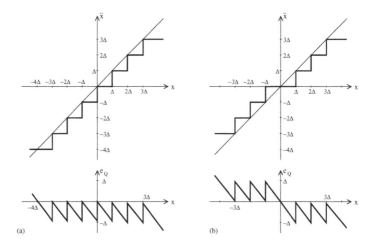

FIGURE 12.106

Characteristics and errors of a truncation quantizer: (a) Two's complement and (b) sign-magnitude representations.

by

$$
p(e_Q) = \begin{cases} \dfrac{1}{\triangle}, & \text{if} \quad -\triangle < e_Q \leq 0 \\ 0, & \text{otherwise.} \end{cases} \tag{12.314}
$$

Show that

$$
E(e_Q) = \int_{-\infty}^{+\infty} e_Q p(e_Q) de_Q = \int_{-\triangle}^{0} e_Q p(e_Q) de_Q = -\frac{\triangle}{2} \tag{12.315}
$$

and

$$
E(e_Q^2) = \int_{-\infty}^{+\infty} e_Q^2 p(e_Q) de_Q = \int_{-\triangle}^{0} e_Q p(e_Q) de_Q = \frac{\triangle^2}{3}. \tag{12.316}
$$

Deduce that the variance or power of the quantization noise is of the form

$$
\sigma_Q^2 = E(e_Q^2) - [E(e_Q)]^2 = \frac{\triangle^2}{12} \tag{12.317}
$$

where $\triangle$ is the quantizer step size.

− Suppose now that a sign-magnitude representation is adopted. The characteristic and quantization error of the truncation quantizer are depicted in Figure 12.106(b) and the probability density function can be obtained as

$$
p(e_Q) = \begin{cases} \dfrac{1}{2\triangle}, & 0 \leq e_Q < \triangle \quad \text{if} \quad x < 0 \\ \dfrac{1}{2\triangle}, & -\triangle < e_Q \leq 0 \quad \text{if} \quad x \geq 0 \\ 0, & \text{otherwise.} \end{cases} \tag{12.318}
$$

Show that

$$E(e_Q) = E(e_Q)|_{x<0} + E(e_Q)|_{x\geq0} = 0 \qquad (12.319)$$

and

$$E(e_Q^2) = E(e_Q^2)|_{x<0} + E(e_Q^2)|_{x\geq0} = \frac{\triangle^2}{3}. \qquad (12.320)$$

Deduce that the variance of the quantization noise is of the form

$$\sigma_Q^2 = \{E(e_Q^2) - [E(e_Q)]^2\}|_{x<0} + \{E(e_Q^2) - [E(e_Q)]^2\}|_{x\geq0} = \frac{5\triangle^2}{2\,12}, \qquad (12.321)$$

where $\triangle$ is the quantizer step size.

3. **Analysis of a second-order DT $\Delta\Sigma$ modulator**

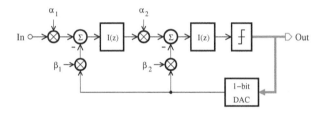

FIGURE 12.107
Block diagram of a second-order $\Delta\Sigma$ modulator.

Consider the block diagram of the second-order $\Delta\Sigma$ modulator depicted in Figure 12.107. The integrator can be implemented such that $I(z) = z^{-1}/(1 - z^{-1})$.

Assuming a linear model for the comparator, that is, $Y(z) = qX(z) + E_Q(z)$, where Y is the comparator output, X is the comparator input, E_Q is the quantization noise, and q is the comparator gain, show that

$$Y(z) = H_S(z)S(z) + H_Q(z)E_Q(z), \qquad (12.322)$$

where

$$H_S(z) = \frac{q\alpha_1\alpha_1 z^{-2}}{1 + (q\beta_2 - 2)z^{-1} + [1 + q(\alpha_2\beta_1 - \beta_2)]z^{-2}} \qquad (12.323)$$

and

$$H_Q(z) = \frac{(1 - z^{-1})^2}{1 + (q\beta_2 - 2)z^{-1} + [1 + q(\alpha_2\beta_1 - \beta_2)]z^{-2}}. \qquad (12.324)$$

Let $z = e^{j\omega T}$, where T is the clock signal period. Evaluate the coefficients α_1, α_2, β_1, and β_2 in terms of q so that the stability criterion, $|H_Q(\omega)| < 1.5$ for $0 \leq \omega T \leq \pi$, is fulfilled.

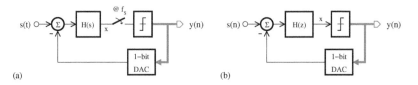

FIGURE 12.108

Block diagram of first-order $\Delta\Sigma$ modulators.

4. Discrete-time to continuous-time transformation

A $\Delta\Sigma$ modulator can be designed to process either a continuous-time or discrete-time signal, as shown in Figures 12.108(a) and (b), respectively. The feedback digital-to-analog converter (DAC) is characterized by the transfer function $H_{DAC}(s) = (1 - e^{sT})/s$, where $T = 1/f_s$ is the sampling period, and f_s is the sampling frequency.

Suppose that

$$H(s) = \frac{1}{sT} \tag{12.325}$$

and use s-transform tables to show that

$$h(t) = \mathcal{L}^{-1}\big[H(s)H_{DAC}(s)\big] = \frac{t}{T}u(t) - \frac{t-T}{T}u(t-T). \tag{12.326}$$

With reference to z-transform tables, show that

$$H(z) = \mathcal{Z}\big[h(t)|_{t=nT}\big] = \frac{z^{-1}}{1 - z^{-1}}. \tag{12.327}$$

Now consider that

$$H(s) = \frac{s_k}{sT + s_k}, \tag{12.328}$$

where $\omega_c = s_k/T$ is the 3-dB bandwidth frequency of the analog lowpass filter, and use s-transform tables to show that

$$h(t) = \mathcal{L}^{-1}\big[H(s)H_{DAC}(s)\big] = \big(1 - e^{-s_k t}\big)u(t) - \big[1 - e^{-s_k(t-T)}\big]u(t-T). \tag{12.329}$$

Show that the equivalent transfer function in the z-domain can be put into the form

$$H(z) = \mathcal{Z}\big[h(t)|_{t=nT}\big] = \frac{(1 - z_k)z^{-1}}{1 - z_k z^{-1}}, \tag{12.330}$$

where $z_k = e^{-s_k T}$.

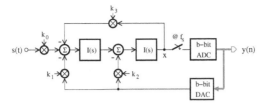

FIGURE 12.109
Block diagram of a second-order CT $\Delta\Sigma$ modulator.

5. **Analysis of a second-order CT $\Delta\Sigma$ modulator**

   Consider the second-order CT $\Delta\Sigma$ modulator shown in Figure 12.20.

   Assuming that $I(s) = 1/(sT)$, where T denotes the clock signal period, show that

   $$H(s) = \frac{X(s)}{Y(s)} = -\frac{k_2 sT + k_1}{s^2 T^2 + k_3}. \qquad (12.331)$$

   In the case where the digital-to-analog converter (DAC) is of the half-return-to-zero (HRZ) type, that is, $H_{DAC}(s) = (e^{-sT/2} - e^{-sT})/s$, find the equivalent discrete-time (DT) transfer function $H(z)$.

   Generate the pole-zero plot of the DT modulator noise transfer function based on the following coefficient values: $k_0 = 1$, $k_1 = 2$, $k_2 = 3.5$, and $k_3 = 0.015$.

6. **Discrete-time and continuous-time 2-1 cascaded $\Delta\Sigma$ modulators**

   For the 2-1 discrete-time (DT) cascaded $\Delta\Sigma$ modulator depicted in Figure 12.110(a), verify that

   $$H_1(z) = \frac{X_2(z)}{Y_1(z)} = -\frac{\alpha_1 \alpha_2 z^{-2}}{(1 - z^{-1})^2} - \frac{2\alpha_1 \alpha_2 z^{-1}}{1 - z^{-1}}, \qquad (12.332)$$

   $$H_2(z) = \frac{X_3(z)}{Y_2(z)} = -\frac{\alpha_3 z^{-1}}{1 - z^{-1}}, \qquad (12.333)$$

   and

   $$H_3(z) = \frac{X_3(z)}{Y_1(z)} = \left[\frac{\kappa_2}{\alpha_1 \alpha_2} H_1(z) - \kappa_1 \kappa_2 \right] \frac{\alpha_3 z^{-1}}{1 - z^{-1}} \qquad (12.334)$$

   $$= -\frac{\kappa_2 \alpha_3 z^{-3}}{(1 - z^{-1})^3} - \frac{2\kappa_2 \alpha_3 z^{-2}}{(1 - z^{-1})^2} - \frac{\kappa_1 \kappa_2 \alpha_3 z^{-1}}{1 - z^{-1}}. \qquad (12.335)$$

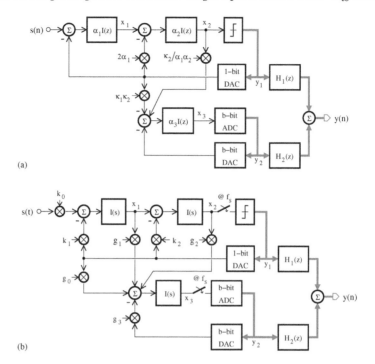

FIGURE 12.110

Block diagrams of 2-1 cascaded $\Delta\Sigma$ modulators.

TABLE 12.5

Equivalent CT Transfer Functions of Some DT Transfer Functions

$$\frac{z^{-1}}{1-z^{-1}} \to \frac{1}{sT} \qquad \frac{z^{-2}}{(1-z^{-1})^2} \to \frac{1}{s^2T^2} - \frac{1}{2}\frac{1}{sT}$$

$$\frac{z^{-3}}{(1-z^{-1})^3} \to \frac{1}{s^3T^3} - \frac{1}{s^2T^2} + \frac{1}{3}\frac{1}{sT}$$

Use the impulse invariant transform relations of Table 12.5, where it was assumed that the digital-to-analog converter (DAC) is modeled by a transfer function of the form $H_{DAC}(s) = (1 - e^{-sT})/s$, where T is the period of the clock signal, to show that the equivalent s-domain transfer functions of the z-domain transfer functions, H_p, $(p = 1, 2, 3)$, are given by

$$H_1(s) = -\frac{\alpha_1 \alpha_2}{s^2 T^2} - \frac{3\alpha_1 \alpha_2/2}{sT}, \tag{12.336}$$

$$H_2(s) = -\frac{\alpha_3}{sT}, \tag{12.337}$$

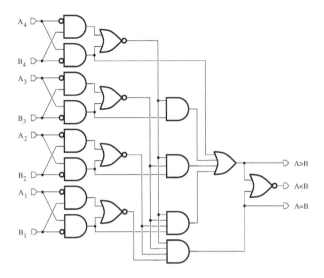

FIGURE 12.111
Block diagram of a 3-bit digital comparator with parallel inputs.

and

$$H_3(s) = -\frac{\kappa_2\alpha_3}{s^3T^3} - \frac{\kappa_2\alpha_3}{s^2T^2} + \frac{(2\kappa_2\alpha_3/3)(1 - 3\kappa_1/2)}{sT}. \qquad (12.338)$$

With reference to the continuous-time (CT) 2-1 cascaded $\Delta\Sigma$ modulator shown in Figure 12.110(b), where $k_0 = k_1$, determine the transfer functions $H_1(s) = X_2(s)/Y_1(s)$, $H_2(s) = X_3(s)/Y_2(s)$, and $H_3(s) = X_3(s)/Y_1(s)$.

Find k_i, $(i = 0, 1, 2)$ and g_j, $(j = 0, 1, 2, 3)$ in terms of α_p $(p = 1, 2, 3)$ and κ_q $(q = 1, 2)$.

With the assumption that $\alpha_1 = 1/2$, $\alpha_2 = 1/2$, $\alpha_3 = 1$, $\kappa_1\kappa_2 = 1$, and $\kappa_2 = 1/2$, generate the pole-zero plot of each of the modulator noise transfer functions.

7. **Implementation of a selection unit based on noise-shaping dynamic element matching**

   Let $A = A_4A_3A_2A_1$ and $B = B_4B_3B_2B_1$. The logic diagram of a 4-bit comparator with parallel inputs is shown in Figure 12.111. Verify that this circuit will generate a logic signal if one of the following conditions, $A > B$, $A = B$, or $A < B$, is satisfied.

   The selection unit of Figure 12.112 was proposed in [44] in order to simplify the hardware required in the implementation of the vector-feedback noise shaping scheme for high-resolution DACs. The partial sorting is implemented here by a tree structure and $F(z) = (2z^{-1} - z^{-2})/(1 - z^{-1})^2$. Digital comparators, as shown in Figure 12.111, are used to evaluate the difference between the

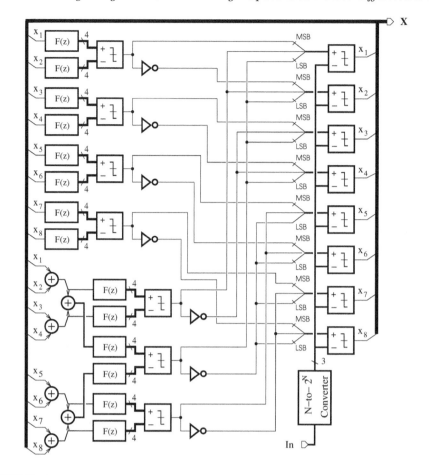

FIGURE 12.112
Block diagram of the DAC selection unit. (From [44], ©1998 IEEE.)

outputs of two adjacent filters, $F(z)$, and for the realization of the partial sorting. Each signal X_i will assume a value of 1 or 0, depending on whether or not the associated DAC unit is selected.

Use simulation results to show that the overall (mismatch and quantization) error is shaped by a function of the form, $1/[1 - F(z)]$.

Bibliography

[1] L. R. Carey, "A noise shaping coder topology for 15+ bit converters," *IEEE J. of Solid-State Circuits*, vol. 24, pp. 267–273, April 1989.

[2] A. Keady and C. Lyden, "Tree structure for mismatch noise-shaping multibit DAC," *Electronics Letters*, vol. 33, pp. 1431–1432, Aug. 1997.

[3] T. Shui, R. Schreier, and F. Hudson, "Mismatch shaping for a current-mode multibit delta-sigma DAC," *IEEE J. of Solid-State Circuits*, vol. 34, pp. 331–338, March 1999.

[4] D. B. Ribner, "A comparison of modulator networks for high-order over-sampled $\Sigma\Delta$ analog-to-digital converters," *IEEE Trans. on Circuits and Systems*, vol. 38, pp. 145–159, Feb. 1991.

[5] K. C.-H. Chao, S. Nadeem, W. L. Lee, and C. G. Sodini, "A higher order topology for interpolative modulators for oversampling A/D converters," *IEEE Trans. on Circuits and Systems*, vol. 37, pp. 309–318, March 1990.

[6] R. Schreier, "An empirical study of high-order single-bit delta-sigma modulators," *IEEE Trans. on Circuits and Systems–II*, vol. 40, pp. 461–466, Aug. 1993.

[7] J. A. Cherry, and W. M. Snelgrove, "Excess loop delay in continuous-time delta-sigma modulators," *IEEE Trans. on Circuits and Systems–II*, vol. 46, pp. 376–389, April 1999.

[8] C.-L. Lin, "Dithering noise cancellation for a delta-sigma modulator," U.S. Patent 7,301,489, filed Dec. 6, 2005; issued Nov. 27, 2007.

[9] T. Ndjountche and R. Unbehauen, "Adaptive calibration techniques for time-interleaved ADCs," *Electronics Letters*, vol. 37, pp. 412–414, March 2001.

[10] M. Sarhang-Nejad and G. C. Temes, "A high-resolution multibit sigma-delta ADC with digital correction and relaxed amplifier requirements," *IEEE J. of Solid-State Circuits*, vol. 28, No 6, pp. 648–660, June 1993.

[11] B. P. Brandt and B. A. Woley, "A 50-MHz multibit sigma-delta modulator for 12-b 2-MHz A/D conversion," *IEEE J. of Solid-State Circuits*, vol. 26, pp. 1746–1756, Dec. 1991.

[12] E. B. Hogenauer, "An economical class of digital filters for decimation and interpolation," *IEEE Trans. on Acoustics. Speech, and Signal Processing*, vol. 29, pp. 155–162, April 1981.

[13] J. C. Candy, "Decimation for sigma delta modulation," *IEEE Trans. on Communications*, vol. 34, pp. 72–76, Jan. 1986.

[14] S. Chu and C. S. Burrus, "Multirate filter designs using comb filters," *IEEE Trans. on Circuits Systems*, vol. 31, pp. 913–924, Nov. 1984.

[15] B. P. Brandt and B. A. Wooley, "A low-power, area-efficient digital filter for decimation and interpolation," *IEEE J. of Solid-State Circuits*, vol. 29, pp. 679–687, June 1994.

[16] R. E. Crochiere and L. R. Rabiner, "Interpolation and decimation of digital signals - A tutorial review," *Proc. of the IEEE*, vol. 69, pp. 300–331, March 1981.

[17] S. K. Mitra, A. Mahalanobis, and T. Saramäki, "Generalized structural subband decomposition of FIR filters and its application in efficient FIR filter design and implementation," *IEEE Trans. on Circuits Systems–II*, vol. 40, pp. 363–374, June 1993.

[18] P. P. Vaidyanathan and T. Q. Nguyen, "A trick for the design of FIR halfband filters," *IEEE Trans. on Circuits and Systems*, vol. 34, pp. 297–300, March 1987.

[19] F. Mintzer, "On half-band, third-band and Nth band FIR filters and their design," *IEEE Trans. on Acoustics, Speech, Signal Processing*, vol. 30, pp. 734–738, Oct. 1982.

[20] G. W. Reitwiesner, "Binary arithmetics," *Advances in Computers*, vol. 1, pp. 231–308, 1960.

[21] A. Avizienis, "Signed-digit number representations for fast parallel arithmetic," *IRE Trans. on Electronic Computers*, vol. 10, pp. 389–400, Sept. 1961.

[22] A. G. Dempster and M. D. Macleod, "Use of minimum-adder multiplier blocks in FIR digital filters," *IEEE Trans. on Circuits and Systems–II*, vol. 42, pp. 569–577, Sept. 1995.

[23] R. I. Hartley, "Subexpression sharing in filters using canonic signed digit multipliers," *IEEE Trans. on Circuits and Systems–II*, vol. 43, pp. 677–688, Oct. 1996.

[24] M. Martínez-Peiró, E. I. Boemo, and L. Wanhammar, "Design of high-speed multiplierless filters using a nonrecursive signed common subexpression algorithm," *IEEE Trans. on Circuits and Systems–II*, vol. 49, pp. 196–203, March 2002.

[25] S. R. Norsworthy, "Optimal nonrecursive noise shaping filters for over-sampling data converters, part 1: Theory," *Proc. of 1993 IEEE Int. Symp. on Circuits and Systems*, vol. 2, pp. 1353–1356, May 1993.

[26] Z.-J. (Alex) Mou, "Symmetry exploitation in digital interpolators/decimators," *IEEE Trans. on Signal Processing*, vol. 44, pp. 2611–2615, Oct. 1996.

[27] I. Løkken, A. Vinje, T. Sæther, and B. Hernes, "Quantizer nonoverload criteria in sigma-delta modulators," *IEEE Trans. on Circuits and Systems–II*, vol. 53, pp. 1383–1387, Dec. 2006.

[28] H. T. Jensen and I. Galton, "A reduced-complexity mismatch shaping DAC for delta-sigma data converter," *Proc. of 1998 IEEE Int. Symp. on Circuits and Systems*, vol. I, pp. 504–507, May 31-June 3, 1998.

[29] E. Fogleman, I. Galton, W. Huff, and H. Jensen, "A 3.3-V single-poly CMOS audio ADC deltasigma modulator with 98-dB peak SINAD and 105-dB peak SFDR," *IEEE J. of Solid-State Circuits*, vol. 35, pp. 297–307, March 2000.

[30] J. Welz, I. Galton, and E. Fogleman, "Simplified logic for first-order and second-order mismatch-shaping digital-to-analog converters," *IEEE Trans. on Circuits and Systems–II*, vol. 48, pp. 1014–1027, Nov. 2001.

[31] T.-H. Kuo, K.-D. Chen, and H.-R. Yeng, "A wideband CMOS sigma-delta modulator with incremental data weighted averaging," *IEEE J. of Solid-State Circuits*, vol. 37, pp. 11–17, Jan. 2002.

[32] J. Doernberg, H.-S. Lee, and D. A. Hodges, "Full-speed testing of A/D converters," *IEEE J. of Solid-State Circuits*, vol. 19, pp. 820–827, Dec. 1984.

[33] A. Brandolini and A. Gandelli, "Testing methodologies for analog-to-digital converters," *IEEE Trans. on Instrumentation and Measurement.*, vol. 41, pp. 595–603, Oct. 1992.

[34] Y.-C. Jenq, "Discrete-time method for signal-to-noise power ratio measurement," *IEEE Trans. on Instrumentation and Measurement*, vol. 45, pp. 431–434, April 1996.

[35] A. K. Lu, G. W. Roberts, and D. A. Johns, "A high-quality analog oscillator using oversampling D/A conversion techniques," *IEEE Trans. on Circuits and Systems–II*, vol. 41, pp. 437–444, July 1994.

[36] B. R. Veillette and G. W. Roberts, "High frequency sinusoidal generation using delta-sigma modulation," *Proc. of 1995 IEEE Int. Symp. on Circuits and Systems*, pp. 637–640, April 30-May 3, 1995.

[37] A. K. Lu and G. W. Roberts, "An analog multi-tone signal generator for built-in self-test applications," *Proc. of the IEEE International Test Conference*, pp. 650–659, Washington, Oct. 1994.

[38] H.-W. Huang, *MC68HC12 An introduction: Software and hardware interfacing*, Clifton Park, NY: Thompson Delmar Learning, 2003.

[39] Jim Ryan, "Interfacing DSPs with high performance analog converters," *DSP Workshops, ICSPAT*, Orlando, FL, 1999.

[40] J.-L. Huang, C.-K. Ong, and K.-T. Cheng, "A BIST scheme for on-chip ADC and DAC testing," *Proc. of the Design, Automation and Test in Europe (DATE) Conference*, pp. 216–220, Paris, 2000.

[41] Y.-C. Wen and K.-J. Lee, "An on chip ADC test structure," *Proc. of the Design, Automation and Test in Europe (DATE) Conference*, pp. 221–225, Paris, 2000.

[42] S. Rabii and B. A. Wooley, "A 1.8-V digital-audio sigma-delta modulator in 0.8 μm CMOS," *IEEE J. of Solid-State Circuits*, vol. 32, pp. 783–796, June 1997.

[43] A. Yasuda and H. Tanimoto, "Noise shaping dynamic element matching method using tree structure," *Electronics Letters*, vol. 33, pp. 130–131, Jan. 1997.

[44] A. Yasuda, H. Tanimoto, and T. Iida, "A third-order Δ-Σ modulator using second-order noise-shaping dynamic element matching," *IEEE J. of Solid-State Circuits*, vol. 33, pp. 1879–1886, Dec. 1998.

13

Circuits for Clock Signal Generation and Synchronization

CONTENTS

A circuit for the clock signal generation or recovery is often required to achieve accurate data transfer between the different building blocks (see Figure 13.1) of very large-scale ICs operating at high speed. In the case of transmission systems, as shown in Figure 13.2, the multiplexer converts the input data into a serial stream of non-return-to-zero data, which then drives a high-speed buffer. At the receiver, the signal level is determined by an amplifier and the clock signal is recovered from the transmitted data and used to control the demultiplexer. The resulting data synchronization determines the accuracy of the information regeneration. The rising edges of the clock signal, whose frequency is set equal to the data rate, should coincide with the midpoint of each data bit, such that the sampling occurs farthest from the preceding and following transitions, yielding a maximum tolerance margin for the jitter and other timing uncertainties.

Precision timing circuits are generally based on a phase-locked loop (PLL)

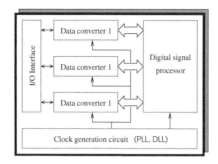

FIGURE 13.1
A typical integrated-circuit floorplan.

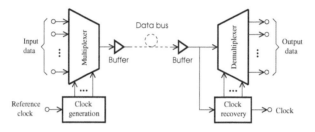

FIGURE 13.2
Transmission system.

or delay-locked loop (DLL). They should typically feature a low sensitivity to process and temperature variations, and generate a clock signal with very low skew and jitter. However, achieving these requirements can be difficult due to a number of design trade-offs to be made between the circuit characteristics. Furthermore, a high level of integration will only be achieved if the objectives of reducing the area and power consumption are met.

13.1 Generation of clock signals with nonoverlapping phases

FIGURE 13.3
(a) Circuit diagram of a two-phase nonoverlapping clock signal generator; (b) clock signals.

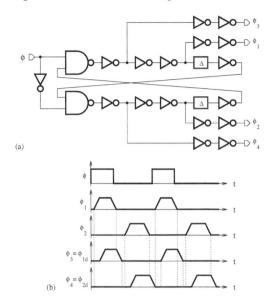

(a)

(b)

FIGURE 13.4
(a) Circuit diagram of a four-phase nonoverlapping clock signal generator; (b) plot of output signals.

To keep negligible the charge leakage, the duration of the sampling phase and hold phase required in the operation of switched-capacitor circuits is controlled by nonoverlapping clock signals. The circuit diagram of a two-phase nonoverlapping clock signal generator is shown in Figure 13.3(a). It consists of NAND gates and inverters. Here, the generator is designed to provide two outputs, which are not allowed to be in the high state during the same time. The nonoverlapping time can be increased by augmenting the number of inverters included between each output and the NAND gate. Each output node is buffered with an inverter sized to drive the on-chip clock bus. Figure 13.3(b) shows the plot of input and output signals. By applying a 50% duty-cycle reference clock signal at the input, the rising edge occurs at one output after the falling edge is produced at the other output. The main advantage of the aforementioned signal generator is its simplicity. However, the propagation delay introduced by the input inverter and changes in clock waveforms due to variations in load conditions can become critical in high-speed applications.

Charge injection errors can be minimized using a four-phase clock signal generator [4]. In the case of the structure shown in Figure 13.4(a), the delay Δ blocks, which can be implemented by an even number of inverters connected in series, are purposely introduced to increase the nonoverlap time between the clock phases ϕ_1 and ϕ_2. The plot of the input and output signals is depicted in Figure 13.4(b), where ϕ_3 and ϕ_4 represent the delayed versions of ϕ_1 and ϕ_2, respectively. The propagation delay of the NAND gate and inverters, which determines the nonoverlap time, is a critical design parameter. If it is chosen with a very small value, clock skew may affect the accuracy of the clock timing. Conversely, if it is sized to be excessively large, the effective time period of

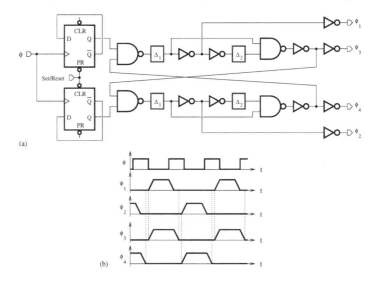

FIGURE 13.5
(a) Circuit diagram of a four-phase nonoverlapping clock signal generator with equal pulse-width complementary signals; (b) plot of output signals.

the clock signals will be considerably reduced. Under these conditions, the clock speed may be reduced. Furthermore, the inverter required at the input of the clock generator delivering nonoverlapping clock signals with the same periodicity as the reference clock introduces a difference in the duty cycle of complementary clock signals.

A solution to improve the performance of the clock generator can rely on forcing the rising edges of a clock phase and its delayed version to occur simultaneously. The circuit diagram of the four-phase nonoverlapping clock signal generator [5] with equal pulse-width complementary signals is shown in Figure 13.5(a). To make the pulse widths of complementary clock signals equal, the input reference clock is applied to two divide-by-2 circuits based on D flip-flops initialized to the high state and low state, respectively. The resulting signals are then applied to a cross-coupled section including a NAND gate, a delay block Δ_1, and two signal-edge synchronization structures, each of which is composed of a series connection of two inverters, a delay block Δ_2, a NAND gate, and an inverter. The synchronization is achieved within a specified maximum amount of time equal to the propagation delay introduced by the series of two inverters and the delay block Δ_2. Each of the delay blocks, Δ_1 and Δ_2, should be realized by an even number of inverters connected in series. Figure 13.4(b) shows the plot of the input and output signals. The rising edges of ϕ_1 and ϕ_2 are aligned to the ones of ϕ_3 and ϕ_4, respectively, while the falling edges of ϕ_3 and ϕ_4 occur after the ones of ϕ_1 and ϕ_2, respectively. In this approach, the clock generator has the advantage of not introducing a difference of pulse width between both complementary clock phases and is then suitable for the control of double-sampled or time-interleaved switched-capacitor circuits.

13.2   Phase-locked loop

The block diagram of the PLL is shown in Figure 13.6. The PLL is a feedback

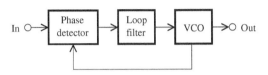

FIGURE 13.6
Phase-locked loop.

system, that operates by generating an oscillation signal, whose frequency has to match the one of the input signal. It consists of a phase detector (PD), loop filter, and voltage-controlled oscillator (VCO). The PD output waveform is proportional to the phase difference between the input and VCO output signals. It is then smoothed by the loop filter and the resulting dc signal is applied to the VCO control node. The VCO can then be driven to minimize the phase difference.

During the initial transient, the PLL operates in the nonlinear region as the VCO tries to find the correct frequency. The PLL linear model is valid when the locked condition is obtained. In this case, the phases of the input and VCO output signals are relatively equal.

Note that the choice of PD architecture has an impact on the overall PLL performance, such as the lock-in range and static phase error. The linear range of a PD detector, which can consist of a simple XOR gate, spreads from $-\pi$ to π. On the other hand, an alternative structure known as a phase and frequency detector (PFD), which generally exhibits the advantage of indicating the lead or lag relation between input waveforms, can handle differences between the clock signals in the range of -2π to 2π.

13.2.1   PLL linear model

Even if the PLL exhibits a nonlinear behavior, its design often starts with the linear model. Figure 13.7 shows the linear model of the PLL. Let Θ_i and Θ_0 be the phase angles associated with the input and output signals, respectively; the output of the phase detector has the form

$$V_p(s) = K_p(\Theta_i(s) - \Theta_0(s)),\tag{13.1}$$

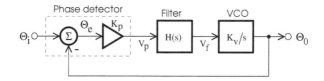

FIGURE 13.7
PLL linear model.

where K_p is the PD conversion gain in units of volts per radian. The transfer function of the filter is denoted by $H(s)$ and that of the VCO is K_v/s, where K_v is the VCO gain in units of radians per volt·second, because the frequency is related to the time derivative of the phase. The phase and error transfer functions can be computed as

$$T(s) = \frac{\Theta_0(s)}{\Theta_i(s)} = \frac{K_v K_p H(s)}{s + K_v K_p H(s)} \qquad (13.2)$$

and

$$T_e(s) = \frac{\Theta_e(s)}{\Theta_i(s)} = \frac{s}{s + K_v K_p H(s)}, \qquad (13.3)$$

respectively. The frequency and transient responses of the loop appear to be affected by the choice of filter characteristic. Due to the contribution of the VCO first-order transfer function, the order of the PLL is equal to that of the filter plus 1. In the special case of a second-order loop, the PLL transfer function contains two poles at the origin due to the VCO and the loop filter implemented by an integrator. To counteract the effect of these poles, the loop transfer function must include a stabilizing zero, which is implemented by connecting a resistor in series with the integrating capacitor.

13.2.2 Charge-pump PLL

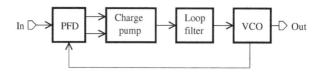

FIGURE 13.8
Charge-pump phase-locked loop.

To achieve an extended tracking range, the PLL can be implemented as shown in Figure 13.8. This structure includes a PFD, a charge-pump circuit, a loop filter, and a VCO. The PFD has the advantage of also being sensitive to frequency error. It acts as an extended-range phase detector and generates a signal, which is indicative of the difference between its input signals. The purpose of the charge-pump circuit is to convert the logic states of the PFD into analog signals, which are appropriate for the VCO control.

13.3 Charge-pump PLL building blocks

A charge-pump PLL system generally consists of a phase and frequency detector (PFD) or phase detector (PD), a charge-pump circuit, a lowpass filter, and a VCO. The PFD or PD compares the phase of the input data and that of the recovered clock signal generated by the VCO, and produces an error signal, typically consisting of Up and Dn signals used to drive a charge-pump circuit differentially. The output signal delivered by the charge-pump circuit is dependent on the phase difference between the data and clock signals. The resulting average signal, which is provided by the loop filter operating as an integrator, is applied to the VCO control input in order to appropriately change the frequency of the clock signal.

13.3.1 Phase and frequency detector

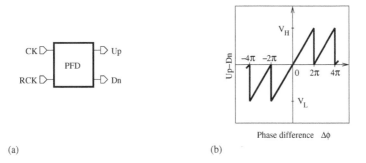

(a)   (b)

FIGURE 13.9
(a) Phase and frequency detector symbol; (b) ideal characteristic of a three-state PFD.

The ideal characteristic of a three-state PFD, as shown in Figure 13.9, is linear for the range of input phase differences from -2π to 2π. The gain of the PFD can be defined as $K_P = (V_H - V_L)/2\pi$, where V_H and V_L denote the highest and lowest output levels, respectively. When the PFD is followed by a charge-pump circuit, the threshold levels V_H and V_L are used to control the value of the charge-pump current, I_P, and the combination of the PFD and charge-pump circuit then exhibits a gain of the form, $K_P = I_P/2\pi$. For the absolute value of the phase difference not exceeding 2π, the PFD is said to be in the lock state.

 In the case of a comparison of two input signals with the same period

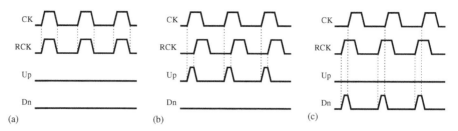

FIGURE 13.10

(a)–(c) Timing diagrams illustrating the three states of a PFD.

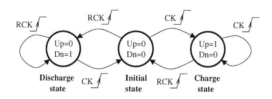

FIGURE 13.11

State graph of the three-state PFD.

and amplitude, the PFD operation principle can be illustrated by the timing diagrams shown in Figure 13.10. If CK leads RCK, pulses will be generated at the output Up while the Dn signal will remain at zero. On the other hand, if CK lags RCK, the Dn signal will be pulsed while the Up signal level will remain low. By applying input signals with different frequencies to the PFD, one output signal is more often set to the high level than the other. As a result, the average value of the output will be positive or negative, depending on the sign of the frequency difference.

The PFD operation can be described using finite state machine, as shown in Figure 13.11, where CK and RCK represent the two periodic input signals. There are only three allowed combinations for the outputs Up and Dn. From the initial state, Up = 0 and Dn = 0, a rising edge of CK causes a transition to the charge state, Up = 1 and Dn = 0, while a rising edge of RCK causes a transition to the discharge state, Up = 0 and Dn = 1. If the PFD is in the charge state, a rising edge of CK will cause no state change, but a rising edge of RCK will cause a change to the initial state. From the discharge state, the PFD can change state only in response to a rising edge of CK. The charge, initial, and discharge states can be respectively associated with the three different values, I_P, 0, and $-I_P$, of the current to be generated by a charge-pump circuit.

At the gate level, a PFD can be realized as shown in Figure 13.12 [6]. Each of the logic states Q_1 and Q_2 are generated by an $\overline{R}\,\overline{S}$ latch, including a pair of cross-coupled two-input NAND gates. From the initial state where both outputs are in the low state, the PFD can move to either the state Up (high

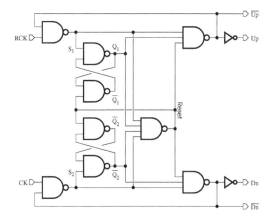

FIGURE 13.12
Gate implementation of the phase and frequency detector.

at the output Up and low at the output Dn) or state Dn (high at the output Dn and low at the output Up) after the detection of the rising edge of one of the input signals. It remains in this last state until the second input goes high, causing the generation of a reset signal, which enables the return to the initial state.

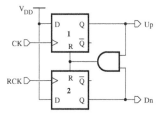

FIGURE 13.13
Circuit diagram of a D flip-flop-based PFD.

An implementation of the PFD based on two resettable D flip-flops, which are clocked by the input signals, is depicted in Figure 13.13. The input terminal D is connected to the positive supply voltage, V_{DD}, and the CK terminals serve respectively as inputs for the signals to be compared. The Dn and Up output pulses are applied to the AND gate, which generates the flip-flop reset signal. When both output signals are high, the AND gate output is set to the high level and this in turn resets both D flip-flops. The reset time is about several gate delays and determines the circuit speed.

In practice, due to the nonzero gate delays in the reset path and violations of the setup and hold time, the PFD characteristic, as shown in Figure 13.14 [7], may exhibit a reduced linear range and a dead zone, where small changes in the input signals are not detected. The dead zone effect can be significantly reduced by introducing an extra delay in the reset path to increase the overlapping time between the Up and Dn output signals. However,

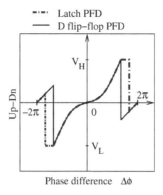

FIGURE 13.14

Characteristics of PFD circuits based respectively on the $\overline{R}\,\overline{S}$ latch and D flip-flop.

the efficiency of this approach appears to be limited by the fact that the delay size is generally dependent on variations in the IC process. For instance, the power consumption will increase in the lock state if the predicted value of the delay time is somewhat high. For a given PFD architecture, the dead zone, the operation frequency range, the power dissipation, and the phase noise are dependent on the design technique (standard logic gates, true single-phase clock logic circuits, differential logic circuits). To extend the operating frequency range, it may be necessary to use a PFD with more than three logic states.

13.3.2 Phase detector

A PD circuit, which can exhibit either a linear or binary transfer characteristic [3], is required for the generation of the phase error signal. Linear PDs, such as the Hogge phase detector, deliver a continuous error signal that is responsible for a linear behavior in the tracking characteristic of the acquisition loop, while binary PDs, also known as Alexander (or bang-bang) phase detectors, generate a quantized phase error signal that contributes to a nonlinear tracking characteristic. The input dynamic range of a phase detector is smaller than that of a PFD.

13.3.2.1 Linear phase detector

The circuit diagram of a Hogge PD [8] and its transfer characteristic are depicted in Figure 13.15. The Hogge PD consists of two flip-flops, which are respectively enabled at the rising and falling edge of the clock signal, and two XOR gates. The input data, D_i, and the output signal of the first flip-flop are processed by the first XOR gate to generate Up phase error signal, while the output signal of the first flip-flop and the output signal of the second flip-flop are applied to the second XOR gate to produce a Dn phase error signal. The retimed data signal is available at the D_0 terminal.

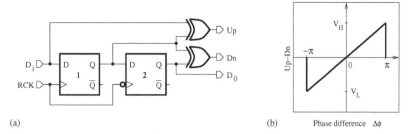

(a)　　　　　　　　　　　　　　　　　　　(b)　　　Phase difference　Δφ

FIGURE 13.15

(a) Circuit diagram and (b) characteristic of a linear (Hogge) PD.

For each data transition, a pulse, whose width varies with the phase difference between the reference clock signal, RCK, and the input data, D_i, is first generated at the Up terminal, and a pulse with a fixed width of half of the clock signal period, or $T_{CK}/2$, is then produced at the Dn terminal. The ideal sampling points of the data correspond to instants where the rising edge of the clock occurs near the center of the data sample, thus yielding the maximum noise margin. The width of the Up pulse will be $T_{CK}/2$ if the rising edge of the signal RCK is nearly aligned with the data center; otherwise, the Up pulse will become smaller or larger than the Dn pulse for early or late clock signals. The average value of the difference between the Up and Dn signals is a linear function of the phase error.

In practical implementations of the Hogge PD, two of the three signals applied to the XOR gates are affected by the RCK-to-Q delay of flip-flops, resulting in an increased width of the Up pulses. These delay variations, which are particularly critical at high frequencies, can cause an increase in the static phase error, and a reduction of the clock phase margin and jitter tolerance. A design solution for the equalization of the RCK-to-Q delay can consist of placing extra delay elements on the path of the data signal to the XOR gate input.

13.3.2.2　Binary phase detector

The binary PD, which is also known as the Alexander or bang-bang PD [9], is shown in Figure 13.16 and is generally used in high-speed clock and data recovery circuits. It consists of four flip-flops and a pair of XOR gates. The input signal is received at the D terminals of the first and third flip-flops. The first, second, and fourth flip-flops are enabled by the rising edge of the clock pulse, while the sampling instants of the third flip-flop correspond to the falling edges of the clock pulse. In addition to the retimed data that can be obtained at the output of the second or fourth flip-flop, the binary PD generates the Up and Dn pulses indicating whether the clock signal is leading or lagging the input data signal.

By sampling two adjacent input data bits and the in-between data transition, the binary PD can deliver the early, late, or no transition information.

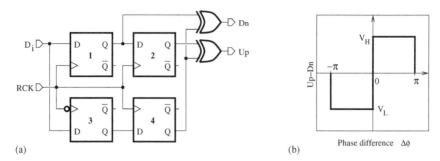

(a) (b)

FIGURE 13.16

(a) Circuit diagram and (b) characteristic of a binary PD.

If the logical states of the first data bit and data transition are identical but differ from that of the second data bit, the clock is early; and if the logical states of the signal transition and the second data bit are equal but differ from that of the first data bit, the clock is late. On the other hand, if all logical states are similar or if the logical states of the first and second data bits are identical but differ from that of the data transition, there is no data transition and a zero dc output is generated.

With Q_1, Q_2, and Q_4 denoting the output of the first, second, and fourth flip-flops, it can be deduced that

- CK is early (E),   if $Q_2 = Q_4 \neq Q_1$
- CK is late (L),   if $Q_2 \neq Q_4 = Q_1$
- There is no transition (X),   if $Q_2 = Q_4 = Q_1$ or $Q_2 = Q_1 \neq Q_4$

While the binary PD can effectively align the clock with the data signal, the average value of its Up and Dn output signals is not proportional to the magnitude of the phase difference. Instead, the output characteristics is a discrete function, which can assume only one of two logical states depending on the result of each phase comparison. By detecting the phase information only at the zero-crossings, the performance of the binary PD can be sensitive to the data transition density.

13.3.2.3   Half-rate phase detector

The PDs described previously are assumed to operate at the full rate or to use a clock frequency equal to the baud rate of the input data. Generally, half-rate PDs are employed to enable the operation at a higher speed with a clock frequency equal to half the input data rate, as they can relax the requirements set for the acquisition loop components. This is in contrast to full-rate architectures, which require more system components to work at higher frequencies and then quickly reach the operating limit of the IC manufacturing process.

- Half-rate linear PD

The circuit diagram of a half-rate linear PD is depicted in Figure 13.17(a).

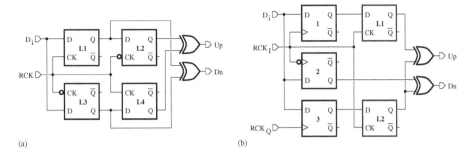

(a) (b)

FIGURE 13.17

Circuit diagram of half-rate PDs with the (a) linear and (b) binary characteristics.

This PD circuit, which is the half-rate version of the Hogge's detector, consists of four latches and two XOR gates [10].

The outputs of the first and third latches follow the data at the D inputs whenever the clock signal is in the high and low state, respectively. They are processed by the XOR gate to generate the signal Dn, whose width is dependent on the phase difference between the half-rate clock and the input data. The retimed data at half rate can be obtained at the outputs of the second and fourth latches, which are respectively enabled on the low and high levels of the clock signal. A full-rate output can be obtained by interleaving the two retimed data streams using a multiplexer controlled by the half-rate clock signal.

The two latches included in each PD path operate as a master-slave flip-flop. Hence, the outputs of the second and fourth latches change on both edges of the clock signal, and the Up signal provided by the XOR gate is a pulse with a constant width of half the clock period, or $T_{CK}/2$, in the case where a data transition is detected. In the locked state, the width of the Dn becomes $T_{CK}/4$, while that of the Up pulse is $T_{CK}/2$. To equalize the effect of both PD outputs, the Up signal can be scaled down by a factor of 2. This is realized in practice by sizing the charge pump circuit such that the current source controlled by the Up signal is two times smaller than the one steered by the Dn signal.

• Half-rate binary PD

The circuit diagram of a half-rate binary PD is shown in Figure 13.17(b) [11]. The incoming data are applied to flip-flops enabled by the in-phase clock signal, the complement of the in-phase clock signal, and the quadrature clock signal, respectively. The output signals of the first and second flip-flops are then compared to that of the third flip-flop using XOR gates to generate the Up and Dn signals. Note that the latches L1 and L2 are introduced to align in time the flip-flop output signals. If a data transition occurs between the rising edge of the in-phase and quadrature clock signals, a pulse will be generated at the Up or Dn terminal indicating whether the clock signal is leading or lagging the input data; otherwise, both the Up and Dn signals remain at the low level.

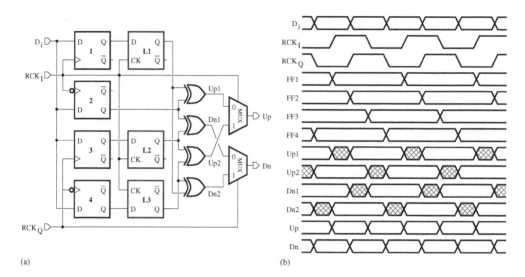

(a)   (b)

FIGURE 13.18

(a) Circuit diagram of a half-rate binary interleaved PD; (b) timing diagram.

The aforementioned PD architecture samples the input data at the rising and falling edges of the in-phase clock, while the quadrature clock is used to track the data transition. Its jitter tolerance is degraded as the operating frequency is increased. A solution can consist of increasing the number of sampling instants [12]. This approach is exploited in the half-rate PD implementation shown in Figure 13.18(a) [13, 14]. The symmetric architecture of this PD circuit helps ensure the matching of delays in the signal paths.

The input data are sampled at $0°$, $90°$, $180°$, and $270°$ of the clock phases using four flip-flops in parallel. The synchronization of data samples, which is particularly useful at high frequencies, is performed using three latches controlled by the in-phase clock signal. The decoding logic for the Up and Dn signals is based on XOR gates and 2-to-1 multiplexers. Each XOR gate produces either an Up1/Up2 or Dn1/Dn2 signal by comparing data samples of two adjacent signal paths. The two multiplexers with the select line controlled by the in-phase and quadrature clock signals, respectively, keep the transfer of the input signals to the Up and Dn terminals within a phase angle of $180°$. The timing diagram of the half-rate binary interleaved PD is depicted in Figure 13.18(b), where the shaded sections represent the invalid time intervals of the Up1/Up2 and Dn1/Dn2 waveforms.

In the locked state, the quadrature clock edges are aligned with the data transitions and the retimed data can be obtained at the output of the first or third flip-flop. The multiplexer selection signals can be delayed by the total amount of signal propagation delay up to the multiplexer input and the multiplexer setup time to improve the timing margin.

13.3.3   Charge-pump circuit

In practice, the PFD or PD incorporated in PLLs does not provide sufficient drive currents to achieve an adequate loop bandwidth. A charge-pump circuit is therefore required to convert the logic pulses generated by the PFD or PD into current signals that are used to drive the loop filter providing the VCO control voltage. The associated current amplification contributes to an increase in the loop bandwidth. The transfer characteristic of the charge pump is generally determined in accordance with the operation principle of the PFD or PD.

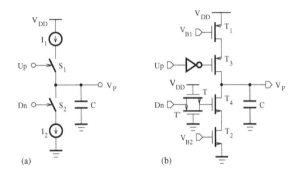

FIGURE 13.19
(a) Conceptual diagram and (b) implementation of a charge-pump circuit.

The conceptual diagram of a charge-pump circuit is shown in Figure 13.19(a). The current sources I_1 and I_2 should be identical. Their connections to the output node are controlled by the Up and Dn signals. The current obtained at the output node will be ideally zero, if the the Up and Dn signals are identical. Otherwise, the average output current over a cycle is $I_p \triangle \phi / 2\pi$, where I_p is the maximum output current and $\triangle \phi$ is the phase difference between the input signals.

A charge-pump circuit can be implemented as shown in Figure 13.19(b). The complementary transistor pair, $T - T'$, is used to equalize the delay of the inverter. Due to the difference in the electron mobility of nMOS and pMOS transistors and the asymmetry in rise and fall times of the Up and Dn signals, a dynamic mismatch between the output currents corresponding respectively to the Up and Dn control voltages can be observed during the operation of this charge-pump circuit. The design of the charge-pump circuit can be improved by minimizing the charge injection errors due to switches and the charge sharing from parasitic capacitances. These charge errors are known to result in a phase offset when a PLL is in the lock state.

For the charge-pump circuit of Figure 13.20(a) [15], two transistors, T_5

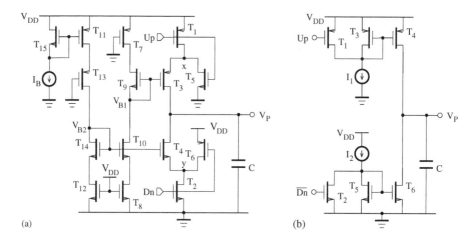

(a) (b)

FIGURE 13.20
Circuit diagrams of charge-pump structures with reduced charge sharing due to the use of (a) compensation switches and (b) nonoverlapping switching pulses.

and T_6, are used to remove the residual charge from the nodes, x and y, during the inactive phases of the Up and Dn signals, thereby mitigating the charge-sharing problem caused by parasitic capacitances at these nodes. The charge-pump current flowing through the transistors, T_3 and T_4, is defined by the voltages, V_{B1} and V_{B2}, set by the biasing circuit section, $T_7 - T_{15}$. A drawback to this charge-pump circuit is the limited dynamic range available at the output node, due to the overdrive voltages required to maintain the transistors, T_3 and T_4, in the saturation region.

An alternative charge-pump structure is shown in Figure 13.20(b) [16, 17]. The current I_1 applied to the current mirror $T_3 - T_4$ will be directed to the output if the signal Up is high, while the input current of the current mirror $T_5 - T_6$ will be transferred to the output if the signal Dn is low. Due to the fact that this charge-pump circuit operates without switching pulse overlap, the charge redistribution associated with overlap capacitances of switches and parasitic capacitances of current sources is reduced.

For the aforementioned charge-pump circuits with switched currents I_1 and I_2, the switches controlled by the Up and Dn pulses can be assumed to be respectively closed for the time periods T_{Up} and T_{Dn}. The output charge can then be expressed as

$$\triangle Q = I_1 T_{Up} - I_2 T_{Dn} . \tag{13.4}$$

Ideally, no charge should be transferred to the output in the lock state because $I_1 = I_2$ and $T_{Up} = T_{Dn}$. In practice, the mismatch between the current levels and the difference in arrival times of the Up and Dn pulses can introduce a steady-state phase offset and increase frequency spurs in the acquisition loop. Furthermore, the transient glitch caused by parasitic capacitors can also contribute to the increase of the spur and jitter levels.

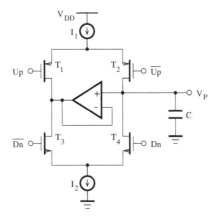

FIGURE 13.21
Charge-pump circuit with a reduced charge-sharing error.

In the case of the charge-pump circuit structure shown in Figure 13.21, the effect of the charge sharing between the parasitic and output capacitors is attenuated using an operational amplifier configured as a unity-gain buffer [18]. The charge pump is implemented as switched current source and sink I_1 and I_2, which can be connected to the capacitor C defining the output voltage, V_P. The upper switches consist of p-channel transistors, while the lower switches are realized with n-channel transistors. For Up and Dn pulses associated with a phase difference, the capacitor C should ideally be charged by either the current source or the current sink. In the lock state, no current should be supplied to the capacitor C by the charge-pump circuit so that the output voltage will remain unchanged.

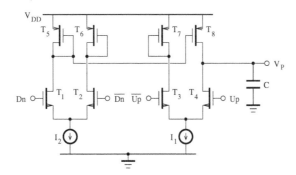

FIGURE 13.22
Charge-pump circuit with an improved current matching.

The charge-sharing errors that can cause mismatch between the current source and sink are reduced by maintaining the output node of all switches connected to the output voltage. In this way, the nonideal voltage variation applied to parasitic capacitors is reduced to the small level of the amplifier offset voltage. However, the accuracy of the nonideal charge cancelation, especially in the lock state, can be limited by the inherent mismatching between the n-channel and p-channel transistors.

The inherent mismatch between the nMOS and pMOS transistors can be eliminated using the charge-pump circuit of Figure 13.22 [19]. Here, only nMOS switches are required. The currents I_1 and I_2 are transferred to the output when either the Up signal or the Dn signal assumes the high level. However, the overall circuit performance can be limited by the dynamic range of the current mirror $T_5 - T_6$. In general, it should also be noted that performance improvement is achieved at the price of a power consumption increase.

13.3.4 Loop filter

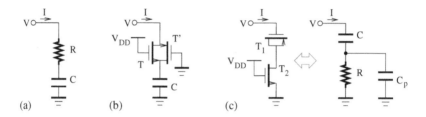

FIGURE 13.23
(a) Circuit diagram and (b) CMOS implementation of a first loop filter; (c) alternative MOS implementation and its equivalent circuit.

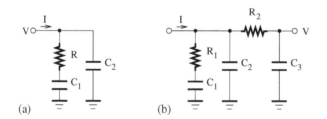

FIGURE 13.24
Circuit diagrams of (a) second-order and (b) third-order loop filters.

The loop filter plays an important role in the determination of PLL characteristics. Generally, it is designed based on the trade-off to be achieved between the lock time, the phase noise, and the residual level of reference spurs. A large loop filter bandwidth helps reduce the lock time, but is also associated with a low attenuation of the phase noise and reference spurs, whereas a narrow bandwidth leads to an improved suppression of the phase noise and reference spurs while increasing the lock time. In practice, the order of the loop filter is increased to satisfy the high attenuation requirement of all unwanted frequencies beyond the cutoff frequency. However, the calculation of the loop filter components then becomes cumbersome, and computer methods must be employed.

The circuit diagram of a first-order filter is shown in Figure 13.23(a). It consists of a resistor connected in series with a capacitor, and can be characterized by a transfer function of the form

$$H(s) = \frac{V(s)}{I(s)} = \frac{R\left(s + \dfrac{1}{RC}\right)}{s} = \frac{1 + \dfrac{s}{\omega_z}}{sC}, \tag{13.5}$$

where $\omega_z = 1/RC$. The closed loop of a PLL using a first-order filter is characterized by a second-order transfer function due to the extra single pole introduced by the VCO. Because the damping factor of the closed-loop transfer function is inversely proportional to the zero at ω_z, the loop bandwidth is enlarged by decreasing the zero frequency. In the s-domain, the stability of a second-order linear system is not affected by the loop gain.

Due to the fact that the loop filter generally requires a resistor with a high resistance, the resistor R can be implemented using a CMOS transistor pair, as illustrated in Figure 13.23(b). However, this simple approach may be limited at low supply voltages, as the effective resistance of the CMOS transistor pair becomes a function of the input node voltage (or VCO control voltage). An alternative implementation of a first-order filter is shown in Figure 13.23(c). To reduce the resistance dependence on the voltage, the resistor R, which is realized by an nMOS transistor, is permutated with the pMOS transistor based capacitor. But, the equivalent model of the filter should take into account the well-substrate parasitic capacitance, C_p, associated here with the MOS capacitor structure.

For a second-order filter, as depicted in Figure 13.24(a), the transfer function is given by

$$H(s) = \frac{V(s)}{I(s)} = \frac{s + \dfrac{1}{RC_1}}{sC_2\left(s + \dfrac{C_1 + C_2}{RC_1C_2}\right)} = \frac{1 + \dfrac{s}{\omega_z}}{sC_T\left(1 + \dfrac{s}{\omega_p}\right)}, \tag{13.6}$$

where $\omega_z = 1/RC_1$, $\omega_p = (C_1 + C_2)/RC_1C_2$, and $C_T = C_1 + C_2$. In practice, the filter components are designed to satisfy the constraints of robust stability, and noise and spur rejection while keeping the component sizes as small as possible.

To improve the suppression of the reference spurs while keeping the bandwidth sufficiently large to meet the requirement of a high lock speed, a third-order filter, as shown in Figure 13.24(b), can be used. This is achieved by setting the pole frequency due to R_2 and C_3 to be lower than the reference frequency, but higher than the loop bandwidth. Using the voltage division principle, the voltage across the R_1C_1 or C_2 branch is of the form $(sR_2C_3+1)V(s)$.

The input current I is then given by

$$I(s) = \frac{(sR_2C_3 + 1)V(s)}{Z(s)} + \frac{(sR_2C_3 + 1)V(s) - V(s)}{R_2}, \tag{13.7}$$

where

$$Z(s) = \left(R_1 + \frac{1}{sC_1}\right) \Big\| \frac{1}{sC_2} = \frac{R_1C_1s + 1}{R_1C_1C_2s^2 + (C_1 + C_2)s}. \tag{13.8}$$

Hence,

$$H(s) = \frac{V(s)}{I(s)} = \frac{R_1C_1s + 1}{D(s)}, \tag{13.9}$$

where

$$D(s) = R_1R_2C_1C_2C_3s^3 + [R_1C_1(C_2 + C_3) + R_2C_3(C_1 + C_2)]s^2 + (C_1 + C_2 + C_3)s. \tag{13.10}$$

Finally, the transfer function, $H(s)$, can be put into the form

$$H(s) = \frac{V(s)}{I(s)} = \frac{1 + \dfrac{s}{\omega_z}}{sC_T\left(1 + \dfrac{s}{\omega_p Q_p} + \dfrac{s^2}{\omega_p^2}\right)}, \tag{13.11}$$

where

$$\omega_z = \frac{1}{R_1C_1} \tag{13.12}$$

$$\omega_p^2 = \frac{C_1 + C_2 + C_3}{R_1R_2C_1C_2C_3} \tag{13.13}$$

$$\omega_p Q_p = \frac{C_1 + C_2 + C_3}{R_1C_1(C_2 + C_3) + R_2C_3(C_1 + C_2)} \tag{13.14}$$

and

$$C_T = C_1 + C_2 + C_3. \tag{13.15}$$

Note that the use of a third-order loop filter has a considerable impact on the design requirements, as the PLL can be prone to instability.

13.3.5 Voltage-controlled oscillator

Generally, an oscillator can be considered a feedback system similar to the one in Figure 13.25(a).

The oscillator closed-loop transfer function can obtained as

$$\left.\frac{V_0(s)}{V_i(s)}\right|_{s=j\omega} = \frac{A(j\omega)}{1 + A(j\omega)}. \tag{13.16}$$

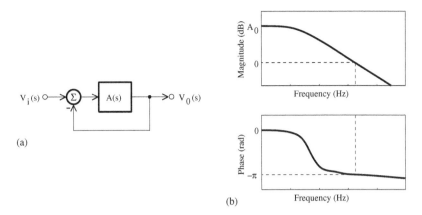

FIGURE 13.25
(a) Linear model of an oscillator; (b) open-loop frequency response.

The system will oscillate at the frequency ω_0 if $A(j\omega_0) = -1$. The oscillation condition or Barkhausen criteria (see Figure 13.25(b)) can then be summarized as

$$|A(j\omega_0)| = 1, \tag{13.17}$$
$$\arg[A(j\omega_0)] = -\pi. \tag{13.18}$$

The oscillatory feature of the system relies on the fact that the feedback signal is added in phase to the forward one. A phase shift of π is introduced by the negative feedback and the overall phase shift is 2π. In practice, the use of the oscillation condition can be limited due to the component imperfections. Then, an open-loop gain larger than unity at the desired oscillation may be required in order to ensure the normal circuit operation.

A voltage-controlled oscillator (VCO) structure with a differential architecture is shown in Figure 13.26 [20]. It consists of a loop of N delay stages with a wire inversion. The ring will oscillate with a period of $2N$ times the stage delay. The differential delay stage using a replica biasing circuit is depicted in Fig 13.27(a) [20]. Fig 13.27(b) shows the circuit diagram of a differential delay stage with symmetric loads. The tail current, which is applied to the differential transistor pair $T_1 - T_2$, is driven by T_7. The voltages V_c and V_B are generated by the replica biasing circuit, which should adjust the bias currents of the delay stage to provide a wide tuning range over the temperature and process variations.

An alternative structure of the delay cell, which also achieves a good power-supply noise rejection, is shown in Figure 13.28 [15]. Transistors $T_5 - T_6$ should fix the output voltage at the minimum value of $V_{DD} - V_T$, where V_T is the transistor threshold voltage, and set an output swing and a common-mode level without the requirement for a replica biasing circuit. That is, the bias current I_B must be greater than the current, I_L, flowing through the pMOS transistor loads. A common practice is to have $I_B = 2I_L$. It should be noted

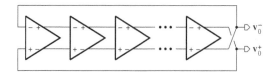

FIGURE 13.26
Oscillator using differential delay stages.

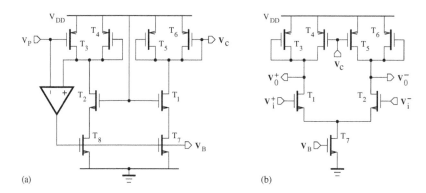

(a)

(b)

FIGURE 13.27
(a) Replica biasing circuit; (b) delay buffer.

that the parasitic capacitance introduced at the output nodes by $T_5 - T_6$ can limit the operating frequency range.

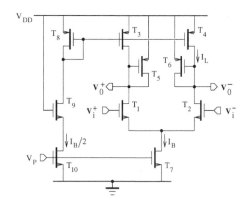

FIGURE 13.28
Delay buffer using pMOS transistor diodes.

Let us consider a delay buffer consisting of a differential pair loaded by transistors in the triode region. The related equivalent model is shown in Figure 13.29.

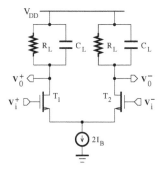

FIGURE 13.29
Equivalent model of the delay buffer.

The differential output voltage can be written as

$$\frac{dV_0(t)}{dt} = -\frac{V_0(t)}{R_L C_L} + \frac{I_B}{C_L}, \tag{13.19}$$

where R_L is the output load capacitor and C_L denotes the output load resistor. An expression of V_0 is then given by

$$V_0(t) = R_L I_B (1 - 2e^{-t/\tau}), \tag{13.20}$$

where $\tau = R_L C_L$ is the time constant. It was assumed that the output voltage initially takes the value $-I_B R_L$ and can increase up to $I_B R_L$. The delay of the buffer can be defined as the time required for the change of V_0 from the initial value to zero, that is,

$$t_d = \ln(2)\tau. \tag{13.21}$$

In practical circuits, the parameter T_d involves a variable contribution due to transistor noise. This delay uncertainty gives rise to clock jitter. Using the relation

$$\left.\frac{dV_0(t)}{dt}\right|_{t=t_d} = \frac{I_B}{C_L}, \tag{13.22}$$

the average jitter component can be obtained as

$$\overline{\triangle t_d^2} = \frac{\overline{v_n^2}}{[(dV_0(t)/dt)|_{t=t_d}]^2} = \overline{v_n^2}\frac{C_L^2}{I_B^2}, \tag{13.23}$$

where $\overline{v_n^2}$ denotes the voltage variance of the total noise.

It should be noted that an additional buffer may be required to provide the single-ended version of the output signal [22]. Figure 13.30 shows the circuit diagram of a differential-to-single-ended converter. Two source followers, a differential stage, and two inverters are required in this design. The input impedance of the buffer is increased by the source followers $T_6 - T_7$ and $T_8 - T_9$, while the output drive capability is improved by the two inverters $T_{10} - T_{11}$ and $T_{12} - T_{13}$.

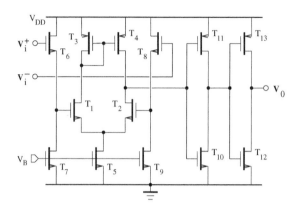

FIGURE 13.30
Differential-to-single-ended buffer.

Using a VCO, the jitter of the output clock is only affected by that of the reference signal, because the loop acts as a lowpass filter. The periodicity of the signal delivered by a VCO is useful in PLL applications such as the clock and data recovery. Furthermore, by inserting a frequency divider in the loop, the output clock period can be a fraction of the reference signal to meet the frequency synthesizer specifications.

13.4 Applications

PLLs find use in a wide variety of applications, including, but not limited to, frequency synthesizers, clock, and data recovery circuits.

13.4.1 Frequency synthesizer

In general, a frequency synthesis consists of generating a desired frequency from one or more reference signals, each at a given frequency and generated by a precise crystal oscillator. The block diagram of a frequency synthesizer based on the PLL is shown in Figure 13.31. It includes an integer-N programmable divider in the feedback path. The output frequency is given by

$$f_0 = N \cdot f_{ref}, \tag{13.24}$$

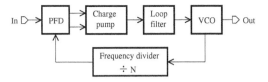

FIGURE 13.31
Block diagram of a frequency synthesizer based on the PLL.

where N is the division ratio and f_{ref} is the frequency of the input reference signal.

Based on the continuous-time linear model of the PLL, the relationship between the phase angles of the error, input, and output signals is of the form

$$\Theta_e = \Theta_i - \Theta_0/N. \tag{13.25}$$

Assuming that the transfer functions of the filter and VCO are $H(s)$ and K_v/s, respectively, the close-loop transfer function can be expressed as

$$T(s) = \frac{\Theta_0(s)}{\Theta_i(s)} = \frac{G(s)}{1 + G(s)}, \tag{13.26}$$

and the error transfer function is

$$T_e(s) = \frac{\Theta_e(s)}{\Theta_i(s)} = \frac{1}{1 + G(s)}, \tag{13.27}$$

where $G(s)$ denotes the open-loop transfer function given by

$$G(s) = \frac{K_p K_v H(s)}{sN}. \tag{13.28}$$

By reducing $H(s)$ to a constant, the frequency synthesizer becomes a first-order system that is unconditionally stable. However, in practice, the transfer function of the loop filter should include a pole/zero pair, which is used to increase the frequency range.

Although the linear model in the s-domain provides a helpful set of design equations, a charge-pump-based PLL should be accurately described and optimized in the z-domain. The direct conversion of the PLL continuous-time equations into the z-domain can be computationally intensive. For this reason, the method based on the impulse invariant transformation[1] is generally

---

[1] Let $F(s)$ be the transfer function of a system in the s-domain. Using the impulse invariant transformation, its z-domain version can obtained as

$$\widehat{F}(z) = \mathcal{Z}\left\{ \frac{1 - e^{-Ts}}{s} F(s) \right\} = (1 - z^{-1}) \cdot \mathcal{Z}\left\{ \mathcal{L}^{-1}\left(\frac{F(s)}{s} \right) \Big|_{t=kT} \right\}. \tag{13.29}$$

It is assumed that the equivalent discrete representation is provided by the series connection of an ideal sampler with the sampling period T, a zero-order hold stage and the analog model of the system.

adopted. The closed-loop transfer function is written as

$$\widehat{T}(z) = \frac{\widehat{\Theta}_0(z)}{\widehat{\Theta}_i(z)} = \frac{\widehat{G}(z)}{1 + \widehat{G}(z)}, \tag{13.30}$$

where

$$\widehat{G}(z) = \frac{K_p K_v \widehat{H}(z)}{N} \tag{13.31}$$

and

$$\widehat{H}(z) = (1 - z^{-1})\mathcal{Z}\left\{ \mathcal{L}^{-1}\left(\frac{H(s)}{s^2} \right) \Big|_{t=nT/N} \right\}. \tag{13.32}$$

Here, T is the sampling frequency, and $\mathcal{Z}$ and $\mathcal{L}^{-1}$ denote the z-transform (or modified z-transform) and inverse Laplace transform, respectively. Note that the correspondence between the models is ensured only at the time nT/N. For the evaluation of $\widehat{H}$, the value of the time-domain function should be zero at the initial time instant.

Let us consider the third-order frequency synthesizer shown in

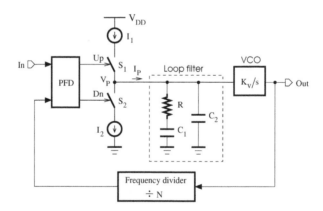

FIGURE 13.32
Block diagram of a third-order frequency synthesizer.

Figure 13.32. The loop filter has a second-order transfer function of the form

$$H(s) = \frac{V_P(s)}{I_P(s)} = \frac{1 + \dfrac{s}{\omega_z}}{sC_T\left(1 + \dfrac{s}{\omega_p}\right)}, \tag{13.33}$$

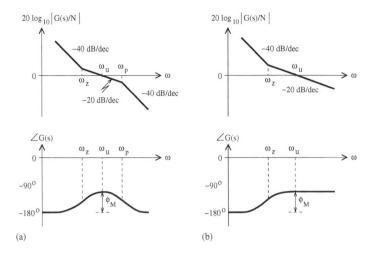

FIGURE 13.33

Magnitude and phase of the open-loop transfer function: (a) $C_2 \neq 0$, (b) $C_2 = 0$.

where $\omega_z = 1/RC_1$, $\omega_p = (C_1 + C_2)/RC_1 C_2$, and $C_T = C_1 + C_2$. The open-loop transfer function is given by

$$G(s) = \frac{K_p K_v H(s)}{sN} \tag{13.34}$$

$$= K_p K_v \frac{1 + \dfrac{s}{\omega_z}}{s^2 N C_T \left(1 + \dfrac{s}{\omega_p}\right)}. \tag{13.35}$$

The combination of the PFD and charge-pump circuit can be characterized by the gain factor, $K_p = I_p/2\pi$. The magnitude and phase of the open-loop transfer function are depicted in Figure 13.33. At the frequency, ω_u, the magnitude of the open-loop transfer function is equal to 1 (or 0 dB).

To determine the zero and pole frequencies needed to obtain the desired phase margin, we compute

$$\phi_M = 180^\circ - \angle G(j\omega_u) = \arctan\left(\frac{\omega_u}{\omega_z}\right) - \arctan\left(\frac{\omega_u}{\omega_p}\right), \tag{13.36}$$

where ω_u is the unity-gain frequency. The phase margin is maximum for a value of ω_u, which can be obtained by setting the derivative of ϕ_M with respect to ω_u equal to zero. That is,

$$\frac{d\phi_M}{d\omega_u} = \frac{1}{\omega_z} \cdot \frac{1}{1 + \left(\dfrac{\omega_u}{\omega_z}\right)^2} - \frac{1}{\omega_p} \cdot \frac{1}{1 + \left(\dfrac{\omega_u}{\omega_p}\right)^2} = 0 \tag{13.37}$$

and

$$\omega_u = \sqrt{\omega_z \omega_p} = \omega_z \sqrt{1 + \eta}, \tag{13.38}$$

where

$$\eta = C_1/C_2. \tag{13.39}$$

By substituting Equation (13.38) into (13.36), the maximum value of ϕ_M is obtained as

$$\phi_M = \arctan\left(\sqrt{1 + \eta}\right) - \arctan\left(\frac{1}{\sqrt{1 + \eta}}\right). \tag{13.40}$$

Recalling that $\arctan x + \arctan y = \arctan[(x - y)/(1 + xy)]$ and solving Equation (13.40) for η gives

$$\eta = 2 \tan^2 \phi_M + 2 \tan \phi_M \sqrt{1 + \tan^2 \phi_M}. \tag{13.41}$$

The stability of the loop in the s-domain is guaranteed, provided the capacitors are chosen such that Equation (13.41) is satisfied.

Given ω_u and ϕ_M, the initial values of the filter components are determined as follows. By definition, we have

$$|G(j\omega_u)| = \frac{K_p K_v}{N C_T \omega_u^2} \cdot \frac{\sqrt{1 + \left(\frac{\omega_u}{\omega_z}\right)^2}}{\sqrt{1 + \left(\frac{\omega_u}{\omega_p}\right)^2}} = 1. \tag{13.42}$$

By combining Equations (13.37) and (13.42), and recalling that $\omega_p/\omega_z = 1 + \eta$ and $C_T = C_1(1 + 1/\eta)$, it can be shown that

$$\frac{K_p K_v}{N C_T \omega_u^2} \sqrt{\frac{\omega_p}{\omega_z}} = \frac{K_p K_v}{N C_1 \omega_u^2} \cdot \frac{\eta}{\sqrt{1 + \eta}} = 1. \tag{13.43}$$

Therefore, according to Equation (13.43), we find that

$$C_1 = \frac{K_p K_v}{N \omega_u^2} \cdot \frac{\eta}{\sqrt{1 + \eta}}. \tag{13.44}$$

From Equations (13.38) and (13.39), we respectively obtain

$$R = \frac{\sqrt{1 + \eta}}{\omega_u C_1} \tag{13.45}$$

and

$$C_2 = \frac{C_1}{\eta}. \tag{13.46}$$

The *s*-domain analysis is limited by the fact that it does not take into account the sampling nature of the loop. It is then not suitable for the prediction of jitter performance and nonlinear acquisition process. Simulations are necessary to fine-tune the values of the filter components.

As a rule-of-thumb, the closed-loop bandwidth of a charge-pump PLL should be chosen to be less than approximately one-tenth of the reference frequency. Otherwise, the stability, speed, and phase noise of a charge-pump PLL will be affected by the sampling process. To take into account the sampling effects, it is necessary to perform the *z*-domain analysis of the loop [23, 24]. The open-loop transfer function, $G(s)$, can be converted from the *s*-domain to the *z*-domain using the impulse invariant transformation. By performing the partial expansion of $G(s)/s$, we obtain

$$\frac{G(s)}{s} = \frac{K_p K_v}{N C_T} \left[\frac{1}{s^3} + \frac{\omega_p - \omega_z}{\omega_z \omega_p} \cdot \frac{1}{s^2} - \frac{\omega_p - \omega_z}{\omega_z \omega_p^2} \cdot \frac{1}{s(1 + s/\omega_p)} \right]. \tag{13.47}$$

The inverse Laplace transform of $G(s)/s$ can then be written as

$$\mathcal{L}^{-1}\left(\frac{G(s)}{s} \right) = \frac{K_p K_v}{N C_T} \left[\frac{t^2}{2} + \frac{\omega_p - \omega_z}{\omega_z \omega_p} t - \frac{\omega_p - \omega_z}{\omega_z \omega_p^2}(1 - e^{-\omega_p t}) \right]. \tag{13.48}$$

The equivalent transfer function in the *z*-domain is given by

$$\widehat{G}(z) = \frac{z-1}{z} \cdot \mathcal{Z}\left\{ \mathcal{L}^{-1}\left(\frac{G(s)}{s} \right)\Big|_{t=kT} \right\}. \tag{13.49}$$

Using *z* transform tables, it can be shown that

$$\widehat{G}(z) = \frac{K_p K_v}{N C_T} \times$$
$$\left[\frac{T^2}{2} \cdot \frac{z+1}{(z-1)^2} + \frac{\omega_p - \omega_z}{\omega_z \omega_p} \cdot \frac{T}{z-1} - \frac{\omega_p - \omega_z}{\omega_z \omega_p^2} \cdot \frac{1 - e^{-\omega_p T}}{z - e^{-\omega_p T}} \right], \tag{13.50}$$

or equivalently,

$$\widehat{G}(z) = \frac{K_p K_v}{N C_T} \frac{p z^2 + q z + r}{z^3 - (2+\alpha)z^2 + (1+2\alpha)z - \alpha}, \tag{13.51}$$

where

$$\alpha = e^{-\omega_p T},\tag{13.52}$$

$$p = \frac{T^2}{2} + \left(\frac{T}{\omega_z} - \frac{1-\alpha}{\omega_z \omega_p}\right)\left(1 - \frac{\omega_z}{\omega_p}\right),\tag{13.53}$$

$$q = \frac{T^2(1-\alpha)}{2} - \left(\frac{T(1+\alpha)}{\omega_z} - \frac{2(1-\alpha)}{\omega_z \omega_p}\right)\left(1 - \frac{\omega_z}{\omega_p}\right),\tag{13.54}$$

$$r = -\frac{T^2\alpha}{2} + \left(\frac{T\alpha}{\omega_z} - \frac{1-\alpha}{\omega_z \omega_p}\right)\left(1 - \frac{\omega_z}{\omega_p}\right),\tag{13.55}$$

and T denotes the period of the clock signal. The parameter of the charge-pump PLL should then be chosen such that the roots of the characteristic equation, $1 + \widehat{G}(z)$, remain inside the unit circle defined by $|z| = 1$ in the z-plane. In general, the stability condition in the z-domain is more constraining than that in the s-domain.

The operation or acquisition process of a charge-pump PLL can be accurately modeled using a time-domain analysis based on difference equations and state-space representation. The input phase, θ_i, and the output phase, θ_0, are related by difference equations of the form

$$\theta_i(t) = \theta_i(0) + \omega_i t,\tag{13.56}$$

$$\theta_0(t) = \theta_0(0) + \omega_v t + K_v \int_0^t v_p(\tau)d\tau,\tag{13.57}$$

$$\theta_e(t) = \theta_i(t) - \theta_0(t),\tag{13.58}$$

where θ_e is the phase error, ω_i represents the input frequency, ω_v is the free running frequency of the oscillator, K_v is the VCO gain, and v_p denotes the output voltage of the loop filter or the VCO control voltage, which is identical to the voltage across the capacitor C_2. It was assumed that the initial conditions for the input and the output phases are $\omega_i(0)$ and $\omega_0(0)$, respectively. The state-space representation of the filter can be written as

$$\frac{dv_p}{dt} = -\frac{v_p}{RC_2} + \frac{v_{C_1}}{RC_2} + \frac{i_p}{C_2},\tag{13.59}$$

$$\frac{v_{C_1}}{dt} = \frac{v_p}{RC_1} - \frac{v_{C_1}}{RC_1},\tag{13.60}$$

where v_{C_1} represents the voltage across the capacitor C_1 and i_p is the charge-pump output current, which is used to drive the loop filter. Assuming that the output capacitor of the charge-pump circuit is either charged for a positive phase error or discharged for a negative phase error, the current i_p is given by

$$i_p = \begin{cases} I_P \cdot \mathrm{sign}(\theta_e), & \text{if} \quad 0 \le t \le t_p \\ 0, & \text{if} \quad t_p < t < t_r, \end{cases} \tag{13.61}$$

where $\mathrm{sign}(\theta_e)$ denotes the polarity (i.e., 1 or -1) of the phase error, θ_e, t_p represents the turn-on duration of the current I_P, and t_r is the time at which the next rising edge of either the VCO or the reference signal occurs. Tools for symbolic analysis can be used to solve the system of equations characterizing the charge-pump PLL. By using linear models, the analysis is valid only near locked states. However, due to the fact that the PFD, for instance, is actually a nonlinear and time-variant component, the loop may be affected by nonlinear mechanisms such as cycle slip, which is caused by a large frequency difference between the reference signal and the feedback signal. In practice, this difference is minimized by setting the initial control voltage appropriately, and thereby facilitating the acquisition process.

Note that the synthesis of a frequency, which is N/M times the reference frequency, can simply be performed by adding a divider with a ratio of M at the input of the PLL.

The aforementioned topology is commonly used due to its simplicity. However, the output frequency can only change by integer multiples of the reference frequency. Furthermore, the reference spurs, which appear centered on the reference frequency and its harmonics, are related to the amount of jitter in the retimed signals. Hence, the achievable resolution of the output frequency is limited because the loop bandwidth, as set by the loop filter, should be at least ten times smaller than the reference frequency to prevent undesirable signal components caused by the sampling action in the phase detector from reaching the input of the VCO and corrupting the output frequency. This can result in a slow settling (or lock) time for the frequency synthesizer.

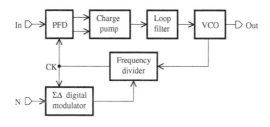

FIGURE 13.34

Block diagram of a $\Delta\Sigma$ fractional-N frequency synthesizer.

To overcome the resolution-bandwidth trade-off of integer-N frequency synthesizers, $\Delta\Sigma$ fractional-N architecture can be used [25]. This last approach is capable of generating frequencies over wide bandwidths with a very fine frequency resolution to accommodate the narrow channel spacing of wireless telephony applications. The block diagram of a $\Delta\Sigma$ fractional-N frequency synthesizer is depicted in Fig 13.34. It is generally based on the concept of division ratio averaging, which is implemented using a PLL whose feedback path includes a multi-modulus divider controlled by a $\Delta\Sigma$ modulator. The division ratio is dynamically switched between two or more values to realize the fractional frequency division.

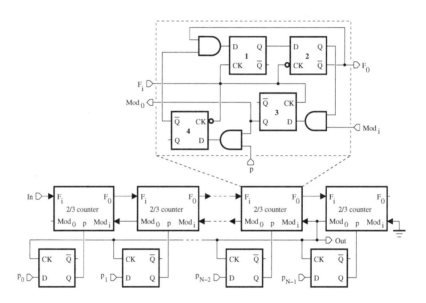

FIGURE 13.35
Block diagram of a dual-modulus frequency divider.

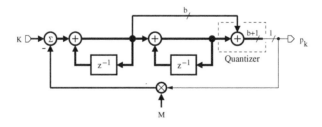

FIGURE 13.36
Block diagram of a second-order $\Delta\Sigma$ modulator.

An dual-modulus frequency divider can be implemented as shown in Figure 13.35. This modular structure consists of a plurality of 2/3 counters arranged in a cascade combination [26]. Each 2/3 counter realizes a division by a factor of 2 or 3, depending on whether the control signal p_k is set to a logic state 0 or 1.

The basic principle of the 2/3 counter cell can be extended to any $P/(P+1)$ cell. Even if the division ratio of each counter cell of the frequency divider is an integer at any instant, its repetitive switching between P and $P+1$ by the modulator will give rise to a fractional division with a ratio comprised between these last two integer values. With the division ratio equal to P during $(1-\eta)T$ and $P+1$ during ηT over a time period of T, the average output frequency is given by

$$F_0 = [\eta(P+1)T + (1-\eta)PT]f_{ref}/T = (P+\eta)Fi, \qquad (13.62)$$

where F_i is the input frequency of the $P/(P+1)$ counter and η is determined by the modulator output.

In operation, the prescaler logic block of each 2/3 counter first divides the frequency of the incoming signal, F_i, and the resulting signal, F_0, is applied to the following counter cell. The division ratio is determined by the logic state of the control signals applied to the end-of-cycle logic block of each 2/3 counter cell. Upon completion of a division cycle, the end-of-cycle logic block of the last 2/3 counter cell in the divider chain generates a signal, Mod_0, which is transferred successively to the preceding 2/3 counter cells after being retimed by each cell. The division cycle corresponds to the clock period of the F_0 signal available at the output of the last 2/3 counter, whose Mod_0 terminal can serve as the divider output. To avoid the perturbation of the current division operation, the updated control signals for the division ratio should transit through latches, which are synchronized with the respective division cycle, on their way to each 2/3 counter cell. A new division ratio can be set only when a division cycle is completed.

The overall division ratio of the frequency divider can be put into the form

$$N = 2^n + 2^{n-1}p_{n-1} + 2^{n-2}p_{n-2} + \cdots + 2p_1 + p_0, \qquad (13.63)$$

where $p_0, p_1, \cdots, p_{n-1}$ denote the control signal logic states from the first to the last 2/3 counter. Using n 2/3 counter cells in cascade, the possible division ratios range from 2^n (if all $p_k = 0$) to $2^{n+1} - 1$ (if all $p_k = 1$).

The block diagram of a second-order $\Delta\Sigma$ modulator is depicted in Figure 13.36. Generally, the data word-length in the modulator should be long enough to prevent the occurrence of an overflow in the first accumulator. Based on simulations, it was shown in [25] that the parameters K and M can be adequately chosen as $-0.5M < K < 0.5M$ and $M < 2^b/2.5$, where b is the accumulator word-length. The computation resolution, which is initially set to b bits, is increased to $b+1$ bits in certain steps to accommodate possible numerical overflows in the adders. The quantization operation is reduced to the overflow of the second adder, the sign bit of which is used as the modulator output.

Using a linear model of the modulator, the z-transform of the output can be expressed as

$$P_k(z) = \frac{H(z)}{1 + M \cdot H(z)} K(z) + \frac{1}{1 + M \cdot H(z)} E_Q(z), \qquad (13.64)$$

where E_Q is the quantization error and

$$H(z) = \frac{2 - z^{-1}}{(1 - z^{-1})^2}. \qquad (13.65)$$

At dc, or $z = 1$, and $M \cdot H(z) \gg 1$, we can obtain

$$P_k \simeq \frac{K}{M} + \frac{E_Q}{M \cdot H}. \qquad (13.66)$$

For a slow-varying input signal, the time average of the binary output sequence produced by the modulator is a high-resolution representation of the value K/M.

A problem generally associated with $\Delta\Sigma$ fractional-N frequency synthesizers is the increased level of in-band spurs. Ideally, the PFD and charge-pump circuits should deliver an output signal that is proportional to the phase difference between the reference and feedback signals. However, their practical characteristics are not fully linear. In the lock condition, the phase difference can still take different values due to the changing division ratio in the feedback path. This in turn stimulates nonlinearities of the PFD and charge-pump circuits, generating a noise floor that boosts the in-band phase noise.

Note that the use of a higher-order $\Delta\Sigma$ modulator helps reduce the in-band quantization noise, but at the price of an increased level of the out-of-band noise.

13.4.2 Clock and data recovery

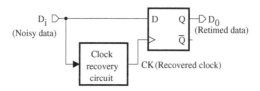

FIGURE 13.37
Principle of a clock and data recovery circuit.

In serial data transmissions, the signal is generally distorted by the transmission channel (coaxial cable, optic fiber). Because it is generally impractical to transmit the necessary sampling clock signal separately from data, the timing information is usually derived from the transmitted data, which are asynchronous and noisy. This is realized using a clock recovery circuit, which can be based on a PLL. With reference to Figure 13.37, the incoming data should be retimed by the D flip-flop, which is synchronized by the recovered clock signal in such a way that the sampling clock edge is aligned with the middle of the data bit period in order to minimize the bit-error rate.

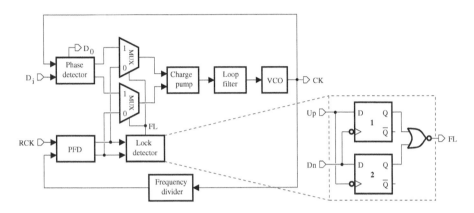

FIGURE 13.38
Block diagram of a dual-loop PLL requiring an external reference clock signal.

A clock and data recovery circuit, which relies on a PLL consisting of a phase detector, a charge-pump circuit, a filter, and a VCO, is generally unable to capture the incoming data if the free-running frequency of the VCO is more than a few hundred parts-per-million (ppm) from the frequency of the input data.

With the VCO free-running frequency exhibiting a wide variation over the IC process, temperature, and supply voltage, it is impossible for the PLL to lock under all situations without relying on a frequency acquisition aid. Hence, the PLL used for clock and data recovery applications can be configured to include a frequency loop and a phase loop. The first one achieves the clock signal frequency acquisition by reducing the difference between the free-running frequency of the VCO and a reference frequency, while the second one is enabled for the phase locking of the clock signal with the incoming data signal. A PLL with a dual-loop configuration may be implemented with or without an external reference clock signal.

Referring to Figure 13.38, a block diagram shows a dual-loop PLL [27] requiring a reference clock signal. Before the closing of the data-acquisition loop by the lock detector, the frequency-acquisition loop should bring the VCO free-running frequency near the desired operating range, if necessary. The data-acquisition loop minimizes the remaining frequency error and aligns the phase of the VCO for an optimal sampling of the incoming data. The lock detector continuously monitors that the difference between the reference signal frequency and the divider output frequency remains within a predetermined range.

An approach to perform the lock detection can consist of monitoring the Up and Dn pulses that are generated by the three-state PFD [28]. This can simply be realized, as shown in the inset of Figure 13.38, by a lock detector structure whose components are two D flip-flops, inverters, and a AND gate. The output signal FL is set to the high logic state indicating the frequency-locked mode only in the case where the Up and Dn pulses are at the low logic state. When the Up and Dn signals are in different logic states due to the fact that one of the PFD input is leading or lagging the other, the output signal FL is maintained to the low logic state. Depending on the signal propagation delay, the resulting error in the detection of phase alignment between the Up and Dn pulses is within 5 and 15% of the full-scale phase deviation of the PFD.

Another scheme for implementing the lock detector consists of using a counter-based frequency comparator to monitor the reference clock frequency and the divider output frequency [29]. The use of Gray counters may help reduce the occurrence of latch metastability due to the fact that the reference clock and the divider output signal can be asynchronous with each other. After a given time period, the lock detector determines whether or not the frequency difference between the two signals is within a predetermined range. The counters should then be reset before the beginning of the next counting interval.

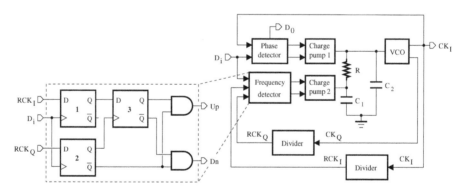

FIGURE 13.39
Block diagram of a dual-loop PLL without a reference clock signal.

A dual-loop PLL can also be designed to operate without a reference clock signal, as shown in Figure 13.39. It requires two separate charge pumps driven by a phase detector and a frequency detector, respectively. In one implementation of the frequency detector, three D flip-flops and two AND gates are necessary [30]. During a frequency acquisition, the frequency detector delivers Up and Dn signals with a frequency that is equal to the difference between the input data frequency and quadrature clock signal frequency. When the free-running frequency of the VCO is set within the loop capture range, the

Up and Dn signals are maintained at a low logic level. It is then not necessary to disable the frequency detector, which can continuously track frequency changes without affecting the operation of the data acquisition loop. Because the VCO is equivalent to a gain element in the frequency domain, a zero is not needed in the transfer function of the filter included in the frequency loop. Hence, the charge-pump circuit controlled by the frequency detector is connected to the node between the resistor R and capacitor C_1.

With V_c being the control voltage delivered by the loop filter in response to the charge-pump current I_p, the filter impedance function can be expressed as

$$Z(s) = \frac{V_c(s)}{I_p(s)} = \frac{N(s)}{sC_2(s + \omega_p)}, \tag{13.67}$$

where $\omega_p = (C_1 + C_2)/RC_1C_2$. Assuming that $\omega_z = 1/RC_1$, the numerator is given by $N(s) = s + \omega_z$ for the data acquisition loop or $N(s) = 1$ for the frequency acquisition loop.

The signal CK_Q is 90° out of phase with the signal CK_I. By implementing the VCO as a four-stage ring oscillator, the quadrature signals CK_I and CK_Q are available at the outputs of two of the stages. The input range of the FD can be numerically predicted to be on the order of 25% of the desired VCO free-running frequency.

13.5 Delay-locked loop

In applications where the desired clock frequency is already available, a voltage-controlled delay line (VCDL) can be used instead of a VCO. Hence, while the PLLs generate a new clock signal, the delay-locked loops (DLLs) actually retard the incoming clock signals in such a way that the delay in the output clock signals is greatly reduced.

The block diagram of a DLL is shown in Figure 13.40. The output signal is generated by delaying the input clock. The phase difference between the VCDL output and reference clock signals is provided by a phase detector whose output is applied to the loop filter, which generates the VCDL control voltage. The delay cell is actually equivalent to a gain stage and the stabilization of the loop can be achieved using a first-order lowpass filter.

The closed-loop phase transfer function is given by

$$T(s) = \frac{\Theta_0(s)}{\Theta_i(s)} = \frac{1}{1 + \dfrac{1}{K_d K_p H(s)}}, \tag{13.68}$$

where K_d is the gain of the VCDL and H is the transfer characteristic of the loop filter, which is realized here by a charge-pump circuit loaded by a capacitor and is then similar to an integrator.

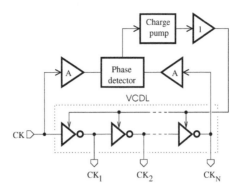

FIGURE 13.40
Delay-locked loop.

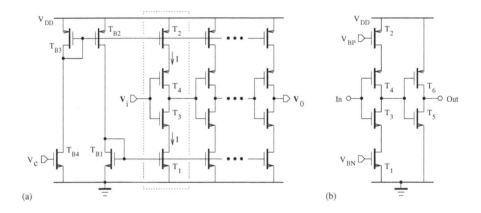

FIGURE 13.41
(a) Delay line-based oscillator using inverters; (b) inverter stage with an improved driving capability.

A VCDL can be implemented as shown in Figure 13.41(a) [18]. The biasing circuit uses a current mirror to source and sink current corresponding to the dc level of the control voltage, V_c. Transistors $T_1 - T_2$ operate as a current source and the inversion of the input signal is achieved by $T_3 - T_4$. The time delay is related to the output load capacitance. The inverter structure depicted in Figure 13.41(b) can be adopted to improve the output driving capability. Generally, single-ended structures like the ones of Figure 13.41 are sensitive to the power supply noise, which affects the phase of the output signal.

A DLL is less prone to stability problems than a PLL whose loop bandwidth is often determined by a high-order narrowband filter requiring large components with a high sensibility to IC process, voltage, and temperature variations. The requirement to use such a loop filter in a PLL also results in a longer acquisition time, usually in the 1/2 to 1 microsecond range, while a DLL can lock to the data rate in just a few clock cycles. By employing components with high values, a PLL occupies a large chip area compared to a

DLL. Because a DLL adjusts the amount of delay or phase without affecting the frequency in order to achieve the desired synchronization, its operating frequency range is severely limited. Furthermore, a DLL may falsely lock into a harmonic frequency of the reference clock signal.

13.6 PLL with a built-in self-test structure

By designing a PLL with a built-in self-test (BIST) structure [31, 32], the number of test devices required after the manufacturing phase is reduced. A possible PLL architecture with a BIST structure, which consists of a multiplexer, and control and register sections, is shown in Figure 13.42(a).

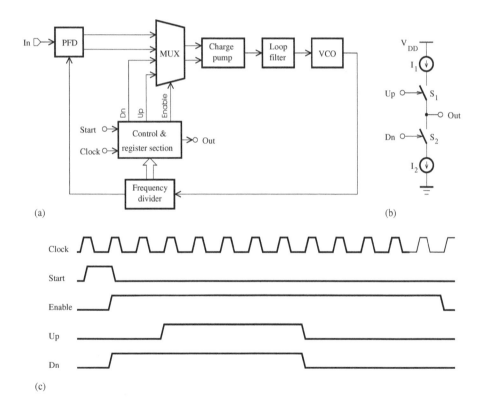

FIGURE 13.42
(a) Block diagram of a PLL including a BIST structure; (b) charge-pump circuit; (c) timing diagram.

The charge-pump circuit of Figure 13.42(b) can be made digitally programmable. It output signal is used as the input stimulus. The divide-by-N section in the frequency counter circuit can provide a digital estimation of the VCO output frequency. A fault in the loop results in a deviation of the oscillation frequency from its nominal value.

The multiplexer connects either the control signal in the test mode or the phase detector output during the normal PLL operation to the charge-pump circuit. It is steered by the output signal of the control and register section. The digital frequencies computed by the divide-by-N circuit are stored in the registers. They can then be scanned out using the IEEE 1149.1 scan path, for instance.

The timing diagram of the BIST section is illustrated in Figure 13.42(b). During a test period, three frequency estimations are achieved according to the next scheme, where the low and high logic states are represented by L and H, respectively.

- Set Enable=H, Up=L and Dn=H and estimate the frequency, f_0.
If f_0 is in the lower half of the output range;
    - Set Enable=H, Up=L and Dn=L, and estimate the frequency, f_1;
    - Set Enable=H, Up=H and Dn=H, and estimate the frequency, f_2;
Otherwise f_0 is in the upper half of the output range;
    - Set Enable=H, Up=H and Dn=H, and estimate the frequency, f_1;
    - Set Enable=H, Up=L and Dn=L, and estimate the frequency, f_2.

The digital output, D, of the divide-by-N section is given by

$$D = Tf_i, \tag{13.69}$$

where f_i is the input frequency and T is the measurement time period. If D_{max} is the maximum value of D, the number of bits used for the number representation will be obtained as

$$B = \lceil \log_2(D_{max}) \rceil + 1, \tag{13.70}$$

where $\lceil x \rceil$ denotes the smallest integer greater than or equal to x.

The deviation between the frequency differences $|f_0 - f_1|$ and $|f_2 - f_1|$ should be greater for a PLL with a fault than the one operating normally.

13.7   PLL specifications

A PLL will operate at the free-running frequency, ω_0, when it is not locked to an input signal. This frequency is determined by the VCO components. A nonlinear model of the PLL is necessary for the characterization of the acquisition process.

A PLL is generally designed to meet a specified lock range, lock time, pull-in range, pull-in time, pull-out range, and loop noise bandwidth.

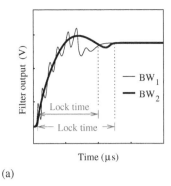

(a)

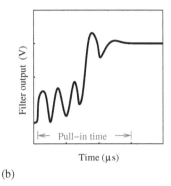

(b)

FIGURE 13.43

Transient response of the VCO control voltage due to a step signal in the PLL input: Illustration of the (a) lock time and (b) pull-in time.

Let the frequency range of the PLL input signal be in the frequency range from $\omega_0 - \triangle\omega$ to $\omega_0 + \triangle\omega$.

Lock range, $\pm\triangle\omega_L$ — It is the angular frequency range over which the PLL acquires lock within the first cycle of the input signal. The lock range is affected by the difference between the input and free-running frequencies. Note that $\triangle\omega_L \geq \triangle\omega$. The parameter $\triangle\omega_L$ is centered at the free-running angular frequency and determines the PLL operating frequency range.

Lock time, t_L — This is the time after which the PLL can lock in one single-beat note between the input reference frequency and the output frequency. The lock time, as illustrated in Figure 13.43(a)) for two loop bandwidths, is primarily a function of the loop characteristics.

Pull-in range, $\pm\triangle\omega_{pi}$ — It is the angular frequency range over which the PLL would naturally become locked after the cycle slipping occurrence, as shown in Figure 13.43(b), where the frequency of the input step is assumed to lie outside the lock range. Cycle slipping occurs when the PFD is unable to accurately follow the phase error variations caused by large frequency offsets, for instance. In the event of a cycle slip, the PLL momentarily loses the frequency lock before settling at the actual output frequency.

Pull-in time, t_p — This is the transient time required by the PLL to always become locked. The pull-in process has a nonlinear nature and can be slow and unreliable. Hence, the pull-in time is dependent on the PLL initial conditions (frequency and phase errors) and loop characteristics.

Pull-out range, $\pm\triangle\omega_{po}$ — It is the angular frequency range over which the PLL operation is stable. The pull-out range is smaller than or equal to the pull-in range.

Loop noise bandwidth, B_n — This parameter is the one-sided bandwidth of a unity-gain and ideal lowpass filter, which transmits as much noise power as does a linear PLL model. The signal components located outside B_n are greatly attenuated.

After the lock is achieved, the accuracy of the acquisition loop can still be limited by the effects of the jitter, phase offset, and step-size errors. The jitter is a random variation of the clock signal transitions due to noise. The phase offset, which is constant in nature, is essentially caused by mismatches between circuit components and timing misalignment. The step-size errors are associated with the minimum resolution of the delay or VCO control signal.

13.8 Summary

Circuits for clock signal generation and synchronization were described at the behavioral and architectural level. Specifically, they are used to reduce the voltage and timing errors in the clocking or data transmission networks.

In addition to a low jitter performance, it is important to achieve a low power consumption and die area in portable equipment. Either passive or active methods can be used to reduce the clock jitter induced by the power supply noise. The first method employs filters on the power supply lines, while the second one regulates the power supply by relying on a feedback control loop that keeps the voltage constant.

13.9 Circuit design assessment

1. **Phase and frequency detector**
   The phase and frequency detector of Figure 13.44 [2] possess the advantage of not having a dead zone. Use SPICE simulations to obtain the transient response of the circuit and show that the error detection range is limited in the range from $-\pi$ to π.

2. **Single-ended VCO based on inverters**
   Consider the VCO structure shown in Figure 13.45(a). It consists of N inverters (see Figure 13.45(b)), capacitors and transistors operating as switches controlled by V_c.

   Why must the number of stages, N, be odd?
   With the assumption that each stage can be modeled with a first-order transfer function of the form

   $$\frac{V_0}{V_i}(s) = \frac{A_0}{1 + \dfrac{s}{\omega_0}}, \tag{13.71}$$

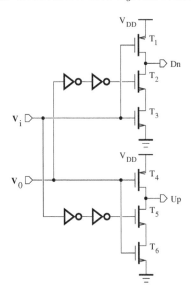

FIGURE 13.44
Phase and frequency detector.

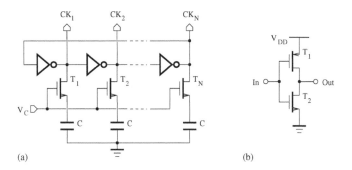

FIGURE 13.45
(a) VCO structure; (b) circuit diagram of the inverter.

where A_0 is the dc gain and ω_0 is the -3 dB bandwidth, find the oscillation frequency of the VCO.

3. **Analysis of a second-order PLL**
   Consider the linear equivalent model of the second-order PLL shown in Figure 13.46.

   Estimate the closed-loop phase transfer function, T, of the PLL.

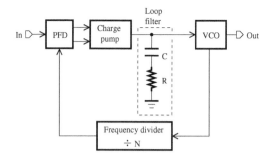

FIGURE 13.46
Block diagram of a second-order PLL.

Show that T can be put into the following form:

$$T(s) = \frac{\Theta_0(s)}{\Theta_i(s)} = \frac{N\left(1 + 2\dfrac{\zeta}{\omega_n}s\right)}{1 + 2\dfrac{\zeta}{\omega_n}s + \left(\dfrac{s}{\omega_n}\right)^2}, \qquad (13.72)$$

where

$$\omega_n = \sqrt{\frac{K_v I_P}{NC}}, \qquad (13.73)$$

$$\zeta = \frac{1}{2}\sqrt{\frac{K_v I_P C R^2}{N}}. \qquad (13.74)$$

By applying an angular frequency step to the PLL, the input phase angle, Θ_i, can be written as

$$\Theta_i(t) = \begin{cases} 0, & \text{for } t < 0, \\ \triangle\omega t, & \text{for } t > 0. \end{cases} \qquad (13.75)$$

Prove that the phase error is given in the s-domain and time domain by

$$\Theta_e(s) = (1 - T(s))\frac{\triangle\omega}{s^2} \qquad (13.76)$$

and

$$\Theta_e(t) = \begin{cases} \dfrac{N\triangle\omega}{\omega_n\sqrt{1-\zeta^2}}\sin(\sqrt{1-\zeta^2}\omega_n t)\exp(-\zeta\omega_n t), & \text{for } \zeta < 1, \\ N\triangle\omega t\exp(-\omega_n t), & \text{for } \zeta = 1, \\ \dfrac{N\triangle\omega}{\omega_n\sqrt{\zeta^2-1}}\sinh(\sqrt{\zeta^2-1}\omega_n t)\exp(-\zeta\omega_n t), & \text{for } \zeta > 1, \end{cases}$$

$$(13.77)$$

respectively.

The noise bandwidth of the PLL is defined as

$$B_n = \int_0^{+\infty} |T(j\omega)|^2 d\omega. \tag{13.78}$$

Verify that

$$B_n = N^2 \omega_n \frac{1 + 4\zeta^2}{8\zeta}. \tag{13.79}$$

The parameter ζ is generally chosen around the value ζ_m, which minimizes B_n. Find ζ_m.

4. **Design of a PLL for the clock recover application**

Consider the phase-locked loop (PLL) shown in Figure 13.47 [21]. It includes a phase and frequency detector (PFD) based on two XOR gates, a charge-pump (CP) circuit, a loop filter, a voltage-controlled oscillator (VCO) with four differential delay stages, and a D flip-flip used for retiming the recovered clock signal.

With the assumption that I_1 and I_2 are nominally equal to I_P, the CP provides the charge $\pm I_P$ to the capacitor C_1.

Show that the transfer function of the section including the PFD, CP, and loop filter is given by

$$\frac{V_c}{\triangle \phi}(s) = \frac{I_P}{2\pi} \frac{1 + RC_1 s}{(C_1 + C_2)s + RC_1 C_2 s^2}, \tag{13.80}$$

where V_c is the VCO control voltage and $\triangle \phi$ is the phase difference between the PD inputs.

Determine the closed-loop transfer function of the PLL in the s-domain and z-domain.

Use the SPICE program to perform the transient analysis of the PLL.

5. **Fourth-order charge-pump PLL**

Consider the fourth-order charge-pump PLL shown in Figure 13.48. The frequency division by N is performed by cascading six divide-by-two stages, each of which is based on a D flip-flop whose inverted output node is connected back to the data input node. The circuit diagrams of a charge-pump circuit and a VCO are depicted in Figures 13.49(a) and (b), respectively. Assuming that that the pole frequency due to R_2 and C_3 is at least ten times higher than the loop bandwidth, and $C_3 \leq C_2/10$, the third-order loop filter can be considered a second-order filter section followed by a first-order filter section providing an attenuation α of the spurious sidebands

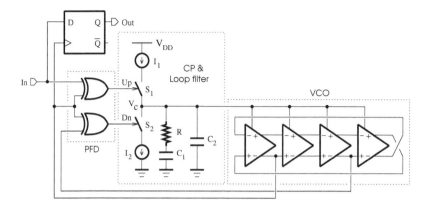

FIGURE 13.47

Circuit diagram of a third-order charge-pump PLL.

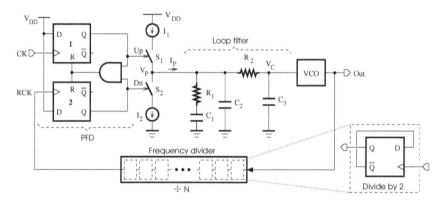

FIGURE 13.48

Circuit diagram of a fourth-order charge-pump PLL.

at multiple of the reference frequency. Hence,

$$C_1 \simeq \frac{K_p K_v}{N \omega_u^2} \cdot \frac{\eta}{\sqrt{1 + \eta}}, \tag{13.81}$$

$$R_1 \simeq \frac{\sqrt{1 + \eta}}{\omega_u C_1}, \tag{13.82}$$

$$C_2 \simeq \frac{C_1}{\eta}, \tag{13.83}$$

and

$$R_2 \simeq \frac{\sqrt{10^{\alpha/10} - 1}}{\omega_{ref} C_3}, \tag{13.84}$$

where

$$\eta = 2 \tan^2 \phi_M + 2 \tan \phi_M \sqrt{1 + \tan^2 \phi_M}. \tag{13.85}$$

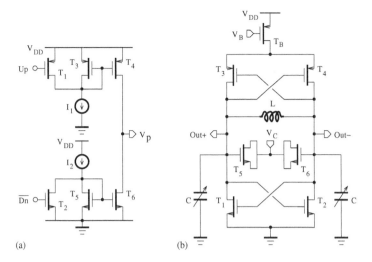

FIGURE 13.49
(a) Charge-pump circuit; (b) VCO circuit.

Here, $K_p = I_P/(2\pi)$, $\omega_{ref} = 2\pi f_{ref}$, and f_{ref} is the reference frequency.

For a reference frequency of 37.5 MHz, a VCO free-running frequency of 2, 4 GHz, $\omega_u/(2\pi) = 200$ kHz, $\phi_M = 65°$, $\alpha = 10$ dB, $K_v = 100$ MHz/V, $N = 64$, and $I_P = 200$ µA, first determine the initial values of the filter components and then use MATLAB and SPICE simulations to size and fine-tune the values of the loop components in a given CMOS technology such that the charge-pump PLL can exhibit a loop bandwidth of 200 kHz.

Bibliography

[1] F. M. Gardner, "Charge-pump phase-lock loops," *IEEE Trans. on Communications*, vol. 28, pp. 1849–1858, Nov. 1980.

[2] H. O. Johansson, "A simple precharged CMOS phase frequency detector," *IEEE J. of Solid-State Circuits*, vol. 33, pp. 295–299, Feb. 1998.

[3] B. Razavi, "Challenges in the design of high-speed clock and data recovery circuits," *IEEE Communications Magazine*, vol. 40, pp. 94–101, Aug. 2002.

[4] D. W. Cline, *Noise, speed, and power trade-offs in pipelined analog to digital converters*, PhD thesis, University of California, Berkeley, 1995.

[5] Seng-Pan U, R. P. Martins, and J. E. Franca, "A 2.5-V 57-MHz 15-tap SC bandpass interpolating filter with 320-Ms/s output for DDFS system in 0.35-μm CMOS," *IEEE J. of Solid-State Circuits*, vol. 39, pp. 87–99, Jan. 2004.

[6] I. W. Young, J. K. Greason, and K. L. Wong, "A PLL clock generator with 5 to 110 MHz of lock range for microprocessors," *IEEE J. Solid-State Circuits*, vol. 27, pp. 1599-1607, Nov. 1992.

[7] M. Mansuri, D. Liu, and C.-K. K. Yang, "Fast frequency acquisition phase-frequency detectors for Gsamples/s phase-locked loops," *IEEE J. of Solid-State Circuits*, vol. 37, pp. 1331–1334, Oct. 2002.

[8] C. R. Hogge Jr., "Signal detection apparatus," U.S. Patent 4,535,459, filed May 26, 1983; issued August 13, 1985.

[9] J. D. H. Alexander, "Clock recovery from random binary signals," *Electronics Letters*, vol. 11, pp. 541–542, Oct. 1975.

[10] J. Savoj and B. Razavi, "A 10-Gb/s CMOS clock and data recovery circuit with a half rate linear phase detector," *IEEE Journal of Solid-State Circuits*, vol. 36, pp. 761–768, May 2001.

[11] M. Fukaishi, K. Nakamura, H. Heiuchi, Y. Hirota, Y. Nakazawa, H. Ikeno, H. Hayama, and M. Yotsuyanagi, "A 20-Gb/s CMOS multichannel transmitter and receiver chip set for ultra-high-resolution digital displays," *IEEE J. of Solid-State Circuits*, vol. 35, pp. 1611–1618, Nov. 2000.

[12] C.-C. Huang, "Phase detector," U.S. Patent 6,259,278, filed June 14, 1999; issued July 10, 2001.

[13] A. Rezayee and K. Martin, "A 9–16 Gb/s clock and data recovery circuit with three-state phase detector and dual-path loop architecture," *Proc. of the 2003 European Solid-State Circuits Conf.*, pp. 683–686, Estoril, Portugal, Sept. 2003.

[14] A. Ong, S. Benyamin, J. Cancio, V. Condito, T. Labrie, Q. Lee, J. P. Mattia, D. K. Shaeffer, A. Shahani, X. Si, H. Tao, M. Tarsia, W. Wong, and M. Xu, "A 4043-Gb/s clock and data recovery IC with integrated SFI-5 1:16 demultiplexer in SiGe technology," *IEEE J. of Solid-State Circuits*, vol. 38, pp. 2155–2168, Dec. 2003.

[15] P. Larsson, "A 2-1600-MHz CMOS clock recovery PLL with low-V_{dd} capability," *IEEE J. of Solid-State Circuits*, vol. 34, pp. 1951–1960, Dec. 1999.

[16] A. Waizman, "A delay line loop for frequency synthesis of de-skewed clock," *1994 IEEE ISSCC Digest of Technical Papers*, pp. 298-299, Feb. 1994.

[17] M. Rau, T. Oberst, R. Lares, A. Rothermel, R. Schweer, and N. Menoux, "Clock/data recovery PLL using half-frequency clock," *IEEE J. Solid-State Circuits*, vol. 32, pp. 1156–1159, July 1997.

[18] M. G. Johnson and E. L. Hudson, "A variable delay line PLL for CPU coprocessor synchronization," *IEEE J. Solid-State Circuits*, vol. 23, pp. 1218–1223, Oct. 1988.

[19] J. G. Maneatis, "Low-jitter process-independent DLL and PLL based on self-biased techniques," *IEEE J. of Solid-State Circuits*, vol. 31, pp. 1723–1732, Nov. 1996.

[20] J. G. Maneatis and M. A. Horowitz, "Precise delay generation using coupled oscillators," *IEEE J. of Solid-State Circuits*, vol. 28, pp. 1273–1282, Dec. 1993.

[21] D.-H. Kim and J.-K. Kang, "Clock and data recovery circuit with two exclusive-OR phase frequency detector," *Electronics Letters*, vol. 36, pp. 1347–1349, Aug. 2000.

[22] I. I. Novof, J. Austin, R. Kelkar, D. Strayer, and S. Wyatt, "Fully integrated CMOS phase-locked loop with 15 to 240 MHz locking range and ±50 ps jitter," *IEEE J. of Solid-State Circuits*, vol. 30, pp. 1259–1266, Nov. 1995.

[23] P. K. Hanumolu, M. Brownlee, K. Mayaram, and U. K. Moon, "Analysis of charge-pump phase-locked loops," *IEEE Trans. on Circuits and Systems–I*, vol. 51, pp. 1665–1674, Sept. 2004.

[24] Z. Wang, "An analysis of charge-pump phase-locked loops," *IEEE Trans. on Circuits and Systems–I*, vol. 52, pp. 2128–2138, Oct. 2005.

[25] T. A. D. Riley, M. A. Copeland, and T. A. Kwasniewski, "Delta-sigma modulation in fractional-N frequency synthesis," *IEEE J. of Solid-State Circuits*, vol. 28, pp. 553–559, May 1993.

[26] C. S. Vaucher, I. Ferencic, M. Locher, S. Sedvallson, U. Voegeli, and Z. Wang, "A family of low-power truly modular programmable dividers in standard 0.35-μm CMOS technology," *IEEE J. of Solid-State Circuits*, vol. 35, pp. 1039–1045, July 2000.

[27] M. Meghelli, B. D. Parker, H. A. Ainspan, and M. Soyuer, "SiGe BiCMOS 3.3-V clock and data recovery circuits foe 10-Gb/s serial transmission systems," *IEEE J. of Solid-State Circuits*, vol. 35, pp. 1992–1995, Dec. 2000.

[28] H. S. Hakkal and J. J. Hughes, "Circuit and method of a three state phase frequency lock detector," U.S. Patent 6,404,240, filed Oct. 30, 2000; issued June 11, 2002.

[29] M. B. Ghaderi and V. W. S. Tso, "Phase-frequency lock detector," U.S. Patent 5,870,002, filed June 23, 1997; issued Feb. 9, 1999.

[30] R. C. H. van de Beek, C. S. Vaucher, D. M. W. Leenaerts, E. A. M. Klumperink, and B. Nauta, "A 2.510-GHz clock multiplier unit with 0.22-ps rms jitter in standard 0.18-μm CMOS," *IEEE J. of Solid-State Circuits*, vol. 39, pp. 1862–1872, Nov. 2004.

[31] S. Kim and M. Soma, "An all-digital built-in self-test for high-speed phase-locked loops," *IEEE Trans. on Circuits and Systems*, vol. 48, pp. 141–150, Feb. 2001.

[32] C.-L. Hsu, Y. Lai, and S.-W. Wang, "Built-in self-test for phase-locked loops," *IEEE Trans. on Instrumentation and Measurement*, vol. 54, pp. 996–1002, June 2005.

A

Logic Building Blocks

CONTENTS

A.1   Boolean algebra

Boolean algebra is used to describe the relations between inputs and outputs of a digital circuit. The input and output signals are considered to be Boolean variables (X, Y, Z) whose values are either 0 (logic low level) or 1 (logic high level).

A.1.1   Basic operations

NOT	AND	OR
$\overline{0} = 1$	$0 \cdot X = 0$	$0 + X = X$
$\overline{1} = 0$	$1 \cdot X = X$	$1 + X = 1$
$\overline{\overline{X}} = X$	$X \cdot X = X$	$X + X = X$
	$X \cdot \overline{X} = 0$	$X + \overline{X} = 1$

- $X + X \cdot Y = X$
- $X \cdot Y + \overline{X} \cdot Z + Y \cdot Z = X \cdot Y + \overline{X} \cdot Z$
- $X + \overline{X} \cdot Y = X + Y$
- $(X + Y)(X + \overline{Y}) = X$
- $X(X + Y) = X$
- $(X + Y)(\overline{X} + Z) = X \cdot Z + \overline{X} \cdot Y$
- $X(\overline{X} + Y) = X \cdot Y$
- $(X+Y)(\overline{X}+Z)(Y+Z) = (X+Y)(\overline{X}+Z)$

A.1.2   Exclusive-OR and equivalence operations

- $X \oplus Y = X \cdot \overline{Y} + \overline{X} \cdot Y$

- $X(Y \oplus Z) = X \cdot Y \oplus X \cdot Z$ $(X \oplus Y) \oplus Z = X \oplus (Y \oplus Z) = X \oplus Y \oplus Z$

- $X \odot Y = \overline{X \oplus Y} = X \cdot Y + \overline{X} \cdot \overline{Y} = \overline{X} \oplus Y = X \oplus \overline{Y}$

- $X \oplus 0 = X \qquad X \oplus X = 0$

- $X \oplus 1 = \overline{X} \qquad X \oplus \overline{X} = 1$

- $\overline{\overline{X} \oplus Y} = \overline{X \oplus \overline{Y}} = \overline{X} \oplus \overline{Y} = X \oplus Y$

- $(X \cdot Y) \oplus (X + Y) = X \oplus Y$ $(X \cdot Y) \oplus (\overline{X} \cdot \overline{Y}) = \overline{X \oplus Y} = X \odot Y$

- $X + Y = (X \oplus Y) \oplus (X \cdot Y)$

- $X \cdot Y = (X \oplus Y) \oplus (X + Y)$

- $X \cdot (Y \oplus Z) = (X \cdot Y) \oplus (X \cdot Z)$

- If $X \cdot Y = 0$, then $X + Y = X \oplus Y$

A.2   Combinational logic circuits

A.2.1   Basic gates

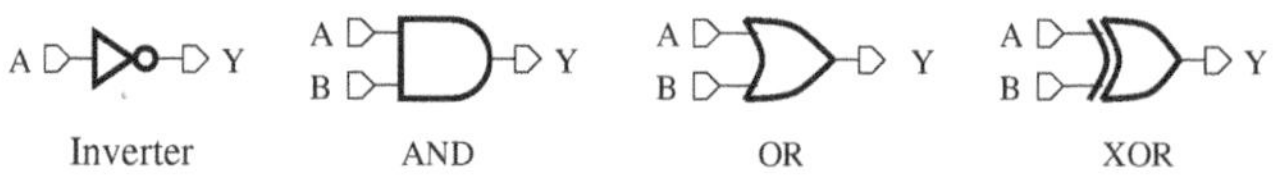

FIGURE A.1
NOT, AND, OR, and XOR gates.

Combinational circuits are digital circuits whose outputs depend only on the current states of inputs. Figure A.1 shows the NOT, AND, OR and XOR gates. Boolean expressions that relate the state of the output to the ones of

TABLE A.1
Logic Equations of NOT, AND, OR, and XOR Gates

Gate type	Inverter	AND	OR	XOR
Boolean function	$Y = \overline{A}$	$Y = A \cdot B$	$Y = A + B$	$Y = A \oplus B$

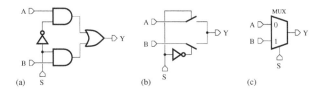

(a)   S   (b)   S   (c)

FIGURE A.2
2-1 Multiplexer: (a) Circuit diagram, (b) principle, (c) symbol.

inputs are given in Table A.1.

A multiplexer permits the transmission of only one of many digital inputs to the output. In general, a multiplexer with 2^k inputs and one output has k select bits that are used to select the input to be connected to the output node. The circuit diagram, operation principle, and symbol of a 2-1 multiplexer are illustrated in Figure A.2. The Boolean expression for the output of the 2-1 multiplexer can be written as

$$Y = \begin{cases} A & \text{if} \quad S = 0 \\ B & \text{if} \quad S = 1, \end{cases} \tag{A.1}$$

or equivalently,

$$Y = A \cdot \overline{S} + B \cdot S, \tag{A.2}$$

where S denotes the selection signal.

Decoding is necessary in applications such as interfacing, where it is often necessary to convert one digital format to another. In general, a binary decoder asserts one of 2^k output lines for each combination of the k input signals, provided it is enabled. It is implemented using k inverters and 2^k AND gates with $k + 1$ inputs. Figure A.3 shows the circuit diagram and symbol of a 2-4 decoder, which consists of two inverters and four AND gates. The Boolean equations of the outputs can be derived as

$$Z_0 = E \cdot \overline{X} \cdot \overline{Y}, \tag{A.3}$$

$$Z_1 = E \cdot \overline{X} \cdot Y, \tag{A.4}$$

$$Z_2 = E \cdot X \cdot \overline{Y}, \tag{A.5}$$

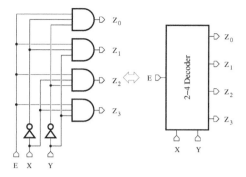

FIGURE A.3
2-4 Decoder: (a) Circuit diagram, (b) symbol.

and

$$Z_3 = E \cdot X \cdot Y, \tag{A.6}$$

where E is the enable signal, and X and Y represent the bits of the input code. For each input code, a different output line is asserted by the decoder. Note that each output line is numbered in accordance with the decimal equivalent of the input binary code.

A.2.2 CMOS implementation

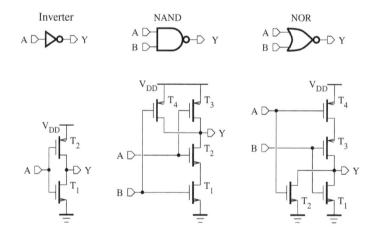

FIGURE A.4
CMOS implementations of (a) inverter, (b) NAND and (c) NOR gates.

A static CMOS inverter consisting of an nMOS transistor and pMOS transistor is shown in Figure A.4(a). It operates according to Table A.2. When the input is at V_{DD}, the nMOS transistor is on while the pMOS transistor is off. The output is then set to 0 V. On the other hand, when the input takes the

TABLE A.2
Truth Table of CMOS Inverter

A	T_1	T_2	Y
0	OFF	ON	V_{DD}
V_{DD}	ON	OFF	0

value of 0 V, the nMOS transistor is off while the pMOS transistor is on. As a result, the output level equals V_{DD}. Assuming that the leakage current is negligible, an unloaded inverter should not dissipate a static power because the path between the supply voltage and ground is interrupted in the steady state.

TABLE A.3
Truth Table of CMOS NAND Gate

A	B	T_1	T_2	T_3	T_4	Y
0	0	OFF	OFF	ON	ON	V_{DD}
0	V_{DD}	ON	OFF	ON	OFF	V_{DD}
V_{DD}	0	OFF	ON	OFF	ON	V_{DD}
V_{DD}	V_{DD}	ON	ON	OFF	OFF	0

The circuit diagram of a two-input NAND gate is depicted in Figure A.4(b). By connecting two nMOS transistors in series between the output node and ground, the output will be set to 0 V only if the level of both inputs is V_{DD}. Due to the two parallel pMOS transistors connected between the supply voltage and the output node, the output level will equal V_{DD} if either the input A or B is 0 V. This is summarized in Table. A.3.

TABLE A.4
Truth Table of CMOS NOR Gate

A	B	T_1	T_2	T_3	T_4	Y
0	0	OFF	OFF	ON	ON	V_{DD}
0	V_{DD}	ON	OFF	OFF	ON	0
V_{DD}	0	OFF	ON	ON	OFF	0
V_{DD}	V_{DD}	ON	ON	OFF	OFF	0

A CMOS NOR gate can be implemented as shown in Figure A.4(c). Its operation is illustrated by Table. A.4. By inserting two nMOS transistors in parallel between the output node and ground, the output will be set to 0 V if either the input A or B is at V_{DD}. The connection of two pMOS transistors in series between the supply voltage and ground forces the output to be at V_{DD} only if the level of both inputs is equal to 0 V.

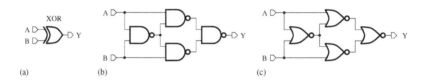

(a)   (b)   (c)

FIGURE A.5

(a) Inverter and its implementations based on (b) NAND and (c) NOR gates.

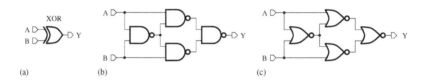

FIGURE A.6

(a) XOR gate and its implementations based on (b) NAND and (c) NOR gates.

The NAND and NOR gates are universal logic elements. Each of them can be used for the implementation of any logic function, as illustrated in Figures A.5 and A.6 in the case of the inverter and XOR gate, respectively.

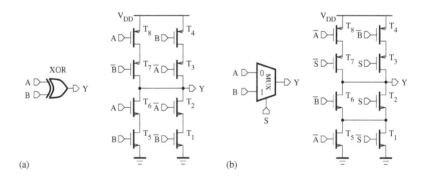

FIGURE A.7

CMOS implementation of (a) XOR gate and (b) 2-1 multiplexer.

Assuming that complementary signals are available, the XOR gate and multiplexer can also be realized as shown in Figure A.7. For the XOR gate, the pull-up transistors are configured based on the function $A \oplus B$, while the structure of the pull-down transistors is determined by the function $\overline{A \oplus B}$.

The output is at the logic high level when only one of the inputs is set to the logic high level. When the logic level of both inputs is either high or low, that of the output is low. Considering the circuit diagram of the 2-1 multiplexer, which is also composed of pull-up and pull-down transistors, it can be shown that the implemented logic function is of the form $A \cdot S + B \cdot \overline{S}$. For any given logic state of the selection signal, S, the logic level of only one of the inputs is transferred to the output.

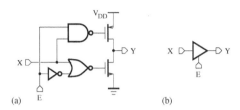

FIGURE A.8
(a) Tri-state buffer and (b) its symbol.

Tri-state buffers allow multiple logic circuits to share a common data bus, provided only one of the logic circuits is active at any given time. Figure A.8 shows the circuit diagram and symbol of a tri-state buffer. The output signal of a tri-state buffer can be written as

$$Y = \begin{cases} X, & \text{if } E = 1 \\ Z, & \text{if } E = 0, \end{cases} \tag{A.7}$$

where Z denotes the high-impedance state. It can then assume one of three states: logic low, logic high, or high impedance state. In the transfer mode, that is, when $E = 1$, the tri-state buffer is equivalent to a closed switch. In the disconnect mode, that is, when $E = 0$, the tri-state buffer output is isolated from the input by a high impedance.

A.3 Sequential logic circuits

Sequential circuits are digital circuits whose outputs depend not only on the current state of inputs, but also on the previous state of outputs. Basic sequential circuits include latch and flip-flop. A latch is sensitive to either the high level or the low level of its inputs and it is said to be level-sensitive or transparent, while a flip-flop can change its output state only at an edge of the clock signal and it is said to be edge-triggered.

A.3.1 Asynchronous SR latch

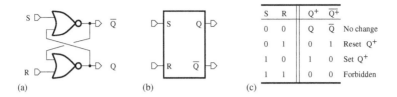

S	R	Q^+	$\overline{Q^+}$	
0	0	Q	$\overline{Q}$	No change
0	1	0	1	Reset Q^+
1	0	1	0	Set Q^+
1	1	0	0	Forbidden

(a) (b) (c)

FIGURE A.9
SR Latch: (a) Circuit diagram, (b) symbol, (c) truth table.

The asynchronous set-reset (SR) latch is one of the simplest sequential circuits. Figure A.9 shows the circuit diagram, symbol and truth table of an SR latch implemented with two NOR gates. The characteristic equation of the SR latch is given by

$$Q^+ = \overline{R} \cdot S + \overline{R} \cdot Q = \overline{R} \cdot (S + Q), \tag{A.8}$$

where Q and Q^+ denote the current and next states of the output, respectively. In the case where the condition $(R = S = 1)$ is not supposed to occur, we have

$$Q^+ = S + \overline{R} \cdot Q, \quad \text{with the assumption that} \quad S \cdot R = 0. \tag{A.9}$$

A.3.2 Asynchronous $\overline{S}\,\overline{R}$ latch

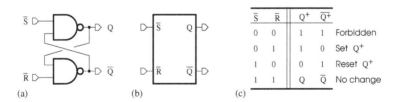

$\overline{S}$	$\overline{R}$	Q^+	$\overline{Q^+}$	
0	0	1	1	Forbidden
0	1	1	0	Set Q^+
1	0	0	1	Reset Q^+
1	1	Q	$\overline{Q}$	No change

(a) (b) (c)

FIGURE A.10
$\overline{S}\,\overline{R}$ latch: (a) Circuit diagram, (b) symbol, (c) truth table.

Another simplest form of the latch circuit can be implemented using two NAND gates. Figure A.10 shows the circuit diagram, symbol, and truth table of the $\overline{S}\,\overline{R}$ latch, whose characteristic equation is of the form

$$Q^+ = S + \overline{R} \cdot Q. \tag{A.10}$$

A.3.3 D latch

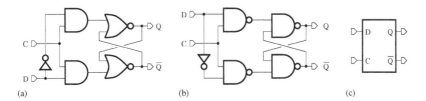

(a)   (b)   (c)

FIGURE A.11

D latch: (a) Implementation based on SR latch, (b) implementation based on $\overline{S}\,\overline{R}$ latch, (c) symbol.

A D (data) latch is used to capture the logic level of the signal at the data input when it is enabled by the control signal. Figure A.11 shows the circuit diagrams of D latches based on SR latch and $\overline{S}\,\overline{R}$ latch, respectively, and the D-latch symbol. It can be found that the characteristic equation is

$$Q^+ = D \cdot C + \overline{C} \cdot Q. \tag{A.11}$$

Hence, the latch output follows the D input when the control signal C is at the logic high level. When the C input is low, the latch output is maintained at the state previously acquired when the C input was high. Note that the outputs Q and $\overline{Q}$ of the D flip-flop are complementary.

A.3.4 D flip-flops

The operation of flip-flops is synchronized by either the rising or falling edge of a clock signal (CK). D flip-flops, whose outputs can only change at one edge of the clock signal, exhibit a characteristic equation of the form

$$Q^+ = D. \tag{A.12}$$

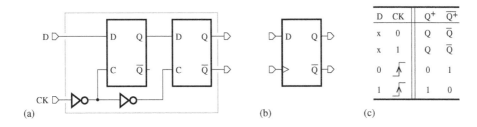

(a)   (b)   (c)

D	CK	Q^+	$\overline{Q^+}$
x	0	Q	$\overline{Q}$
x	1	Q	$\overline{Q}$
0	↑	0	1
1	↑	1	0

FIGURE A.12

D flip-flop with master-slave configuration: (a) Circuit diagram, (b) symbol, (c) truth table.

A D flip-flop can be implemented using a master-slave configuration, which consists of two D latches that are connected in series and controlled by inverted phases of the clock signal. Figure A.12 shows the circuit diagram, symbol, and truth table of the master-slave D flip-flop. The input data are captured when the clock signal is low, but its state is transferred to the output only at the beginning of the next clock phase. Provided the setup and hold requirements are met, the aforementioned master-slave D flip-flop is referred to as a device triggered by the rising edge of the clock signal.

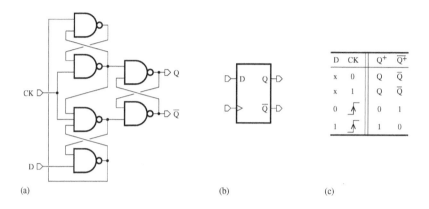

D	CK	Q^+	$\overline{Q^+}$
x	0	Q	$\overline{Q}$
x	1	Q	$\overline{Q}$
0	↑	0	1
1	↑	1	0

(a) (b) (c)

FIGURE A.13
D flip-flop: (a) Circuit diagram, (b) symbol, (c) truth table.

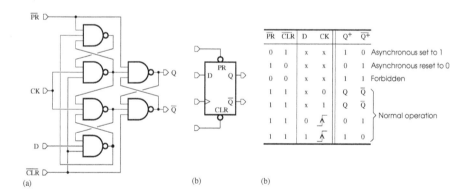

$\overline{PR}$	$\overline{CLR}$	D	CK	Q^+	$\overline{Q^+}$	
0	1	x	x	1	0	Asynchronous set to 1
1	0	x	x	0	1	Asynchronous reset to 0
0	0	x	x	1	1	Forbidden
1	1	x	0	Q	$\overline{Q}$	
1	1	x	1	Q	$\overline{Q}$	
1	1	0	↑	0	1	Normal operation
1	1	1	↑	1	0	

(a) (b) (b)

FIGURE A.14
D flip-flop with asynchronous (preset and clear) inputs: (a) Circuit diagram, (b) symbol, (c) truth table.

A D flip-flop can also be realized by relying on the use of an edge detector so that the input pulse can occur synchronously with a transition of the clock signal. Figure A.13 shows the circuit diagram, symbol, and truth table of a D flip-flop triggered by the rising edge of the clock signal. Six NAND gates are required in this implementation of the D flip-flop. The state of the data input is captured at the rising edge of the clock signal. It is then transferred to the output a short time after the edge occurrence due to the gate propagation delay.

In some applications, it may be necessary to asynchronously drive the flip-flop, thereby bypassing the clock signal control. Figure A.14 shows the circuit diagram, symbol, and truth table of a D flip-flop with asynchronous inputs. The D input is synchronous, while the preset and clear inputs, which are used to respectively set or reset the outputs regardless of the signal levels on the other input nodes (and especially the clock signal node), are asynchronous. The asynchronous set and reset functions are activated by the low level of signals, which are then denoted as $\overline{PR}$ and $\overline{CLR}$. Asynchronous inputs are required to determine the initial state of the flip-flop and are not used during the normal operation.

A.3.5 CMOS implementation

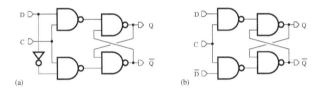

FIGURE A.15
Circuit diagrams of conventional D latches (a) with a single data input and (b) with complementary data inputs.

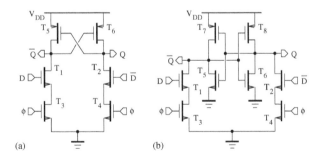

FIGURE A.16
CMOS implementations of (a) dynamic and (b) static D latches.

D latches can be designed by substituting each logic gate of the circuit diagrams shown in Figure A.15 with its CMOS implementation. However, the resulting maximum operating frequency appears to be limited by the long delay of the critical path, especially at low supply voltages. An approach to reduce the critical path delay is to use latches with differential cascode structures [1], such as the ones depicted in Figure A.16. The first latch in Figure A.16(a)

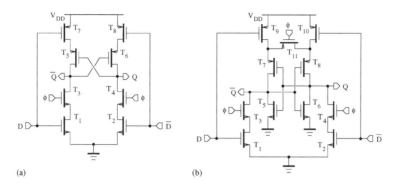

FIGURE A.17
CMOS implementations of ratio-insensitive (a) dynamic and (b) static D latches.

is a dynamic logic circuit based on cross-coupled transistors, which can track the changes of the input data only during one phase of the clock signal, while the second latch in Figure A.16(b), which uses cross-coupled inverters acting as a storage element, can be considered a static logic circuit. A static latch possesses an output storage node that remains connected to either the supply voltage or ground, while a dynamic latch exhibits a conducting path to the supply voltage or ground only during the evaluation of the input data. In general, dynamic CMOS logic circuits require less silicon area due to the number of transistors needed than their static counterparts. However, they can be affected by charge sharing at the output nodes. One solution to prevent variations in the logic level consists of using sufficiently large static inverters to isolate the output nodes of the dynamic latch stage.

The latches of Figure A.16 have the drawback of being sensitive to the aspect ratio between the nMOS and pMOS transistors, especially in the case where the input transistor is of the pMOS type. This is due to the fact that the current delivered by input section should overcome that from the cross-coupled transistors in order for the latch to switch, thereby increasing the length of the switching period.

Ratio-insensitive latches can be designed as shown in Figure A.17. Extra transistors are used so that the input node can be formed by connecting together the gate terminals of nMOS and pMOS transistors.

Flip-flops can be realized using two latches in a master-slave configuration, as shown in Figure A.18. The state of the input is acquired by the master latch when the clock signal becomes low. It is then transferred to the outputs of the slave latch when the clock signal goes high. The master latch is transparent, while the slave latch remains opaque — and vice versa. This flip-flop has the advantage of using only one clock signal, but it can be prone to a substantial voltage drop at the outputs due to the capacitive coupling effect between the

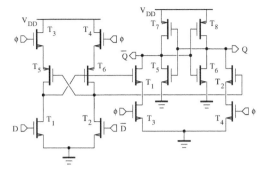

FIGURE A.18
CMOS implementation of a semi-static D flip-flop.

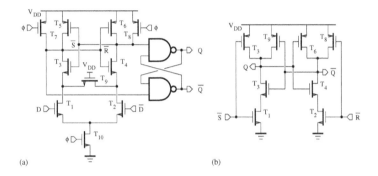

FIGURE A.19
CMOS implementations of (a) D flip-flop and (b) $\overline{S}\,\overline{R}$ latch.

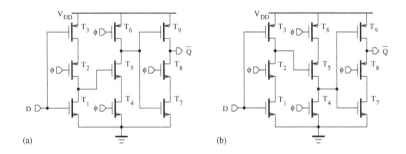

FIGURE A.20
CMOS implementations of (a) rising-edge and (b) falling-edge triggered D flip-flops.

master and slave latches.

Another approach to design flip-flop relies on the structure depicted in Figure A.19(a) [2, 3]. The first stage is based on a sense amplifier and the second stage is an $\overline{S}\,\overline{R}$ latch, which can be implemented as shown in Figure A.19(b). When the clock signal becomes low, the input differential transistor pair is in the cutoff region and the output nodes of the sense amplifier are precharged through transistors T_7 and T_8 to the high logic state. The $\overline{S}\,\overline{R}$ latch then holds its previous logic state. When the rising edge of the clock signal occurs, the sense amplifier is enabled and can track the input data to produce a transition from the high state to the low state on one of the outputs. The $\overline{S}\,\overline{R}$ latch updates its outputs according to the actual state of the sense amplifier. By applying the high state of the input data to the node D, T_3, T_1, and T_{10} provide a discharge path to the node $\overline{R}$, while T_4 and T_6 are biased in the cutoff and conduction region, respectively. On the other hand, when the high state of the input data occurs at the node $\overline{D}$, a discharge path is supplied to the node $\overline{S}$ by T_4, T_2, and T_{10}, while T_3 and T_5 operate in the cutoff and conduction region, respectively. After the completion of the state transition at one of the output nodes, the inputs become decoupled from the outputs and the sense amplifier then remains unaffected by any subsequent change of input data during the active clock phase. If the state of the input data changes, the discharge path will be interrupted, leaving the node at the low logic level floating. In order to prevent this floating node from being charged by the leakage or coupling currents, another discharge path should be provided by using the nMOS pass transistor T_9. This pass transistor also helps equalize the voltage values of the two differential branches during the pre-charge and reset the output nodes prior to the next sensing of the input data. The primary feature of the sense-amplifier flip-flop is the provision of high input impedance. However, the operation speed can be limited by the propagation delay introduced by the output latch.

More efficient circuits for D flip-flops can be derived using true single phase clocking (TSPC) techniques [4]. TSPC D flip-flops operate with only one clock signal, and offer advantages such as a small circuit area for clock lines, a reduced clock skew, and high-speed operation. Figure A.20 shows the circuit diagrams of the rising-edge and falling-edge triggered D-flip-flops, which consist of three clocked inverting stages. The output signal is available in an inverted version. When the clock slope is not sufficiently steep, both nMOS and pMOS clocked transistors simultaneously operate in the conduction region during the clock transition. As a result, the logic level of internal signals may become undefined and a race condition may occur.

A.4   Bibliography

[1] J. Yuan and C. Svensson, "New single-clock CMOS latchs and flipflops with improved speed and power savings," *IEEE J. of Solid-State Circuits*, vol. 32, pp. 62–69, Jan. 1997.

[2] M. Matsui, H. Hara, Y. Uetani, L.-S. Kim, T. Nagamatsu, Y. Watanabe, A. Chiba, K. Matsuda, and T. Sakurai, "A 200 MHz 13 mm^2 2-D DCT macro-cell using sense-amplifying pipeline flip-flop scheme," *IEEE J. of Solid-State Circuits*, vol. 29, pp. 1482–1490, Dec. 1994.

[3] J. Montanaro, R. T. Witek, K. Anne, A. J. Black, E. M. Cooper, D. W. Dobberpuhl, P. M. Donahue, J. Eno, G. W. Hoeppner, D. Kruckemyer, T. H. Lee, P. C. M. Lin, L. Madden, D. Murray, M. H. Pearce, S. Santhanam, K. J. Snyder, R. Stephany, and S. C. Thierauf, "A 160-MHz, 32-b, 0.5-W CMOS RISC microprocessor," *IEEE J. of Solid-State Circuits*, vol. 31, pp. 1703–1714, Nov. 1996.

[4] J. Yuan and C. Svensson, "High-speed CMOS circuit technique," *IEEE J. of Solid-State Circuits*, vol. 24, pp. 62–70, Jan. 1989.

B

Transistor sizing in building blocks

CONTENTS

Transistor sizing affects the performance characteristics (speed, power consumption, chip area, etc.) of integrated circuits. It is an iterative process that is generally achieved using SPICE-based simulation tools. Because the design parameters such as transistor sizes and bias currents depend on the performance metrics that are nonlinear functions, their optimal values may be derived only using computation-intensive programs. The effects of manufacturing yields are minimized by taking into account the process and operating-condition variations

The design variables are the aspect ratios (width over length) of transistors, the value of the passive components (capacitors and resistors), and the value of bias currents and bias voltages. The optimization of the design variables for a component can only improve the operation characteristics (speed, precision, power dissipation, and area) to an extent allowed by the chosen topology.

For any given IC process, the determination of transistor sizes may require transistor matching data extracted from previous wafer fabrications. The minimum transistor area, WL, is set by the matching requirements, while parasitic capacitances are increased by scaling up transistor dimensions. The W/L ratio is determined by the drain current level and the operating region. For low power consumption, the bias currents should be chosen to be as small as possible while satisfying the noise and speed requirements.

B.1 MOS transistor

Two different types of MOS transistors are generally available for circuit design. In nMOS transistors, electrons are majority carriers in the source, which should operate at a voltage lower than the one of the drain, while in pMOS transistors, holes are generated in the source, which should be biased at a voltage higher than that of the drain. The operation of MOS transistors can be described as follows:

nMOS	pMOS
Cutoff region	
$V_{GS} < V_{T_n}$	$V_{SG} < -V_{T_p}$
$I_D = 0$	$I_D = 0$
Linear region	
$V_{GS} > V_{T_n}, V_{DS} < V_{DS(sat)}$	$V_{SG} > -V_{T_p}, V_{SD} < V_{SD(sat)}$
$I_D = K_n[2(V_{GS} - V_{T_n})V_{DS} - V_{DS}^2]$	$I_D = K_p[2(V_{SG} + V_{T_p})V_{SD} - V_{SD}^2]$
Saturation region	
$V_{GS} > V_{T_n}, V_{DS} \geq V_{DS(sat)}$	$V_{SG} > -V_{T_p}, V_{SD} \geq V_{SD(sat)}$
$I_D = K_n(V_{GS} - V_{T_n})^2(1 + \lambda_n V_{DS})$	$I_D = K_p(V_{SG} + V_{T_p})^2(1 + \lambda_p V_{SD})$

Note that, for the nMOS transistor,

$$K_n = \mu_n(C_{ox}/2)(W/L), \qquad V_{DS(sat)} = V_{GS} - V_{T_n}, \quad \text{and} \quad V_{T_n} > 0;$$

and for the pMOS transistor,

$$K_p = \mu_p(C_{ox}/2)(W/L), \qquad V_{SD(sat)} = V_{SG} + V_{T_p}, \quad \text{and} \quad V_{T_p} < 0.$$

The saturation voltage, $V_{DS(sat)}$ or $V_{SD(sat)}$, is also referred as the overdrive. The linear region is also known as the ohmic or triode region.

In the saturation region, the transconductance and conductance can be derived as

$$g_m = \frac{\partial I_D}{\partial V_{GS}} = 2K_n(V_{GS} - V_{T_n})(1 + \lambda_n V_{DS})$$
$$= 2\sqrt{K_n I_D(1 + \lambda_n V_{DS})} = \frac{2I_D}{V_{GS} - V_{T_n}} \tag{B.1}$$

and

$$g_{ds} = \frac{\partial I_D}{\partial V_{DS}} = \frac{\lambda_n I_D}{1 + \lambda_n V_{DS}} \tag{B.2}$$

for nMOS transistors, and

$$g_m = \frac{\partial I_D}{\partial V_{SG}} = 2K_p(V_{SG} + V_{T_p})(1 + \lambda_p V_{SD})$$
$$= 2\sqrt{K_p I_D(1 + \lambda_p V_{SD})} = \frac{2I_D}{V_{SG} + V_{T_p}} \tag{B.3}$$

and

$$g_{sd} = \frac{\partial I_D}{\partial V_{SD}} = \frac{\lambda_p I_D}{1 + \lambda_p V_{SD}} \qquad (B.4)$$

for pMOS transistors. To simplify the circuit analysis, the effect of λ can at first be neglected in g_m expressions.

In submicron CMOS technology, transistors are characterized using Berkeley short-channel IGFET model (BSIM), which is supposed to better take into account short-channel effects. Let us consider an nMOS transistor with a known (W/L) aspect ratio. Given a value of the gate-source voltage, the values of the drain current can be obtained by running SPICE simulations with BSIM parameters for at least two values of the drain-source voltage. These values can then be inserted into an $I - V$ characteristic of the form

$$I_D = K'_n \frac{W_{eff}}{L_{eff}} (V_{GS} - V_{T_n})^2 (1 + \lambda_n V_{DS}) \qquad (B.5)$$

to determine the parameters K'_n and λ_n required for hand calculations. For a better fit, this must be achieved in a region around the drain-source saturation voltage, which defines the operating point of the transistor. Here, the effective channel width and length[1], W_{eff} and L_{eff}, are used due to the fact that the mask-defined sizes (W and L) are reduced by the amount of lateral diffusions, which are specified in the BSIM model card. Similarly, the values of K'_p and λ_p can be determined using the expression

$$I_D = K'_p \frac{W_{eff}}{L_{eff}} (V_{SG} + V_{T_p})^2 (1 + \lambda_p V_{SD}). \qquad (B.6)$$

TABLE B.1
Simulated Drain Currents

nMOS Transistor, $V_{GS} = 0.95$ V		pMOS Transistor, $V_{SG} = 0.95$ V	
V_{DS} (V) 0.50	1.10	V_{SD} (V) 0.50	1.10
I_D (mA) 11.96	12.45	I_D (mA) 23.29	26.15

Parameter extraction example

Using BSIM models for a 0.13 μm CMOS process, the simulated drain currents for both nMOS and pMOS transistors with the effective aspect ratio of 20 μm/0.12 μm can be obtained as shown in Table B.1. Use the values of the threshold voltages, $V_{T_n} = 0.46$ V and $V_{T_p} = -0.44$ V, to determine the transistor parameters K'_n, λ_n, K'_p, and λ_p.

---

[1]Based on BSIM3v3 parameters, the effective channel width and length are of the form $W_{eff} = W - 2 \cdot WINT$ and $L_{eff} = L - 2 \cdot LINT$, where $WINT$ and $LINT$ are specified in the BSIM model card.

BSIM SPICE Model Card Example

```
*
* SPICE BSIM3 VERSION 3.1 PARAMETERS
* SPICE 3f5 Level 8, Star-HSPICE Level 49, UTMOST Level 8
* 0.13 μm CMOS process
*
* Temperature_ parameters=Default
*
.model CMOS NMOS (
```

+VERSION = 3.1	TNOM=27	TOX= 3.2E-9
+XJ=1E-7	NCH=2.3549E17	VTH0=0.0458681
+K1=0.3661767	K2=-0.0334177	K3=1E-3
+K3B=4.0506568	W0=1E-7	NLX=1E-6
+DVT0W=0	DVT1W=0	DVT2W=0
+DVT0=1.4508861	DVT1=0.1491907	DVT2=0.2337763
+U0=436.7862785	UA=-3.86228E-10	UB=3.278288E-18
+UC=4.781785E-10	VSAT=1.929894E5	A0=1.9927058
+AGS=0.751416	B0=1.840348E-6	B1=5E-6
+KETA=0.05	A1=7.776166E-4	A2=0.3
+RDSW=150	PRWG=0.3498753	PRWB=0.1103551
+WR=1	WINT=4.847999E-9	LINT=1.039837E-8
+DWG=1.179843E-8	DWB=8.997945E-9	VOFF=-0.0270176
+NFACTOR=2.5	CIT=0	CDSC=2.4E-4
+CDSCD=0	CDSCB=0	ETA0=2.751524E-6
+ETAB=-0.0111499	DSUB=4.060052E-6	PCLM=1.9774199
+PDIBLC1=0.9702431	PDIBLC2=0.01	PDIBLCB=0.1
+DROUT=0.9994828	PSCBE1=7.965102E10	PSCBE2=5.021019E-10
+PVAG=0.5368546	DELTA=0.01	RSH=7
+MOBMOD=1	PRT=0	UTE=-1.5
+KT1=-0.11	KT1L=0	KT2=0.022
+UA1=4.31E-9	UB1=-7.61E-18	UC1=-5.6E-11
+AT=3.3E4	WL=0	WLN=1
+WW=0	WWN=1	WWL=0
+LL=0	LLN=1	LW=0
+LWN=1	LWL=0	CAPMOD=2
+XPART=0.5	CGDO=3.74E-10	CGSO=3.74E-10
+CGBO=1E-12	CJ=9.581273E-4	PB=0.9758836
+MJ=0.4044874	CJSW=1E-10	PBSW=0.8002027
+MJSW=0.6	CJSWG=3.3E-10	PBSWG=0.8002027
+MJSWG=0.6	CF=0	PVTH0=2.009264E-4
+PRDSW=0	PK2=1.30501E-3	WKETA=0.013236
+LKETA=0.0327523	PU0=4.4729531	PUA=1.66833E-11
+PUB=0	PVSAT=653.2294237	PETA0=1E-4
+PKETA=-9.655097E-3	)	

*

.MODEL CMOS PMOS (

+VERSION=3.1	TNOM=27	TOX=3.2E-9
+XJ=1E-7	NCH=4.1589E17	VTH0=-0.2219851
+K1=0.2770146	K2=5.044386E-3	K3=0.0971898
+K3B=6.5020562	W0=1E-6	NLX=2.628685E-7
+DVT0W=0	DVT1W=0	DVT2W=0
+DVT0=9.146632E-3	DVT1=1	DVT2=0.1
+U0=111.7597102	UA=1.237083E-9	UB=1.90335E-21
+UC=-1.69849E-11	VSAT=1.22678E5	A0=2
+AGS=0.4944995	B0=5.266819E-6	B1=5E-6
+KETA=0.0118456	A1=0.4157385	A2=0.9596542
+RDSW=109.2955948	PRWG=-0.4797803	PRWB=0.5
+WR=1	WINT=0	LINT=7.536533E-9
+DWG=5.119137E-9	DWB=-1.84021E-8	VOFF=-0.1022829
+NFACTOR=1.5332272	CIT=0	CDSC=2.4E-4
+CDSCD=0	CDSCB=0	ETA0=0.0125544
+ETAB=-6.066043E-3	DSUB=2.751452E-3	PCLM=0.3090456
+PDIBLC1=0	PDIBLC2=-1.27526E-13	PDIBLCB=0.1
+DROUT=1.003724E-3	PSCBE1=2.606086E9	PSCBE2=8.052708E-10
+PVAG=0.0181389	DELTA=0.01	RSH=7
+MOBMOD=1	PRT=0	UTE=-1.5
+KT1=-0.11	KT1L=0	KT2=0.022
+UA1=4.31E-9	UB1=-7.61E-18	UC1=-5.6E-11
+AT=3.3E4	WL=0	WLN=1
+WW=0	WWN=1	WWL=0
+LL=0	LLN=1	LW=0
+LWN=1	LWL=0	CAPMOD=2
+XPART=0.5	CGDO=3.42E-10	CGSO=3.42E-10
+CGBO=1E-12	CJ=1.156238E-3	PB=0.8
+MJ=0.4407762	CJSW=1.125225E-10	PBSW=0.8
+MJSW=0.1152909	CJSWG=4.22E-10	PBSWG=0.8
+MJSWG=0.1152909	CF=0	PVTH0=4.284016E-4
+PRDSW=60.4471984	PK2=2.405903E-3	WKETA=0.0352518
+LKETA=0.0207754	PU0=-1.4797175	PUA=-5.65562E-11
+PUB =7.212046E-25	PVSAT=-50	PETA0 = 1.069996E-5
+PKETA=-4.427073E-3	)	

Solution

Knowing the values of the threshold voltages, V_{T_n} and V_{T_p}, we can solve Equations (B.5) and (B.6) for

$$K'_n = 288.4 \ \mu A/V^2 \quad \lambda_n = 0.072 \ V^{-1}\,,$$
$$K'_p = 48.2 \ \mu A/V^2 \quad \lambda_p = 0.229 \ V^{-1}\,.$$

Note that due to a further reduction of the transistor channel length, the value of the channel-length modulation parameter is increased, yielding a diminution of the small-signal drain-source resistance.

B.2   Amplifier

Basically, there are two structure types for the design of amplifiers: single-stage and multistage structures. Using a given IC process, an amplifier can be designed to meet the following specifications:

Supply voltage	V_{DD} (V)
Capacitive load	C_L (pF)
Slew rate	SR (V/µs)
Settling time	t_s (µs)
DC voltage gain	A_0 (dB)
Gain-bandwidth product	GBW (MHz)
Phase margin	ϕ_M (°)
Power consumption	P (mW)
Input-referred noise	$\overline{v_{n,i}^2}$ (nV/$\sqrt{\text{Hz}}$ at f Hz)

For a given amplifier, the same dc common-mode voltage is applied to each input node. Its value can vary from the maximum $V_{ICM,max}$ to the minimum $V_{ICM,min}$. Similarly, the output dc voltage is assumed to be bounded by $V_{0,max}$ and $V_{0,min}$.

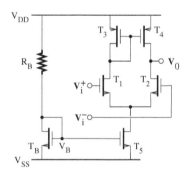

FIGURE B.1
Circuit diagram of a single-stage differential amplifier.

One of the simplest single-stage differential amplifiers is depicted in Figure B.1. It consists of a differential transistor pair, a current mirror, and a biasing circuit. The slew rate can be expressed as

$$SR = I_5/C_L, \tag{B.7}$$

where I_5 is the drain current of the transistor T_5. Hence,

$$I_5 = SR \cdot C_L. \tag{B.8}$$

Ideally, the current I_5 should be equal to the bias current I_B. Because the accuracy achieved for the value of I_5 can be limited by the biasing circuit, it is good practice to set the bias current value with a safe margin, that is, $I_B = I_5(1 + x)$, where x is a fractional number denoting the worst-case variation of the current.

The gain-bandwidth product or unity-gain frequency is of the form

$$GBW = \frac{g_{m1}}{C_L}. \tag{B.9}$$

The transcondutance of the transistor T_1 can be derived as

$$g_{m1} = 2K_n(V_{GS_1} - V_{T_n}) = GBW \cdot C_L. \tag{B.10}$$

The aspect ratio of the transistor T_1 is then given by

$$\frac{W_1}{L_1} = \frac{GBW \cdot C_L}{\mu_n C_{ox}(V_{GS_1} - V_{T_n})} = \frac{GBW \cdot C_L}{\mu_n C_{ox} V_{DS_1(sat)}}, \tag{B.11}$$

where

$$V_{DS_1(sat)} = \frac{2I_{D_1}}{g_{m1}} = \frac{I_B}{GBW \cdot C_L}. \tag{B.12}$$

Note that the transistors T_1 and T_2 are matched, that is, $W_1/L_1 = W_2/L_2$.

Because the dc current flowing through the transistors T_3 and T_4 is $I_B/2$, we can obtain

$$\frac{W_3}{L_3} = \frac{2I_{D_3}}{\mu_p C_{ox}(V_{SG_3} + V_{T_p})^2} = \frac{I_B}{\mu_p C_{ox}(V_{SG_3} + V_{T_p})^2}, \tag{B.13}$$

where

$$V_{SG_3} = V_{DD} - V_{ICM,max} + V_{T_n} \tag{B.14}$$

and

$$\frac{W_4}{L_4} = \frac{W_3}{L_3}. \tag{B.15}$$

The bias current flows through the transistors T_5 and T_6 and it can be found that

$$\frac{W_5}{L_5} = \frac{2I_{D_5}}{\mu_n C_{ox}(V_{GS_5} - V_{T_n})^2} = \frac{2I_B}{\mu_n C_{ox}(V_{GS_5} - V_{T_n})^2}, \tag{B.16}$$

where

$$V_{GS_5} - V_{T_n} = V_{DS_5(sat)} = V_{ICM,min} - V_{SS} - V_{GS_1} \tag{B.17}$$

and

$$\frac{W_6}{L_6} = \frac{W_5}{L_5}. \tag{B.18}$$

The value of the resistor R_B is given by

$$R_B = \frac{V_{DD} - V_{GS_6} - V_{SS}}{I_B}, \tag{B.19}$$

where $V_{GS_6} = V_{DS_6(sat)} + V_{T_n}$.

The power consumption of the differential stage is estimated to be

$$P = (V_{DD} - V_{SS})I_B. \qquad (B.20)$$

For the above differential stage, the dc gain is typically about 35 dB, and the phase margin should be at least $60°$ to meet the stability requirement.

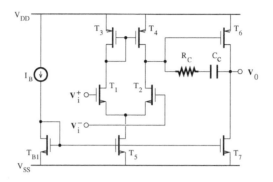

FIGURE B.2

Circuit diagram of a two-stage amplifier.

Let us consider the two-stage amplifier [1,2] shown in Figure B.2. The input stage consists of a differential transistor pair loaded by a current mirror, while the output stage is a common source amplifier with an active load. The frequency stabilization is achieved using a compensation network.

All transistors are biased to operate in the saturation region. Applying Kirchhoff's voltage law between the positive input node and the positive supply voltage, we can obtain

$$V_{DD} - V_i^+ = V_{SG_3} + V_{DS_1} - V_{GS_1}, \qquad (B.21)$$

where $V_i^+ = V_{ICM,max}$ and $V_{DS_1} = V_{GS_1} - V_{T_n}$. The transistor T_1 operates in the saturation region provided

$$\sqrt{\frac{2I_{D_3}}{\mu_n C_{ox}(W_3/L_3)}} \leq V_{DD} - V_{ICM,max} + V_{T_p} + V_{T_n}. \qquad (B.22)$$

On the other hand, the application of Kirchhoff's voltage law between the positive input node and the negative supply voltage gives

$$V_i^+ - V_{SS} = V_{GS_1} + V_{DS_5}, \qquad (B.23)$$

where $V_i^+ = V_{ICM,min}$ and $V_{GS_1} = V_{DS_1} + V_{T_n}$. The operation of the transistor T_5 in the saturation region then requires that

$$\sqrt{\frac{2I_{D_1}}{\mu_n C_{ox}(W_1/L_1)}} + \sqrt{\frac{2I_{D_5}}{\mu_n C_{ox}(W_5/L_5)}} \leq V_{ICM,min} - V_{T_n} - V_{SS}. \qquad (B.24)$$

The source-drain voltage of the transistor T_6 is given by

$$V_{SD_6} = V_{DD} - V_0 \,. \tag{B.25}$$

To bias T_6 in the saturation region, the following condition must be fulfilled:

$$\sqrt{\frac{2I_{D_6}}{\mu_n C_{ox}(W_6/L_6)}} \leq V_{DD} - V_{0,max} \,. \tag{B.26}$$

In the case of the transistor T_7, the drain-source voltage is of the form

$$V_{DS_7} = V_0 - V_{SS} \,. \tag{B.27}$$

The saturation condition corresponds to

$$\sqrt{\frac{2I_{D_7}}{\mu_n C_{ox}(W_7/L_7)}} \leq V_{0,min} - V_{SS} \,. \tag{B.28}$$

The spectral density of the output noise can be computed as

$$\overline{v_{n,o}^2} = g_{m6}^2 R_2^2 [\overline{v_{n,6}^2} + \overline{v_{n,7}^2} + R_1^2 (g_{m1}^2 \overline{v_{n,1}^2} + g_{m2}^2 \overline{v_{n,2}^2} + g_{m3}^2 \overline{v_{n,3}^2} + g_{m4}^2 \overline{v_{n,4}^2})] \,. \tag{B.29}$$

Assuming that $g_{m1} = g_{m2}$, $g_{m3} = g_{m4}$ $\overline{v_{n,1}^2} = \overline{v_{n,2}^2}$, $\overline{v_{n,3}^2} = \overline{v_{n,4}^2}$, and $\overline{v_{n,6}^2} = \overline{v_{n,7}^2}$, the spectral density of the input-referred noise can be derived as

$$\overline{v_{n,i}^2} = \frac{\overline{v_{n,o}^2}}{g_{m1}^2 g_{m6}^2 R_1^2 R_2^2} = 2 \frac{\overline{v_{n,6}^2}}{g_{m1}^2 R_1^2} + 2\overline{v_{n,1}^2} \left[1 + \left(\frac{g_{m3}}{g_{m1}} \right)^2 \frac{\overline{v_{n,3}^2}}{\overline{v_{n,1}^2}} \right] \,. \tag{B.30}$$

Because $g_{m1}R_1 \gg 1$, $\overline{v_{n,1}^2} = 8kT/(3g_{m1})$, and $\overline{v_{n,3}^2} = 8kT/(3g_{m3})$, we obtain

$$\overline{v_{n,i}^2} \simeq 2\overline{v_{n,1}^2} \left[1 + \left(\frac{g_{m3}}{g_{m1}} \right)^2 \frac{\overline{v_{n,3}^2}}{\overline{v_{n,1}^2}} \right] = \frac{16kT}{3g_{m1}} \left(1 + \frac{g_{m3}}{g_{m1}} \right) \,. \tag{B.31}$$

To reduce the noise contribution due to the transistors acting as loads or current sources, the transconductance g_{m1} (or equivalently g_{m2}) should be made as large as possible. The gain-bandwidth product is given by

$$GBW = \frac{g_{m1}}{C_C} \,, \tag{B.32}$$

where C_C is the compensation capacitor. The slew rate can be written as

$$SR = \frac{I_{D_5}}{C_C} = \frac{I_{D_7} - I_{D_5}}{C_L} \,, \tag{B.33}$$

where C_L is the total load capacitor at the output node. Hence,

$$I_{D_7} = SR(C_C + C_L) \,. \tag{B.34}$$

Because the positive input common-mode range is of the form

$$ICMR^+ = V_{DD} - V_{ICM,max} = V_{SD_3(sat)} - V_{T_n} \qquad \text{(B.35)}$$

and $I_{D_3} = I_{D_1} = I_{D_5}/2$, it can be found that

$$g_{m3} = \frac{2I_{D_3}}{V_{SD_3(sat)}} = \frac{SR \cdot C_C}{ICMR^+ + V_{T_n}}. \qquad \text{(B.36)}$$

Substituting Equations (B.32) and (B.36) into (B.31) gives

$$C_C = \frac{16kT}{3 \cdot GBW \cdot v_{n,i}^2} \left[1 + \frac{SR}{GBW(ICMR^+ + V_{T_n})} \right]. \qquad \text{(B.37)}$$

From Equation (B.32),

$$g_{m1} = \sqrt{2\mu_n C_{ox}(W_1/L_1)I_{D_1}} = GBW \cdot C_C. \qquad \text{(B.38)}$$

Assuming that $I_{D_1} = I_{D_5}/2 = SR \cdot C_C/2$, we have

$$\frac{W_1}{L_1} = \frac{GBW^2 C_C}{\mu_n C_{ox} SR}. \qquad \text{(B.39)}$$

Note that the aspect ratios of T_1 and T_2 are identical. The head room voltage for the negative input common-mode range input can be expressed as

$$ICMR^- = V_{ICM,min} - V_{SS} = V_{DS_5(sat)} + V_{DS_1(sat)} + V_{T_n}, \qquad \text{(B.40)}$$

where $V_{DS_1(sat)} = SR/GBW$. The current I_{D_5} is given by

$$I_{D_5} = \mu_n C_{ox}(W_5/L_5)V_{DS_5(sat)}^2/2 = SR \cdot C_C \qquad \text{(B.41)}$$

Combining Equations (B.40) and (B.41), it can be shown that

$$\frac{W_5}{L_5} = \frac{2 \cdot SR \cdot C_C}{\mu_n C_{ox}(ICMR^- - V_{T_n} - SR/GBW)^2}. \qquad \text{(B.42)}$$

Note that the transistors T_5 and T_{B_1} are designed with the same aspect ratio. The transistors T_5 and T_7 are biased in the saturation region and $V_{GS_5} = V_{GS_7}$. Hence,

$$\frac{I_{D_7}}{I_{D_5}} = \frac{W_7/L_7}{W_5/L_5}. \qquad \text{(B.43)}$$

From Equation (B.33), we find that

$$\frac{I_{D_7}}{I_{D_5}} = 1 + \frac{C_L}{C_C} \qquad \text{(B.44)}$$

and therefore

$$\frac{W_7}{L_7} = \left(1 + \frac{C_L}{C_C}\right)\frac{W_5}{L_5}. \tag{B.45}$$

Based on the small-signal analysis of the amplifier, the zero of the transfer function is rejected at infinity by choosing the compensation resistor such that

$$R_C = \frac{1}{g_{m6}}. \tag{B.46}$$

The second pole of the transfer function is approximately located at

$$\omega_{p2} \simeq -\frac{g_{m6}}{C_L}, \tag{B.47}$$

where C_L is the load capacitor. The amplifier is designed to exhibit the behavior of a two-pole system, that is, $\omega_{p3} \geq 10 \cdot GBW$. That is, its phase margin is of the form

$$\phi_M \simeq 90° - \arctan\frac{GBW}{|\omega_{p2}|}. \tag{B.48}$$

Combining Equations (B.47) and (B.48) gives

$$g_{m6} = GBW \cdot C_L \cdot \tan(\phi_M). \tag{B.49}$$

Note that, for a given angle x, $\tan(90° - x) = 1/\tan(x)$. Because $g_{m6} = \sqrt{2\mu_n C_{ox}(W_6/L_6)I_{D_6}}$ and $I_{D_6} = I_{D_7}$, we obtain

$$\frac{W_6}{L_6} = \frac{[GBW \cdot C_L \tan(\phi_M)]^2}{2\mu_n C_{ox} I_{D_7}}. \tag{B.50}$$

To reduce the offset voltage due to the difference, which can exist between the values of currents I_{D_6} and I_{D_7} when the amplifier input nodes are grounded, we should have $V_{SD_4} = V_{SD_3} = V_{SG_3}$ and $V_{SD_4} = V_{SG_6}$. It can also be deduced that $I_{D_3} = I_{D_4} = I_{D_5}/2$ and $I_{D_6} = I_{D_7}$. The aspect ratios of transistors T_3, T_4, T_5, T_6, and T_7 satisfy the following relation,

$$\frac{W_3/L_3}{W_6/L_6} = \frac{W_4/L_4}{W_6/L_6} = \frac{1}{2}\frac{W_5/L_5}{W_7/L_7}. \tag{B.51}$$

Equation (B.51) can be used for the determination of the aspect ratio of T_3 and T_4.

For the resulting amplifier, the dc gain is approximately given by

$$A_0 = -\frac{g_{m1}}{g_2 + g_4} \cdot \frac{g_{m6}}{g_6 + g_7}, \tag{B.52}$$

where g_2, g_4, g_6, and g_7 denote the drain-source conductance of transistors T_2, T_4, T_6, and T_7, respectively. The power dissipation can be computed as

$$P = (V_{DD} - V_{SS})(2I_{D_5} + I_{D_7}). \qquad (B.53)$$

The results of the initial transistor sizing for the two-stage amplifier may be optimized using appropriate CAD programs [3]. The maximum and minimum sizes of transistors are primarily set by process-related variations and limitations of the IC fabrication technique.

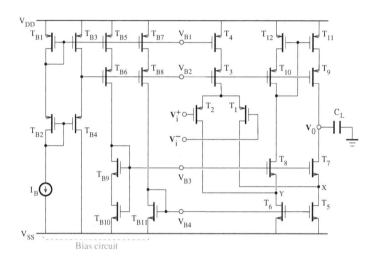

FIGURE B.3
Circuit diagram of a folded-cascode amplifier.

In cases where a large gain is required with a high gain-bandwidth product and a limited power budget, a folded-cascode amplifier [4, 5], as shown in Figure B.3, is the architecture of choice. Due to symmetry, the transistors, T_1 and T_2, T_5 and T_6, T_7 and T_8, T_9 and T_{10}, and T_{11} and T_{12} are matched. As a result of the small-signal analysis, the *dc* gain of the folded-cascode amplifier can be written as

$$A_0 = g_{m1}r_0, \qquad (B.54)$$

where

$$r_0 = \cfrac{1}{\cfrac{(g_1 + g_5)g_7}{g_{m7} + g_{mb7}} + \cfrac{g_{11}g_9}{g_{m9} + g_{mb9}}} \qquad (B.55)$$

and g_k ($k = 1, 5, 9.11$) is the drain-source conductance of transistor T_k. The slew rate is given by

$$SR = \frac{I_{D_3}}{C_L}, \qquad (B.56)$$

where I_{D_3} is equal to the bias current I_{B_1} of the differential transistor pair. The gain-bandwidth product is of the form

$$GBW = \frac{g_{m1}}{C_L} \tag{B.57}$$

Because $g_{m1} = \sqrt{2\mu_p C_{ox}(W_1/L_1)I_{D_1}}$ and $I_{D_1} = I_{D_3}/2$, we obtain

$$\frac{W_1}{L_1} = \frac{GBW^2 C_L}{\mu_p C_{ox} SR}. \tag{B.58}$$

Using Equation (B.56), the aspect ratios of transistors T_3 and T_4 can be computed as

$$\frac{W_3}{L_3} = \frac{2I_{D_3}}{\mu_p C_{ox} V_{SD_3(sat)}^2} = \frac{2 \cdot SR \cdot C_L}{\mu_p C_{ox} V_{SD_3(sat)}^2} \tag{B.59}$$

and

$$\frac{W_4}{L_4} = \frac{2I_{D_4}}{\mu_p C_{ox} V_{SD_4(sat)}^2} = \frac{2 \cdot SR \cdot C_L}{\mu_p C_{ox} V_{SD_4(sat)}^2}, \tag{B.60}$$

where $V_{SD_3(sat)} = V_{SD_4(sat)}$, $V_{SD_3(sat)} = (V_{DD} - V_{SG_1} - V_{ICM,max})/2$, and $V_{SG_1} = V_{SD_1(sat)} - V_{T_p}$. The power consumption of the amplifier is of the form

$$P = (V_{DD} - V_{SS})(I_{B_1} + 2I_{B_2}) \tag{B.61}$$

where $I_{B_1} = I_{D_3} = I_{D_4}$ and $I_{B_2} = I_{D_5}$. Using Equations (B.56) and (B.61), the aspect ratio of the transistor T_{11} is derived as

$$\frac{W_5}{L_5} = \frac{2I_{D_5}}{\mu_n C_{ox} V_{DS_5(sat)}^2} = \frac{\dfrac{P}{V_{DD} - V_{SS}} - SR \cdot C_L}{\mu_n C_{ox} V_{DS_5(sat)}^2}. \tag{B.62}$$

Because

$$V_i^- = V_{SS} + V_{DS_5(sat)} + V_{SD_1(sat)} - V_{SG_1}, \tag{B.63}$$

it can be shown that

$$V_{DS_5(sat)} = V_{ICM,min} - V_{SS} - V_{T_p}, \tag{B.64}$$

where $V_i^- = V_{ICM,min}$. For the size of the transistor T_7, we find

$$\frac{W_7}{L_7} = \frac{2I_{D_7}}{\mu_n C_{ox} V_{DS_7(sat)}^2}, \tag{B.65}$$

where

$$I_{D_7} = I_{D_5} - I_{D_3}/2 = \frac{P}{2(V_{DD} - V_{SS})} - SR \cdot C_L \tag{B.66}$$

and

$$V_{DS_7(sat)} = V_{0,min} - V_{SS} - V_{DS_5(sat)}. \tag{B.67}$$

The small-signal analysis indicates that the transfer function of the folded-cascode amplifier has four poles and two zeros [6]. Because two of the poles are approximately compensated by the zeros, the amplifier phase margin can be expressed as

$$\phi_M \simeq 180° - \arctan \frac{GBW}{|\omega_{p_1}|} - \arctan \frac{GBW}{|\omega_{p_2}|}, \tag{B.68}$$

where $\omega_{p_1} = -GBW/A_0$ and $\omega_{p_2} = -g_{m7}/C_p$. Note that A_0 is the amplifier dc gain and C_p is the total parasitic capacitance at the source of the transistor T_7 ($C_p \simeq C_{gs7} \simeq 2W_7 L_7 C_{ox}/3$). To proceed further, we can simplify Equation (B.68) to

$$\phi_M \simeq 90° - \arctan \frac{GBW}{|\omega_{p_2}|} \tag{B.69}$$

or equivalently

$$g_{m7} = GBW \cdot C_p \cdot \tan \phi_M . \tag{B.70}$$

Here, g_{m7} and C_p are functions of the width and length of the transistor T_7, making difficult the determination of the width-to-length ratio from Equation (B.70) alone.

The aspect ratios of transistors T_9 and T_{11} are respectively given by

$$\frac{W_9}{L_9} = \frac{2I_{D_9}}{\mu_p C_{ox} V_{SD_9(sat)}^2} \tag{B.71}$$

and

$$\frac{W_{11}}{L_{11}} = \frac{2I_{D_{11}}}{\mu_p C_{ox} V_{SD_{11}(sat)}^2}, \tag{B.72}$$

where

$$I_{D_9} = I_{D_{11}} = I_{D_7} = \frac{P}{2(V_{DD} - V_{SS})} - SR \cdot C_L \tag{B.73}$$

and

$$V_{SD_9(sat)} = V_{SD_{11}(sat)} = (V_{DD} - V_{0,max})/2. \tag{B.74}$$

Given the current I_B, the biasing circuit should be sized such that it can generate the required voltages V_{B_1}, V_{B_2}, V_{B_3}, and V_{B_4}. All transistors, except T_{B10}, are biased in the saturation region. Transistors $T_{B1} - T_{B8}$ operate as a current mirror delivering the bias current to each of the load $T_{B9} - T_{B10}$ and T_{B11}. Hence,

$$\begin{aligned} W_3/L_3 = W_{B1}/L_{B1} = 4(W_{B2}/L_{B2}) = W_{B3}/L_{B3} = W_{B4}/L_{B4} \\ = W_{B5}/L_{B5} = W_{B6}/L_{B6} = W_{B9}/L_{B9} = W_{B10}/L_{B10} . \end{aligned} \tag{B.75}$$

Let $I_{DB9} = I_{DB10} = I_B$, $V_{DSB9(sat)} = V_{GSB9} = V_{GS7}$, and $V_{DSB10(sat)} = V_{DS5(sat)}$. The aspect ratio of the transistor T_{B9} can be computed as

$$W_{B9}/L_{B9} = \frac{2I_{DB9}}{\mu_p C_{ox} V_{DSB9(sat)}^2}. \tag{B.76}$$

Because the transistor T_{B10} operates in the triode region, we have

$$W_{B10}/L_{B10} = \frac{2I_{DB10}}{3\mu_p C_{ox} V_{DSB10(sat)}^2}. \tag{B.77}$$

In the case of the transistor T_{B11}, it can be shown that

$$\frac{W_{B11}}{L_{B11}} = \frac{2I_{DB11}}{\mu_p C_{ox} (V_{GSB11} - V_{T_n})^2}, \tag{B.78}$$

where $I_{DB11} = I_B$ and $V_{GSB11} = V_{GS5}$.

In practice, due to IC process variations, a small safety margin is often added to $V_{DS(sat)}$ or $V_{SD(sat)}$ to ensure that the transistors remain in the saturation region. That is,

$$V_{DS} - (V_{GS} - V_{T_n}) \geq V_{DS(sat)} \tag{B.79}$$

and

$$V_{SD} - (V_{SG} + V_{T_p}) \geq V_{SD(sat)}. \tag{B.80}$$

For submicrometer circuit designs, the saturation voltage is generally in the range from 0.2 V to 0.6 V.

B.3   Comparator and latch

One of the structures that can be used for the design of a comparator is depicted in Figure B.4(a). It consists of an input differential stage loaded by cross-coupled transistors. Due to the positive feedback provided by the loading transistors, the comparator transfer characteristic exhibits hysteresis as shown in Figure B.4(b).

Given the values of some specifications (slew rate, gain-bandwidth product, input common-mode voltage, input-referred noise, power dissipation), the sizes of transistors composing the input stage can be determined in the same way as in the case of an amplifier.

The difference between the positive and negative trigger points is the hysteresis band, which is given by

$$V_{HB} = V_{trig+} - V_{trig-}. \tag{B.81}$$

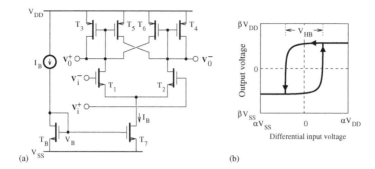

FIGURE B.4

(a) Comparator circuit with hysteresis; (b) comparator characteristics in the noninverting configuration.

Assuming that the transistors T_1 and T_2 are matched, that is, $W_1/L_1 = W_2/L_2$, we obtain

$$V_{trig+} = \sqrt{\frac{I_B}{2K'(W_1/L_1)}} \frac{\sqrt{(W_5/L_5)/(W_3/L_3)} - 1}{\sqrt{1 + (W_5/L_5)/(W_3/L_3)}} \tag{B.82}$$

and

$$V_{trig-} = \sqrt{\frac{I_B}{2K'(W_1/L_1)}} \frac{1 - \sqrt{(W_6/L_6)/(W_4/L_4)}}{\sqrt{1 + (W_6/L_6)/(W_4/L_4)}}. \tag{B.83}$$

For a symmetrical structure, it is required that $V_{trig+} = -V_{trig-}$. Hence, $W_3/L_3 = W_4/L_4$ and $W_5/L_5 = W_6/L_6$. Given the values of I_B, $2K'(W_1/L_1)$ and the specification V_{trig+}, the ratio $x = (W_5/L_5)/(W_3/L_3)$ can be determined by solving the next equation, which is derived from Equation (B.82) and can be written as

$$x + \frac{2}{\Delta^2 - 1}\sqrt{x} + 1 = 0, \tag{B.84}$$

where $\Delta = V_{trig+}\sqrt{2K'(W_1/L_1)/I_B}$. Solving Equation (B.84) gives

$$\sqrt{x} = -\frac{1}{\Delta^2 - 1} \pm \sqrt{\frac{1}{(\Delta^2 - 1)^2} - 1}, \tag{B.85}$$

where $|\Delta| < \sqrt{2}$ and $\Delta \neq 1$. To choose between the plus and minus sign in the valid expression of x, we note that the condition $x > 1$ should be realized in order for the comparator to operate with a positive feedback.

A D latch [7] can be implemented as shown in Figure B.5(a). The input data are applied differentially to a pair of transistors connected through clocked transistors to two cross-coupled inverters.

Ideally, the transition between the low state and high state should occur

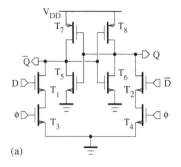
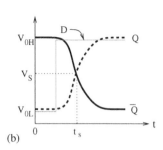

(a)   (b)

FIGURE B.5
(a) Circuit diagram of a D latch; (b) output waveforms.

at the mid-point of the logic swing or $V_{DD}/2$ (see Figure B.5(b)). This is achieved when both transistors are in the saturation region. Hence,

$$\mu_n(W_5/L_5) = \mu_p(W_7/L_7) \tag{B.86}$$

and

$$\mu_n(W_6/L_6) = \mu_p(W_8/L_8). \tag{B.87}$$

Provided $\mu_n/\mu_p = r$, the difference in charge mobilities is compensated in an inverter by choosing the aspect ratio (W/L) of the pMOS transistor to be r times the one of the nMOS transistor.

To switch the latch from the low state to the high state, the low level of the circuit section $T_3 - T_1 - T_7$ should be below the switching threshold of the inverter $T_5 - T_7$. Assuming that the transistors are biased in the triode region, and transistors T_1 and T_3 can be reduced to a single transistor with the aspect ratio W_k/L_k, it can be shown that

$$K_{p7}[2(V_{SG_7} + V_{T_p})V_{SD_7} - V_{SD_7}^2] = K_{n_k}[2(V_{GS_k} - V_{T_n})V_{DS_k} - V_{DS_k}^2], \tag{B.88}$$

where $V_{SG_7} = V_{GS_k} = V_{DD}$ and $V_{SD_7} = V_{DS_k} = V_{DD}/2$. With $V_{T_n} \leq |V_{T_p}|$, we arrive at the following condition:

$$K_{n_k} \geq K_{p7}. \tag{B.89}$$

Substituting $K_{n_k} = \mu_n C_{ox}(W_k/L_k)$ and $K_{p7} = \mu_p C_{ox}(W_7/L_7)$ into Equation (B.89) gives

$$(W_k/L_k) \geq (\mu_p/\mu_n)(W_7/L_7). \tag{B.90}$$

Because $(W_1/L_1) = (W_3/L_3) = 2(W_k/L_k)$, we can obtain

$$(W_1/L_1) = (W_3/L_3) \geq 2(\mu_p/\mu_n)(W_7/L_7) = 2(W_5/L_5). \tag{B.91}$$

Similarly, due to the circuit symmetry, it can be found that

$$(W_2/L_2) = (W_4/L_4) \geq 2(\mu_p/\mu_n)(W_8/L_8) = 2(W_6/L_6). \tag{B.92}$$

Because the above derivation of transistor sizes is based on first-order transistor models and simplified circuit operation, it may still be necessary to use simulation results to adequately adjust the transistor widths and lengths.

B.4 Bibliography

[1] P. R. Gray and R. G. Meyer, "MOS operational amplifier design — A tutorial overview," *IEEE J. of Solid-State Circuits*, vol. 17, pp. 969–982, Dec. 1982.

[2] G. Palmisano, G. Palumbo, and S. Pennisi, "Design procedure for two-stage transconductance operational amplifiers: a tutorial," *Analog Integrated Circuits and Signal Processing*, vol. 27, pp. 179–189, June 2001.

[3] M. del M. Hershenson, S. P. Boyd, and T. H. Lee, "Optimal design of a CMOS op-amp via geometric programming," *IEEE Trans. on CAD of Integrated Circuits and Systems*, vol. 20, pp. 1–21, Jan. 2001.

[4] C.-C. Shih, and P. E. Gray, "Reference refreshing cyclic analog-to-digital and digital-to-analog converters," *IEEE J. of Solid-State Circuits*, vol. 21, pp. 544–554, Aug. 1986.

[5] S. M. Mallya and J. H. Nevin, "Design procedures for a fully differential folded-cascode amplifier," *IEEE J. of Solid-State Circuits*, vol. 24, pp. 1737–1740, Dec. 1989.

[6] H. C. Yang, M. A. Abu-Dayeh, and D. J. Allstot, "Small-signal analysis and minimum setling time design of a one-stage folded-cascode CMOS operational amplifier," *IEEE Trans. on Circuits and Systems*, vol. 38, pp. 804–807, July 1991.

[7] J. Yuan and C. Svensson, "New single-clock CMOS latchs and flipflops with improved speed and power savings," *IEEE J. of Solid-State Circuits*, vol. 32, pp. 62–69, Jan. 1997.

C

Signal-Flow Graph

CONTENTS

Signal-flow graphs (SFGs) [1] are suitable for the representation of continuous-time and discrete-time systems in the s-domain and z-domain, respectively. They can be considered a network diagram in which nodes are connected by unidirectional branches. Each node represents a system variable, and a scaling factor is indicated along each branch. The analysis and design of small and medium-size systems is often facilitated by the use of signal-flow graph techniques.

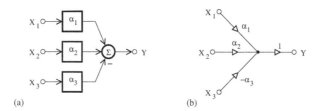

(a)   (b)

FIGURE C.1

(a) Block diagram and (b) signal-flow graph representations of a system.

A block diagram and an SFG contain the same network information. They can represent a system of equations or an equation. The SFG representation is then not unique for a network, which can be characterized by equations in different forms. In the special case of Figure C.1, where the input nodes are labeled X_1, X_2, and X_3, the variable Y at the output node can be expressed as

$$Y = \alpha_1 X_1 + \alpha_2 X_2 - \alpha_3 X_3. \tag{C.1}$$

For the SFG, the summer is implemented by a node. Hence, a mixed node, that is, a node that receives both incoming and outgoing branches, sums the weighted signals of all incoming branches and the result is transmitted to all outgoing branches.

C.1 SFG reduction rules

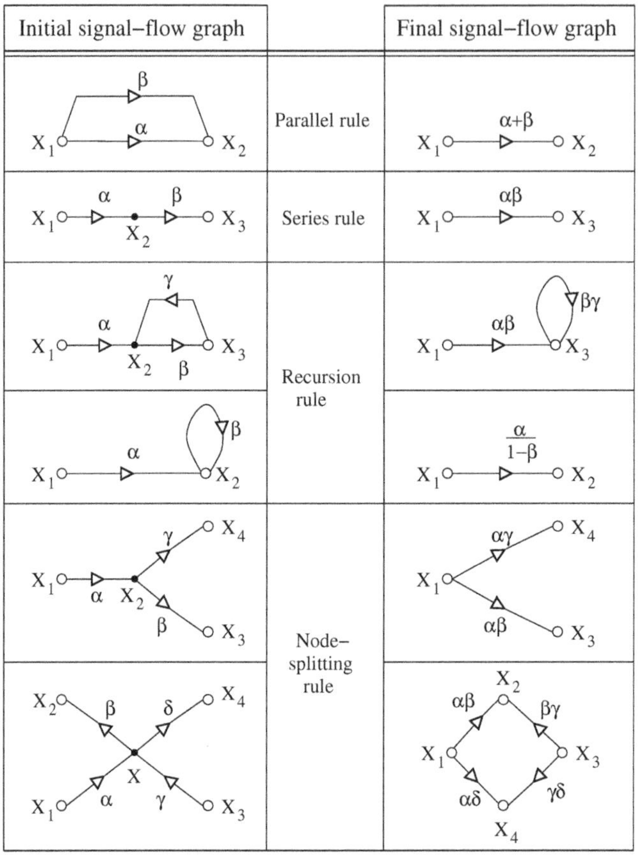

FIGURE C.2
Basic SFG reduction rules.

SFGs are often used in system analysis. Generally, the objective is to determine the relationship or transfer function between various variables. This can be achieved by the application of a set of rules such that nodes are absorbed and branches are combined to form new branches with the equivalent scaling factors. The reduction process is repeated until an SFG with the desired degree of complexity, or a residual graph in which only one path exists between the input node and the selected output node, is obtained. Basic SFG reduction rules are summarized in Figure C.2. However, the reduction procedure can become complicated as the system complexity increases. An alternative method for finding the relationship between system variables can be based on Mason's gain formula [2].

C.2 Mason's gain formula

To proceed further, it is necessary to give the following definitions:

- A path is a series connection of branches following the direction of the signal flow from one node to a given node.

- A forward path is a path from the input node to the output node that does not go through the same node twice.

- A loop is a path that starts and ends at the same node, without passing through any other node more than once.

- A path gain is the product of the gains or scaling factors of all its branches.

- Two loops are *nontouching* if they do not have any node in common.

Mason's gain formula gives the transfer function T from an input (or source) node to an output (or sink) node in the form

$$T = \frac{\sum_{k}^{P} T_k \triangle_k}{\triangle}, \tag{C.2}$$

where

$\triangle = 1 -$ (Sum of all individual loop gains) + (Sum of the product of the loop gains of all possible combinations of nontouching loops taken two at a time) $-$ (Sum of the product of the loop gains of all possible combinations of nontouching loops taken three at a time) + (Sum of the product of the loop gains of all possible combinations of nontouching loops taken four at a time) $- \cdots + (-1)^r$ (Sum of the product of the loop gains of all possible combinations of nontouching loops taken r at a time),

P is the total number of forward paths between the input and output nodes,

T_k is the gain of the k-th forward path between the input and output nodes,

$\triangle_k$ is the value of $\triangle$ for the part of the graph not touching the k-th forward path.

The term $\triangle$ can be considered the determinant of the graph, while $\triangle_k$ represents the cofactor of the k-th forward path.

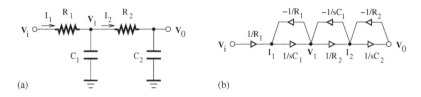

(a) (b)

FIGURE C.3
(a) RC network; (b) equivalent SFG representation.

Consider the circuit diagram of the RC network depicted in Figure C.3(a). Using Kirchhoff's current law and Kirchhoff's voltage law gives

$$I_1 = \frac{V_i - V_1}{R_1}, \tag{C.3}$$

$$V_1 = \frac{I_1 - I_2}{sC_1}, \tag{C.4}$$

$$I_2 = \frac{V_1 - V_0}{R_2}, \tag{C.5}$$

and

$$V_0 = \frac{I_2}{sC_2}. \tag{C.6}$$

An SFG of the RC network can then be derived as shown in Figure C.3(b). There are three individual loops with the gains $P_{11} = -1/sC_1R_1$, $P_{21} = -1/sC_1R_2$, and $P_{31} = -1/sC_2R_2$. Because the first and third loops are non-touching, we have $P_{12} = 1/s^2C_1C_2R_1R_2$. Hence, the determinant $\triangle$ is given by

$$\triangle = 1 - (P_{11} + P_{21} + P_{31}) + P_{12} \tag{C.7}$$

$$= 1 + 1/sC_1R_1 + 1/sC_1R_2 + 1/sC_2R_2 + 1/s^2C_1C_2R_1R_2. \tag{C.8}$$

The gain of the forward path joining the input node and output node is

$$T_1 = 1/s^2C_1C_2R_1R_2. \tag{C.9}$$

Because the forward path touches all three loops, it can be shown that

$$\triangle_1 = 1. \tag{C.10}$$

Note that $\triangle_1$ is obtained from $\triangle$ by removing the contributions due to the loops that touch the forward path. The circuit transfer function is then of the form

$$T = \frac{V_0}{V_i} = \frac{T_1\triangle_1}{\triangle} \tag{C.11}$$

$$= \frac{1}{s^2C_1C_2R_1R_2 + s(C_1R_1 + C_2R_1 + C_2R_2) + 1}. \tag{C.12}$$

It can be deduced that the aforementioned RC circuit is a second-order low-pass filter, whose transfer function only exhibits finite poles.

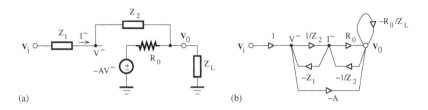

(a)   (b)

FIGURE C.4
(a) Equivalent model of an amplifier and (b) its SFG representation.

As a second example of circuit analysis using Mason's gain formula, consider the equivalent model of an amplifier shown in Figure C.4(a). Using Kirchhoff's laws, the following equations can be derived:

$$V^- = V_i - Z_1 I^- \tag{C.13}$$

$$I^- = \frac{V^- - V_0}{Z_2} \tag{C.14}$$

$$V_0 = -AV^- + R_0 \left(I^- - \frac{V_0}{Z_L} \right). \tag{C.15}$$

The corresponding SFG representation is obtained as illustrated in Figure C.4(b). Four individual loops can be identified. Their gains are of the form $P_{11} = -Z_1/Z_2$, $P_{21} = -R_0/Z_2$, $P_{31} = -AZ_1/Z_2$, and $P_{41} = -R_0/Z_L$. We also have $P_{12} = Z_1 R_0/Z_2 Z_L$ because the loops P_{11} and P_{41} are nontouching. The term $\triangle$ is then given by

$$\triangle = 1 - (P_{11} + P_{21} + P_{31} + P_{41}) + P_{12} \tag{C.16}$$
$$= 1 + Z_1/Z_2 + R_0/Z_2 + AZ_1/Z_2 + R_0/Z_L + Z_1 R_0/Z_2 Z_L . \tag{C.17}$$

The SFG exhibits two forward paths, whose gains can be written as, $T_1 = -A$ and $T_2 = R_0/Z_2$. Each forward path has a common node with all loops, and it can then be found that $\triangle_1 = 1$ and $\triangle_2 = 1$. Therefore, the transfer function between the input and output nodes reads

$$T = \frac{V_0}{V_i} = \frac{T_1 \triangle_1 + T_2 \triangle_2}{\triangle} \tag{C.18}$$

$$= -\frac{Z_2}{Z_1} \cdot \frac{1 - \dfrac{1}{A} \cdot \dfrac{R_0}{Z_2}}{1 + \dfrac{1}{A}\left(1 + \dfrac{Z_2}{Z_1} + \dfrac{R_0}{Z_1} + \dfrac{R_0}{Z_L} + \dfrac{Z_2 R_0}{Z_1 Z_L}\right)}. \tag{C.19}$$

For a high gain, A, the transfer function is reduced to the ratio $-Z_2/Z_1$ and an inverting amplifier is realized.

C.3   Bibliography

[1] S. J. Mason, "Feedback theory — Some properties of signal flow graphs," *Proc. of the IRE*, vol. 41, pp. 1144–1156, Sept. 1953.

[2] S. J. Mason, "Feedback theory — Further properties of signal flow graph," *Proc. of the IRE*, vol. 44, pp. 920–926, July 1956.

Index

Index 873